Engineering Analyses and Finite Element Methods

GUANGMING ZHANG

Department of Mechanical Engineering
and the Institute for Systems Research

The University of Maryland @ College Park

COLLEGE HOUSE ENTERPRISES, LLC
Knoxville, TN

USE AT YOUR OWN RISK

The author and the publisher have expended every effort to avoid errors in preparing this textbook. However, we are human and errors are inevitable. Readers and users of this book are cautioned to verify the results and the designs suggested in the text. The author and the publisher cannot guarantee, and do not assume responsibility for any aspect of any development based on the information contained herein.

THANKS FOR PERMISSIONS

The author and the publisher thank the copyright and trademark owners providing permission for use of copyrighted materials and trademarks. We apologize for any errors or omissions in obtaining permissions. Where appropriate, we referenced similar treatments and gave credit for prior work. Errors or omissions in obtaining permissions or in giving proper references are not intentional. We will correct them at the earliest opportunity after the error or omission is brought to our attention. Please contact the publisher at the address given below.

The manuscript was prepared in Microsoft Word 2000 using 11 point Times New Roman font. The book was printed from camera ready copy by Publishing and Printing Inc, Knoxville, TN.

College House Enterprises, LLC.
5713 Glen Cove Drive
Knoxville, TN 37919-8611, U.S.A.
Phone: (865) 558 6111
FAX (865) 558 6111

ISBN 0-9762413-1-5

To my wife Jinyu
my children Zumei and Haowei, and my granddaughter Chianchian

who have always been with me especially at those difficult times. Their love and understanding have supported me in a journey for survival and success.

ABOUT THE AUTHOR

Guangming Zhang obtained a bachelor degree and a master degree both in mechanical engineering from Tianjin University, the People's Republic of China. He obtained a master degree and a Ph. D. degree in mechanical engineering from the University of Illinois at Urbana-Champaign. He is currently an associate professor in the Department of Mechanical Engineering and Director of the Advanced Design and Manufacturing Laboratory at the University of Maryland at College Park. He holds a joint appointment with the Institute for Systems Research.

Professor Zhang worked at the Northwest Medical Surgical Instruments Factory in China where he served as a principal engineer to design surgical instruments and dental equipment. He also taught at the Beijing Institute of Printing and received the National Award for Outstanding Teaching from the Press and Publication Administration of the People's Republic of China in 1987. In 1992 he was selected by his peers at the University of Maryland to receive the Outstanding Systems Engineering Faculty Award. He was the recipient of the E. Robert Kent Outstanding Teaching Award of the College of Engineering in 1993, and the Poole and Kent Company Senior Faculty Teaching Award in 2004. He was a recipient of the 1992 Blackall Machine Tool & Gage Award of the American Society of Mechanical Engineering. In 1995, 1997, 1998 and 1999, he received the Award of Commendation, Award of Member of the Year, and the Award of Appreciation from the Society of Manufacturing Engineers, Region 3, for his outstanding service as the faculty advisor to the SME Student Chapter at College Park.

Professor Zhang actively participated in the NSF sponsored ECSEL grant between 1990 and 2000. He served as the principal investigator for this grant at the University of Maryland between 1997 and 2000, and coordinated the ECSEL sponsored projects on integration of design, on active learning and hands-on experiences, and on developing methods for team learning. He has written 3 textbooks, co-authored 2 textbooks, about 70 technical papers, and holds one patent.

PREFACE

This book presents an introduction to the fundamentals of finite element methods with computer applications. Finite element methods are numerical methods for solving problems of engineering and science. Those problems involve geometry, material properties, loading conditions and boundary conditions. Seeking analytical based solutions are generally difficult if not impossible. With capabilities of computers expanding at an unthinkable pace, the importance of using unbeatable computing power to perform finite element analyses must be emphasized. The book is written as an introductory text for undergraduate students in engineering. The book should also be useful to those engaged in the engineering analyses.

The book is organized into 10 chapters. The first chapter provides a fundamental coverage of finite element methods. It stresses the mathematical and mechanical fundamentals, which support the development of finite element methods. Chapter 2 introduces the Pro/ENGINEER Wildfire design system and Pro/Mechanica, which is engineering simulation software to carry out finite element analyses. Chapter 3 focused on modeling a variety of load conditions. Chapter 4 focuses on modeling the boundary conditions of mechanical structures in general, illustrating the procedure of dealing the physical entities both at the component level and at the assembly level. Chapter 5 presents a review of the stress condition so as to understand the mechanism of the engineering simulation software in the computing process. Chapter 6 presents an important branch of finite element methods. The branch is shell modeling through idealization. In Chapter 7, we focus on applications. The three case studies deal with engineering vibration analyses and buckling analyses. Chapter 8 introduces a specific concept called design optimization. The engineering simulation system of ALGOR is introduced in Chapter 9. We focus on performing FEA for truss and beam structures. In Chapter 10, we introduce the engineering simulation software of ANSYS, which is the most widely used to perform FEA, not only nationally, but worldwide as well.

The material assembled in this book is an outgrowth of a senior-level design elective course taught by the author at the University of Maryland at College Park for the past three years. Moreover, this book is written in a style adaptable for self-study and reference. This book is the first edition. Suggestions for improving the contents are welcome and the author deeply appreciate the efforts made by the readers in this regard.

It is a pleasure to extend special thanks to Dr. James W. Dally, President of the College House Enterprises, LLC. His support has been invaluable to the author, not only in the academic area, but also in many aspects of the author's life and career development.

Guangming Zhang
College Park, MD
December 2004

CONTENTS

CHAPTER 6 IDEALIZATION: MODELNG WITH SHELLS

CHAPTER 7 APPLICATIONS ON MODAL AND BUCKLING ANALYSES

CHAPTER 8 EVALUATION OF ENGINEERING DESIGN AND DESIGN OPTIMIZATION

CHAPTER 9 IDEALIZATION: MODELING WITH STRUSS AND BEAM STRUCTURES AND ALGOR

CHAPTER 10 FEA USING ANSYS

CHAPTER 1

INTRODUCTION TO FINITE ELEMENT METHODS

1.1 Introduction

Engineering design is an activity which demands the responsibility of the designer or the design team to meet the customer needs in terms of the design specifications. When an initial design is completed, a critical step to follow is to evaluate the performance of the designed component under a specified working environment. Results obtained from the evaluation lead to accepting the initial design or directing to an improvement through a redesign process. Finite element analysis is such a process, which has been used to carry out those evaluations of the initial designs through simulation under a computer based environment.

Finite element methods originated from an intuitive thinking, in which people viewed an entire body or an entire structure or a system (we consider the structure, the materials, the boundary conditions and the load condition as an integrated system) as composed of elements. As a result, the response of a system when it was subjected to an external load could be viewed as an integration of the responses of those elements under the loading conditions. To carry out such an integration process, numerical methods were introduced to facilitate the evaluation process, which was element based. It was clear that the results obtained from such an evaluation process were in the nature of approximation. The accuracy of the obtained results from the finite element method based evaluation was dependent on the number of elements used, the algorithm applied and the computing accuracy which could be achieved.

Finite element methods made their progress due to a great number of engineers and scientists devoted their efforts. Engineering analyses using finite element methods were gradually accepted in engineering applications because the methods had its unique capability of dealing with complex geometry and complex loading conditions by offering approximate solutions. Under those circumstances, obtaining a solution simply solely based on an analytical approach seemed very difficult, if not impossible. On the other hand, directly using an experimental method to evaluate the performance of a structural design would be extremely costly in general. In 1970s and 1980s, a significant amount of progress had been made in developing new computation algorithms, which significantly improved the computing efficiency and accuracy. While those progresses were being made, there were some difficulties in application implementations. All those difficulties were clustered at the limitation of the computation capability. As the number of elements was increased significantly in problem formulation, the requirement for data storage and distribution in computing was also increased accordingly. It was not uncommon that the computer time demanded to obtain a solution through the use of FEM was with day's time even weeks to complete.

Thanks to the rapid advancement of computer technology in past two decades, the capability of data storage and computing speed a computer system can offer not only has reached the needs of design engineers, but also has opened a new dimension to the design engineers for seeking solutions to those structures and systems, which are beyond their imagination to deal with during the past. Nowadays, finite element methods have already gained their incomputable popularity in engineering applications from the field of continuum mechanics, to heat transfer, flow dynamics, and even to medical fields where approximate solutions are needed for treatments and disease control. Finite element methods have already revolutionized to finite element analyses, which are a branch of engineering and science. Finite element analyses focus on obtaining approximate solutions to a wide variety of engineering and scientific problems, based on the integration of scientific fundamentals and computer based numerical analyses. Such a unique integration creates a favorable, stimulating environment for design evaluations. Today, software systems based on finite element methods to carry out finite element analyses for seeking solutions to real life challenges have been available. These software systems provide design engineers and scientists with a common platform for fast, efficient and cost-conscious product development, from design concept to final-stage testing and validation.

In this book, we will use three (3) of those FEA software systems available in the market to demonstrate the procedure(s) to perform FEA. These 3 software systems are Pro/Mechanica, ALGOR and ANSYS. We introduce Pro/Mechanica first in this book because we want to utilize the Pro/ENGINEER design system in the process of creating 3D solid models, and Pro/Mechanica and Pro/ENGINEER are an integrated system. We hope that the learning process related to the applications of Pro/Mechanica forms a basis for readers to understand the fundamentals related to the general framework embedded in the FEA software systems commercially available in the market.

In section 1.1, we introduce the finite element method in a one-dimensional space. In section 1.2, we introduce the finite element method in a two-dimensional space. In section 1.3, we introduce the finite element method in a three-dimensional space. In section 1.4, we emphasize the importance of setting a unit system when using a FEA software system.

1.2 Finite Element Method in a One-Dimensional Space

In this section, we use the concept of idealization applied in modeling a mechanical structure in one-dimensional space to introduce readers the basic concepts involved in using finite element methods. In the field of engineering, components such as springs are used in structural design. For the purpose of analyzing the response of a mechanical structure when subjected to loading, springs are widely used in so-called idealization. For example, when dealing with the displacement of a cylindrical bar, which is subjected to an axial load, the stress and strain can be evaluated based on the following constitution equations:

$$\sigma = \frac{F}{A}$$

$$\varepsilon = E\sigma$$

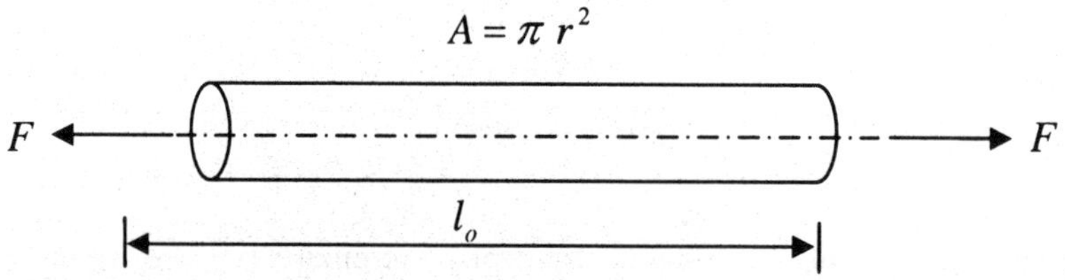

Where E = Young's modulus of elasticity, or elastic modulus, and ε is evaluated by

$$\varepsilon = \frac{\Delta l}{l_o} = E\sigma$$

By substitution, we will have the following relation:

$$F = \sigma\, A = (E\varepsilon)A = (E\frac{\Delta l}{l_o})A = (\frac{AE}{l_o})\Delta l$$

In the field of theoretical mechanics, we have the following constitution relation for a spring component where parameter *k* is the spring constant or the stiffness of the spring.

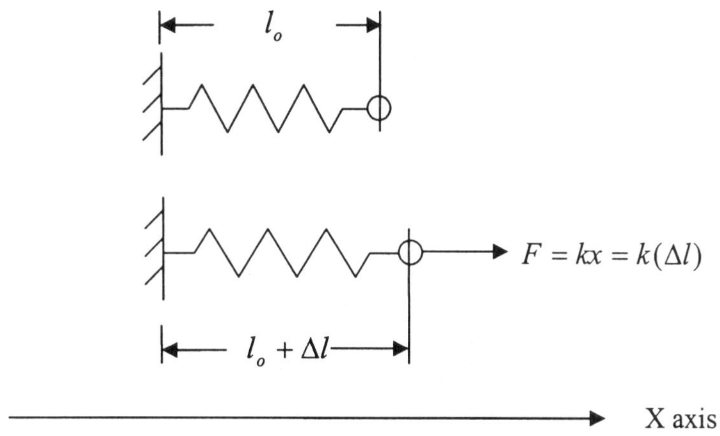

When we use a spring component to represent the cylindrical bar for evaluating the displacement, we may use a concept called an equivalent stiffness of the cylindrical bar, which is given by

$$k = \frac{AE}{l_o}$$

Now let us introduce a concept of shape function. The need to have a shape function comes from quantitatively and systematically representing the displacement at each node.

For example, when using FEA, a spring is modeled as an element. In a 2D space, each element has 2 nodes, as shown below.

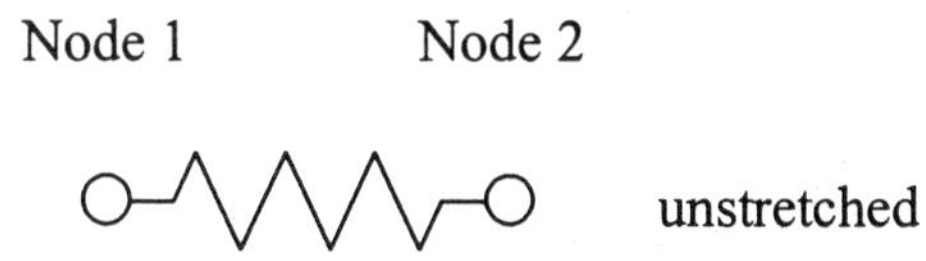

Under a loading condition, the spring element can be stretched or compressed. The displacement at each node is used to quantify the deformation of the spring component. Let us assume that the displacement at node 1 is d_{1x} and the displacement at node 2 is d_{2x}.

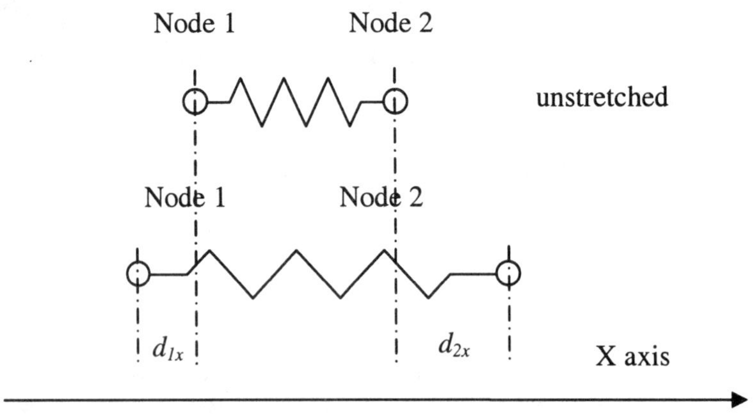

Mathematically, we are able to use the following relation to represent the displacement:

$$u(x) = \begin{cases} a_1 + a_2 x & 0 =< x <= l \\ d_{1x} & x = 0 \\ d_{2x} & x = l \end{cases}$$

where $a_1 = d_{1x}$ and $a_2 = \frac{d_{2x} - d_{1x}}{l}$.

To evaluate the displacement at a location within the element, the following relation applies:

$$u(x) = d_{1x} + \frac{d_{2x} - d_{1x}}{l} x \qquad 0 =< x <= l$$

In preparing a computer code to evaluate the displacement of an element, a matrix format is always used. To represent the displacement at a particular location within an element, the following matrix notation is introduced:

$$u(x) = \begin{bmatrix} (1 - \frac{x}{l}) & \frac{x}{l} \end{bmatrix} \begin{Bmatrix} d_{1x} \\ d_{2x} \end{Bmatrix} = \begin{bmatrix} N_1 & N_2 \end{bmatrix} \begin{Bmatrix} d_{1x} \\ d_{2x} \end{Bmatrix}$$

where $N_1 = 1 - \frac{x}{l}$ and $N_2 = \frac{x}{l}$ are called the shape function of the element to characterize the displacement at specific location within the element based on the known nodal displacements.

The shape function at node 1 is given by $N_1 = 1$ and $N_2 = 0$.
The shape function at node 2 is given by $N_1 = 0$ and $N_2 = 1$.

It is evident that N1 and N2 function as weighting factors. These weighting factors can be a zero-order polynomial function, such as the one in the current case. They can also be a first-order or second-order or higher-order polynomial functions, depending on the need in evaluation.

Now let us introduce a concept of stiffness matrix of an element. To characterize the system response of a spring component when subjected to loading, we have the following relations:

$$f_{1x} = k(d_{1x} - d_{2x})$$

$$f_{2x} = k(d_{2x} - d_{1x})$$

Pay a special attention to the sign of each entity with respect to the position direction of the X-axis defined, as shown below:

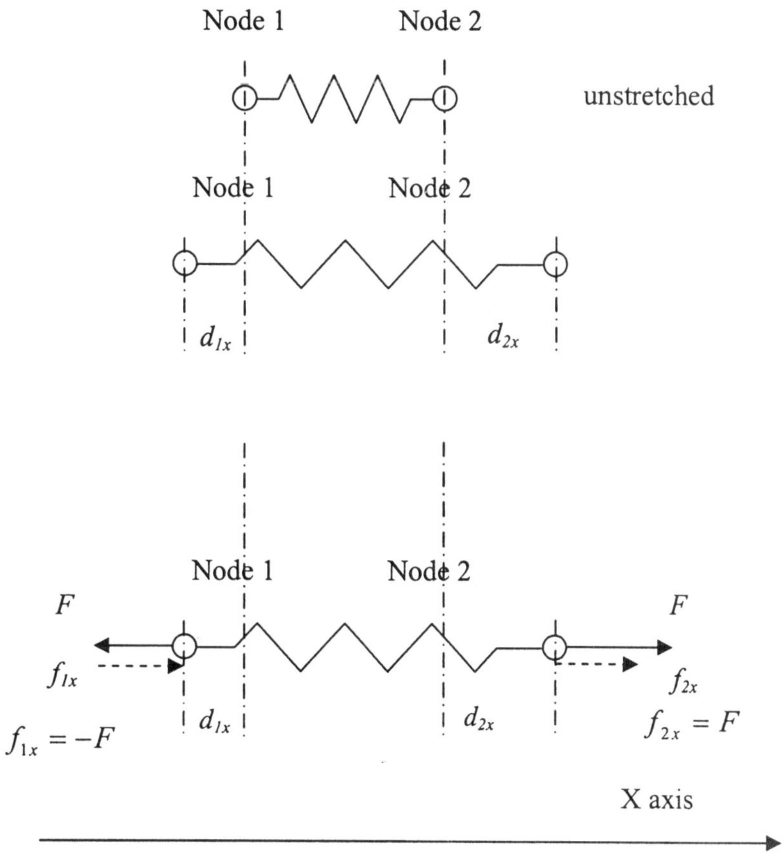

In preparing a computer code to evaluate the force acting on each node, a matrix format is always used. To represent the forces at the 2 nodes, the following matrix notation is introduced:

$$\begin{Bmatrix} f_{1x} \\ f_{2x} \end{Bmatrix} = \begin{bmatrix} k & -k \\ -k & k \end{bmatrix} \begin{Bmatrix} d_{1x} \\ d_{2x} \end{Bmatrix} = \begin{bmatrix} k_{11} & k_{12} \\ k_{21} & k_{22} \end{bmatrix} \begin{Bmatrix} d_{1x} \\ d_{2x} \end{Bmatrix} = [N_1 \quad N_2] \begin{Bmatrix} d_{1x} \\ d_{2x} \end{Bmatrix}$$

Where f_{1x} and f_{2x} are local nodal forces, and d_{1x} and d_{2x} are local nodal displacements. The 4 parameters, k_{11}, k_{12}, k_{21} and k_{22} are coefficients of the spring stiffness. These coefficients can be identified as

$$k_{11} = k$$

$$k_{12} = -k$$

$$k_{21} = -k$$

$$k_{22} = k$$

$$f_{1x} = k_{11}d_{1x} + k_{12}d_{2x}.$$

$$f_{2x} = k_{21}d_{1x} + k_{22}d_{2x}$$

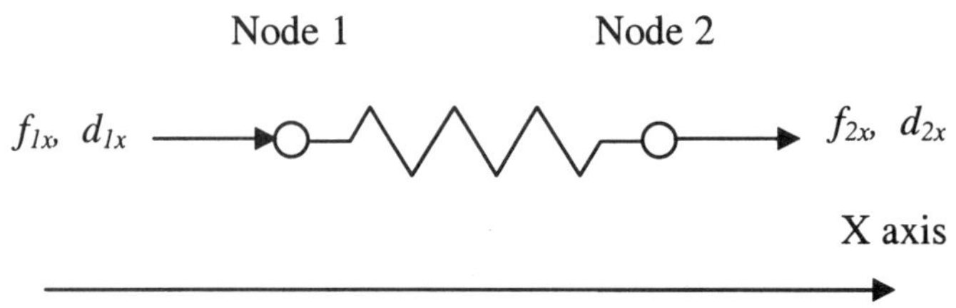

When a stiffness matrix for each element is constructed, we need to assemble all of them so that we will have a global stiffness matrix, which characterizes the entire system under study. We demonstrate the procedure to perform FEA in a one-dimensional space through a case study. The case study is a lumped mass system with 3 springs. An external force acts on the mass, as shown below. In this case, we view the lumped mass system consisting of three (3) elements. Each element is treated as a spring component. The system is in a static equilibrium status. Therefore, we model this system, using 3 elements and 4 nodes. Note that node 2 is shared by all 3 elements, as show below:

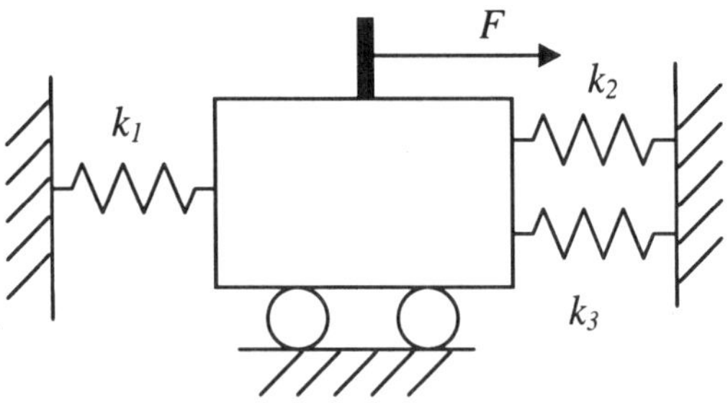

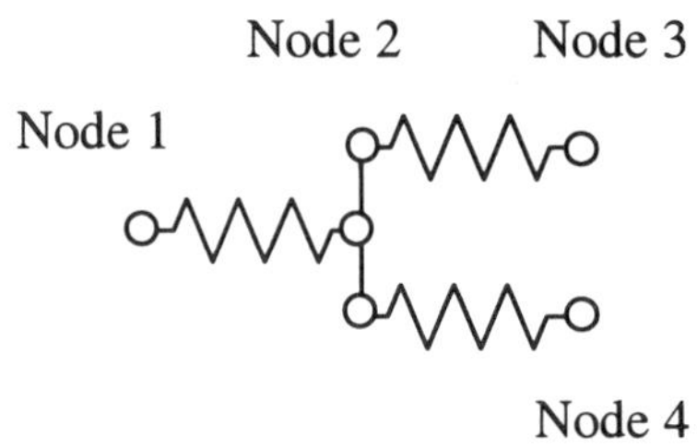

Secondly, we formulate a stiffness matrix for each of the 3 elements. The 3 stiffness matrices for the 3 elements are shown below:

$$\begin{Bmatrix} f_{2x}^{(2)} \\ f_{3x}^{(2)} \end{Bmatrix} = \begin{bmatrix} k_{22}^{(2)} & k_{23}^{(2)} \\ k_{32}^{(2)} & k_{33}^{(2)} \end{bmatrix} \begin{Bmatrix} d_{2x}^{(2)} \\ d_{3x}^{(2)} \end{Bmatrix}$$

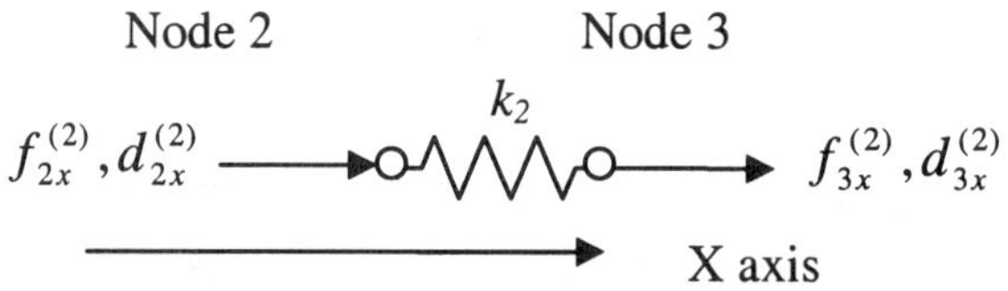

$$\begin{Bmatrix} f_{1x}^{(1)} \\ f_{2x}^{(1)} \end{Bmatrix} = \begin{bmatrix} k_{11}^{(1)} & k_{12}^{(1)} \\ k_{21}^{(1)} & k_{22}^{(1)} \end{bmatrix} \begin{Bmatrix} d_{1x}^{(1)} \\ d_{2x}^{(1)} \end{Bmatrix}$$

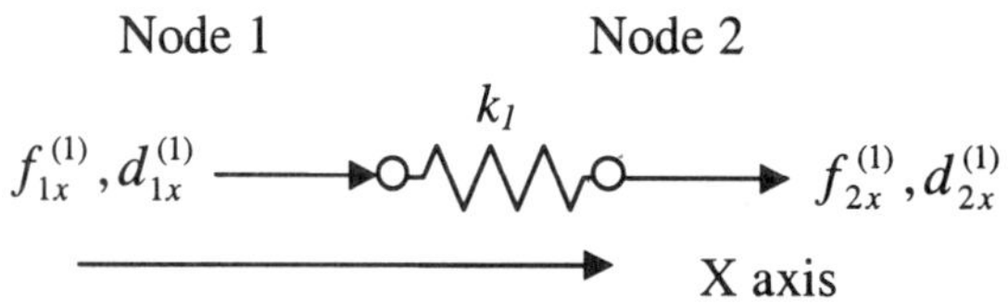

$$\begin{Bmatrix} f_{2x}^{(3)} \\ f_{4x}^{(3)} \end{Bmatrix} = \begin{bmatrix} k_{22}^{(3)} & k_{24}^{(3)} \\ k_{42}^{(3)} & k_{44}^{(3)} \end{bmatrix} \begin{Bmatrix} d_{2x}^{(3)} \\ d_{4x}^{(3)} \end{Bmatrix}$$

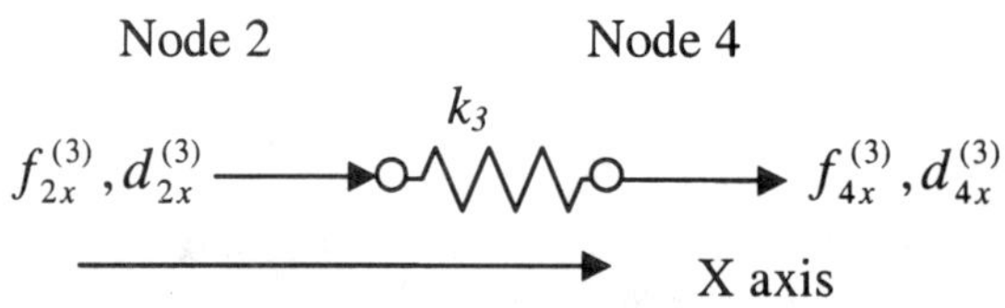

Now let us assemble these 3 stiffness matrices to obtain the global stiffness matrix.

$$\begin{Bmatrix} f_{1x}^{(1)} \\ f_{2x}^{(1)} + f_{2x}^{(2)} + f_{2x}^{(3)} \\ f_{3x}^{(2)} \\ f_{4x}^{(3)} \end{Bmatrix} = \begin{bmatrix} k_{11}^{(1)} & k_{12}^{(1)} & 0 & 0 \\ k_{21}^{(1)} & k_{22}^{(1)} + k_{22}^{(2)} + k_{22}^{(3)} & k_{23}^{(2)} & k_{24}^{(3)} \\ 0 & k_{32}^{(2)} & k_{33}^{(2)} & 0 \\ 0 & k_{42}^{(3)} & 0 & k_{44}^{(3)} \end{bmatrix} \begin{Bmatrix} d_{1x}^{(1)} \\ d_{2x} \\ d_{3x}^{(2)} \\ d_{4x}^{(3)} \end{Bmatrix}$$

where $d_{2x} = d_{2x}^{(1)} = d_{2x}^{(2)} = d_{2x}^{(3)}$.

It is important that we need to have a specific set of boundary conditions for the Assembled System. Those boundary conditions are also called constraint conditions, widely adopted in the FEA software systems.

Referring this specific case study, we have 2 constraint conditions. The first constraint condition has already been used in the process of assembling the 3 stiffness matrices of the 3 elements. The first constraint condition is "node 2 is shared by all 3 elements". The second constraint condition is $d_{1x}^{(1)} = d_{3x}^{(2)} = d_{4x}^{(3)} = 0$, or

$$\begin{Bmatrix} d_{1x}^{(1)} \\ d_{2x} \\ d_{3x}^{(2)} \\ d_{4x}^{(3)} \end{Bmatrix} = \begin{Bmatrix} 0 \\ d_{2x} \\ 0 \\ 0 \end{Bmatrix}$$

To determine the displacement of each node under a static equilibrium condition, we need a specific load condition, which is acting on the assembled system.

In general, we call $f_{1x}^{(1)}$ the reaction force at the joint between the spring of k_1 and the wall structure, $f_{3x}^{(2)}$ the reaction force at the joint between the spring of k_2 and the wall structure, and $f_{4x}^{(3)}$ the reaction force at the joint between the spring of k_1 and the wall structure. Note that we use UPPER CASE to represent the load condition or the reaction force because they characterize the load condition acting on the assembly system, not on a node of a single element. Certainly, the load acting on the system could be the nodal force under some circumstances.

$$\begin{Bmatrix} F_{1x} \\ F_{2x} \\ F_{3x} \\ F_{4x} \end{Bmatrix} = \begin{Bmatrix} f_{1x}^{(1)} \\ f_{2x}^{(1)} + f_{2x}^{(2)} + f_{2x}^{(3)} \\ f_{3x}^{(2)} \\ f_{4x}^{(3)} \end{Bmatrix}$$

After applying the boundary conditions and the load conditions, the mathematical form of FEA for this system is represented by:

$$\begin{Bmatrix} F_{1x} \\ F_{2x} \\ F_{3x} \\ F_{4x} \end{Bmatrix} = \begin{bmatrix} k_{11}^{(1)} & k_{12}^{(1)} & 0 & 0 \\ k_{21}^{(1)} & k_{22}^{(1)} + k_{22}^{(2)} + k_{22}^{(3)} & k_{23}^{(2)} & k_{24}^{(3)} \\ 0 & k_{32}^{(2)} & k_{33}^{(2)} & 0 \\ 0 & k_{42}^{(3)} & 0 & k_{44}^{(3)} \end{bmatrix} \begin{Bmatrix} 0 \\ d_{2x} \\ 0 \\ 0 \end{Bmatrix}$$

$$\begin{Bmatrix} F_{1x} \\ F_{2x} \\ F_{3x} \\ F_{4x} \end{Bmatrix} = \begin{bmatrix} k_1 & -k_1 & 0 & 0 \\ -k_1 & k_1 + k_2 + k_3 & -k_2 & -k_3 \\ 0 & -k_2 & k_2 & 0 \\ 0 & -k_3 & 0 & k_3 \end{bmatrix} \begin{Bmatrix} 0 \\ d_{2x} \\ 0 \\ 0 \end{Bmatrix}$$

Now let us the FEA formula to obtain solutions to 2 questions, namely, Example 1-1 and Example 1-2.

Example 1-1: Assume that k_1 = 10 Newton/mm, k_2 = 8 Newton/mm and k_3 = 6 Newton/mm. If we need to place the body 0.5 mm away from its original position, what is the load condition to keep the body a distance of 0.5 mm stationary? What are the reaction forces at the 3 joints?

$$\begin{Bmatrix} F_{1x} \\ F_{2x} \\ F_{3x} \\ F_{4x} \end{Bmatrix} = \begin{bmatrix} k_1 & -k_1 & 0 & 0 \\ -k_1 & k_1 + k_2 + k_3 & -k_2 & -k_3 \\ 0 & -k_2 & k_2 & 0 \\ 0 & -k_3 & 0 & k_3 \end{bmatrix} \begin{Bmatrix} 0 \\ d_{2x} \\ 0 \\ 0 \end{Bmatrix}$$

$$\begin{Bmatrix} F_{1x} \\ F_{2x} \\ F_{3x} \\ F_{4x} \end{Bmatrix} = \begin{bmatrix} 10 & -10 & 0 & 0 \\ -10 & 24 & -8 & -6 \\ 0 & -8 & 8 & 0 \\ 0 & -6 & 0 & 6 \end{bmatrix} \begin{Bmatrix} 0 \\ 0.5 \\ 0 \\ 0 \end{Bmatrix} = \begin{Bmatrix} -5 \\ 12 \\ -4 \\ -3 \end{Bmatrix}$$

Example 1-2: Assume that k_1 = 10 Newton/mm, k_2 = 8 Newton/mm and k_3 = 6 Newton/mm. There is a lumped force acting on the body. Its magnitude is 48 Newton. What is the displacement of the body with respect to its original equilibrium condition? What are the reaction forces at the 3 joints?

$$\begin{Bmatrix} F_{1x} \\ F_{2x} \\ F_{3x} \\ F_{4x} \end{Bmatrix} = \begin{bmatrix} k_1 & -k_1 & 0 & 0 \\ -k_1 & k_1 + k_2 + k_3 & -k_2 & -k_3 \\ 0 & -k_2 & k_2 & 0 \\ 0 & -k_3 & 0 & k_3 \end{bmatrix} \begin{Bmatrix} 0 \\ d_{2x} \\ 0 \\ 0 \end{Bmatrix}$$

$$\begin{Bmatrix} F_{1x} \\ 48 \\ F_{3x} \\ F_{4x} \end{Bmatrix} = \begin{bmatrix} 10 & -10 & 0 & 0 \\ -10 & 24 & -8 & -6 \\ 0 & -8 & 8 & 0 \\ 0 & -6 & 0 & 6 \end{bmatrix} \begin{Bmatrix} 0 \\ d_{2x} \\ 0 \\ 0 \end{Bmatrix}$$

We have 48 = 24*d_{2x}, and d_{2x} = 2 mm. Using the matrix multiplication, we have the 3 reaction forces listed below:

$$\begin{Bmatrix} F_{1x} \\ 48 \\ F_{3x} \\ F_{4x} \end{Bmatrix} = \begin{bmatrix} 10 & -10 & 0 & 0 \\ -10 & 24 & -8 & -6 \\ 0 & -8 & 8 & 0 \\ 0 & -6 & 0 & 6 \end{bmatrix} \begin{Bmatrix} 0 \\ 2 \\ 0 \\ 0 \end{Bmatrix} = \begin{Bmatrix} -20 \\ 48 \\ 16 \\ 12 \end{Bmatrix}$$

1.3 Finite Element Method in a Two-Dimensional Space

In the previous section, we dealt with structures which can be represented in a one-dimensional space. In this section, we will deal with structures, which have to be represented in a two-dimensional space. For example, a truss structure in real life is generally described in a two-dimensional space. In the following, we us a plane pin-jointed truss structure to demonstrate the procedure to obtain the stiffness matrix of an element in the 2-dimensional space. We assume that the loads act only at the joints, the axial force in each member is constant throughout the length of the member.

In FEA, we view each rod within a truss structure as an element. As illustrated, a rod is in an inclined position. The inclined angle with respect to the horizontal axis is 30 degrees. Under a load condition, there is a displacement along the x-axis, and there is no displacement along the y-axis because the force acting on the rod is along the x-axis. Assume that $d_{1x} = 1$.

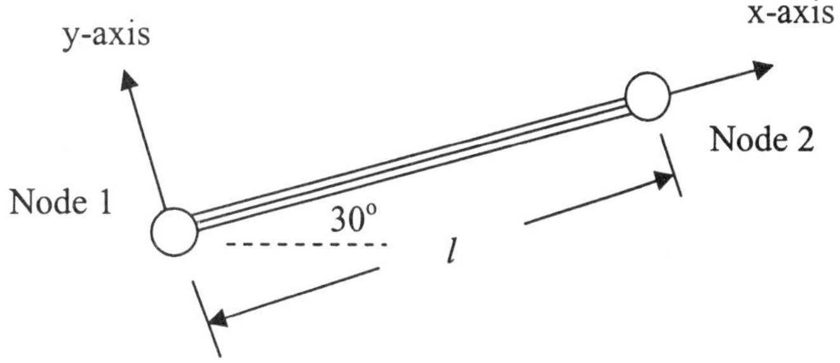

In general, a truss structure has more than 1 rod, and each rod has its own orientation with respect to the horizontal direction and/or the vertical direction. In order to effectively represent the system performance indices, such as displacement or load, we generally use a global coordinate system to define the boundary condition(s) and the load condition(s). This is extremely important when the assemblage of stiffness matrices is performed.

A displacement associated with an individual rod is also reflected into the displacement observed in the global coordinate system. The relationship between the displacements evaluated based on a local coordinate system and the displacements observed from the global coordinate system is established through the following transformation.

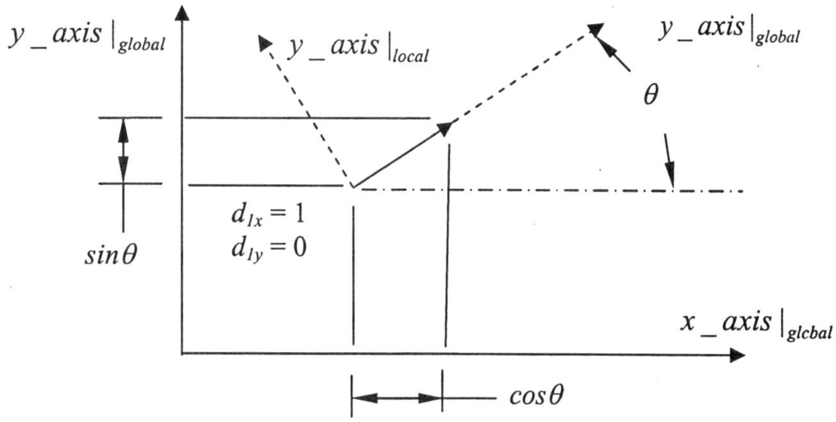

$$d_{1x\_global} = d_{1x\_local} \cos\theta \vec{i} - d_{1y\_local} \sin\theta \vec{j}$$

$$d_{1y\_global} = d_{1x\_local} \sin\theta \vec{i} + d_{1y\_local} \cos\theta \vec{j}$$

$$\begin{Bmatrix} d_{1x} \\ d_{1y} \end{Bmatrix}_{global} = \begin{bmatrix} \cos\theta & -\sin\theta \\ \sin\theta & \cos\theta \end{bmatrix} \begin{Bmatrix} d_{1x} \\ d_{1y} \end{Bmatrix}_{local} = \begin{bmatrix} C & -S \\ S & C \end{bmatrix} \begin{Bmatrix} d_{1x} \\ d_{1y} \end{Bmatrix}_{local}$$

$$\begin{Bmatrix} d_{1x} \\ d_{1y} \end{Bmatrix}_{local} = \begin{bmatrix} \cos\theta & \sin\theta \\ -\sin\theta & \cos\theta \end{bmatrix} \begin{Bmatrix} d_{1x} \\ d_{1y} \end{Bmatrix}_{global} = \begin{bmatrix} C & S \\ -S & C \end{bmatrix} \begin{Bmatrix} d_{1x} \\ d_{1y} \end{Bmatrix}_{global}$$

When we consider both nodes, or both Node 1 and Node 2, we have the following relations:

$$\begin{Bmatrix} d_{1x} \\ d_{1y} \\ d_{2x} \\ d_{2y} \end{Bmatrix}_{global} = \begin{bmatrix} \cos\theta & -\sin\theta & 0 & 0 \\ \sin\theta & \cos\theta & 0 & 0 \\ 0 & 0 & \cos\theta & -\sin\theta \\ 0 & 0 & \sin\theta & \cos\theta \end{bmatrix} \begin{Bmatrix} d_{1x} \\ d_{1y} \\ d_{2x} \\ d_{2y} \end{Bmatrix}_{local}$$

$$\begin{Bmatrix} d_{1x} \\ d_{1y} \\ d_{2x} \\ d_{2y} \end{Bmatrix}_{local} = \begin{bmatrix} \cos\theta & \sin\theta & 0 & 0 \\ -\sin\theta & \cos\theta & 0 & 0 \\ 0 & 0 & \cos\theta & \sin\theta \\ 0 & 0 & -\sin\theta & \cos\theta \end{bmatrix} \begin{Bmatrix} d_{1x} \\ d_{1y} \\ d_{2x} \\ d_{2y} \end{Bmatrix}_{global}$$

When considering the load conditions, it is natural to have the following relationship to interpret the mapping of a vector representing a force component from the global coordinate system to a local coordinate system and vice versa.

$$\begin{Bmatrix} F_{1x} \\ F_{1y} \\ F_{2x} \\ F_{2y} \end{Bmatrix}_{global} = \begin{bmatrix} \cos\theta & -\sin\theta & 0 & 0 \\ \sin\theta & \cos\theta & 0 & 0 \\ 0 & 0 & \cos\theta & -\sin\theta \\ 0 & 0 & \sin\theta & \cos\theta \end{bmatrix} \begin{Bmatrix} f_{1x} \\ f_{1y} \\ f_{2x} \\ f_{2y} \end{Bmatrix}_{local}$$

$$\begin{Bmatrix} f_{1x} \\ f_{1y} \\ f_{2x} \\ f_{2y} \end{Bmatrix}_{local} = \begin{bmatrix} \cos\theta & \sin\theta & 0 & 0 \\ -\sin\theta & \cos\theta & 0 & 0 \\ 0 & 0 & \cos\theta & \sin\theta \\ 0 & 0 & -\sin\theta & \cos\theta \end{bmatrix} \begin{Bmatrix} F_{1x} \\ F_{1y} \\ F_{2x} \\ F_{2y} \end{Bmatrix}_{global}$$

Now let us consider the stiffness matrix of a rod in the 2-dimensional space. We start from the stiffness matrix defined locally.

$$\begin{Bmatrix} f_{1x} \\ f_{1y} \\ f_{2x} \\ f_{2y} \end{Bmatrix}_{local} = \begin{bmatrix} k_1 & 0 & -k_1 & 0 \\ 0 & 0 & 0 & 0 \\ -k_1 & 0 & k_1 & 0 \\ 0 & 0 & 0 & 0 \end{bmatrix} \begin{Bmatrix} d_{1x} \\ d_{1y} \\ d_{2x} \\ d_{2y} \end{Bmatrix}_{local}$$

We substitute the local displacement vector by the global displacement vector. We have

$$\begin{Bmatrix} f_{1x} \\ f_{1y} \\ f_{2x} \\ f_{2y} \end{Bmatrix}_{local} = \begin{bmatrix} k_1 & 0 & -k_1 & 0 \\ 0 & 0 & 0 & 0 \\ -k_1 & 0 & k_1 & 0 \\ 0 & 0 & 0 & 0 \end{bmatrix} \begin{Bmatrix} d_{1x} \\ d_{1y} \\ d_{2x} \\ d_{2y} \end{Bmatrix}_{local} = \begin{bmatrix} k_1 & 0 & -k_1 & 0 \\ 0 & 0 & 0 & 0 \\ -k_1 & 0 & k_1 & 0 \\ 0 & 0 & 0 & 0 \end{bmatrix} \begin{bmatrix} C & S & 0 & 0 \\ -S & C & 0 & 0 \\ 0 & 0 & C & S \\ 0 & 0 & -S & C \end{bmatrix} \begin{Bmatrix} d_{1x} \\ d_{1y} \\ d_{2x} \\ d_{2y} \end{Bmatrix}_{global}$$

$$\begin{Bmatrix} f_{1x} \\ f_{1y} \\ f_{2x} \\ f_{2y} \end{Bmatrix}_{local} = \begin{bmatrix} k_1 & 0 & -k_1 & 0 \\ 0 & 0 & 0 & 0 \\ -k_1 & 0 & k_1 & 0 \\ 0 & 0 & 0 & 0 \end{bmatrix} [T] \begin{Bmatrix} d_{1x} \\ d_{1y} \\ d_{2x} \\ d_{2y} \end{Bmatrix}_{global}$$

$$[T] = \begin{bmatrix} C & S & 0 & 0 \\ -S & C & 0 & 0 \\ 0 & 0 & C & S \\ 0 & 0 & -S & C \end{bmatrix}$$

We substitute the local load vector by the global load vector. We have

$$\begin{bmatrix} C & S & 0 & 0 \\ -S & C & 0 & 0 \\ 0 & 0 & C & S \\ 0 & 0 & -S & C \end{bmatrix} \begin{Bmatrix} F_{1x} \\ F_{1y} \\ F_{2x} \\ F_{2y} \end{Bmatrix}_{local} = \begin{bmatrix} k_1 & 0 & -k_1 & 0 \\ 0 & 0 & 0 & 0 \\ -k_1 & 0 & k_1 & 0 \\ 0 & 0 & 0 & 0 \end{bmatrix} [T] \begin{Bmatrix} d_{1x} \\ d_{1y} \\ d_{2x} \\ d_{2y} \end{Bmatrix}_{global}$$

$$\begin{Bmatrix} F_{1x} \\ F_{1y} \\ F_{2x} \\ F_{2y} \end{Bmatrix}_{local} = [T]^T \begin{bmatrix} k_1 & 0 & -k_1 & 0 \\ 0 & 0 & 0 & 0 \\ -k_1 & 0 & k_1 & 0 \\ 0 & 0 & 0 & 0 \end{bmatrix} [T] \begin{Bmatrix} d_{1x} \\ d_{1y} \\ d_{2x} \\ d_{2y} \end{Bmatrix}_{global}$$

$$[T]^T = \begin{bmatrix} C & -S & 0 & 0 \\ S & C & 0 & 0 \\ 0 & 0 & C & -S \\ 0 & 0 & S & C \end{bmatrix}$$

$$\begin{Bmatrix} F_{1x} \\ F_{1y} \\ F_{2x} \\ F_{2y} \end{Bmatrix}_{local} = [K]\begin{Bmatrix} d_{1x} \\ d_{1y} \\ d_{2x} \\ d_{2y} \end{Bmatrix}_{global}$$

$$[K] = [T]^T \begin{bmatrix} k_1 & 0 & -k_1 & 0 \\ 0 & 0 & 0 & 0 \\ -k_1 & 0 & k_1 & 0 \\ 0 & 0 & 0 & 0 \end{bmatrix} [T]$$

Now let us the FEM formula to obtain the required solutions to the following truss structure. As illustrated, the truss structure has 3 bars. Their lengths are 300 mm, 400 mm and 500 mm, as shown. If we assign three nodes to the three joints, we will have 3 elements. Among the 3 joints, 2 joints are pin connections. The other is a roller support. Each element represents each bar. The load condition is a horizontal force acts on node 2.

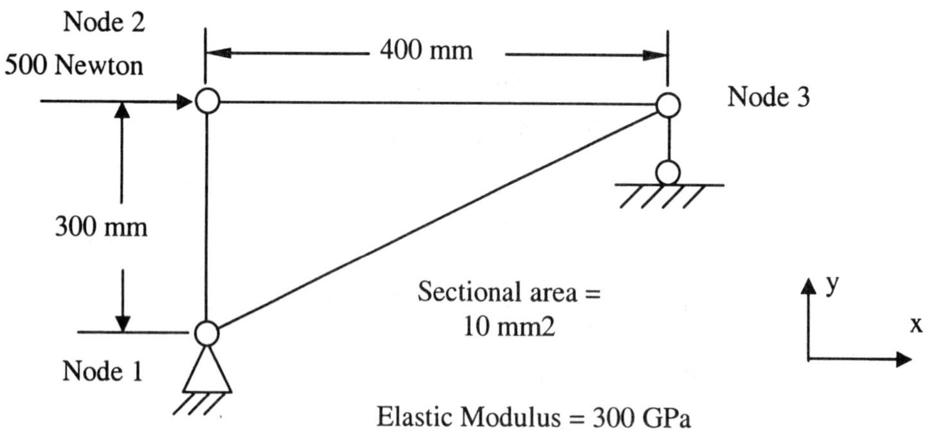

Our objectives of this study are

(1) Obtain the stiffness matrix for each of the 3 elements.
(2) Obtain the global stiffness matrix of the truss structure.
(3) Determine the displacements of Node 2 and Node 3.
(4) Determine the reaction forces on Node 1 and Node 3.

First let us examine the characteristics of this truss structure. There is no normal displacement of node 1. There is no normal displacement in node 3. As a result, there is no normal displacement in node 2. Therefore, under the global coordinate system, we have

$$\begin{Bmatrix} d_{1x} \\ d_{1y} \\ d_{2x} \\ d_{2y} \\ d_{3x} \\ d_{3y} \end{Bmatrix}_{global} = \begin{Bmatrix} d_{1x} \\ 0 \\ d_{2x} \\ 0 \\ d_{3x} \\ 0 \end{Bmatrix}_{global}$$

It is important to emphasize that in the local coordinate system the x direction is along the length of the element and the y direction is normal to the x direction.

Let us consider the first element, which will be the vertical bar. In this orientation θ=90°. The local stiffness matrix has the following form:

$$\begin{bmatrix} f_{1x} \\ f_{1y} \\ f_{2x} \\ f_{2y} \end{bmatrix}_{local} = \begin{bmatrix} k_1 & 0 & -k_1 & 0 \\ 0 & 0 & 0 & 0 \\ -k_1 & 0 & k_1 & 0 \\ 0 & 0 & 0 & 0 \end{bmatrix} \begin{bmatrix} d_{1x} \\ d_{1y} \\ d_{2x} \\ d_{2y} \end{bmatrix}_{local}$$

In order to get the local coordinate system into the global coordinate system the force and displacement matrices must be multiplied by a matrix T.

$$[T] = \begin{bmatrix} \cos\theta & \sin\theta & 0 & 0 \\ -\sin\theta & \cos\theta & 0 & 0 \\ 0 & 0 & \cos\theta & \sin\theta \\ 0 & 0 & -\sin\theta & \cos\theta \end{bmatrix}$$

Because the force matrix is on the other side of the equation, the inverse of the T matrix is also required.

$$[T]^T = \begin{bmatrix} \cos\theta & -\sin\theta & 0 & 0 \\ \sin\theta & \cos\theta & 0 & 0 \\ 0 & 0 & \cos\theta & -\sin\theta \\ 0 & 0 & \sin\theta & \cos\theta \end{bmatrix}$$

When the T matrix is used to convert the local coordinate system into the global coordinate system the stiffness matrix takes the form

$$\begin{bmatrix} F_{1x} \\ F_{1y} \\ F_{2x} \\ F_{2y} \end{bmatrix}_{Global} = [T]^T \begin{bmatrix} k_1 & 0 & -k_1 & 0 \\ 0 & 0 & 0 & 0 \\ -k_1 & 0 & k_1 & 0 \\ 0 & 0 & 0 & 0 \end{bmatrix} [T] \begin{bmatrix} d_{1x} \\ d_{1y} \\ d_{2x} \\ d_{2y} \end{bmatrix}_{global}$$

K_1 can be determined from the equation

$$K_1 = \frac{AE}{l_1} = \frac{(1*10^{-5})(300*10^9)}{.3} = 1*10^7\, N/m$$

The values can be inserted into the stiffness matrix

$$\begin{bmatrix} F_{1x} \\ F_{1y} \\ F_{2x} \\ F_{2y} \end{bmatrix}_{Global} = \begin{bmatrix} 0 & -1 & 0 & 0 \\ 1 & 0 & 0 & 0 \\ 0 & 0 & 0 & -1 \\ 0 & 0 & 1 & 0 \end{bmatrix} \begin{bmatrix} k_1 & 0 & -k_1 & 0 \\ 0 & 0 & 0 & 0 \\ -k_1 & 0 & k_1 & 0 \\ 0 & 0 & 0 & 0 \end{bmatrix} \begin{bmatrix} 0 & 1 & 0 & 0 \\ -1 & 0 & 0 & 0 \\ 0 & 0 & 0 & 1 \\ 0 & 0 & -1 & 0 \end{bmatrix} \begin{bmatrix} d_{1x} \\ d_{1y} \\ d_{2x} \\ d_{2y} \end{bmatrix}_{global}$$

The matrix multiplication can be evaluated through the use of a software system, such as Matlab.

$$\begin{bmatrix} F_{1x} \\ F_{1y} \\ F_{2x} \\ F_{2y} \end{bmatrix}_{Global} = \begin{bmatrix} 0 & 0 & 0 & 0 \\ 0 & 1e7 & 0 & -1e7 \\ 0 & 0 & 0 & 0 \\ 0 & -1e7 & 0 & 1e7 \end{bmatrix} \begin{bmatrix} d_{1x} \\ d_{1y} \\ d_{2x} \\ d_{2y} \end{bmatrix}_{global}$$

Now let us work with the next element, which is the horizontal bar. Let us name element 2. The orientation is θ=0°. The local stiffness matrix takes the form

$$\begin{bmatrix} f_{2x} \\ f_{2y} \\ f_{3x} \\ f_{3y} \end{bmatrix}_{local} = \begin{bmatrix} k_2 & 0 & -k_2 & 0 \\ 0 & 0 & 0 & 0 \\ -k_2 & 0 & k_2 & 0 \\ 0 & 0 & 0 & 0 \end{bmatrix} \begin{bmatrix} d_{2x} \\ d_{2y} \\ d_{3x} \\ d_{3y} \end{bmatrix}_{local}$$

The matrix is the same form as the matrix for element one because it is again assumed that there is no motion normal to the local x coordinate. The T matrix is needed to transform the local coordinate system into the global coordinate system. The T matrix was already presented so will not be defined again. The result of the local to global transformation is

$$\begin{bmatrix} f_{2x} \\ f_{2y} \\ f_{3x} \\ f_{3y} \end{bmatrix}_{global} = \left[T^T\right] \begin{bmatrix} k_2 & 0 & -k_2 & 0 \\ 0 & 0 & 0 & 0 \\ -k_2 & 0 & k_2 & 0 \\ 0 & 0 & 0 & 0 \end{bmatrix} [T] \begin{bmatrix} d_{2x} \\ d_{2y} \\ d_{3x} \\ d_{3y} \end{bmatrix}_{global}$$

The value of k_2 can be determined the same way as K_1

$$[k_2] = \frac{AE}{l_2} = \frac{(1*10^{-5})(300*10^9)}{.4} = 7.5*10^6\, N/m$$

The values can now be used in the matrix

$$\begin{bmatrix} f_{2x} \\ f_{2y} \\ f_{3x} \\ f_{3y} \end{bmatrix}_{global} = \begin{bmatrix} 1 & 0 & 0 & 0 \\ 0 & 1 & 0 & 0 \\ 0 & 0 & 1 & 0 \\ 0 & 0 & 0 & 1 \end{bmatrix} \begin{bmatrix} 7.5e6 & 0 & -7.5e6 & 0 \\ 0 & 0 & 0 & 0 \\ -7.5e6 & 0 & 7.5e6 & 0 \\ 0 & 0 & 0 & 0 \end{bmatrix} \begin{bmatrix} 1 & 0 & 0 & 0 \\ 0 & 1 & 0 & 0 \\ 0 & 0 & 1 & 0 \\ 0 & 0 & 0 & 1 \end{bmatrix} \begin{bmatrix} d_{2x} \\ d_{2y} \\ d_{3x} \\ d_{3y} \end{bmatrix}_{global}$$

After the matrix multiplication, we have

$$\begin{bmatrix} f_{2x} \\ f_{2y} \\ f_{3x} \\ f_{3y} \end{bmatrix}_{global} = \begin{bmatrix} 7.5e6 & 0 & -7.5e6 & 0 \\ 0 & 0 & 0 & 0 \\ -7.5e6 & 0 & 7.5e6 & 0 \\ 0 & 0 & 0 & 0 \end{bmatrix} \begin{bmatrix} d_{2x} \\ d_{2y} \\ d_{3x} \\ d_{3y} \end{bmatrix}_{global}$$

The last element is the diagonal element. This is element 3. The orientation angle can be determined by the following calculation.

$$\theta = \sin^{-1} \frac{300}{500} = 36.87°$$

The local coordinate system for element 3 is the x direction along the beam and the y direction normal to the x direction. The local stiffness matrix is

$$\begin{bmatrix} f_{3x} \\ f_{3y} \\ f_{1x} \\ f_{1y} \end{bmatrix}_{local} = \begin{bmatrix} k_3 & 0 & -k_3 & 0 \\ 0 & 0 & 0 & 0 \\ -k_3 & 0 & k_3 & 0 \\ 0 & 0 & 0 & 0 \end{bmatrix} \begin{bmatrix} d_{3x} \\ d_{3y} \\ d_{1x} \\ d_{1y} \end{bmatrix}_{local}$$

By transforming the local matrix into the global matrix

$$\begin{bmatrix} f_{3x} \\ f_{3y} \\ f_{1x} \\ f_{1y} \end{bmatrix}_{global} = [T]^T \begin{bmatrix} k_3 & 0 & -k_3 & 0 \\ 0 & 0 & 0 & 0 \\ -k_3 & 0 & k_3 & 0 \\ 0 & 0 & 0 & 0 \end{bmatrix} [T] \begin{bmatrix} d_{3x} \\ d_{3y} \\ d_{1x} \\ d_{1y} \end{bmatrix}_{global}$$

K_3 can be determined from the equation

$$k_3 = \frac{AE}{l_3} = \frac{(1*10^{-5})(300*10^9)}{.5} = 6*10^6\, N/m$$

The values can be used in the global coordinate stiffness matrix to get

$$\begin{bmatrix} F_{3x} \\ F_{3y} \\ F_{1x} \\ F_{1y} \end{bmatrix}_{global} = \begin{bmatrix} .8 & .6 & 0 & 0 \\ -.6 & .8 & 0 & 0 \\ 0 & 0 & .8 & .6 \\ 0 & 0 & -.6 & .8 \end{bmatrix} \begin{bmatrix} 6e6 & 0 & -6e6 & 0 \\ 0 & 0 & 0 & 0 \\ -6e6 & 0 & 6e6 & 0 \\ 0 & 0 & 0 & 0 \end{bmatrix} \begin{bmatrix} .8 & -.6 & 0 & 0 \\ -.6 & .8 & 0 & 0 \\ 0 & 0 & .8 & -.6 \\ 0 & 0 & .6 & .8 \end{bmatrix} \begin{bmatrix} d_{3x} \\ d_{3y} \\ d_{1x} \\ d_{1y} \end{bmatrix}_{global}$$

This matrix can be evaluated through matrix multiplication.

$$\begin{bmatrix} F_{3x} \\ F_{3y} \\ F_{1x} \\ F_{1y} \end{bmatrix}_{global} = \begin{bmatrix} 3.84e6 & -.288e6 & -3.84e6 & 2.88e6 \\ -2.88e6 & 2.16e6 & 2.88e6 & -2.16e6 \\ -3.84e6 & 2.88e6 & 3.84e6 & -2.88e6 \\ 2.88e6 & -2.16e6 & -2.88e6 & 2.16e6 \end{bmatrix} \begin{bmatrix} d_{3x} \\ d_{3y} \\ d_{1x} \\ d_{1y} \end{bmatrix}_{global}$$

The global stiffness matrix can be assembled by summing the parts of the individual stiffness matrices. The global stiffness matrix is

$$\begin{bmatrix} F_{1x} \\ F_{1y} \\ 500 \\ F_{2y} \\ F_{3x} \\ F_{3y} \end{bmatrix} = \begin{bmatrix} 3.84 & -2.88 & 0 & 0 & 3.84 & -2.88 \\ -2.88 & 12.16 & 0 & -10 & 2.88 & -2.16 \\ 0 & 0 & 7.5 & 0 & -7.5 & 0 \\ 0 & -10 & 0 & 10 & 0 & 0 \\ -3.84 & 2.88 & -7.5 & 0 & 11.34 & -2.88 \\ 2.88 & -2.16 & 0 & 0 & -2.88 & 2.16 \end{bmatrix} *10^6 \begin{bmatrix} 0 \\ 0 \\ d_{2x} \\ 0 \\ d_{3x} \\ 0 \end{bmatrix}$$

Let us apply the load condition, which is given by

$$\begin{bmatrix} F_{1x} \\ F_{1y} \\ 500 \\ F_{2y} \\ F_{3x} \\ F_{3y} \end{bmatrix}$$

Let us apply the boundary condition or the constraint condition, which is given by

$$\begin{bmatrix} 0 \\ 0 \\ d_{2x} \\ 0 \\ d_{3x} \\ 0 \end{bmatrix}$$

The results obtained from the calculations are listed below:

$$\begin{bmatrix} F_{1x} \\ F_{1y} \\ 500 \\ F_{2y} \\ F_{3x} \\ F_{3y} \end{bmatrix} = \begin{bmatrix} -500 \\ -375 \\ 500 \\ 0 \\ 0 \\ 375 \end{bmatrix} (Newton)$$

$$\begin{bmatrix} d_{1x} \\ d_{1y} \\ d_{2x} \\ d_{2y} \\ d_{3x} \\ d_{3y} \end{bmatrix} = \begin{bmatrix} 0 \\ 0 \\ 0.137 \\ 0 \\ 0.130 \\ 0 \end{bmatrix} (mm)$$

1.4 Finite Element Method in a Three-Dimensional Space

In this section, we deal with a plate structure. The plate structure is shown below. Its width or height varies along the x axis. The height at the left is 200 mm, and the height at the right is 60 mm.

The thickness of the plate is uniform and the thickness dimension is 20 mm. We assume that the material of the plate is aluminum. The modulus of elasticity is 70 GPa. The Poisson's ratio is 0.33.

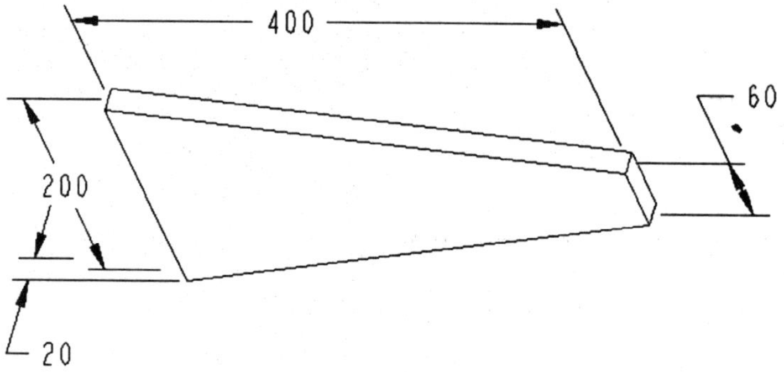

Assume that the left side of the plate is fixed, and the end on the right side is free to move. A tensile force is acting on the right side. The magnitude of the force is 1000 Newton.

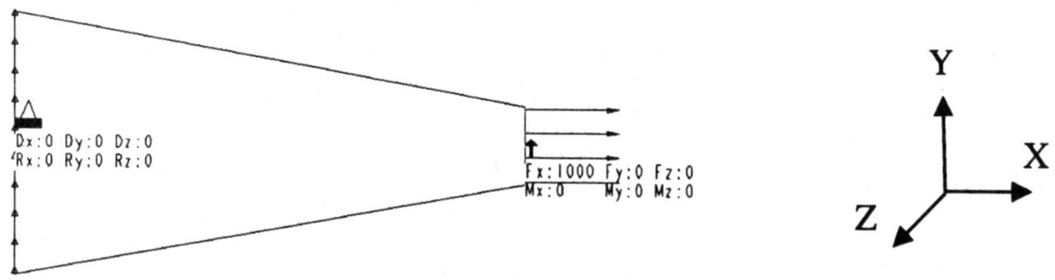

Step 1: divide this plate into four elements with an equal length of 100 mm, as shown below:

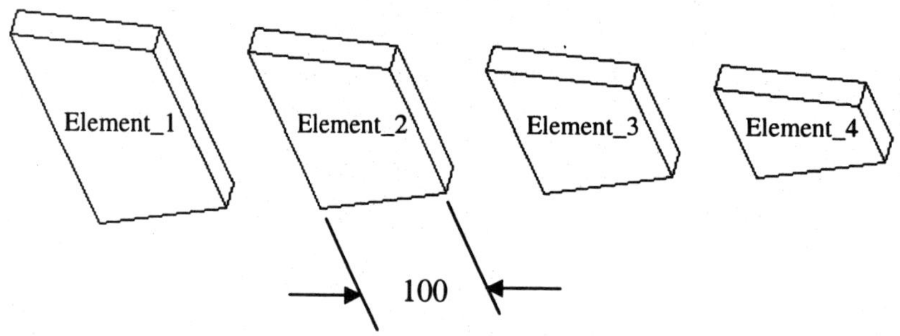

We assume that, under the loading, the stress and/or deformation are uniformly distributed on the cross sections normal to the horizontal axis. We define five nodal points, as shown below:

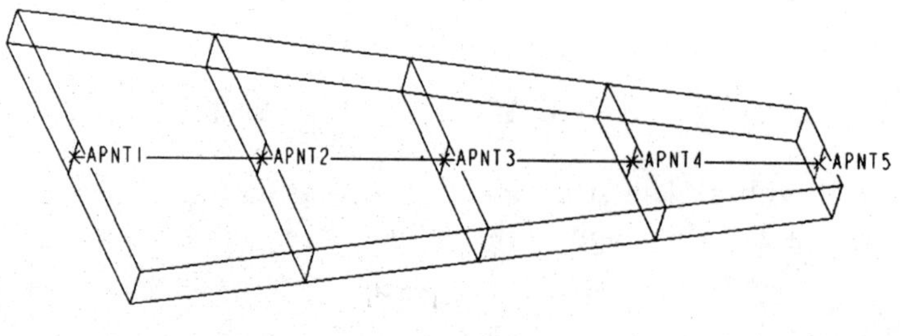

Step 2: We use a shape function to represent the physical behavior of an element so that we are able to interpolate entities of interest quantitatively. The interpolation can be linear, or quadratic or a specific form as needed. The following case illustrates a linear interpolation

Middle Section

$\sigma_{left} = 0.38(MPa)$

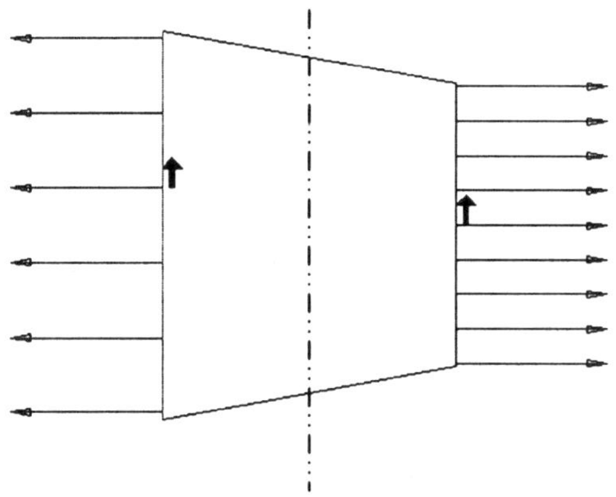

$\sigma_{right} = 0.53(MPa)$

$$\sigma_{middle} = \frac{\sigma_{left} + \sigma_{right}}{2}$$

Step 3: Develop an equation for an element so that an interpolation defined by the shape function can be carried out.

In the solid mechanics, we have the relations among stress, strain and elongation, as follows:

$$\sigma = \frac{F}{A} \qquad \sigma = E\varepsilon \qquad \varepsilon = \frac{\Delta l}{l_0}$$

When combining these relations, we may come up with

$$F = \left(\frac{AE}{l}\right)\Delta l$$

We may call the term *AE/l* as an equivalent stiffness of the element if we view the element is a spring component.

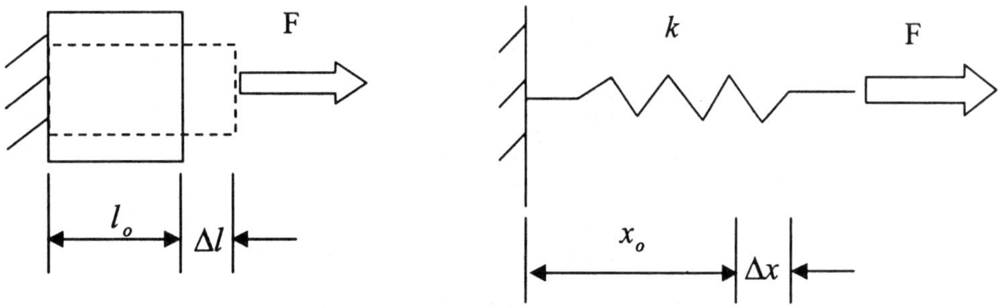

$$k_{equivalent} = \frac{A_{average}E}{l_o} = \frac{(A_{i+1} + A_i)E}{2l_o}$$

$$\Delta x_{equivalent} = u_{i+1} - u_i$$

Step 4: Derive the constituent equation for each element, using a free-body diagram.

Element 1

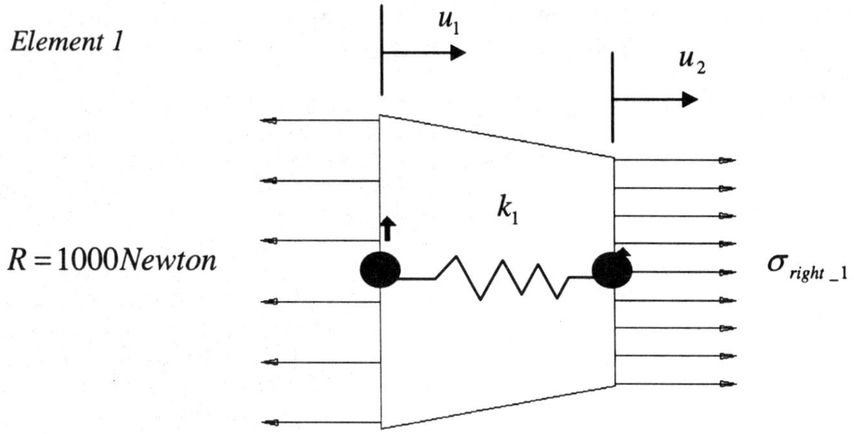

$$node1: R - k_1(u_2 - u_1) = 0$$

Element 2

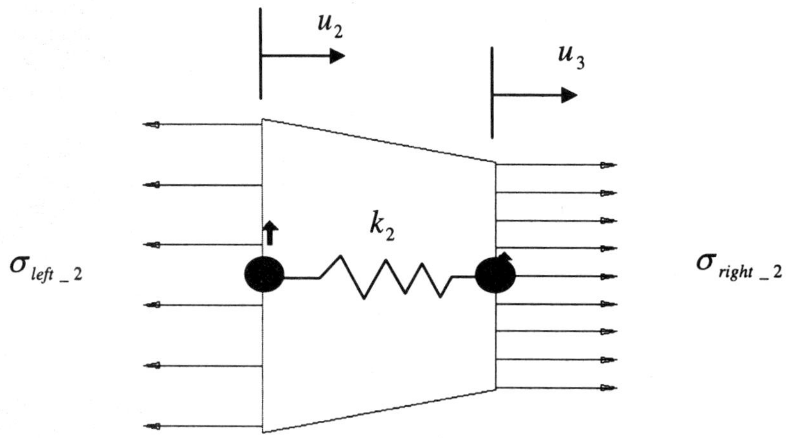

$$node2: k_1(u_2 - u_1) - k_2(u_3 - u_2) = 0$$

Element 3

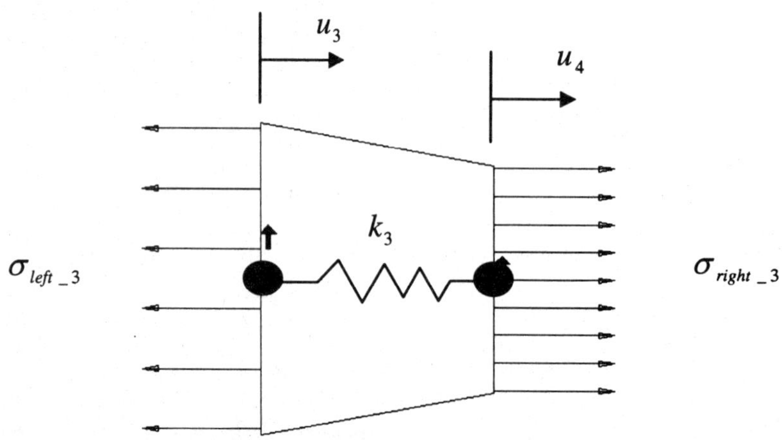

$$node3: k_2(u_3 - u_2) - k_3(u_4 - u_3) = 0$$

Element 4

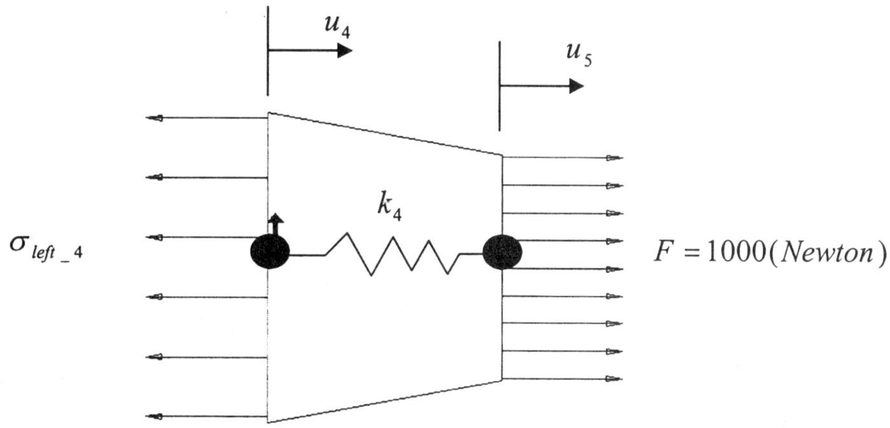

$$node4: k_3(u_4 - u_3) - k_4(u_5 - u_4) = 0$$
$$node5: k_4(u_5 - u_4) - F = 0$$

We arrange the above 5 equilibrium equations in a matrix form as follows:

$$\begin{bmatrix} k_1 & -k_1 & 0 & 0 & 0 \\ -k_1 & k_1+k_2 & -k_2 & 0 & 0 \\ 0 & -k_2 & k_2+k_3 & -k_3 & 0 \\ 0 & 0 & -k_3 & k_3+k_4 & -k_4 \\ 0 & 0 & 0 & -k_4 & k_4 \end{bmatrix} \begin{Bmatrix} u_1 \\ u_2 \\ u_3 \\ u_4 \\ u_5 \end{Bmatrix} = \begin{Bmatrix} -R \\ 0 \\ 0 \\ 0 \\ F \end{Bmatrix}$$

If we decompose the force vector to a load vector and a reaction vector, we have

$$\begin{bmatrix} k_1 & -k_1 & 0 & 0 & 0 \\ -k_1 & k_1+k_2 & -k_2 & 0 & 0 \\ 0 & -k_2 & k_2+k_3 & -k_3 & 0 \\ 0 & 0 & -k_3 & k_3+k_4 & -k_4 \\ 0 & 0 & 0 & -k_4 & k_4 \end{bmatrix} \begin{Bmatrix} u_1 \\ u_2 \\ u_3 \\ u_4 \\ u_5 \end{Bmatrix} = \begin{Bmatrix} 0 \\ 0 \\ 0 \\ 0 \\ F \end{Bmatrix} + \begin{Bmatrix} -R \\ 0 \\ 0 \\ 0 \\ 0 \end{Bmatrix}$$

Recognizing that the left side of the plate is fixed, the displacement of the left side is constrained to be zero, or

$$u_1 = 0$$
$$node1: R - k_1(u_2 - u_1) = 0$$
$$node1: R - k_1 u_2 = 0$$

When we move the reaction force vector to the left side of the above matrix equation, we have

$$\begin{bmatrix} k_1 & -k_1 & 0 & 0 & 0 \\ -k_1 & k_1+k_2 & -k_2 & 0 & 0 \\ 0 & -k_2 & k_2+k_3 & -k_3 & 0 \\ 0 & 0 & -k_3 & k_3+k_4 & -k_4 \\ 0 & 0 & 0 & -k_4 & k_4 \end{bmatrix} \begin{Bmatrix} u_1 \\ u_2 \\ u_3 \\ u_4 \\ u_5 \end{Bmatrix} + \begin{Bmatrix} R \\ 0 \\ 0 \\ 0 \\ 0 \end{Bmatrix} = \begin{Bmatrix} 0 \\ 0 \\ 0 \\ 0 \\ F \end{Bmatrix}$$

To eliminate the reaction vector, in the first row of the matrix, we replace k_1 by 1, and $-k_1$ by 0, and we have an explicit form of the stiffness matrix:

$$[K]\{u\} = \{F\}$$

$$\begin{bmatrix} 1 & 0 & 0 & 0 & 0 \\ -k_1 & k_1+k_2 & -k_2 & 0 & 0 \\ 0 & -k_2 & k_2+k_3 & -k_3 & 0 \\ 0 & 0 & -k_3 & k_3+k_4 & -k_4 \\ 0 & 0 & 0 & -k_4 & k_4 \end{bmatrix} \begin{Bmatrix} u_1 \\ u_2 \\ u_3 \\ u_4 \\ u_5 \end{Bmatrix} = \begin{Bmatrix} 0 \\ 0 \\ 0 \\ 0 \\ F \end{Bmatrix}$$

The stiffness matrix [K] represents the characteristics of the plate system when it is subjected to a load or loads.

Step 5: Determine the deformation of the plate when subjected to the uniform load acting on the free end of the plate.

<u>Calculations of the equivalent stiffness values of the four (4) elements</u>

(1) Calculate the areas of the cross sections at the five (5) nodal points.

$A_1 = 200 \times 20 = 4000\ (\text{mm}^2)$

$A_2 = [\ 200 - \left(\frac{200-60}{400}\right) * 100\ ] \times 20 = 3300\ (\text{mm}^2)$

$A_3 = [\ 200 - \left(\frac{200-60}{400}\right) * 200\ ] \times 20 = 2600\ (\text{mm}^2)$

$A_4 = [\ 200 - \left(\frac{200-60}{400}\right) * 300\ ] \times 20 = 1900\ (\text{mm}^2)$

$A_5 = [\ 200 - \left(\frac{200-60}{400}\right) * 400\ ] \times 20 = 1200\ (\text{mm}^2)$

(2) Calculate the equivalent stiffness for the four elements, using the formula

$$k_{equivalent} = \frac{A_{average} E}{l_o} = \frac{(A_{i+1} + A_i)E}{2l_o}$$

$$k_1 = \frac{10^{-6} * (4000 + 3300) * 70 * 10^9}{2 * 100 * 10^{-3}} = 1.28x10^9 (pascal)$$

$$k_2 = \frac{10^{-6} * (3300 + 2600) * 70 * 10^9}{2 * 100 * 10^{-3}} = 1.03x10^9 (pascal)$$

$$k_3 = \frac{10^{-6} * (2600 + 1900) * 70 * 10^9}{2 * 100 * 10^{-3}} = 0.788x10^9 (pascal)$$

$$k_4 = \frac{10^{-6} * (1900 + 1200) * 70 * 10^9}{2 * 100 * 10^{-3}} = 0.543x10^9 (pascal)$$

Stiffness matrix for the 4 elements

Element 1

$$[K]^1 = \begin{bmatrix} k_1 & -k_1 \\ -k_1 & k_1 \end{bmatrix} = 10^9 \begin{bmatrix} 1.28 & -1.28 \\ -1.28 & 1.28 \end{bmatrix}$$

Element 2

$$[K]^2 = \begin{bmatrix} k_2 & -k_2 \\ -k_2 & k_2 \end{bmatrix} = 10^9 \begin{bmatrix} 1.03 & -1.03 \\ -1.03 & 1.03 \end{bmatrix}$$

Element 3

$$[K]^3 = \begin{bmatrix} k_3 & -k_3 \\ -k_3 & k_3 \end{bmatrix} = 10^9 \begin{bmatrix} 0.788 & -0.788 \\ -0.788 & 0.788 \end{bmatrix}$$

Element 4

$$[K]^4 = \begin{bmatrix} k_4 & -k_4 \\ -k_4 & k_4 \end{bmatrix} = 10^9 \begin{bmatrix} 0.543 & -0.543 \\ -0.543 & 0.543 \end{bmatrix}$$

Assemble the 4 stiffness matrices into a global stiffness matrix

$$10^9 \begin{bmatrix} 1.28 & -1.28 & 0 & 0 & 0 \\ -1.28 & 1.28 + 1.03 & -1.03 & 0 & 0 \\ 0 & -1.03 & 1.03 + 0.788 & -0.788 & 0 \\ 0 & 0 & -0.788 & 0.788 + 0.543 & -0.543 \\ 0 & 0 & 0 & -0.543 & 0.543 \end{bmatrix}$$

Apply the load condition (F=1000 Newton) and the boundary condition (u_1=0), we have the static equilibrium equation of the plate system in its matrix form, as follows:

$$10^9\begin{bmatrix} 1.00 & 0 & 0 & 0 & 0 \\ -1.28 & 1.28+1.03 & -1.03 & 0 & 0 \\ 0 & -1.03 & 1.03+0.788 & -0.788 & 0 \\ 0 & 0 & -0.788 & 0.788+0.543 & -0.543 \\ 0 & 0 & 0 & -0.543 & 0.543 \end{bmatrix}\begin{Bmatrix} u_1 \\ u_2 \\ u_3 \\ u_4 \\ u_5 \end{Bmatrix} = \begin{Bmatrix} 0 \\ 0 \\ 0 \\ 0 \\ 1000 \end{Bmatrix}$$

Solving the above equation (for example, using Matlab), we have

$$\begin{Bmatrix} u_1 \\ u_2 \\ u_3 \\ u_4 \\ u_5 \end{Bmatrix} = 10^{-9}\begin{bmatrix} 1.00 & 0 & 0 & 0 & 0 \\ -1.28 & 1.28+1.03 & -1.03 & 0 & 0 \\ 0 & -1.03 & 1.03+0.788 & -0.788 & 0 \\ 0 & 0 & -0.788 & 0.788+0.543 & -0.543 \\ 0 & 0 & 0 & -0.543 & 0.543 \end{bmatrix}^{-1}\begin{Bmatrix} 0 \\ 0 \\ 0 \\ 0 \\ 1000 \end{Bmatrix}$$

$$\begin{Bmatrix} u_1 \\ u_2 \\ u_3 \\ u_4 \\ u_5 \end{Bmatrix} = 10^{-9}\begin{bmatrix} 1.00 & 0 & 0 & 0 & 0 \\ 1.00 & 0.7813 & 0.7813 & 0.7813 & 0.7813 \\ 1.00 & 0.7813 & 1.7521 & 1.7521 & 1.7521 \\ 1.00 & 0.7813 & 1.7521 & 3.0212 & 3.0212 \\ 1.00 & 0.7813 & 1.7521 & 3.0212 & 4.8628 \end{bmatrix}\begin{Bmatrix} 0 \\ 0 \\ 0 \\ 0 \\ 1000 \end{Bmatrix} = 10^{-6}\begin{Bmatrix} 0 \\ 0.7813 \\ 1.7521 \\ 3.0212 \\ 4.8628 \end{Bmatrix}(meter)$$

$$\begin{Bmatrix} u_1 \\ u_2 \\ u_3 \\ u_4 \\ u_5 \end{Bmatrix} = \begin{Bmatrix} 0 \\ 0.7813 \\ 1.7521 \\ 3.0212 \\ 4.8628 \end{Bmatrix}(\mu m)$$

Step 6: Post-processing information

Determine the normal strain in each of the four (4) elements, using the following formula

$$\varepsilon_i = \frac{u_{i+1} - u_i}{l_i}$$

Element 1

$$\varepsilon_1 = \frac{u_2 - u_1}{l_1} = \frac{0.7813 - 0}{100 * 10^{-3}} x10^{-6} = 7.813x10^{-6}$$

Element 2

$$\varepsilon_2 = \frac{u_3 - u_2}{l_2} = \frac{1.7521 - 0.7813}{100 * 10^{-3}} x10^{-6} = 9.708x10^{-6}$$

Element 3

$$\varepsilon_3 = \frac{u_4 - u_3}{l_3} = \frac{3.0212 - 1.7521}{100 * 10^{-3}} x10^{-6} = 12.691x10^{-6}$$

Element 4

$$\varepsilon_4 = \frac{u_5 - u_4}{l_4} = \frac{4.8628 - 3.0212}{100 * 10^{-3}} x10^{-6} = 18.416x10^{-6}$$

Determine the normal stress in each of the four (4) elements, using the following formula

$$\sigma_i = E\varepsilon_i$$

Element 1

$$\sigma_1 = E\varepsilon_1 = (70x10^9)(7.813x10^{-6}) = 0.546x10^6(pascal) = 0.546(MPa)$$

Element 2

$$\sigma_2 = E\varepsilon_2 = (70x10^9)(9.708x10^{-6}) = 0.680x10^6(pascal) = 0.680(MPa)$$

Element 3

$$\sigma_3 = E\varepsilon_3 = (70x10^9)(12.691x10^{-6}) = 0.888x10^6(pascal) = 0.888(MPa)$$

Element 4

$$\sigma_4 = E\varepsilon_4 = (70x10^9)(18.416x10^{-6}) = 1.289x10^6(pascal) = 1.289(MPa)$$

Determine the reaction force R

$$R = k_1(u_2 - u_1) = (1.28x10^9)(0.7813x10^6 - 0) = 1000(Newton)$$

Determine the strain in the Y-direction and the strain in the Z-direction

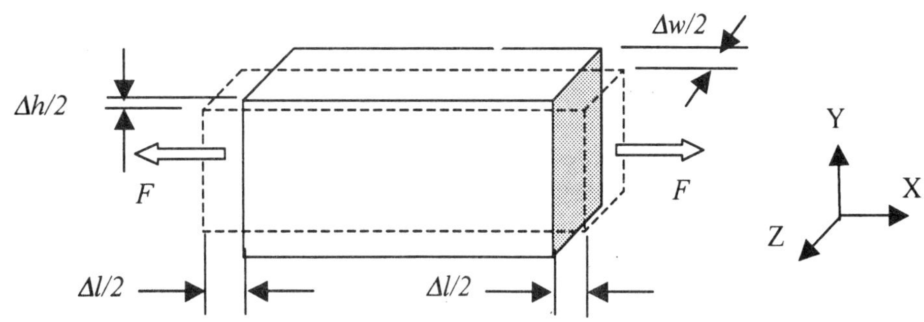

A parameter is used to characterize the relation between the axial elongation and the lateral contraction, either in the Y-direction or in the Z-direction. The parameter is called Poisson's ratio. It is defined as follows:

$$\nu = -\frac{lateral\_strain}{axial\_strain}$$

In this example, we assume that the material has the same properties in all directions, or the material is of an isotropic type.

Element 1

$$\varepsilon_{y1} = \nu\varepsilon_{x1} = 0.33*7.813x10^{-6} = 2.58x10^{-6}$$

$$\varepsilon_{z1} = \nu\varepsilon_{x1} = 0.33*7.813x10^{-6} = 2.58x10^{-6}$$

Element 2

$$\varepsilon_{y2} = \nu\varepsilon_{x2} = 0.33*9.708x10^{-6} = 3.20x10^{-6}$$

$$\varepsilon_{z2} = \nu\varepsilon_{x2} = 0.33*9.708x10^{-6} = 3.20x10^{-6}$$

Element 3

$$\varepsilon_{y3} = \nu\varepsilon_{x3} = 0.33*12.691x10^{-6} = 4.19x10^{-6}$$

$$\varepsilon_{z3} = \nu\varepsilon_{x3} = 0.33*12.691x10^{-6} = 4.19x10^{-6}$$

Element 4

$$\varepsilon_{y4} = \nu\varepsilon_{x4} = 0.33*18.416x10^{-6} = 6.08x10^{-6}$$

$$\varepsilon_{z4} = \nu\varepsilon_{x4} = 0.33*18.416x10^{-6} = 6.08x10^{-6}$$

<u>Determine the stress in the Y-direction and the stress in the Z-direction</u>

$$\sigma_{yi} = E\varepsilon_{yi} \qquad\qquad \sigma_{zi} = E\varepsilon_{zi}$$

Element 1

$$\sigma_{y1} = E\varepsilon_{y1} = (70x10^9)(2.58x10^{-6}) = 0.181x10^6(pascal) = 0.181(MPa)$$

$$\sigma_{z1} = E\varepsilon_{z1} = (70x10^9)(2.58x10^{-6}) = 0.181x10^6(pascal) = 0.181(MPa)$$

Element 2

$$\sigma_{y2} = E\varepsilon_{y2} = (70x10^9)(3.20x10^{-6}) = 0.224x10^6(pascal) = 0.224(MPa)$$

$$\sigma_{z2} = E\varepsilon_{z2} = (70x10^9)(3.20x10^{-6}) = 0.224x10^6(pascal) = 0.224(MPa)$$

Element 3

$$\sigma_{y3} = E\varepsilon_{y3} = (70x10^9)(4.19x10^{-6}) = 0.293x10^6(pascal) = 0.293(MPa)$$

$$\sigma_{z3} = E\varepsilon_{z3} = (70x10^9)(4.19x10^{-6}) = 0.293x10^6(pascal) = 0.293(MPa)$$

Element 4

$$\sigma_{y4} = E\varepsilon_{y4} = (70x10^9)(6.08x10^{-6}) = 0.426x10^6(pascal) = 0.426(MPa)$$

$$\sigma_{z4} = E\varepsilon_{z4} = (70x10^9)(6.08x10^{-6}) = 0.426x10^6(pascal) = 0.426(MPa)$$

Step 7: Graphical Representations

Use a set of 4 colors and characterize the strain distribution in the Y-direction or in the Z-direction. It is important to note that the numerical values represent the strains in the 4 middle cross sections.

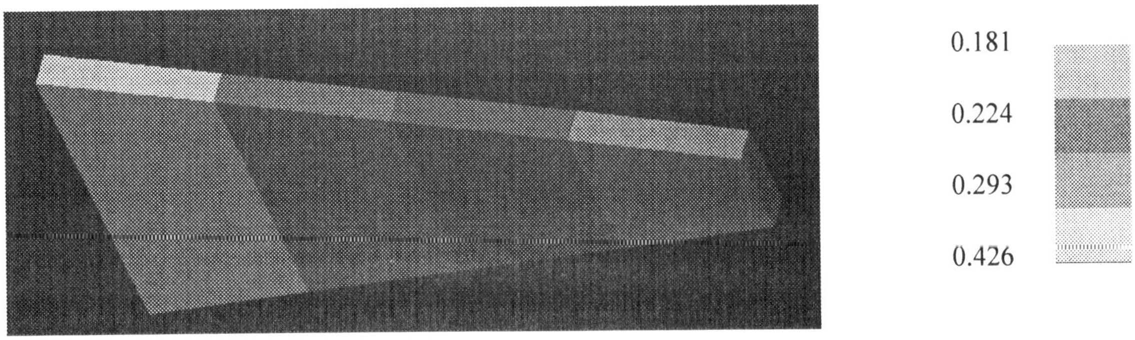

To verify the results obtained from this evaluation, we run **Pro/MECHANICA**, an FEA software system you will learn in Chapter 2.

Step 1: Setting the material properties

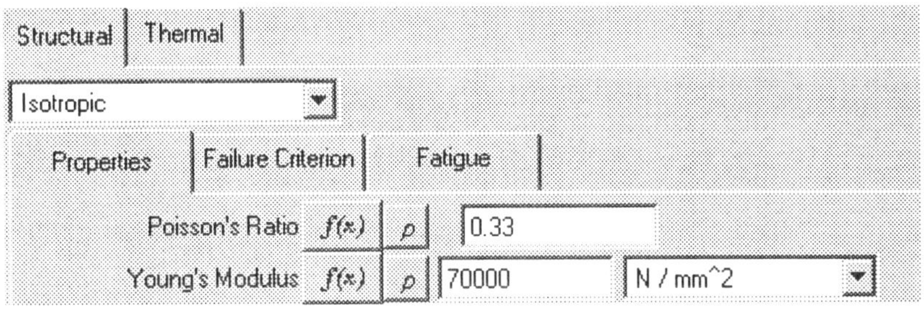

Step 2: Setting the constraint condition and the load condition

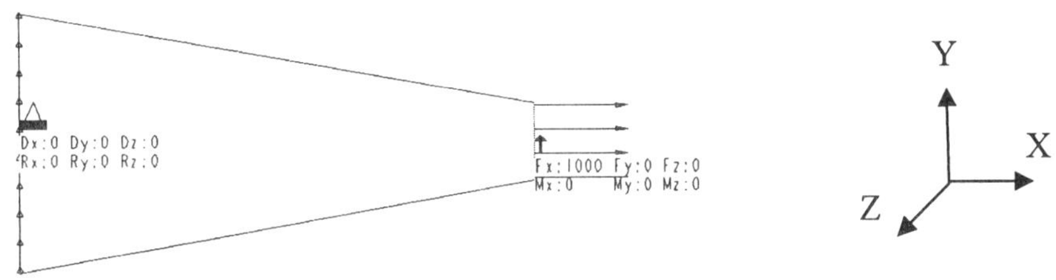

Step 3: Results obtained from running Pro/MECHANICA

<u>Type and number of elements</u>

```
Model Type: Three Dimensional

Points:                16
Edges:                 47
Faces:                 50

Springs:                0
Masses:                 0
Beams:                  0
Shells:                 0
Solids:                18

Elements:              18
```

The maximum stress and displacement in the X direction

max_beam_bending:	0.000000e+00	
max_beam_tensile:	0.000000e+00	
max_beam_torsion:	0.000000e+00	
max_beam_total:	0.000000e+00	Compare to
max_disp_mag:	2.534972e-03	
max_disp_x:	2.530680e-03	4.8628 μm
max_disp_y:	1.282444e-04	
max_disp_z:	-1.501406e-04	
max_prin_mag:	8.830912e-01	
max_rot_mag:	0.000000e+00	
max_rot_x:	0.000000e+00	
max_rot_y:	0.000000e+00	
max_rot_z:	0.000000e+00	Compare to
max_stress_prin:	8.830912e-01	
max_stress_vm:	8.430616e-01	1.289 MPa
max_stress_xx:	8.825928e-01	
max_stress_xy:	1.486066e-01	
max_stress_xz:	-8.724412e-02	
max_stress_yy:	1.833535e-01	
max_stress_yz:	-8.150371e-02	
max_stress_zz:	2.279374e-01	

Plot of the distribution of strain in the Y direction

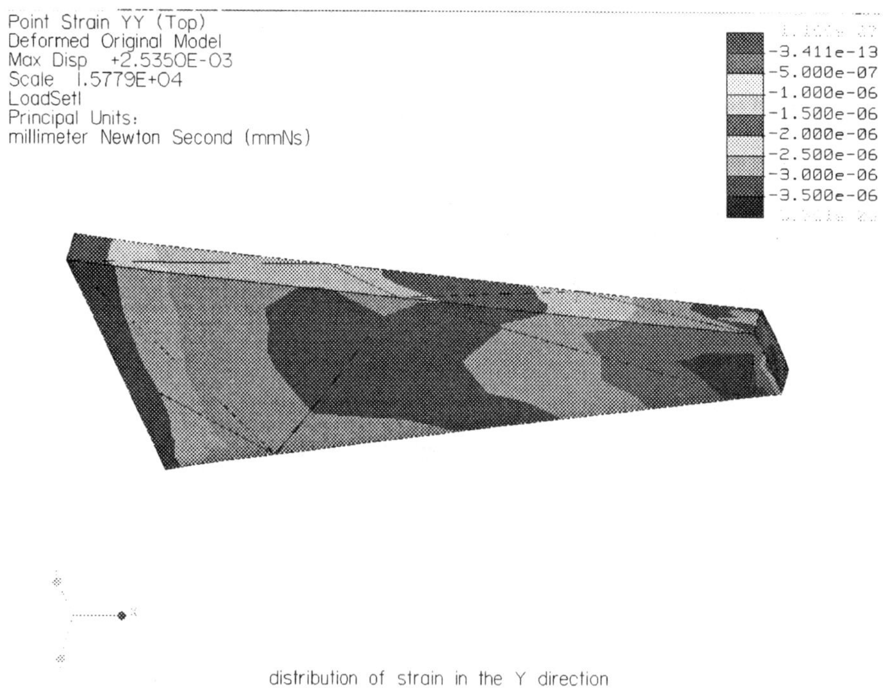

It is evident that the results obtained from the use of Pro/MECHANICA are better in terms of clarification and clearness to represent the deformation and stress distributions. This is one of the reasons why people are using those software systems commercially available to perform engineering analyses, such as FEA.

1.5 Unit Systems Used in the FEA Software Systems

Although there are about 10 FEA software systems commercially available in the market, the unit systems used in those software systems are almost idiential among them. In the following, we illustrate the unit systems used by Pro/Mechanica, ALGOR and ANSYS. It is important that users have a full understanding the basics of each of them so that the numerical values displayed on computer screen, or saved into the data files can be correctly interpreted.

Let us first review the unit systems used by Pro/Mechanica. From the set up menu of the Pro/Mechanica system, the 7 unit systems available under Pro/Mechanica are listed. Let us first examine Meter Kilogram Second (MKS) system, which is the SI unit system. There are 5 basic quantities, namely, length, mass, force, time and temperature. Besides, the gravity constant is set at 9.80665 m/second^2.

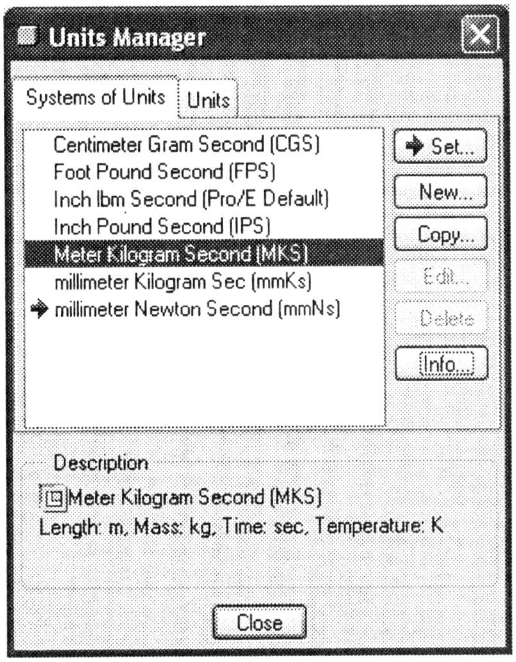

Basic quantities:

Length	m
Mass	kg
Force	N
Time	sec
Temperature	K

Gravity 9.80665 m / sec^2

Based on these 5 basic quantities, the units for other parameters can be derived. The following table shows a set of parameters, such as area, volume, velocity, acceleration, augular velocity, etc.

Derived quantities:

Area	m^2
Volume	m^3
Velocity	m / sec
Acceleration	m / sec^2
Angular Velocity	rad / sec
Angular Acceleration	rad / sec^2
Frequency	1 / sec
Density	kg / m^3
Torque/Moment	m N
Distributed Force	N / m
Distributed Moment	N
Areal Distr Force	N / m^2

In the later chapters, we will use Pro/Mechanica to perform FEA. The unit system used in those chapters is the mmNs system. The 5 basic quantities are mm for the unit of length, tonne for the unit of mass, Newton for the unit of force, second for the unit of time and Celsius (deg C) for the unit of temperature. The gravity constant is set at 9806.65 mm/second^2.

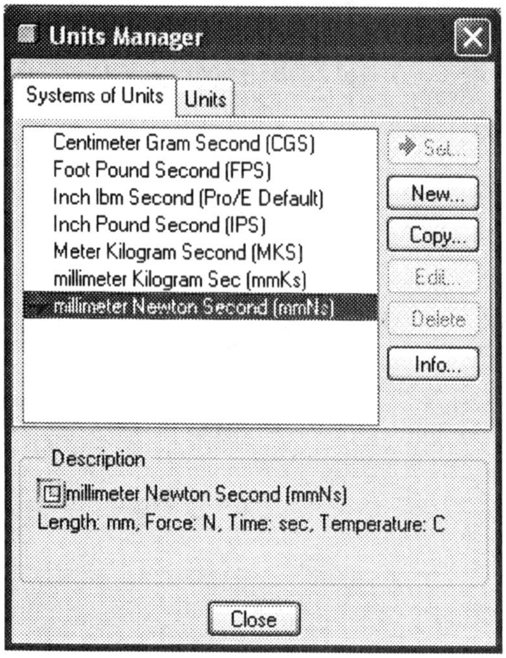

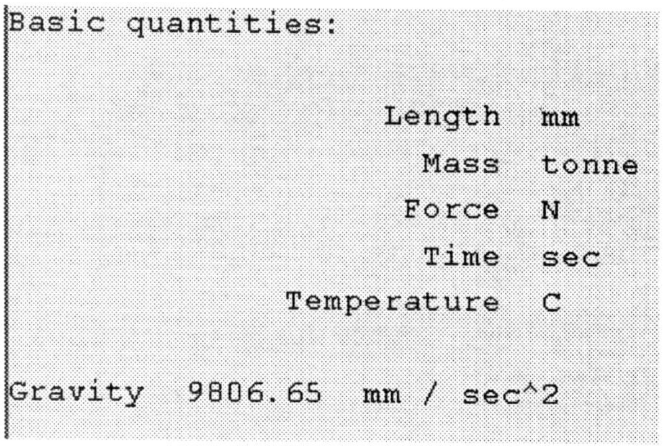

It is extremely important to remember that the unit of stress is

$$unit\_of\_stress = \frac{Newton}{mm^2} = \frac{Newton}{m^2} = 10^6(pascal) = 1(MPa)$$

It is extremely important to remember that the unit of displacement is

$$unit\_of\_deformation = mm$$

Now let us review the unit systems used by ALGOR. From the unit menu of the ALGOR system, the five unit systems available under ALGOR are shown below. They are Metric mks (SI), Metric cgs, English (in), English (ft) and Custom, which allows the user to set up the unit system on his/her own, which is not recommended strongly by the author).

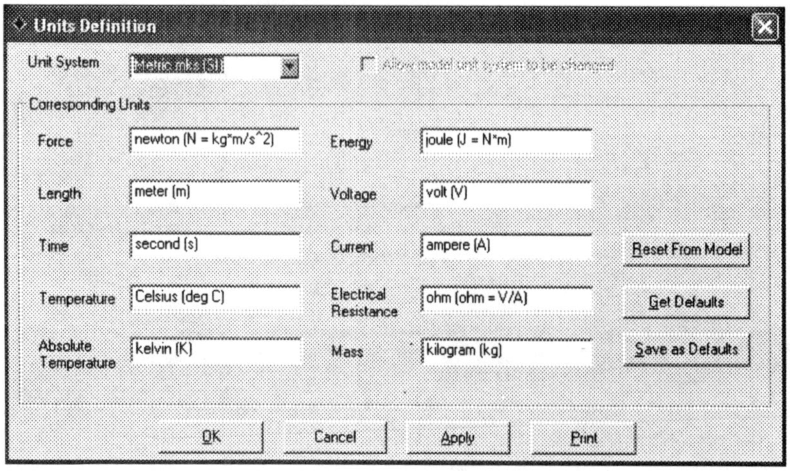

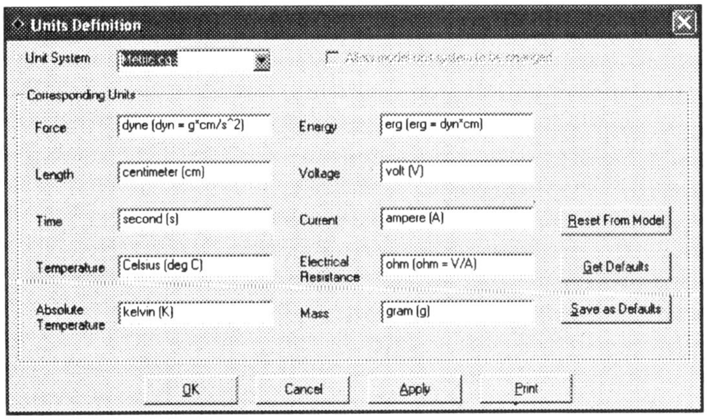
Units Definition
Unit System
Corresponding Units
Force
dyne (dyn = g*cm/s^2)
Energy
erg (erg = dyn*cm)
Length
centimeter (cm)
Voltage
volt (V)
Time
second (s)
Current
ampere (A)
Reset From Model
Temperature
Celsius (deg C)
Electrical Resistance
ohm (ohm = V/A)
Get Defaults
Absolute Temperature
kelvin (K)
Mass
gram (g)
Save as Defaults
OK
Cancel
Apply
Print

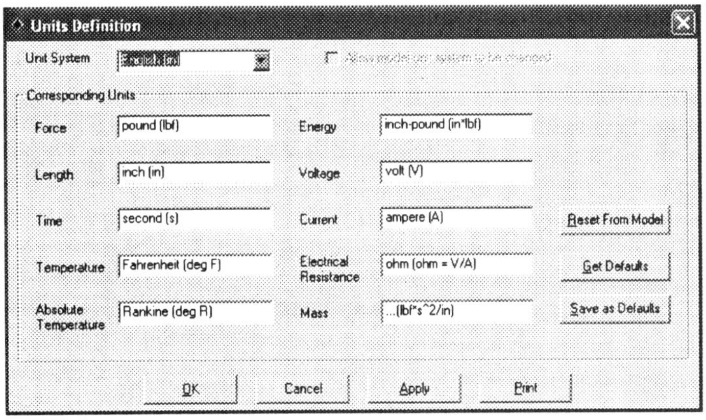
Units Definition
Unit System
Corresponding Units
Force
pound (lbf)
Energy
inch-pound (in*lbf)
Length
inch (in)
Voltage
volt (V)
Time
second (s)
Current
ampere (A)
Reset From Model
Temperature
Fahrenheit (deg F)
Electrical Resistance
ohm (ohm = V/A)
Get Defaults
Absolute Temperature
Rankine (deg R)
Mass
...(lbf*s^2/in)
Save as Defaults
OK
Cancel
Apply
Print

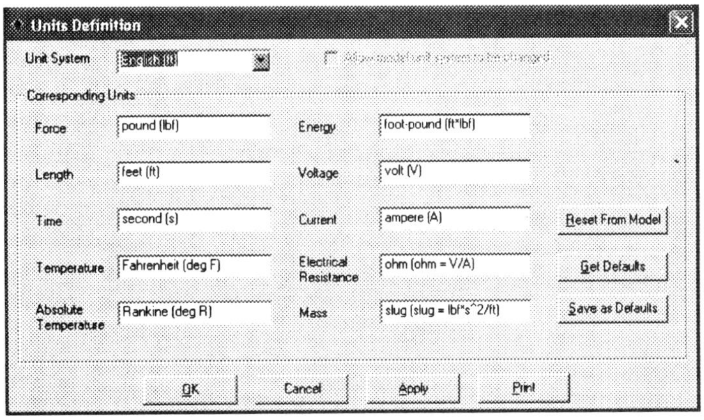
Units Definition
Unit System
Corresponding Units
Force
pound (lbf)
Energy
foot-pound (ft*lbf)
Length
feet (ft)
Voltage
volt (V)
Time
second (s)
Current
ampere (A)
Reset From Model
Temperature
Fahrenheit (deg F)
Electrical Resistance
ohm (ohm = V/A)
Get Defaults
Absolute Temperature
Rankine (deg R)
Mass
slug (slug = lbf*s^2/ft)
Save as Defaults
OK
Cancel
Apply
Print

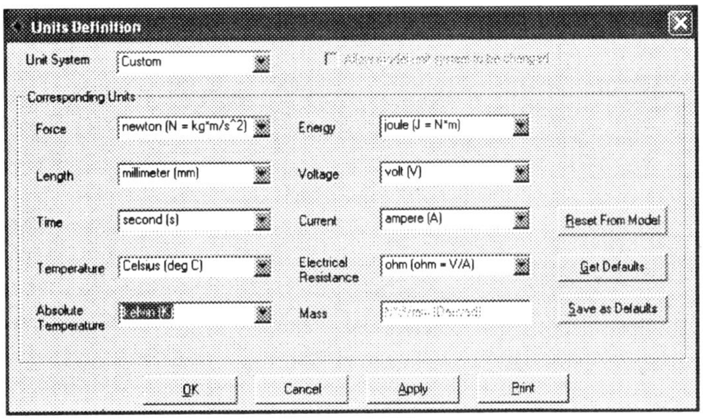
Units Definition
Unit System
Custom
Corresponding Units
Force
newton (N = kg*m/s^2)
Energy
joule (J = N*m)
Length
millimeter (mm)
Voltage
volt (V)
Time
second (s)
Current
ampere (A)
Reset From Model
Temperature
Celsius (deg C)
Electrical Resistance
ohm (ohm = V/A)
Get Defaults
Absolute Temperature
Mass
Save as Defaults
OK
Cancel
Apply
Print

Now let us review the unit systems used by ANSYS. From the menu of Material Library > Select Unit. There are 4 unit systems used by ANSYS. For example, the currently selected unit system, as illustrated, is MKS or the SI unit system.

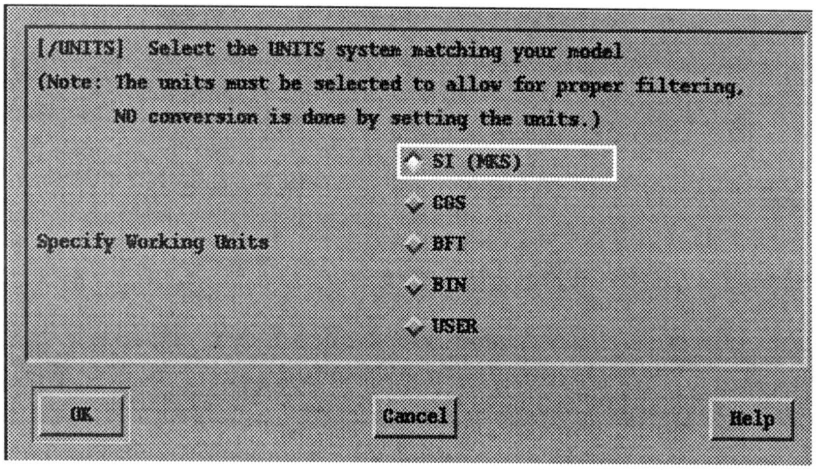

1.6 References

1. Algor Processor Reference Manual, Algor Interactive Systems, 150 Alpha Drive, Pittsburgh, PA 15238
2. O. Axelsson, Iterative Solution Methods, Cambridge University Press, New York, 1994.
3. K. J. Bathe, Finite Element Procedure, Prentice Hall, Upper Saddle River, New Jersey, 1996.
4. K. Bell, A Refined Triangular Plate Bending Finite Element, Int. J. Numer. Methods Eng., Vol. 1, No. 1 January 1969, pp. 101-122.
5. B. Carnahan, H. A. Luther, and J. O. Wikes, Applied Numerical Methods, Wiley New York, NY, 1969.
6. R. W. Clough, The Finite Element in Plane Stress Analysis, Proceedings of 2nd ASCE Conference on Electronic Computation, Pittsburgh, PA. September 1960.
7. R. D. Cook, D. S. Maikus and M. E. Plesha, Concepts and Applications of Finite Element Analysis, 3rd edition, Wiley, New York, NY, 1989.
8. J. W. Dally and W. F. Riley, Experimental Stress Analysis, McGraw Hill, 1965.
9. J. W. Dally, Design Analysis of Structural Elements, Goodington, New York, NY, 2003.
10. C. S. Desal and J. F. Abel, Introduction to the Finite Element Method, Van Nostrand Reinhold, New York, NY, 1972.
11. J. H. Earle, Graphics for Engineers, AutoCAD Release 13, Addision-Wesley, Reading, Massachusetts, 1996.
12. R. H. Gallagher, Finite Element Analysis Fundamentals, Prentice-Hall, Englewoood Cliffs, NJ, 1975.
13. J. M. Gere and S. P. Timoshenko, Mechanics of Materials, 3rd edition, PWS-KENT Publishing Company, Boston, 1990.
14. W. W. Hager, Applied Numerical Linear Algebra, Prentice-Hall, Englewood Cliffs, NJ, 1988.
15. Holland, The Finite Element Method in Plane Stress Analysis, in the Finite Element Method in Stress Analysis, I. Holland and K. Bell (eds), Tapir Press, Trondheim, Norway, 1969, Chapter 2.
16. Y. Y. Hsieh, Elementary Theory of Structure, 2nd edition, Prentice-Hall, Englewood Cliffs, NJ, 1982.
17. M. A. N. Hendriks, H Jongedijk, J. G. Rots and W. J. E. van Spange (eds), Finite Elements in Engineering and Science, DIANA Computational Mechanics, 1997.

18. K. H. Huebner, D. L. Dewhirst, D. E. Smith, and T. G. Byrom, The Finite Element Method for Engineers, 4th edition, John Wiley & Sons, Inc., 2001.
19. C.S. Krishnamoorthy, Finite Element Analysis, Theory and Programming, 2nd Ed., 1995.
20. D. L. Logan, A First Course in the Finite Element Method Using ALGOR, PWS Publishing, Boston, 1997.
21. H. C. Martin, Introduction to Matrix Methods of Structural Analysis, McGraw-Hill, New York, NY, 1966.
22. R. L. Norton, Machine Design: An Integrated Approach, Prentice Hall, Upper Saddle River, New Jersey, 1996.
23. M. Ortiz, Y. Leroy and A. Needleman, A Finite Element Method for Localized Failure Analysis, Comp. Meth. Applied Mech. Enging, 61, pp. 189-214, 1987.
24. M. Petyl, Introduction to Finite Element Vibration Analysis, Cambridge University Press, 1990.
25. T. H. H. Pian and P. Tong, Basis of Finite Element Methods for Solid Continua, Int. J. Numer. Methods Eng., Vol. 2, No. 7, April 1964, pp. 1333-1336.
26. P. G. Plockner, Symmetry in Structural Mechanics, J. of the Structural Division, American Society of Civil Engineers, Vol. 99, No. ST1, pp. 71-89, 1973.
27. J. N. Reddy, An Introduction to the Finite Element Method, McGraw-Hill Publishing Company, New York, NY, 1984.
28. T. Ross, T. Johns and Emile, Dynamic Buckling of Thin Walled Domes under External Water Pressure, Applied Solid Mechanics 2nd Conference, Ed. Tooth & Spence, Elsevier, pp. 211-224, 1987.
29. T. Ross, Finite Element Program in Structural Engineering and Continuum Mechanics, Horwood Publishing Limited, Chichester, England, 1996.
30. T. Ross, Advanced Applied Finite Element Methods, Horwood Publishing Limited, Chichester, England, 1998.
31. L. Stasa, Applied Finite Element Analysis for Engineers, Holt, Rinehart and Winston, New York, NY, 1985, pp. 101-157.
32. Y. Saad, Iterative Methods for Sparse Linear Systems, PWS Publishing Company, Boston, 1996.
33. P. Silvester, Higher-Order Polynomial Triangular Finite Element for Potential Problems, Int. J. Eng. Soc., Vol. 7, No. 8, pp. 849-861.
34. Y. Tada and G. C. Lee, Finite Element Solution to an Elastic Problem of Beams, Int. J. Numer. Methods Eng., Vol. 2, No. 2, April 1970, pp. 229-241.
35. R. L. Taylor, On Completeness of Shape Functions for Finite Element Analysis, Int. J. Numer., Methods Eng., Vol. 4., No. 1, 1972, pp. 17-22.
36. S. P. Timoshenko and D. H. Young, Theory of Structure, McGraw-Hill, Kogaksha Limited, 1965.
37. M. J. Turner, R. W. Clough, H. C. Martin, and L. J. Topp, Stiffness and Deflection Analysis of Complex Structures, J. of the Aeronautical Sciences, Vol. 23, No. 9, pp. 805-824, Sept. 1956.
38. O. C. Zienkiewicz, The Finite Element Method, third edition, McGraw-Hill, New York, NY, 1977.
39. O. C. Zienkiewicz and R. L. Taylor, The Finite Element Method, 4th edition, Vol. 1, McGraw-Hill (UK),London, 1989.
40. Web Addresses

ALGOR	http://www.algor.com
ANSYS	http://www.ansys.com
AutoDesk	http://www.autodesk.com
CATIA	http://www.catia.com
PTC	http://www.ptc.com

1.7 Exercises

1. Use your personal experience to describe the essential steps involved in engineering design

and engineering analysis. Make sure that the following information is presented.

(1) What was the product you designed?
(2) Was the design process a team effort or your own effort?
(3) What was the design objective(s)?
(4) How did you and/or your design team perform an engineering analysis?

2. Present any experiences you have had using CAD and computer aided engineering software systems. Have you used Pro/Mechanica, ALGOR and/or ANSYS? What type or types of the problem(s) did you deal with at those times?

3. What is your expectation from taking a course in computer aided engineering analyses?

4. Search the Internet with the following URLs to get the latest information about the software systems used in this book.

CATIA	http://www.catia.com
I-DEAS	http://www.sdrc.com
Pro/ENGINEER	http://www.ptc.com
UNIGRAPHICS	http://www.unigraphics.com
AutoCAD	http://www.autodesk.com
SOLIDWORKS	http://www.solidworks.com

5. A truss structure is shown below.

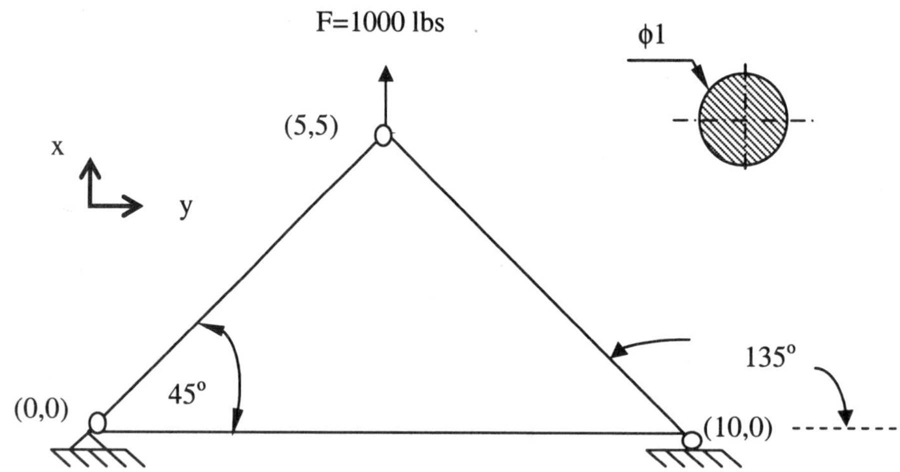

(1) As illustrated, 2 elements associated with 3 nodes are needed. Calculate k_1 and k_2 using $\frac{AE}{l}$ where E = 10.1 x 10^6 lbf/in^2.
(2) Determine the 2 stiffness matrices individually.
(3) Assemble the 2 stiffness matrices and obtain the global stiffness matrix.
(4) Assume that a vertical force is acting on the upper node, as shown. Determine the displacements at each of the 3 joints.

6. A truss structure is shown below.

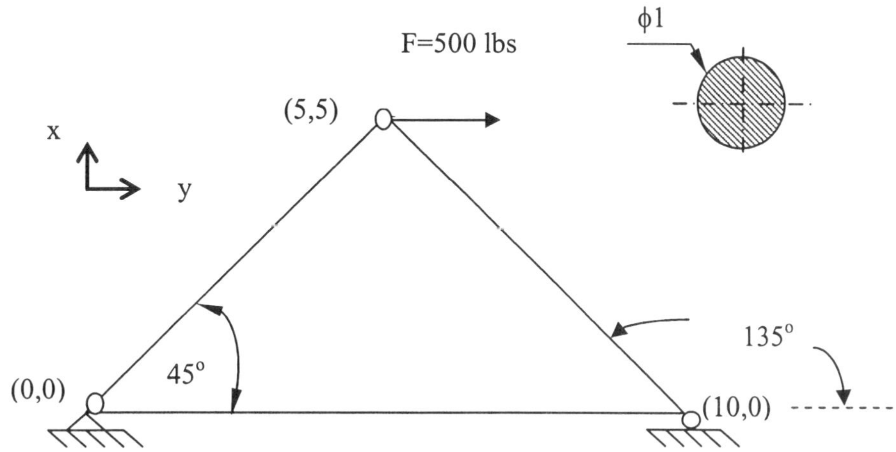

(1) As illustrated, 2 elements associated with 3 nodes are needed. Calculate k_1 and k_2 using $\frac{AE}{l}$ where $E = 10.1 \times 10^6$ lbf/in^2.
(2) Determine the 2 stiffness matrices individually.
(3) Assemble the 2 stiffness matrices and obtain the global stiffness matrix.
(4) Assume that a vertical force is acting on the upper node, as shown. Determine the displacements at each of the 3 joints.

7. The following figure illustrates a truss structure, which consists of 3 bars. The section area of each bar is circular. The diameter is 100 mm. The geometrical information of the truss structure is shown in the figure. The material type is determined by the student.

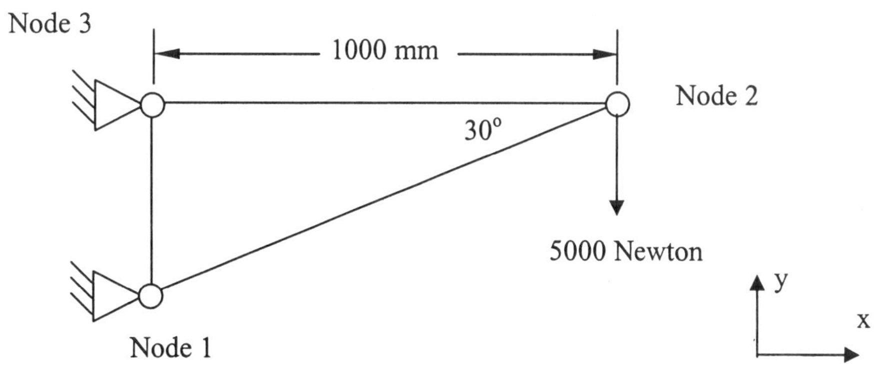

(1) Obtain the stiffness matrix for each of the 3 elements.
(2) Obtain the global stiffness matrix of the truss structure.
(3) Under the shown load condition, determine the displacement of Node 2 in the horizontal and vertical directions.
(4) Determine the reaction forces on Node 1 and Node 3.

8. The following figure illustrates a truss structure, which consists of 3 bars. The dimension of the section area is 100 mm^2. The geometrical information of the truss structure is shown in the figure. The material type is determined by the student team.

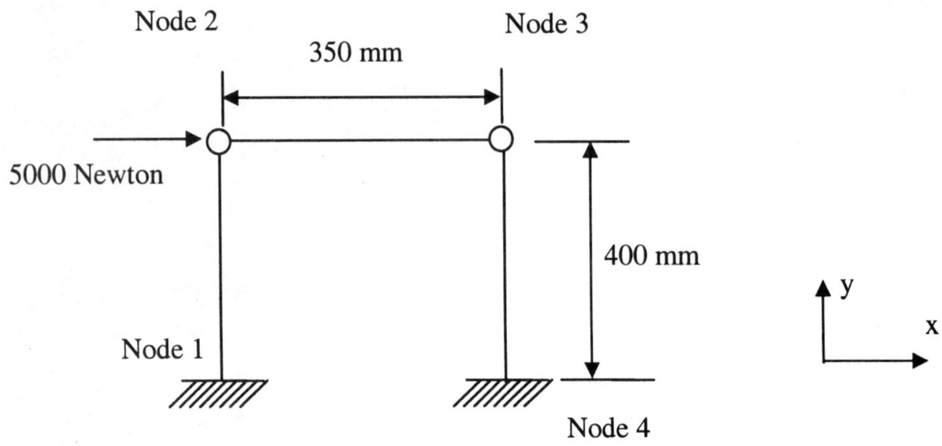
Node 2
Node 3
350 mm
5000 Newton
400 mm
y
x
Node 1
Node 4

CHAPTER 2

FEA USING Pro/Mechanica

2.1 Introduction

As discussed in Chapter 1, in order to perform a finite element analysis, we need a geometrical model which is a representation of the physical object under study. When the geometrical model is available, the finite element analysis can be carried out. In this chapter, section 2.1 is a brief interlocution to the Pro/ENGINEER Wildfire design system. An example of creating a block component is used to demonstrate the essential steps needed to create a geometrical model. Section 2.2 is an introduction to the Pro/Mechanica system. In section 2.2, an example of a column structure subjected to a compressive load is used to demonstrate the essential steps needed for performing a finite element analysis, such as specify the material type, define a set of constraint conditions, and define a set of loading conditions so that the maximum displacement of the column structure can be numerically evaluated and located, and the maximum principal stress can also be evaluated and its location can be identified. In section 2.3, we present a case study. The case study deals with a cantilever beam subjected to a uniformly distributed load. Our objectives are to evaluate the maximum displacement and the maximum principal stress occurred under the constraint and load conditions.

2.2 Introduction to the Pro/ENGINEER Wildfire Design System

Before performing an analysis, we need a 3D solid model in general. In this book, we focus on using the Pro/ENGINEER Wildfire design system to create 3D solid models. The command to start Pro/ENGINEER Wildfire design system depends on the computer system users are using. (i.e., a PC, a Sun workstation, or a SGI machine.) However, the following procedure is applicable to initiate the design process regardless of the computer system used to run Pro/ENGINEER Wildfire.

Figure 2-1 presents the window on display when launching Pro/ENGINEER Wildfire 2.0. The window offers the following functions:

1. Indicate the current working directory, such as **My Document** highlighted, and allowing the user to make a change to the current working directory.
2. Provide a Main Toolbar or Menu Toolbar, from which a new file can be opened for initiating the design process, such as creating a new model for the object under study.
3. Provide tutorial materials on Pro/ENGINEER Wildfire through the connection to the PTC website.
4. Provide the guidelines for getting help from the PTC website.

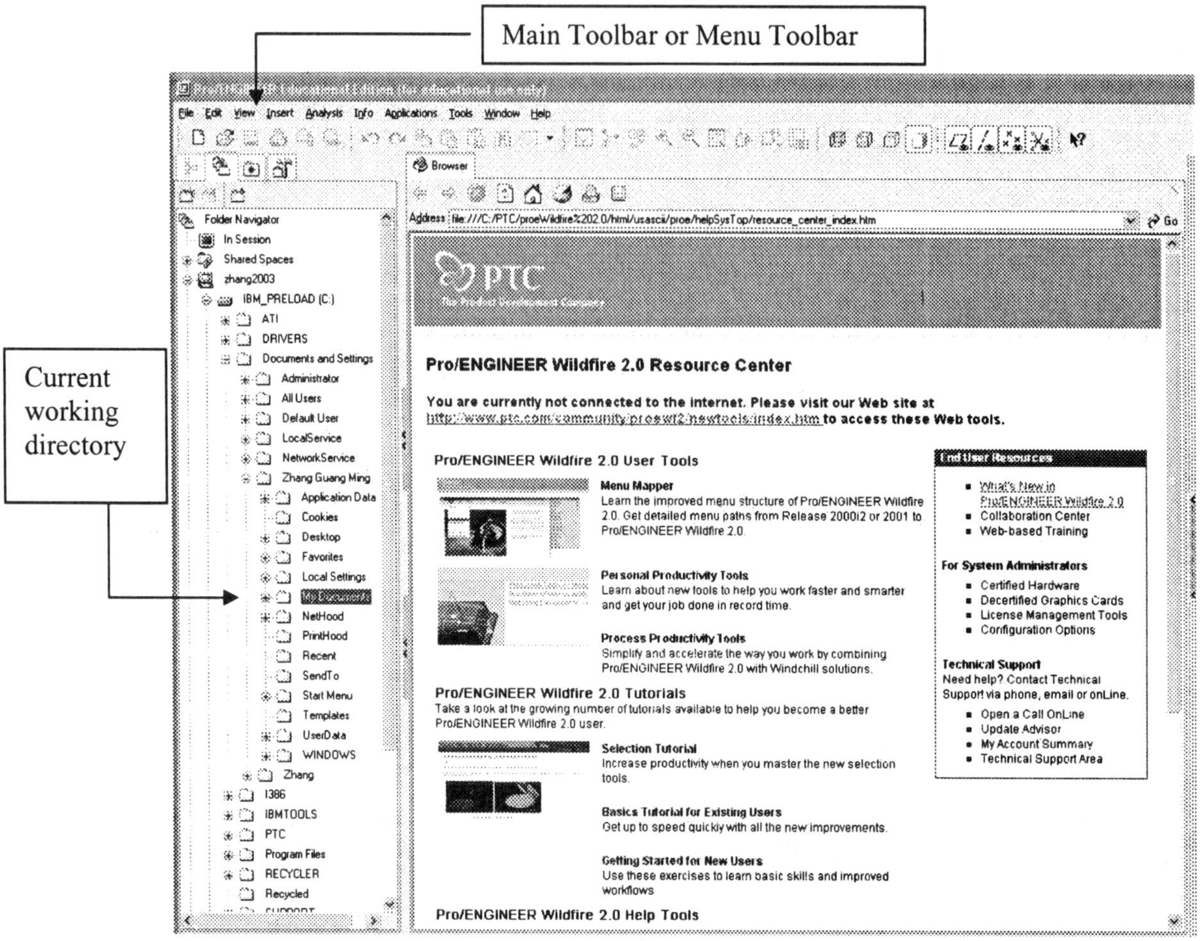

Figure 2-1 Window that Appears when Launching Pro/ENGINEER Wildfire 2.0

In the following, we use an example of creating a block component to demonstrate the process of creating a 3D solid model for an object. The three dimensions of the block are 10 x 6 x 4 inch.

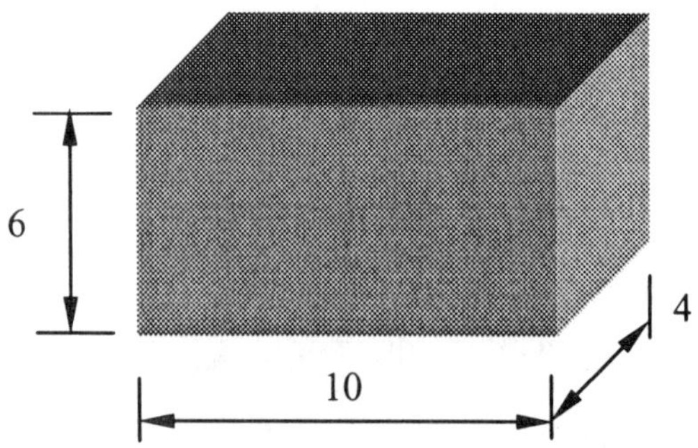

To start a design process using Pro/ENGINEER Wildfire, select the icon, called "**Create a new file**". The icon is displayed on the menu toolbar or the main toolbar, as illustrated below:

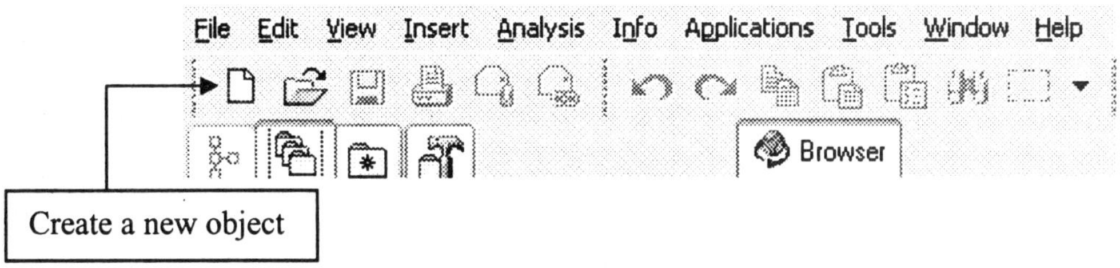

Make sure **"Part"** is selected from the window called **New**. Type the name of the file, or accept the default name, such as *prt0001* shown on display. Click the button of **OK**. A design screen will be on display, as shown in Figure 2-2.

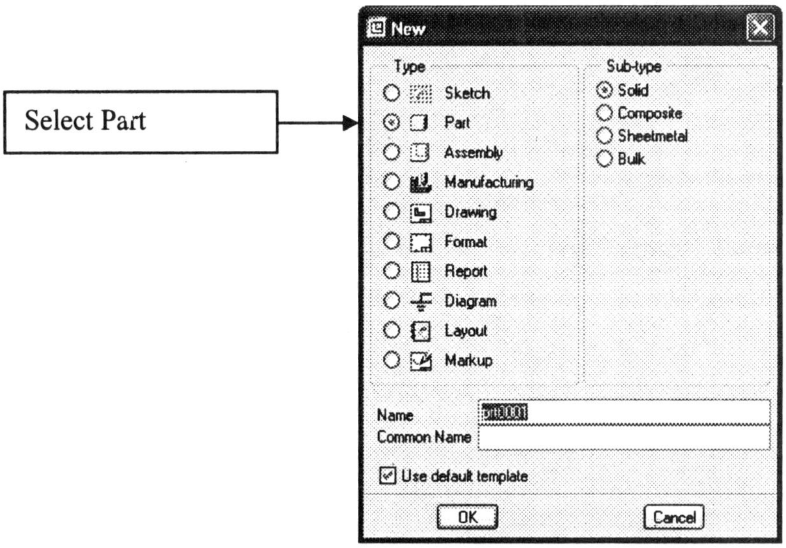

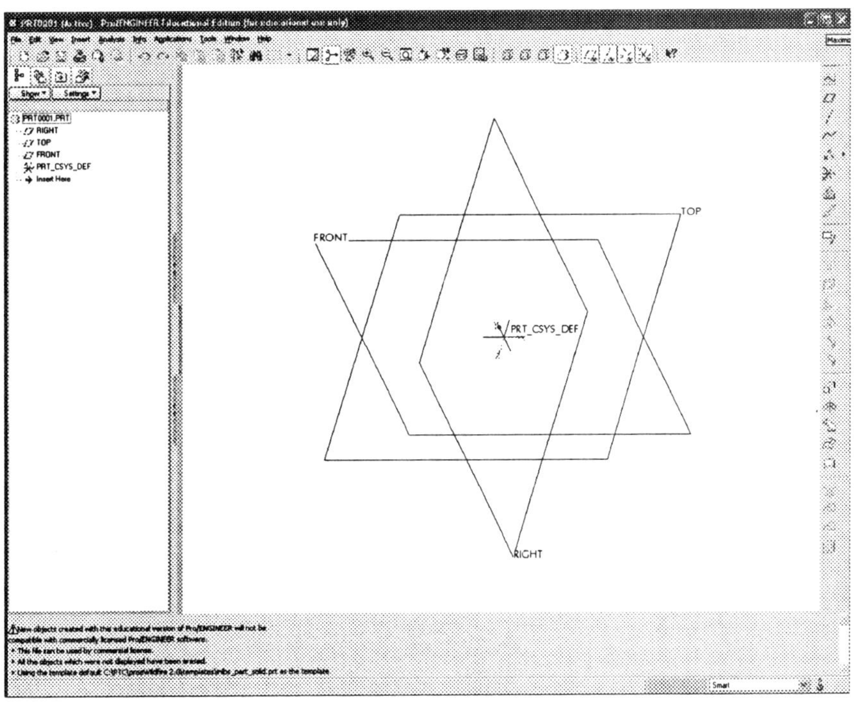

Figure 2-2 Window used for the Design Process

There are three datum planes displayed on screen. They are marked as **FRONT, TOP** and **RIGHT**. These three datum planes are orthogonal to each other. The intersection of these three planes is a point. It is this intersecting point where the origin of a default coordinate system is established. The orientation of the default coordinate system, called **PRT_CSYS_DEF**, is set in the following way, as illustrated in Figure 2-3:

The **FRONT** datum plane represents the X-Y plane.
The **TOP** datum plane represents the X-Z plane.
The **RIGHT** datum plane represents the Y-Z plane.

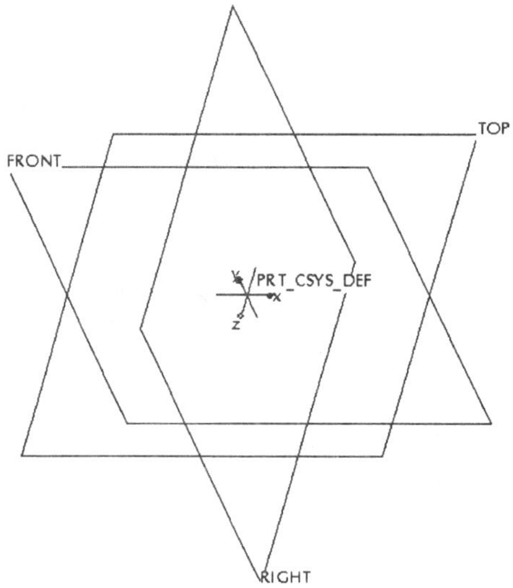

Figure 2-3 The Three Default Datum Planes and the Default Reference System

To start the design process, from the main toolbar, select the icon of **Insert > Extrude**, and note the appearance of a new small window at the bottom of the design screen. The left side of the small window is called "**Dashboard**". The right side of the small window is called "**Feature Control Panel**".

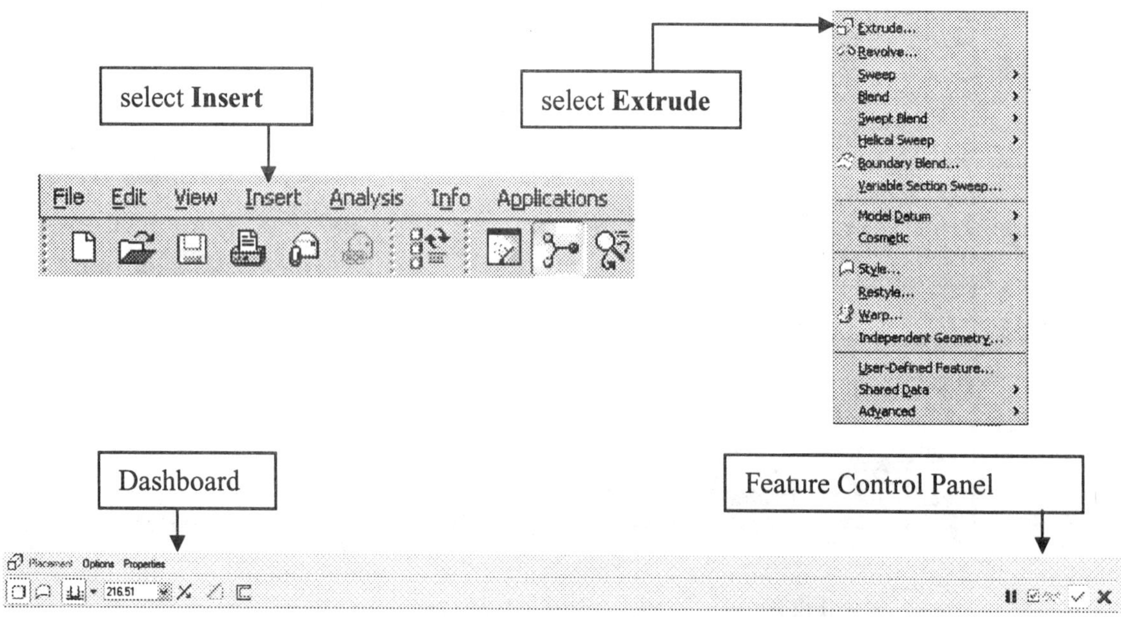

To simplify this procedure to get to the **Dashboard**, a user may just pick up the icon of **Extrude** displayed on the toolbar of feature creation. Make sure that the icon of **Solid** is selected. To initiate the process of sketching, activate **Placement** > **Define.**

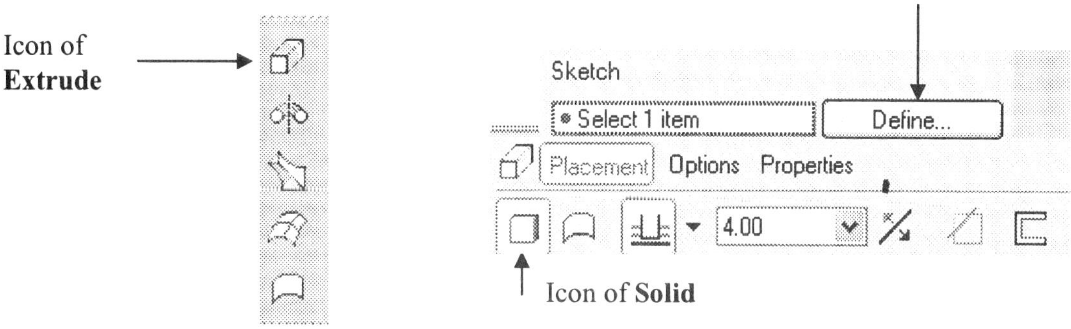

Select the **FRONT** datum plane displayed on screen as the sketch plane. Pick the box of **Sketch** to accept the **RIGHT** datum plane as the default reference to orient the sketch plane, as illustrated below:

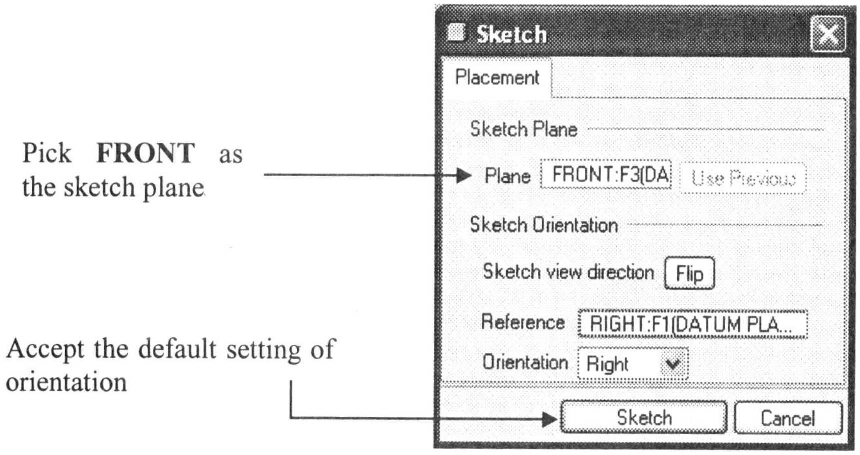

Upon completing this step, the selected sketch plane is on display with the default orientation setting, as shown in Figure 2-4. Let us first look at the small window called References. The Pro/ENGINEER design system, by default, has selected two datum planes as references for making a 2D sketch. These two datum planes are the **RIGHT** datum plane and the **TOP** datum plane. Let us accept these selections by clicking the box of **Close**.

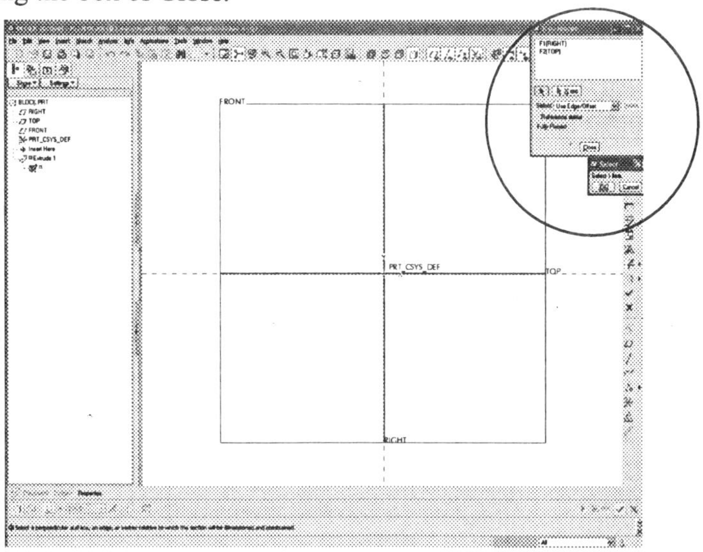

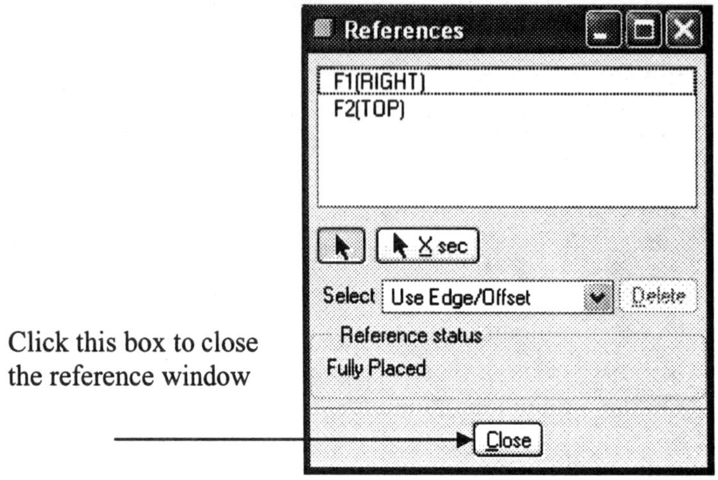

Figure 2-4 Window to Display the Selected References for Making a Sketch

Let us create a 2D sketch. Because we have selected the **FRONT** datum plane as the sketch plane, the 2D sketch will be created in the x-y plane. From the toolbar of sketcher, pick the icon of **Rectangle**, and start the sketch by picking one point on the Y axis and the other point on the X axis, as shown below. To change the horizontal dimension to 10, double click the numerical number, type *10*, and press the **Enter Key**. To change the vertical dimension to 6, double click the numerical number, type *6*, and press the **Enter Key**.

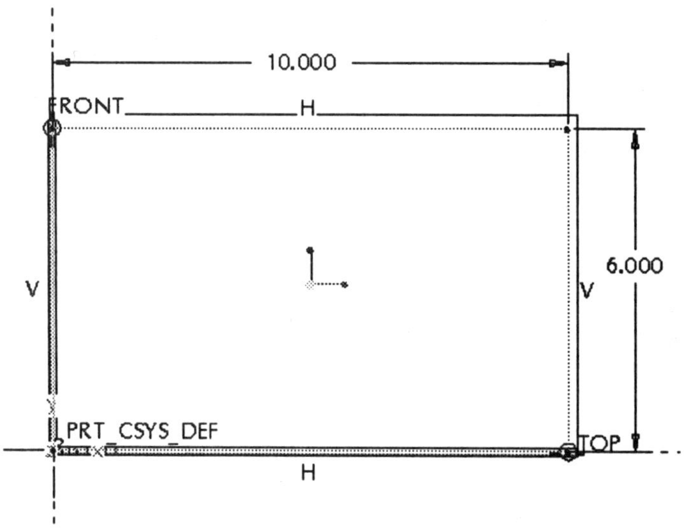

We have now completed the 2D sketch. Pick the icon of **Done** from the toolbar of sketcher. On the **Dashboard**, type *4* as the depth value.

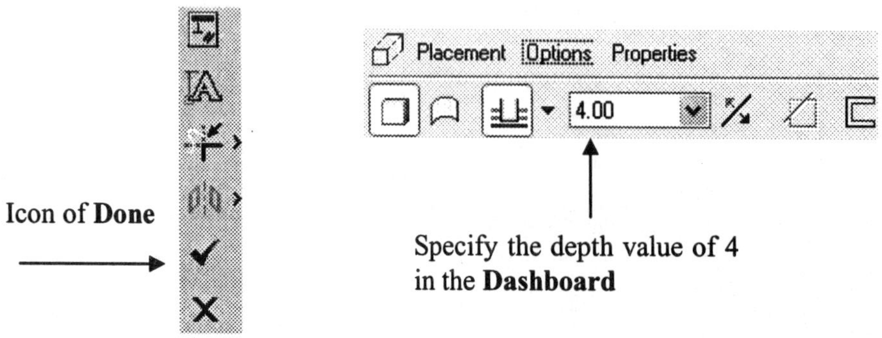

On the **Feature Control Panel**, pick the icon of **Complete** to complete the creation of a 3D solid model, as shown below:

Pick the icon of **Complete** from the **Feature Control Panel**

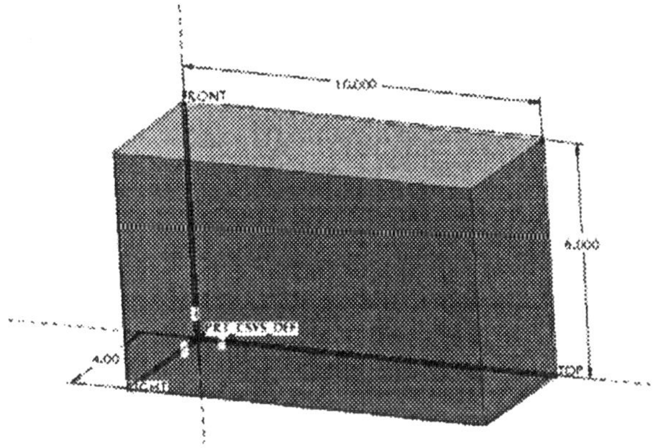

When the creation of a 3D solid model is completed, users are able to view the 3D solid model from different settings in terms of the viewing angle. Select the icon of **Saved view** from the main toolbar > select **Standard Orientation**. An isometric view of the created 3D model is on display.

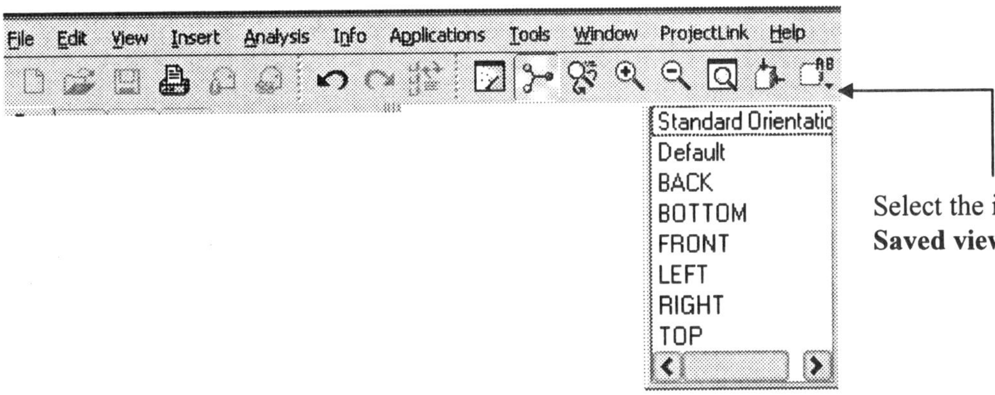

Select the icon of **Saved view list**

Users may also use the buttons of mouse to spin, turn, zoom and pan the 3D solid model. For example, holding down the middle button and moving the pointer will spin the 3D solid model. Information on the mouse controls is available on the Pro/ENGINEER Wildfire Web Tools, and the following information is copied from the Pro/ENGINEER Wildfire Web Tools:

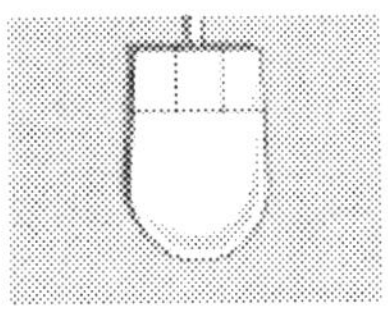

Spin the model by holding down the middle mouse button and moving the pointer.

Turn the model by moving your mouse from left to right while holding down the CTRL key and the middle mouse button.

Zooming takes advantage of the roll wheel, which pulls in the model or pushes it away. You can also zoom by moving your mouse up and down while holding down the CTRL key and middle mouse button.

Panning also complies with the new roll wheel standard by using the SHIFT key and the middle mouse button.

The user may also select the icon of **Shade** from the menu toolbar. The block component displayed on the computer screen will turn to a shaded object. The user may also try other options, such as viewing the object with no hidden line, with hidden line or under wireframe, etc.

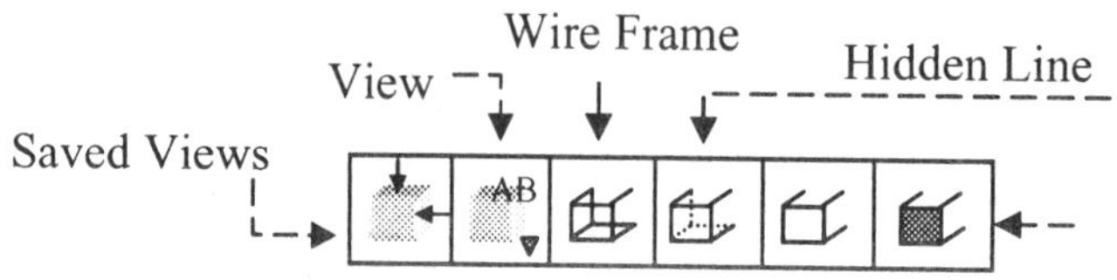

At the time you have successfully completed the block design, remember to save all of your work. You select **Save** from the main toolbar.

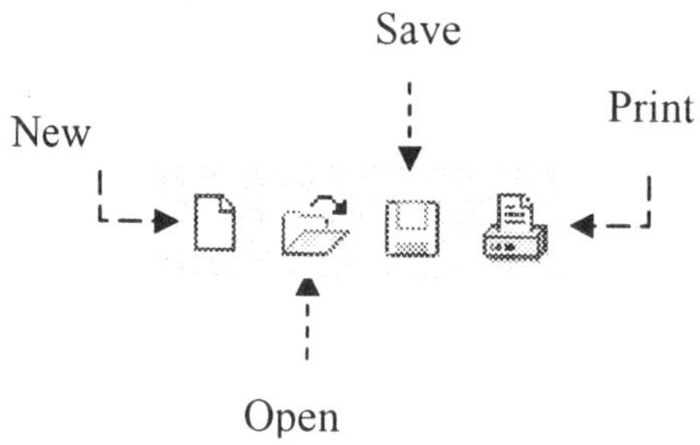

2.2 FEA of a Column Structure Subjected to Compressive Load

In section 2.1, we presented the process of creating a 3D solid model of a block component. In this section, we will get on the process of performing FEA. The example to be used in this section is a column structure, which is subjected to compressive loading. The column structure is shown below:

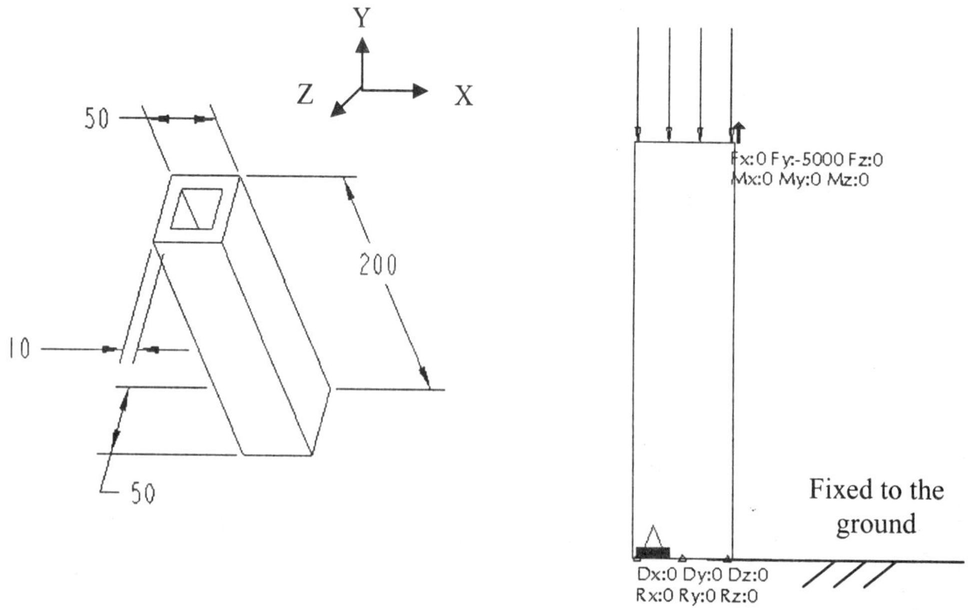

As illustrated, the height of the column is 200 mm. The section area is a square. The two dimensions of the section area are 50 x 50 mm, as shown. It is assumed that the bottom part of the column is fixed to the ground. As illustrated, a compressive load is uniformly acting on the top surface of the column. The magnitude of the load is 5000 Newton. There are 2 objectives to be expected from performing FEA:

(1) Find the location of the maximum displacement on the column when subjected to this compressive load and specify its magnitude.
(2) Find the location of the maximum principal stress and the location of the minimum principal stress on the column when subjected to this compressive load and specify their magnitudes.

To achieve these 2 objectives, in section 2.2.1, we work on the creation of a 3D solid model of the column structure under study. In section 2.2.2, we work on the process of performing FEA.

2.2.1 Creation of a 3D Solid Model of a Column Structure

To begin with the process of geometrical model, we launch the Pro/ENGINEER Wildfire design system. After launching the Pro/ENGINEER Wildfire, select **File** from the menu toolbar or the mail toolbar > **New** to initiate the creation of a 3D solid model of the column structure.

Select **File** from the menu toolbar or the main toolbar

We may directly choose the icon of **Create a new object** from the menu toolbar to initiate the creation of a 3D model.

Select the icon of **Create a new object** from the menu toolbar or the main toolbar

Type *column* as the file name, clear the box of **Use default template** > **OK**. Select the unit of **mmns_part_solid**, type *compression* under the description of the model, and type *student* or *your name* under the **modeled_by** > **OK**. It is important to note the necessity of selecting the unit system. The default unit system used in Pro/ENGINEER Wildfire is the inch_lb_second system. In this textbook, we use the unit system is the mm_Newton_second system. If a unit system other than the mm_Newton_second system is used, a specific note will be made.

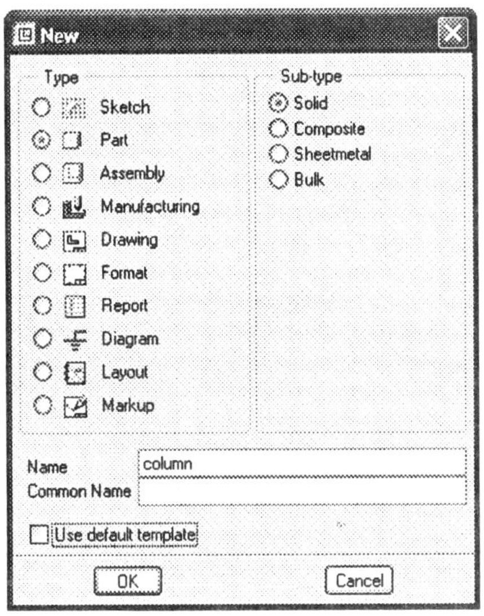

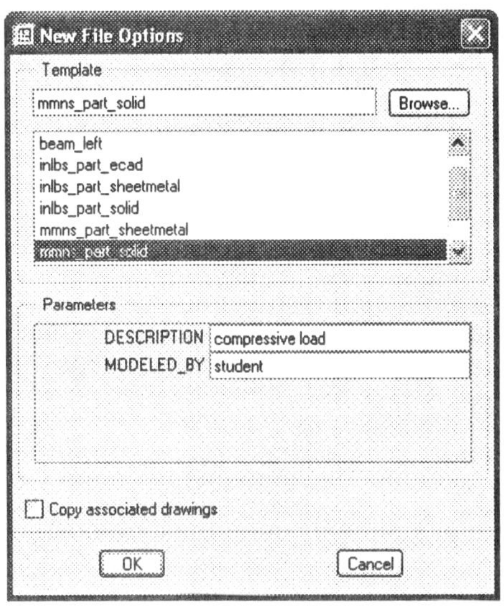

Pick up the icon of **Extrude** displayed on the toolbar of feature creation. Make sure that the icon of **Solid** is selected. To initiate the process of sketching, activate **Placement** > **Define.**

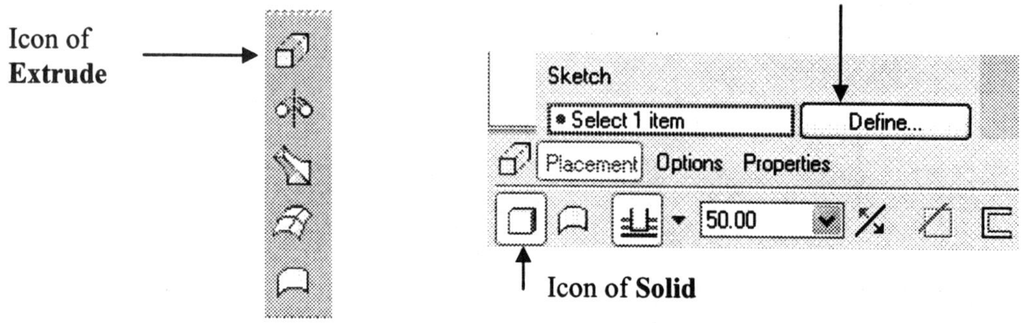

Select the **FRONT** datum plane displayed on screen as the sketch plane. Pick the box of **Sketch** to accept the **RIGHT** datum plane as the default reference to orient the sketch plane, as illustrated below:

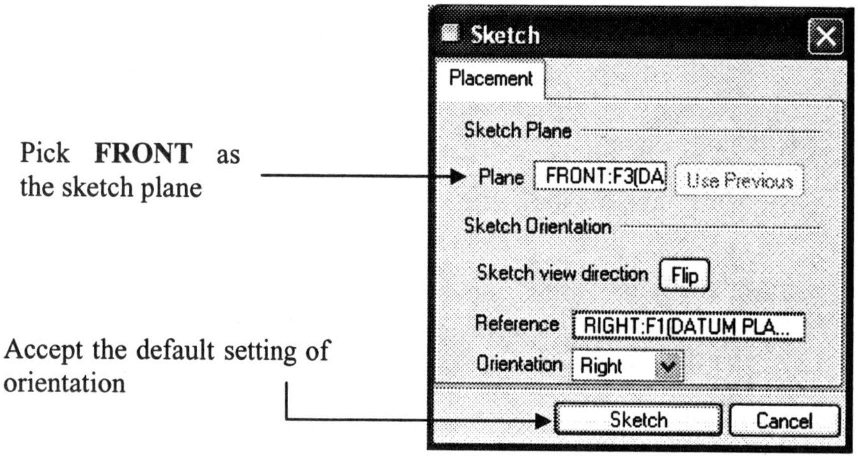

Both the sketch window and the references window appear. Let us first close the reference window. Select the icon of **rectangle** from the toolbar of sketcher and sketch a rectangle. The 2 dimensions are 200 and 50, respectively.

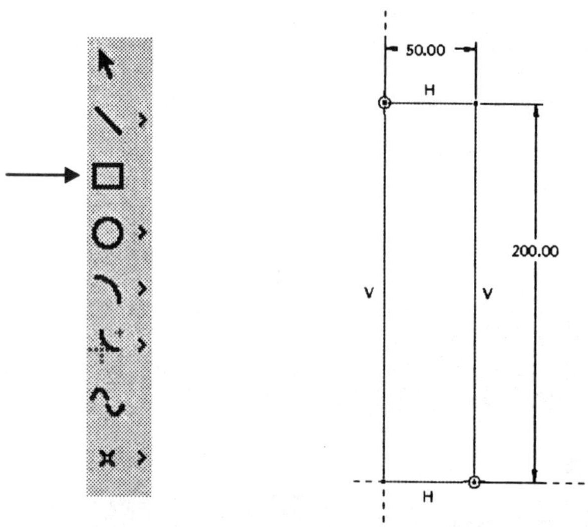

Upon completing the sketch, select the icon of **Done** from the toolbar of sketcher. On the dashboard, type *50* as the thickness value, as shown.

Icon of **Done**

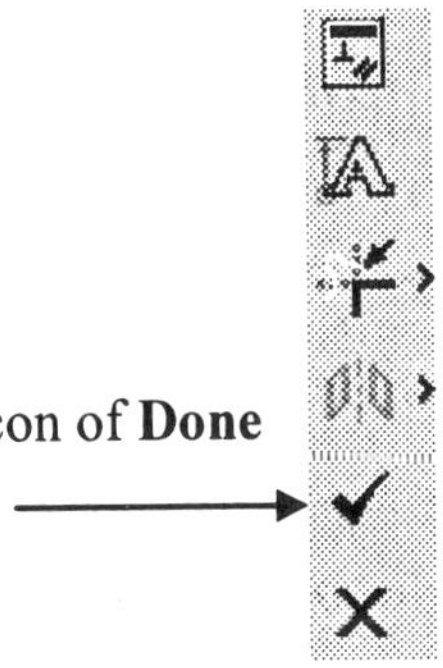

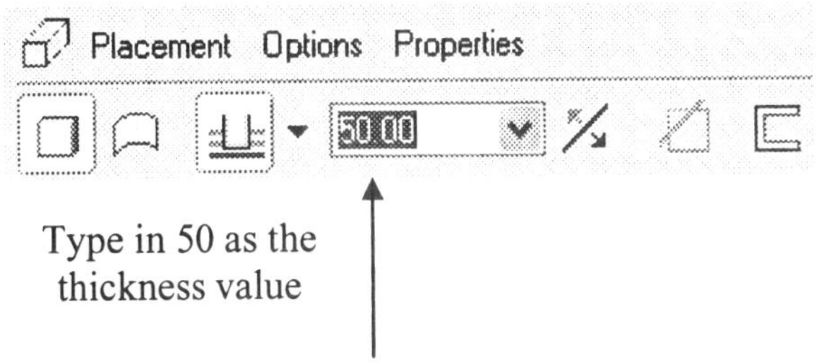

Type in 50 as the thickness value

Select the icon of **Complete** from the feature control panel, completing the creation of a 3D solid model for the column structure.

Pick the icon of **Complete** from the **Feature Control Panel**

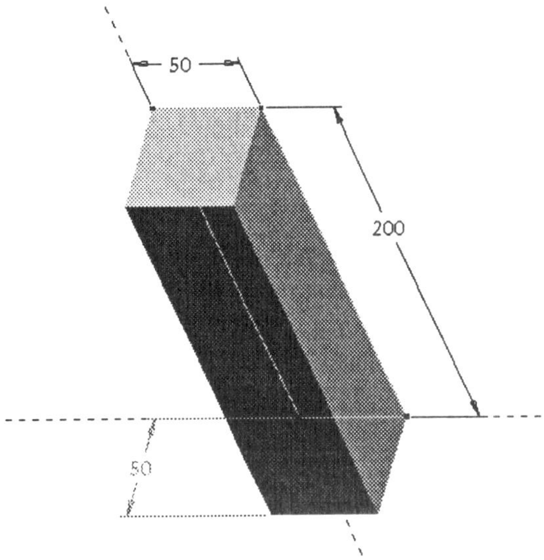

Now let us create a cut feature so that the wall thickness becomes 10 mm.

Select the icon of **extrude** from the toolbar of feature creation. Make sure that the icon of **Cut** is selected. Set the depth choice to **Through All**.

Extrude

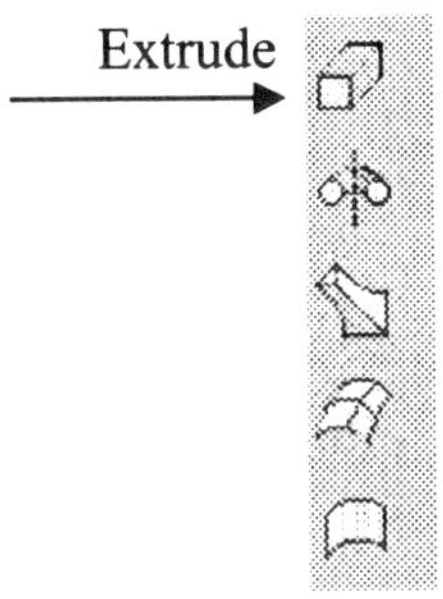

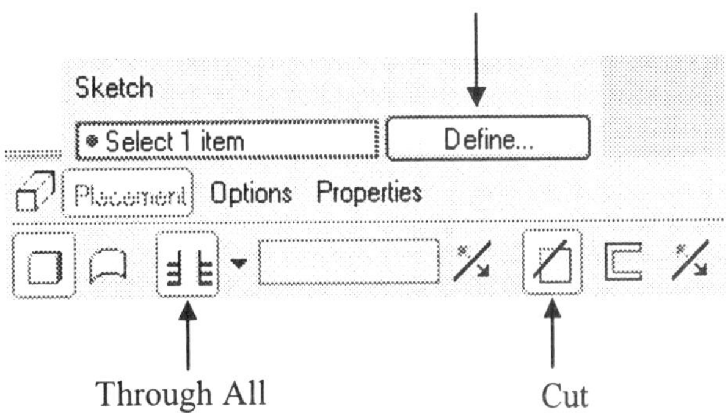

Activate **Placement** > **Define**. Select the top surface of the Column as the sketch plane and use the right-sided surface to be at right to orient the sketch plane > **Sketch**.

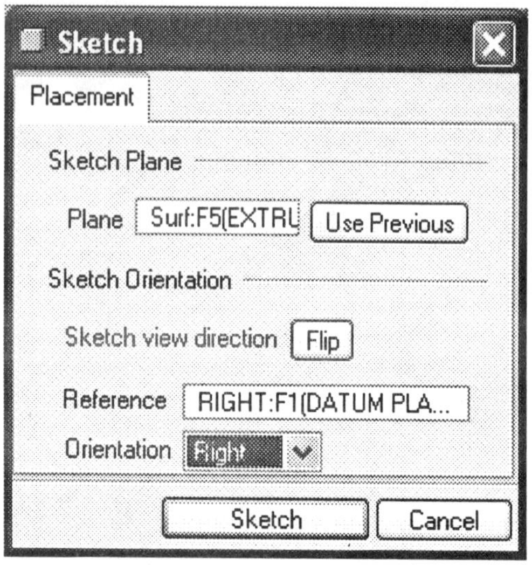

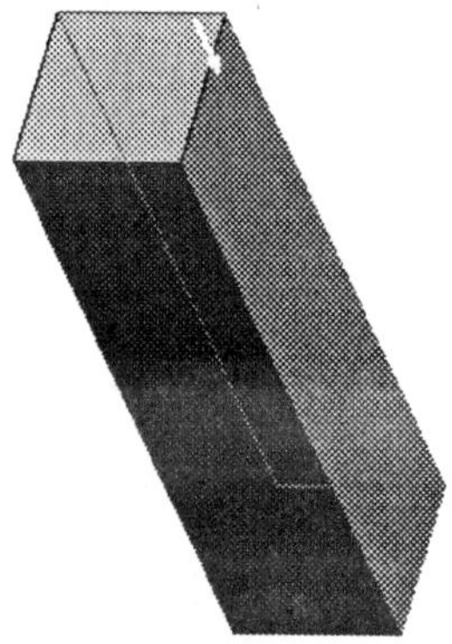

Select the icon of rectangle from the toolbar of sketcher and sketch a rectangle. The 4 dimensions shown below are all set to 10 mm as the wall thickness of the column. Upon completing the sketch, select the icon of **Done** from the toolbar of sketcher.

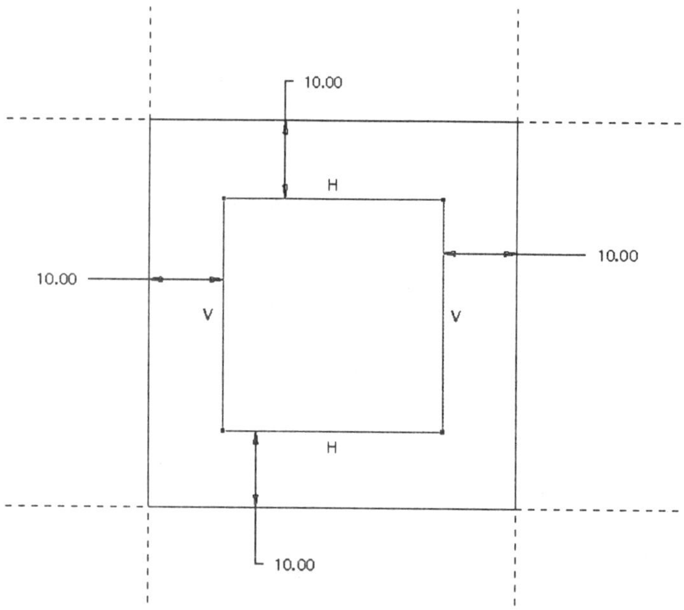

Icon of **Done**

Select the icon of **Complete** from the feature control panel, completing the creation of a cut feature.

Pick the icon of **Complete** from the **Feature Control Panel**

Up to this point, the process of creating a 3D solid model for the column structure is completed. Now let us get on to the process of performing FEA.

2.2.2 Procedure of Performing FEA Using Pro/Mechanica

When the 3D solid model is available, we are able to use Pro/Mechanica to perform FEA for the column structure. There are 5 basic steps involved in such a process.

Step 1: Shift from the Pro/ENGINEER design system to Pro/Mechanica
From the main toolbar or the menu toolbar, select **Applications** > **Mechanica.**

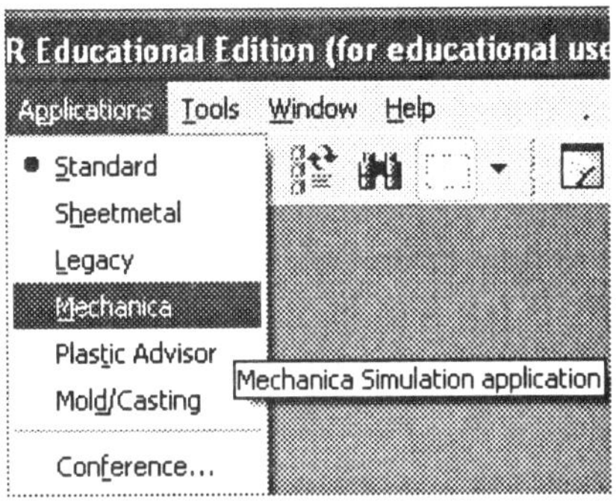

A window called **Unit Info** appears. The displayed information is to verify the unit system used in the process of creating the 3D solid model. As illustrated, the unit system used to create the column model is mm_Newton_second system where the unit of mass is tonne and the unit of temperature is set to Celsius (deg C).

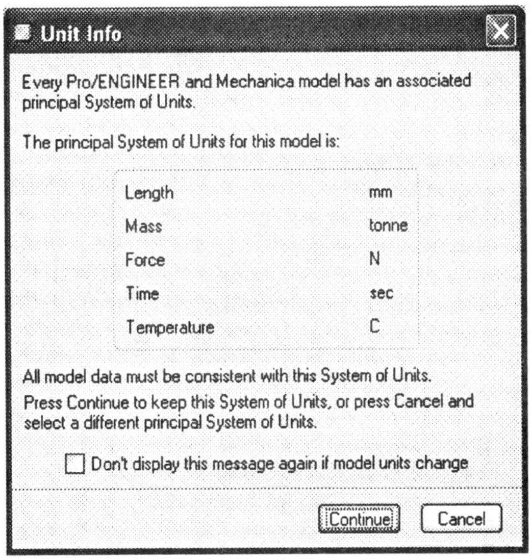

Click the box of **Continue** after checking the unit system. Select **Structure** from the Model Type window > OK.

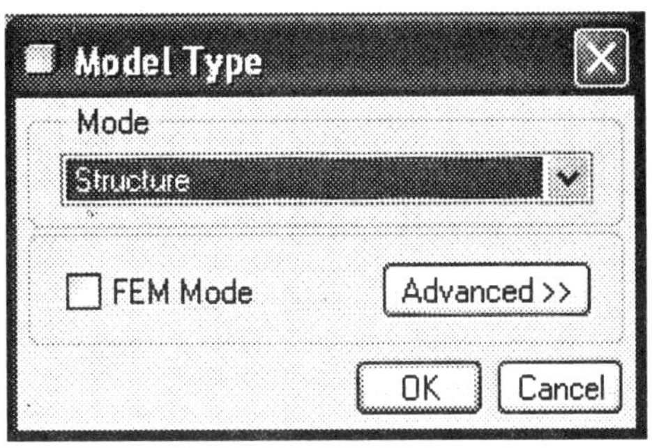

From the toolbar of functions, select the icon of **Materials** > select **STEEL** from the left side called Materials in Library > click the directional arrow so that the selected material type goes to the right side called Materials in Model.

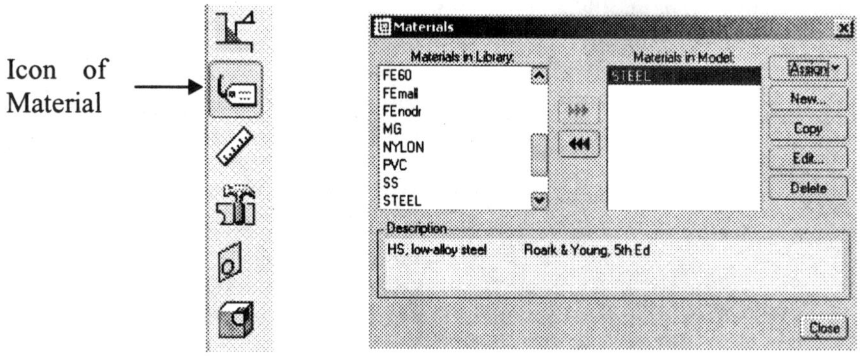

Assign > **Part** > select the column structure from the screen display > **OK** > **Close**.

To define the fixed end constraint on the bottom of the column, select the icon of **New Displacement Constraint** from the toolbar of functions > **Surface(s)** > pick the surface at the bottom of the column > **OK** to accept the surface selection.

Because the constraint condition is "a fixed end", we set all 6 degrees of freedom equal to zero, which is the default setting > **OK**.

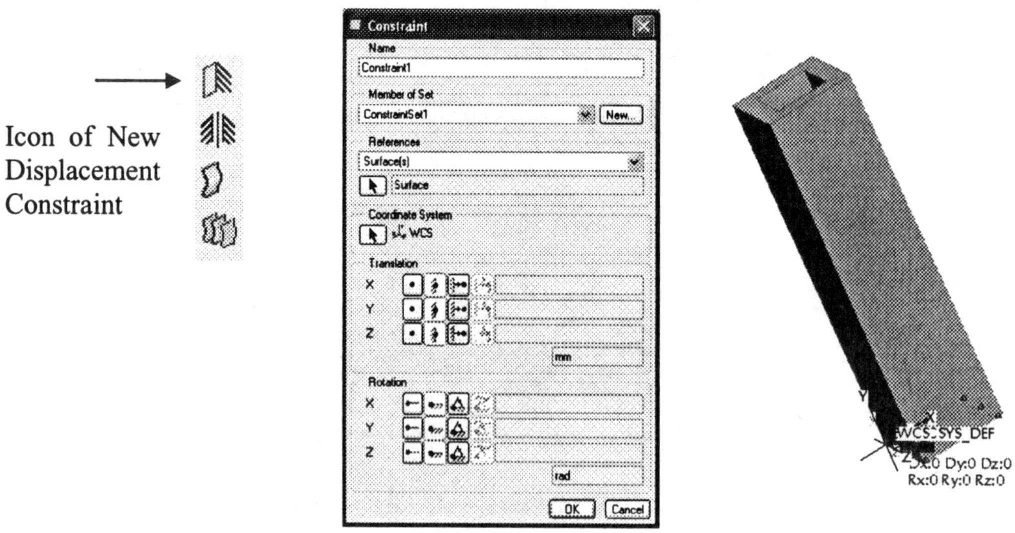

To define the load condition on the top surface of the column, select the icon of New Force/Moment Load from the toolbar of functions > **Surface(s)** > pick the surface at the top of the column > **OK** to accept the surface selection.

Because the load is acting along the negative Y direction, we type -5000 in the Y force component box > **OK**.

Icon of New Force/Moment Load

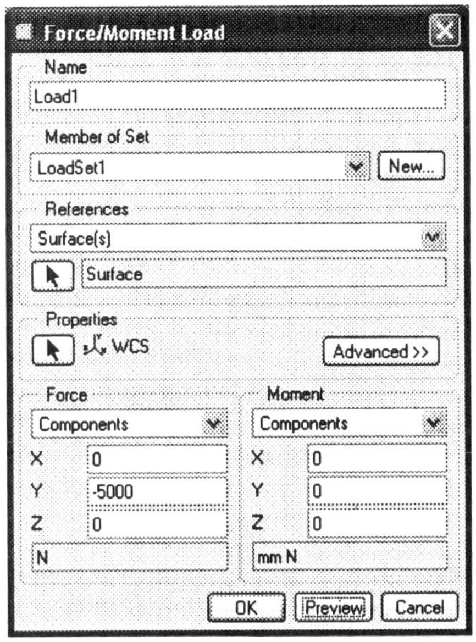

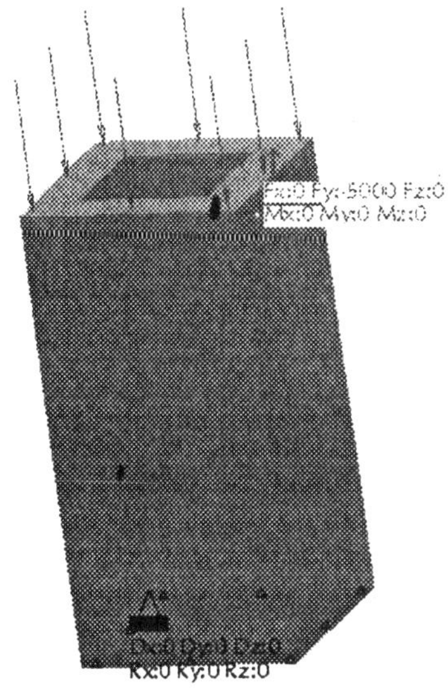

Under the Pro/MECHANICA environment, set up **Analyses** and **Run** it

Select the icon of **Run an Analysis** from the main toolbar > **File** > **New Static** > type *column* as the name of the analysis folder > **OK.** Click the box of **Run.**

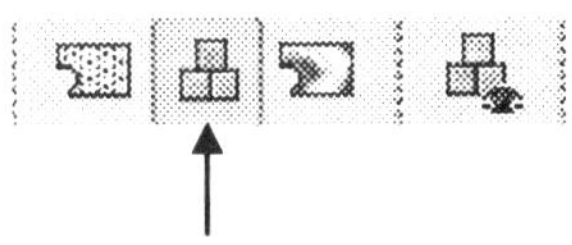

Run an Analysis

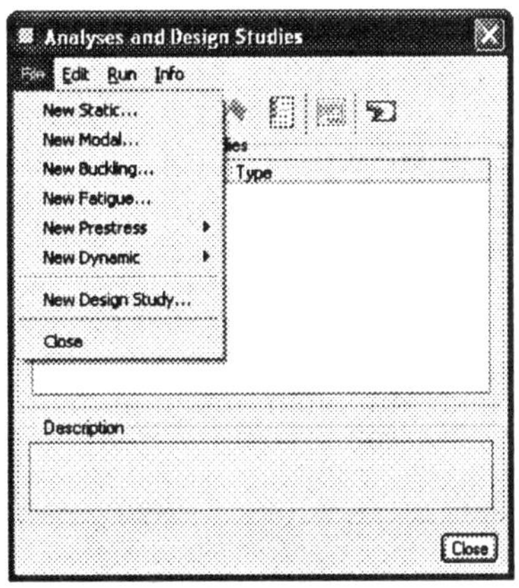

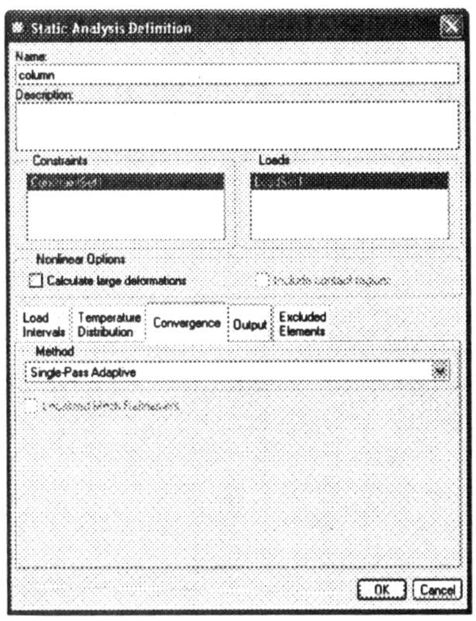

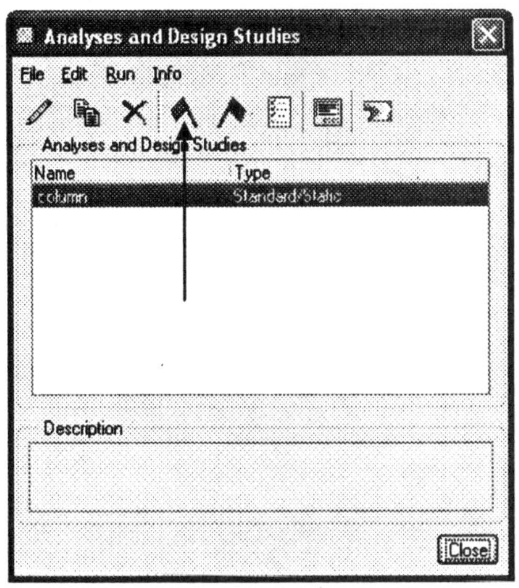

There is always a message displayed on screen. The message is asking the user "Do you want error detection?" Click **Yes**. In this way a warning message will appear if an abnormal operation occurs.

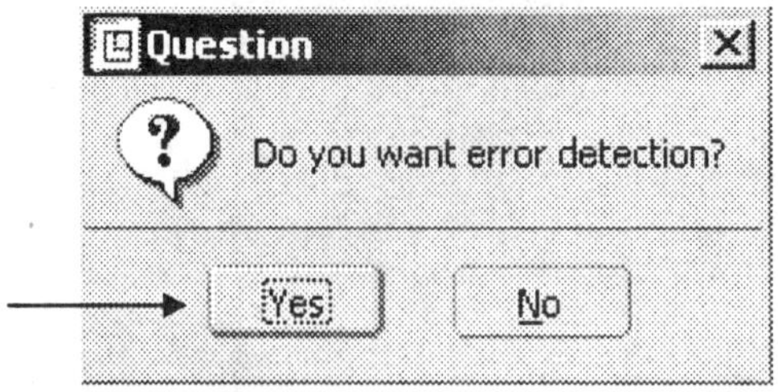

To monitor the computing process, users may click the icon of **Display study status**.

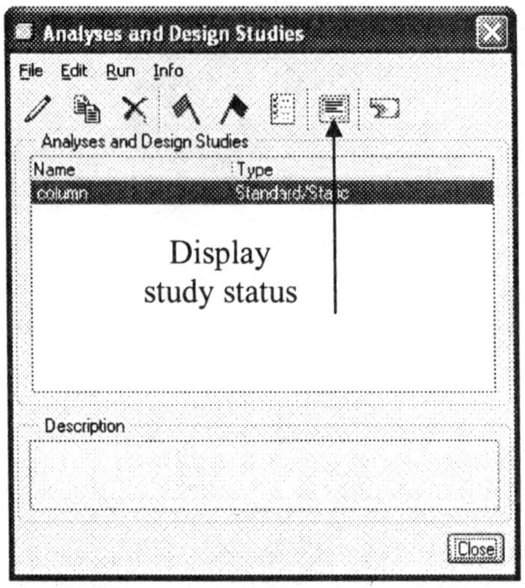

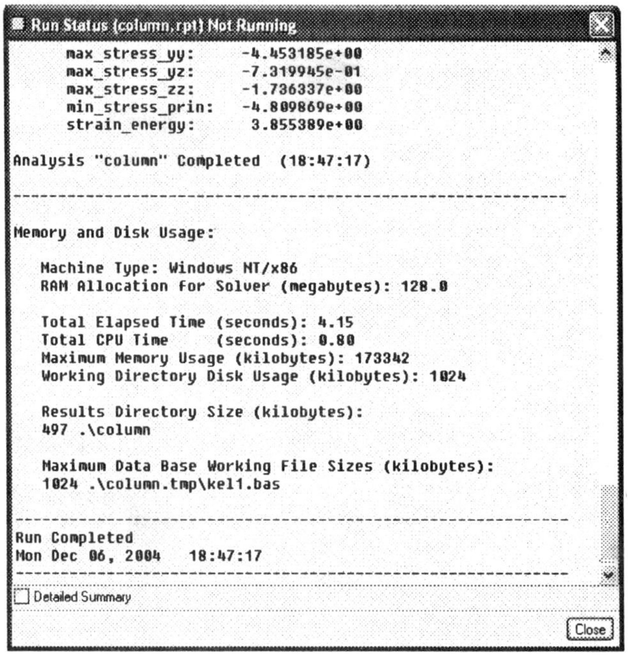

To study the obtained result, say the displacement of the column under the static loading condition, click the icon of **Review results** > **Displacement** > **Magnitude** > **OK and Show.** Note that if the folder called column does not appear in the Result Window Definition, click the icon of folder under Design Study, locate it and open it.

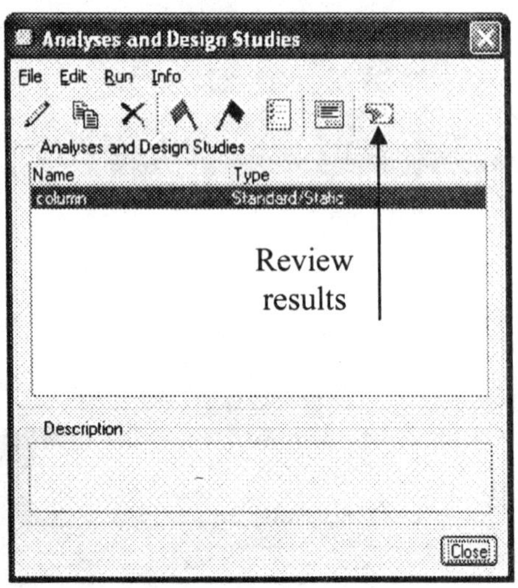

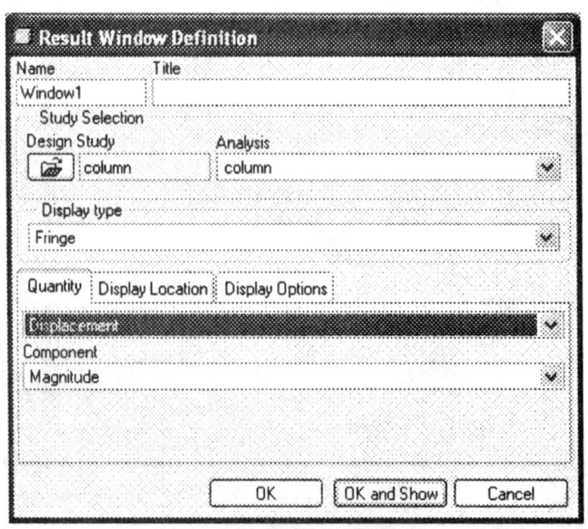

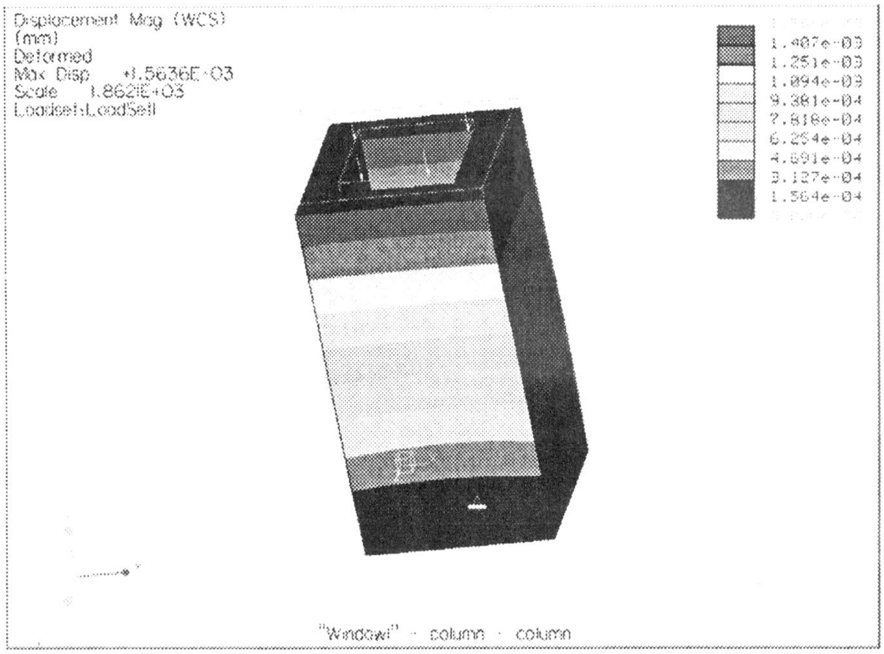

To study the distribution of stress, say the maximum principal stress distribution, select **Edit** from the main toolbar > **Copy** and a new Result Window appears > **Stress** > **Max Principal** > **OK and Show**.

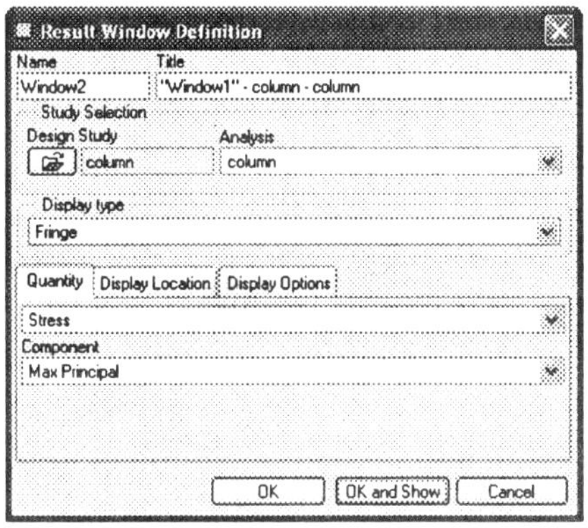

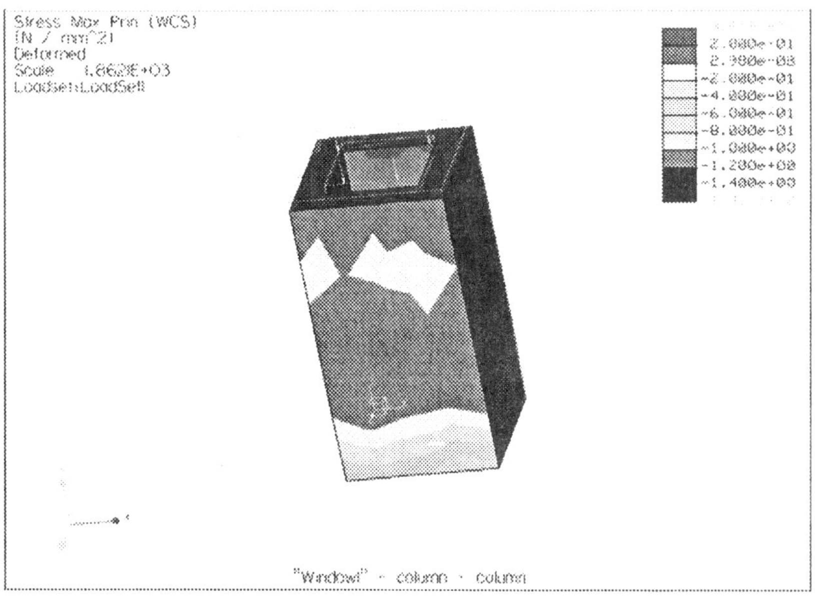

Now let us locate the position of the maximum displacement occurred on the column structure. We need to go back to the window illustrating the displacement distribution. From the main toolbar, select the icon of **Display definition**(s). A new window appears, on which both Window1 and Window2 are highlighted. De-highlighting Window2 and leaving Window1 highlighted will allow the distribution of displacement on display only.

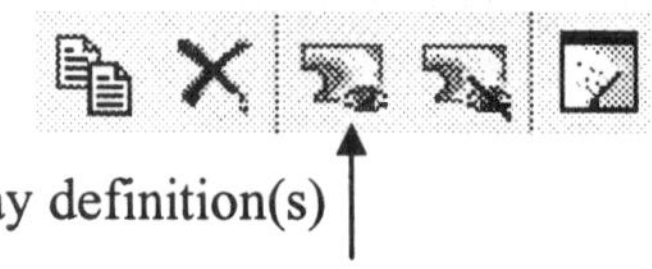

Icon of Display definition(s)

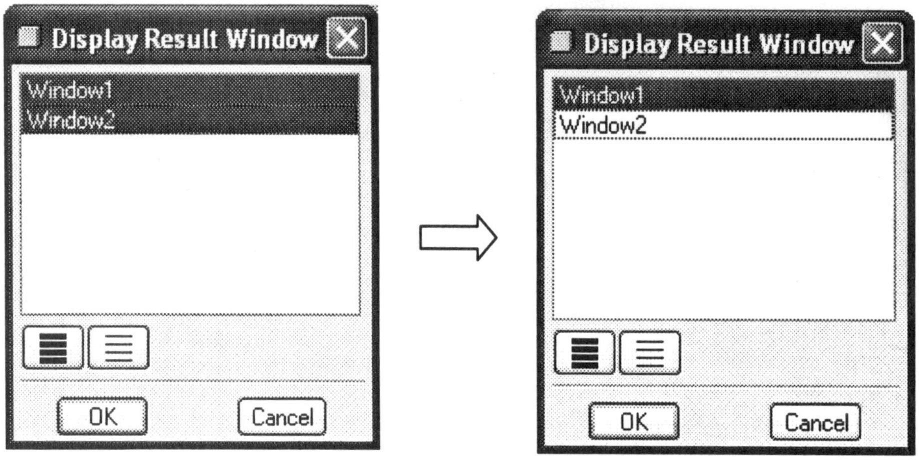

Select **Info** from the main toolbar > **Model Max,** and the magnitude and the location of the maximum displacement appear on the display. As illustrated, the maximum displacement is located at the corner of the top surface (x=0, y=100 and z=50) with the magnitude equal to 0.001564 mm. The minimum displacement is located at the bottom surface, which is fixed to the ground. Users can read the numerical values of those coordinates indicating the model max in the message window, as shown below.

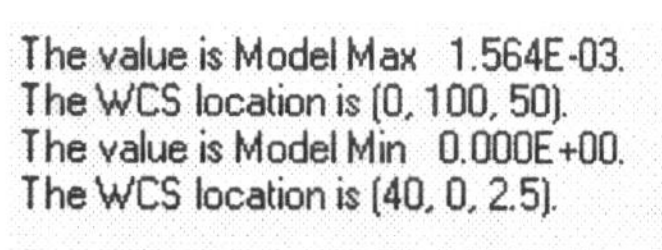

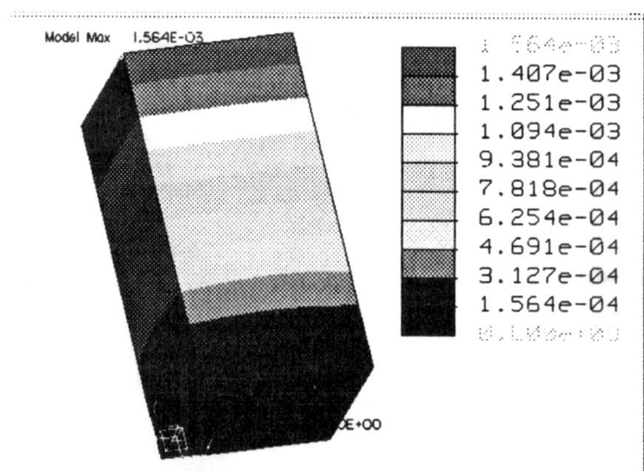

To locate the positions of the maximum and minimum principal stresses occurred on the column structure, follow the same procedure. First go back to the window illustrating the distribution of maximum principal stress. Afterwards, select **Info** from the main toolbar > **Model Max**. Then, select **Info** from the main toolbar again > **Model Min**. As illustrated, the maximum compressive stress is located at the corner of the bottom surface (x=0, y=0 and z=0) with the magnitude equal to 1.421 MPa (10^6 N/mm^2). The maximum tensile stress is located at the point (x=0, y=25 and z=0) with the magnitude equal to 0.3649 MPa (10^6 N/mm^2).

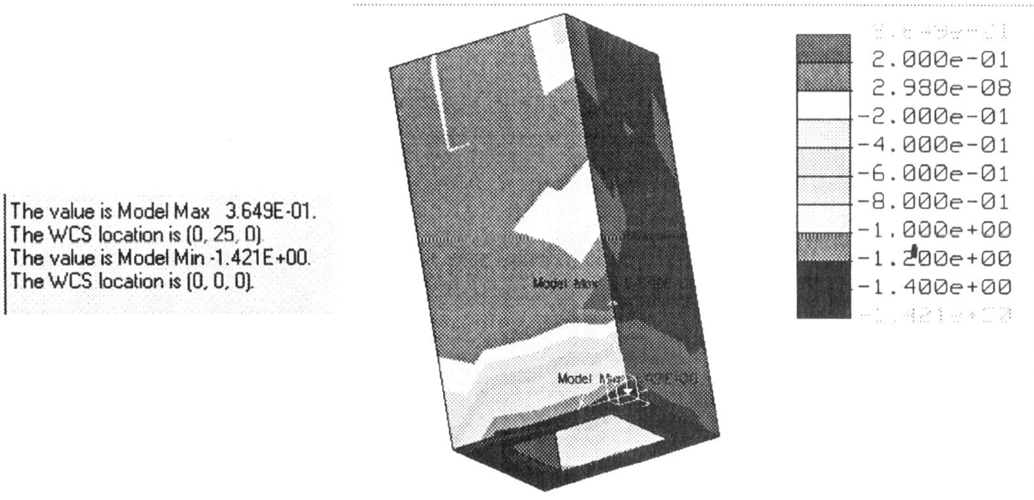

2.3 Beam Structure Subjected to a Uniformly Distributed Load

In section 2.1 and section 2.2, we presented the two examples to demonstrate the procedure of creating a 3D solid model for the physical object under investigation and perform FEA using Pro/Mechanica afterwards. In this section, we will perform a finite element analysis for a beam structure. The geometrical shape of the beam is shown below. The beam by its self is a block. The 3 dimensions are 400 x 80 x 20 mm. There is a hole cut through the beam. The diameter of the hole is 60 mm. The 2 dimensions to locate the hole positions are 200 and 40, respectively.

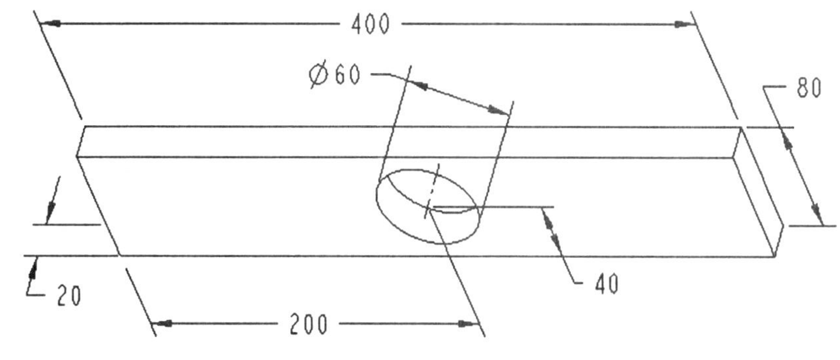

As illustrated below, the left end of the beam is fixed to the wall. Therefore, the structure is a cantilever beam. The load acts on the top surface of the beam. The load is a uniformly distributed load with a total value equal to 1000 Newton. Note the direction of the load is downward. Assume that the material type of the beam is steel.

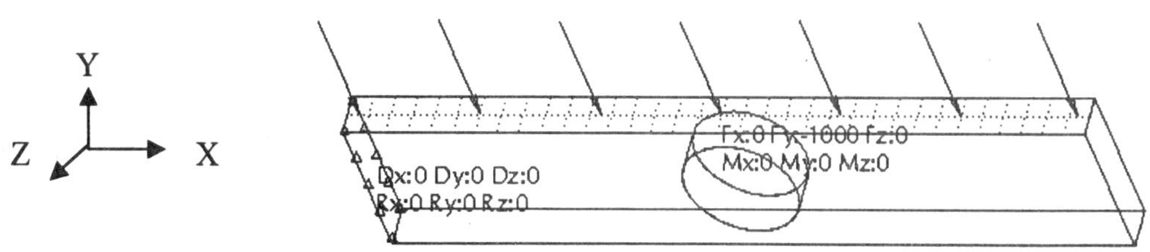

There are 2 objectives for performing FEA:

(1) Find the location of the maximum displacement on the beam when subjected to this uniformly distributed load and specify its magnitude.
(2) Find the location of the maximum principal stress and specify this magnitude.

To achieve these 2 objectives, we follow the procedure discussed in section 2.2. We first work on the creation of a 3D solid model of the beam structure, and afterwards work on performing FEA.

We launch the Pro/ENGINEER Wildfire design system. After launching it, select **File** from the menu toolbar > **New** to initiate the creation of a 3D solid model of the beam.

Select **File** from the menu toolbar or the main toolbar

We may directly choose the icon of **Create a new object** from the menu toolbar to initiate the creation of a 3D model.

Select the icon of **Create a new object** from the menu toolbar or the main toolbar

Type *beam_1* as the file name, clear the box of **Use default template** > **OK**. Select the unit of **mmns_part_solid**, type *cantilever beam* under the description of the model, and type *student* or *your name* under the **modeled_by** > **OK**.

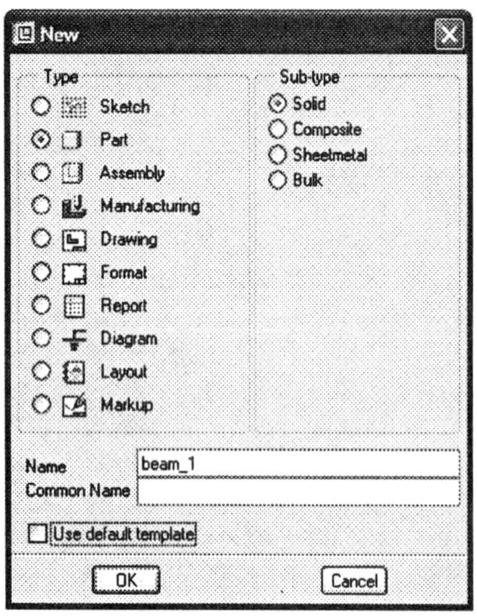

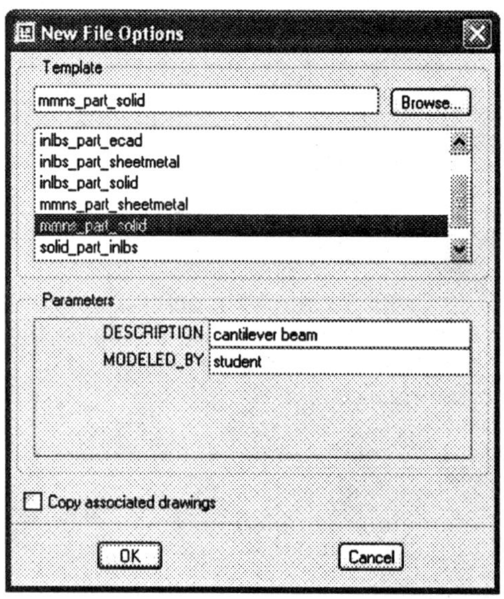

Pick up the icon of **Extrude** displayed on the toolbar of feature creation. Make sure that the icon of **Solid** is selected. Let us set the depth choice by typing *20* (the thickness of the beam). To initiate the process of sketching, activate **Placement** > **Define.**

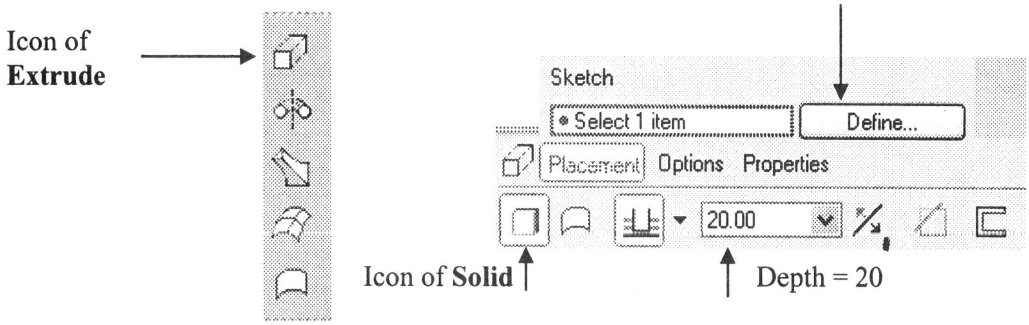

Select the **FRONT** datum plane displayed on screen as the sketch plane. Pick the box of **Sketch** to accept the **RIGHT** datum plane as the default reference to orient the sketch plane, as illustrated below:

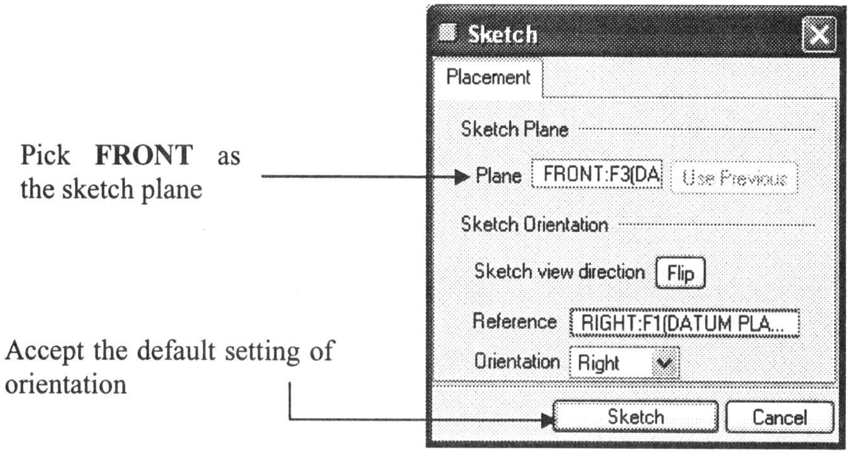

Both the sketch window and the references window appear. Let us first close the reference window. Select the icon of **rectangle** from the toolbar of sketcher and sketch a rectangle. The 2 dimensions are 400 and 80, respectively.

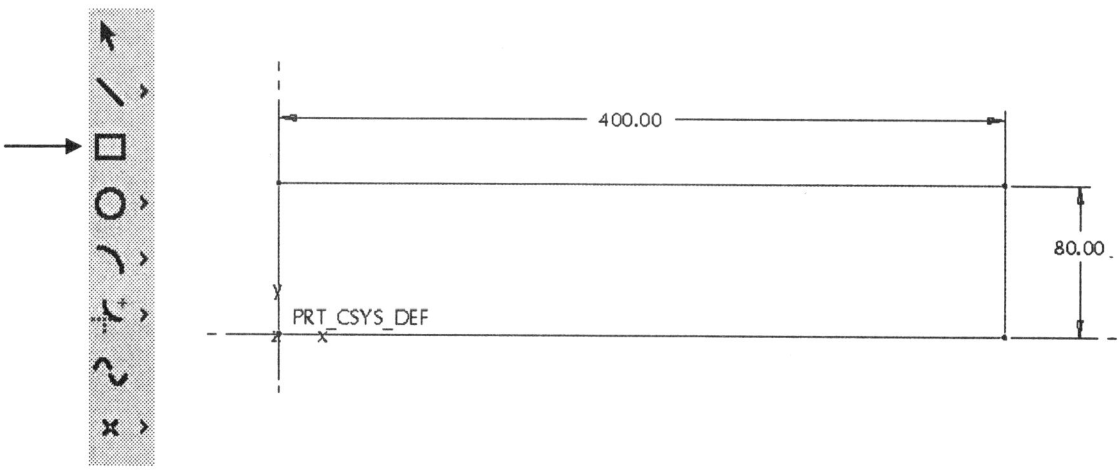

Upon completing the sketch, select the icon of **Done** from the toolbar of sketcher. On the dashboard, type *20* as the thickness value if you have not set the thickness value, as shown.

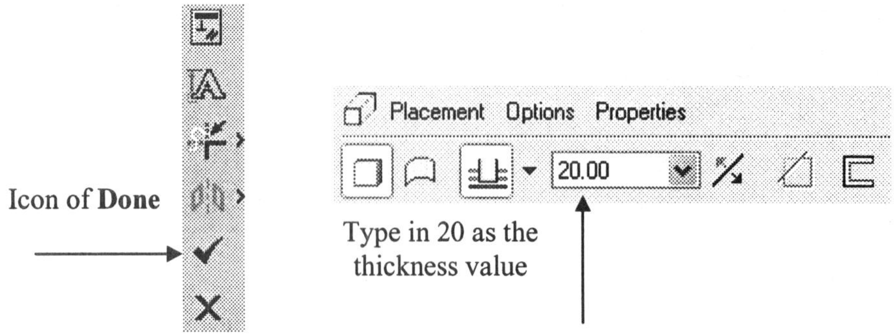

Select the icon of **Complete** from the feature control panel, completing the creation of a 3D solid model for the column structure.

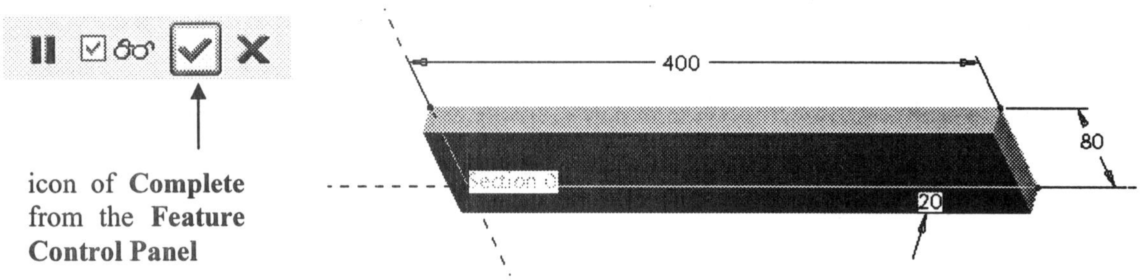

Now let us create a cut feature for the through hole with the diameter equal to 60 mm.

Select the icon of **extrude** from the toolbar of feature creation. Make sure that the icon of **Cut** is selected. Set the depth choice to **Through All**. Activate **Placement** > **Define**.

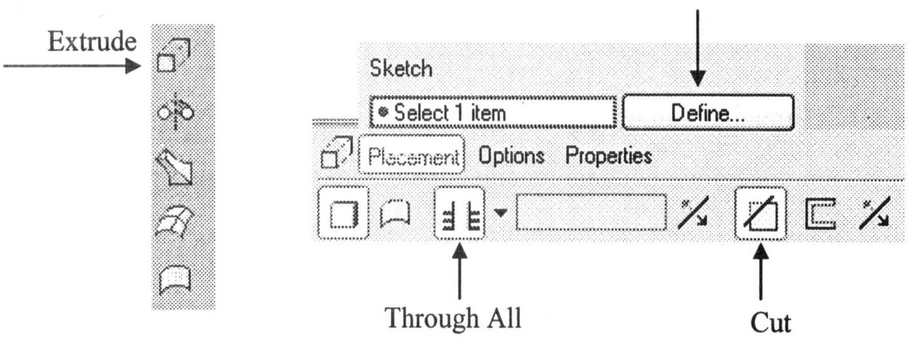

Select the front surface of the beam (do not select the FRONT datum plane) as the sketch plane and use the right-sided surface to be at right to orient the sketch plane > **Sketch**.

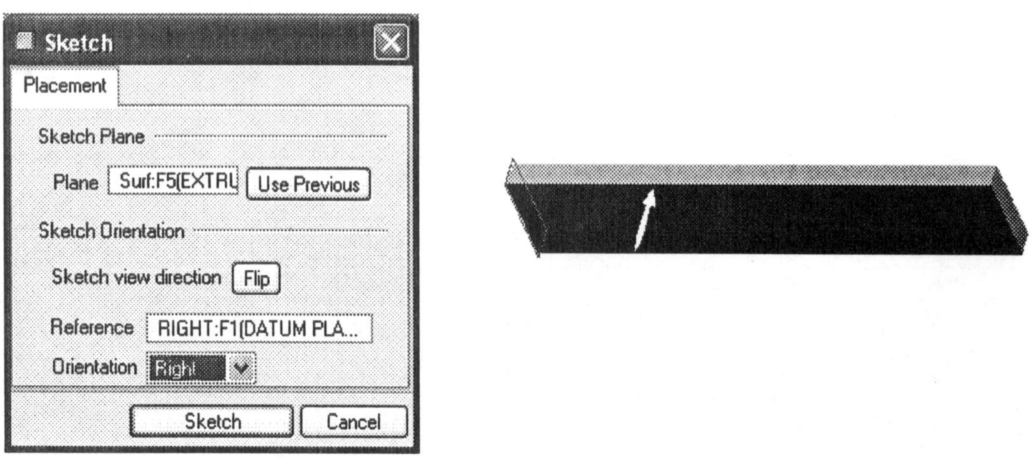

Select the icon of **circle** from the toolbar of sketcher and sketch a circle. Change the diameter dimension to 60, and the two position dimensions to 200 and 40, as shown. Upon completing the sketch, select the icon of **Done** from the toolbar of sketcher.

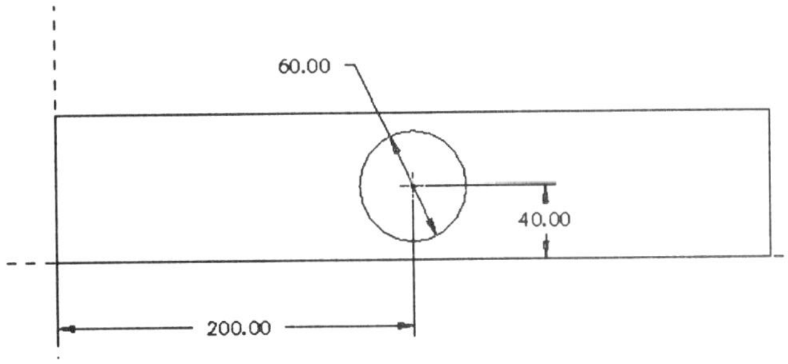

Icon of **Done**

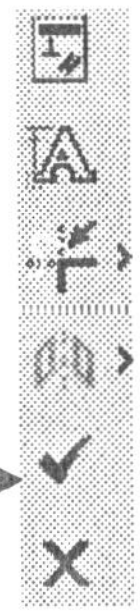

Select the icon of **Complete** from the feature control panel, completing the creation of a cut feature.

Pick the icon of **Complete** from the **Feature Control Panel**

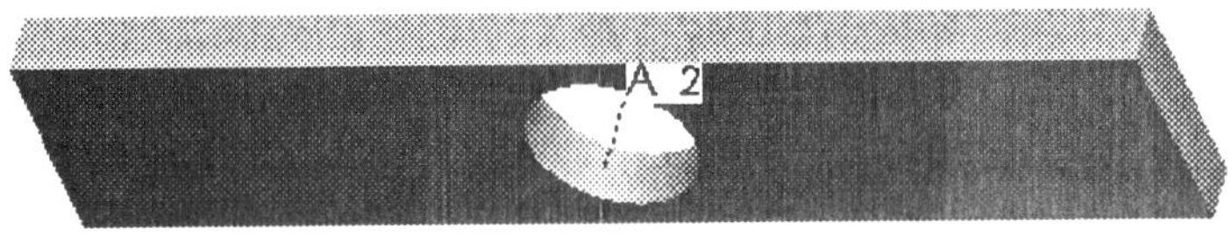

Up to this point, the process of creating a 3D solid model for the beam structure is completed. Now let us get on to the process of performing FEA.

Let us shift from the Pro/ENGINEER design system to Pro/Mechanica. From the main toolbar or the menu toolbar, select **Applications** > **Mechanica.**

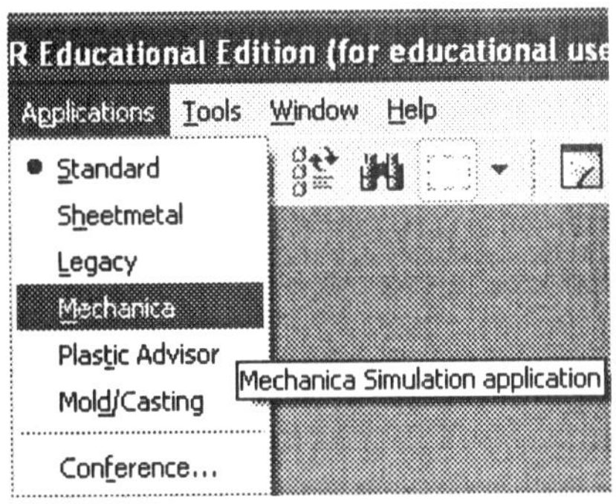

A window called **Unit Info** appears. The displayed information is to verify the unit system used in the process of creating the 3D solid model. As illustrated, the unit system used to create the column model is mm_Newton_second system.

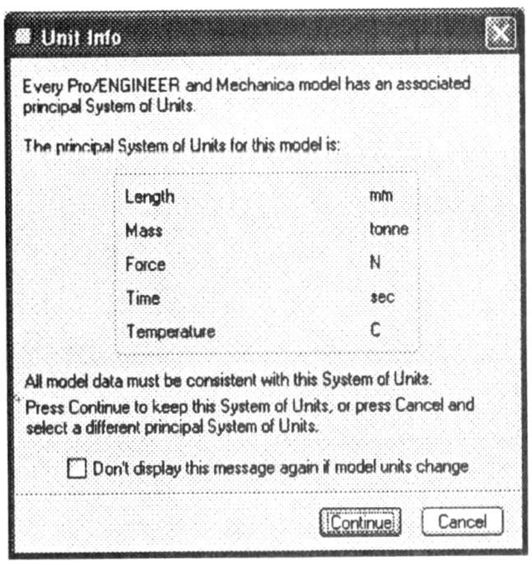

Click the box of **Continue** after checking the unit system. Select **Structure** from the Model Type window > **OK**.

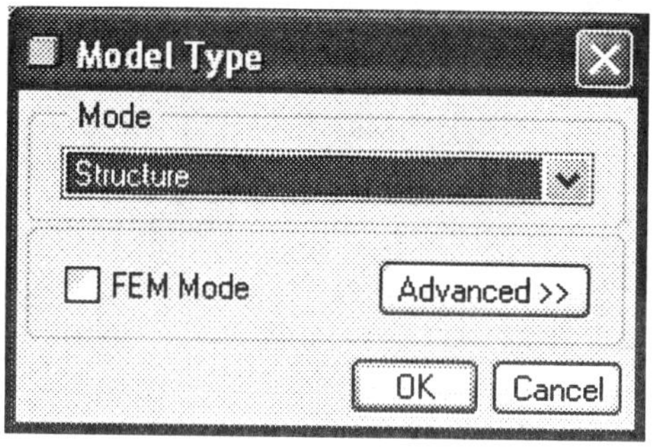

From the toolbar of functions, select the icon of **Materials** > select **STEEL** from the left side called Materials in Library > click the directional arrow so that the selected material type goes to the right side called Materials in Model.

Icon of Material

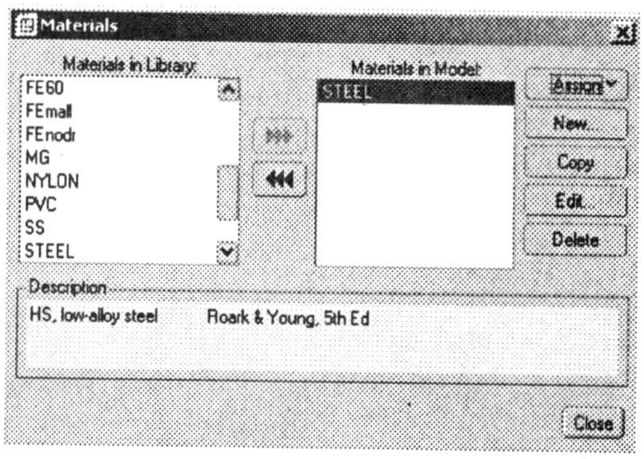

Assign > **Part** > select the beam structure from the screen display > **OK** > **Close.**

To define the fixed end constraint on the left side of the beam, select the icon of **New Displacement Constraint** from the toolbar of functions > **Surface(s)** > pick the surface at the left side of the beam > **OK** to accept the surface selection.

Because the constraint condition is "a fixed end", we set all 6 degrees of freedom equal to zero, which is the default setting > **OK**.

Icon of New Displacement Constraint

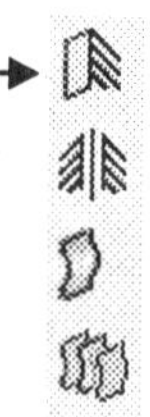

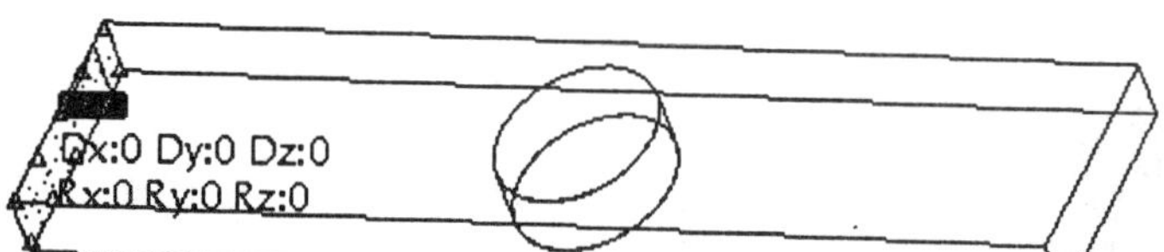

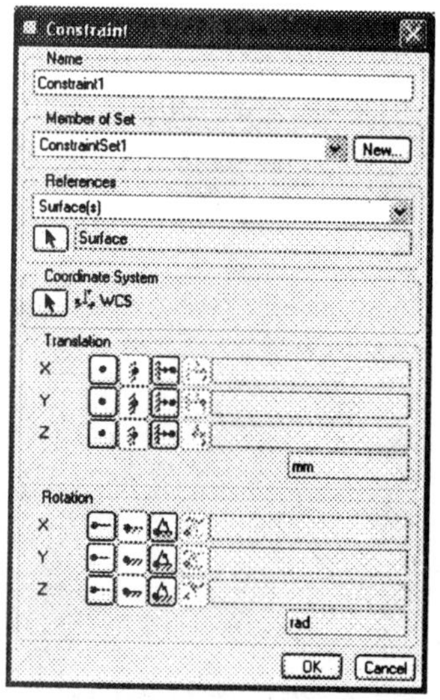

To define the load condition on the top surface of the beam, select the icon of New Force/Moment Load from the toolbar of functions > **Surface(s)** > pick the surface at the top of the beam > **OK** to accept the surface selection.

Icon of New Force/Moment Load

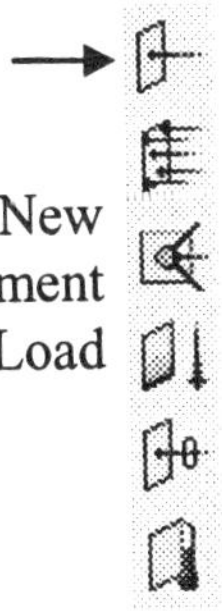

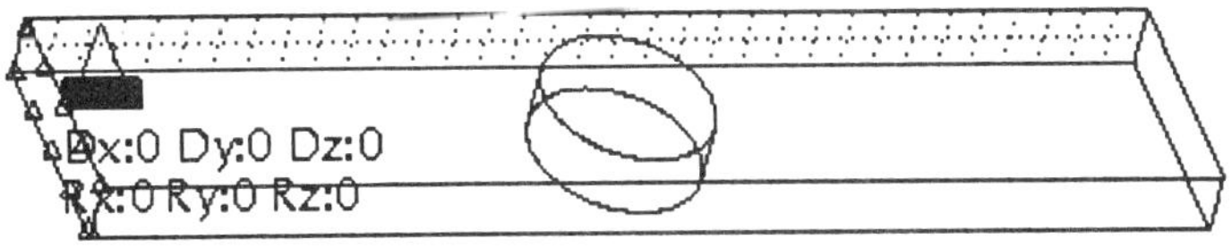

Because the load is acting along the negative Y direction, we type -1000 in the Y force component box > **OK**.

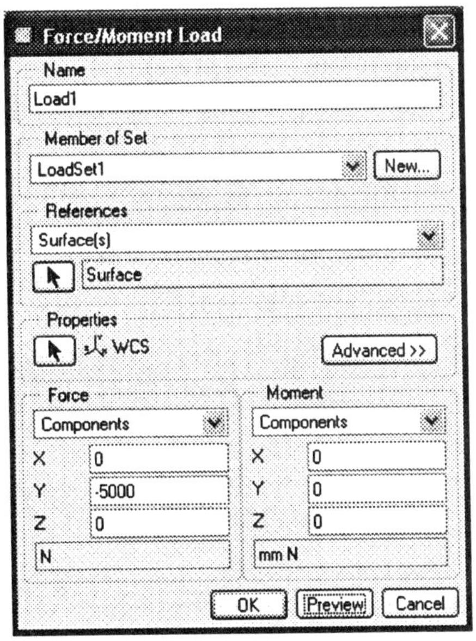

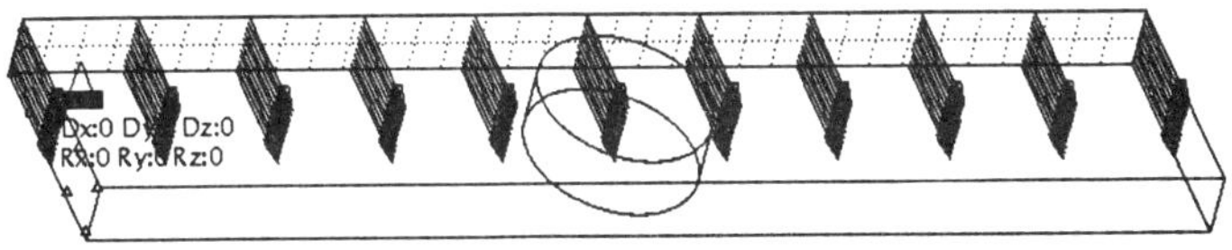

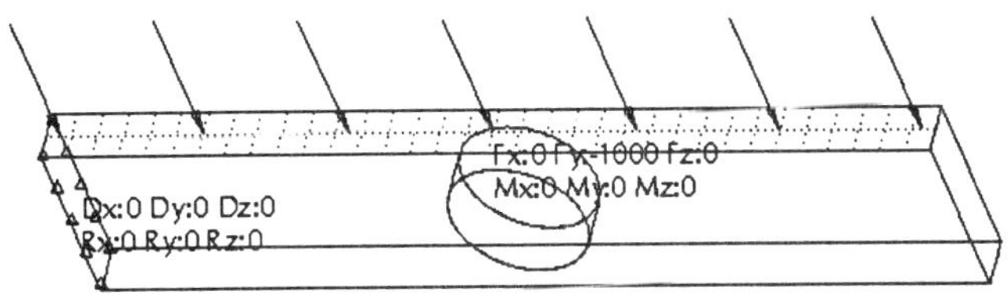

Under the Pro/MECHANICA, set up **Analyses** and **Run** it

Select the icon of **Run an Analysis** from the main toolbar > **File** > **New Static** > type *beam_1* as the name of the analysis folder > **OK.** Click the box of **Run.**

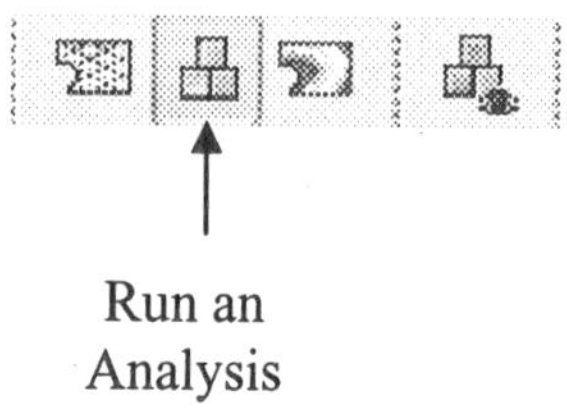

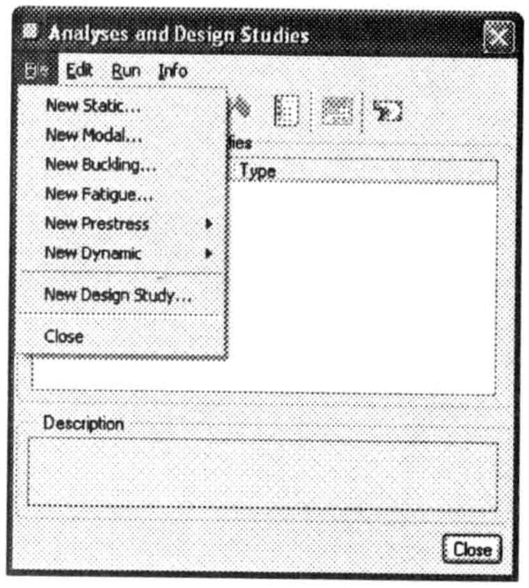

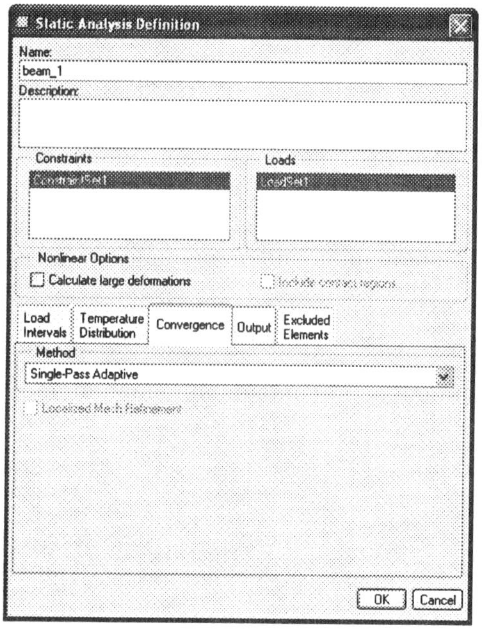

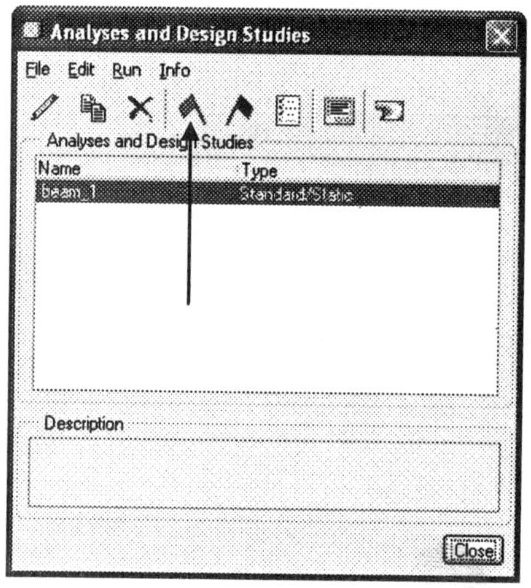

There is always a message displayed on screen. The message is asking the user "Do you want error detection?" Click **Yes**. In this way a warning message will appear if an abnormal operation occurs.

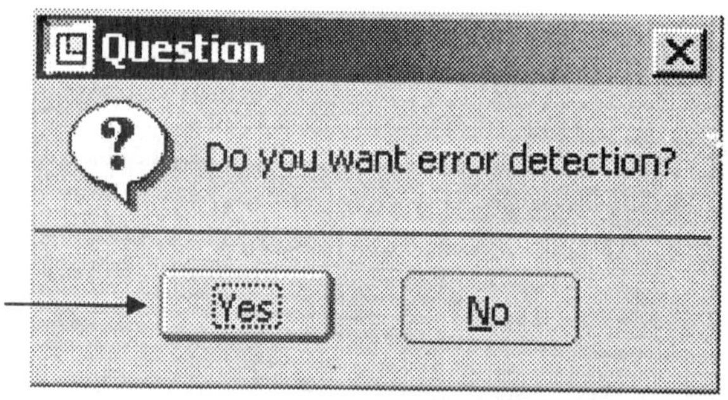

To monitor the computing process, users may click the icon of **Display study status**.

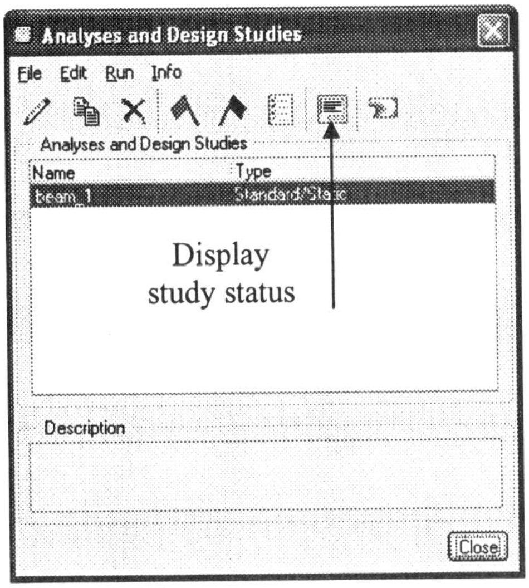

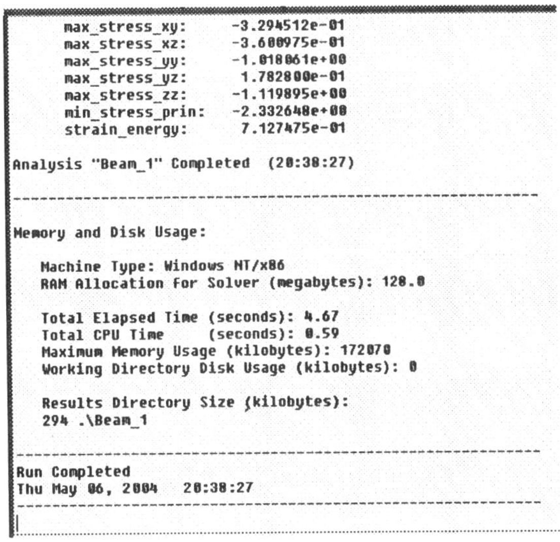

To study the obtained result, say the displacement of the column under the static loading condition, click the icon of **Review results** > **Displacement** > **Magnitude** > **OK and Show**. Note that if the folder called column does not appear in the Result Window Definition, click the icon of folder under Design Study, locate it and open it.

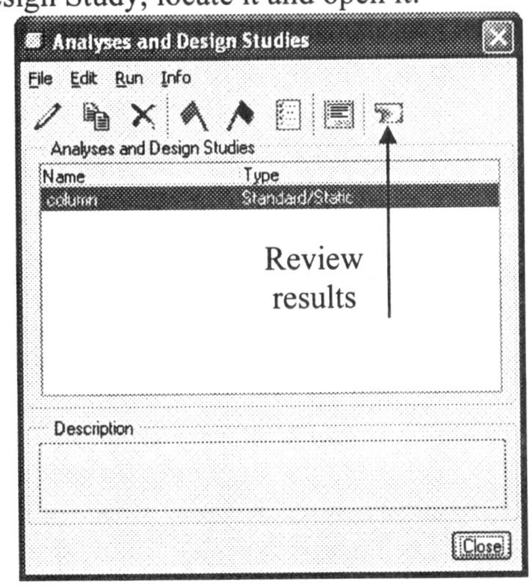

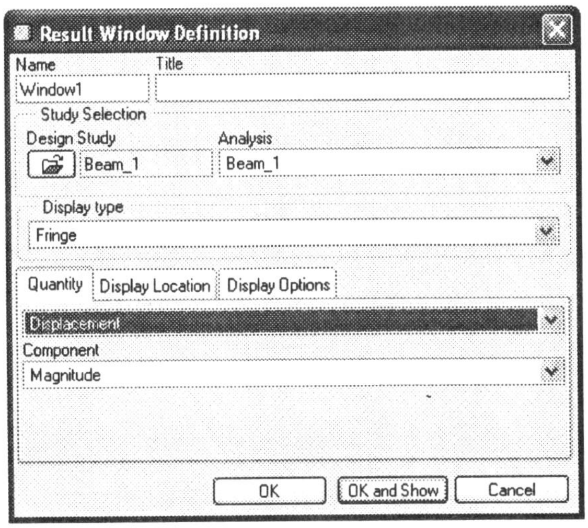

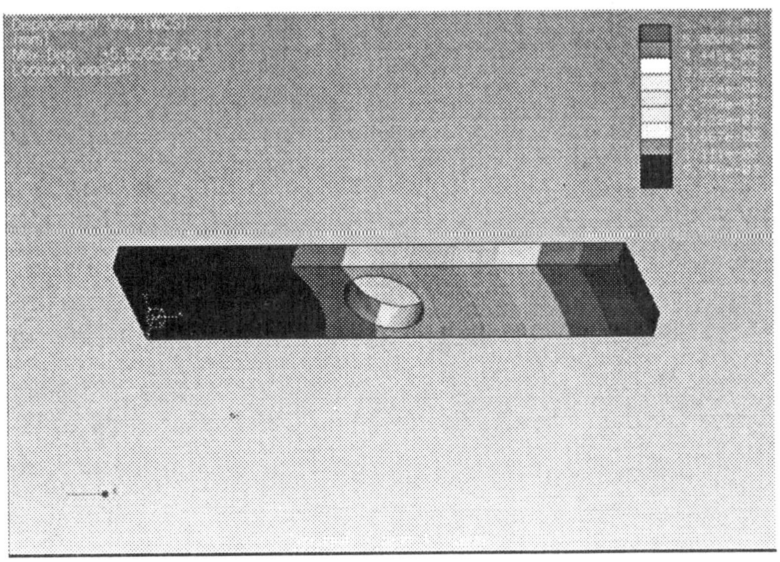

To locate the position of the maximum displacement occurred on the beam structure, select **Info** from the main toolbar, select **Model Max**. The location of the maximum displacement is shown on the display. The location is on the right side of the beam, or the free end of the beam. The magnitude is 0.05556 mm. The numerical values of the coordinates of the location where the maximum displacement occurs are also displayed in the message window, showing x=400, y=80 and z=20.

Select
Model Max

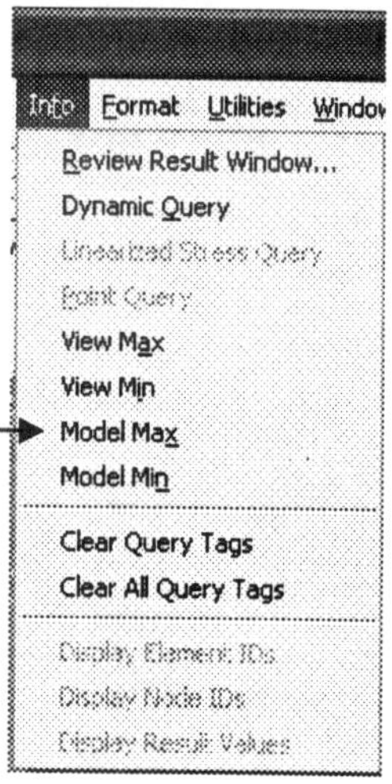

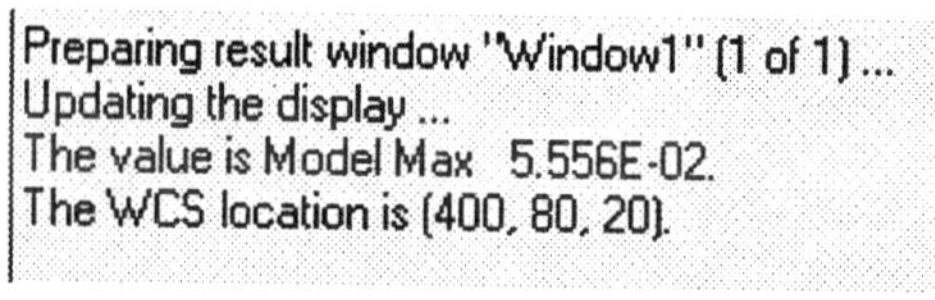

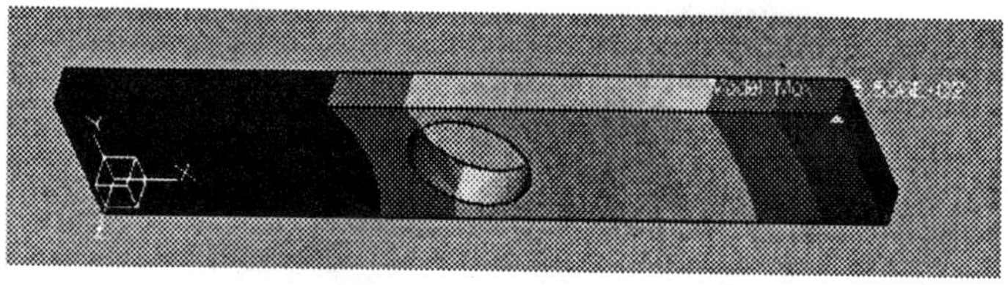

To change the background color to white, select **Format** from the main toolbar > **Result Window** > **White** > **OK**.

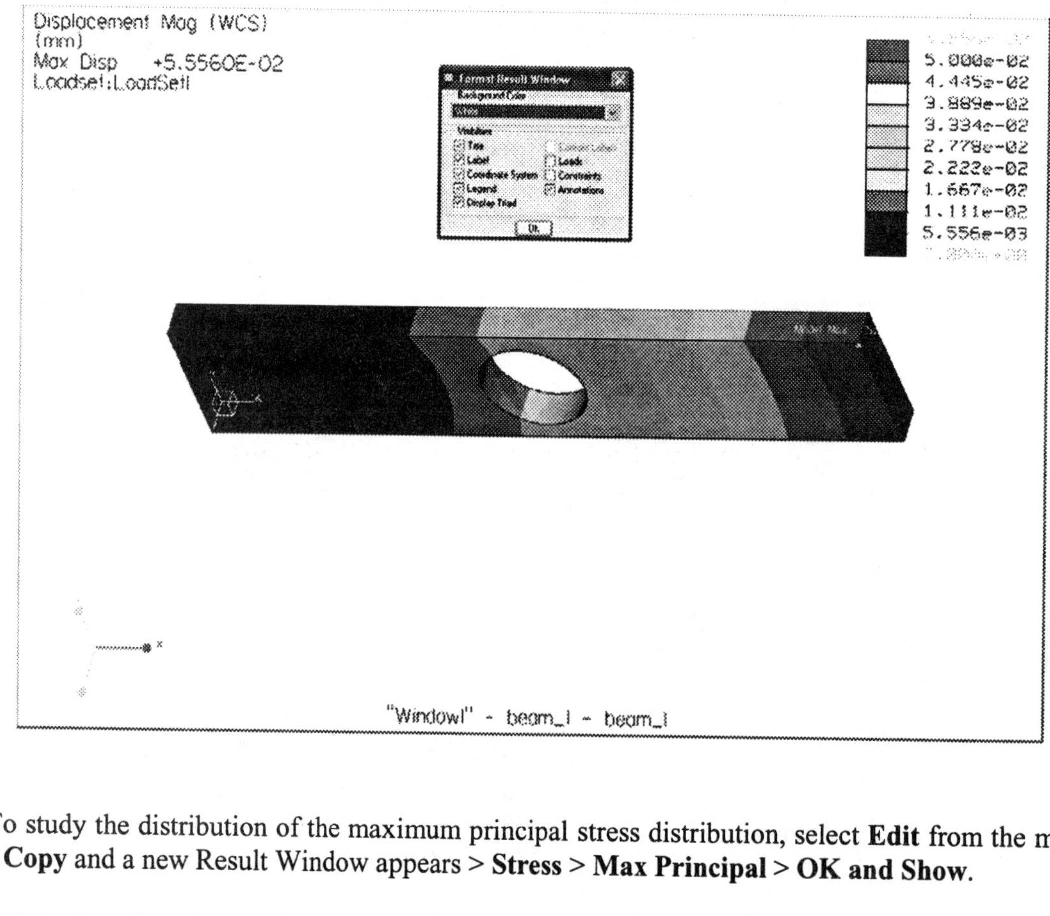

To study the distribution of the maximum principal stress distribution, select **Edit** from the main toolbar > **Copy** and a new Result Window appears > **Stress** > **Max Principal** > **OK and Show.**

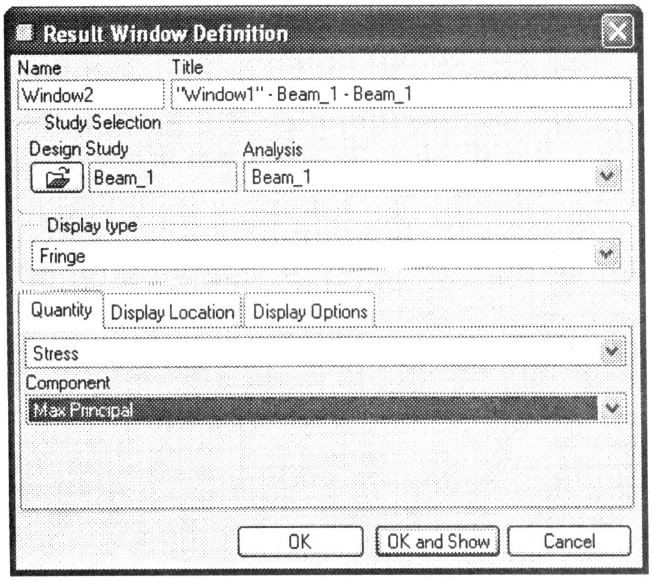

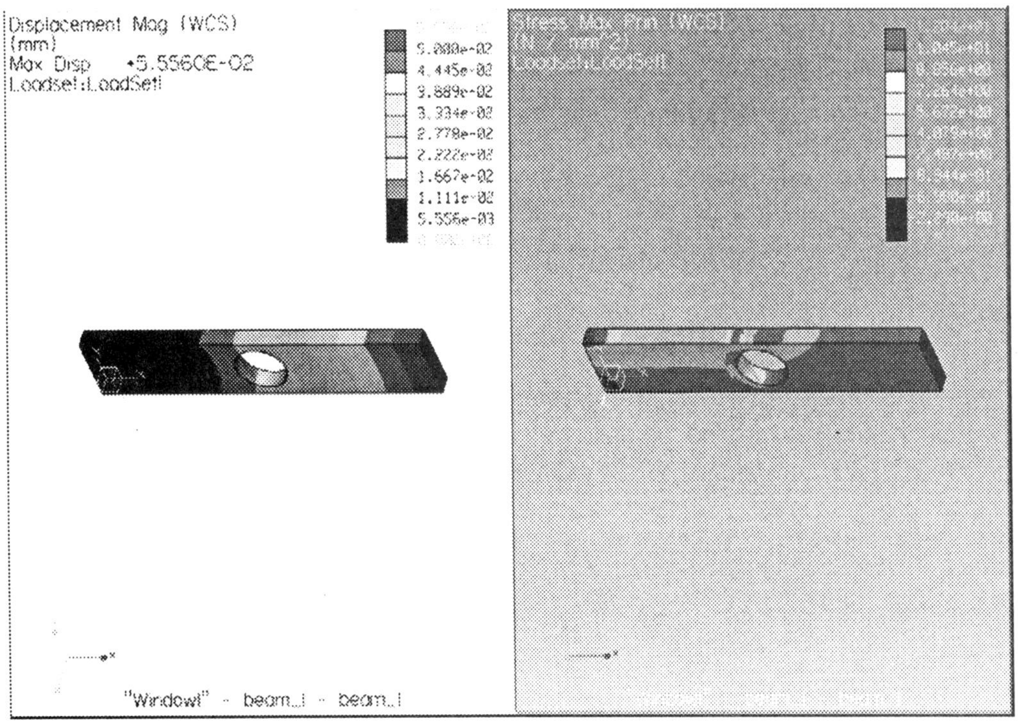

At this moment, we may have two windows or two definitions on display. One definition is the displacement distribution and the other definition is the distribution of the maximum principal stress. To have one definition on display only, click the icon of **Display definition(s)** from the main toolbar, a new window appears, on which both Window1 and Window2 are highlighted. De-highlighting Window1 and leaving Window2 highlighted will allow the distribution of maximum principal stress on display only.

Icon of Display definition(s)

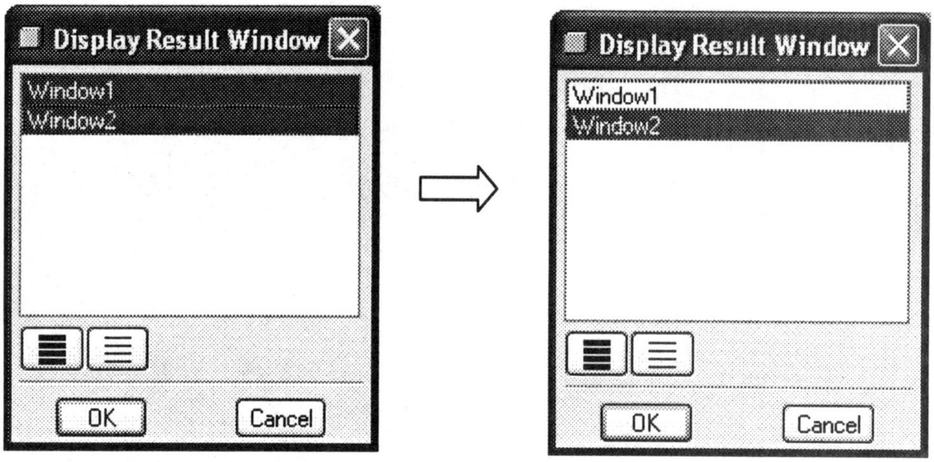

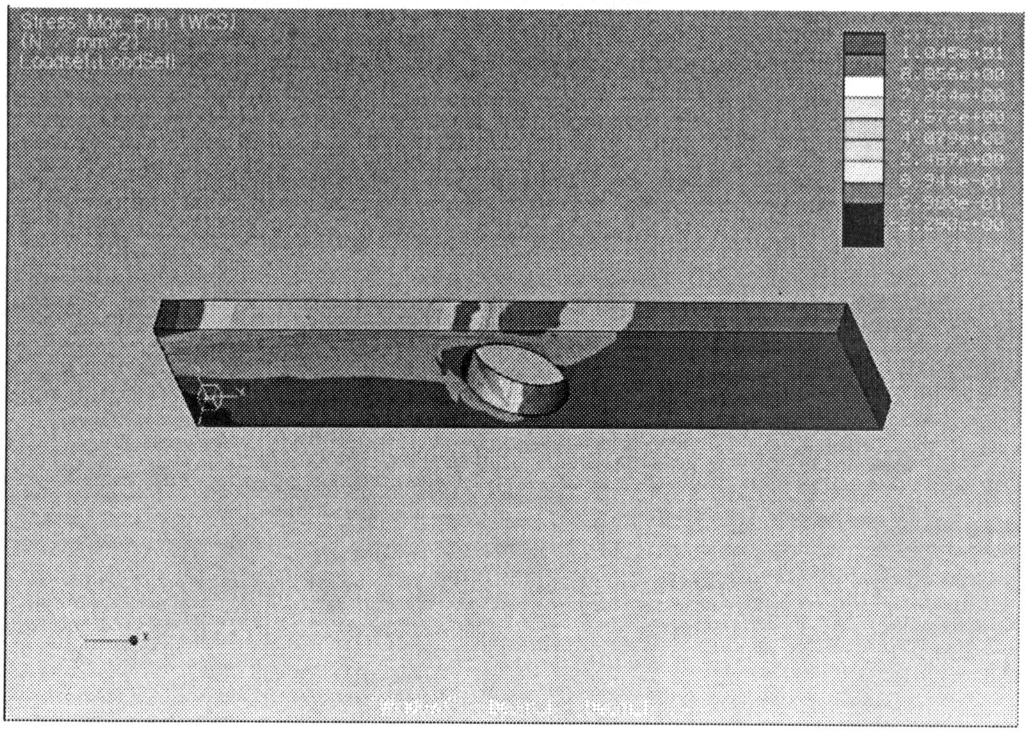

To locate the position of the maximum principal stress occurred on the beam structure, select **Info** from the main toolbar, select **Model Max**. The location of the maximum principal stress is shown on the display is at the left side of the beam, which is the fixed end. The magnitude is 12.04 MPa (10^6 Newton/mm^2). The numerical values of the coordinates of the location where the maximum principal stress occurs are also displayed in the message window, showing x=0, y=80 and z=20.

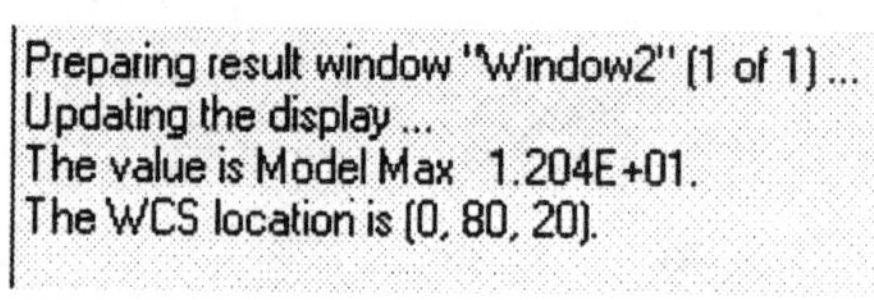
Preparing result window "Window2" (1 of 1)...
Updating the display...
The value is Model Max 1.204E+01.
The WCS location is (0, 80, 20).

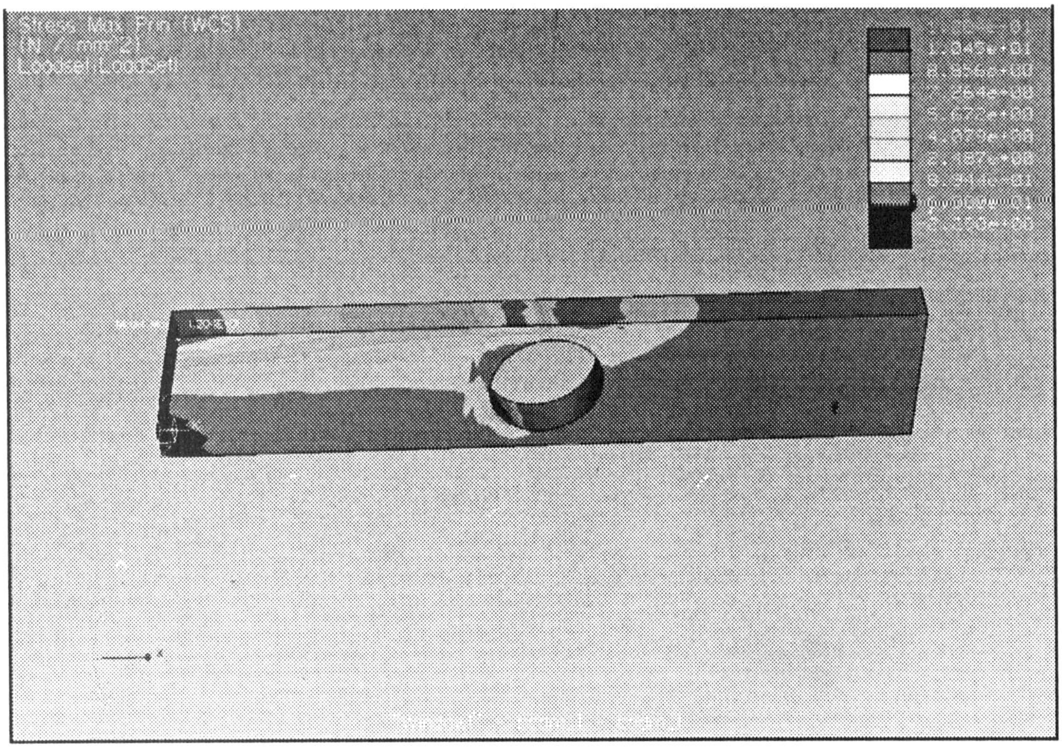

To change the background color to white, select **Format** from the main toolbar > **Result Window** > **White** > **OK**.

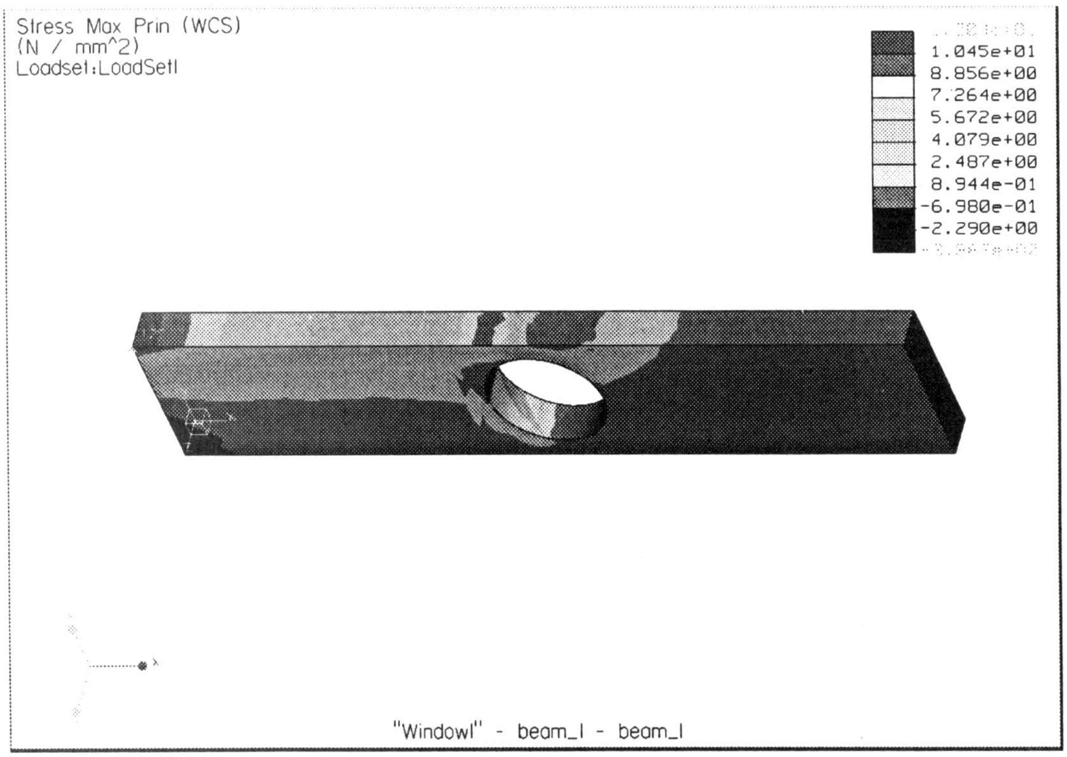

2.4 References

1. O. Axelsson, Iterative Solution Methods, Cambridge University Press, New York, 1994.
2. K. J. Bathe, Finite Element Procedure, Prentice Hall, Upper Saddle River, New Jersey, 1996.
3. K. Bell, A Refined Triangular Plate Bending Finite Element, Int. J. Numer. Methods Eng., Vol. 1, No. 1 January 1969, pp. 101-122.
4. B. Carnahan, H. A. Luther, and J. O. Wikes, Applied Numerical Methods, Wiley New York, NY, 1969.
5. R. W. Clough, The Finite Element in Plane Stress Analysis, Proceedings of 2nd ASCE Conference on Electronic Computation, Pittsburgh, PA. September 1960.
6. R. D. Cook, D. S. Maikus and M. E. Plesha, Concepts and Applications of Finite Element Analysis, 3rd edition, Wiley, New York, NY, 1989.
7. J. W. Dally and W. F. Riley, Experimental Stress Analysis, McGraw Hill, 1965.
8. J. W. Dally, Design Analysis of Structural Elements, Goodington, New York, NY, 2003.
9. C. S. Desal and J. F. Abel, Introduction to the Finite Element Method, Van Nostrand Reinhold, New York, NY, 1972.
10. J. H. Earle, Graphics for Engineers, AutoCAD Release 13, Addision-Wesley, Reading, Massachusetts, 1996.
11. R. H. Gallagher, Finite Element Analysis Fundamentals, Prentice-Hall, Englewoood Cliffs, NJ, 1975.
12. J. M. Gere and S. P. Timoshenko, Mechanics of Materials, 3rd edition, PWS-KENT Publishing Company, Boston, 1990.
13. W. W. Hager, Applied Numerical Linear Algebra, Prentice-Hall, Englewood Cliffs, NJ, 1988.
14. Holland, The Finite Element Method in Plane Stress Analysis, in the Finite Element Method in Stress Analysis, I. Holland and K. Bell (eds), Tapir Press, Trondheim, Norway, 1969, Chapter 2.
15. Y. Y. Hsieh, Elementary Theory of Structure, 2nd edition, Prentice-Hall, Englewood Cliffs, NJ, 1982.
16. M. A. N. Hendriks, H Jongedijk, J. G. Rots and W. J. E. van Spange (eds), Finite Elements in Engineering and Science, DIANA Computational Mechanics, 1997.
17. K. H. Huebner, D. L. Dewhirst, D. E. Smith, and T. G. Byrom, The Finite Element Method for Engineers, 4th edition, John Wiley & Sons, Inc., 2001.
18. C.S. Krishnamoorthy, Finite Element Analysis, Theory and Programming, 2nd Ed., 1995.
19. D. L. Logan, A First Course in the Finite Element Method Using ALGOR, PWS Publishing, Boston, 1997.
20. H. C. Martin, Introduction to Matrix Methods of Structural Analysis, McGraw-Hill, New York, NY, 1966.
21. R. L. Norton, Machine Design: An Integrated Approach, Prentice Hall, Upper Saddle River, New Jersey, 1996.
22. M. Ortiz, Y. Leroy and A. Needleman, A Finite Element Method for Localized Failure Analysis, Comp. Meth. Applied Mech. Enging, 61, pp. 189-214, 1987.
23. M. Petyl, Introduction to Finite Element Vibration Analysis, Cambridge University Press, 1990.
24. T. H. H. Pian and P. Tong, Basis of Finite Element Methods for Solid Continua, Int. J. Numer. Methods Eng., Vol. 2, No. 7, April 1964, pp. 1333-1336.
25. P. G. Plockner, Symmetry in Structural Mechanics, J. of the Structural Division, American Society of Civil Engineers, Vol. 99, No. ST1, pp. 71-89, 1973.
26. J. N. Reddy, An Introduction to the Finite Element Method, McGraw-Hill Publishing Company, New York, NY, 1984.
27. T. Ross, T. Johns and Emile, Dynamic Buckling of Thin Walled Domes under External Water Pressure, Applied Solid Mechanics 2nd Conference, Ed. Tooth & Spence, Elsevier, pp. 211-224, 1987.
28. T. Ross, Finite Element Program in Structural Engineering and Continuum Mechanics, Horwood Publishing Limited, Chichester, England, 1996.
29. T. Ross, Advanced Applied Finite Element Methods, Horwood Publishing Limited, Chichester, England, 1998.
30. L. Stasa, Applied Finite Element Analysis for Engineers, Holt, Rinehart and Winston, New York, NY, 1985, pp. 101-157.

31. Y. Saad, Iterative Methods for Sparse Linear Systems, PWS Publishing Company, Boston, 1996.
32. P. Silvester, Higher-Order Polynomial Triangular Finite Element for Potential Problems, Int. J. Eng. Soc., Vol. 7, No. 8, pp. 849-861.
33. Y. Tada and G. C. Lee, Finite Element Solution to an Elastic Problem of Beams, Int. J. Numer. Methods Eng., Vol. 2, No. 2, April 1970, pp. 229-241.
34. R. L. Taylor, On Completeness of Shape Functions for Finite Element Analysis, Int. J. Numer., Methods Eng., Vol. 4., No. 1, 1972, pp. 17-22.
35. S. P. Timoshenko and D. H. Young, Theory of Structure, McGraw-Hill, Kogaksha Limited, 1965.
36. M. J. Turner, R. W. Clough, H. C. Martin, and L. J. Topp, Stiffness and Deflection Analysis of Complex Structures, J. of the Aeronautical Sciences, Vol. 23, No. 9, pp. 805-824, Sept. 1956.
37. O. C. Zienkiewicz, The Finite Element Method, third edition, McGraw-Hill, New York, NY, 1977.
38. O. C. Zienkiewicz and R. L. Taylor, The Finite Element Method, 4th edition, Vol. 1, McGraw-Hill (UK),London, 1989.

2.5 Exercises

1. The geometry of the cantilever beam is shown below:

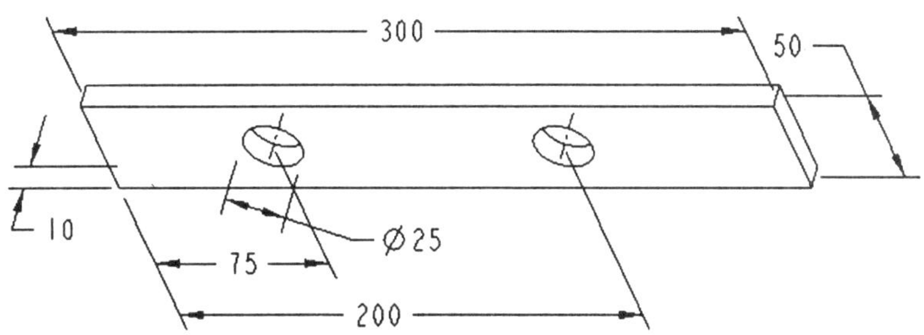

The material of the beam is steel. Analyze the stress and displacement distributions when subjected to

(1). Uniformly distributed pressure load. The magnitude is 10 N/mm^2.

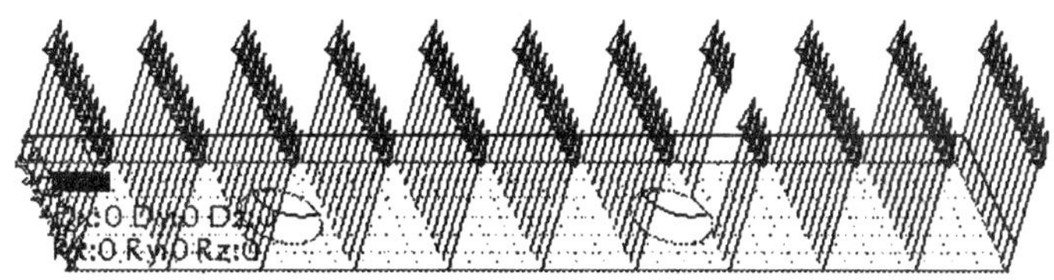

(2). Uniformly distributed load acting on the top surface. The magnitude is 1000 Newton.

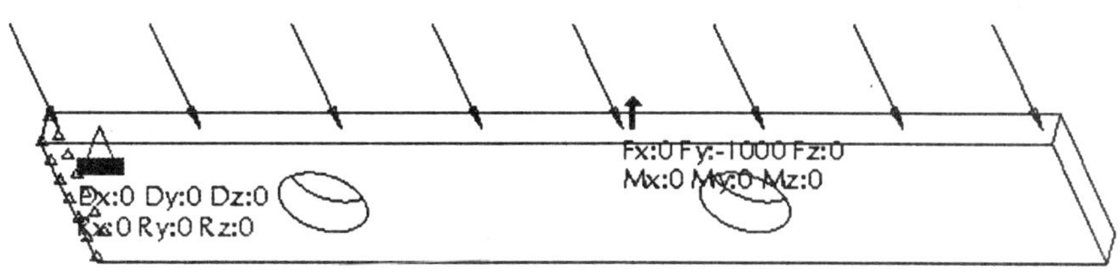

2. The geometry and dimensions of a plate component are shown below. Note that the geometry is symmetric with respect to the horizontal axis. The plate is subjected to an axial load during operation. The material of the plate is aluminum. The modulus of elasticity is 70 GPa. The Poisson's ratio is 0.33.

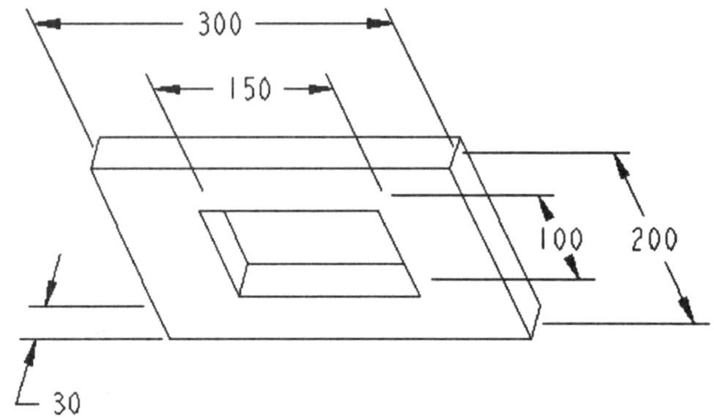

Assume that the left side of the plate is fixed, and the end on the right side is free to move. A tensile force is acting on the right side. The magnitude of the force is 500 Newton.

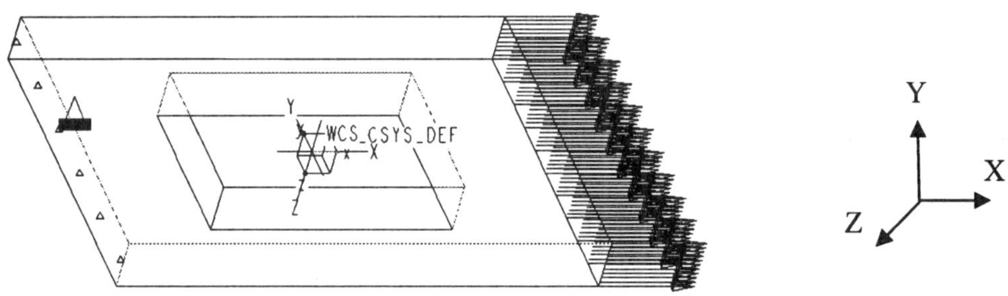

Analyze the stress and displacement distributions when subjected to the load of 500 Newton. The student has the choice to select the type of stress and the displacement in one of the three directions (x, y, z) or the total magnitude of the displacement.

3. The geometry and dimensions of a bracket component are shown below

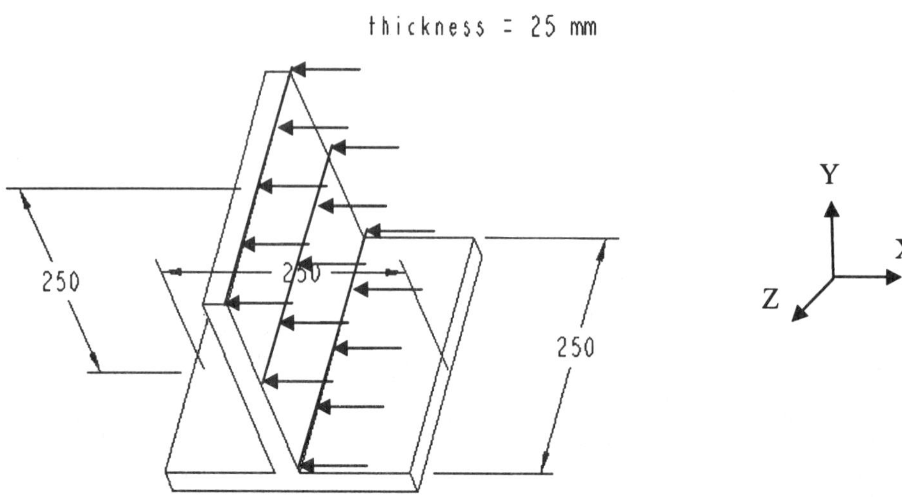

The bracket is subjected to a uniformly distriburted load during operation. The magnitude of the uniformly distributed load is 1000 Newton, acting on the right side of the vertical plate.

The material of the bracket is steel The bracket is fixed to the ground through the bottom surface.

Analyze the stress and displacement distributions when subjected to the load of 1000 Newton. The student is asked to plot the following stress distributions:

(1) The maximum principal stress distribution.
(2) The von Mises stress distribution.
(3) The normal stress distribution in the x direction.
(4) The normal stress distribution in the y direction.
(5) The normal stress distribution in the z direction.
(6) One of the 3 shear stress distributions.
(7) The displacement distribution (magnitude).
(8) The displacement distribution in the y direction.
(9) Identify the location of the maximum von Mises stress.
(10) Calculate the safety factor if the maximum von Mises stress allowable is set at 60 MPa.

CHAPTER 3

MODELING OF LOAD CONDITIONS

3.1 Introduction

In Chapter 2, two examples were used for the purpose of introducing the basic procedure required for performing a finite element analysis. The two examples were a column structure subjected to a compressive load and a cantilever beam structure subjected to a uniformly distributed load acting on the entire top surface of the cantilever beam structure. Readers may have already noticed that the areas, on which the loads were acting, are the entire surfaces. And furthermore, the two loads were uniformly distributed on those entire surfaces. In the engineering field, such load conditions are very common. However, it is not uncommon to observe that a load acts on part of a surface, such as the one shown in Figure 3-1, namely, uniformly distributed load acting on part of the top surface of a beam structure.

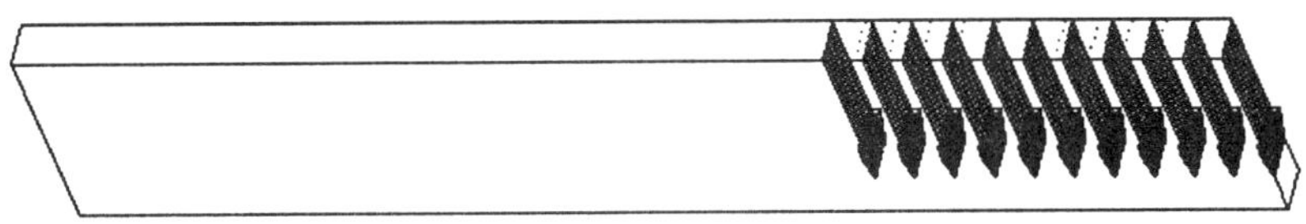

Figure 3-1 A Uniformly Distributed Load Acting on Part of the Top Surface

To model such a load condition, there is need to split the top surface of the beam structure so that the uniformly distributed load only acting on the right portion of the top surface. Sometimes the magnitude of a load varies on the area on which it is acting; a load may act on a curved surface, or load acting on part of a circle. Figure 3-2 presents 3 load cases as just described.

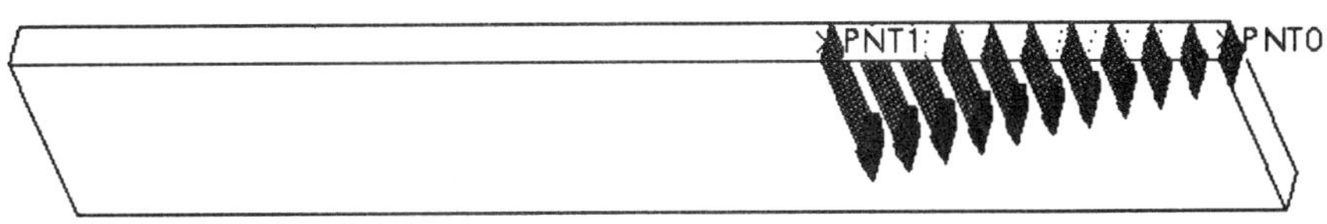

(a) A Non-Uniformly Distributed Load Acting on Part of the Top Surface

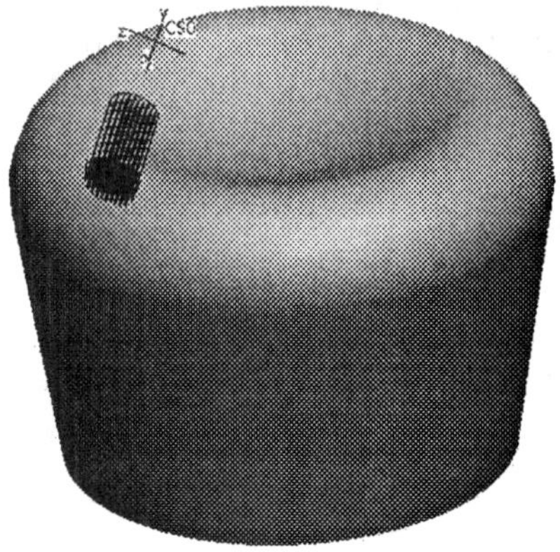

(b) Load Acting on a Curved Surface

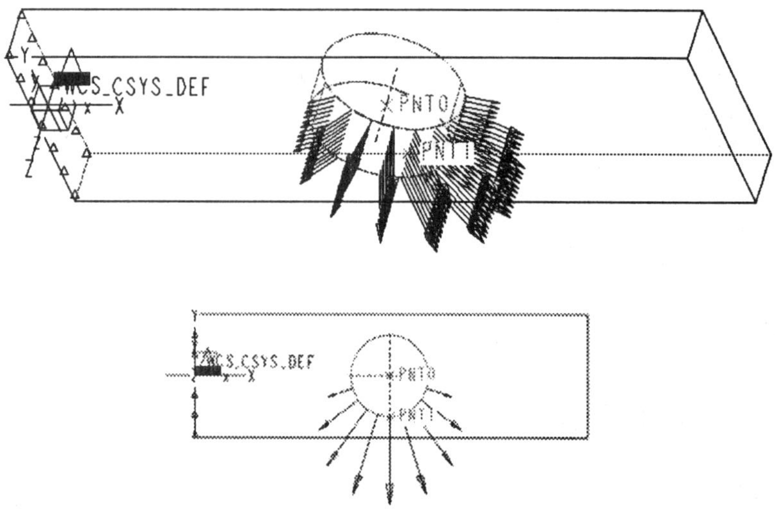

(c) Bearing Load

Figure 3-2 A Variety of Loads Observed in Engineering

In this chapter, section 3.2 presents uniformly distributed load acting on part of a surface. To model this type of load, a new concept called surface region is introduced. Section 3.3 presents non-uniformly distributed load acting on part of a surface. To model the characteristics of a non-uniform distribution, datum points are needed. In section 3.3, the procedure of creating datum points is illustrated. Section 3.4 presents a bearing load where the creation of a directional vector is introduced. Section 3.5 presents a pressure load acting on the wall of a pipe structure. In section 3.5, an example is used to demonstrate how to model load acting on a point. In the later chapters, loads acting on a curved surface and loads such as torque will be introduced as well.

Section 3.5 is devoted to a special discussion. The discussion focuses on the process of modeling a load, which is acting on a point, or a very small area of a cantilever beam structure. The discussion also deals with consideration of the gravity force in the evaluation process where results obtained from FEA are compared with the results obtained from calculations, which are based on well-established beam theory.

3.2 Modeling a Uniformly Distributed Load with a Surface Region

A cantilever beam is shown below. As illustrated, a load is acting on part of the top surface. The load is a uniformly distributed load and the magnitude of the total load is 200 Newton, acting downward.

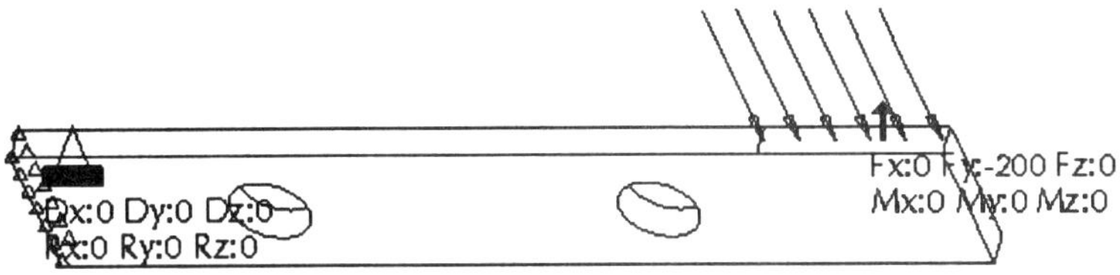

The geometrical shape and dimensions are shown in the following figure. The three dimensions of the beam are 300, 50 and 10 mm, respectively. The diameters of the two through holes are equal. The diameter dimension is 25 mm. The 2 dimensions of the surface area, on which the load is acting, are 60 mm and 10 mm, respectively.

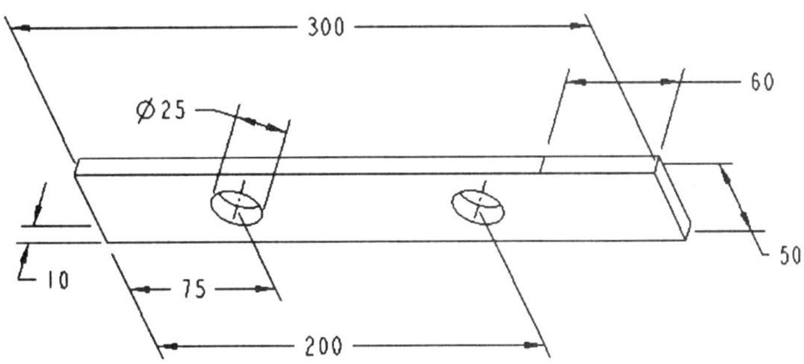

It is assumed that the material of the beam is steel. There are 2 objectives to be expected from performing FEA: the deflection pattern and the tensile stress distribution is the x direction.

Step 1: Create a file for the 3D solid mode.

File > New > Part > type *beam_2* as the file name and clear the icon of **Use default template.** Select **mmns_part_solid** (units: Millimeter, Newton, Second) and type *partial surface* in **DESCRIPTION**, and *student* in **MODELED_BY**, then **OK.**

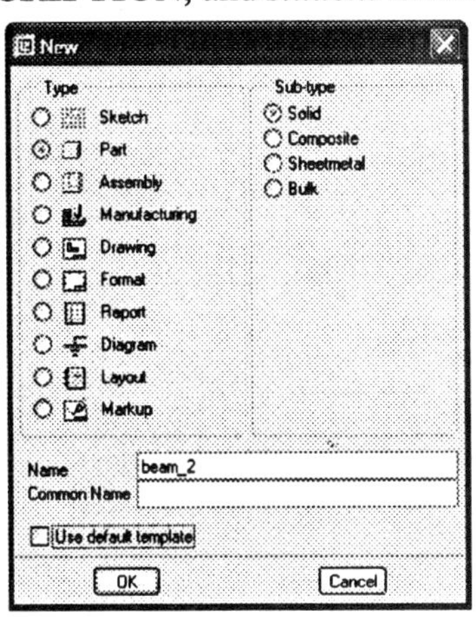

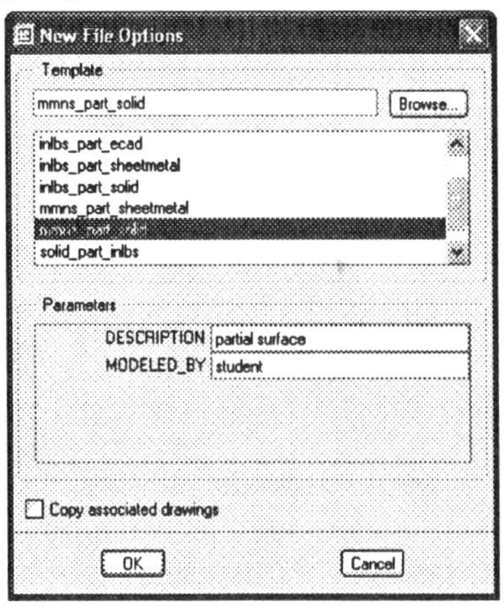

Directly select the icon of **Extrude** from the toolbar of feature creation. Make sure the icon of **Solid** is selected. Click the box of **Placement** > **Define**.

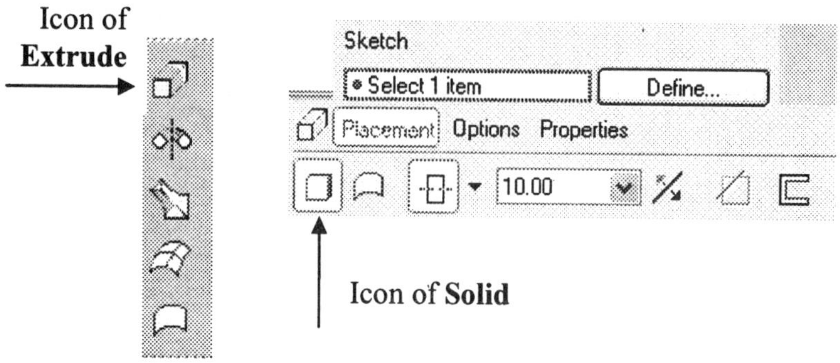

Select the **FRONT** datum plane as the sketch plane. Click the box of **Sketch** to accept the **RIGHT** datum plane as the default reference to orient the sketch plane.

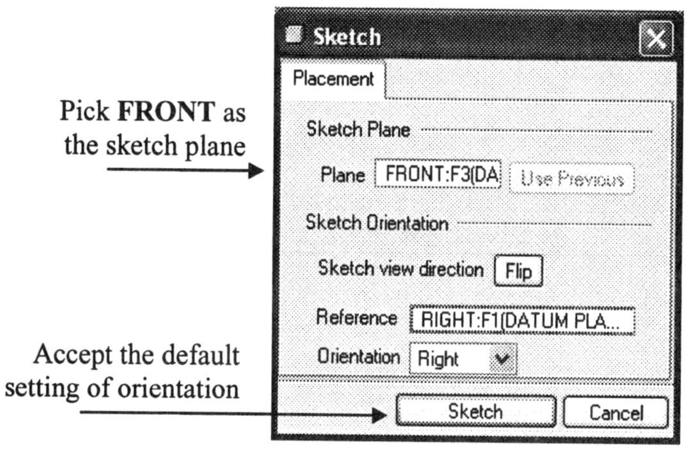

Sketch a horizontal centerline and sketch a rectangle with the dimensions equal to 300 and 50, respectively. Keep the rectangle symmetric about the horizontal centerline.

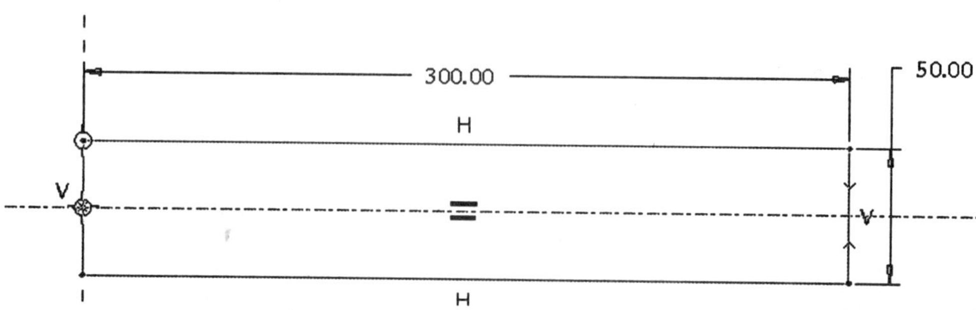

If your sketch is not symmetric about the horizontal centerline, pick the icon of **impose sketcher constraints** from the toolbar of sketcher, and select the icon of **vertices symmetric** from the pop up window called Constraints > click the horizontal centerline and click the 2 vertices so that the 2 vertices are symmetric about the horizontal centerline.

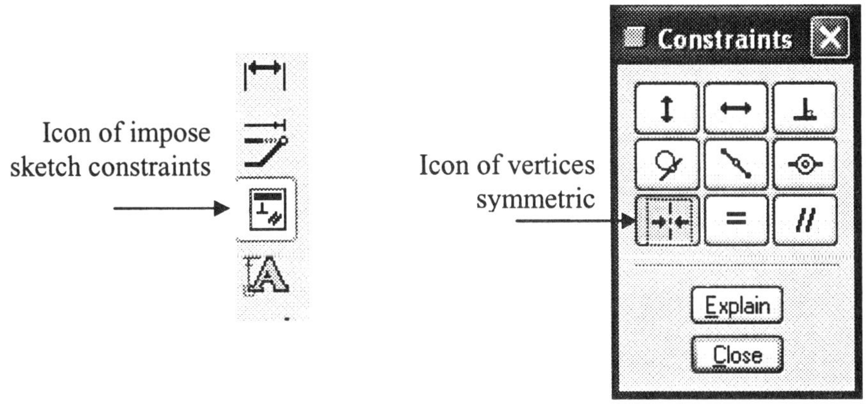

Upon completion, select the icon of **Done.** From the dashboard, select the icon of symmetry and type *10* as the thickness of the block. Select the icon of **Complete** from the feature control panel.

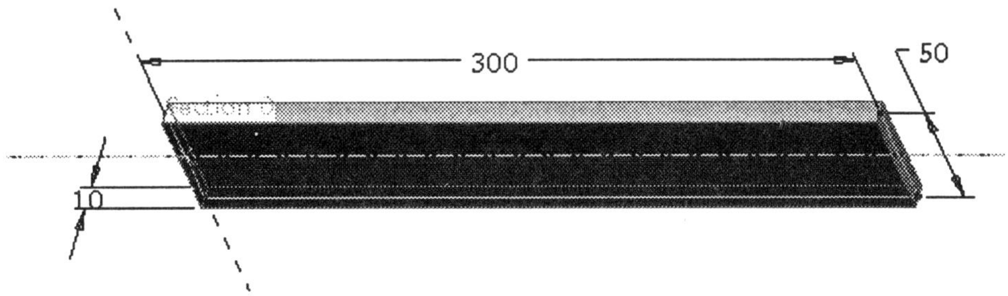

Feature 2: Two holes created through a cut operation

Select the icon of **Extrude** from the toolbar of feature creation. Select the icon of **Cut** and set the depth choice to **Thru All**. Click the box of **Placement** > **Define**. Select the front surface of the block as the sketch plane and accept the default to orient the sketch plane.

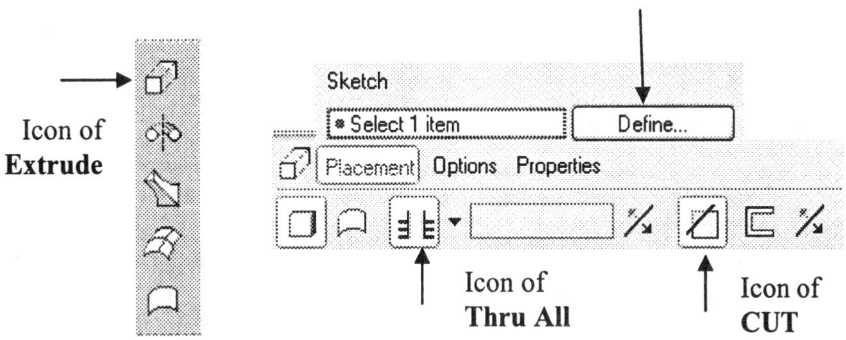

Select the icon of **circle** from the toolbar of sketcher > sketch 2 circles with the dimensions as shown.

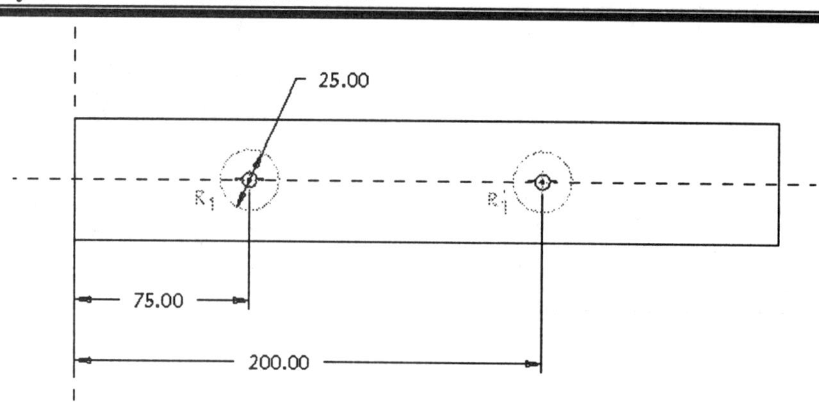

Upon completion, select the icon of **Done** from the sketcher toolbar. Select the icon of **Complete** from the Feature Control Panel to complete the feature creation.

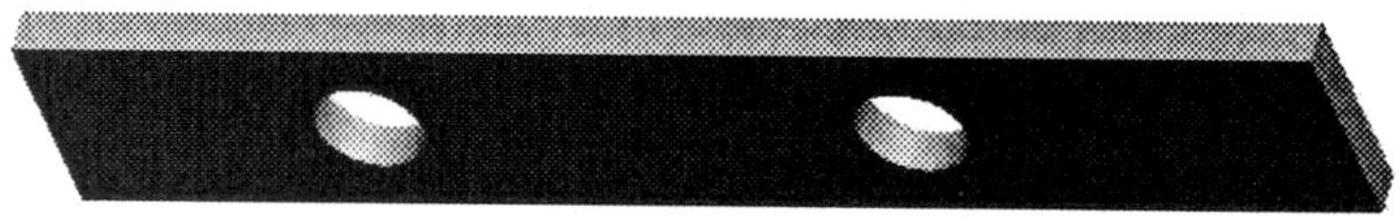

Step 2: Create a datum curve so that a surface region can be specified for loading.

Let us first introduce the concept of a surface region. Assume that the surface is a rectangle shown below. The right portion of the rectangle is the area where a load acts. The left portion of the rectangle is the area where no load acts. As a result, a split operation is needed. After splitting, the right portion becomes a surface region.

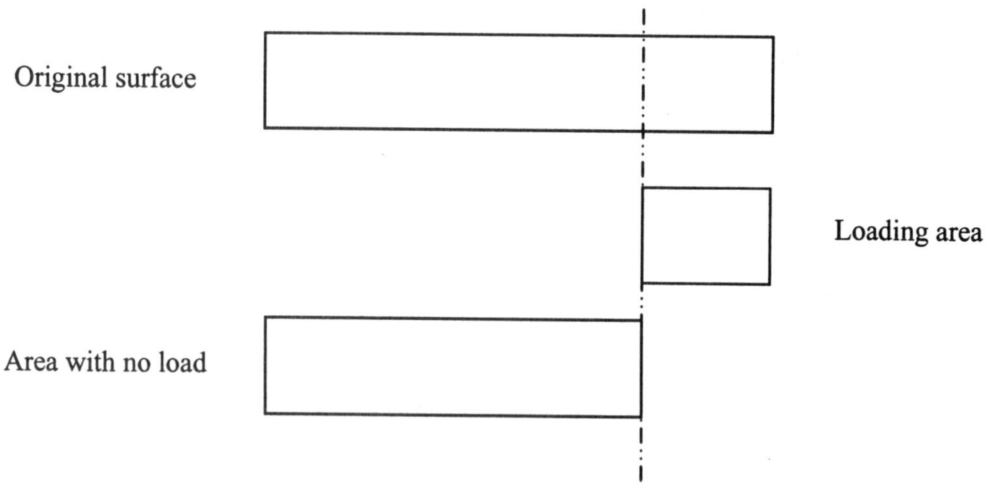

Splitting the original surface
into 2 portions

There are two steps in creating a surface region. The first step is to sketch the boundary of the surface region on the original surface. The second step is to split the original surface into 2 regions based on the sketched boundary.

To define a surface region in this example, select the icon of **Sketch Tool** from the toolbar of datum feature creation > Pick the top surface of the block as the sketch plane > accept the default orientation of the sketch plan

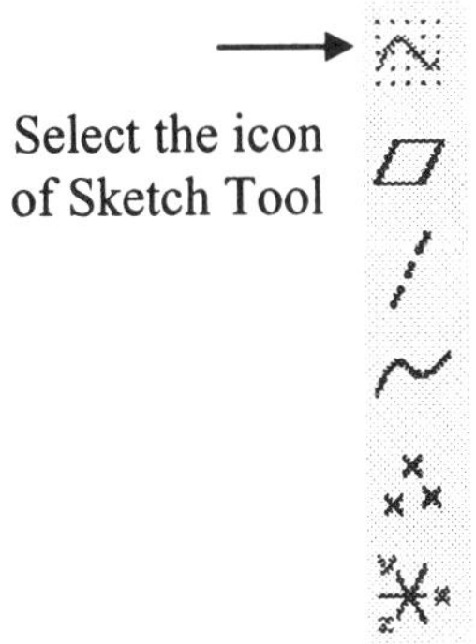

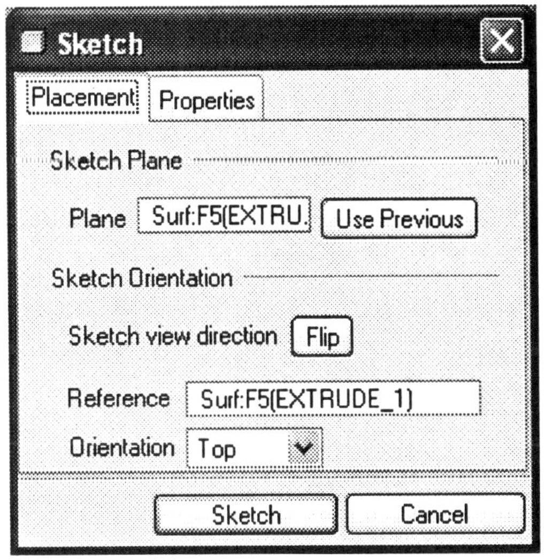

Before making a sketch, let us first add 2 new references. These 2 new references are the surface at the right end and the surface in the front of the block, as shown. Afterwards, close the reference window.

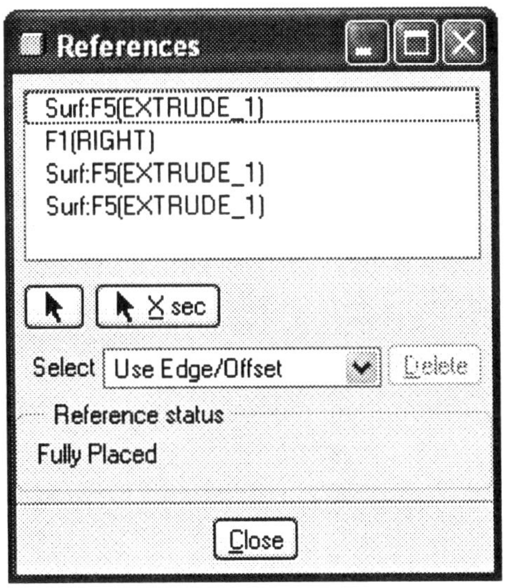

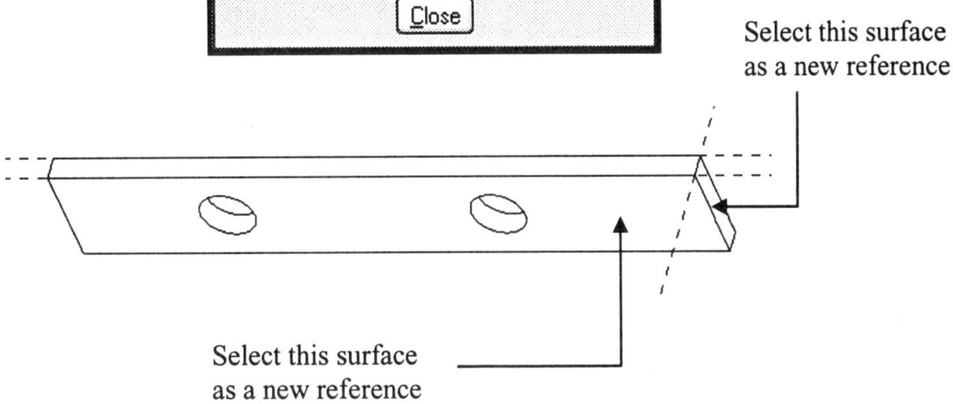

Select the icon of **rectangle** from the toolbar of sketcher and draw a rectangle with the dimension of 60. Only one dimension is needed because the other sides of the sketched rectangle are aligned with the selected references. Pick the icon of Done from the toolbar of sketcher.

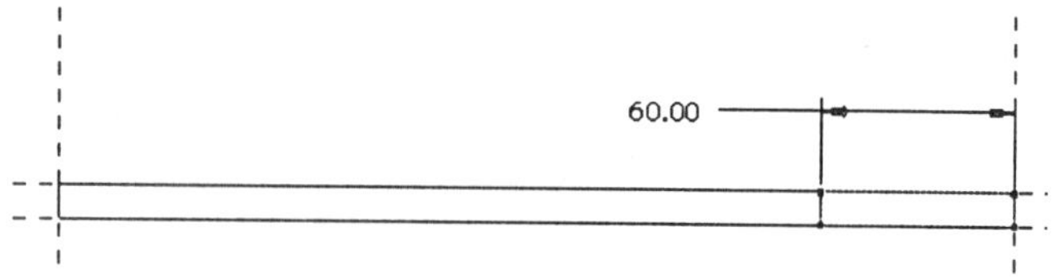

Step 4: Under the Pro/MECHANICA, perform FEA.

From the main toolbar or the menu toolbar, select **Applications** > **Mechanica.**

Continue after checking the unit system > Structure > OK.

From the toolbar of functions, select the icon of **Materials >** select **STEEL** from the left side called Materials in Library > click the directional arrow so that the selected material type goes to the right side called Materials in Model.

Icon of Material

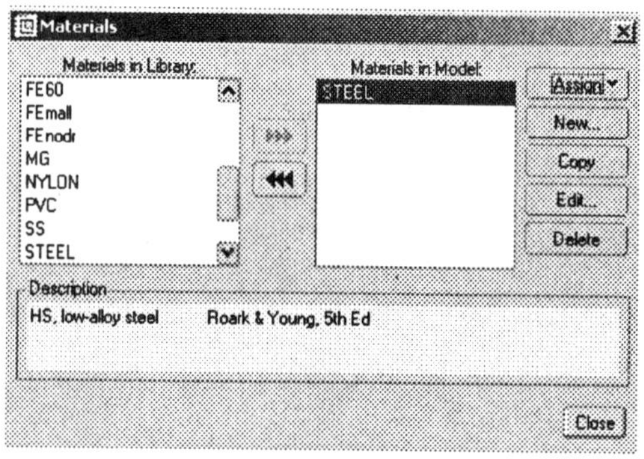

Assign > **Part** > select the column structure from the screen display > **OK** > **Close.**

Step3: Under the Pro/MECHANICA environment, define the fixed end constraint condition.

From the toolbar of functions, select the icon of **Constraints >** select the surface at the left end of the block as the fixed end, and accept the default settings, namely, fixing all six degrees of freedom.

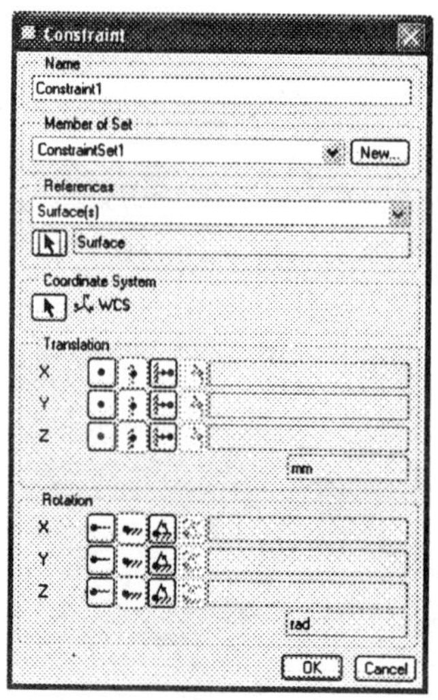

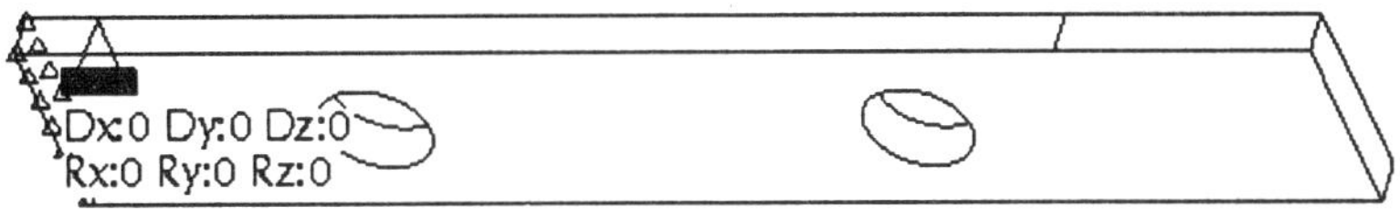

Step 4: Under the Pro/MECHANICA, define the load condition.

We first define a surface region, on which the non-uniformly distributed load will be defined.

From the toolbar of functions, select the icon of surface region > **Select** > **Done**.

Select the datum curve of rectangle > **Done** > select the top surface of the beam for splitting the surface region from the entire surface > **OK** > **Done** > **OK**.

Icon of Surface Region

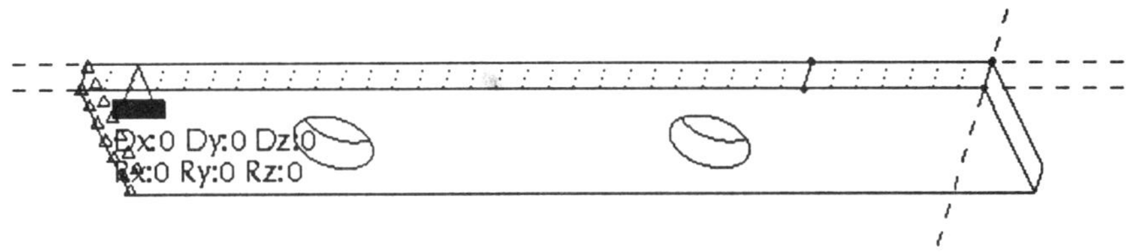

From the toolbar of functions, select the icon of **Force/Moment Loads > Surface(s) >** select the surface region just defined > OK.

Type -200 as the Y Force Component > OK.

Icon of New Force/Moment Load

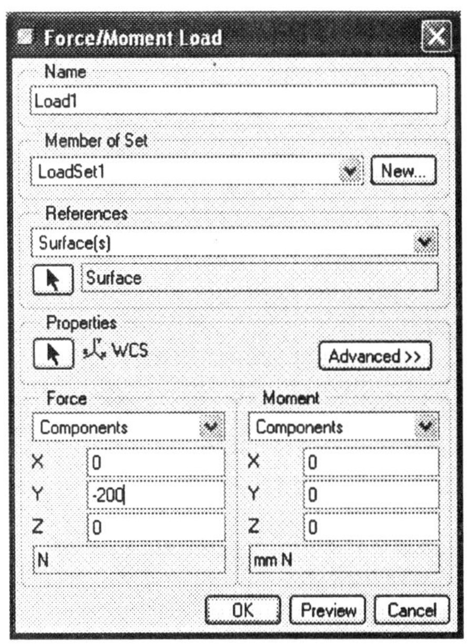

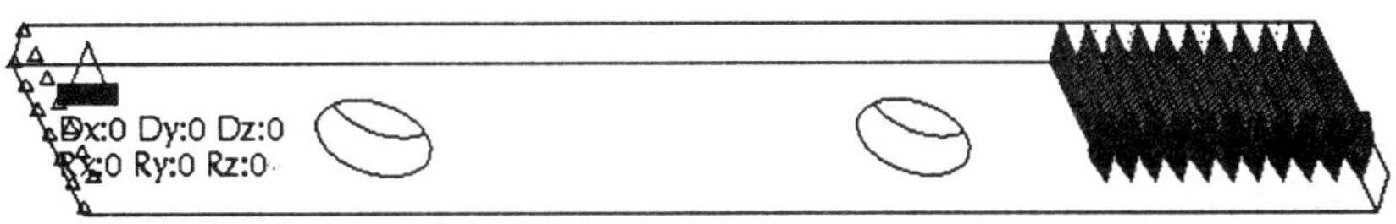

Step 5: Under the Pro/MECHANICA, set up **Analyses** and **Run.**

Select the icon of **Run an Analysis** > **File** > **New Static** > type in beam_2 > **OK**.

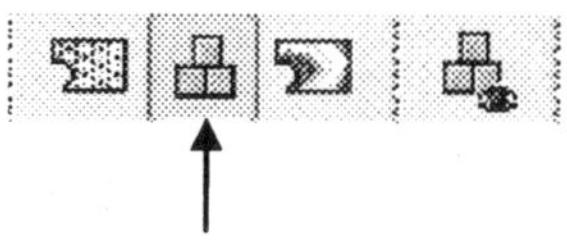

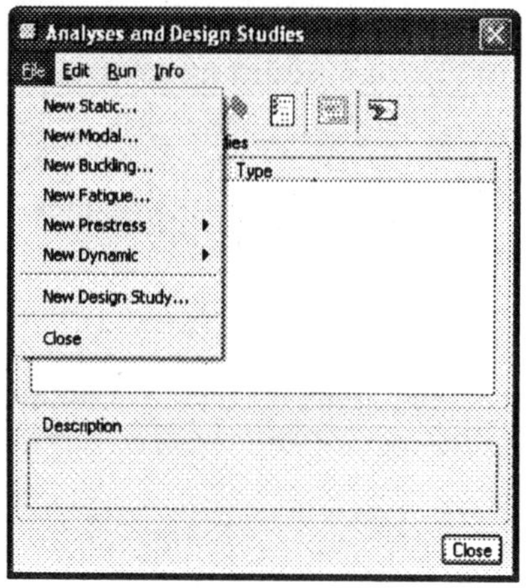

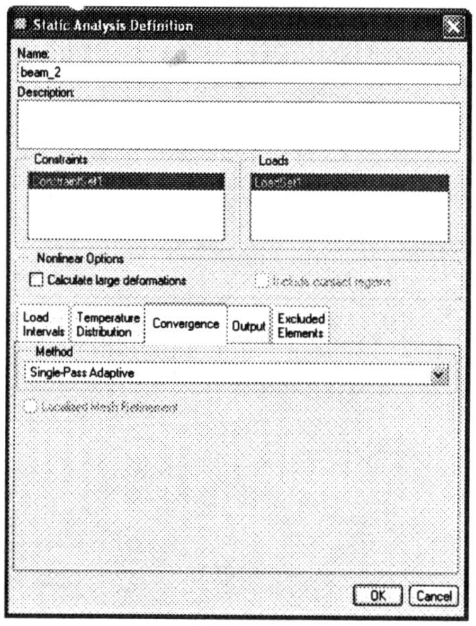

Before running the program, select Settings and specify a sufficient memory space, say 128 Megabytes for RAM Allocation. Run the program

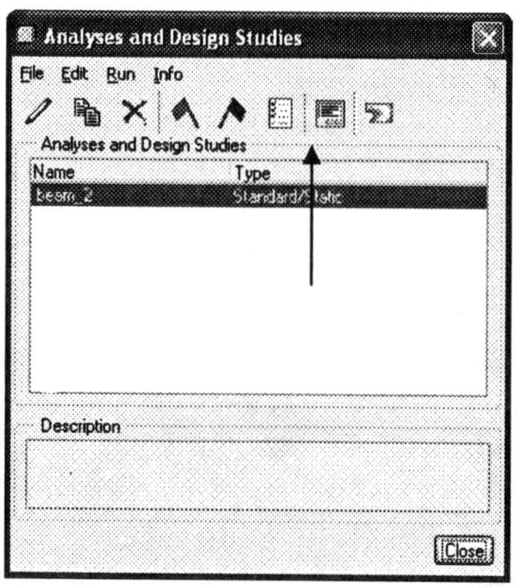

There is always a message displayed on screen. The message is asking the user "Do you want error detection?" Click **Yes**. In this way a warning message will appear if an abnormal operation occurs.

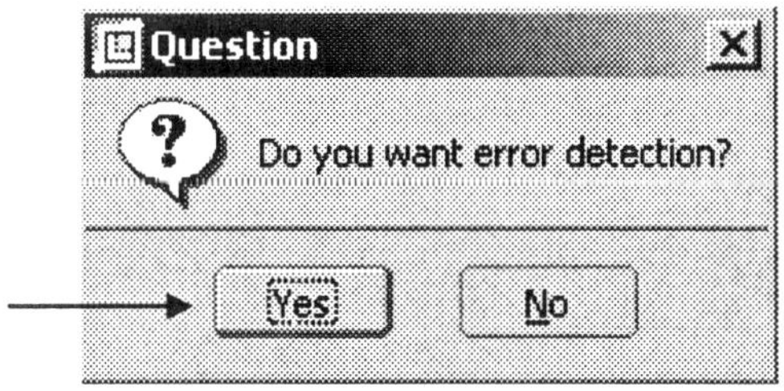

To monitor the computing process, users may click the icon of **Display study status**.

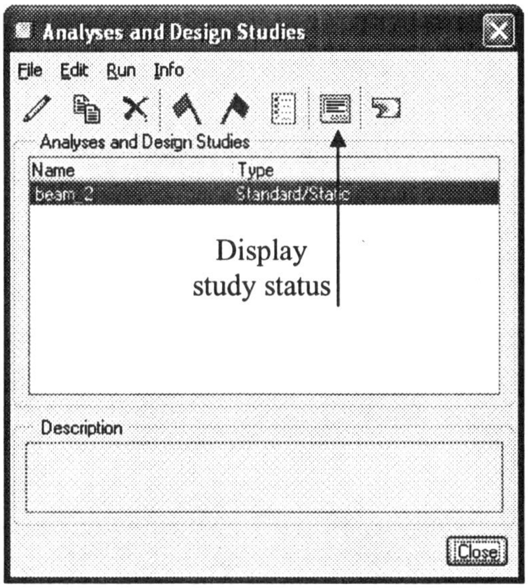

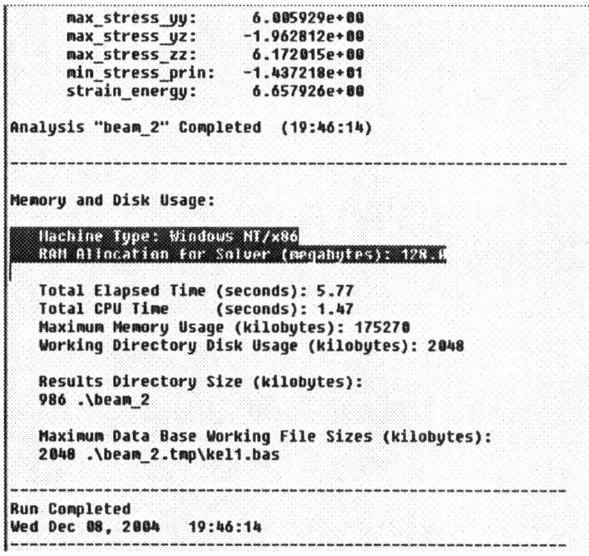

Step 7: Under the Pro/MECHANICA, study the **Results**
Select the icon of **Results**

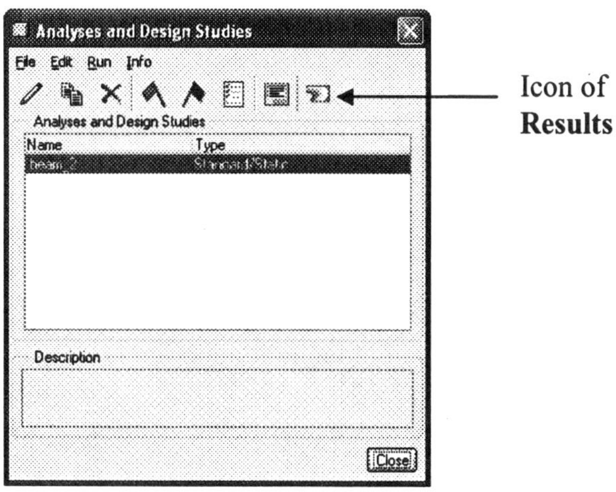

Type in Beam Deflection as the title > locate the Design Study folder called beam_2.
Select Fringe > choose Displacement > Magnitude > OK and Show, indicating the bending effect.

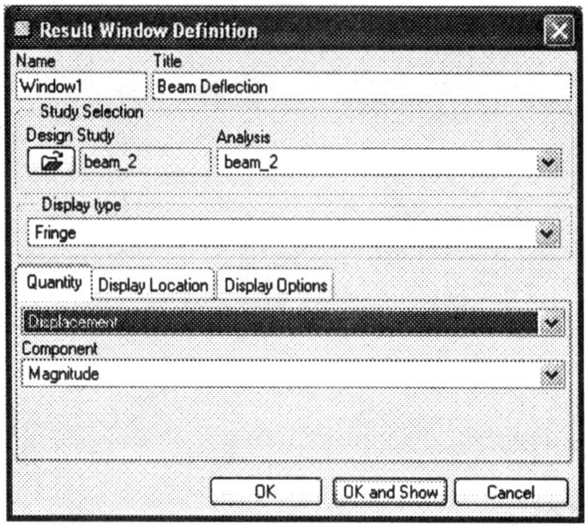

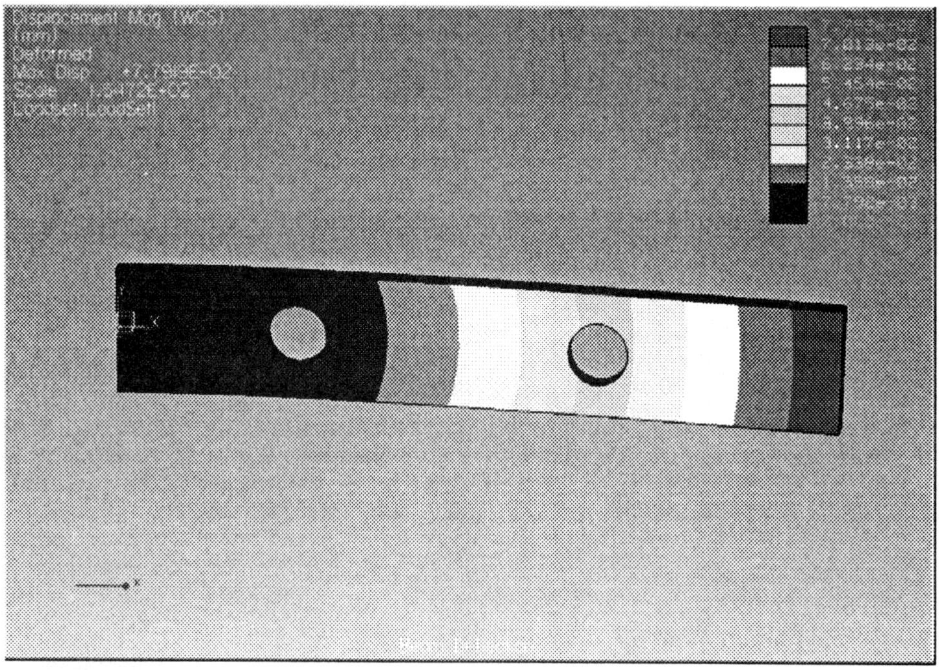

To study the distribution of stress, say the stress distribution in the x direction, select **Edit** from the main toolbar > **Copy** and a new Result Window appears > **Stress** > **XX** > **OK and Show**.

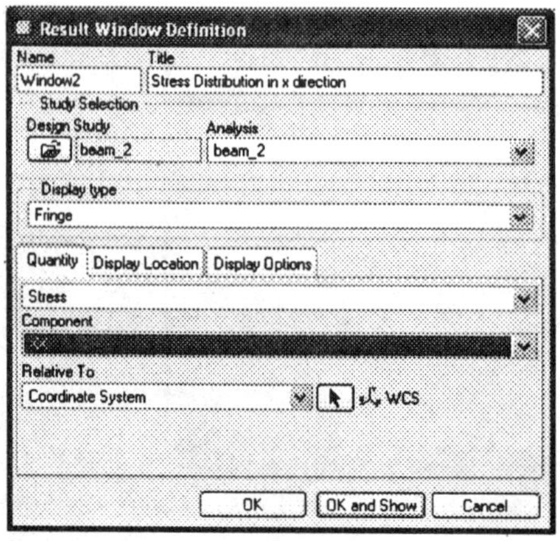

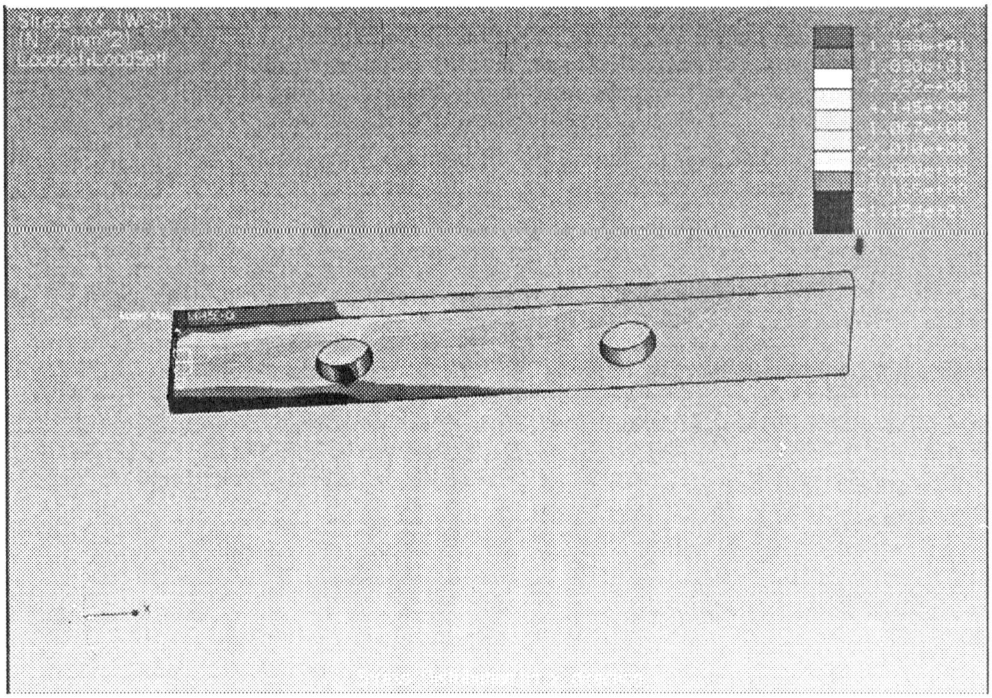

3.3 Modeling a Non-uniformly Distributed Load

In this section, we deal with a non-uniformly distributed load. As shown in the following figure, a non-uniformly distributed load is acting on part of the top surface of a cantilever beam structure.

Magnitude of total load = 200 Newton   Ratio =1:2

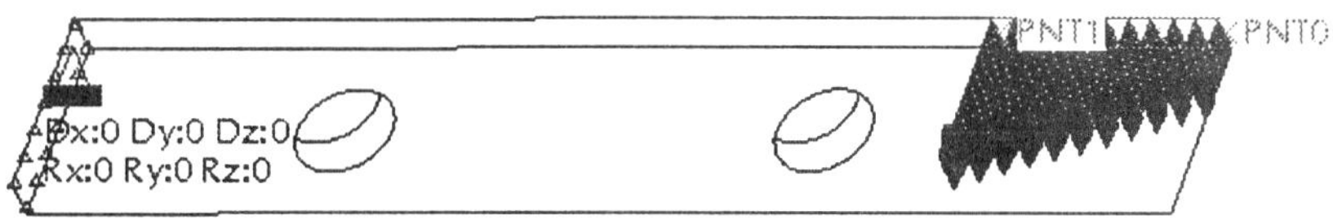

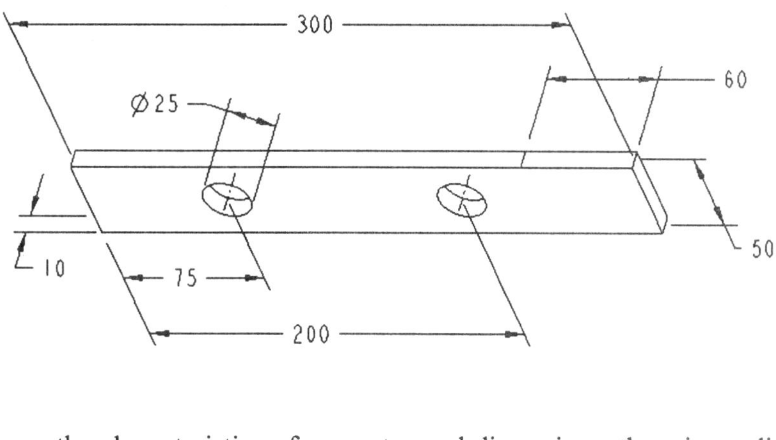

As we compare the characteristics of geometry and dimensions, there is no difference between this example and the example presented in section 3.2. However, there is a difference between this example and the example presented in section 3.2 in terms of the load condition. The distributed load in this example is non-uniformly distributed. As indicated, the ratio is 1:2, indicating that the distributed load at the right edge is smaller than the distributed load at the left edge of the surface region. As we recall, the distributed load in the previous example was uniformly distributed.

Because of the match in geometry and dimensions, there is no need to go through the process of creating a new 3D solid model. Instead of, we directly use the geometrical model created in the previous example. Afterwards, we add 2 datum points to be used for defining the non-uniformly distributed load.

Step 1: Create 2 datum points

To create a datum point, select the icon of **Datum Point** from the toolbar of datum feature creation > click the central location of the right edge, as shown > type 0.50 as the offset value to define **PNT0**, which is the mid-point of the picked edge > **OK**.

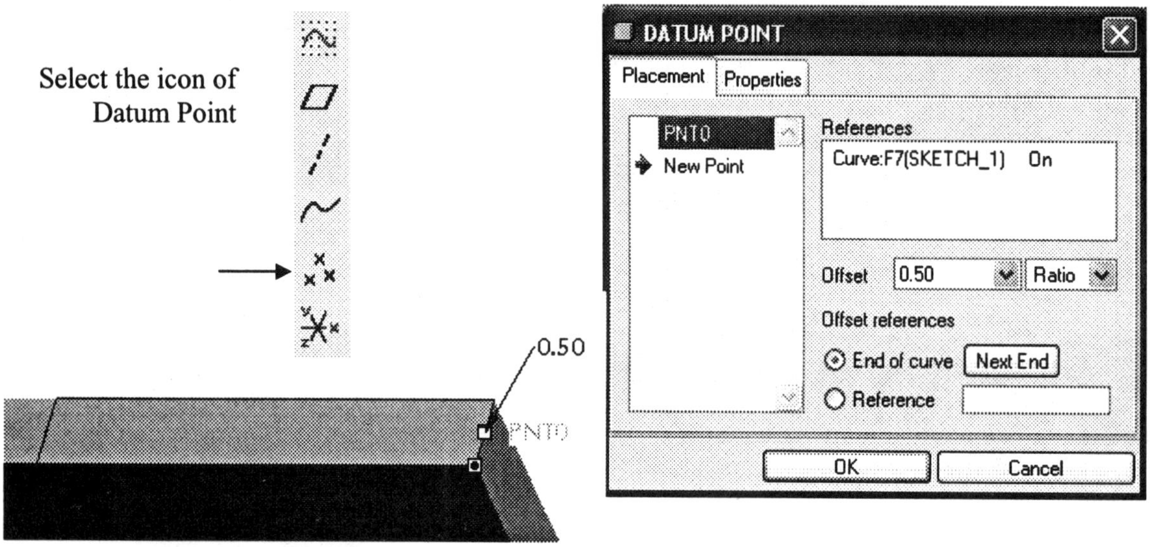

Repeat the above procedure to create another datum point, PNT1. Note that the uniformly distributed load is still presented.

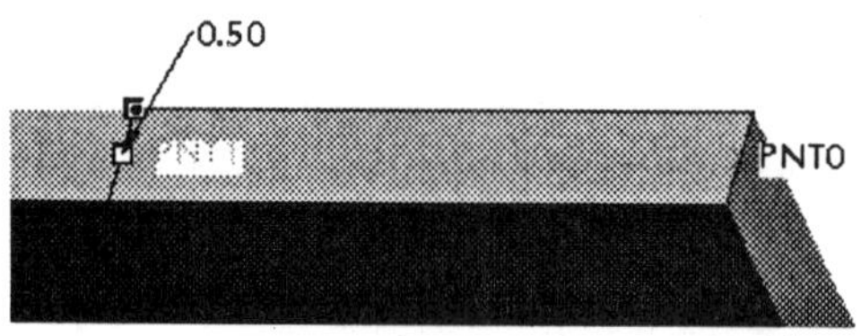

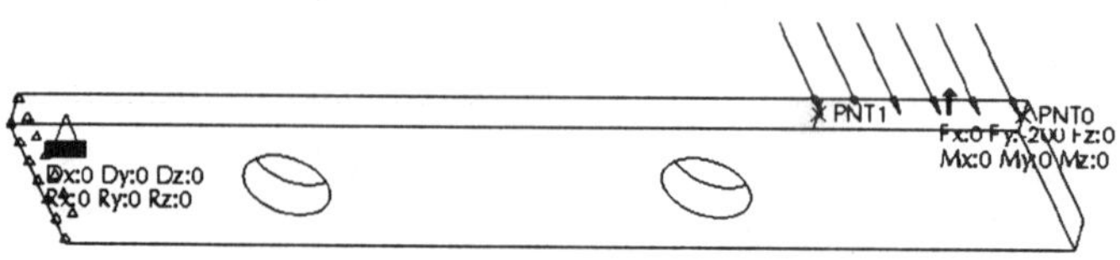

Step 2: Define a non-uniformly distributed load

Now let us define the non-uniformly distributed load. In order to distinguish this load condition, let us call it LoadSet2.

From the toolbar of functions, select the icon of **Force/Moment** > click **New** under Member of Set > a new window called **Load Set Definition** appears and **LoadSet2** is automatically specified > type *non-uniformly distributed load* > **OK** to close this window.

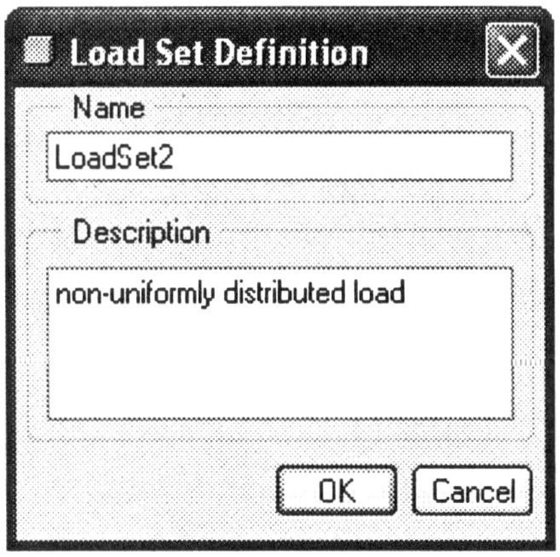

Select **Surface(s) >** select the surface region just defined > **OK**.

Click **Advanced** > change Uniform to **Interpolated Over Entity** > **Define** and a new window called **Interpolation Over Entity** appears.

Add > pick PNT0, while holding down the Ctrl key, pick PNT1 > **OK > Done/Return.**

Type the ratio numbers, namely, 1 and 2, and **OK.** Type -200 as the Y force component > **Preview** > **OK**.

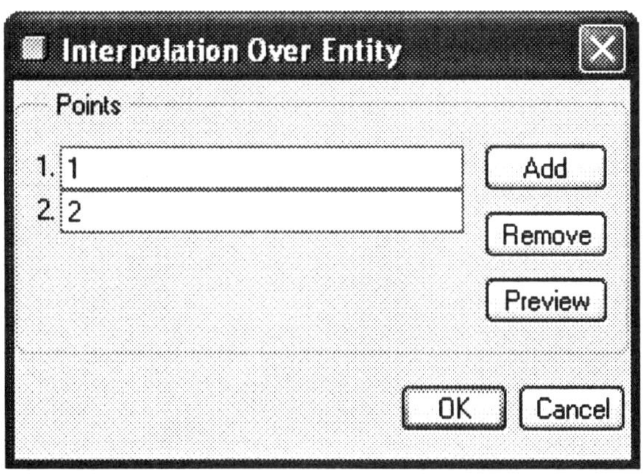

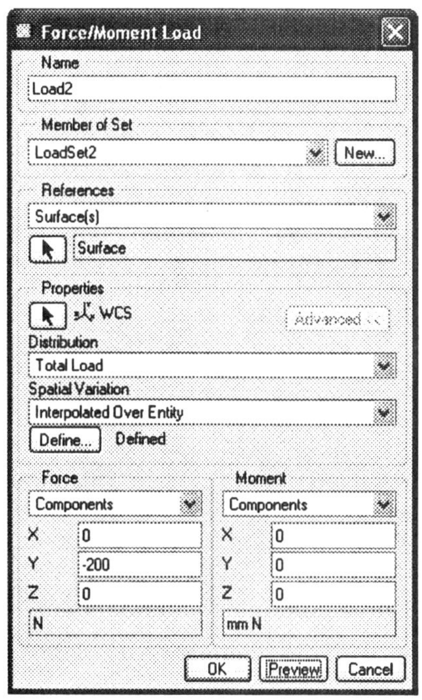

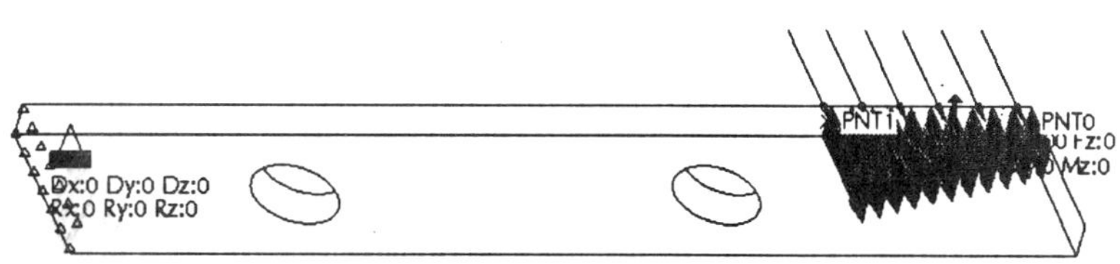

Step 3: Under the Pro/MECHANICA, set up **Analyses** and **Run** it

Select the icon of **Run an Analysis** > **File** > **New Static** > type *beam_3* > It is very important that you select LoadSet2 alone (do not select LoadSet1) > **OK**.

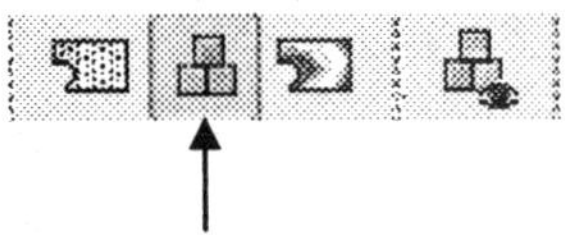

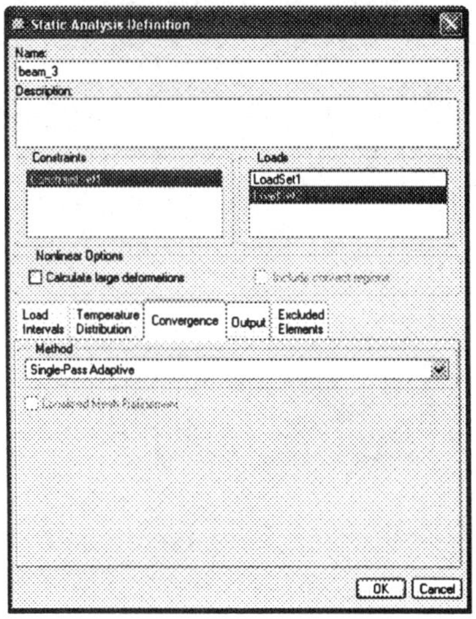

Run the program and afterwards, select the icon of **Results**

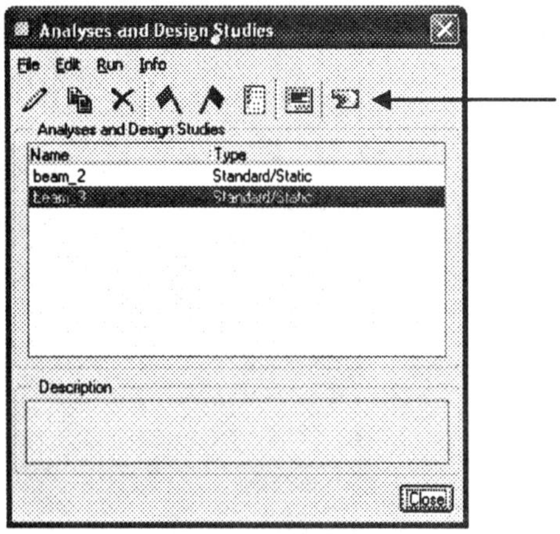

Icon of Results

Type in Beam Deflection as the title > locate the Design Study folder called beam_3. Select Fringe > choose Displacement > OK and Show.

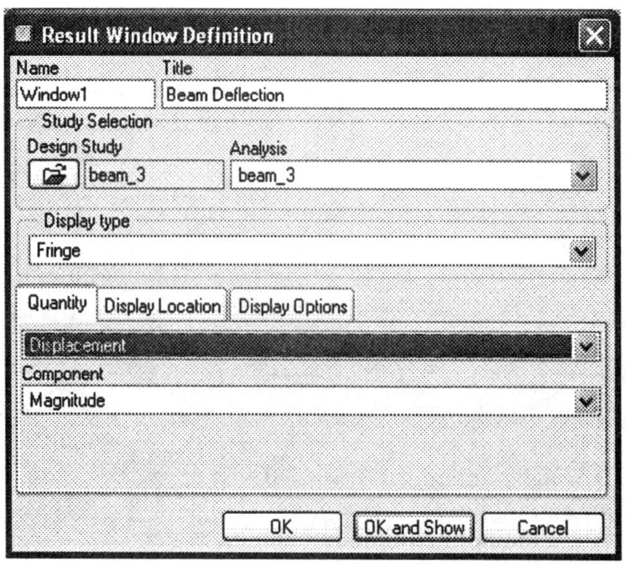

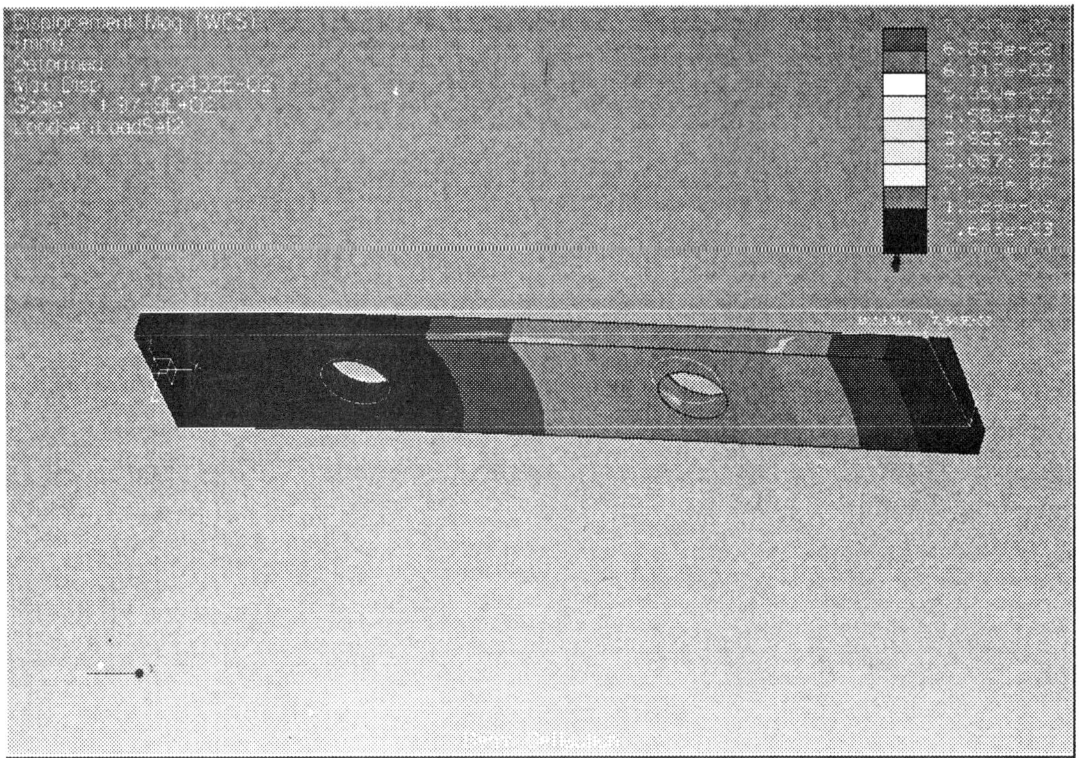

To study the distribution of stress, say the stress distribution in the x direction, select **Edit** from the main toolbar > **Copy** and a new Result Window appears > **Stress** > **XX** > **OK and Show**.

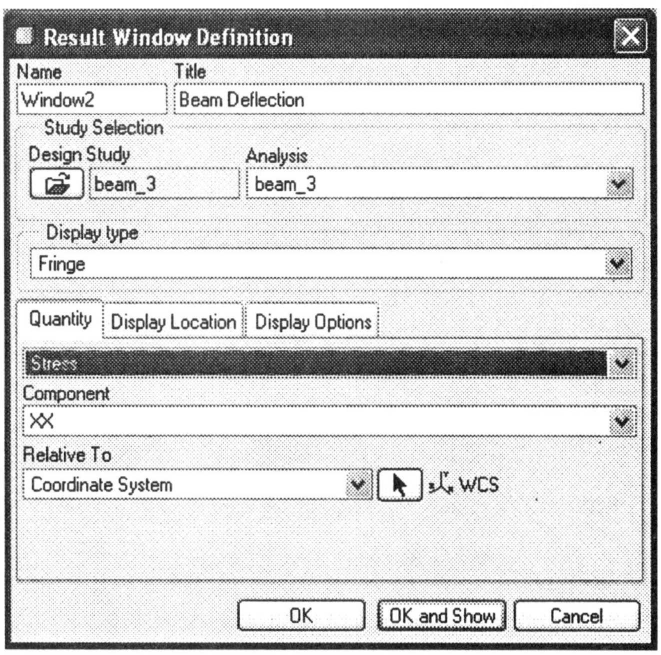

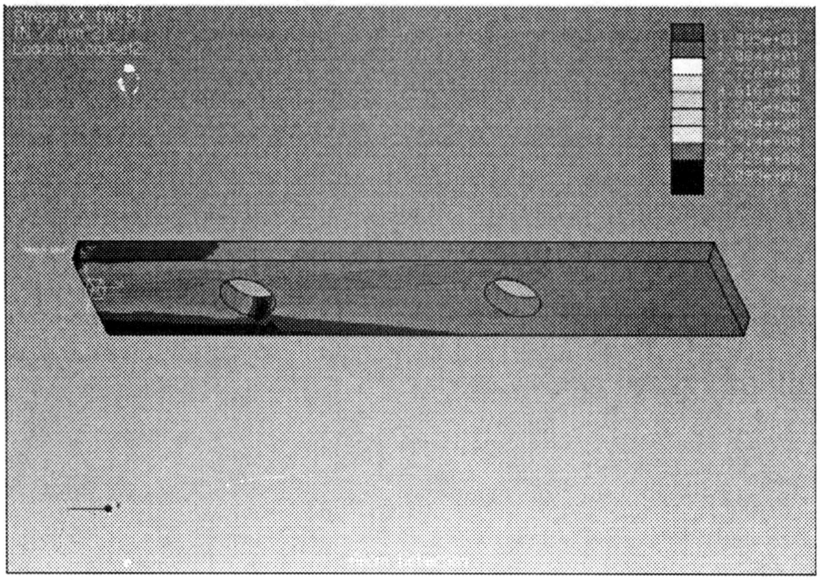

3.4 Modeling of a Bearing Load Condition

Figure 3-4 illustrates that a shaft component is being assembled with a beam structure. The beam structure has a through hole. The diameter dimension of the hole is slightly larger than the diameter dimension of the shaft so that a clearance fit can be assured during the assembly process.

(a) two components before assembling

(b) two components after assembling

Figure 3-4 Clearance Fit of 2 Components during Assembling

Because of the presence of clearance between the two components, the 2 components are in contact partially. As illustrate in the following figure, the lower portion of the shaft component is in contact with the lower portion of the cylindrical surface of the hole. The upper portion of the shaft is not in contact with the cylindrical surface of the hole. When a load is acting on the shaft component, the action transferring from the shaft component to the beam structure is through the contact area. Such a load condition to the beam structure is termed as a bearing load.

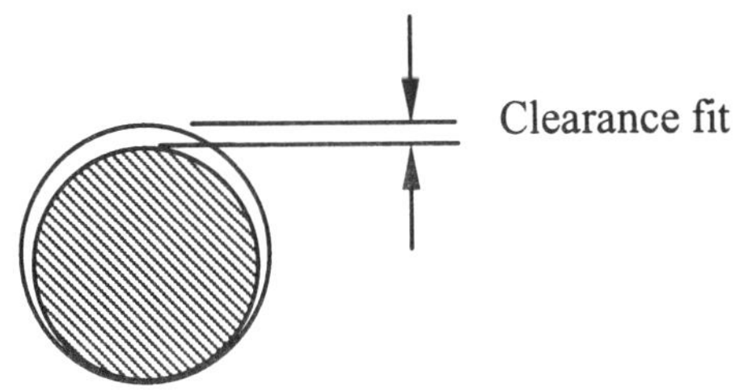

In this section, the example presented is a cantilever beam. There is a through-hole positioned at the central location. The dimensions of the beam and hole are shown in the isometric view of the beam.

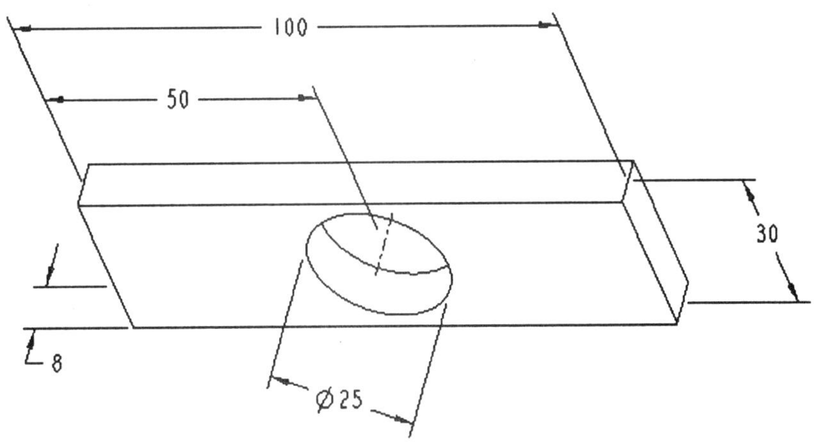

The following constraint and load conditions are assumed:

- The left side of the beam is fixed.
- The right side of the beam is a free end.
- The material of the cantilever is steel.
- The magnitude of the bearing load is 500 Newton, as shown.

The objectives of this study are

(1) Plot the distribution of the deformation of the entire beam.
(2) Plot the distribution of the maximum principal stress of the 3D solid model.
(3) What is the highest value of the maximum principal stress? Where does it occur?

Step 1: Create a 3D solid model to represent the cantilever beam.

File > **New** > **Part** > type *beam_4* as the file name. Clear the box called **Use default template,** and select the **mmns_part_solid**. Fill in **DESCRIPTION** by typing in *bearing load.* Fill in **MODELED_BY** by typing your name, say *student* > **OK.**

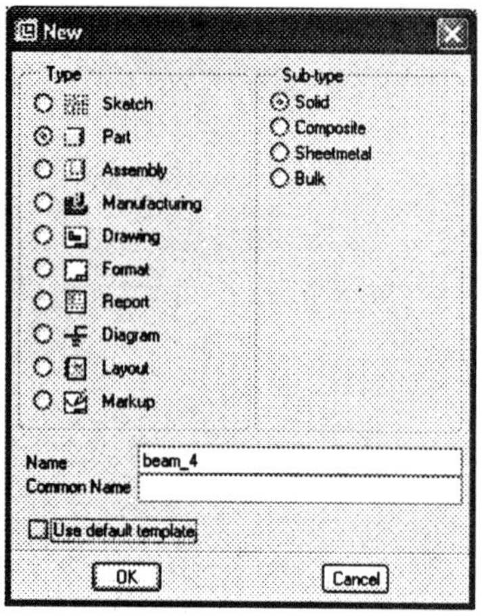

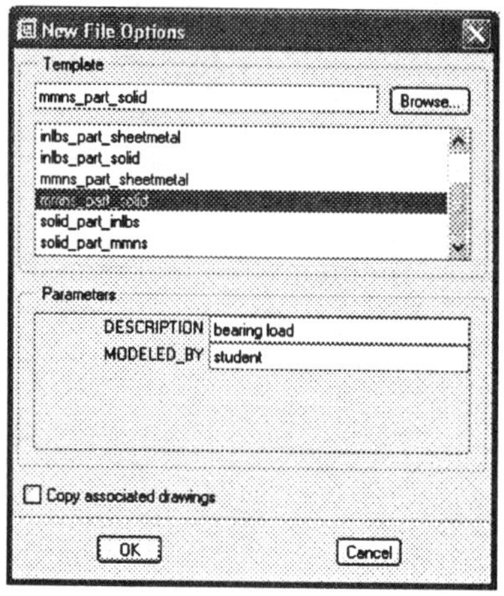

Pick up the icon of **Extrude** displayed on the toolbar of feature creation. Make sure that the icon of **Solid** is selected. Let us set the depth choice by typing *8* (the thickness of the beam). To initiate the process of sketching, activate **Placement** > **Define.**

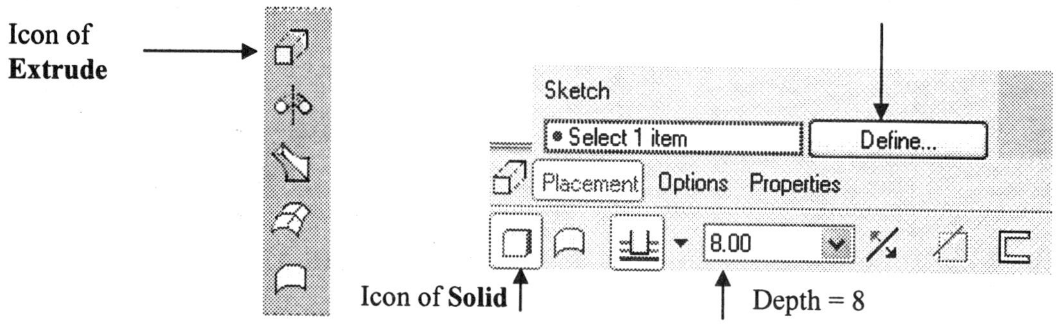

Select the **FRONT** datum plane displayed on screen as the sketch plane. Pick the box of **Sketch** to accept the **RIGHT** datum plane as the default reference to orient the sketch plane, as illustrated below:

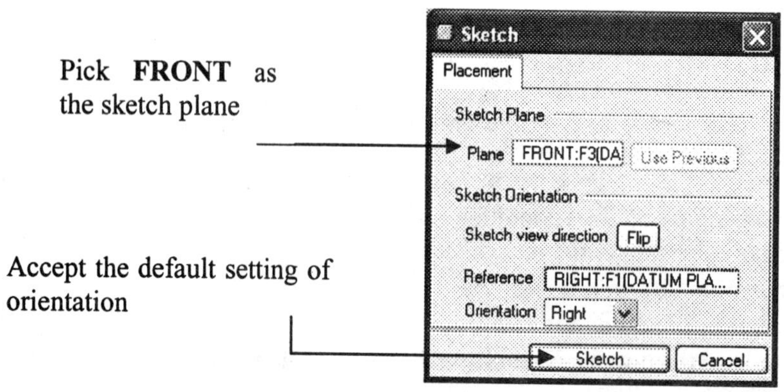

Both the sketch window and the references window appear. Let us first close the reference window. Add a horizontal centerline. Select the icon of **rectangle** from the toolbar of sketcher and sketch a rectangle. The 2 dimensions are 100 and 30, respectively. Make sure that the symmetry about the horizontal centerline is maintained.

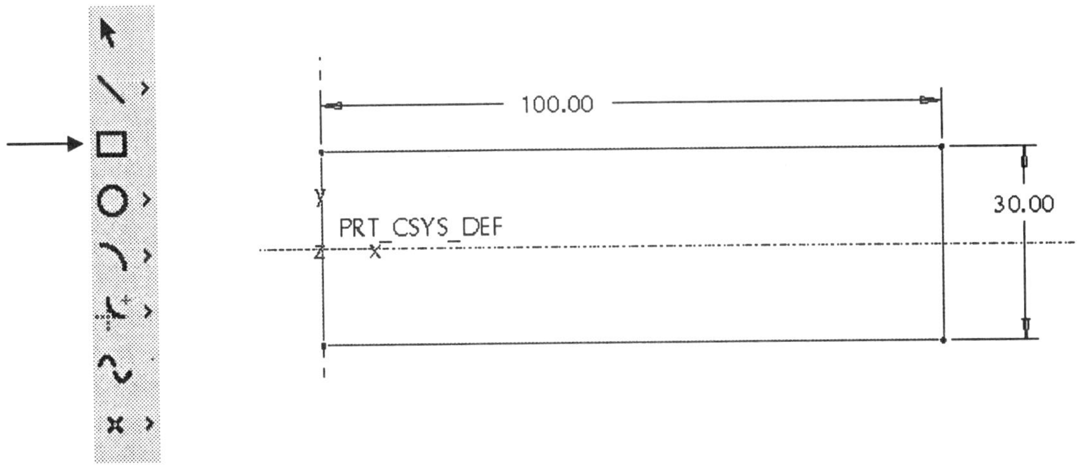

Upon completing the sketch, select the icon of **Done** from the toolbar of sketcher. On the dashboard, type *8* as the thickness value if you have not set the thickness value, as shown.

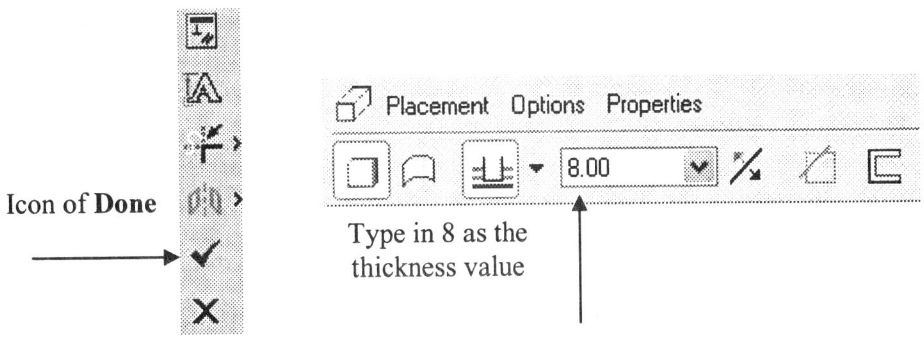

Select the icon of **Complete** from the feature control panel, completing the creation of a 3D solid model for the column structure.

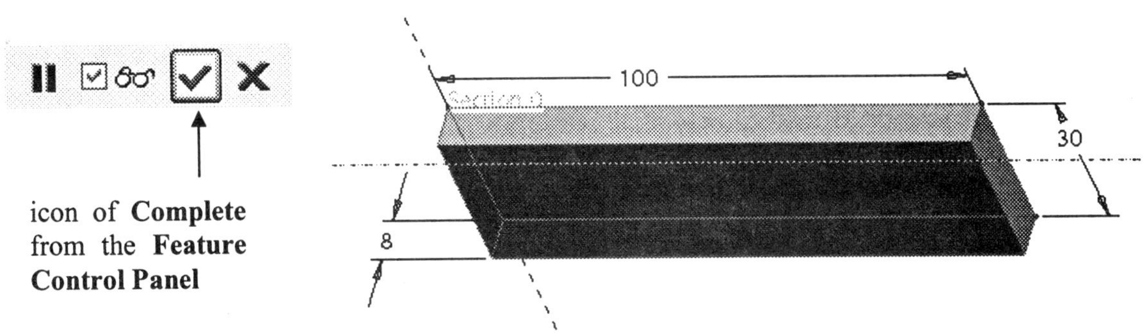

Now let us create a cut feature for the through hole with the diameter equal to 25 mm.

Select the icon of **extrude** from the toolbar of feature creation. Make sure that the icon of **Cut** is selected. Set the depth choice to **Through All**. Activate **Placement** > **Define**.

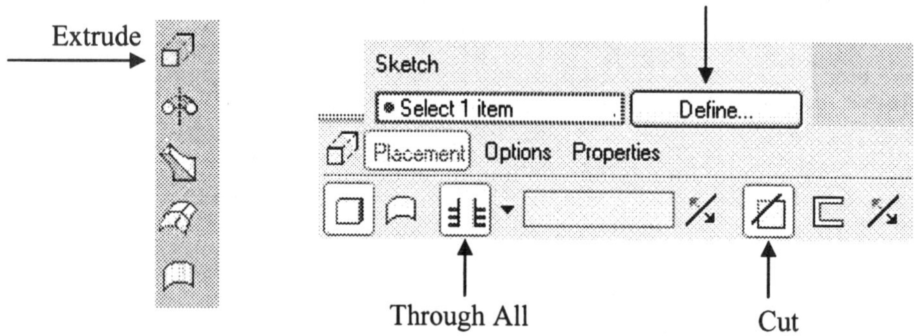

Select the front surface of the beam (do not select the FRONT datum plane) as the sketch plane and use the right-sided surface to be at right to orient the sketch plane > **Sketch**.

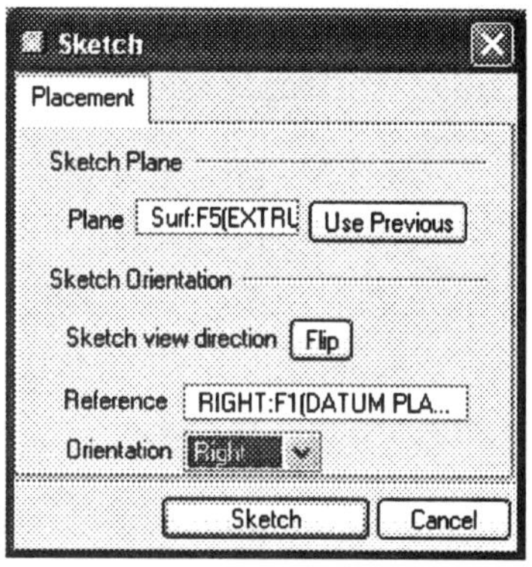

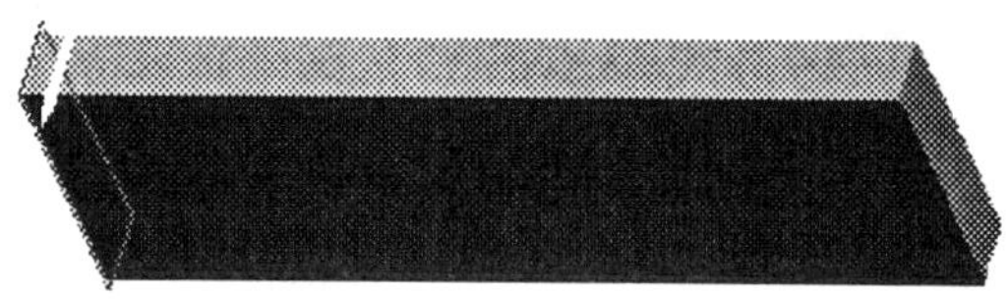

Select the icon of **circle** from the toolbar of sketcher and sketch a circle. Change the diameter dimension to 25, and the two position dimensions to 200 and 40, as shown. Upon completing the sketch, select the icon of **Done** from the toolbar of sketcher.

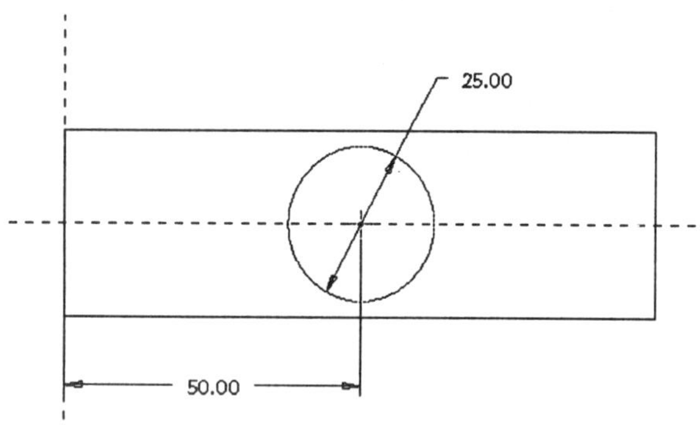

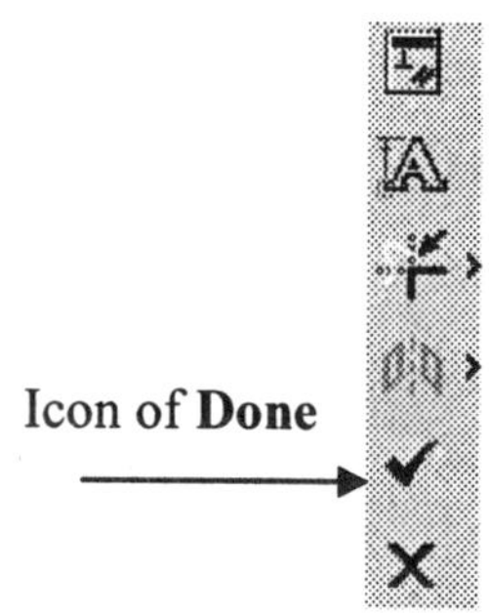

Select the icon of **Complete** from the feature control panel, completing the creation of a cut feature.

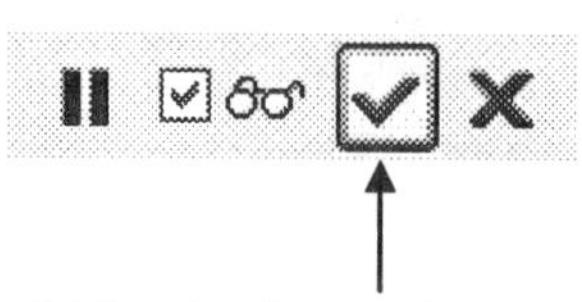

Pick the icon of **Complete** from the **Feature Control Panel**

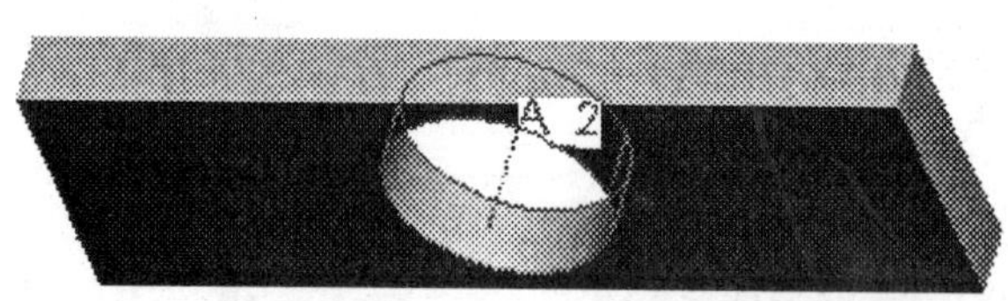

Up to this point, the process of creating a 3D solid model for the beam structure is completed. Now let us get on to the process of performing FEA.

Step 2: Define two datum points to be used to define the bearing load condition.
To create **PNT0**, click the icon of datum point from the toolbar of datum creation.
Pick the datum axis called A_2 from the display. While holding down the Ctrl key, pick the **FRONT** datum plane > OK.

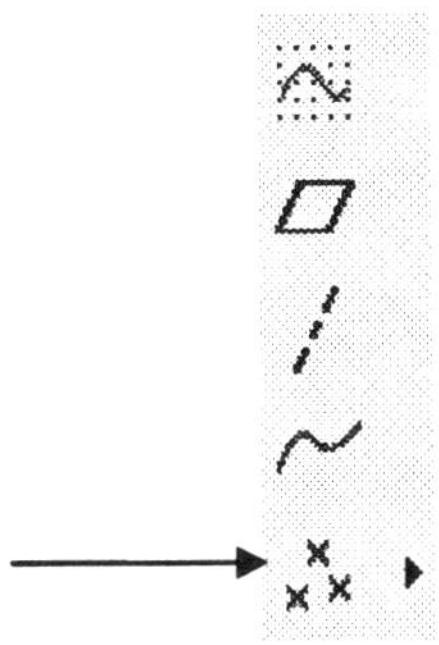

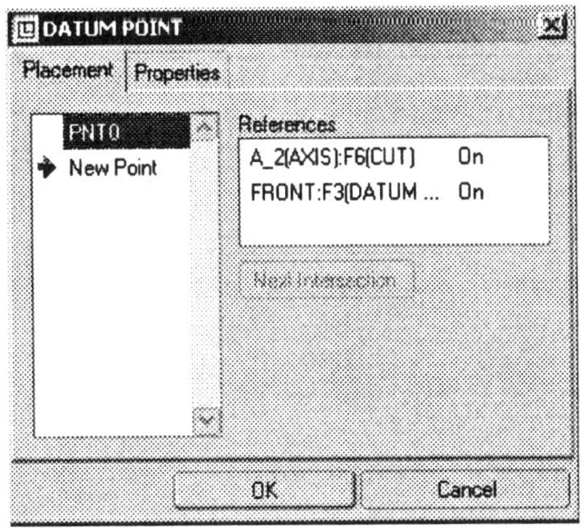

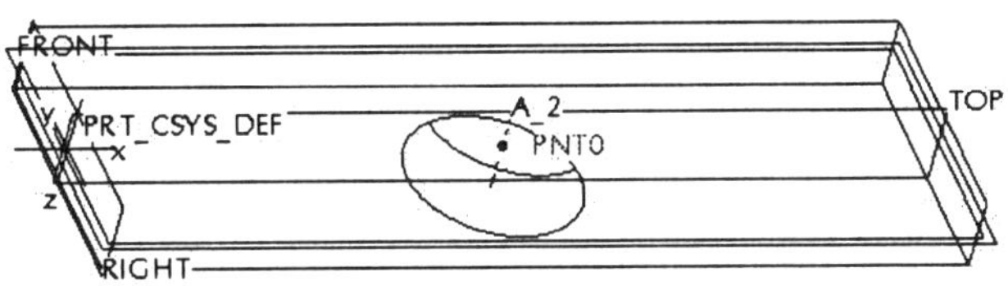

To create **PNT1**, click the icon of datum point > **Offset Csys** > pick the coordinate system on display > **Cartesian** > **Enter Points** > type in (50, -10, 0) as the x, y, z coordinates, respectively > **Done.**

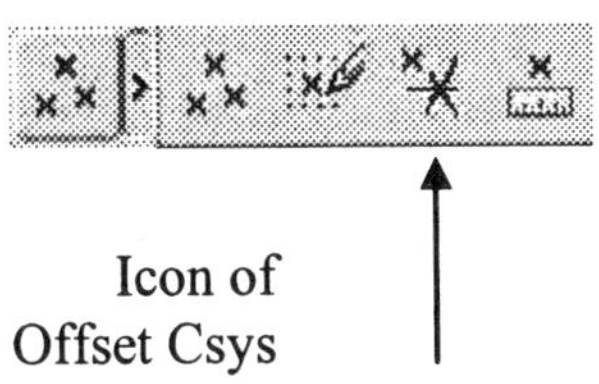

Icon of Offset Csys

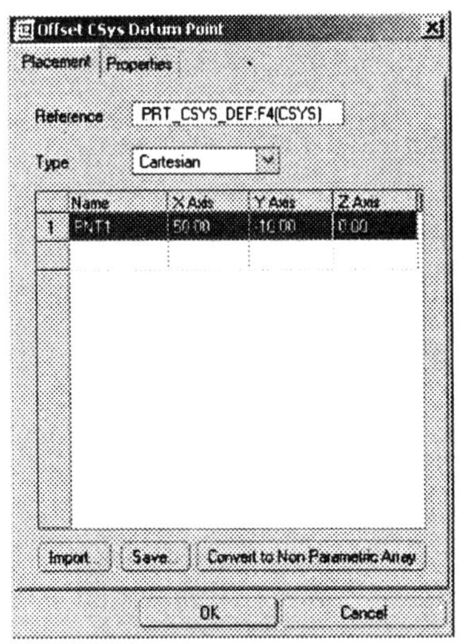

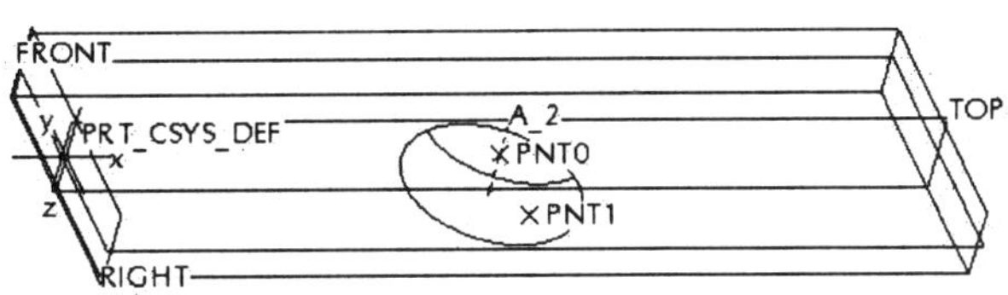

We will use PNT0 and PNT1 to form a directional vector, which is along the vertical direction downward.

Let us shift from the Pro/ENGINEER design system to Pro/Mechanica. From the main toolbar or the menu toolbar, select **Applications > Mechanica > Continue > Structure > OK.**

From the toolbar of functions, select the icon of **Materials >** select **STEEL** from the left side called Materials in Library > click the directional arrow so that the selected material type goes to the right side called Materials in Model.

Icon of Material

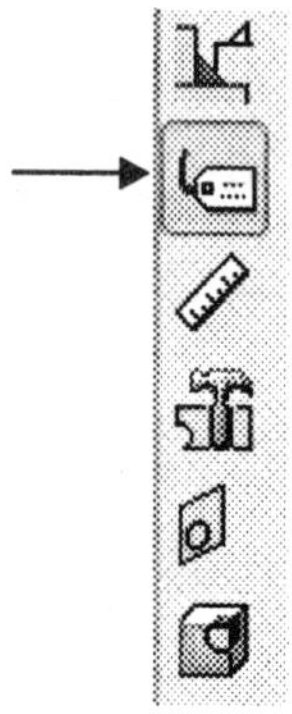

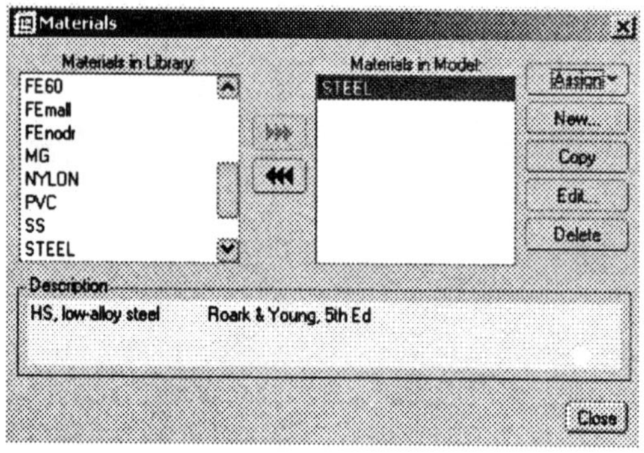

Assign > **Part** > select the beam structure from the screen display > **OK** > **Close.**

To define the fixed end constraint on the left side of the beam, select the icon of **New Displacement Constraint** from the toolbar of functions > **Surface(s)** > pick the surface at the left side of the beam > **OK** to accept the surface selection.

Because the constraint condition is "a fixed end", we set all 6 degrees of freedom equal to zero, which is the default setting > **OK.**

Icon of New Displacement Constraint

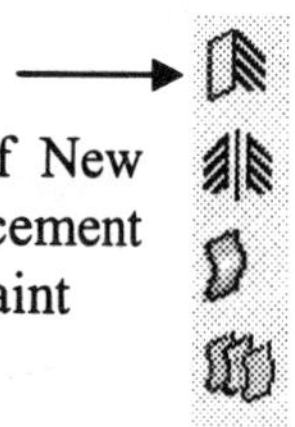

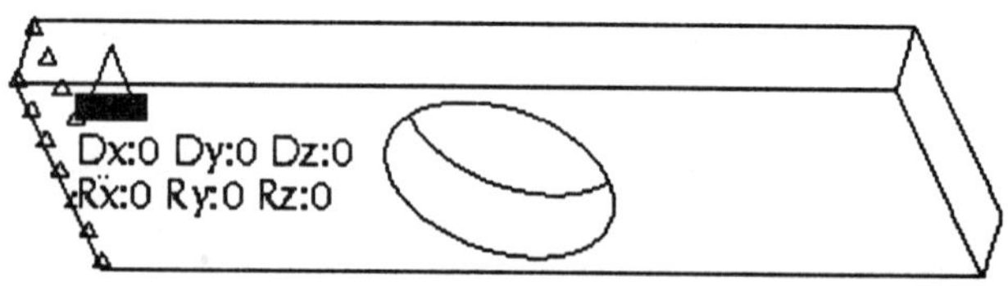

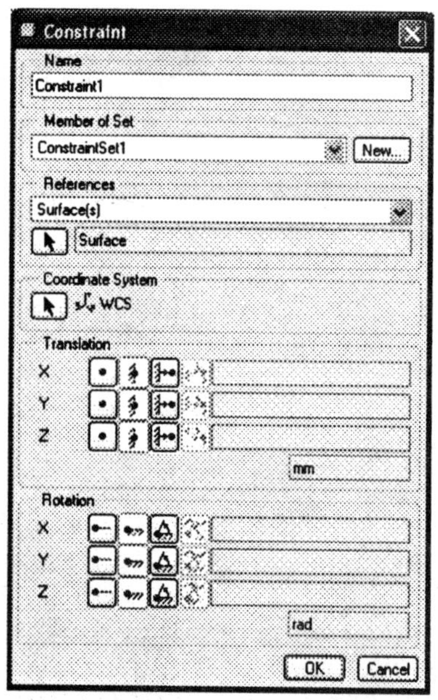

To define the bearing load condition, select the icon of Bearing Load from the toolbar of functions > pick the low half surface of the hole > **OK** to complete the bearing hole selection.

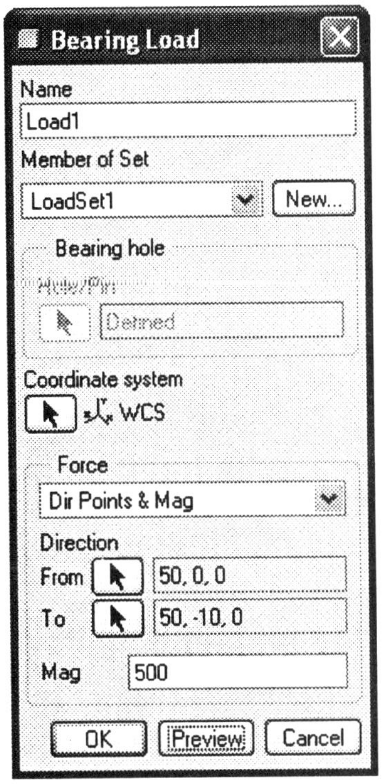

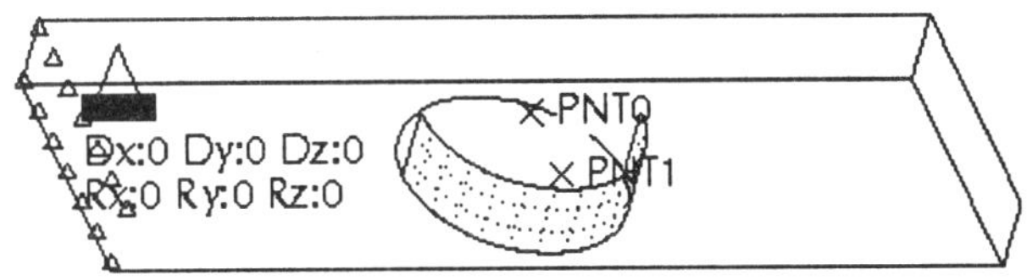

Select Dir **Points & Mag** from the box called **Force.**
Pick PNT0 as the starting point or **From**.
Pick PNT1 as the ending point or **To.**
Type 500 as the magnitude value > **OK**.

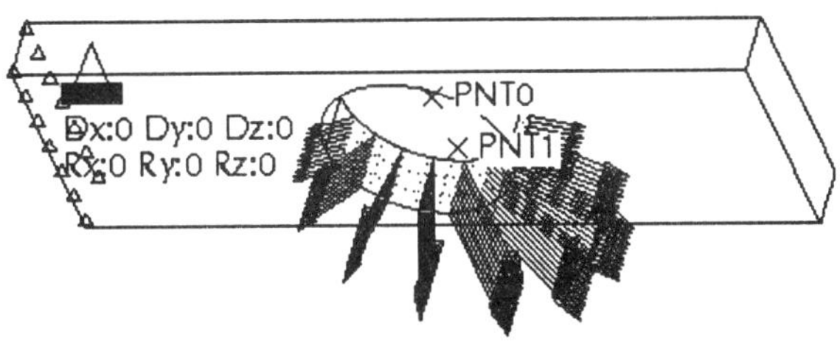

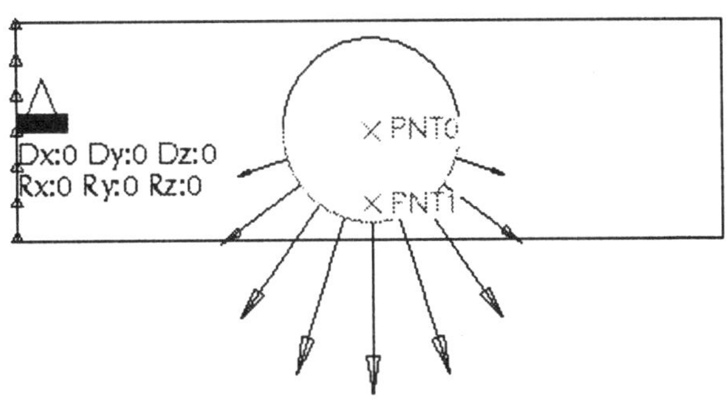

Under the Pro/MECHANICA, set up **Analyses** and **Run** it

Select the icon of **Run an Analysis** from the main toolbar > **File** > **New Static** > type *bearing_load* as the name of the analysis folder > **OK.** Click the box of **Run.**

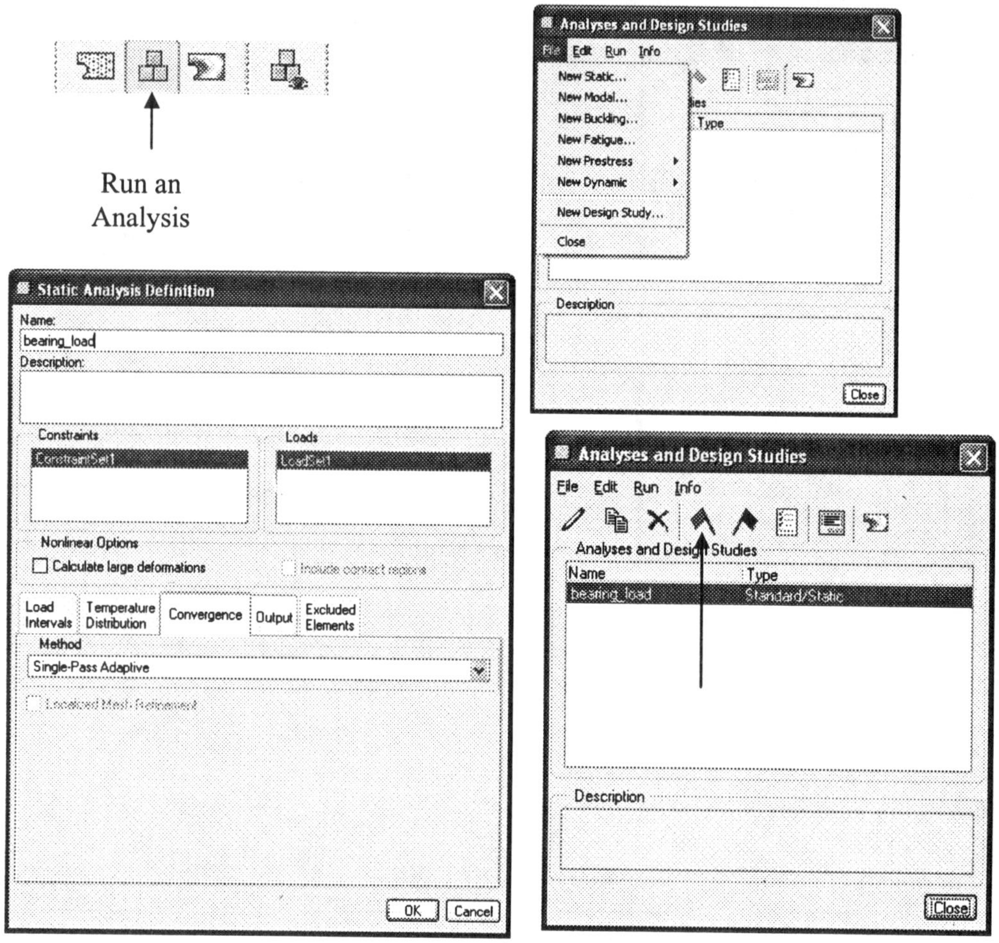

There is always a message displayed on screen. The message is asking the user "Do you want error detection?" Click **Yes**. In this way a warning message will appear if an abnormal operation occurs.

To study the obtained result, say the displacement of the column under the static loading condition, click the icon of **Review results** > **Displacement** > **Magnitude** > **OK and Show**. Note that if the folder called column does not appear in the Result Window Definition, click the icon of folder under Design Study, locate it and open it.

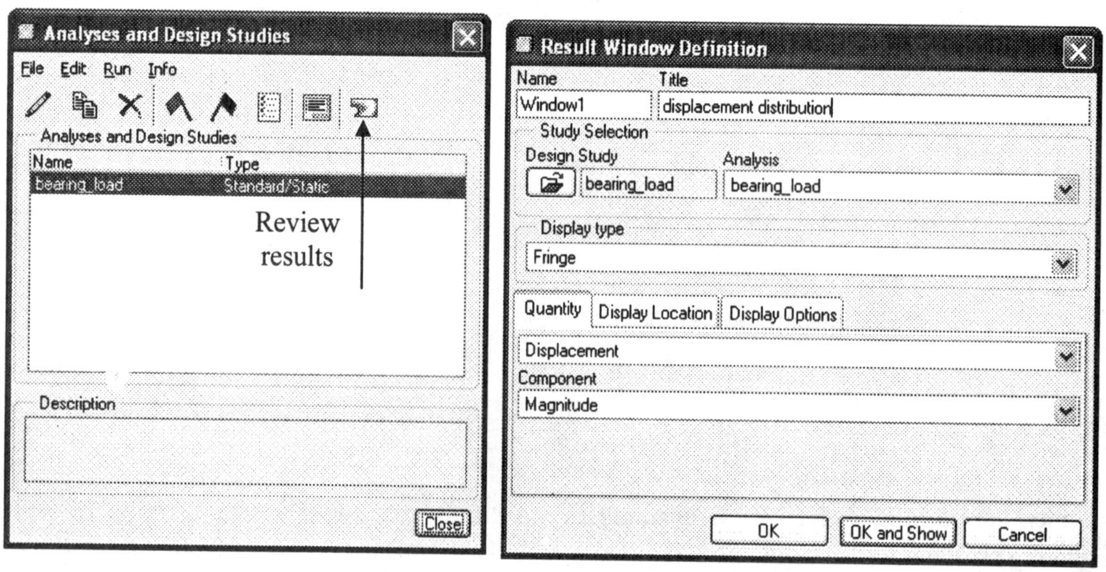

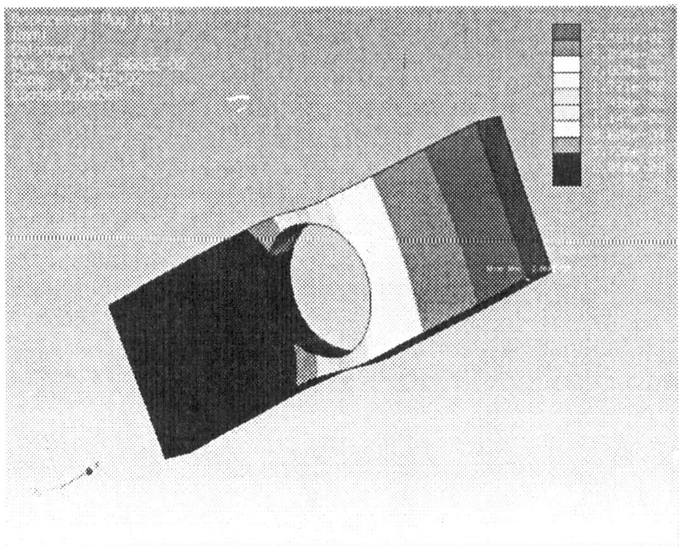

To study the distribution of the maximum principal stress distribution, select **Edit** from the main toolbar > **Copy** and a new Result Window appears > **Stress** > **Max Principal** > **OK and Show**.

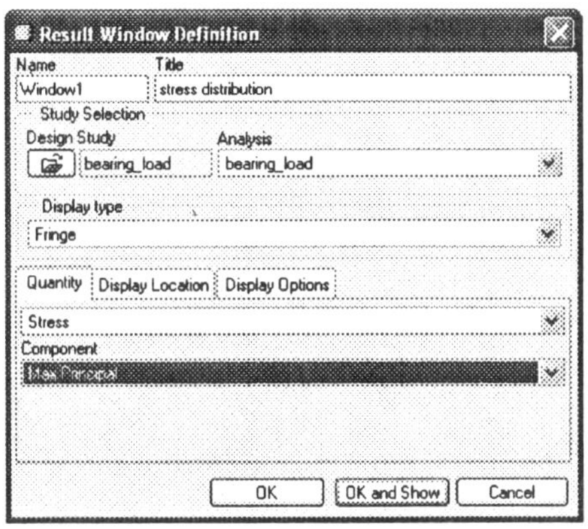

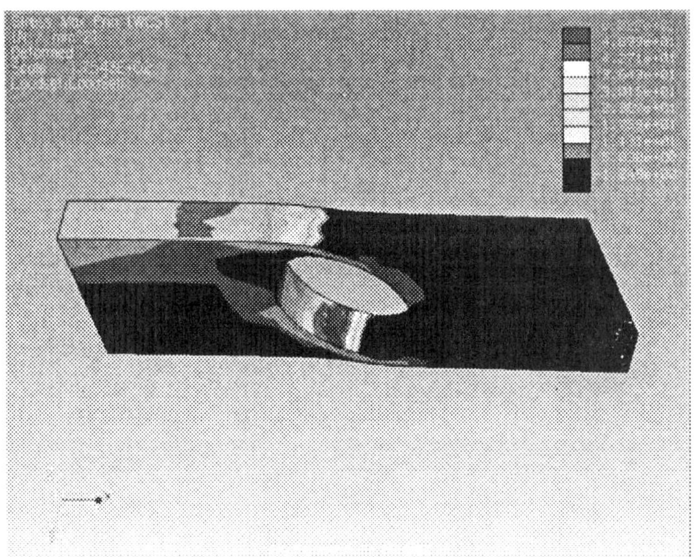

To find the location of the maximum principal stress, select Info from the main toolbar > select **Model Max.** The location of the maximum principal stress is shown on the display. The location is on the inner surface of the hole. The magnitude is 55.27 MPa. The numerical values of the coordinates of the location where the maximum principal stress occurs are also displayed in the message window, showing x=44.7, y=-11.3 and z=2.

```
Preparing result window "Window1" (1 of 1)...
Updating the display...
The value is Model Max  5.654E+01.
The WCS location is (44.73191070557, -11.33558750153, 2.000076532364).
```

Because the location of the maximum principal stress occurred on the beam structure is at the inner surface of the hole, in order to depict the location of the maximum principal stress clearly, let us use a function called Capping Surface. From the main toolbar, select **Insert > Cutting/Capping Surfs.** Pick **Capping Surface** > select **YZ** > **Apply** > **OK.**

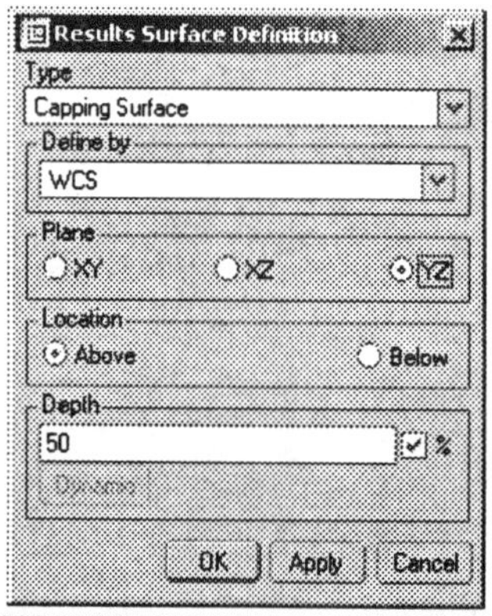

3.5 Modeling of a Pressure Load Condition

A support structure shown below is subjected to pressure loads. There are two pairs of pressure loads, as illustrated. The pressure acting on the walls of the 2 holes at the lower position is 20 N/mm2. The pressure acting on the walls of the 2 holes at the upper position is 10 N/mm2.

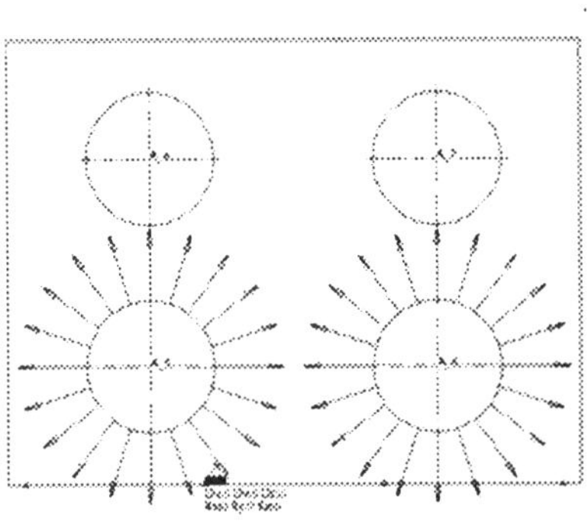

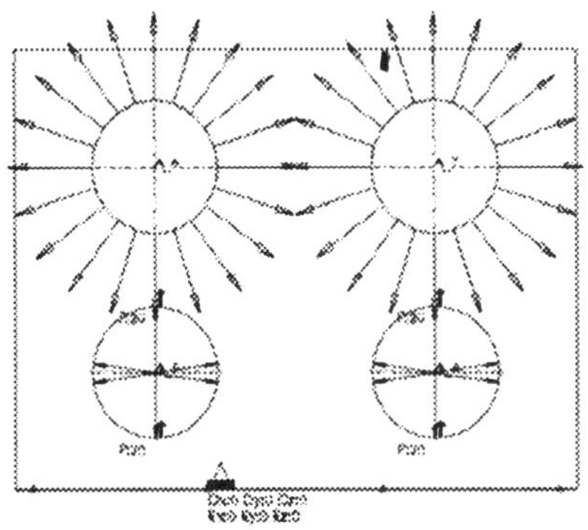

The geometrical shape and dimensions of the support structure are shown below.

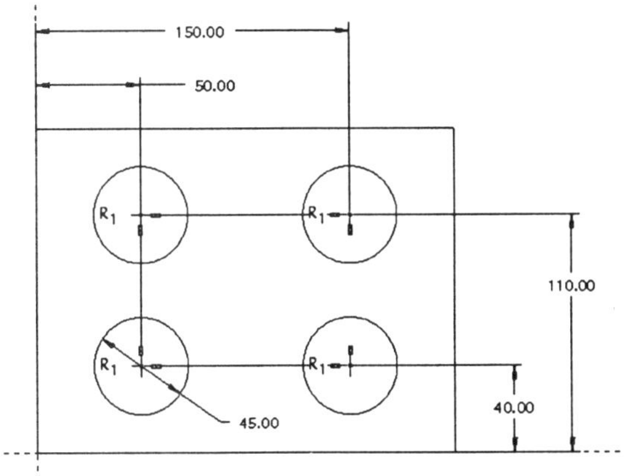

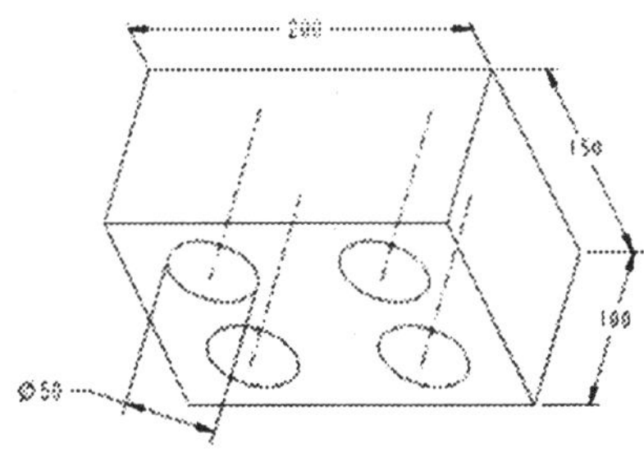

Assume that the bottom surface of the support structure is fixed to the ground. The material type of the support is steel. Our objectives of this study are "investigate the deformation pattern under the pressure loads and the maximum principal stress distribution when subjected to the pressure loads.

Step 1: Create a 3D solid model to represent the support structure.

File > **New** > **Part** > type *pipe_base* as the file name. Clear the box called **Use default template,** and select the **mmns_part_solid**. Fill in **DESCRIPTION** by typing *pressure loads.* Fill in **MODELED_BY** by typing your name, say *student.*> **OK.**

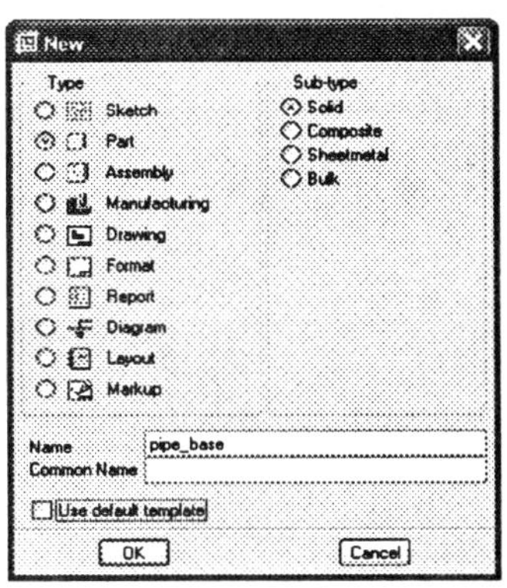

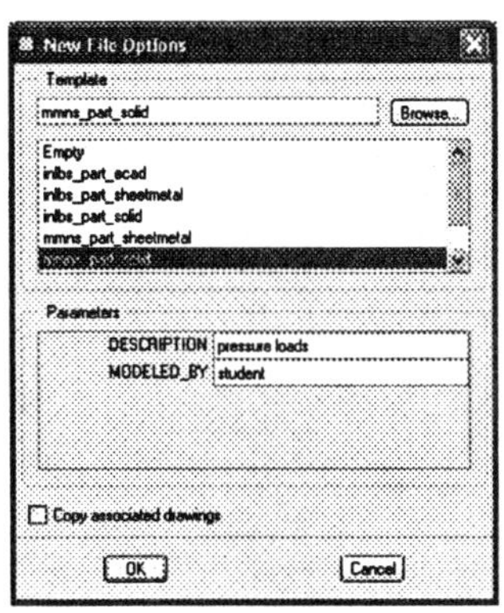

From the toolbar of feature creation, select the icon of **Extrude.** The dashboard appears. Make sure that the icon of **Solid** on the dashboard is selected. Set the depth value to 150. Activate **Placement** to define a sketch plane > **Define.**

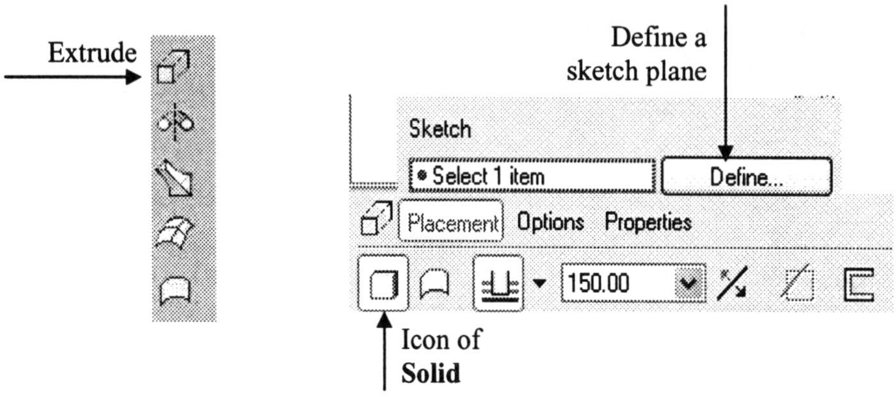

Pick the **TOP** datum plane from the display, and click **Sketch** to accept the orientation of the sketch plane by positioning the **RIGHT** datum plane at the right side. Click the button of **Sketch** to complete the process of defining the sketch plane.

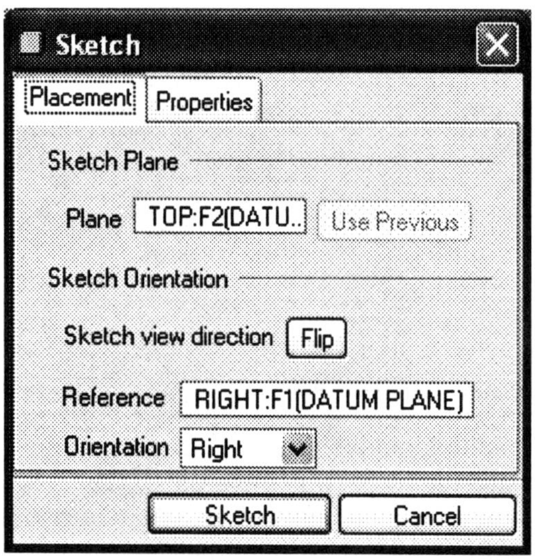

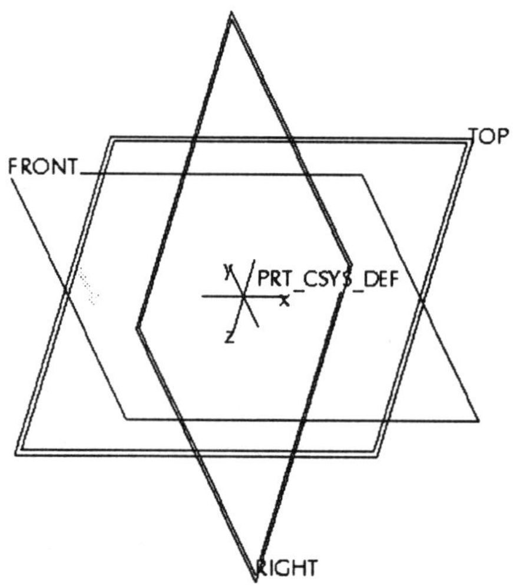

Both the sketch window and the references window appear. Let us first close the reference window. Select the icon of **Rectangle** from the toolbar of sketcher, and draw a rectangle, as shown below.

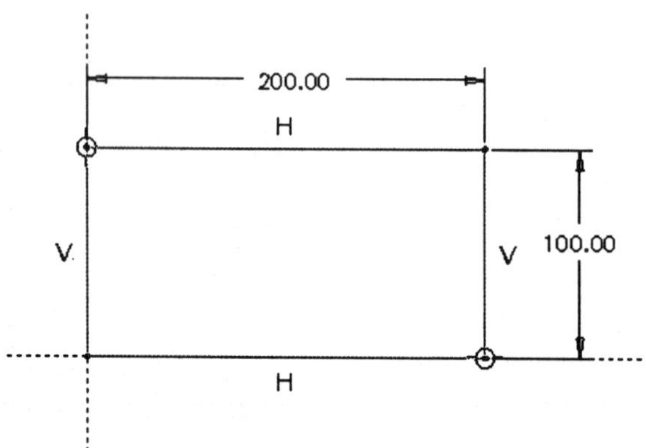

Upon completing the sketch, select the icon of **Done** from the sketcher toolbar. On the dashboard, type in *150* as the height value if this value has not been set, as shown. From the feature control panel, select the icon of **Complete** to complete the block feature creation.

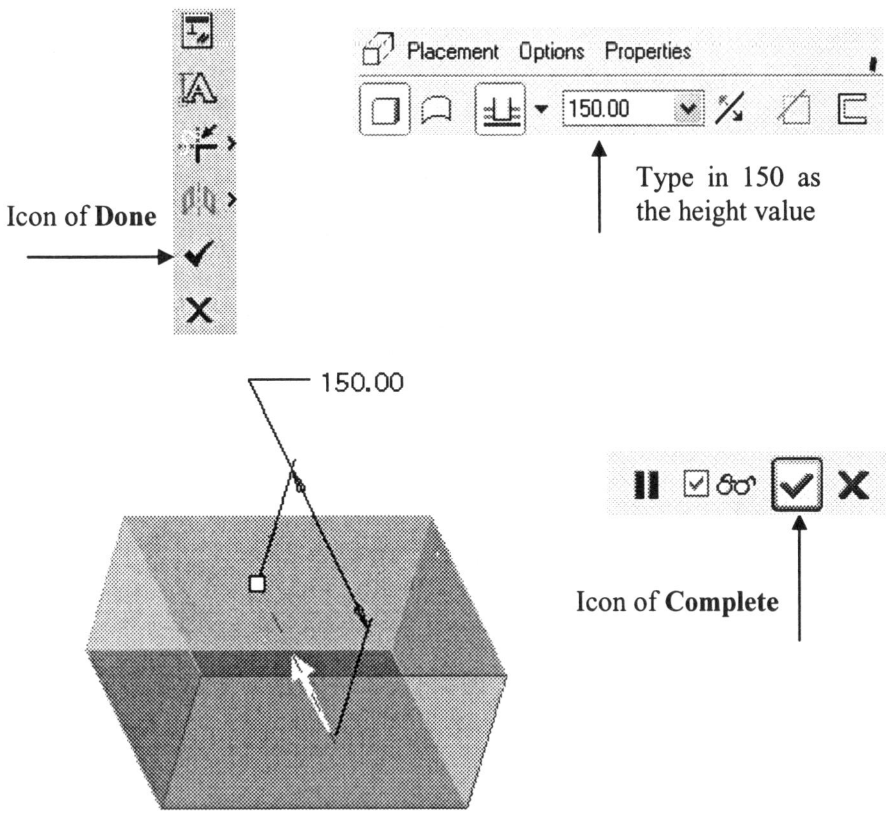

Step 2: Create a feature having 4 through holes using the **Cut** operation.

From the toolbar of feature creation, select the icon of **Extrude.** The dashboard appears. Make sure that the icon of **Solid** on the dashboard is selected. Select the icon of **Cut** and set the depth choice to **Through All**.

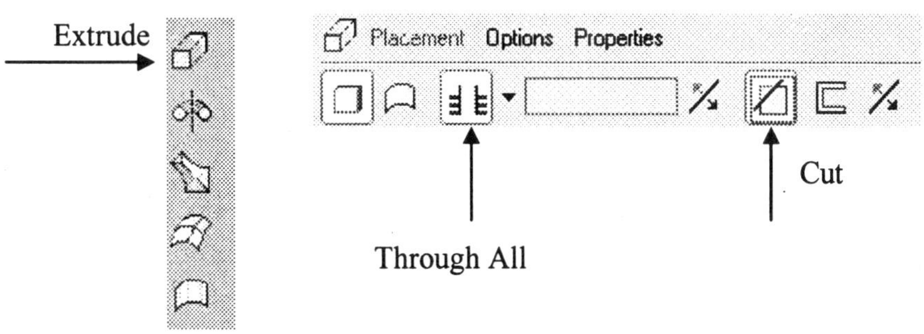

From the dashboard, activate **Placement** to define a sketch plane > **Define** > pick the front surface of the block (not the FRONT datum plane) as the sketch plane > **Sketch.**

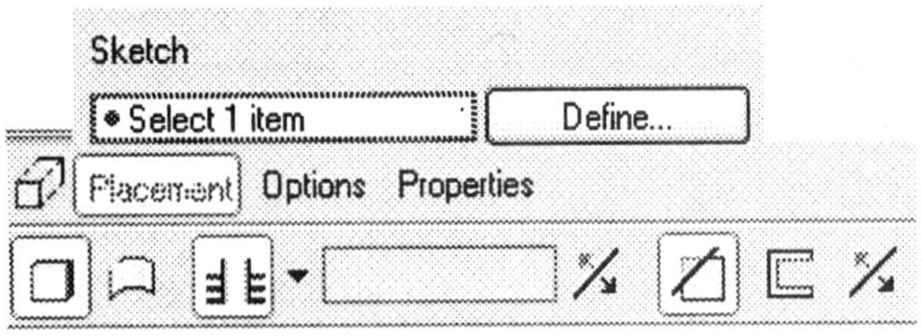

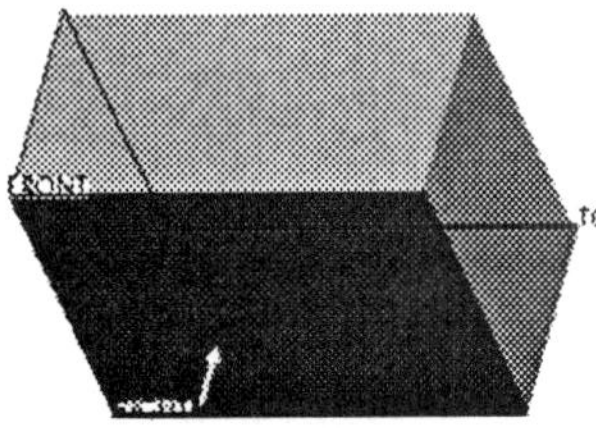

Both the sketch window and the references window appear. Let us first close the reference window. Select the icon of **Circle** from the sketcher toolbar, and draw 4 circles, which are equal to one another in diameter, as shown below. The diameter dimension is 40. Note the 4 position dimensions to locate the 4 centers of the 4 circles. They are 40, 50, 110 and 150, respectively.

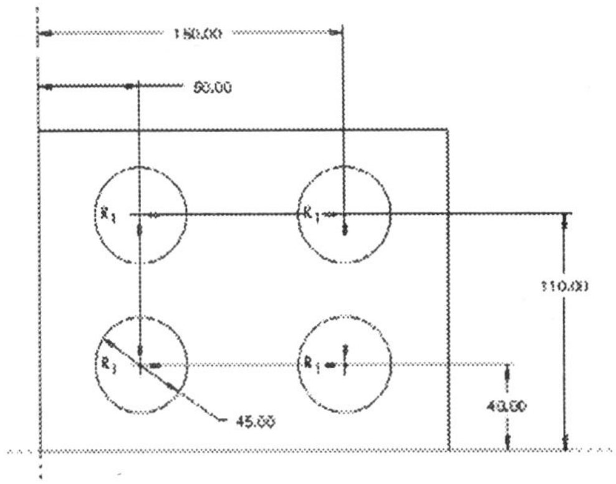

Upon completing the sketch, select the icon of **Done** from the sketcher toolbar. From the feature control panel, select the icon of **Complete** to complete the hole feature creation.

Icon of **Done**

Icon of **Done**

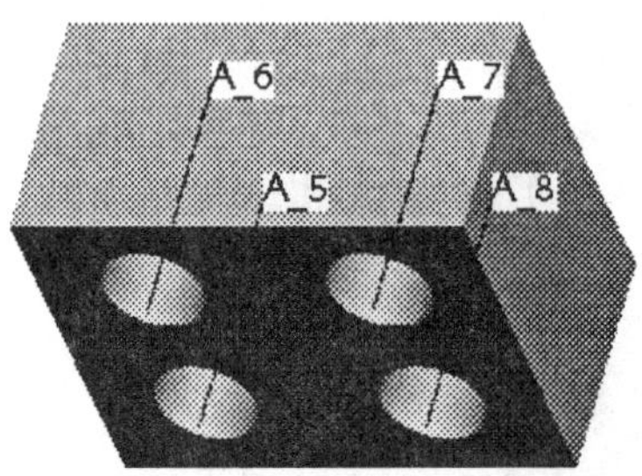

Step 3: FEA under Pro/MECHANICA.

Applications > **Mechanica** > **Continue** > **Structure** > **OK.**

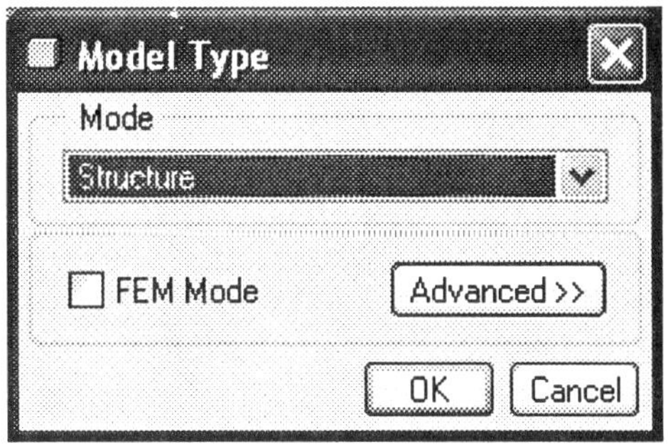

From the toolbar of functions, select the icon of **Materials** > select **STEEL**

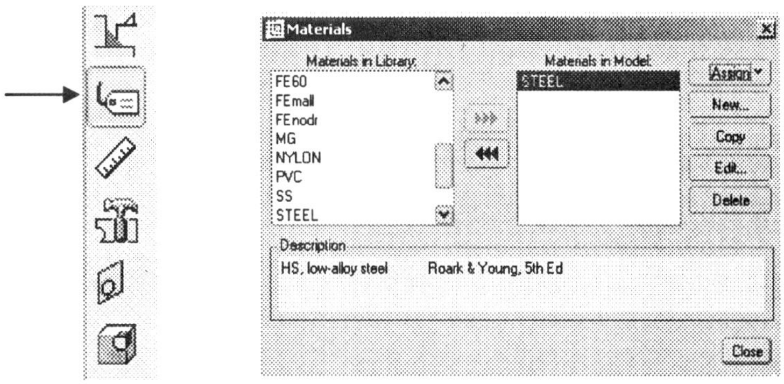

Assign > **Part** > select the pipe base structure from the screen display > **OK** > **Close**

Step 4: Under the Pro/MECHANICA environment, define the fixed end constraint condition.

From the toolbar of functions, select the icon of **Constraints** > select the surface at the bottom of the pipe base component as the fixed end, and accept the default settings, namely, fixing all six degrees of freedom.

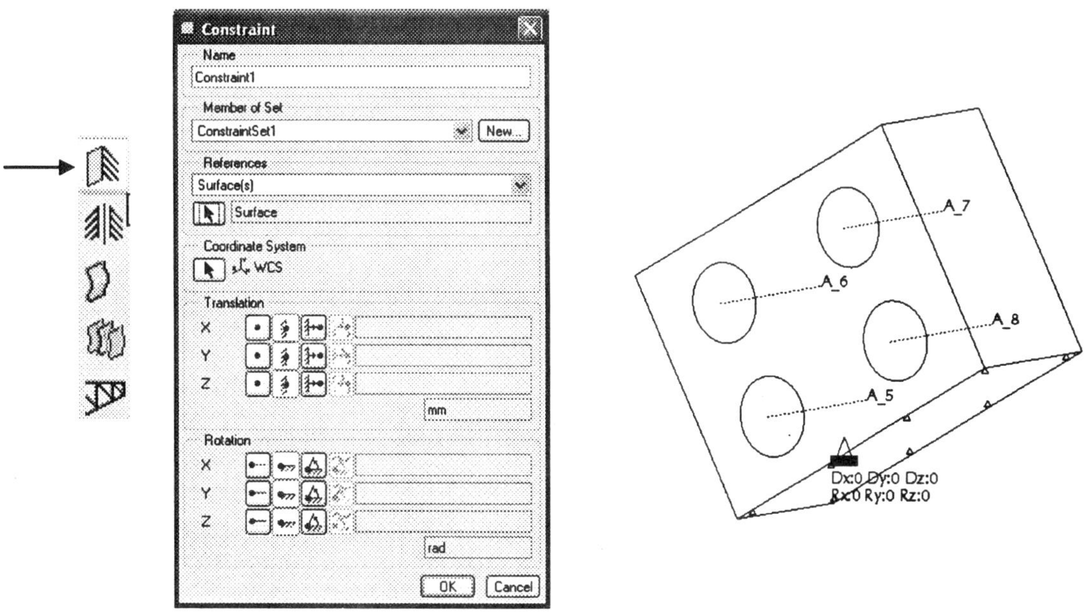

Step 5: Under the Pro/MECHANICA, define the load condition.

Select the icon of New Pressure Load from the toolbar of functions.

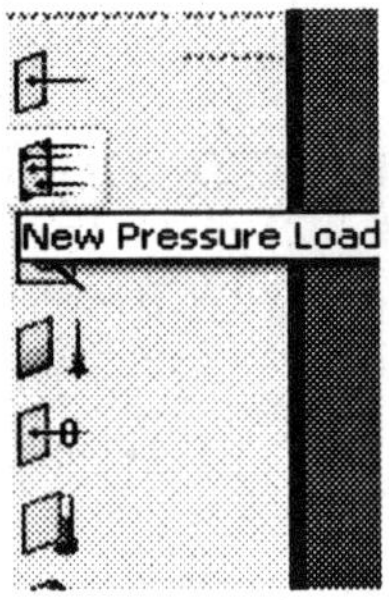

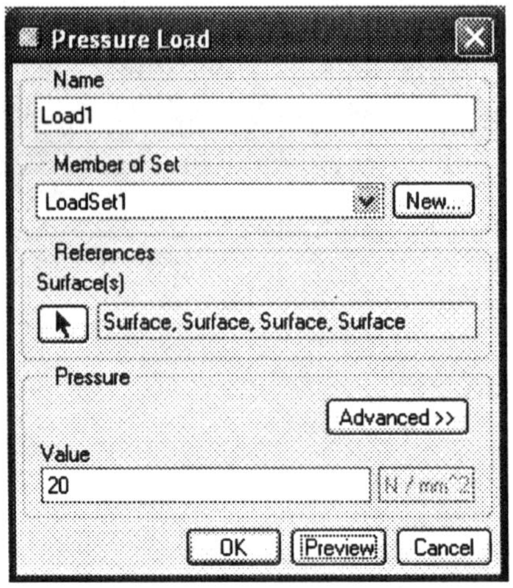

Activate the surface selection > holding down the Ctrl key, pick the 2 internal circular surfaces at the lower position, as shown > **OK.** Set the pressure value to 20 N/mm2 > **OK**.

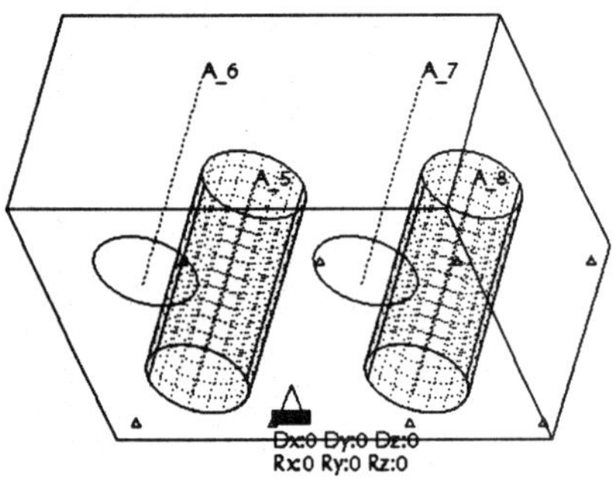

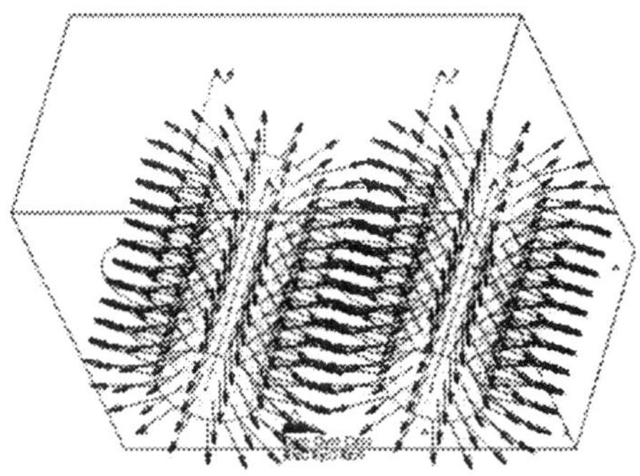

To define the second pressure load, select the icon of **New Pressure Load** from the toolbar of functions, again.

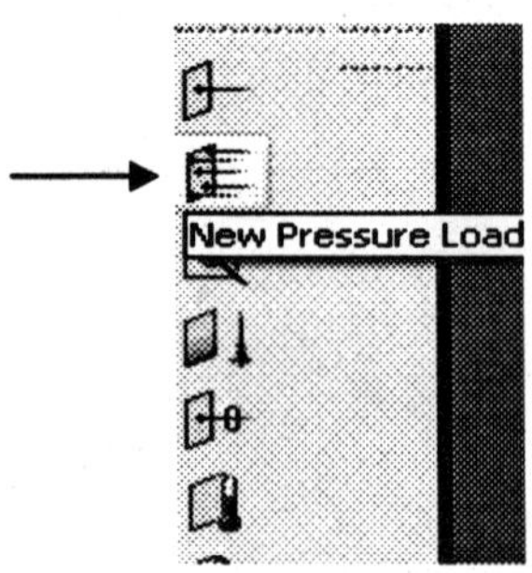

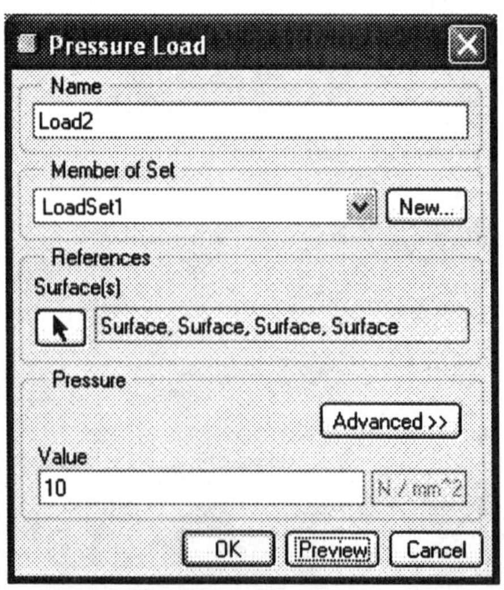

Activate the surface selection > holding down the Ctrl key, pick the 2 internal circular surfaces at the upper position, as shown > **OK.** Set the pressure value to 10 N/mm2 > **OK**.

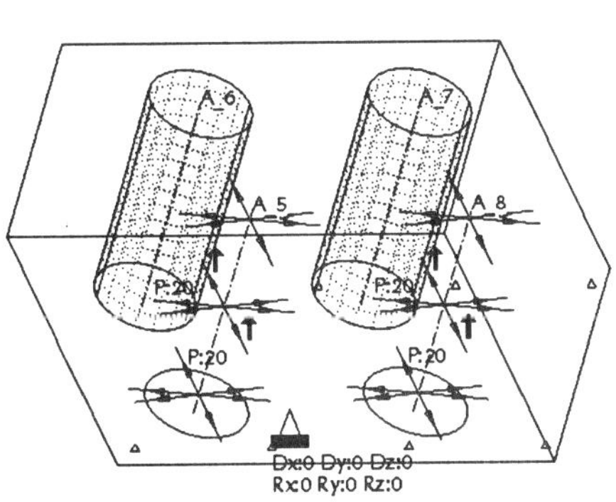

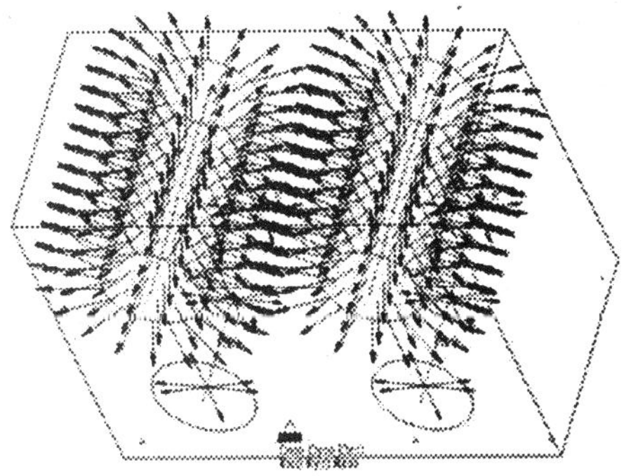

Step 6: Under the Pro/MECHANICA environment, set up **Analyses** and **Run** it
Select the icon of **Run an Analysis** > **File** > **New Static** > type in pipe_base > **OK**.

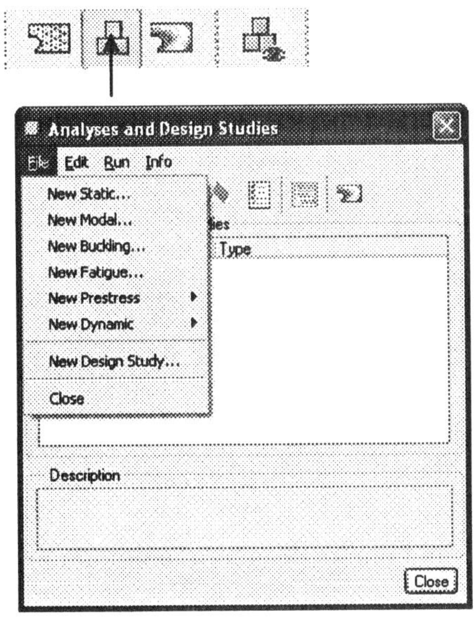

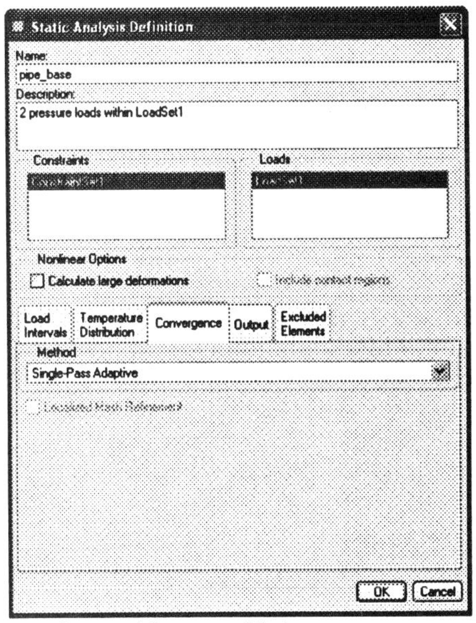

Before running the program, select Settings and specify a sufficient memory space, say 128 Megabytes for RAM Allocation. Run the program

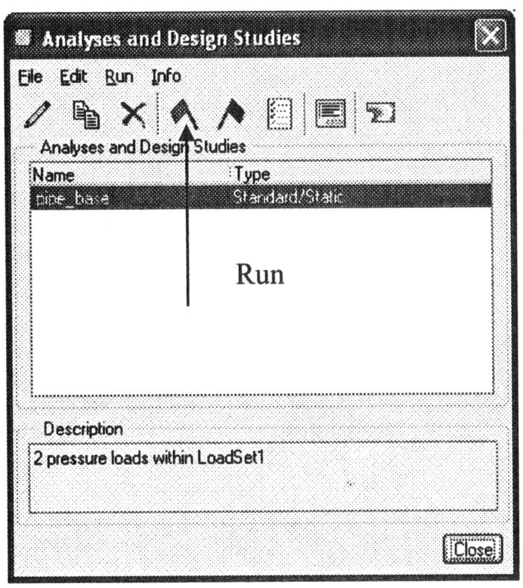

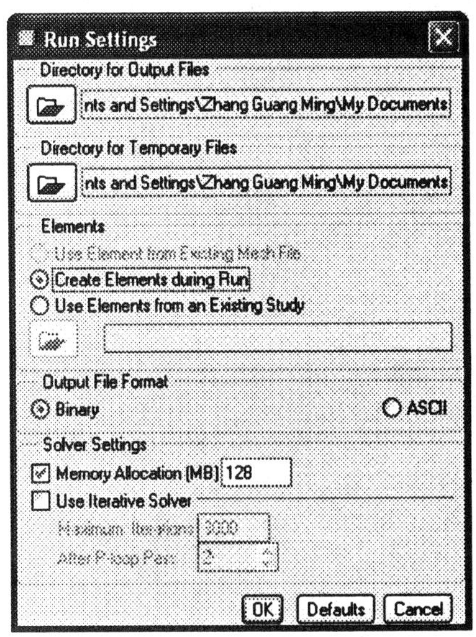

There is always a message disaplayed on screen. The message is asking the user “Do you want error detection?”. Click Yes. In this way, a warning message will appear if an abnormal operation occurs.

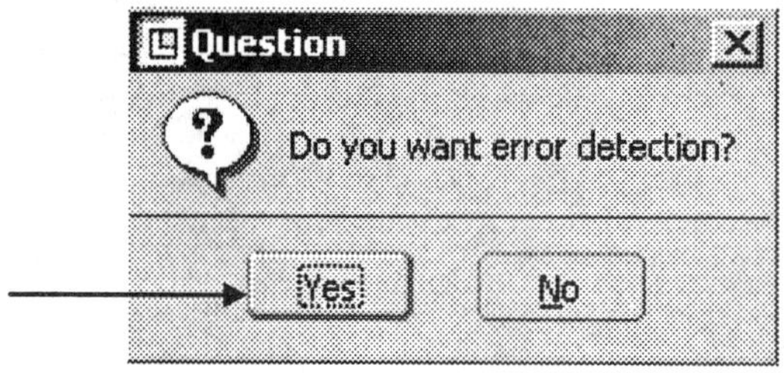

Step 8: Under the Pro/MECHANICA environment, study the **Results**
Select the icon of **Results**

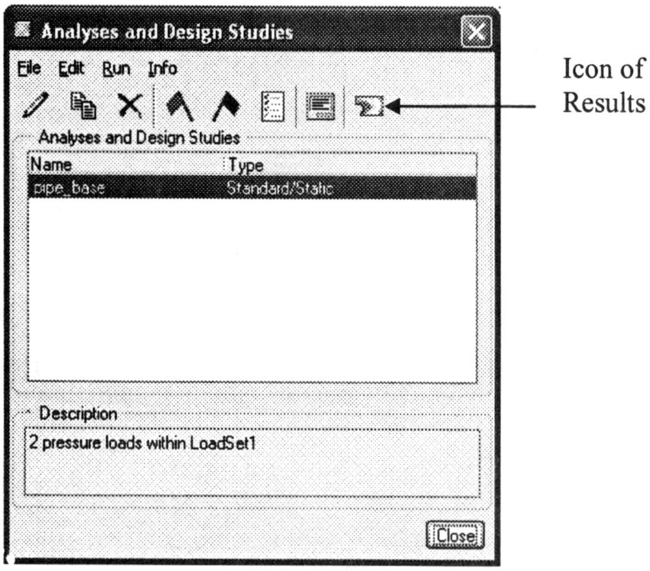

Type in block deflection as the title > locate the Design Study folder called pipe_base.
Select Fringe > choose Displacement > OK and Show.

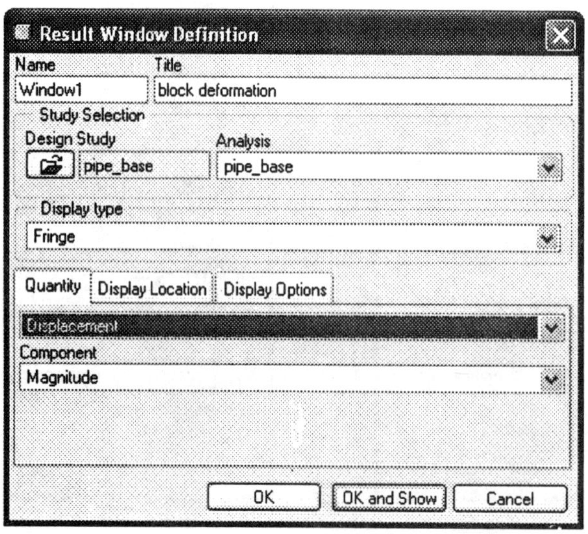

Plot of the displacement distribution.

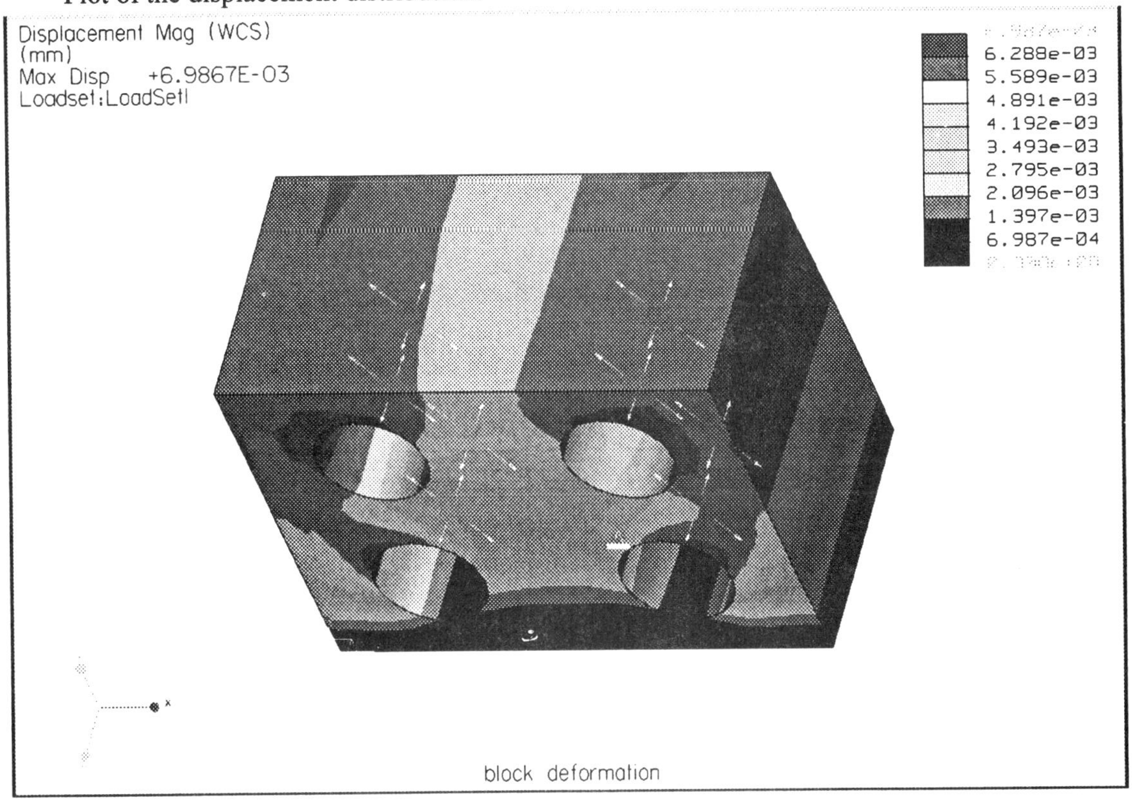

Plot of the displacement distribution with a capping surface crossing the 2 holes at the upper location. From the main toolbar, select **Insert** > **Cutting/Capping Surfs** > select **Capping Surface** > select **XZ** > the user may need to adjust the value of the Depth percentage to 30 or 40 > **Apply** > **OK**.

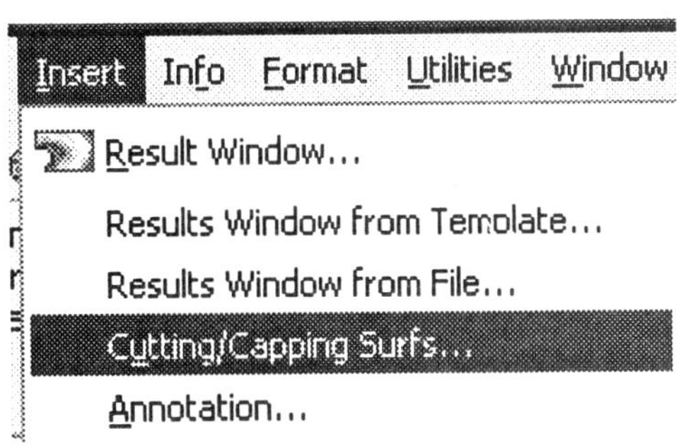

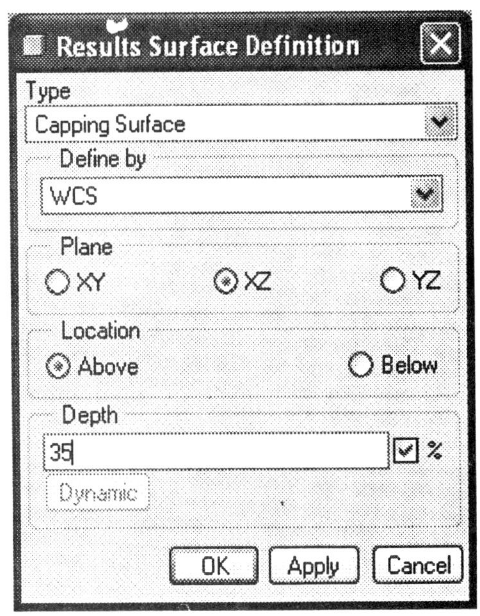

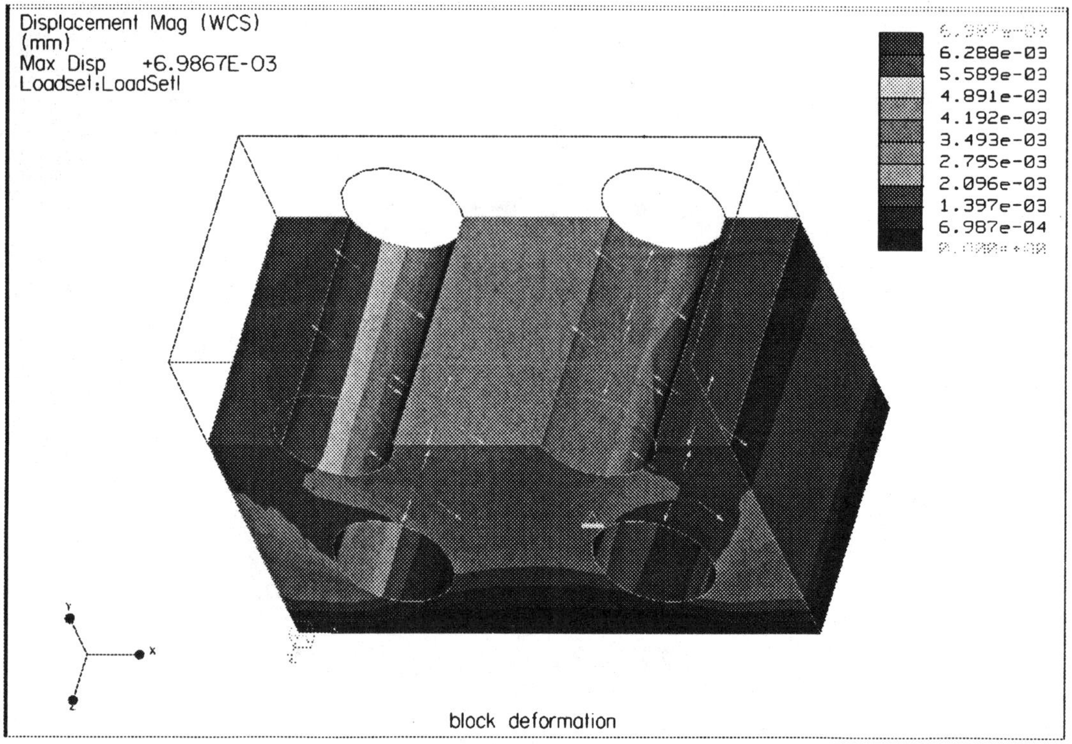

Plot of the displacement distribution with a capping surface crossing the 2 holes at the lower location. In this case, the user may need to adjust the value of the Depth percentage to *70* or *80* > **Apply** > **OK**.

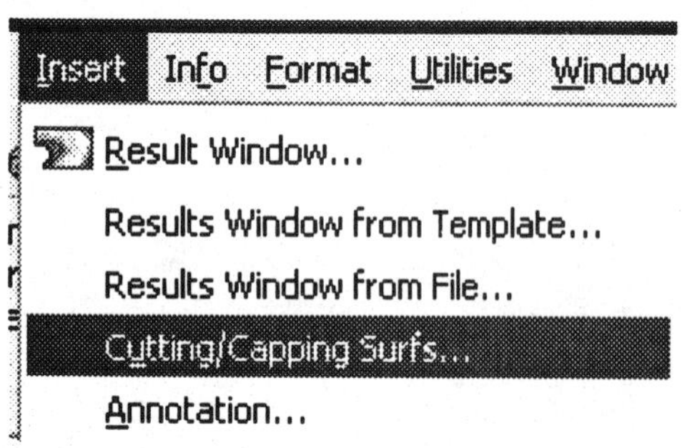

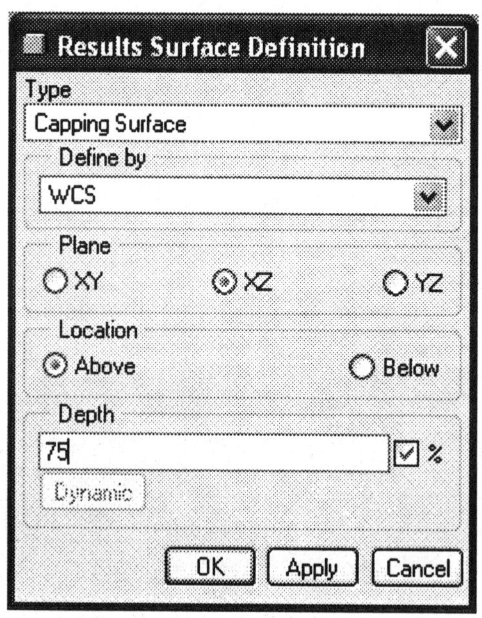

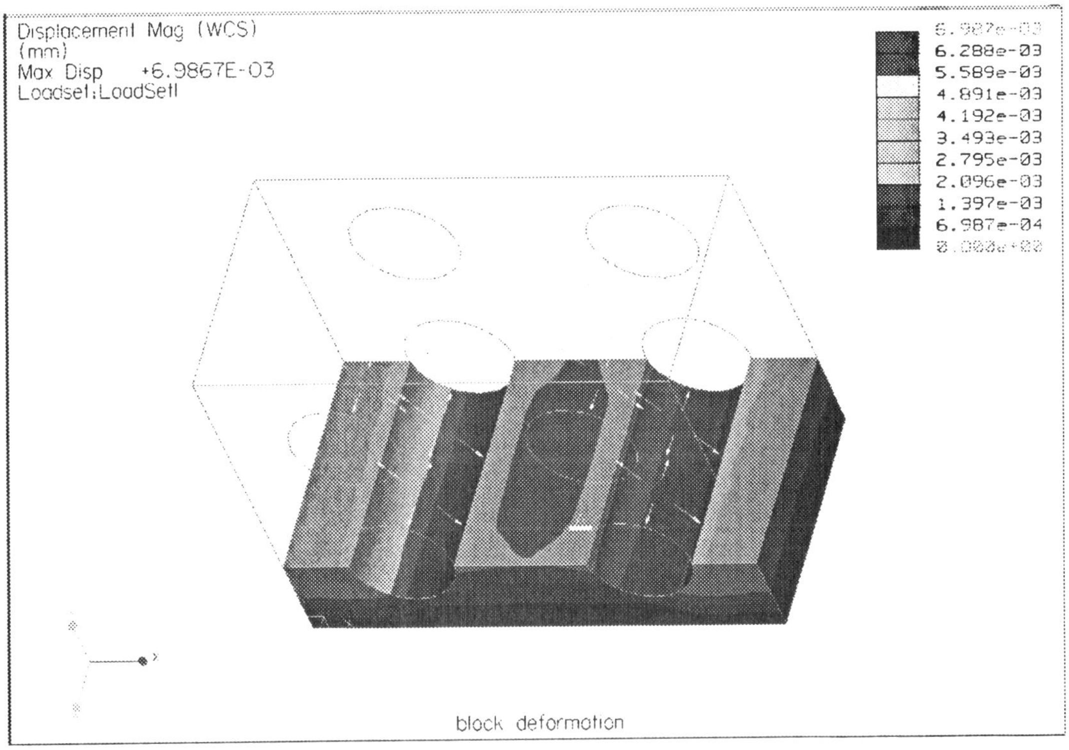

Plot of the maximum principal stress distribution. From the main toolbar, select **Edit** > **Copy** > change the selection of **Displacement** to **Stress** and pick **Max Principal**> **OK and Show**.

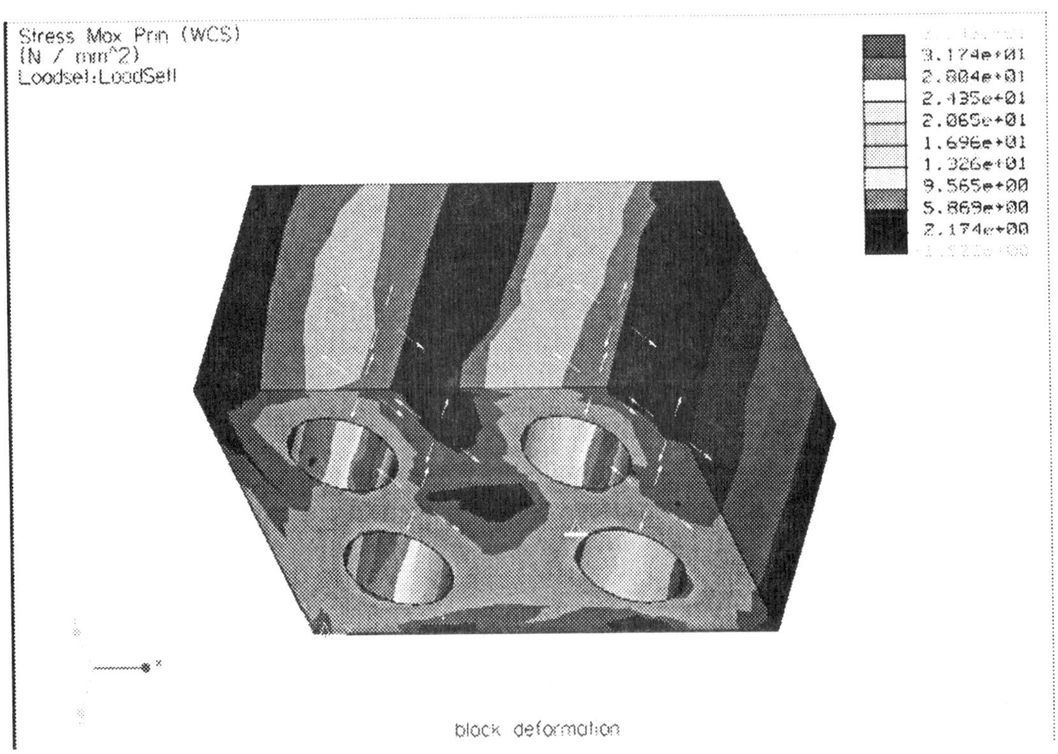

Plot of the maximum principal stress distribution with a horizontal capping surface. Note that, after **Apply** is clicked, a box called **Dynamic** is active. Click the box of **Dynamic** > locate your cursor to the cross-section and move your mouth. The position of capping surface moves following the movement of your mouth. This allows the user to dynamically view the stress distribution > click the middle button of mouth when it is done.

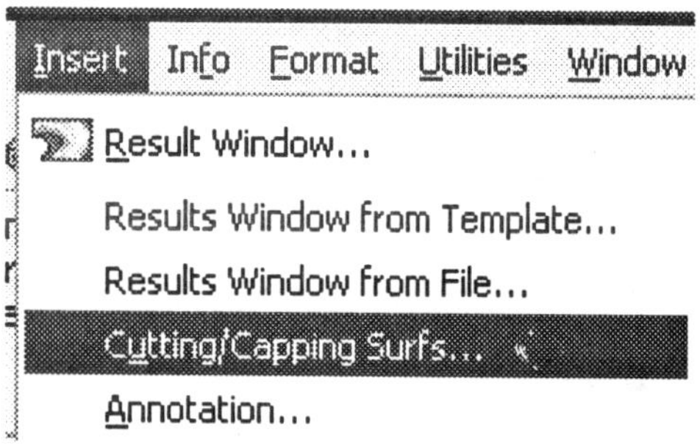

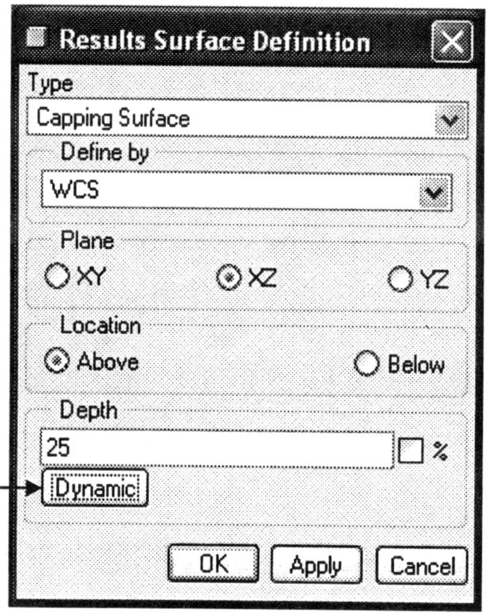

Use Dynamic to adjust the position of capping surface

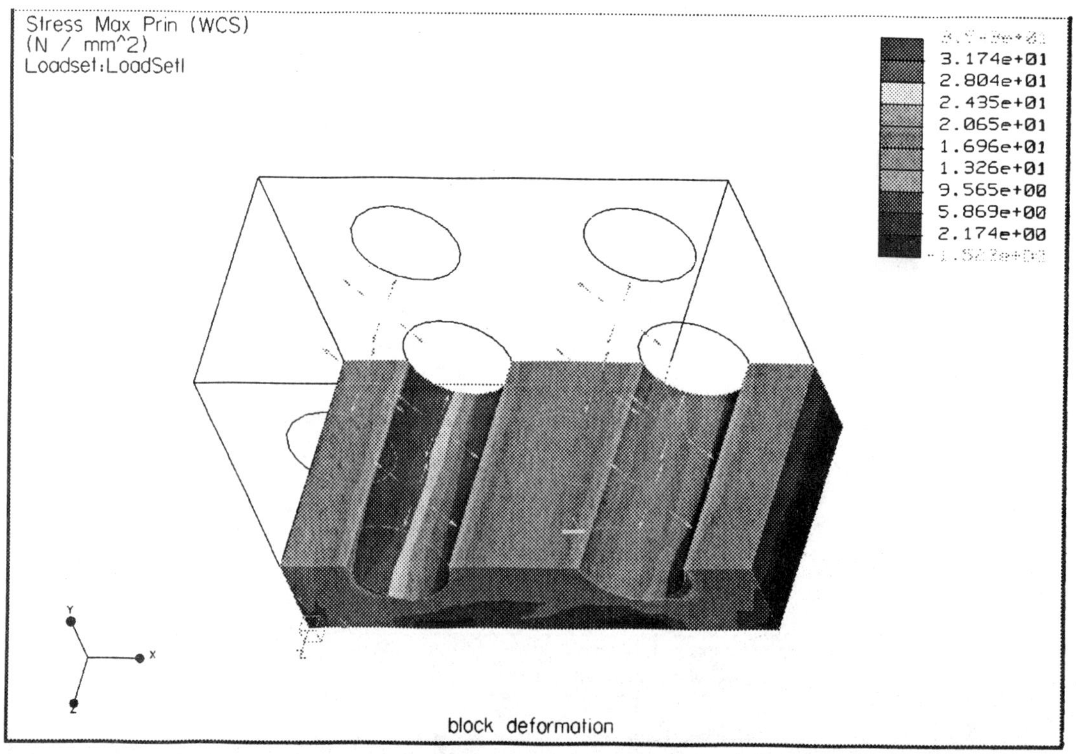

Plot of the maximum principal stress distribution with a vertical capping surface. In this case, select YZ, instead of selecting XZ. Using the function of **Dynamic** to relocate the capping surface.

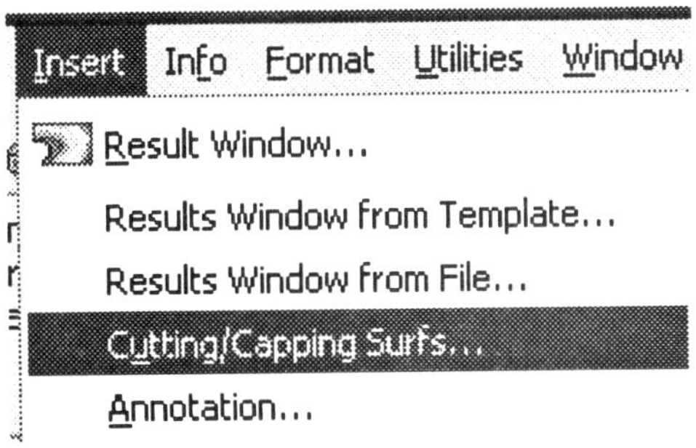

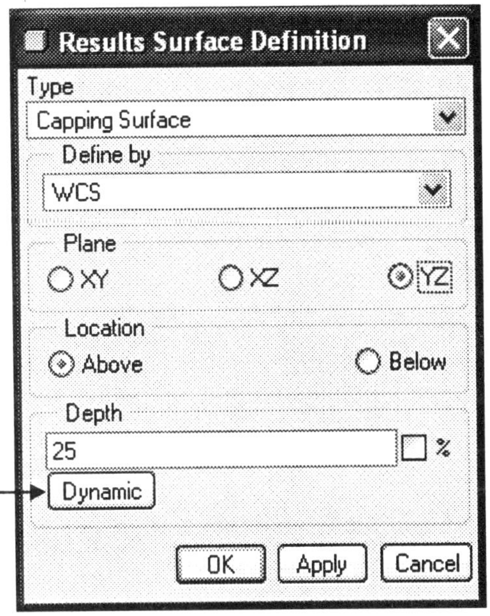

Use Dynamic to adjust the position of capping surface

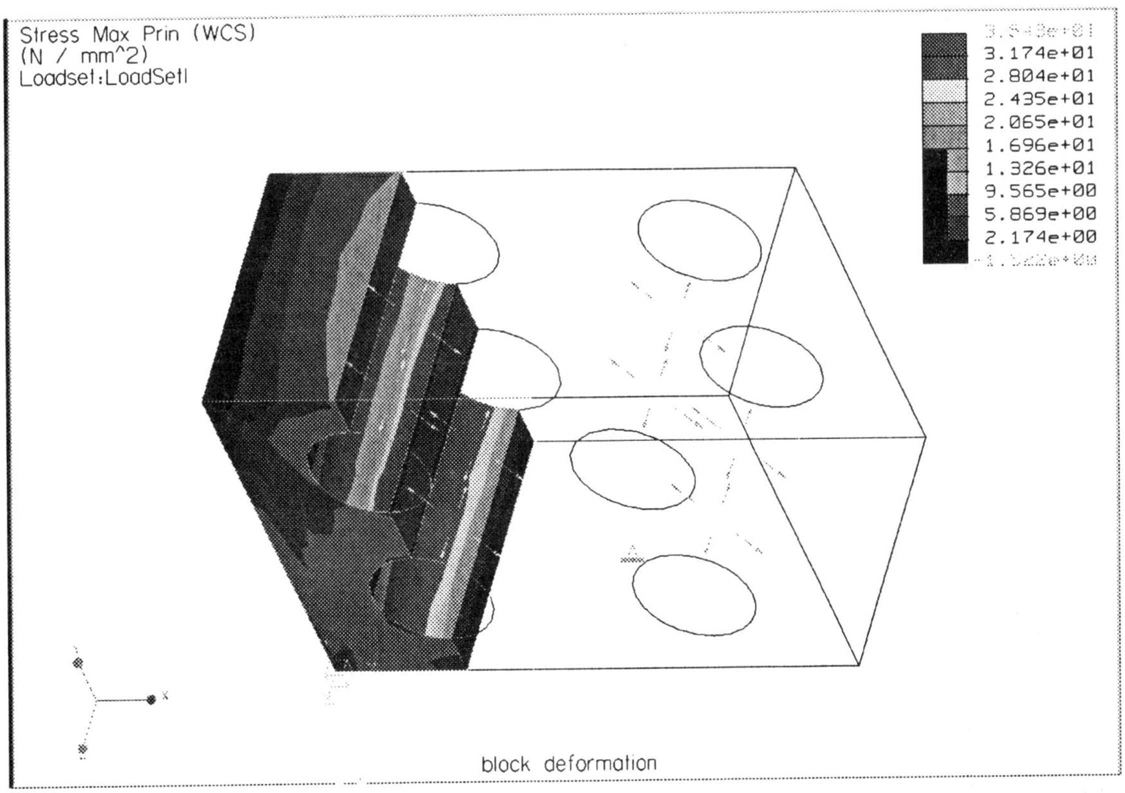

3.6 Modeling of a Point Load Condition and Gravity

When students take a course in mechanics of materials, they draw free-body diagrams when analyzing the forces acting on the object under study. On those free-body diagrams, some forces are assumed to act on points. The following figure presents a simple beam. The force acting on the beam is marked as P. Two forces marked as R1 and R2 are the reaction forces acting on the two supports. All these three forces are assumed to act on three point locations.

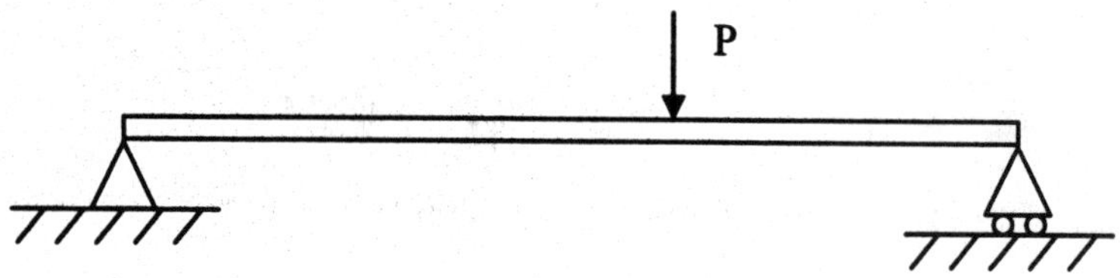

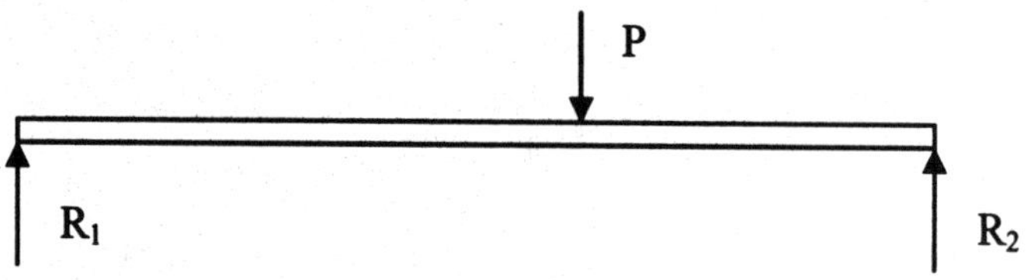

This modeling process used in mechanics of materials may pose difficulties in performing FEA where the stress calculation is related to an area. For example, the stress concentration occurred nearby the location where load P is acting can be viewed as

$$\sigma_{concentration} = \frac{P}{A}$$

A numerical value presenting the size of an area occupied by a point is assumed to be zero. As a result, the FEA software system will send a warning message to the user, indicating that a calculation error has been occurred. To avoid such difficulties, it is a common practice in performing FEA that a surface region is defined at the point location so that the size of an area is specified before calculating the value of stress. In the following example, we demonstrate the procedure of modeling a point load.

The following figure illustrates a cantilever beam. The left end is a fixed end and the right end is a free end. A concentration load is acting on the free end. The magnitude is 200 Newton. The section area is a rectangle. The two dimensions are 50 and 10 mm, respectively. Assume that the material is steel and the Young's modulus is $2x10^{11}$ N/m^2.

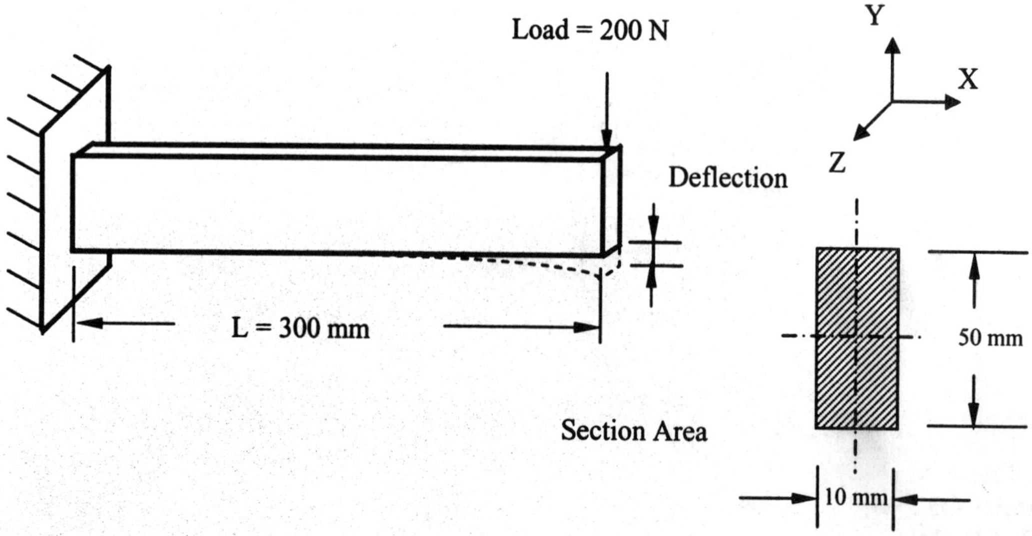

(1) Find the deflection at the free end based on an analytical formula.

The following formula was taken from a textbook titled Mechanics of Materials, authored by Geri and Thimoshenko (publisher: PWS-Kent, 1990). Using this formula to calculate the maximum deflection at the free end, the result is 0.0865 mm.

$$\delta_{free\_end} = \frac{PL^3}{3EI} = \frac{PL^3}{3E(\frac{bh^3}{12})} = \frac{(200)(0.3)^3}{3(2x10^{11})(\frac{(0.01)(.05^3)}{12})} = \frac{5.4}{(6*10^{11})(1.04x10^{-7})} = 0.0865x10^{-3}(meter)$$

(2) Find the deflection at the free end based on the result obtained from FEA analysis. Note that the loading area has to be defined as a surface region.

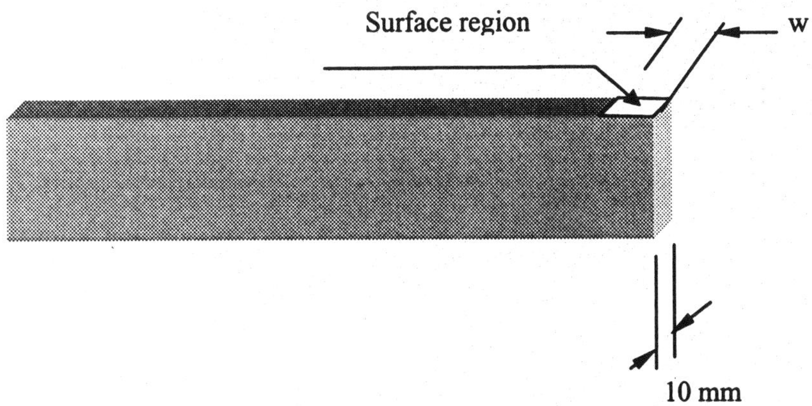

In this example, we will perform FEA through 3 runs. Each run will have a specific surface region. As shown in the following table, the surface region is set to 10 x 1 mm2 for Run 1. The surface region is set for 10 x 1 mm2 for Run 2. The surface region is set for 10 x 0.5 mm2 for Run 3. We will compare the results obtained from the 3 runs to gain a feeling on the sensitivity issue in modeling a point load.

	b	w	deflection (mm)
Run 1	10	1	
Run 2	10	2	
Run 3	10	0.5	

Step 1: Create a 3D solid model for the cantilever beam.

File > **New** > **Part** > type *cantilever_beam* as the file name. Clear the box called **Use default template** > **OK.** Select the **mmns_part_solid**. Fill in **DESCRIPTION** by typing *point load.* Fill in **MODELED_BY** by typing your name, say *student.*> **OK.**

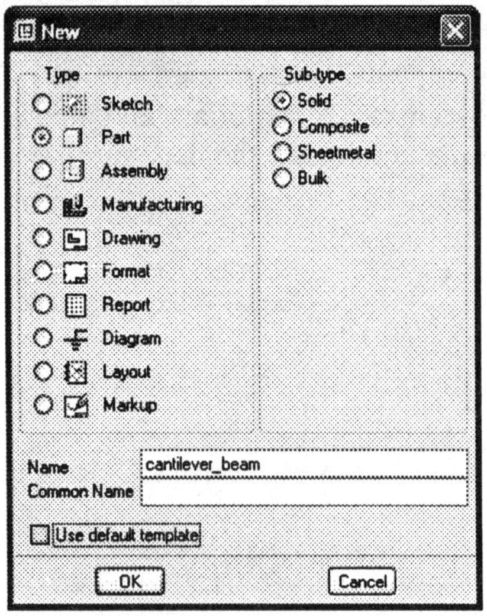

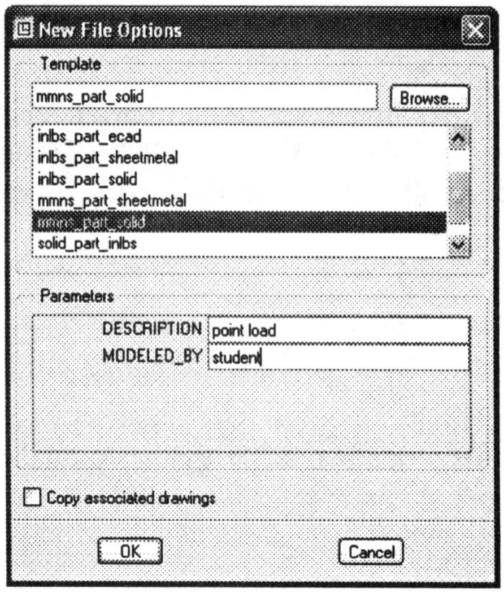

Pick up the icon of **Extrude** displayed on the toolbar of feature creation. Make sure that the icon of **Solid** is selected. Select **Symmetric** before setting the depth value to 10 (the thickness of the beam). To initiate the process of sketching, activate **Placement** > **Define.**

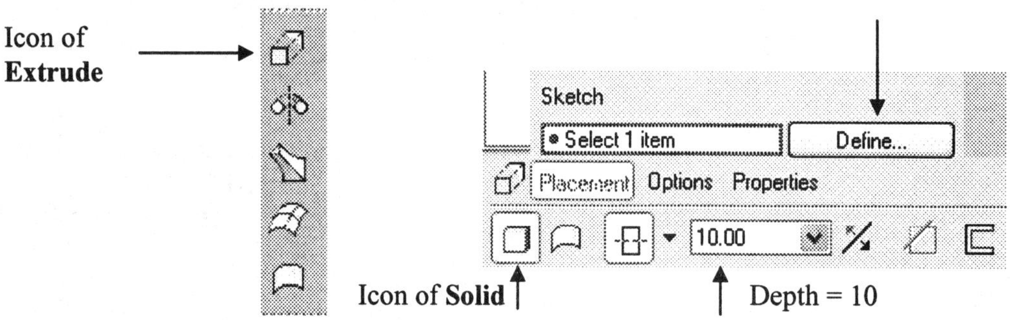

Select the **FRONT** datum plane displayed on screen as the sketch plane. Pick the box of **Sketch** to accept the **RIGHT** datum plane as the default reference to orient the sketch plane, as illustrated below:

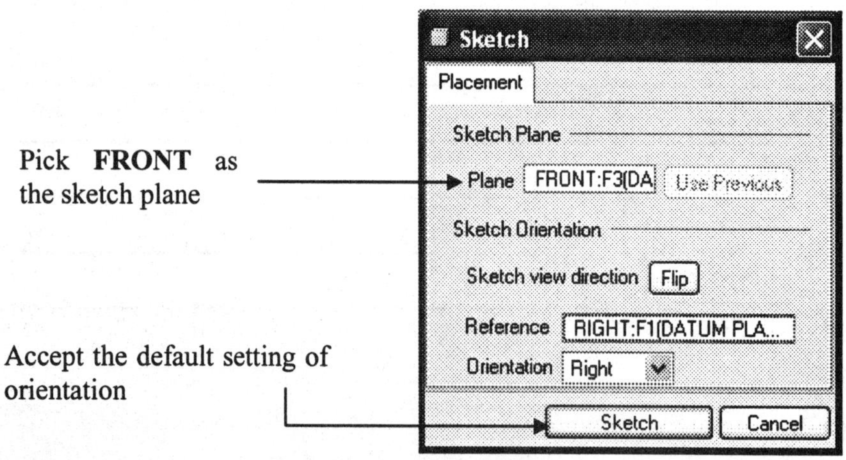

Both the sketch window and the references window appear. Let us first close the reference window. Add a horizontal centerline and a vertical centerline. Select the icon of **rectangle** from the toolbar of sketcher and sketch a rectangle. The 2 dimensions are 300 and 50, respectively. Make sure that the symmetry about the horizontal centerline and the symmetry about the vertical centerline are maintained.

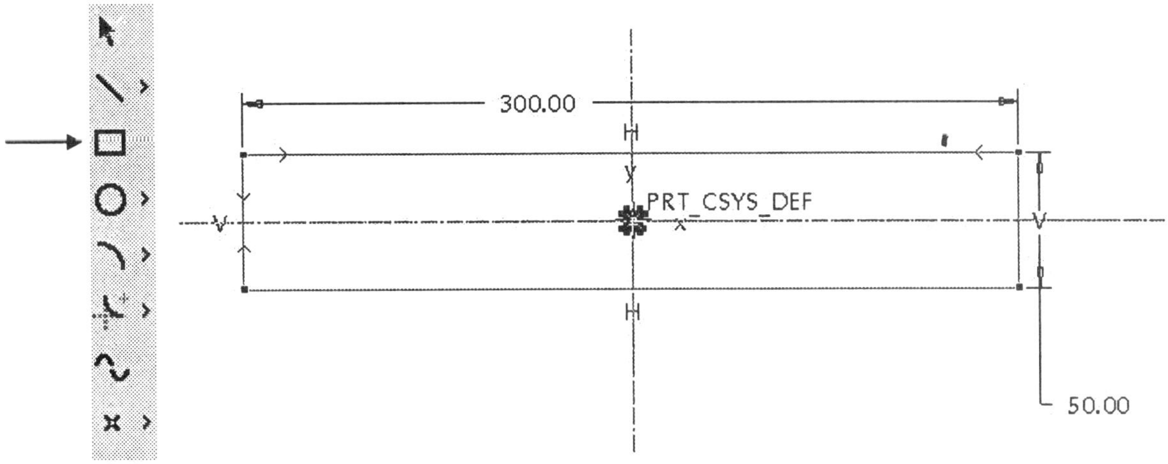

Upon completing the sketch, select the icon of **Done** from the toolbar of sketcher. On the dashboard, select the icon of **Complete** from the feature control panel, completing the creation of the cantilever beam model.

Icon of **Complete**

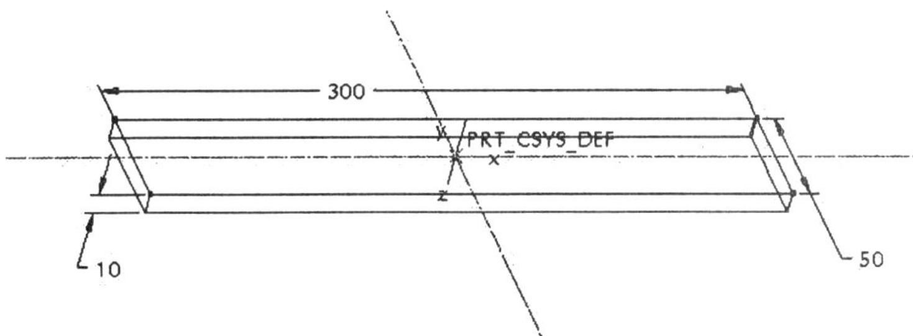

Step 2: Create a datum curve feature so that a surface regions can be define for modeling a point load.

To sketch a datum curve select the icon of **Sketch Tool** from the toolbar of datum feature creation > Pick the top surface of the beam as the sketch plane > accept the default orientation of the sketch plan

Select the icon of Sketch Tool

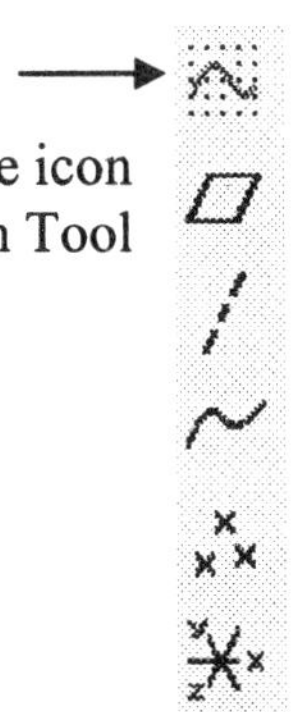

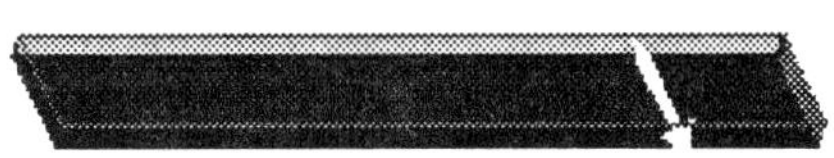

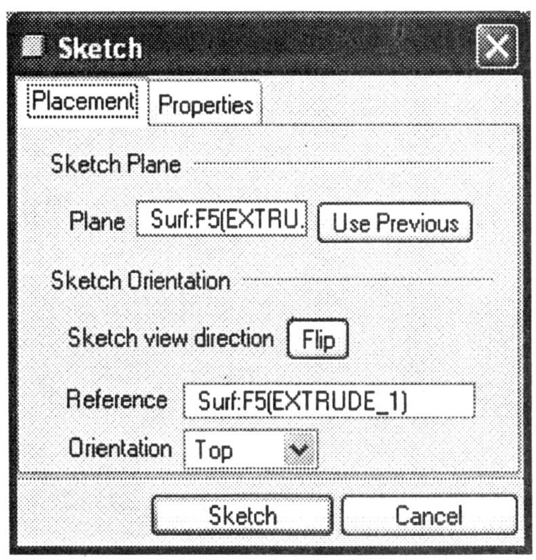

Before making a sketch, let us first add 2 new references. These 2 new references are the surface at the right end and the surface in the front of the block, as shown. Afterwards, close the reference window.

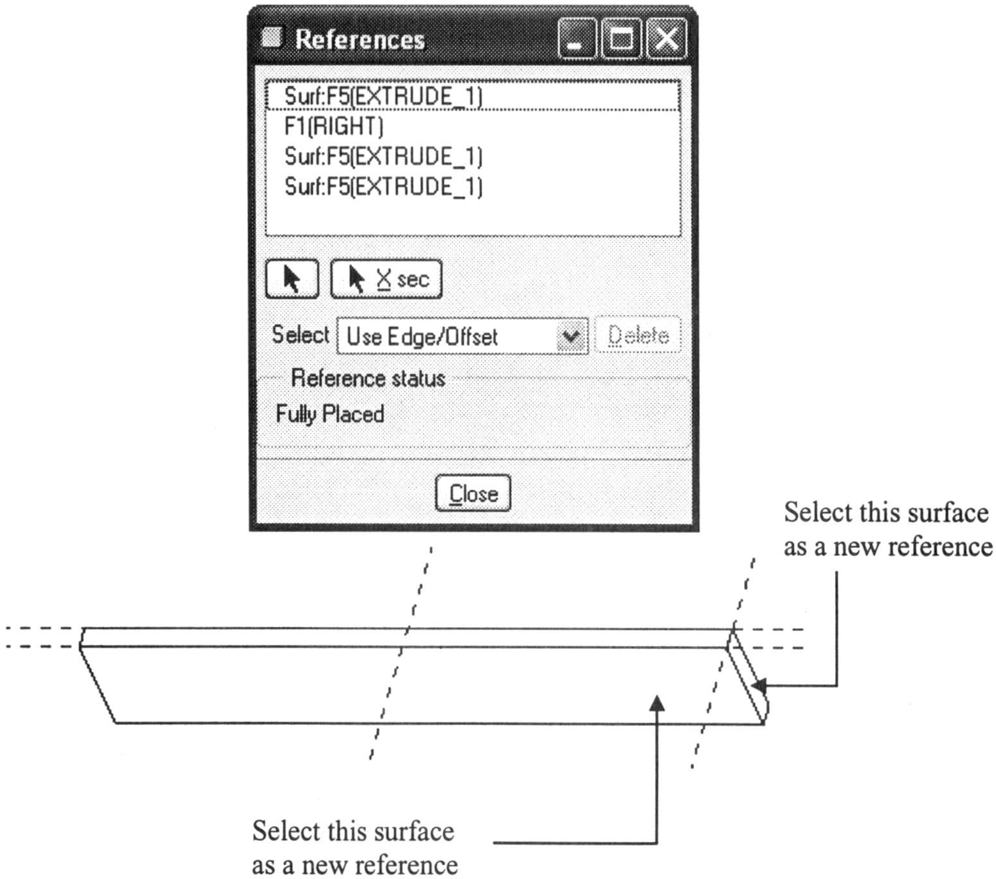

Select the icon of **rectangle** from the toolbar of sketcher and draw a rectangle with the dimension of 1. Only one dimension is needed because the other sides of the sketched rectangle are aligned with the selected references. Pick the icon of Done from the toolbar of sketcher.

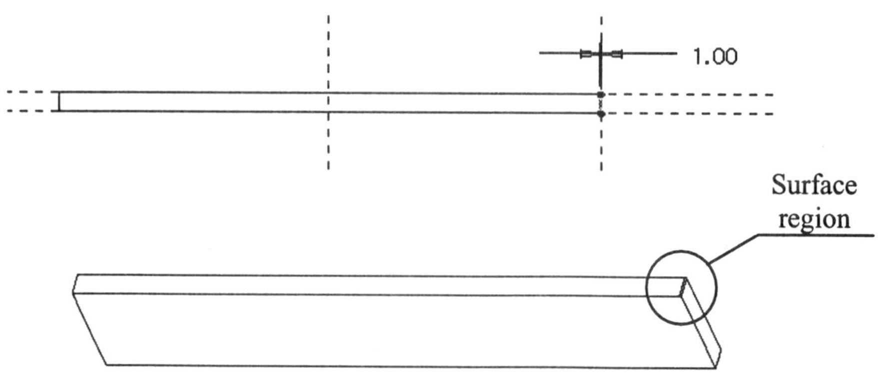

Step 4: Under the Pro/MECHANICA, perform FEA.

From the main toolbar or the menu toolbar, select **Applications** > **Mechanica.**

Continue after checking the unit system > **Structure** > **OK**.

From the toolbar of functions, select the icon of **Materials** > select **STEEL** from the left side called Materials in Library > click the directional arrow so that the selected material type goes to the right side called Materials in Model.

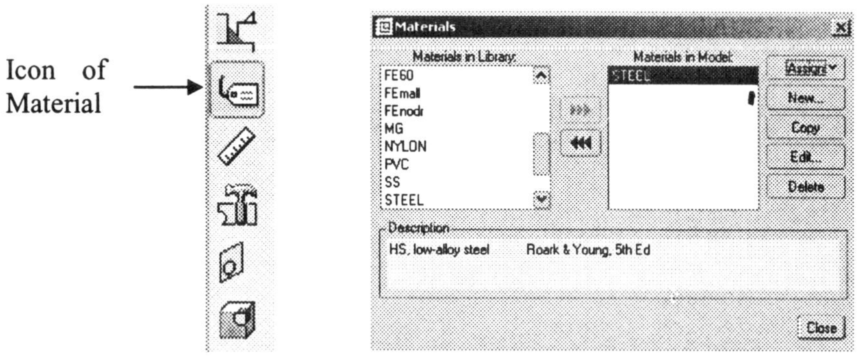

Assign > **Part** > select the column structure from the screen display > **OK** > **Close**.

To define the fixed end constraint condition, from the toolbar of functions, select the icon of **Constraints >** select the surface at the left end of the block as the fixed end > **OK** to accept the surface selection > **OK** to accept the default settings, namely, fixing all six degrees of freedom.

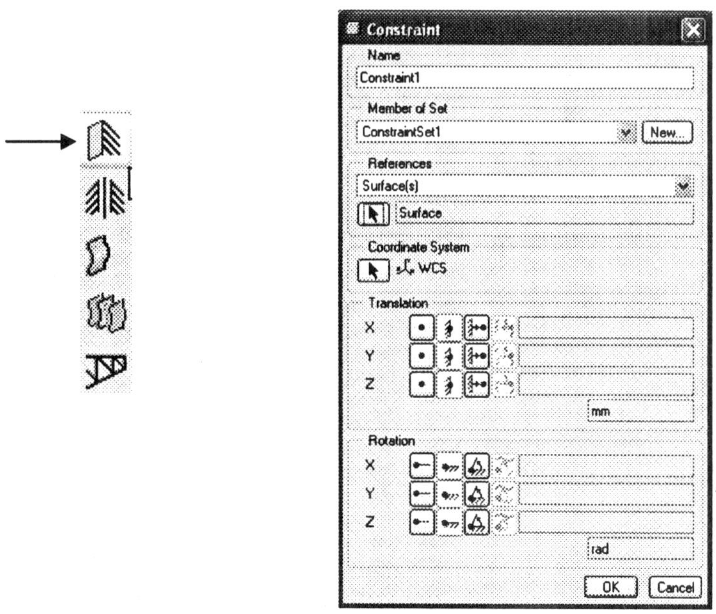

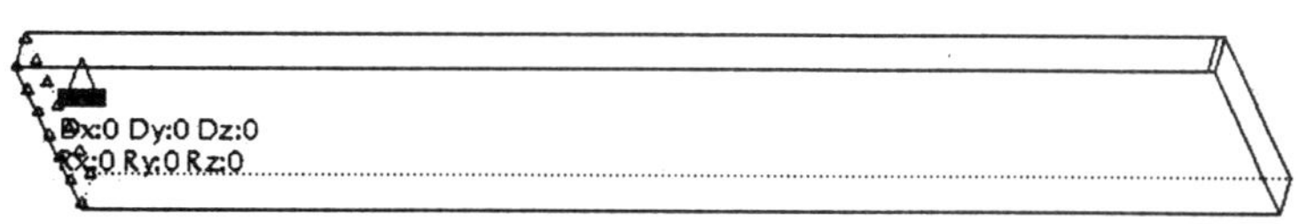

To define the load condition, we first define a surface region, and afterwards to define the point load.

From the toolbar of functions, select the icon of surface region > **Select** > **Done**.

Select the datum curve of rectangle > **Done** > select the top surface of the beam for splitting the surface region from the entire surface > **OK** > **Done** > **OK**.

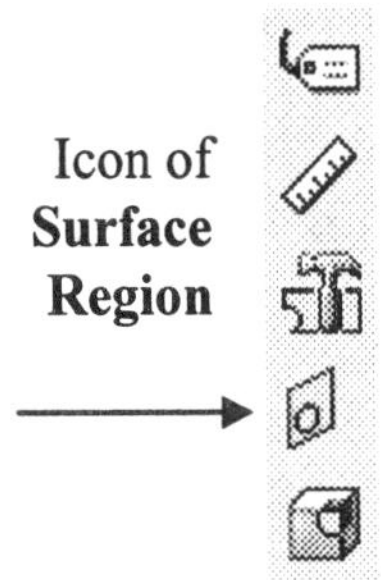

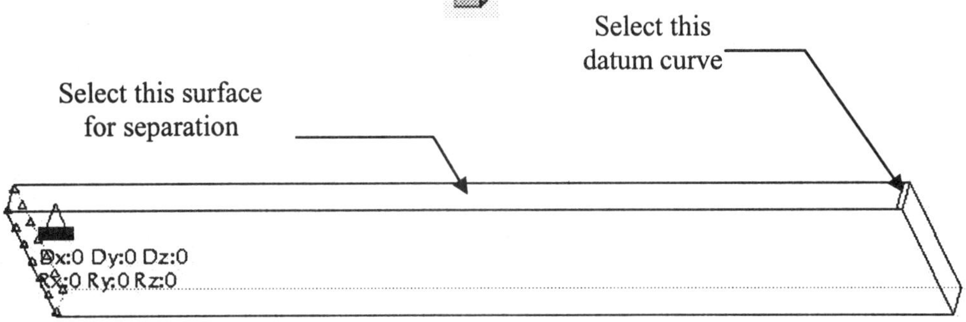

From the toolbar of functions, select the icon of **Force/Moment Loads > Surface(s) >** select the surface region just defined > OK.

Type -200 as the Y Force Component > OK.

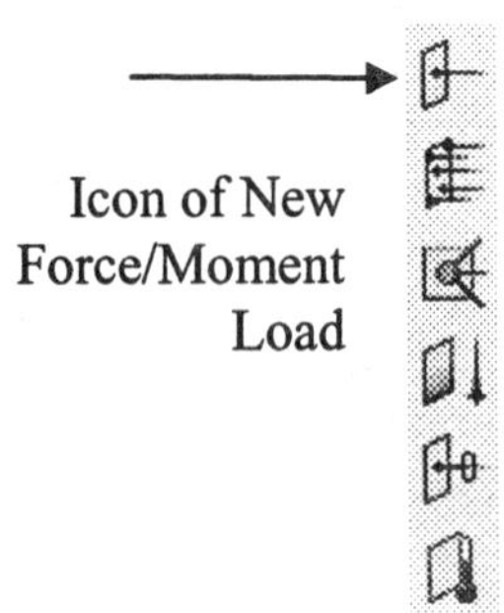

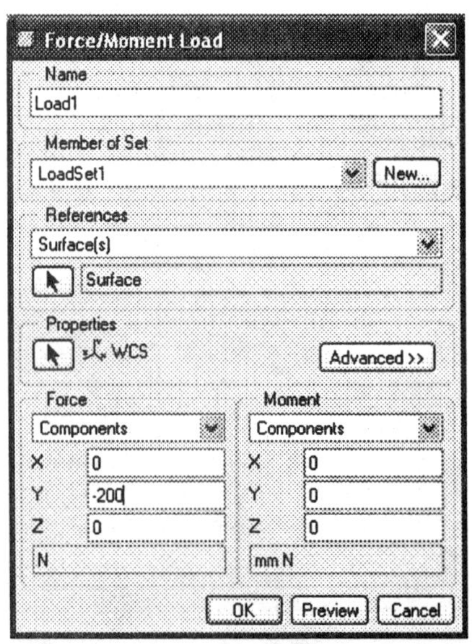

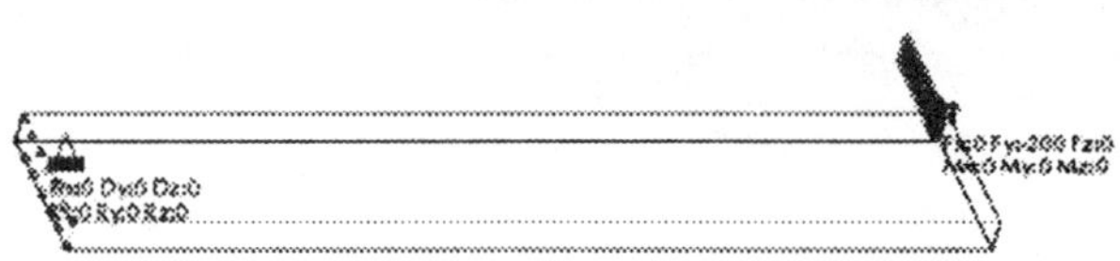

Step 5: Under the Pro/MECHANICA, set up **Analyses** and **Run.**
Select the icon of **Run an Analysis** > **File** > **New Static** > type in *cantilever_beam* > **OK**.

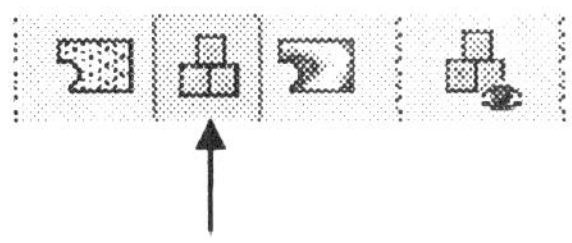

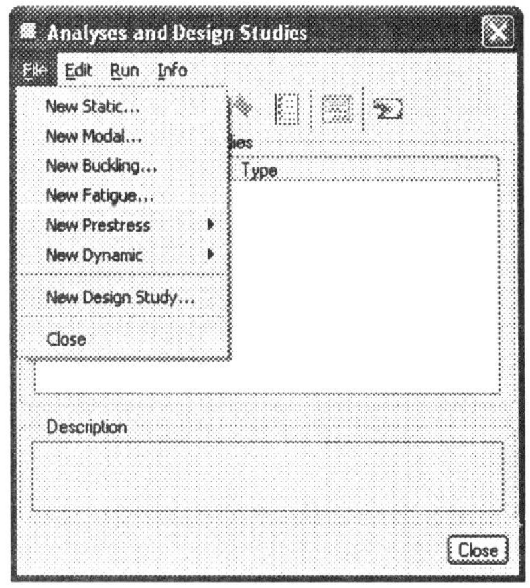

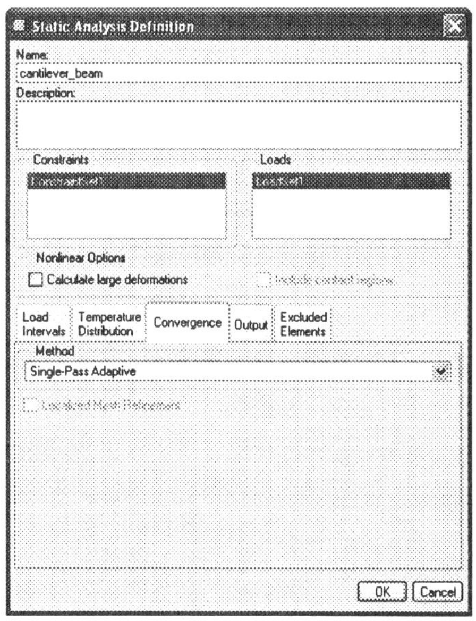

There is always a message displayed on screen. The message is asking the user "Do you want error detection?" Click **Yes**. In this way a warning message will appear if an abnormal operation occurs.

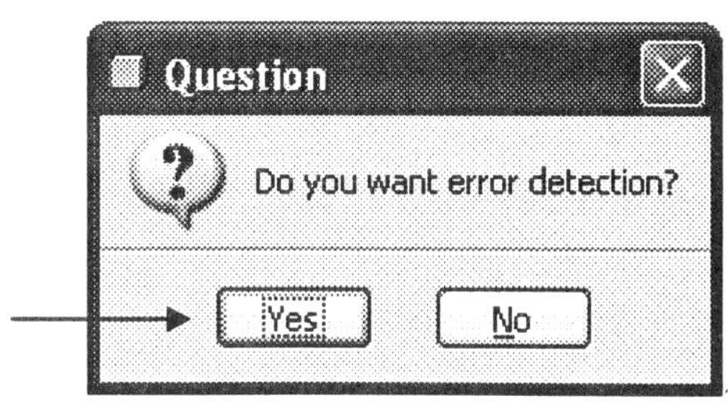

To monitor the computing process, users may click the icon of **Display study status**.

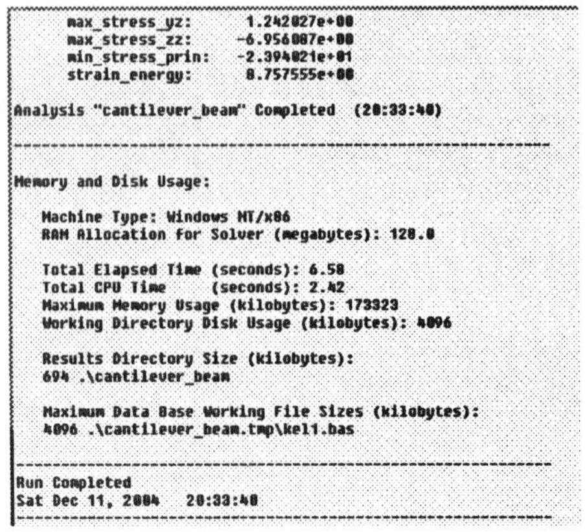

Step 7: Under the Pro/MECHANICA, study the **Results**
Select the icon of **Results**

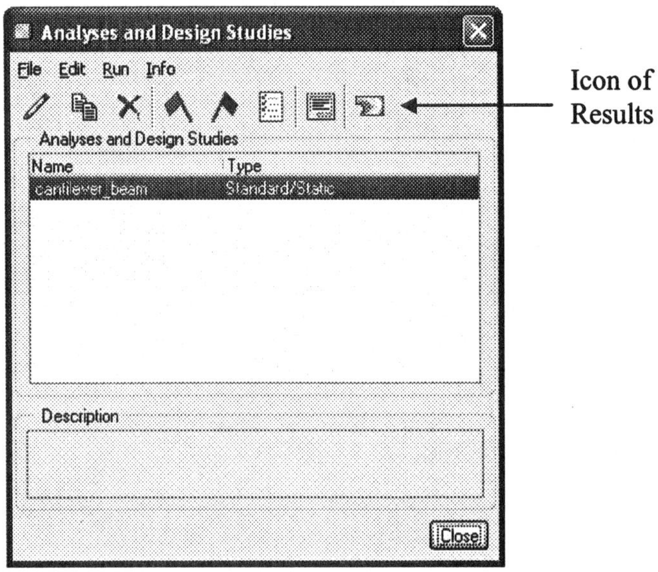

Type in Beam Deflection as the title > locate the Design Study folder called cantilever_beam. Select Fringe > choose Displacement > Magnitude > OK and Show.

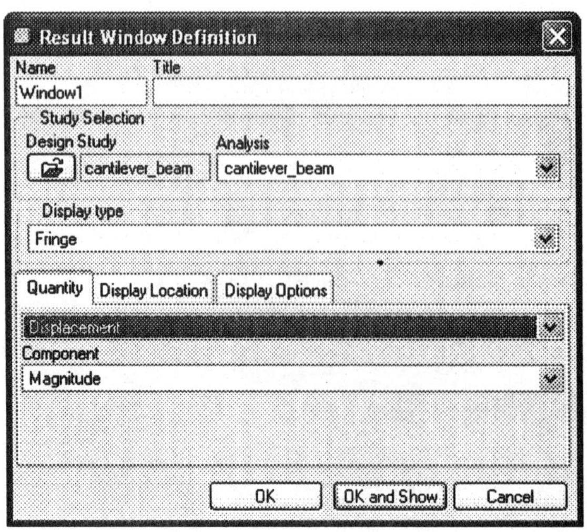

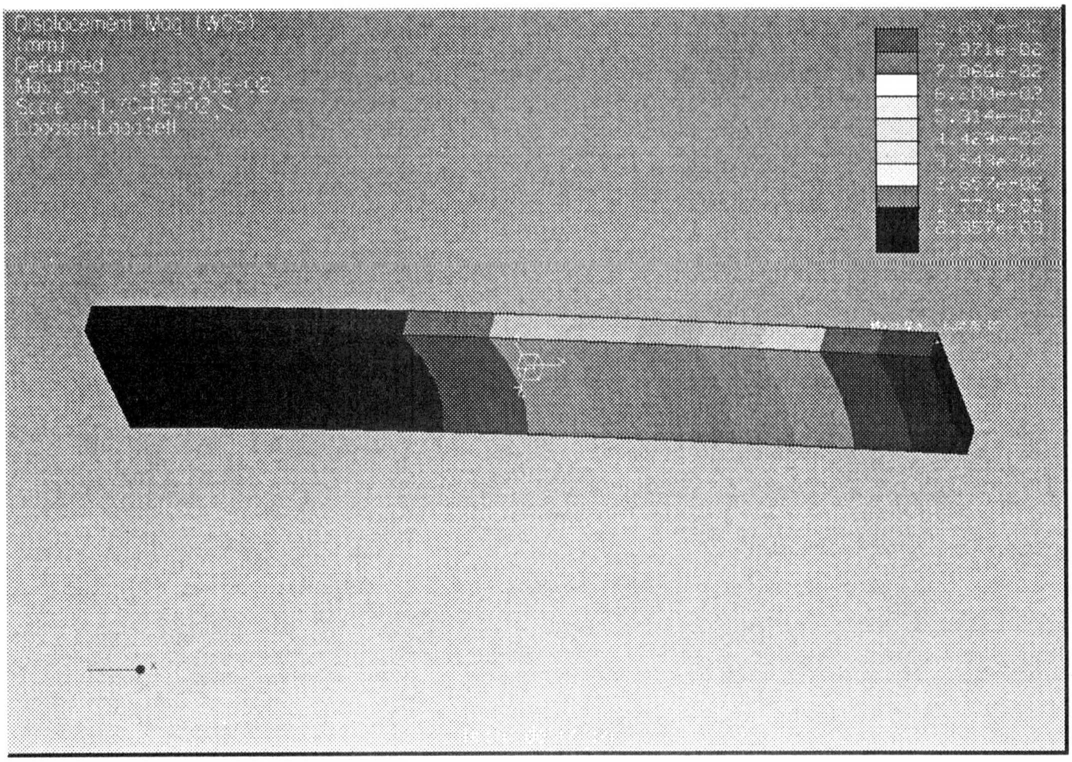

Examining the value of the maximum deflection shown at the free end, the numerical value is 0.08857 mm. As we recall, the numerical value obtained from the calculation using an analytical formula was 0.08650 mm. The difference is 0.00207 mm, or about 2 μm.

Note that the numerical value obtained from the calculation using an analytical formula was the deflection along the Y direction. Let us display the displacement in the Y direction. To do so, select the icon of Edit the selected definition from the main toolbar > change **Magnitude** to **Y** > **OK & Show**.

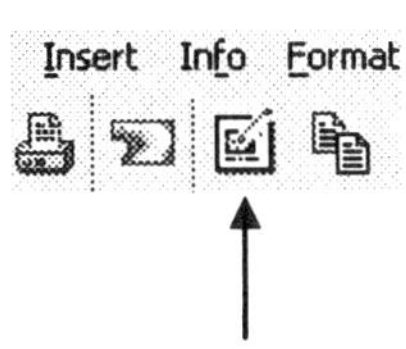

Edit the definition

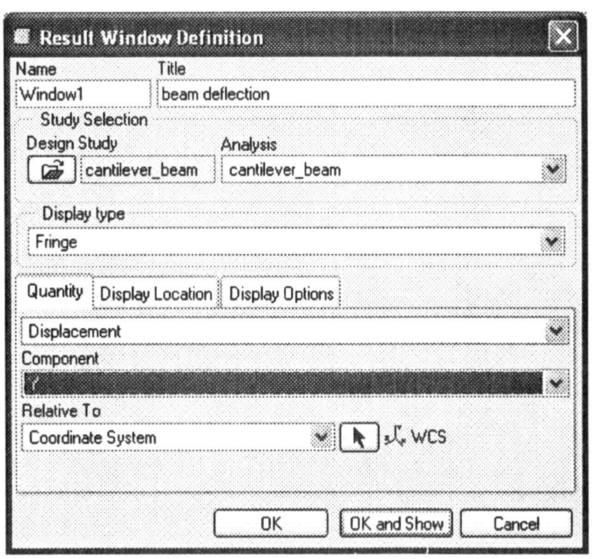

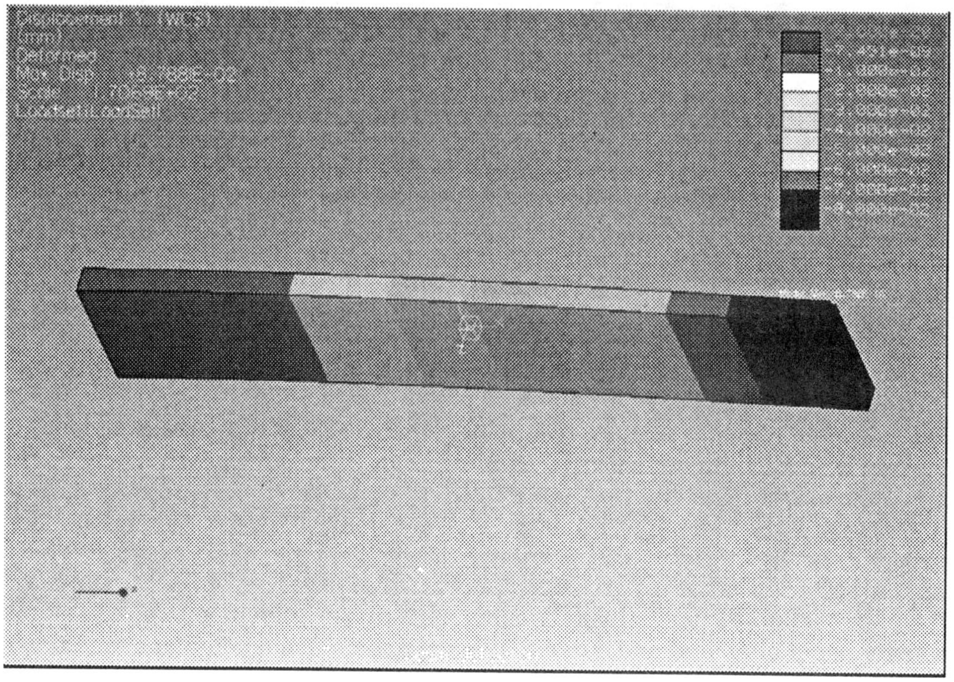

Examining the value of the maximum deflection shown at the free end, the numerical value is 0.08788 mm. As we recall, the numerical value obtained from the calculation using an analytical formula was 0.08650 mm. The difference is 0.00138 mm, or about 1 μm.

Now let us change the area of the surface region from 10 x 1 mm to 10 x 2 mm to perform Run 2. Let us go back to Pro/ENGINEER design system. From the menu bar, select **Applications** > **Standard**.

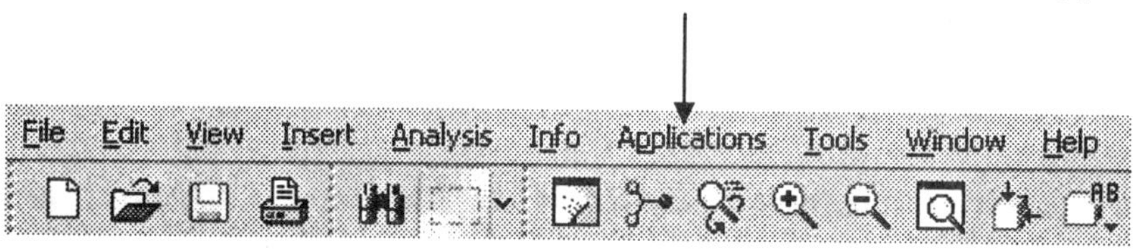

From the model tree, select **Sketch1** > right click on the mouse to select **Edit Definition.**

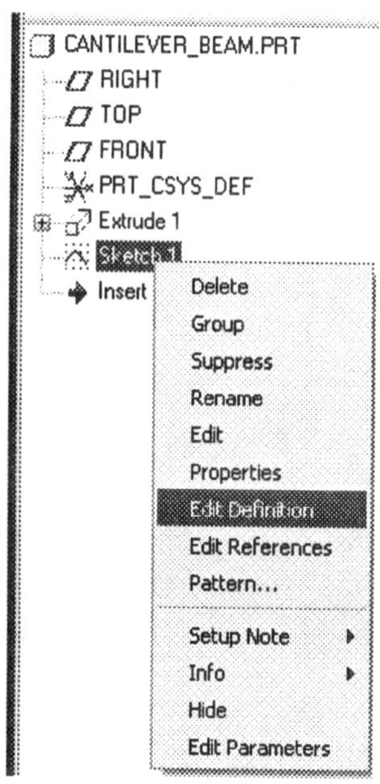

The window called Sketched Datum Curve appears. Select **Use Previous** to go back to the previously selected sketch plane > **Sketch.**

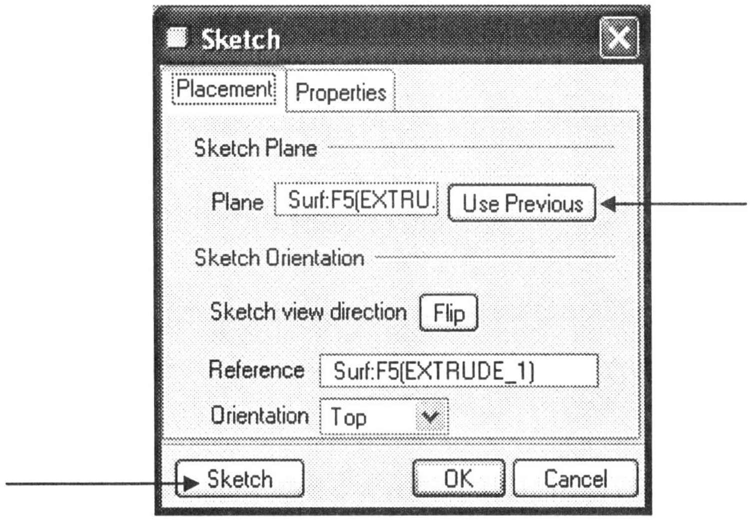

Modify the dimension from 1.00 to 2.00 by double clicking the dimension and typing 2 afterwards > OK to complete the modification process > select the icon of **Done** from the toolbar of **sketcher** > OK from the **Sketch** window.

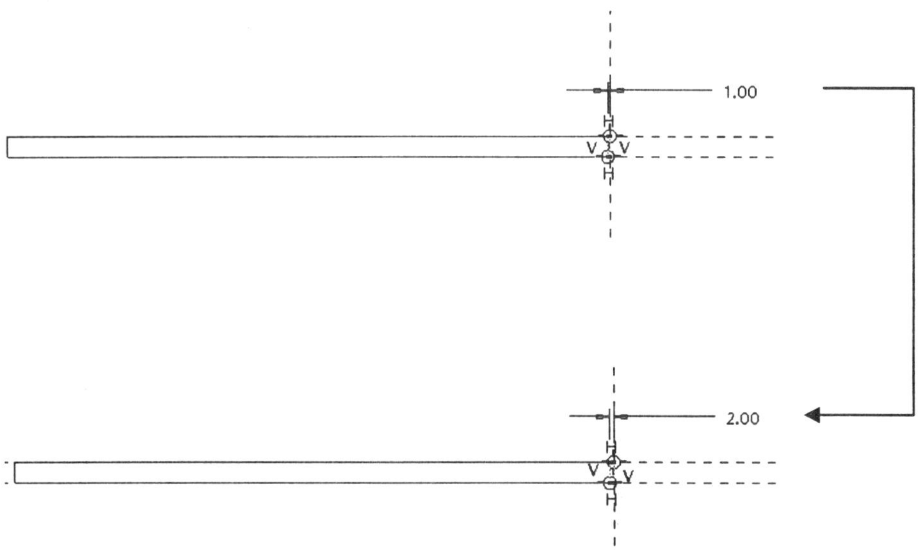

Step 6: Perform the FEA with the loading area equal to 10 x 2 mm^2.
From the menu bar, select **Applications** > **Mechanica**, going back to the Pro/Mechanica system.
Click **Continue** after checking the unit system, if asked > **Structure**

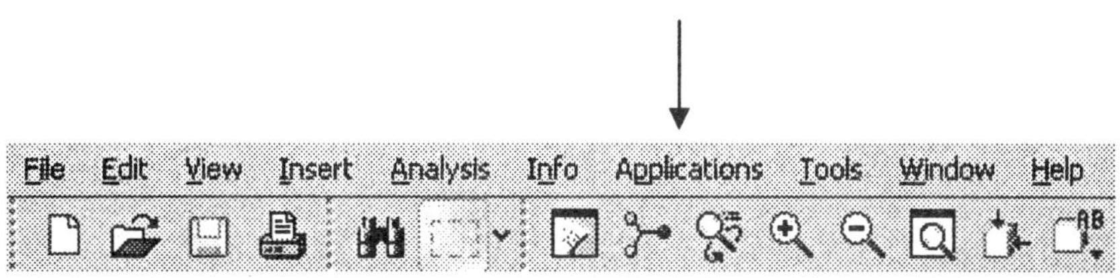

Perform a new static analysis
Select the icon of **Run an Analysis** > **File** > **New Static** > type in *cantilever_beam_run2* > **OK**.

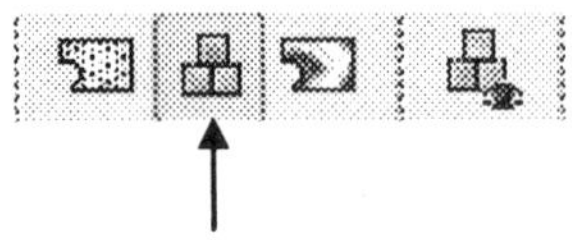

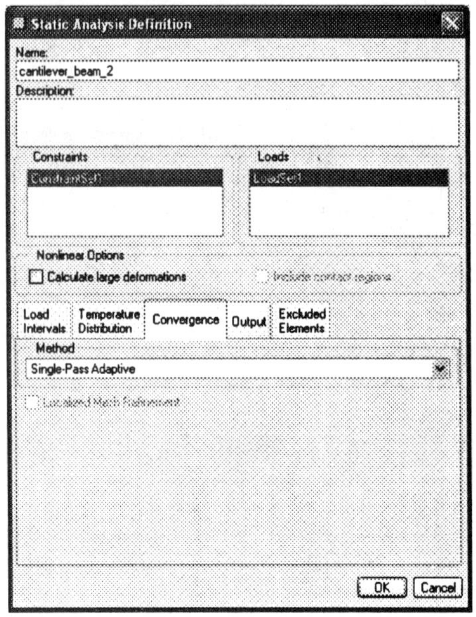

Click the icon of **Run**. There is always a message displayed on screen. The message is asking the user "Do you want error detection?" Click **Yes**. In this way a warning message will appear if an abnormal operation occurs.

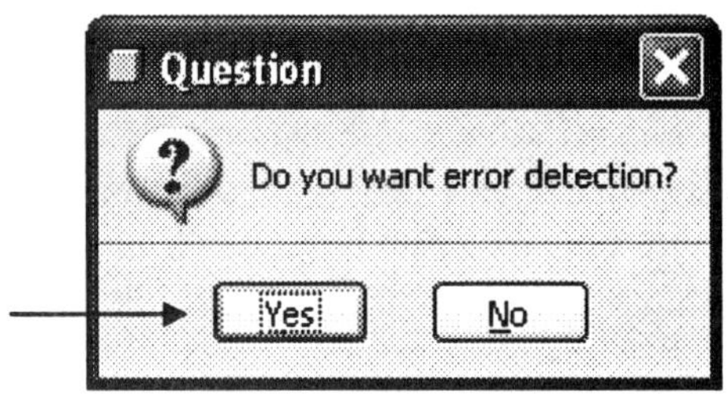

When the run process is completed, select the icon of **Results**

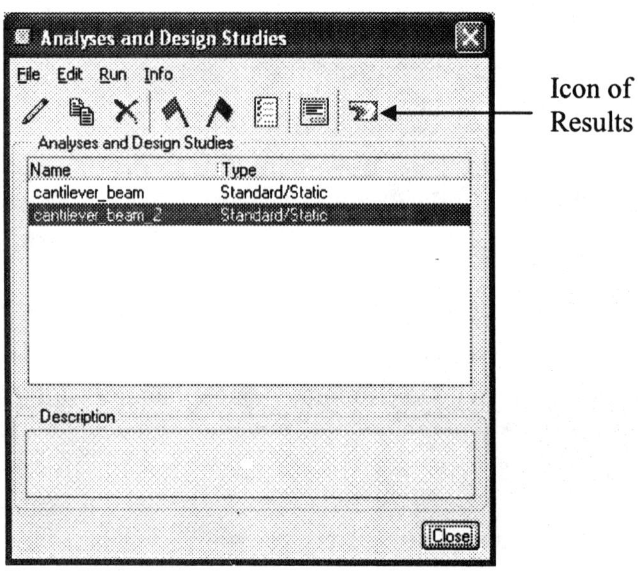

Type in Beam Deflection as the title > locate the Design Study folder called cantilever_beam_run2.

Select **Fringe** > choose **Displacement** > **Y** > **OK and Show**.

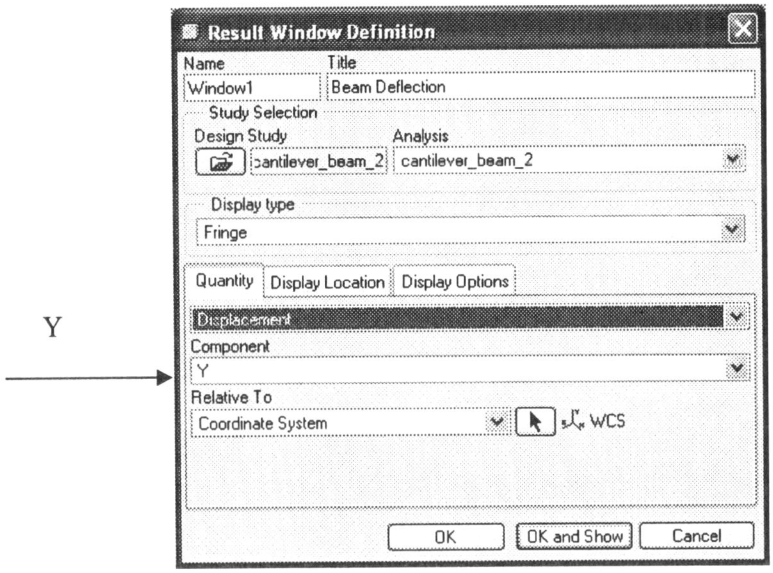

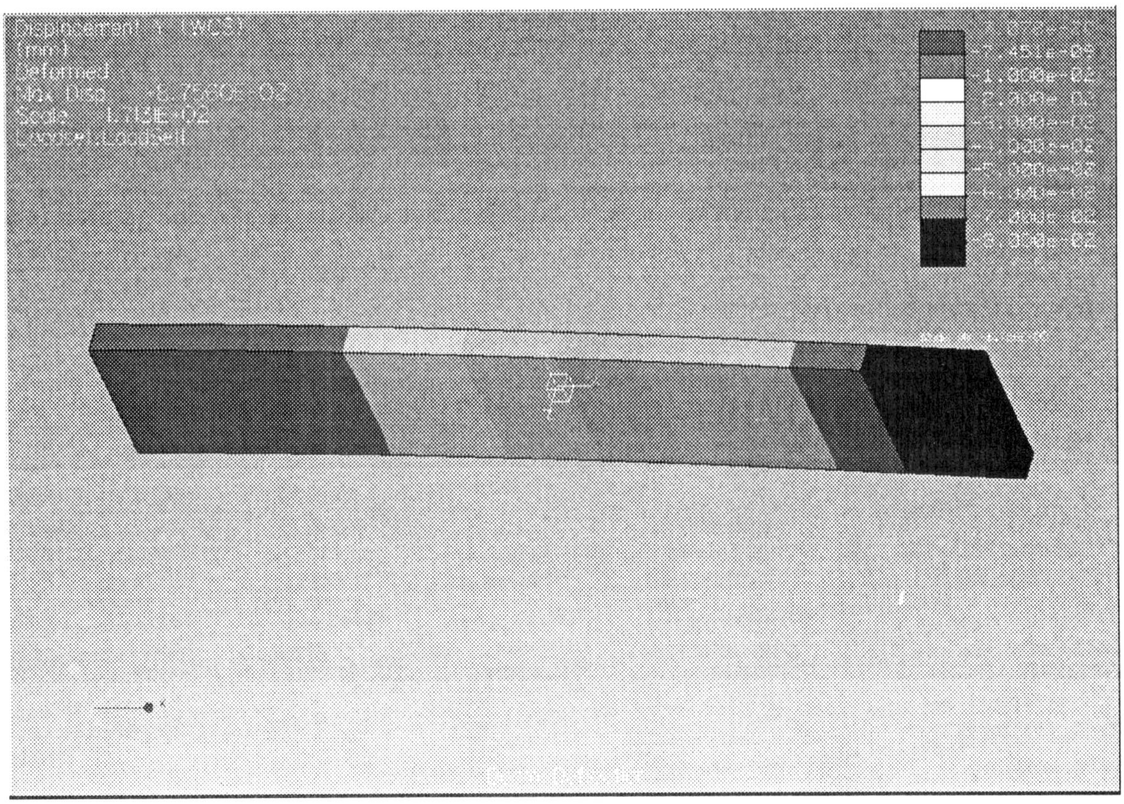

Examining the value of the maximum deflection shown at the free end, the numerical value is 0.08756 mm. As we recall, the numerical value obtained from the calculation using an analytical formula was 0.08650 mm. The difference is 0.00106 mm, or about 1 μm.

Now let us change the area of the surface region from 10 x 2 mm to 10 x 0.5 mm to perform Run 3. Let us go back to Pro/ENGINEER design system. From the menu bar, select **Applications** > **Standard**.

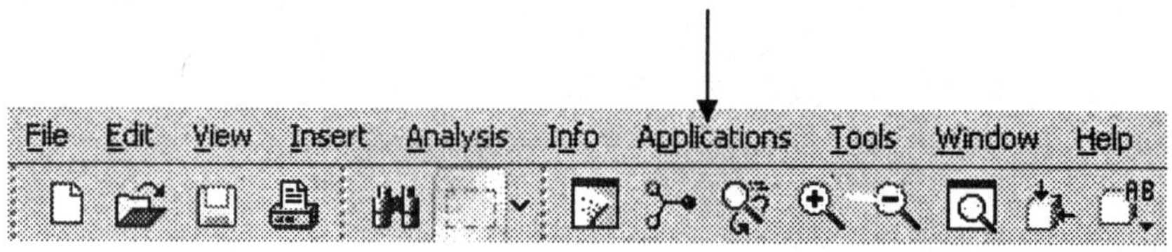

From the model tree, select **Sketch1** > right click on the mouse to select **Edit Definition.**

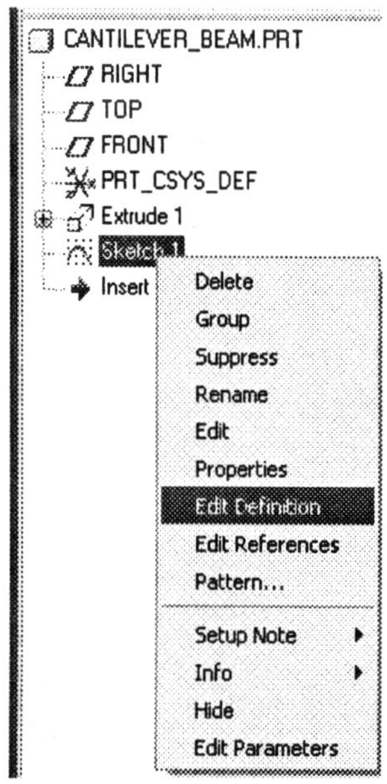

The window called Sketched Datum Curve appears. Select **Use Previous** to go back to the previously selected sketch plane > **Sketch.**

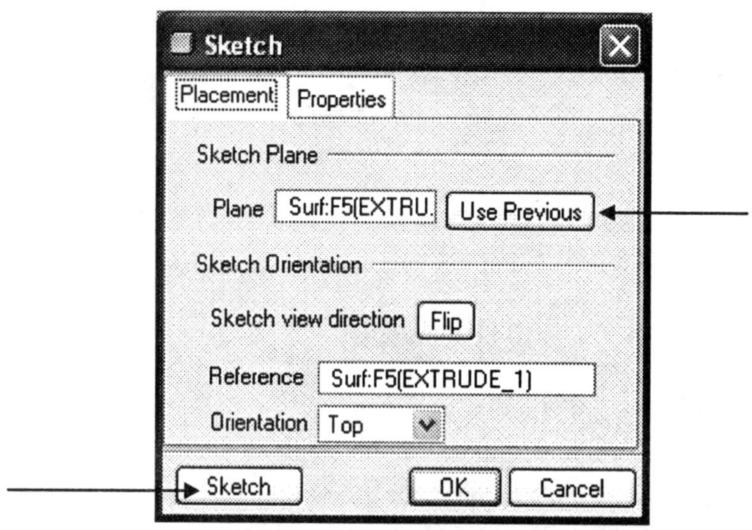

Modify the dimension from 2.00 to 0.5 by double clicking the dimension and typing 0.5 afterwards > press **Enter** to complete the modification process > select the icon of **Done** from the toolbar of **sketcher** > **OK** from the **Sketch** window.

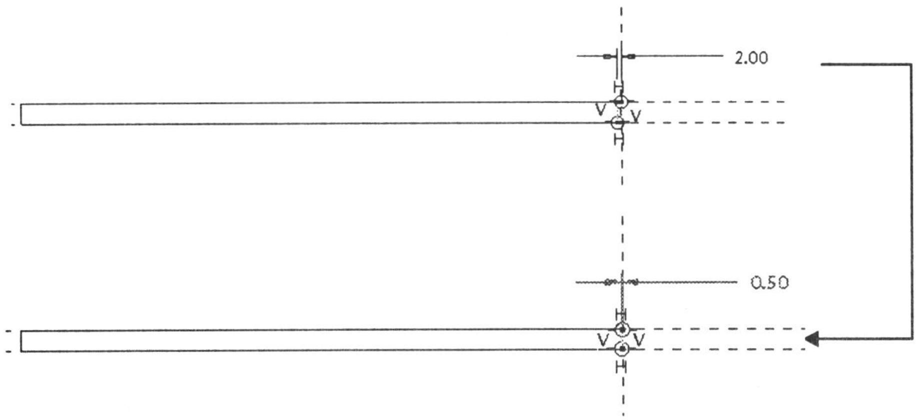

Step 6: Perform the FEA with the loading area equal to 10 x 2 mm^2.

From the menu bar, select **Applications** > **Mechanica**, going back to the Pro/Mechanica system. Click **Continue** after checking the unit system, if asked > **Structure.**

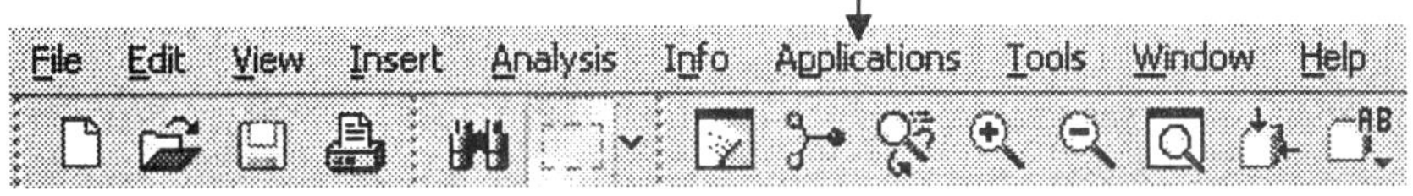

Perform a new static analysis
Select the icon of **Run an Analysis** > **File** > **New Static** > type in *cantilever_beam_run3* > **OK**.

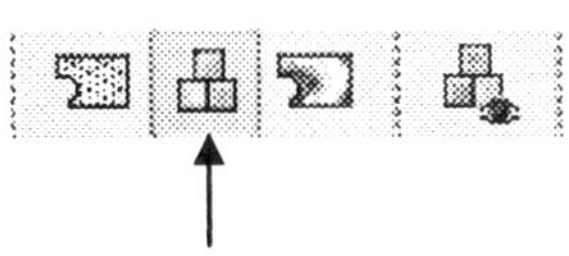

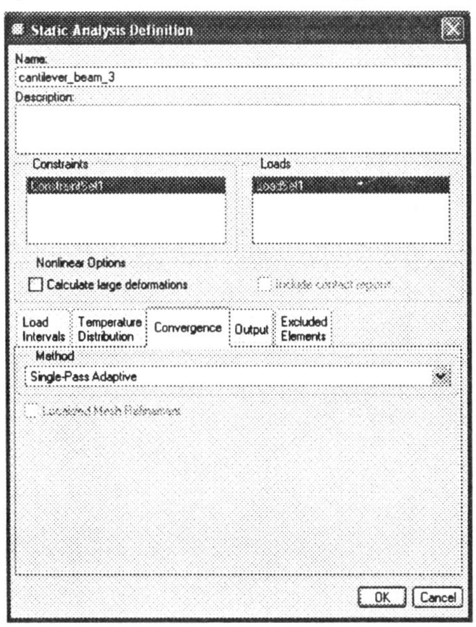

Click the icon of **Run**. There is always a message displayed on screen. The message is asking the user "Do you want error detection?" Click **Yes**. In this way a warning message will appear if an abnormal operation occurs.

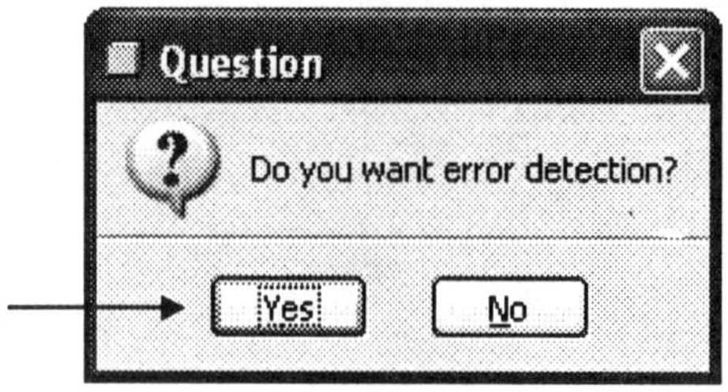

When the run process is completed, select the icon of **Results**

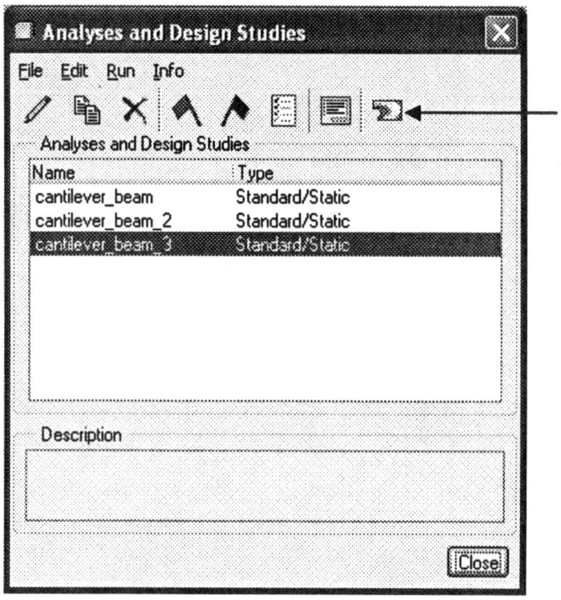

Icon of **Results**

Type in Beam Deflection as the title > locate the Design Study folder called cantilever_beam_run3.

Select **Fringe** > choose **Displacement** > **Y** > **OK and Show**.

Y

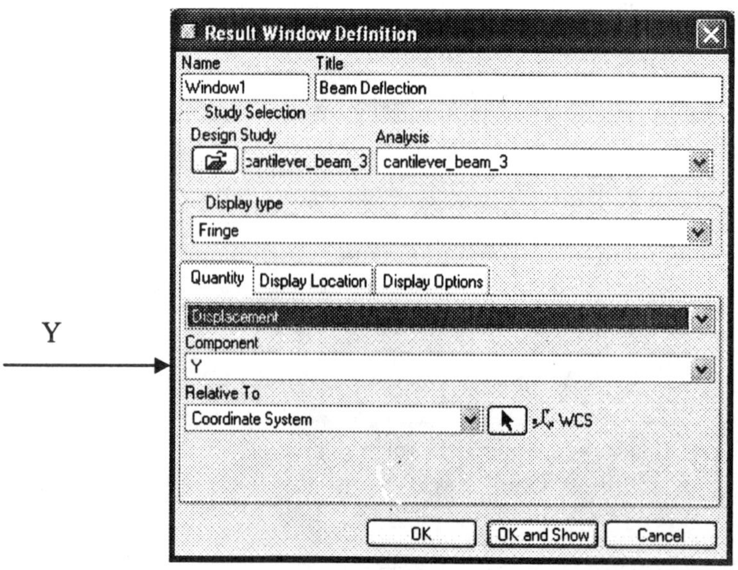

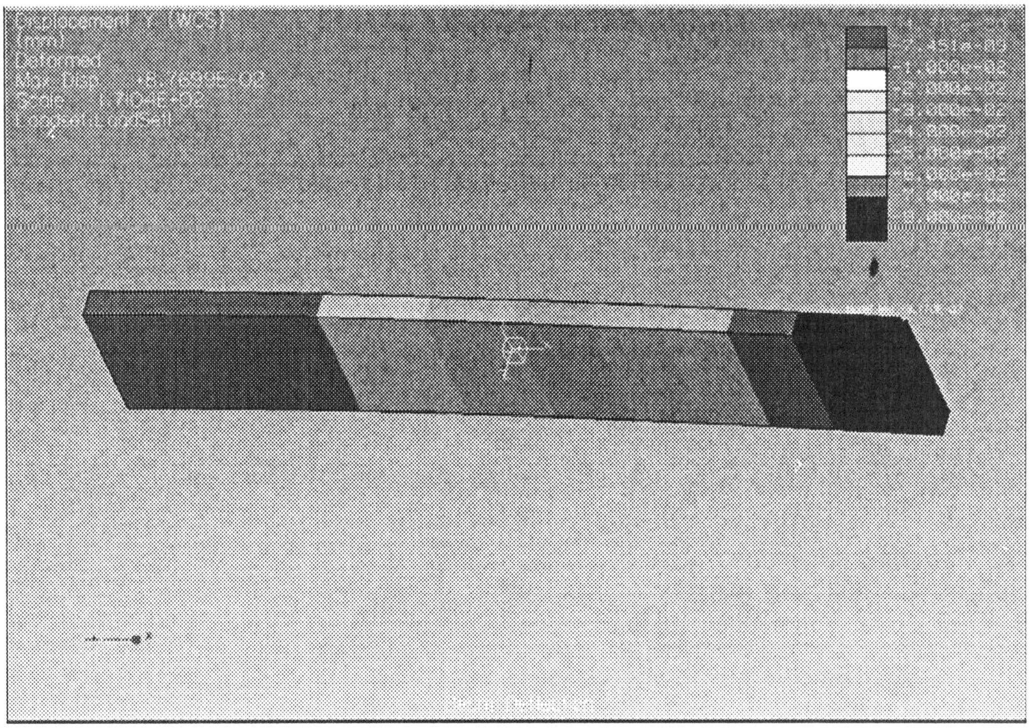

Examining the value of the maximum deflection shown at the free end, the numerical value is 0.08770 mm. As we recall, the numerical value obtained from the calculation using an analytical formula was 0.08650 mm. The difference is 0.00120 mm, or about 1 μm.

Now we list the 3 results obtained from the 3 runs. There is no significant difference among the 3 results in terms of the numerical values. This indicates that the area selected to model a point load is not so sensitive to its numerical value. Certainly it is important to limit the magnitude of the area used so that it represents a point load.

	b	w	deflection (mm)
Run 1	10	1	0.08857
Run 2	10	2	0.08756
Run 3	10	0.5	0.08770

3.7 Consideration of the Gravity Force in Performing FEA

The cantilever beam studied in section 3.6 has its mass. As a result, there is a gravity force uniformly acting on the beam. The action direction is downward. Such a gravity force can be calculated as follows:

$$\text{Volume of the Beam} = (300)(50)(10) = 150000\ (\text{mm}^3) = 150\ (\text{cm}^3)$$

$$\text{Mass of the Beam} = (150)(7.85) = 1177.5\ (\text{g}) = 1.1775\ (\text{kg})$$

$$\text{Gravity Force} = (1.1775)(9.81) = 11.55\ (\text{Newton})$$

Now let us perform Run 4, which considers the effect of the gravity force on the beam deflection. To do so, we need to add a new load set. Let us call it LoadSet 2.

From the toolbar of functions, select the icon of **New Gravity Load** and a new window called **Gravity Load** appears, indicating the gravity force is defined as LoadSet2 > OK.

Type -9810 in the Y direction > **OK**.

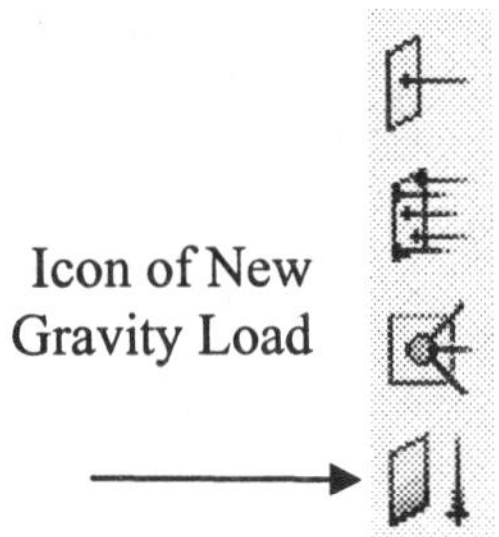

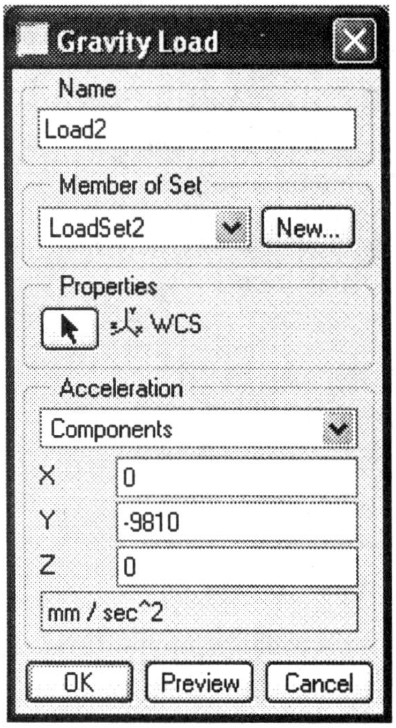

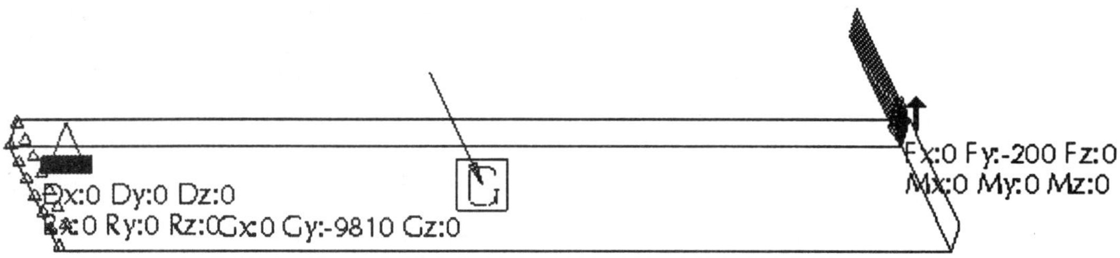

Perform a new static analysis

Select the icon of **Run an Analysis** > **File** > **New Static** > type in *add_gravity_force* > make sure that both Loadset1 and Loadset 2 are selected > **OK**.

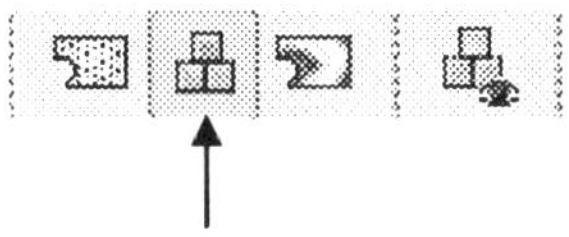

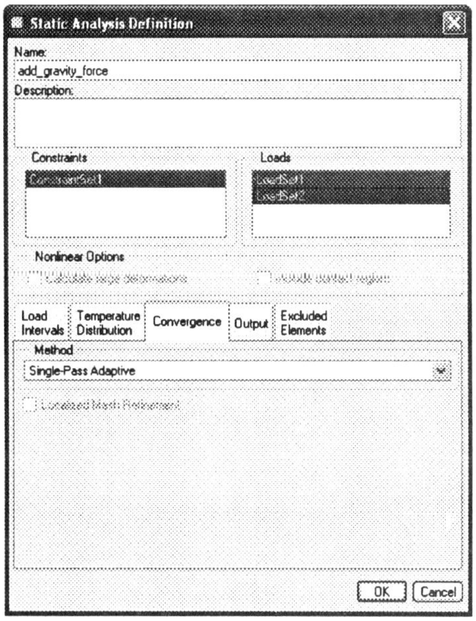

Click the icon of **Run**. There is always a message displayed on screen. The message is asking the user "Do you want error detection?" Click **Yes**. In this way a warning message will appear if an abnormal operation occurs.

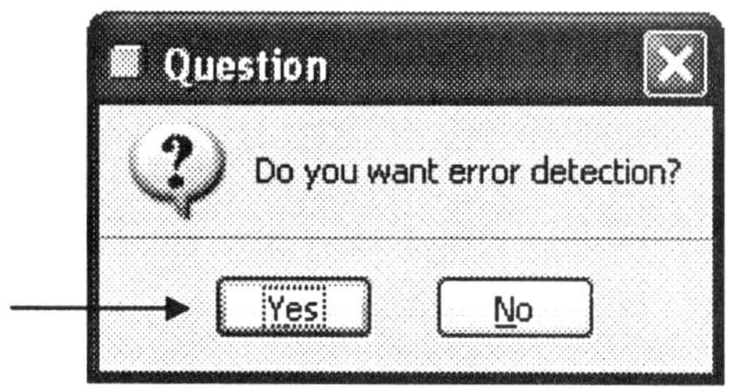

When the run process is completed, select the icon of **Results**

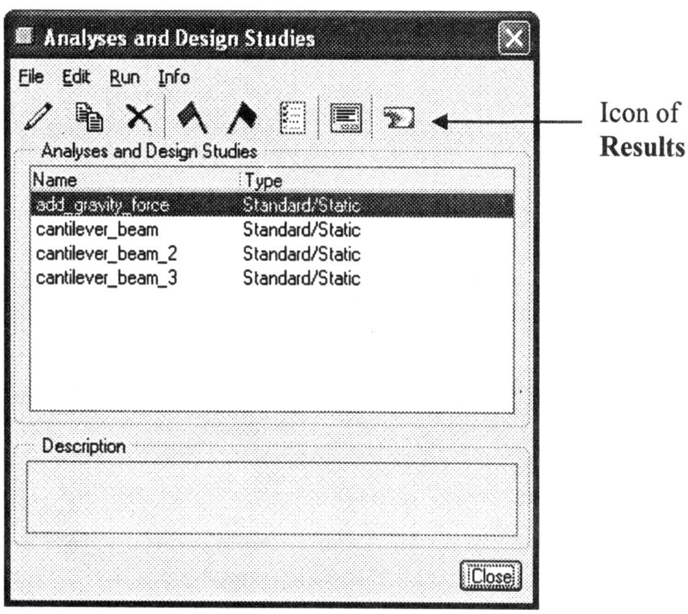

Type *Beam Deflection* as the title > locate the Design Study folder called add_gravity_force. It is important to note that Loadset1 and Loadset2 are shown. Select both LoadSet1 and LoadSet2.

Select **Fringe** > choose **Displacement** > **Y** > **OK and Show**.

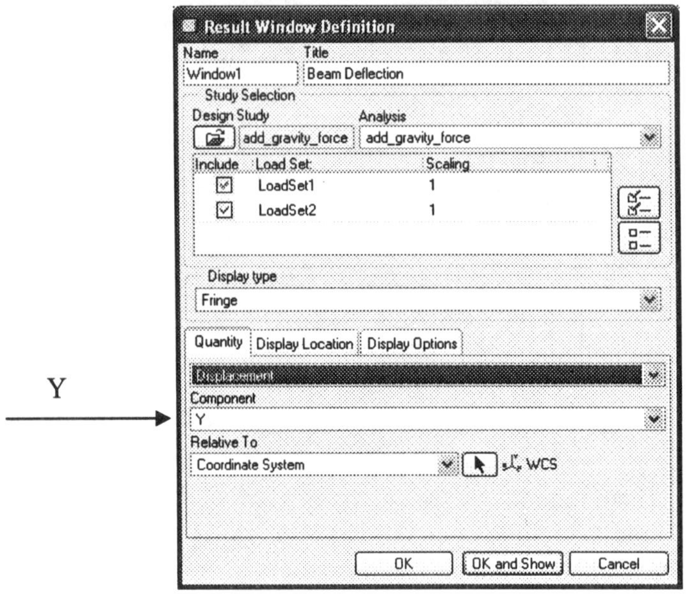

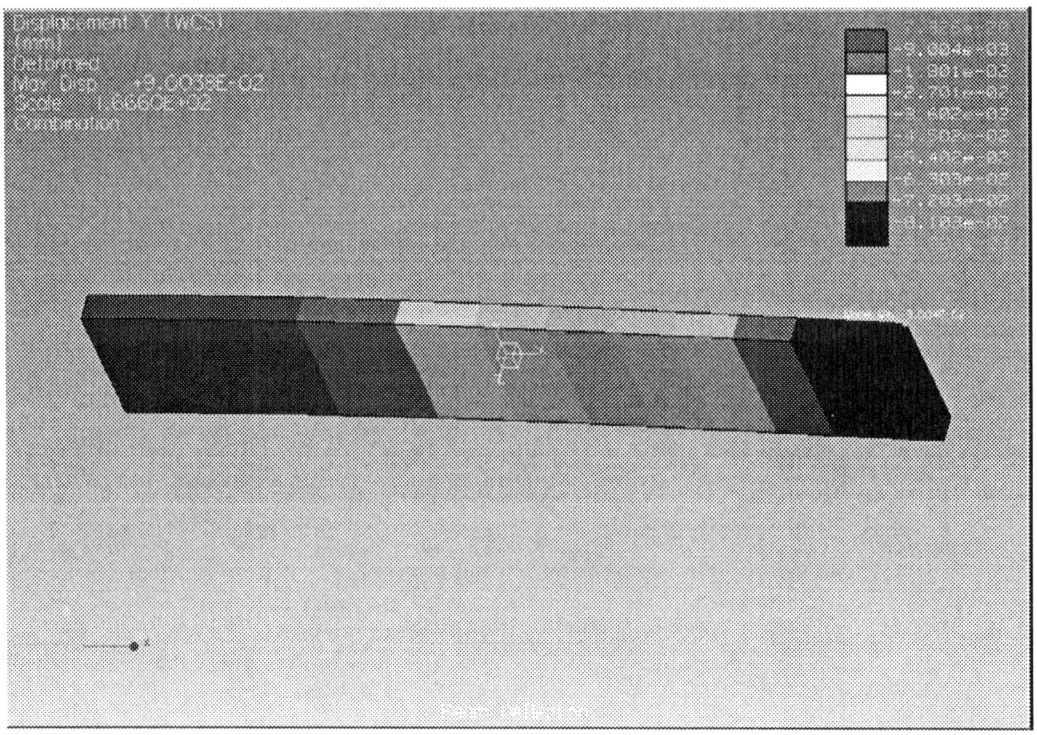

Examining the value of the maximum deflection shown at the free end, the numerical value is 0.0900 mm. As we recall, the numerical value obtained from FEA without considering the gravity force was 0.08770 mm for the surface region set at 10 x 0.5 mm^2. The difference is 0.0023 mm or 2.3 μm, which is rather small as compared with 0.08770 mm.

Now let us examine the contribution by the gravity force only. Select the icon of Edit the Select Definition from the main toolbar > select Loadset 2 only > **OK and Show**.

LoadSet2

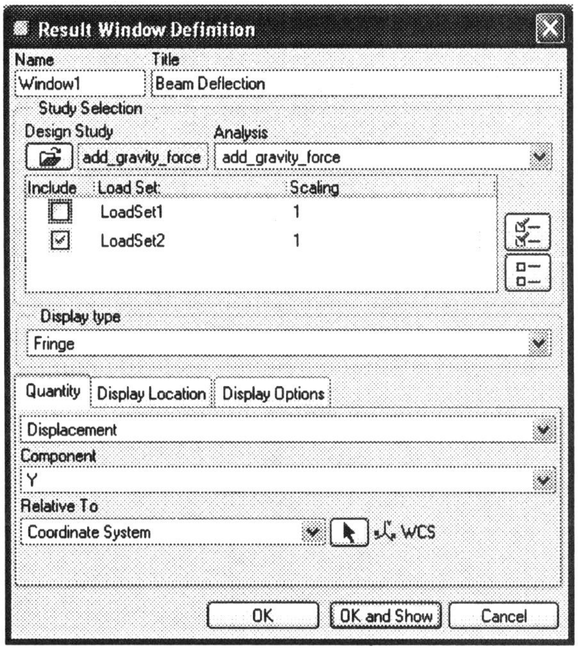

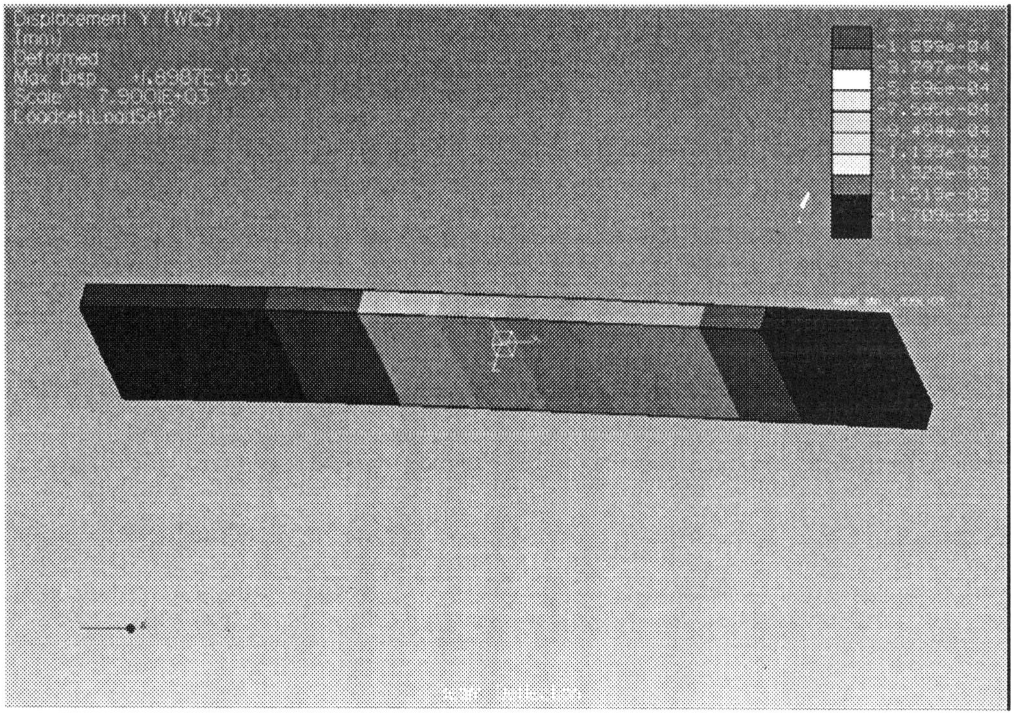

Examining the value of the maximum deflection shown at the free end, the numerical value is 0.0019 mm, which is a value very close to the difference of 0.0023 mm calculated above. In general, the gravity effect is ignored in performing FEA. However, under certain circumstances, we may still need to consider the effect due to the gravity force.

3.8 References

1. Algor Processor Reference Manual, Algor Interactive Systems, 150 Alpha Drive, Pittsburgh, PA 15238
2. O. Axelsson, Iterative Solution Methods, Cambridge University Press, New York, 1994.
3. K. J. Bathe, Finite Element Procedure, Prentice Hall, Upper Saddle River, New Jersey, 1996.

4. K. Bell, A Refined Triangular Plate Bending Finite Element, Int. J. Numer. Methods Eng., Vol. 1, No. 1 January 1969, pp. 101-122.
5. B. Carnahan, H. A. Luther, and J. O. Wikes, Applied Numerical Methods, Wiley New York, NY, 1969.
6. R. W. Clough, The Finite Element in Plane Stress Analysis, Proceedings of 2nd ASCE Conference on Electronic Computation, Pittsburgh, PA. September 1960.
7. R. D. Cook, D. S. Maikus and M. E. Plesha, Concepts and Applications of Finite Element Analysis, 3rd edition, Wiley, New York, NY, 1989.
8. J. W. Dally and W. F. Riley, Experimental Stress Analysis, McGraw Hill, 1965.
9. J. W. Dally, Design Analysis of Structural Elements, Goodington, New York, NY, 2003.
10. C. S. Desal and J. F. Abel, Introduction to the Finite Element Method, Van Nostrand Reinhold, New York, NY, 1972.
11. J. H. Earle, Graphics for Engineers, AutoCAD Release 13, Addision-Wesley, Reading, Massachusetts, 1996.
12. R. H. Gallagher, Finite Element Analysis Fundamentals, Prentice-Hall, Englewoood Cliffs, NJ, 1975.
13. J. M. Gere and S. P. Timoshenko, Mechanics of Materials, 3rd edition, PWS-KENT Publishing Company, Boston, 1990.
14. W. W. Hager, Applied Numerical Linear Algebra, Prentice-Hall, Englewood Cliffs, NJ, 1988.
15. Holland, The Finite Element Method in Plane Stress Analysis, in the Finite Element Method in Stress Analysis, I. Holland and K. Bell (eds), Tapir Press, Trondheim, Norway, 1969, Chapter 2.
16. Y. Y. Hsieh, Elementary Theory of Structure, 2nd edition, Prentice-Hall, Englewood Cliffs, NJ, 1982.
17. M. A. N. Hendriks, H Jongedijk, J. G. Rots and W. J. E. van Spange (eds), Finite Elements in Engineering and Science, DIANA Computational Mechanics, 1997.
18. K. H. Huebner, D. L. Dewhirst, D. E. Smith, and T. G. Byrom, The Finite Element Method for Engineers, 4th edition, John Wiley & Sons, Inc., 2001.
19. C.S. Krishnamoorthy, Finite Element Analysis, Theory and Programming, 2nd Ed., 1995.
20. D. L. Logan, A First Course in the Finite Element Method Using ALGOR, PWS Publishing, Boston, 1997.
21. H. C. Martin, Introduction to Matrix Methods of Structural Analysis, McGraw-Hill, New York, NY, 1966.
22. R. L. Norton, Machine Design: An Integrated Approach, Prentice Hall, Upper Saddle River, New Jersey, 1996.
23. M. Ortiz, Y. Leroy and A. Needleman, A Finite Element Method for Localized Failure Analysis, Comp. Meth. Applied Mech. Enging, 61, pp. 189-214, 1987.
24. M. Petyl, Introduction to Finite Element Vibration Analysis, Cambridge University Press, 1990.
25. T. H. H. Pian and P. Tong, Basis of Finite Element Methods for Solid Continua, Int. J. Numer. Methods Eng., Vol. 2, No. 7, April 1964, pp. 1333-1336.
26. P. G. Plockner, Symmetry in Structural Mechanics, J. of the Structural Division, American Society of Civil Engineers, Vol. 99, No. ST1, pp. 71-89, 1973.
27. J. N. Reddy, An Introduction to the Finite Element Method, McGraw-Hill Publishing Company, New York, NY, 1984.
28. T. Ross, T. Johns and Emile, Dynamic Buckling of Thin Walled Domes under External Water Pressure, Applied Solid Mechanics 2nd Conference, Ed. Tooth & Spence, Elsevier, pp. 211-224, 1987.
29. T. Ross, Finite Element Program in Structural Engineering and Continuum Mechanics, Horwood Publishing Limited, Chichester, England, 1996.
30. T. Ross, Advanced Applied Finite Element Methods, Horwood Publishing Limited, Chichester, England, 1998.
31. L. Stasa, Applied Finite Element Analysis for Engineers, Holt, Rinehart and Winston, New York, NY, 1985, pp. 101-157.
32. Y. Saad, Iterative Methods for Sparse Linear Systems, PWS Publishing Company, Boston, 1996.
33. Y. Tada and G. C. Lee, Finite Element Solution to an Elastic Problem of Beams, Int. J. Numer. Methods Eng., Vol. 2, No. 2, April 1970, pp. 229-241.
34. S. P. Timoshenko and D. H. Young, Theory of Structure, McGraw-Hill, Kogaksha Limited, 1965.

35. M. J. Turner, R. W. Clough, H. C. Martin, and L. J. Topp, Stiffness and Deflection Analysis of Complex Structures, J. of the Aeronautical Sciences, Vol. 23, No. 9, pp. 805-824, Sept. 1956.
36. O. C. Zienkiewicz, The Finite Element Method, third edition, McGraw-Hill, New York, NY, 1977.
37. O. C. Zienkiewicz and R. L. Taylor, The Finite Element Method, 4th edition, Vol. 1, McGraw-Hill (UK),London, 1989.

3.9 Exercises

1. A cantilcvcr bcam is subjected to a non-uniformly distributed load. The total load is 500 Newton, acting downward. The following conditions are assumed:

    - The left side of the beam is fixed.
    - The right side of the beam is free.
    - The material of the cantilever beam is steel.
    - The two weighting factors, which characterize the non-uniform distribution of the load condition, are 1.0 and 0.5 respectively.

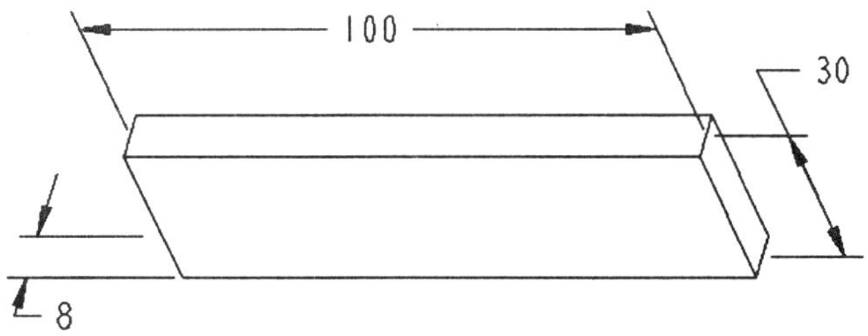

Total load = 500 Newton

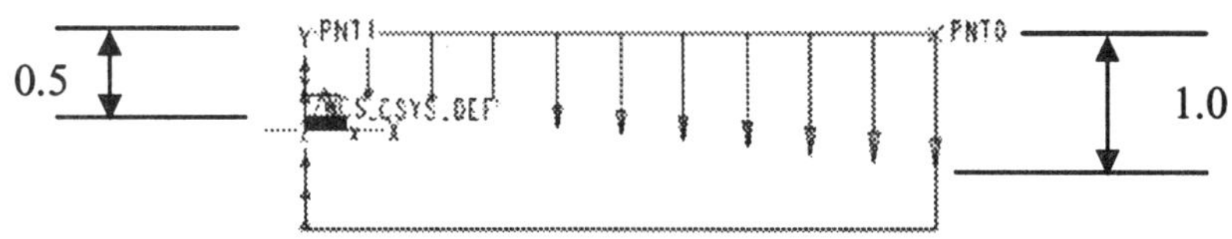

- Plot the distribution of the deformation in the vertical direction.
- Plot the distribution of the maximum principal stress of the 3D solid model.

2. In Problem 1, asssume that the load condition is changed to a partial loaded situation, as shown below:

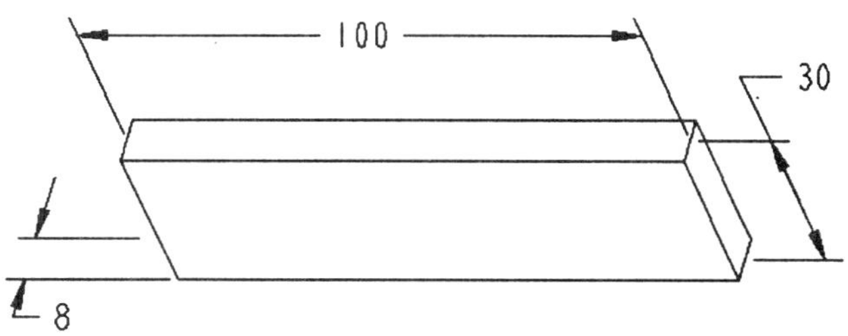

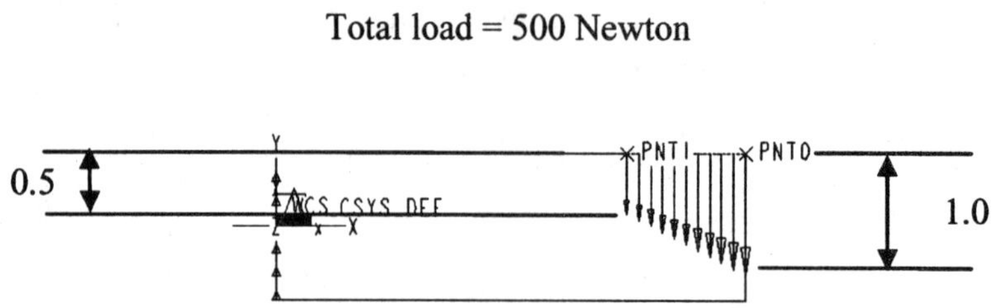

3. A cantilever beam is shown below. The left end is fixed and the right end is set to free. The structure is subjected to a uniform load, which acts on a surface region. The surface region is a rectangle. The two dimensions are 150 mm and 8 mm, respectively.

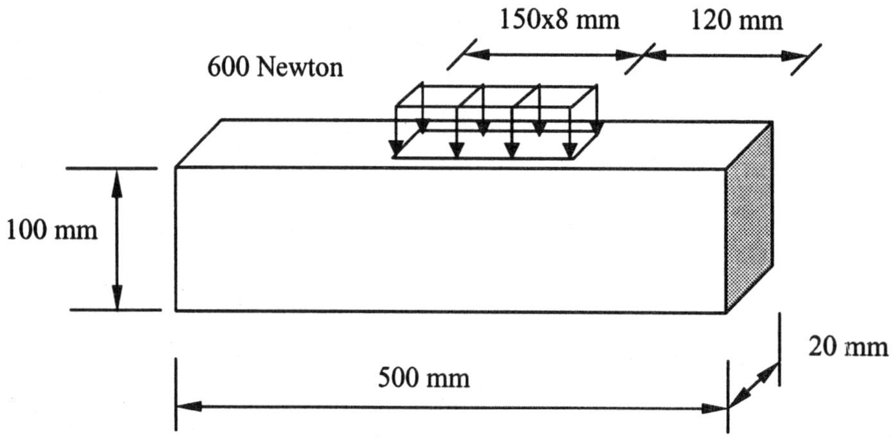

1) A 3D solid model is available and the datum curve for defining the load condition is also available. You are asked to create a surface region so that a uniform load condition can be defined.
2) Assume that the material is STEEL, as listed in the Pro/E material library. Define a constraint set for the fixed end at the left end of the beam, and define uniform load acting on the surface region as specified in (1).
3) Fill in the numerical values in the following table with their units.

	maximum displacement of the beam in Y	max principal stress of the beam	max Von mises stress of the beam
Location (x, y, z)			

4. A pin component is shown below. The surface with 2 arrows associated is fixed to the ground. The surface at the right side is subjected to a uniform load. The magnitude is 300 Newton. Identify the location where the maximum shear stress is present. Identify the magnitude of the maximum shear stress. If the pin component fails during an operation, what is the most likely failure mode? Is it due to the maximum principal stress? Or is it due to the presence of the maximum shear stress? The student is asked to select the material type for the pin component before performing FEA.

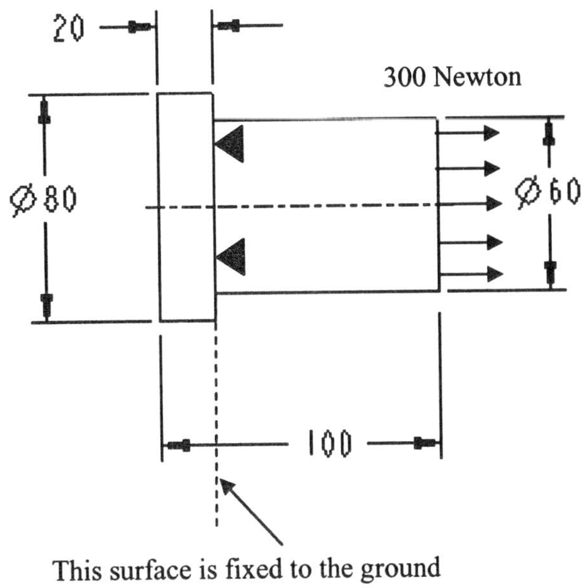
20
300 Newton
Ø80
Ø60
100
This surface is fixed to the ground

CHAPTER 4

MODELING OF CONSTRAINTS AND FEA WITH ASSEMBLIES

4.1 Introduction

In Chapter 3, we discussed the process of modeling a variety of load conditions. There are two focuses in this chapter. The first focus is to model certain types of constraint conditions. The second focus is to perform FEA for assemblies, insteadof components as we have discussed in chapters 2 and 3. In the chapter, we also continue our discussion on modeling the load conditions. The process of modeling load conditions with torque is presented.

4.2 Modeling of Constraint Conditions

We discussed the importance of modeling the constraint condition in chapter 1. When a beam structure is subjected to a load, the status of the beam deflection under the load is also dependent on the boundary conditions of the support conditions. As an example of a cantilever beam, tthe deflection at the free end is large and there is no deflection at the fixed end. On the other hand, if the beam structure is a simple beam, the maximum deflection occurs at a location between the two end supports. In the following example, we demonstrate the importance of modeling constraint conditions in performing FEA.

As illustrated, a support structure is held to the ground through 4 bolts and 4 washers. The load conditions include torque acting on the top surface, and a load uniformly acting on the top surface in the vertical direction. The torsion effect can be visualized from the figure shown on the next page.

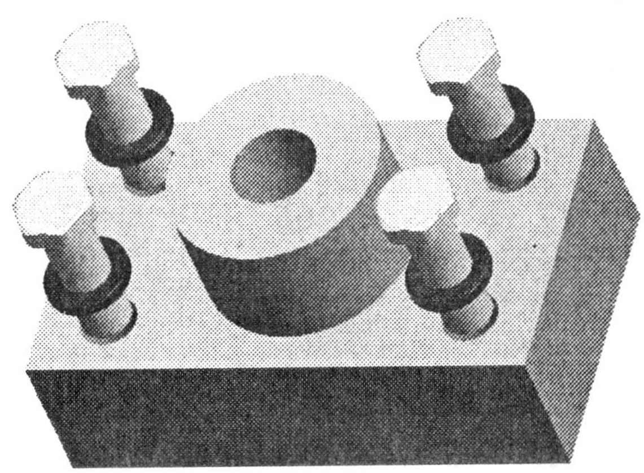

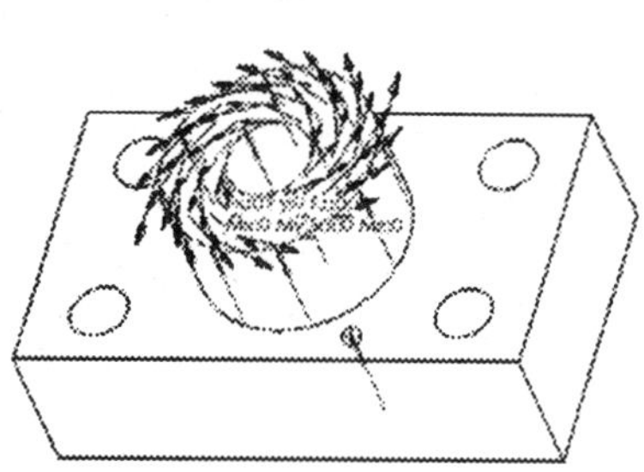

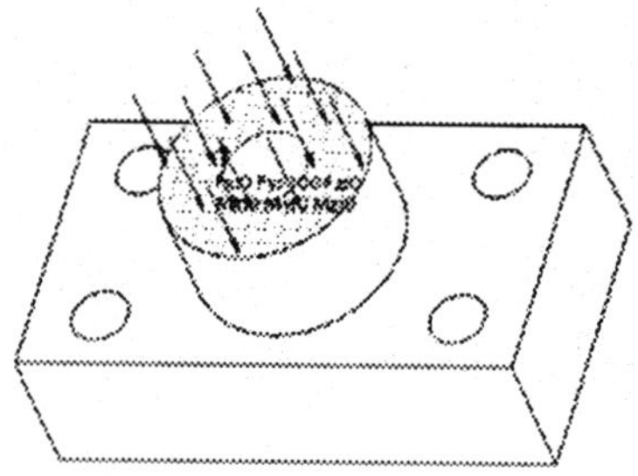

The geometry of the support structure and its dimensions are shown below. We will perform FEA to illustrate the deformation distribution and the distribution of the maximum principal stress. The constraint set consists of the 4 cylindrical surfaces of the 4 holes. In addition, we consider that the bottom surface in being held to the ground.

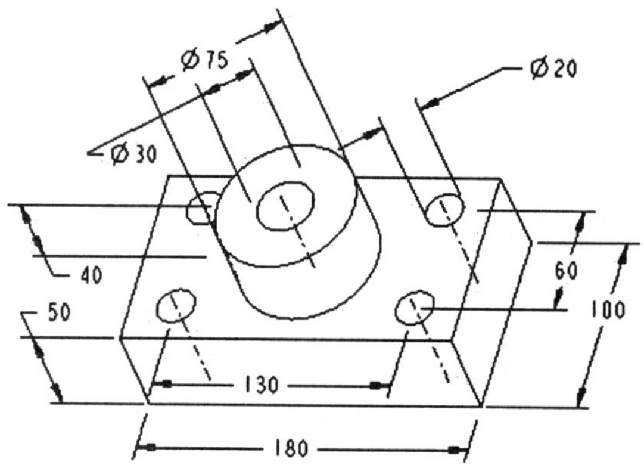

Step 1: Create a file for the 3D solid mode.

File > New > Part > type *support* as the file name and clear the icon of **Use default template > OK.** Select **mmns_part_solid** (units: Millimeter, Newton, Second) and type *torque and constraint* in **DESCRIPTION**, and *student* in **MODELED_BY**, then **OK.**

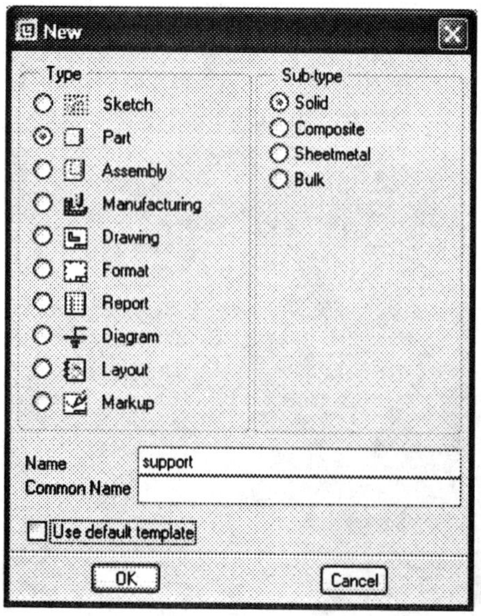

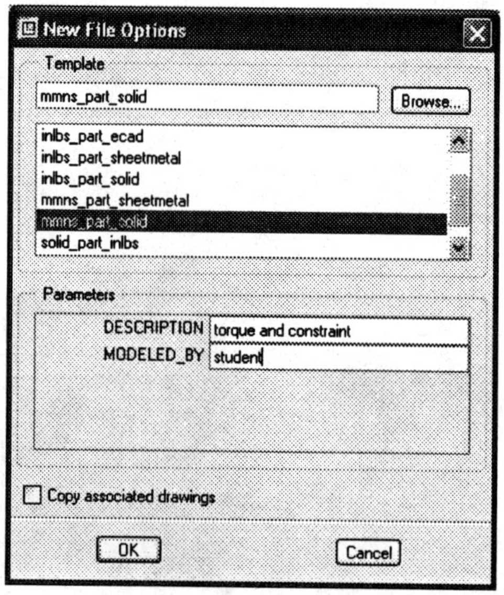

Create a 3D solid model and start with the block feature. The size is 180 x 100 x 50 mm.

From the menu bar, **Insert** > **Extrude.** The dashboard appears. Make sure that the icon of **Solid** on the dashboard is selected. Set the height value to 50. Activate **Placement** to define a sketch plane > **Define.**

Icon of **Extrude**

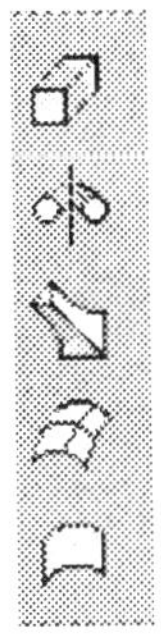

Define a sketch plane

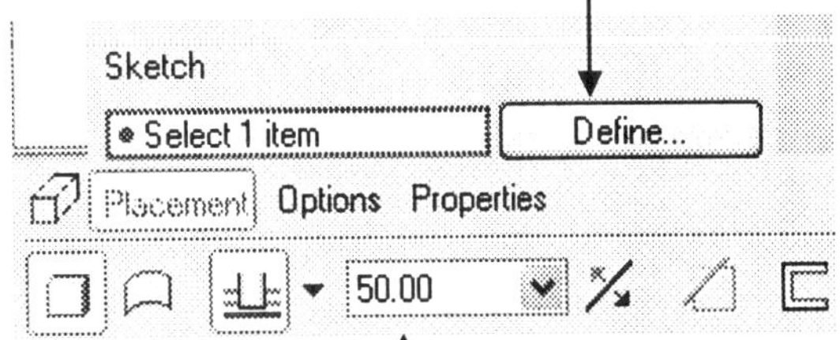

solid

50

Select the **TOP** datum plane as the sketch plan. Pick the box of **Sketch** to accept the **RIGHT** datum plane as the default reference to orient the sketch plane, as illustrated below:

Pick **TOP** as the sketch plane

Accept the default setting of orientation

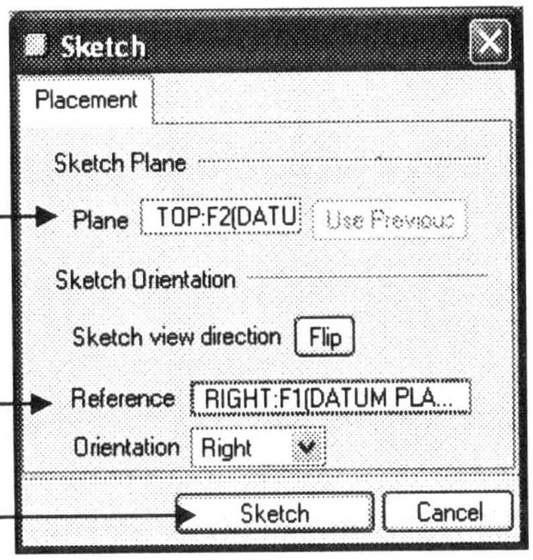

The Pro/ENGINEER design system, by default, has selected two datum planes (**RIGHT** and **FRONT**) as references for making a 2D sketch. Let us accept them, and click the box of **Close** on the **References** window.

Select the icon of **Centerline** (click the black arrow on the Line Types Flyout before selecting) to draw 2 centerlines along the y-axis and the z-axis.

Select the icon of **Rectangle** to draw a rectangle, which is symmetric with respect to the two centerlines, by allowing the rectangle to snap to symmetry.

Modify the two dimensions to 180 and 100, respectively. (if the sketched rectangle is not symmetric about the y axis or z axis or both axes, pick the icon of **Constraints** > select the icon of **Symmetry** > select a centerline, say y axis and, afterwards, pick the two corner points, which are constrained to meet the symmetrical requirement).

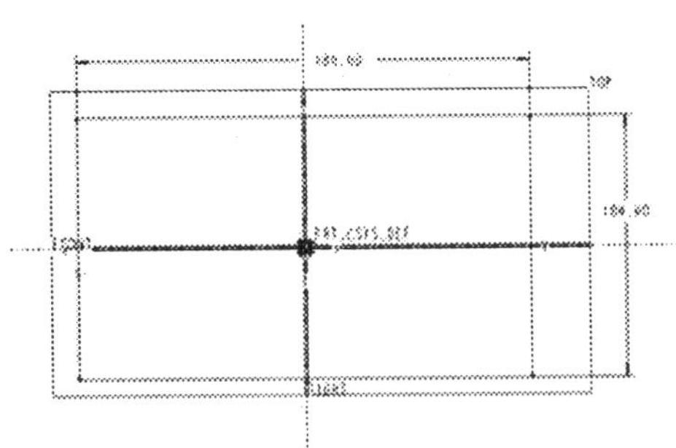

Pick the icon of **Done** from the feature creation toolbar to complete the sketch. On the dashboard, type 50 as the depth value if this value has not been set previously.

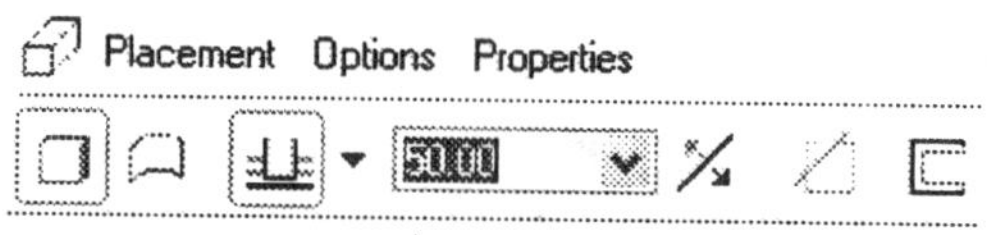

Specify the depth value of 50 in the dashboard

From the Feature Control Panel, pick the icon of **Complete** to complete the creation of the block feature, as shown below:

Pick the icon of **Complete** from the Feature Control Panel

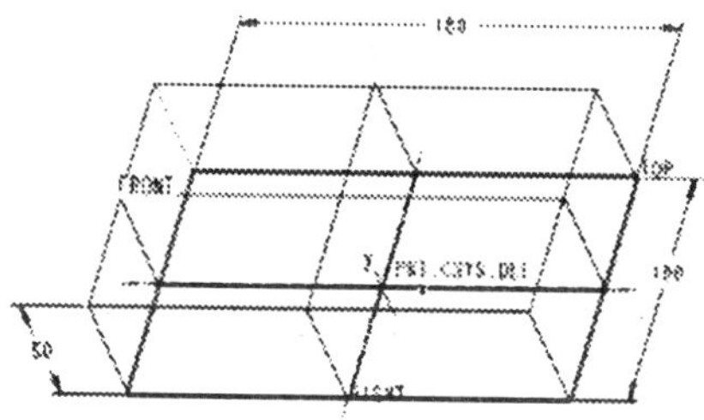

After creating the block feature, we create the cylinder feature. Its size is ϕ75 x 40 mm. Pick the icon of **Extrude** from the toolbar of feature creation. From the dashboard, set the height value to 40. Activate **Placement** to define a sketch plane > **Define**.

Icon of **Extrude**

Icon of **Solid**

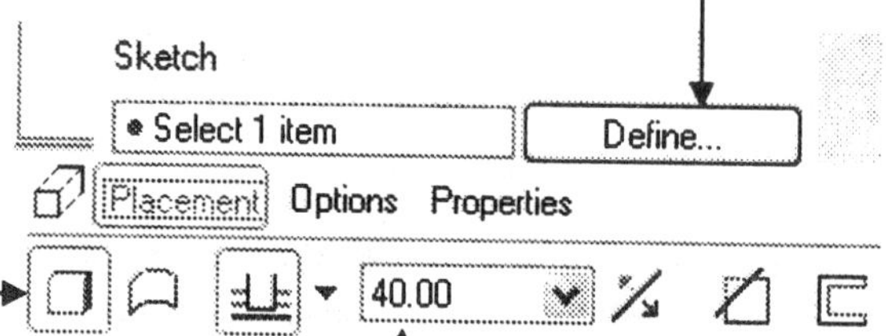

40

Select the top surface of the block (note: do not select the **TOP** datum plane) displayed on screen, and pick the button of **Sketch** to accept the **RIGHT** datum plane as the default reference to orient the sketch plane, as illustrated below:

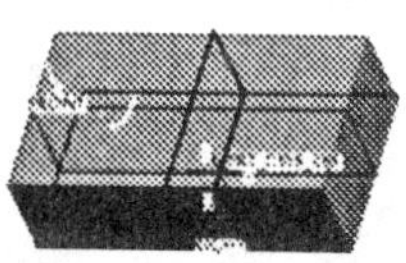

Pick the top surface as the sketch plane

Accept the default setting of orientation

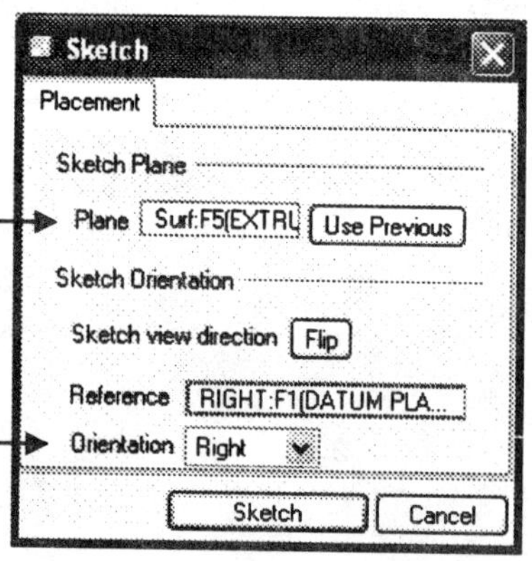

Accept the default references, and select the icon of **Circle** to draw a circle with diameter equal to 75.

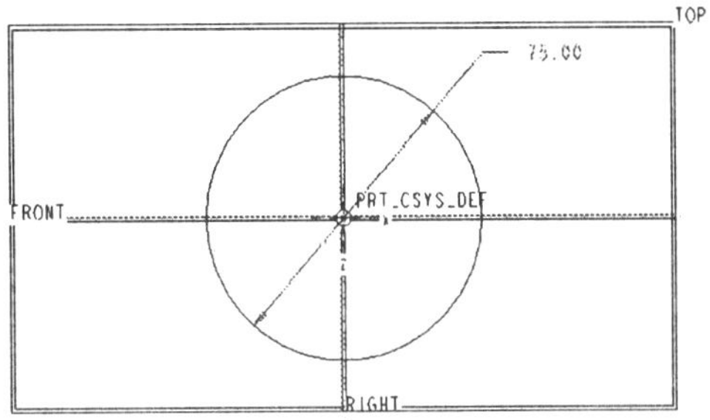

Pick the icon of **Done** from the feature creation toolbar to complete the sketch. On the dashboard, type 40 as the height value if this value has not been set previously.

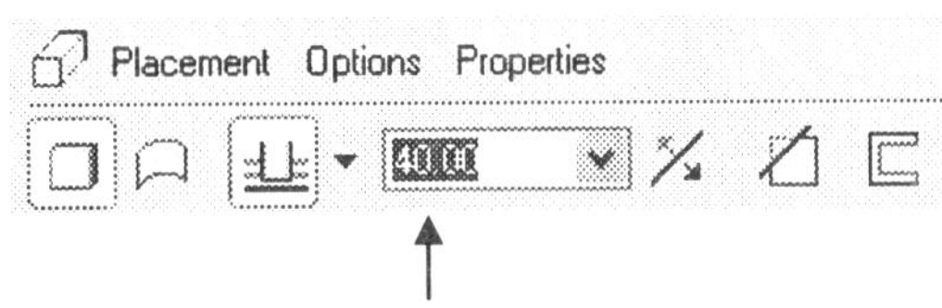

Specify the depth value of 40 in the dashboard

On the Feature Control Panel, pick the icon of **Complete** to complete the creation of the cylinder feature.

Pick the icon of **Complete** from the Feature Control Panel

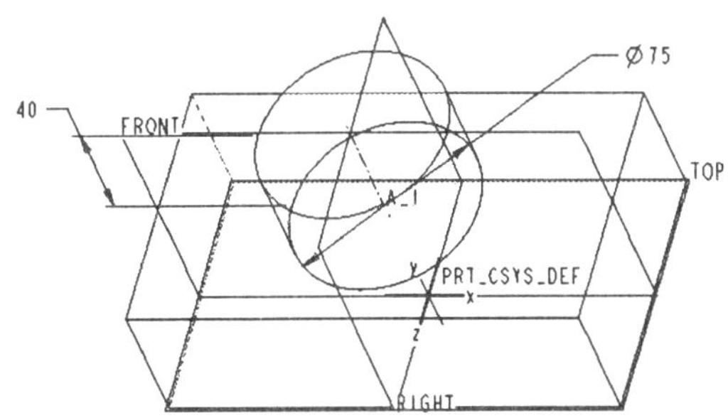

To create a through hole feature of $\phi 30$, from the tool bar of feature creation, pick the icon of **Hole**. On the dashboard, type in *30* as the diameter value and select **Through All** as the depth choice (click the black arrow on the Depth Options Flyout in the dashboard).

Icon of **Hole**

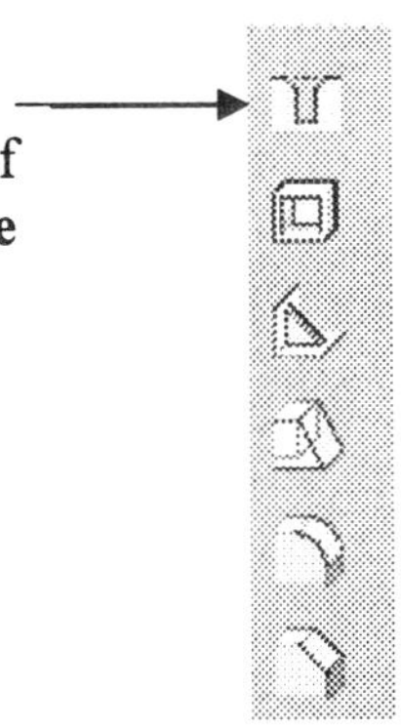

Select Through All

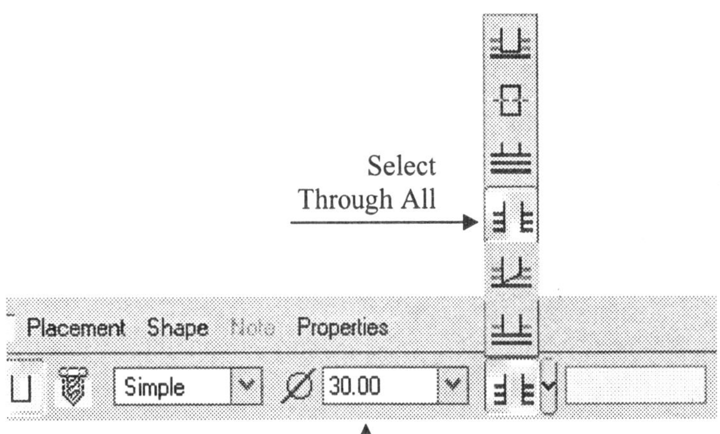

Type 30 as the diameter value

Click the **Primary** box, and select the top surface of the cylinder as the surface for placement. Click the **Secondary Reference** box and select the **FRONT** datum plane. While holding down the Ctrl key on your keyboard, select the **RIGHT** datum plane. Both **FRONT** and **RIGHT** serve as references. Select "Align", indicating that the center of the hole aligns with the two datum planes.

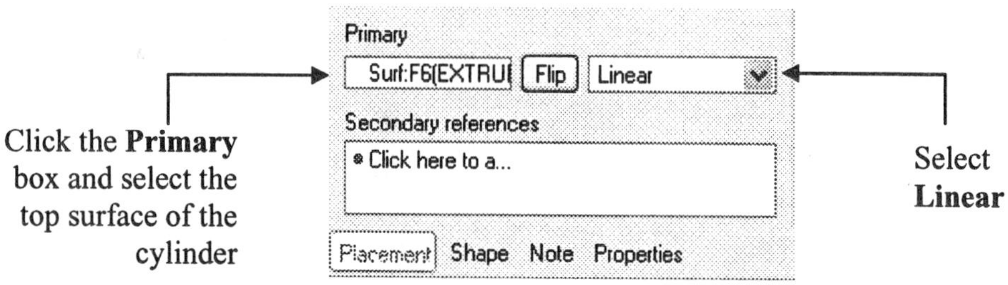

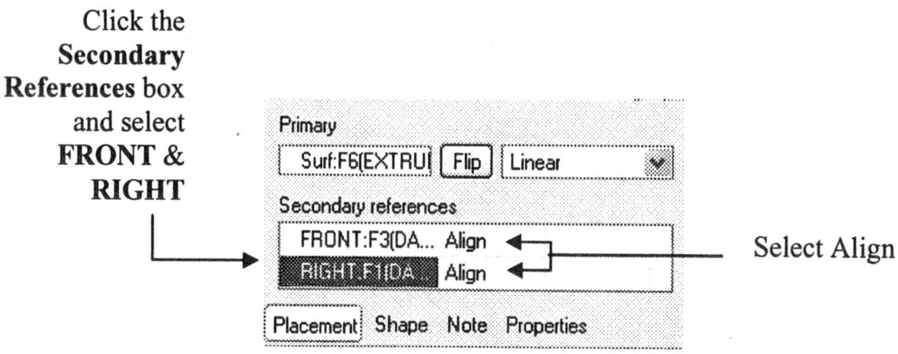

On the Feature Control Panel, pick the icon of **Complete** to complete the creation of the hole feature, as shown below:

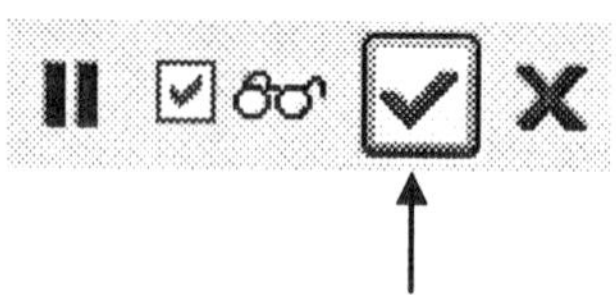

Pick the icon of **Complete** from the Feature Control Panel

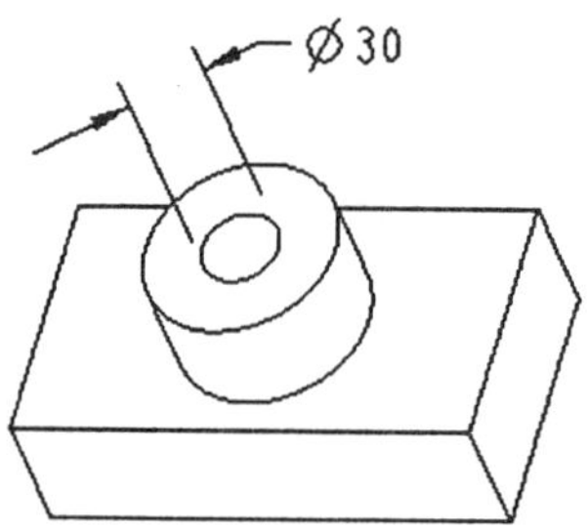

As a summary to the process of creating a hole feature, the Pro/ENGINEER design system provides its users with two ways to construct features:

- Feature construction by sketching as described in the process of creating the block feature.
- Feature construction by picking & placing as described in creating the hole feature

After creating the hole feature, we create the fourth feature, which is a set of 4 though holes with diameter equal to 20. Pick the icon of **Extrude** from the tool bar of feature creation. Select the icon of **Cut** and set the depth choice to **Through All**. Activate **Placement** to define a sketch plane > **Define**.

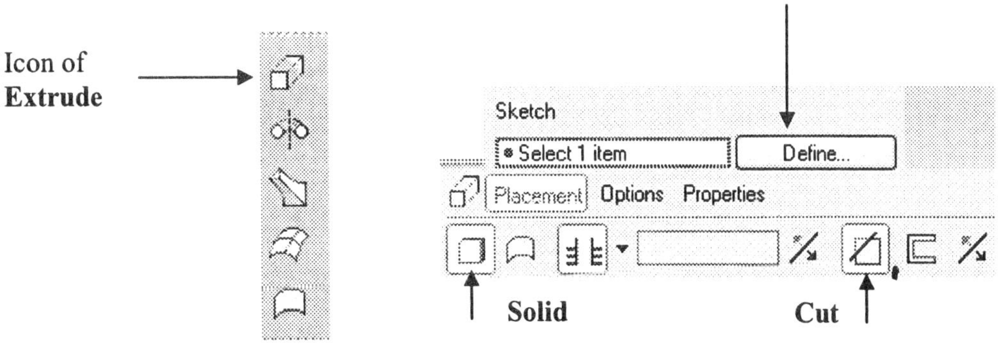

It is important that the user selects the icon of **Cut** because we are going to take material away from the 3D solid model, and set the depth choice to **Through All** because the 4 holes are through holes.

Select the top surface of the block (note: do not select the **TOP** datum plane) displayed on screen, and pick the button of **Sketch** to accept the **RIGHT** datum plane as the default reference to orient the sketch plane, as illustrated below:

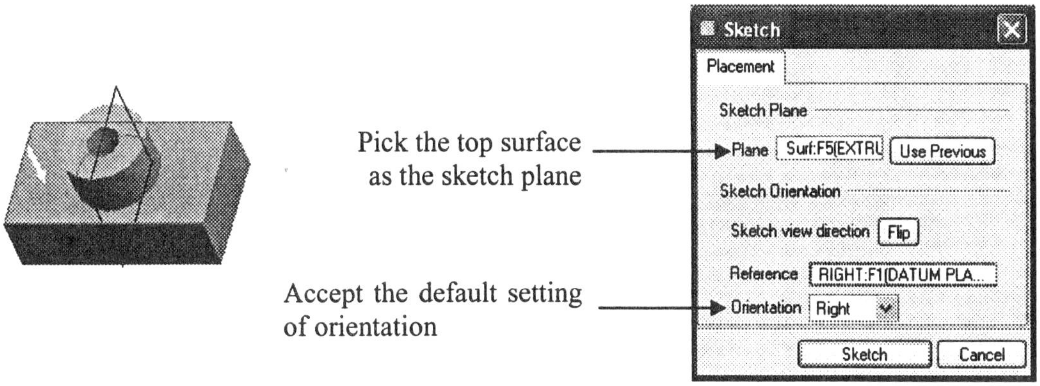

Accept the default references. Select the icon of **Centerline** to draw 2 centerlines along the x-axis and the z-axis, respectively. Select the icon of **Circle** to draw a circle with diameter equal to 20. Make changes so that the two position dimensions are equal to 65 and 30, as shown in the following sketch.

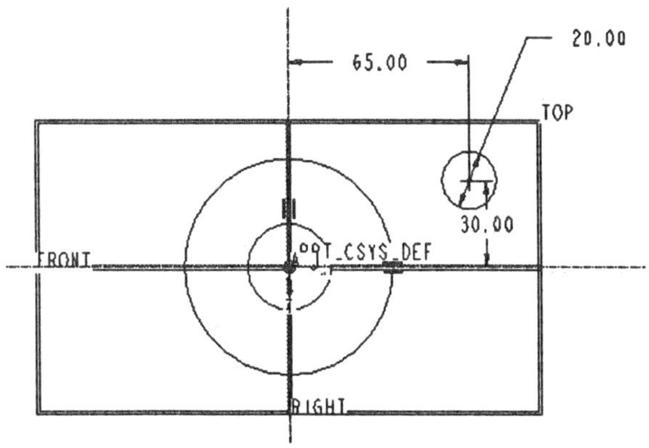

Use symmetry to create the other 3 circles.

First, pick the circle just created > pick the icon of **Symmetry** or **Mapping** > pick the vertical centerline.

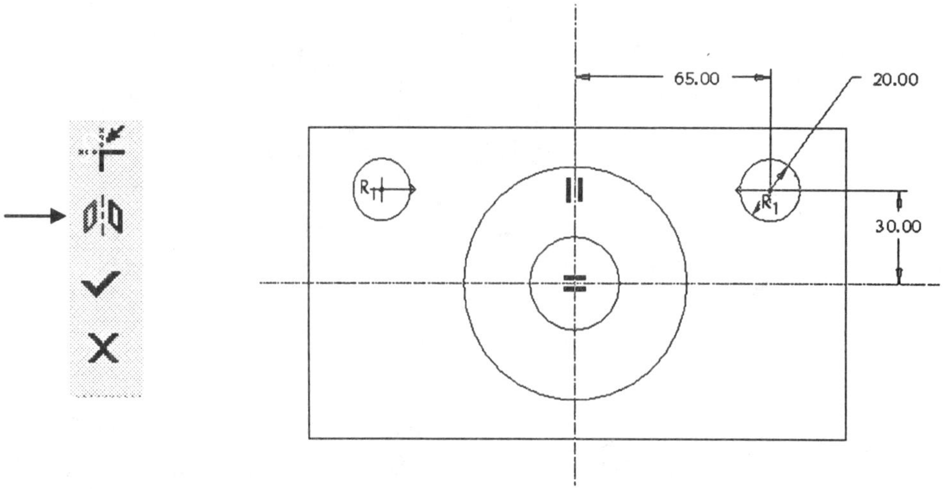

Second, pick the 2 created circles (holding down the Ctrl key) > pick the icon of **Symmetry** > pick the horizontal centerline to complete the creation of the other 2 circles.

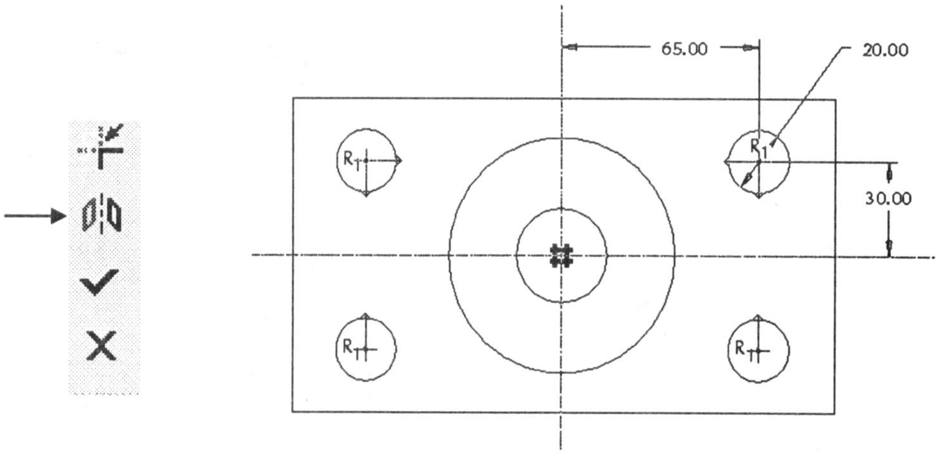

Upon completing the sketch, click the icon of **Done** from the feature creation toolbar. Make sure, on the dashboard, the icon of **Cut** is selected because we are removing material from the current model. Make sure the icon of **Through All** is selected as the depth choice, and check the direction of the cut operation.

To complete the feature creation, pick the icon of **Complete** from the Feature Control Panel.

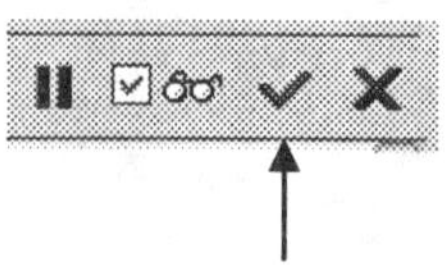

Pick the icon of **Complete** from the Feature Control Panel

Step 2: Perform FEA under **Pro/Mechanica (material, constraints and loads)**

From the menu bar, select **Applications** > **Mechanica** > **Continue** after checking the unit system > **Structure** > **OK**.

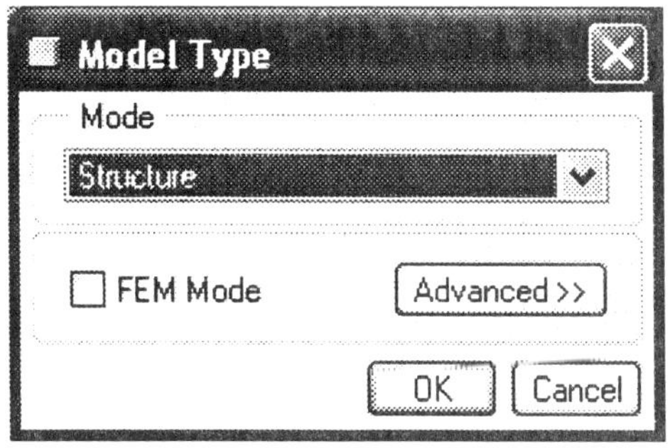

From the toolbar of functions, select the icon of **Materials** > select **STEEL** from the **Materials in Library** > click the arrow so that **STEEL** is transferred to Materials in Model > **Assign** > **Part** > select the 3D solid model on display > **OK** > **Close.**

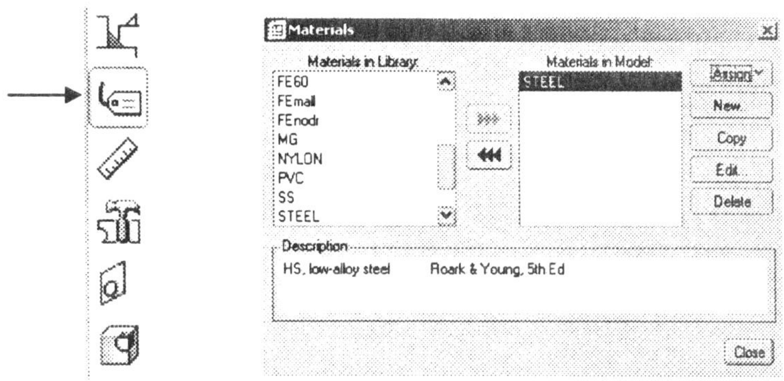

Now let us define Constraint1, simulating the support structure resting on the ground.

From the toolbar of functions, select the icon of **Displacement Constraint** > Activate the surface selection > pick the bottom surface shown below > **OK** > set the translation in X, translation in Z and rotation about Y to free, and fully constrain the other 3 degrees of freedom > **OK.**

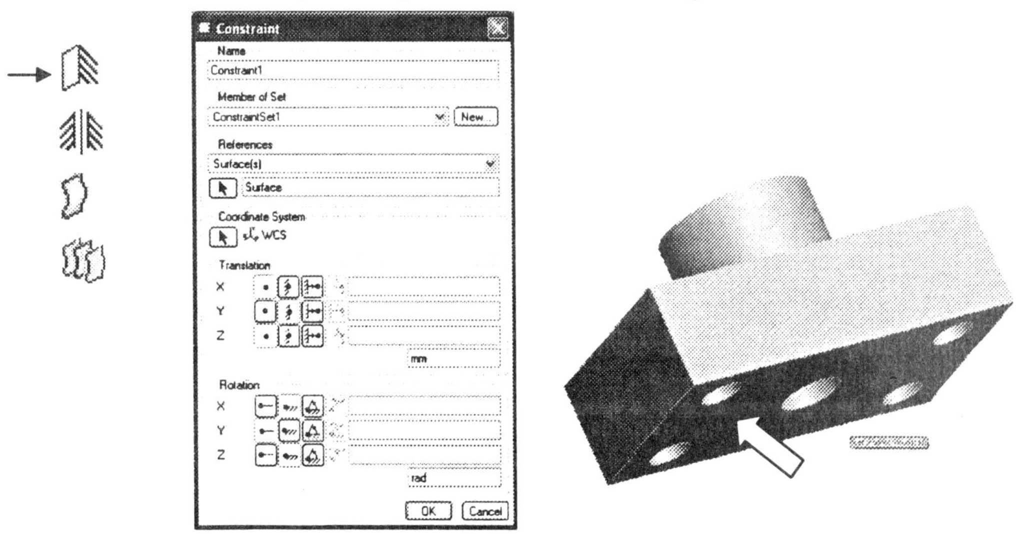

Now let us define Constraint2, simulating the support structure is firmly held to the ground through the use of 4 bolts.

In order to do so, we need to establish 4 local coordinate systems, which are cylindrical coordinate systems.

Let us create the first cylindrical coordinate system. We begin with creating a Cartesian coordinate system. Select the icon of creating coordinate system > pick the axis of the hole and the edge shown below:

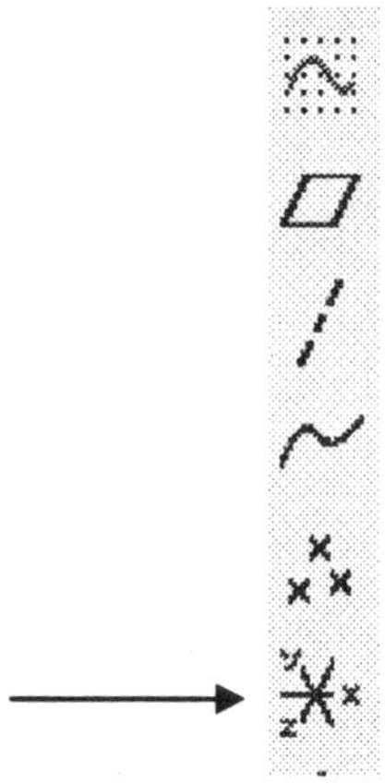

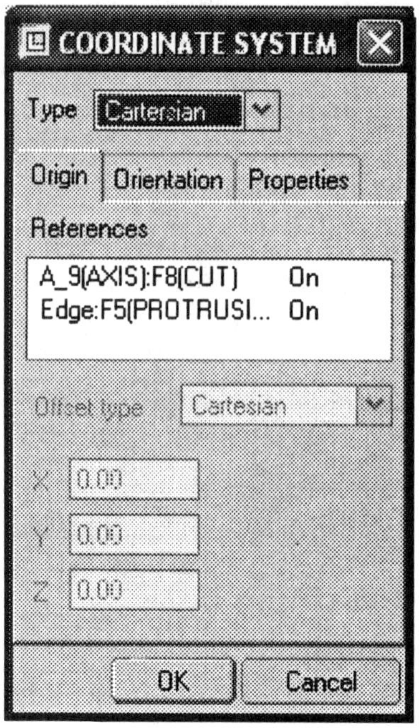

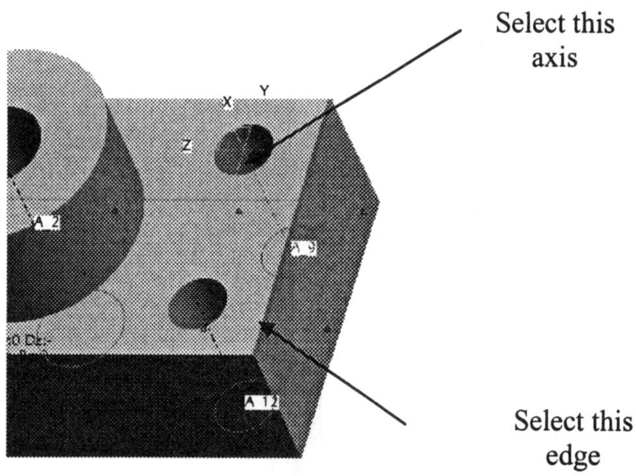

Change the vertical axis from X to Z before converting it to a cylindrical coordinate system.

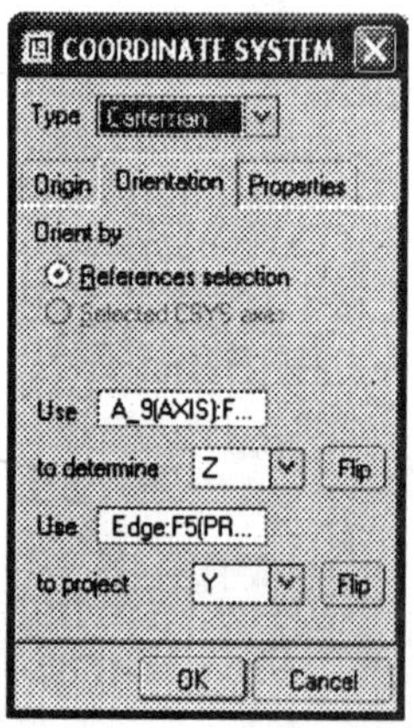

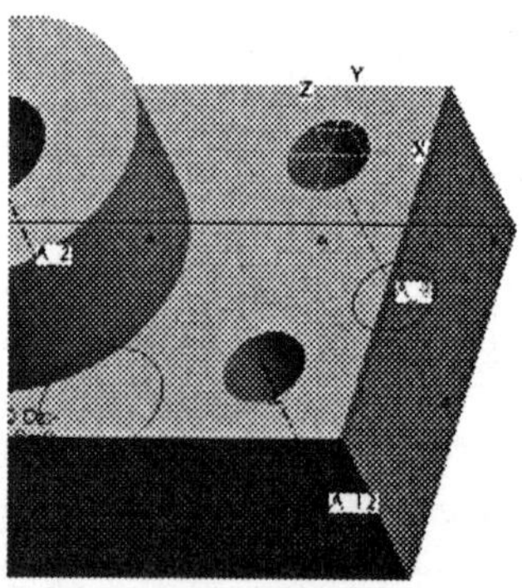

In the box marked as **Type**, change Cartesian to Cylindrical > OK.

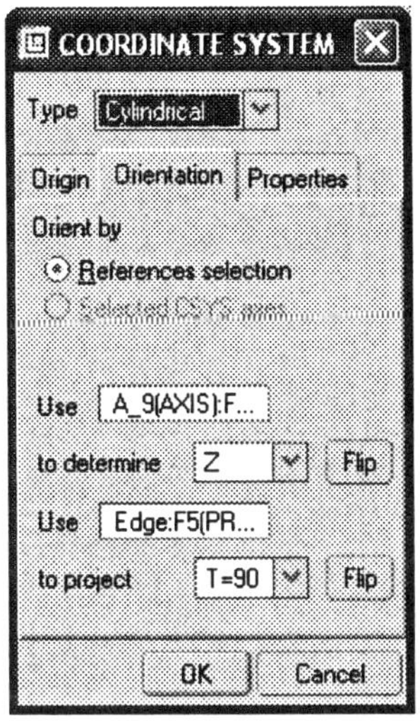

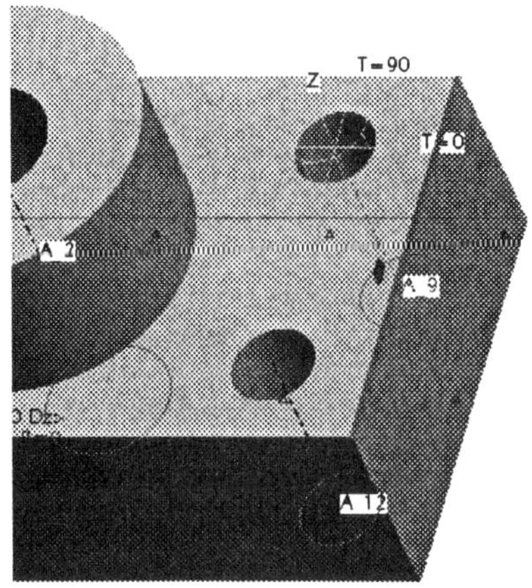

Repeat the procedure three times to create the other three cylindrical coordinate systems, as shown below:

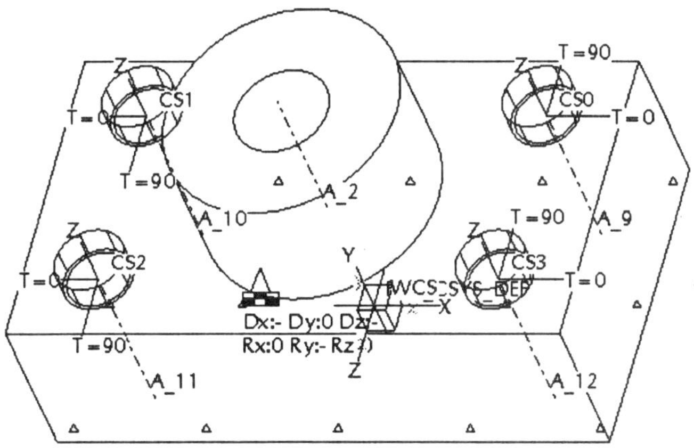

From the toolbar of functions, select the icon of **Displacement Constraint** > Activate the surface selection > pick the cylindrical surface as shown below > **OK.**

Activate the selection of Coordinate system > pick CS0.

Set the translation in R to free, and fully constrain the other 5 degrees of freedom > **OK**

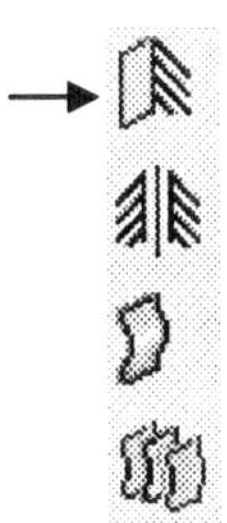

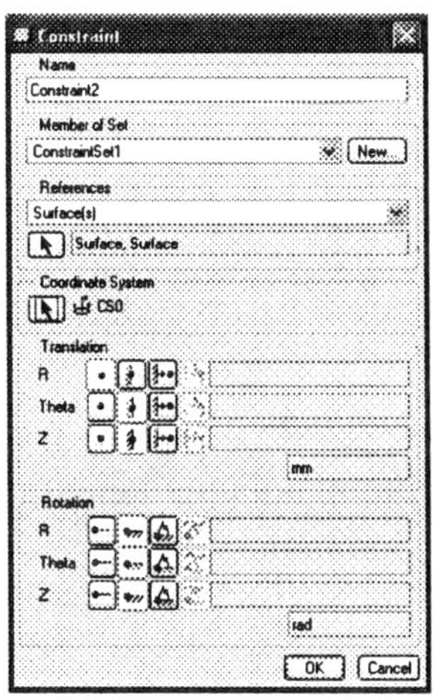

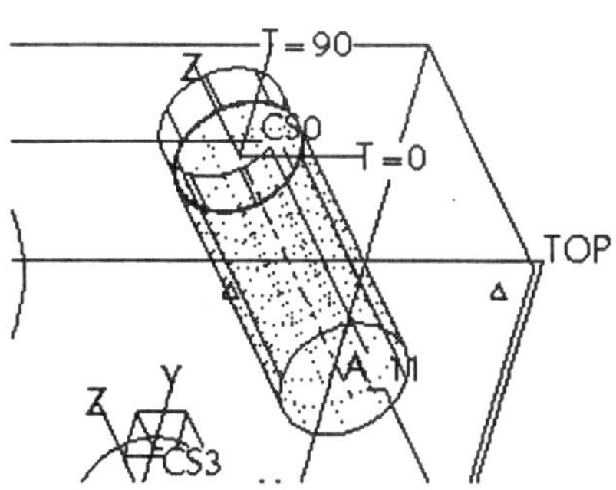

Repeat the procedure three times to define Constraint3, Constraint4 and Constraint5, as shown below:

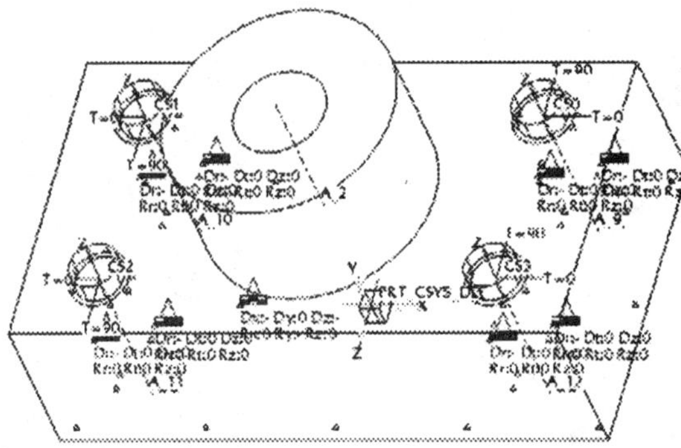

Now let us define the loads. There are two loads. The first load is torque acting on the top surface of the cylinder.

In order to define a torque, we need to define a point, about which the magnitude of the torque can be evaluated. To define a datum point, select the icon of creating datum point from the toolbar of creating datum features. While holding down the Ctrl key, pick the axis of the cylinder and the top surface of the cylinder, PNT0 is defined > **OK**.

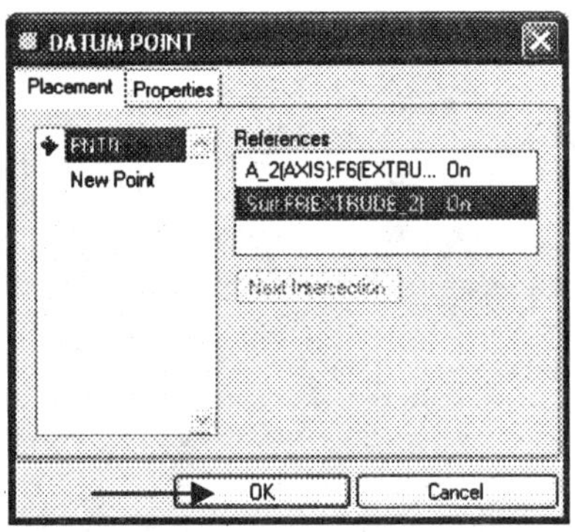

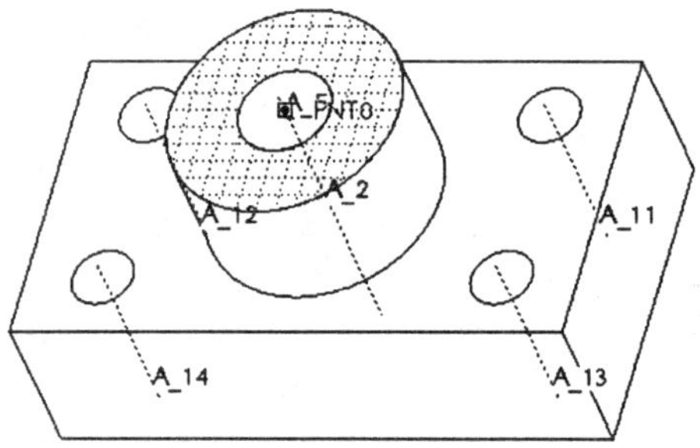

To define the torque load, select the icon of force and moment from the toolbar > Surface(s) > activate the selection of surface and pick the top surface of the cylinder > OK.

Click Advanced > change the distribution from Total Load to Total Load at Point and select PNT0 > type in 5000 in Y of the Moment box > **OK**.

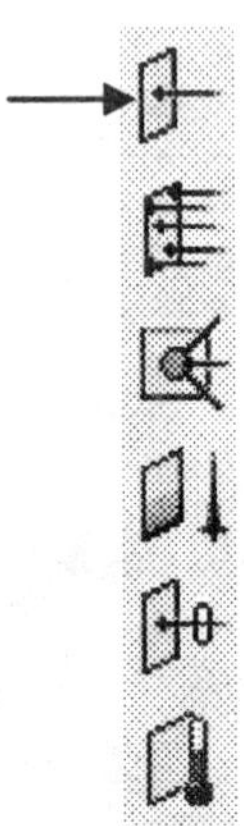

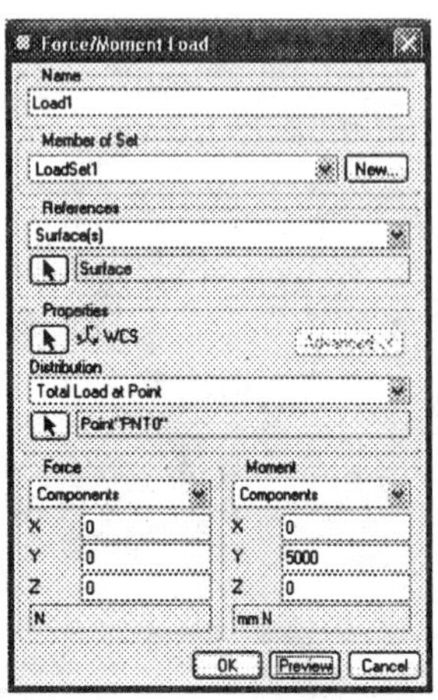

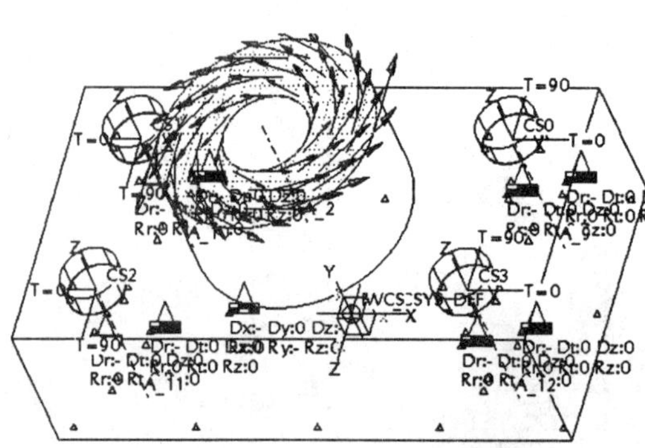

The second load is uniform load acting on the top surface of the cylinder in the Y direction (downward). To define it, select the icon of force and moment from the toolbar > Surface(s) > activate the selection of surface and pick the top surface of the cylinder > OK.

Click Advanced > change the distribution to Total Load and Uniform > type in -100 in Y of the Force box > **OK**.

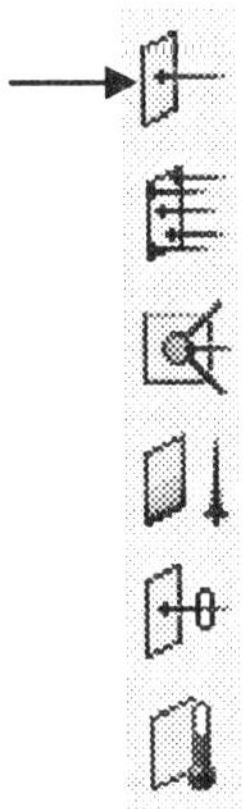

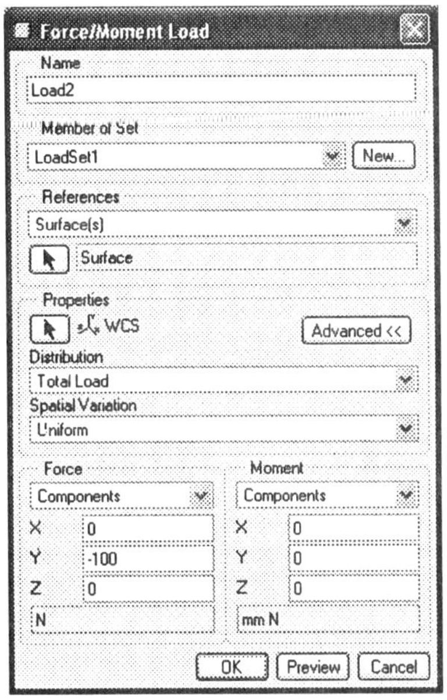

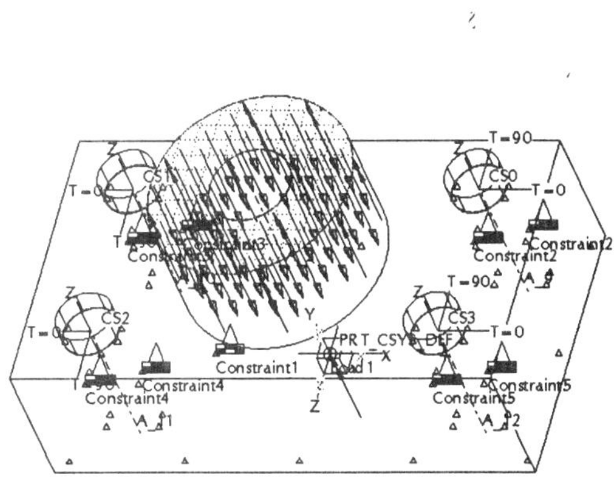

Perform a static analysis

Click the icon of performing a design study from the main toolbar > File > Select New static analysis.

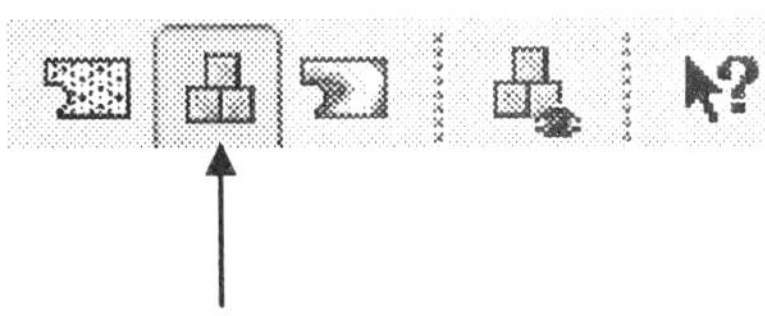

Icon of performing a design study

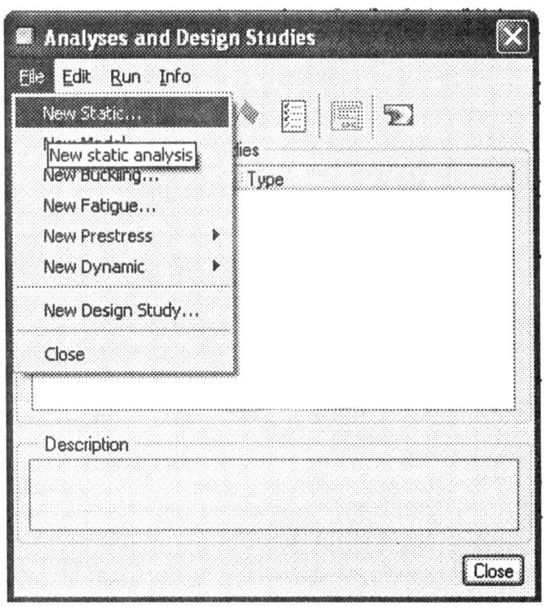

Type in *support_structure_1*

Make sure that ConstraintSet1 and LoadSet1 are highlighted > **Single-Pass Adaptive > OK > Run.**

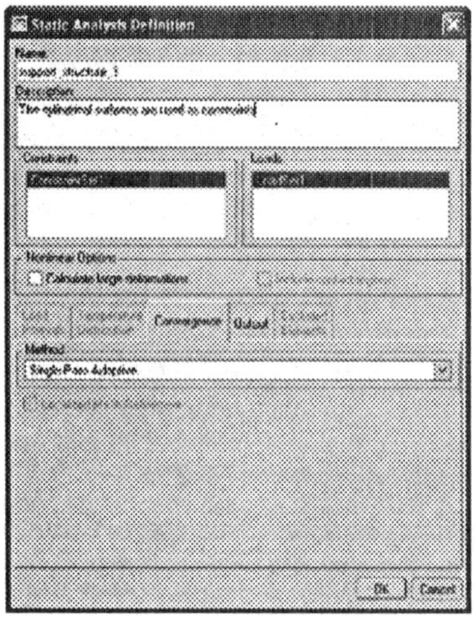

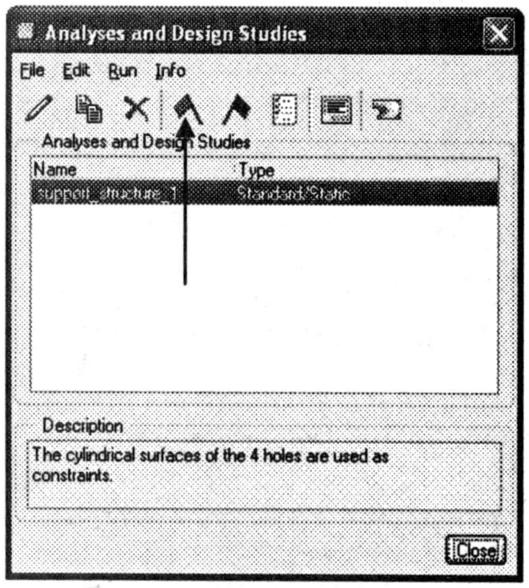

To review the results, click the icon of **review results** > type in deformation distribution (cylindrical surfaces) as the title of the result window > select the folder called support_structure_1 > **Fringe** > **Displacement** > **Magnitude**.

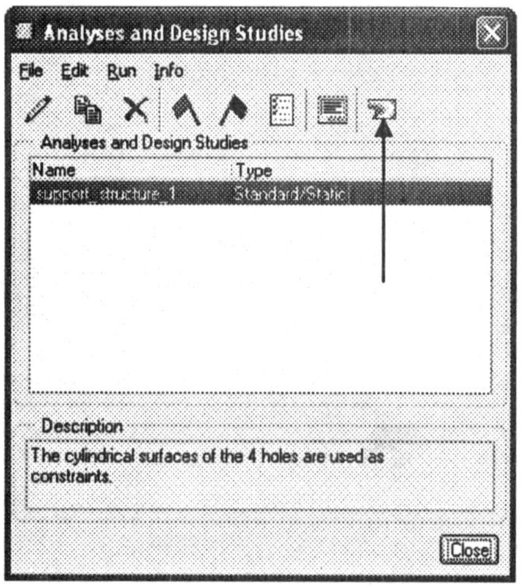

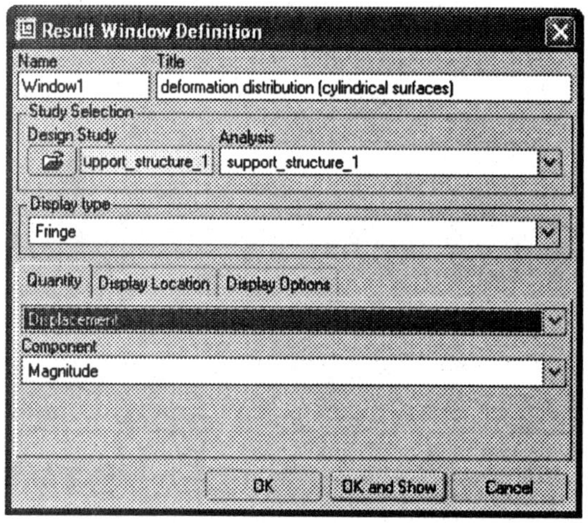

In the Display Option, select **Deformed** and set the scale to *20* > **OK and Show**

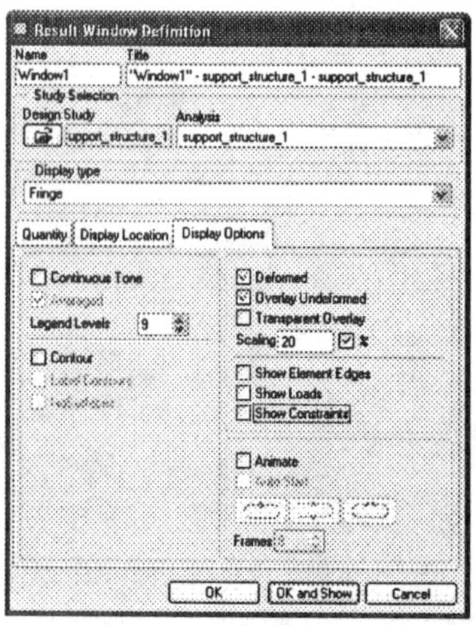

The deformation distribution

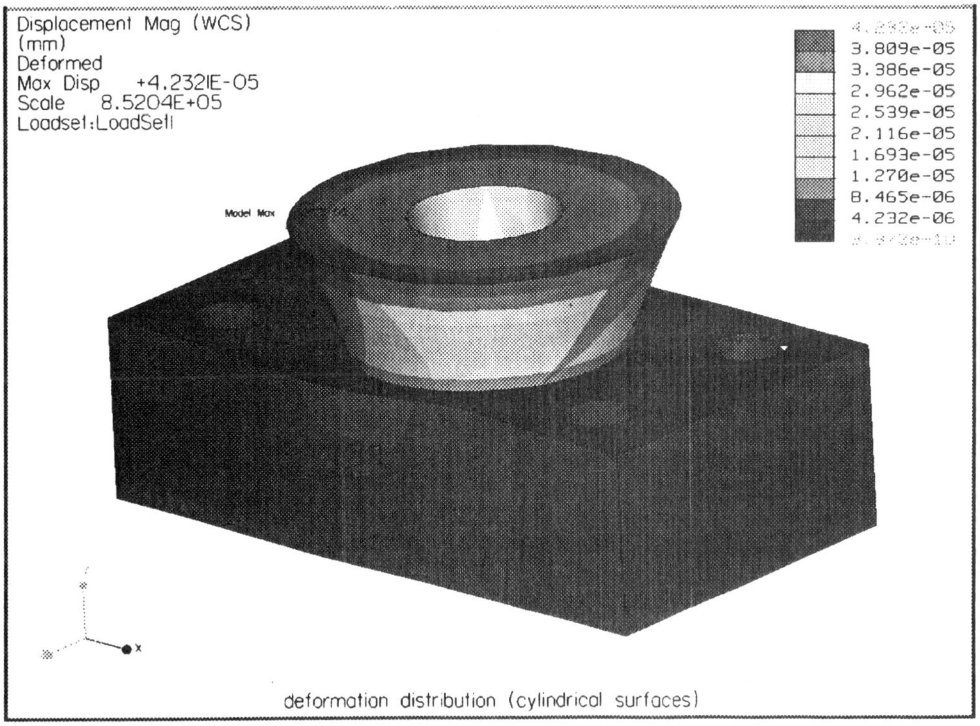

To display the distribution of the maximum principal stress, Edit > Copy > type in distribution of max prin stress (cylindrical surfaces) > change to **Stress** and **Max Principal > OK and Show.**

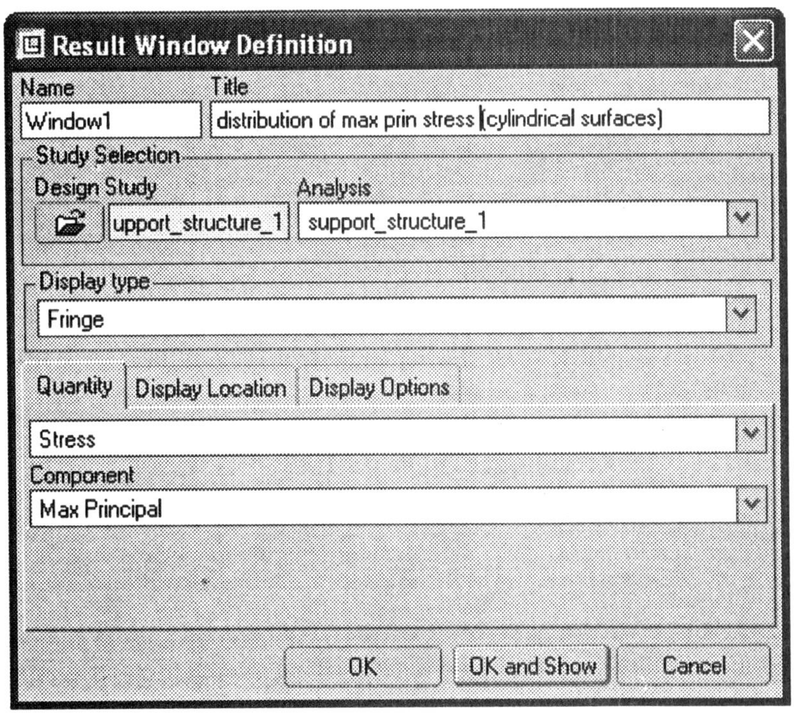

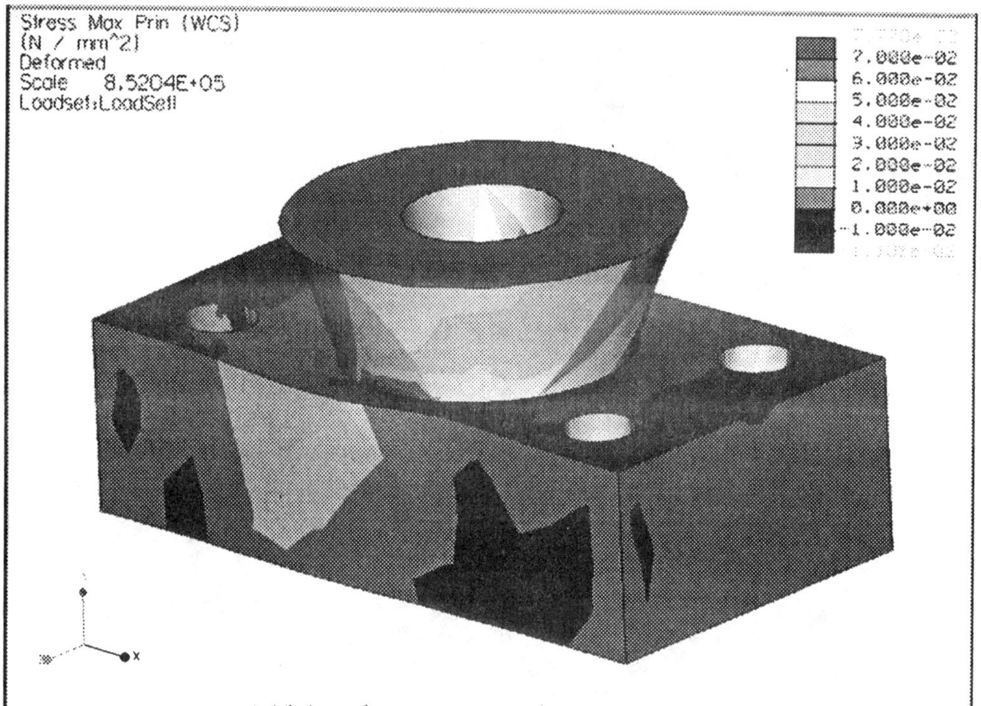

distribution of max prin stress (cylindrical surfaces)

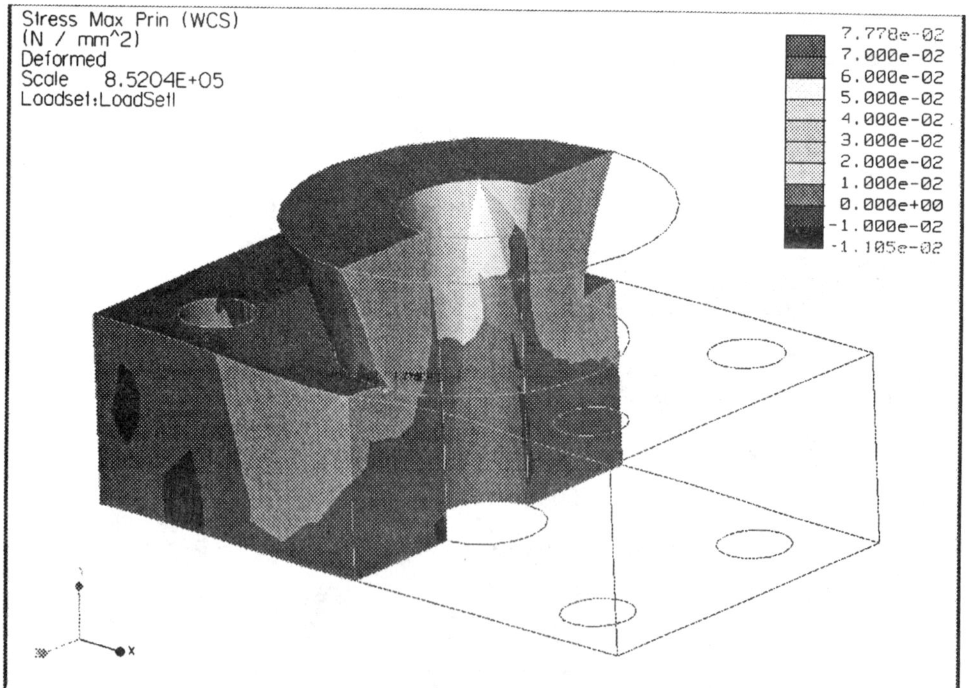

distribution of max prin stress (cylindrical surfaces)

4.3 Modeling of the Constraints Using Surface Regions

In this section, we continue our discussion on the support structure. We want to change the constraint condition from using the cylindrical surfaces to using 4 surface regions, which characterize the use of 4 washers in the real life of engineering. As we recall, we need to create datum curves prior to creating the surface regions.

Let us save the 3D model created in section using another name, say support_2. We continue our discussion with the new model called support_2.

Step 1: create a new file, name it as support_2 and open it.
From the main toolbar, select **File** > **Save a Copy** > type in *support_2* as the New Name > **OK**.

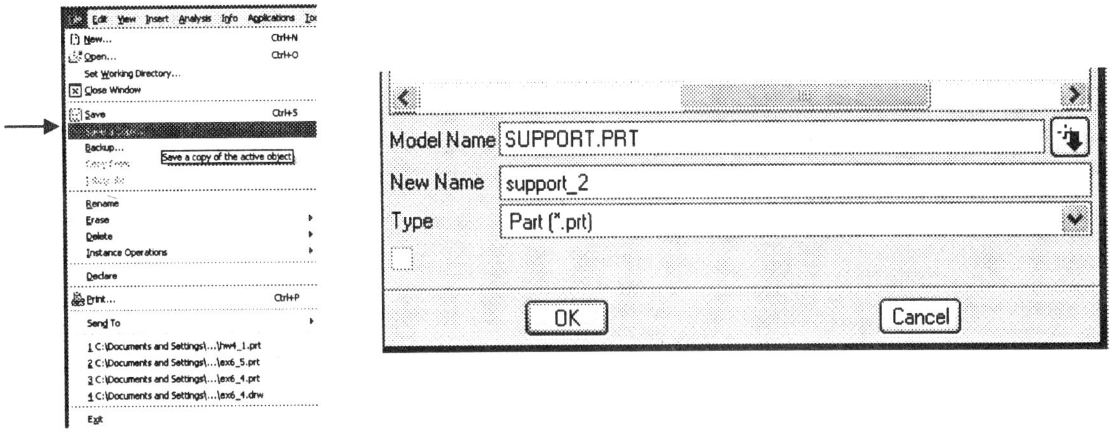

From the main toolbar, select **File > Open** > select *support_2* > **Open**.

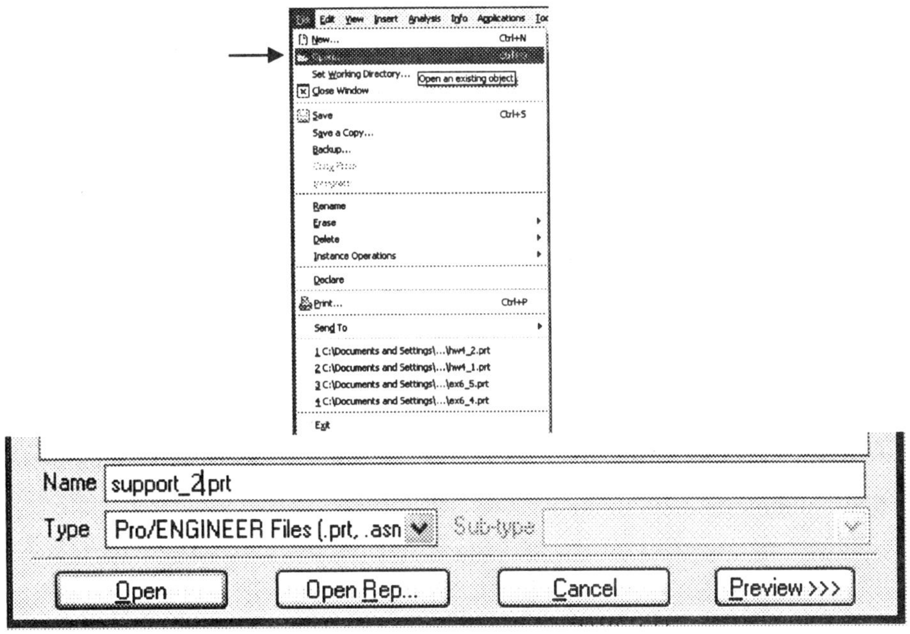

Step 2: create 4 datum curves through sketch. Select the icon of **Sketch** from the toolbar of datum feature creation

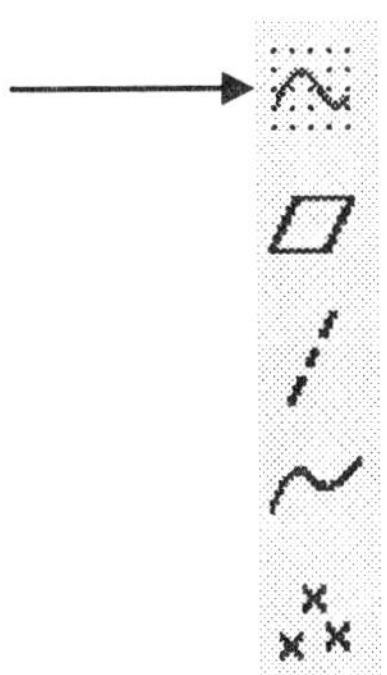

Select the top of the block feature as the sketch plane and accept the default setting for the orientation of the sketch plane > **Sketch**.

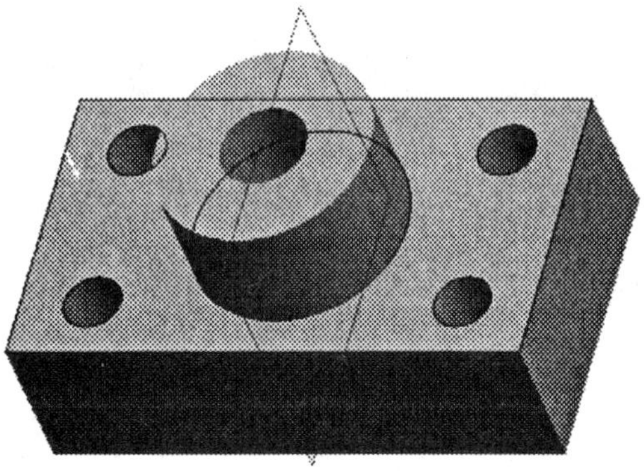

Select the cylindrical surface of the hole as a new reference, as shown > **Close** the window of References.

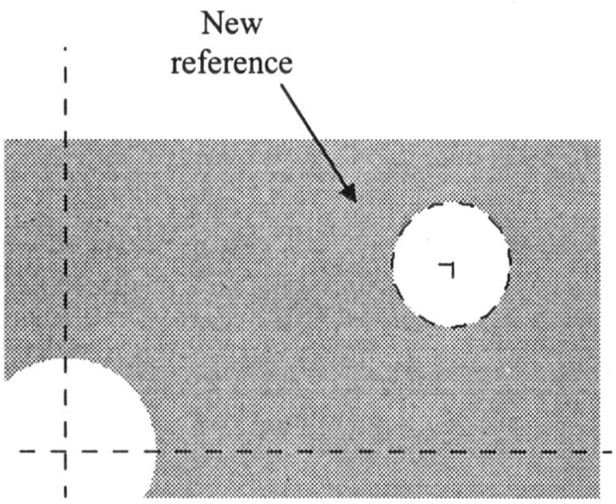

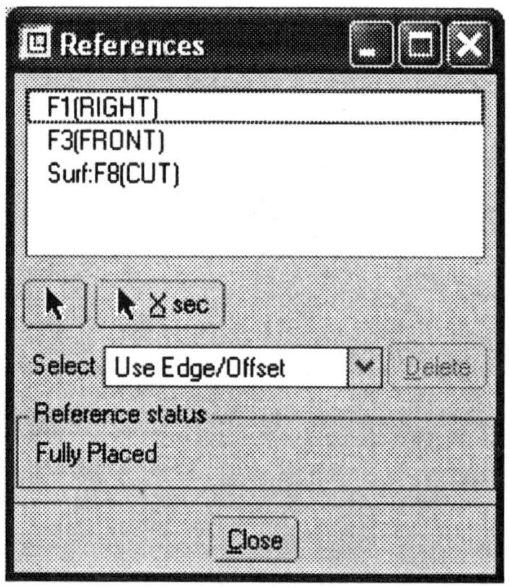

From the toolbar of sketcher, select the icon of **Circle** to sketch a circle. The center of the sketched circle is the center of the referenced circle. The diameter of the circle is 30 mm, which is the outer diameter of the washer used to hold the plate > select the icon of **Done** from the toolbar of sketcher.

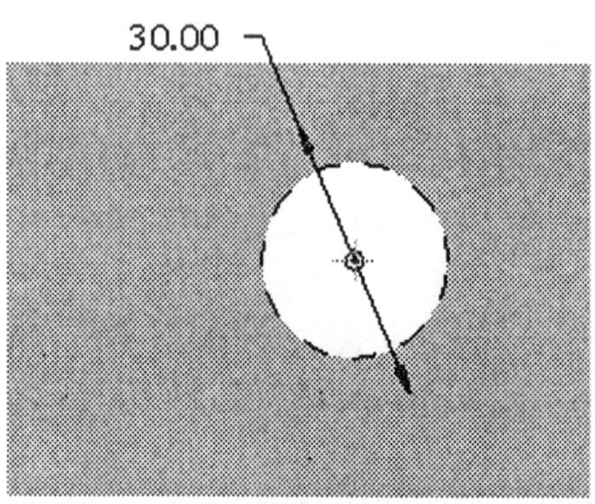

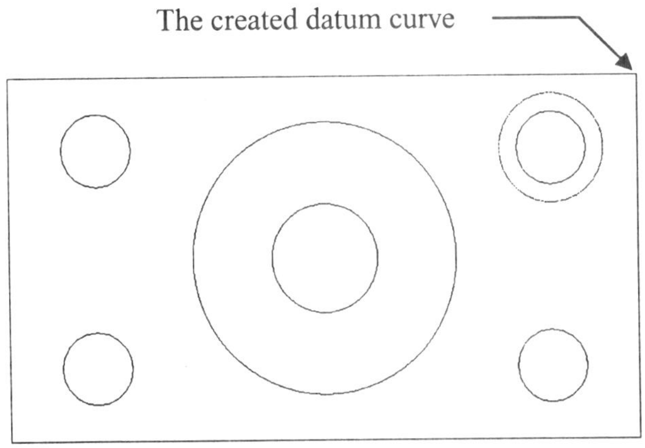

Repeat the above procedure three times to create the other three datum curves. It is important to create a single datum curve each time.

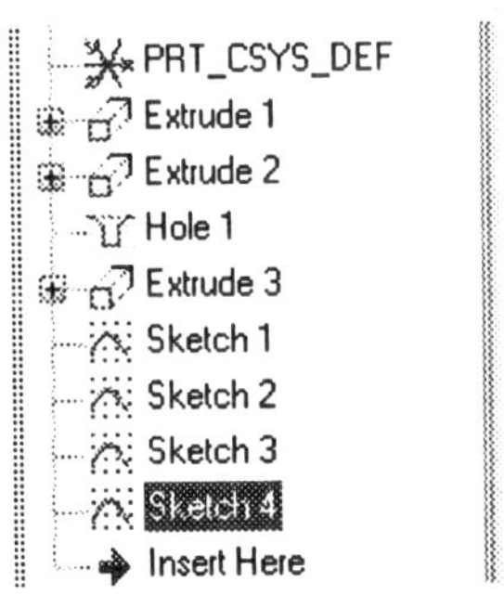

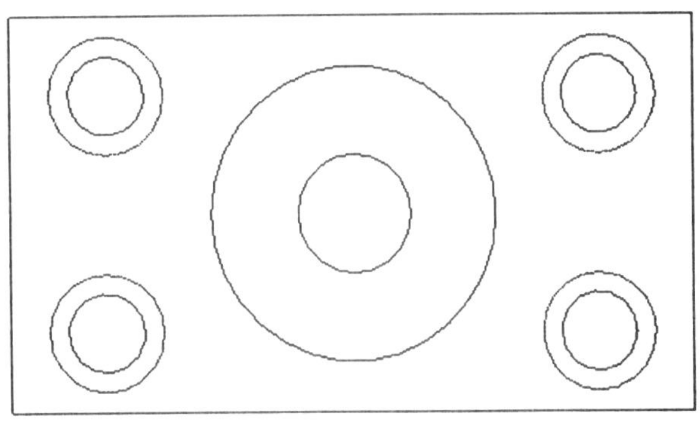

Step 3: Perform FEA under **Pro/Mechanica (material, constraints and loads)**

From the menu bar, select **Applications** > **Mechanica** > **Continue** after checking the unit system > **Structure.**

At this moment, the defined constraint conditions, the load conditions and the material selection are shown. We only need to define a new constraint set in this case.

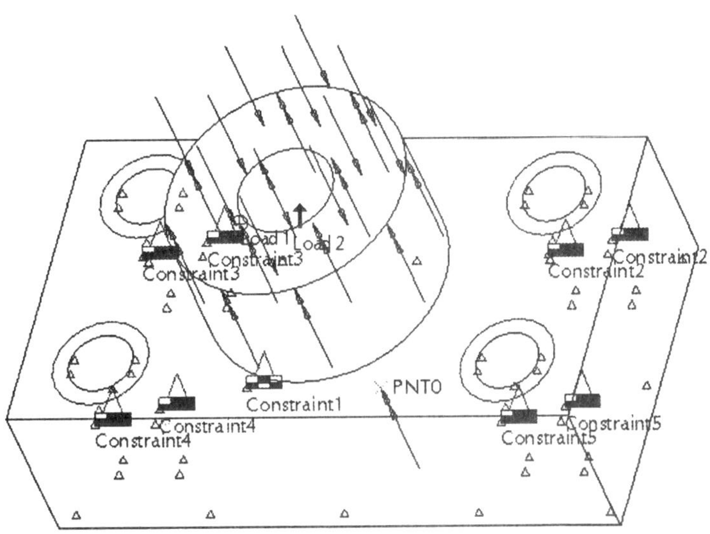

Let us first define 4 surface regions using the four sketched datum curves. To define the first surface region, select the icon of **surface region** from the toolbar of functions.

Select > Done. Pick the datum curve first, as shown > pick the surface to be split afterwards > **OK > Done > OK**.

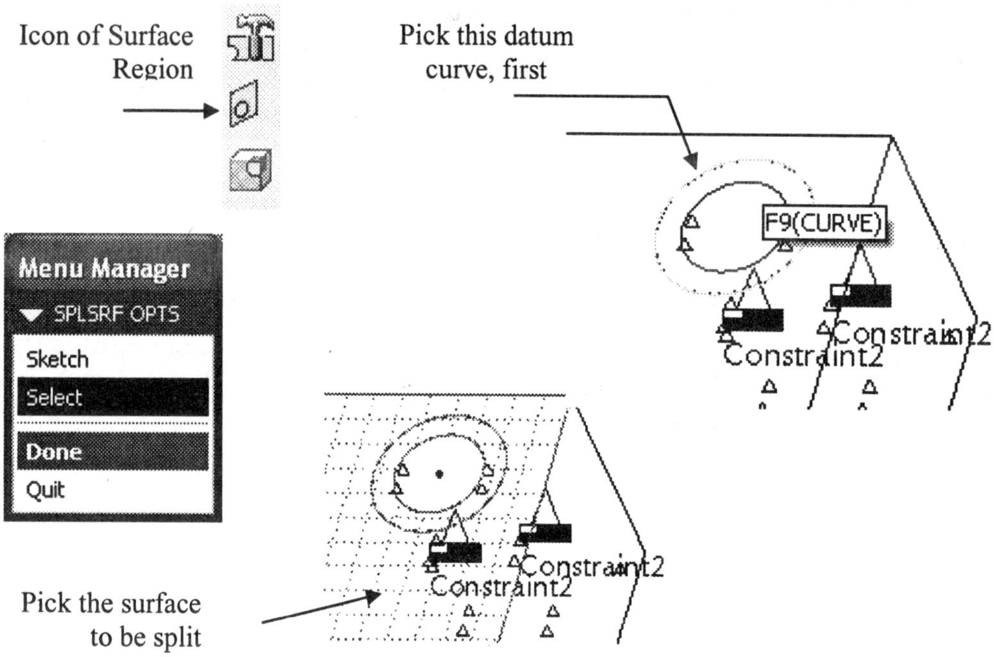

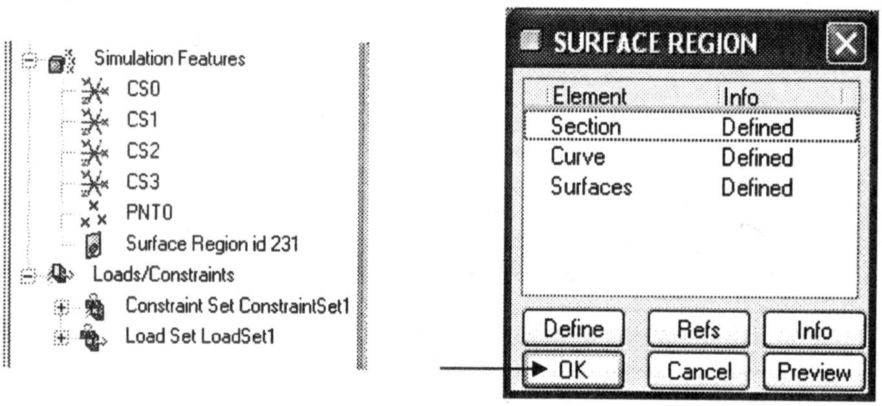

Repeat the above procedure three times to create the other 3 datum curves. It is important to create an individual surface region at a time.

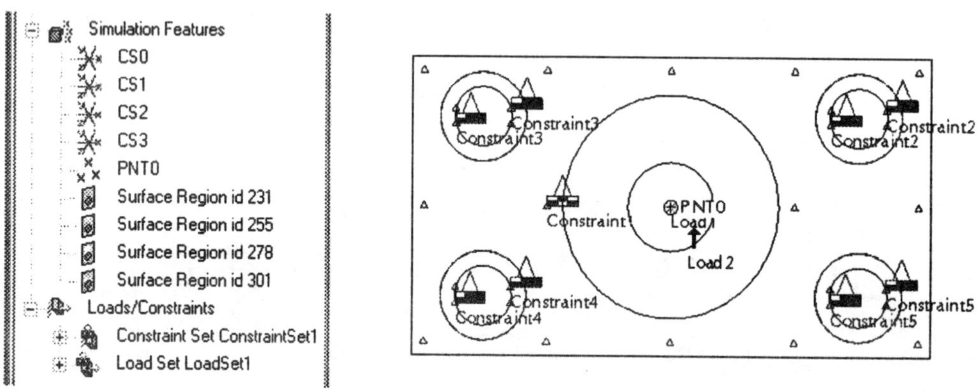

To define a new constraint set, select icon of Constraints > New > Surface > make sure to select New from Member of Set so that ConstraintSet2 appears > type a note on the second constraint set > **OK.**

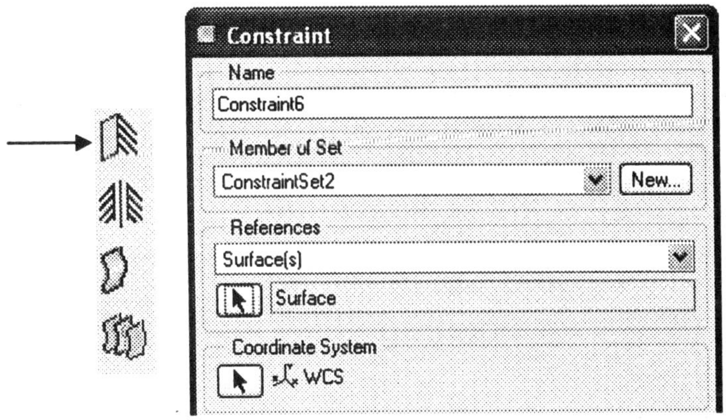

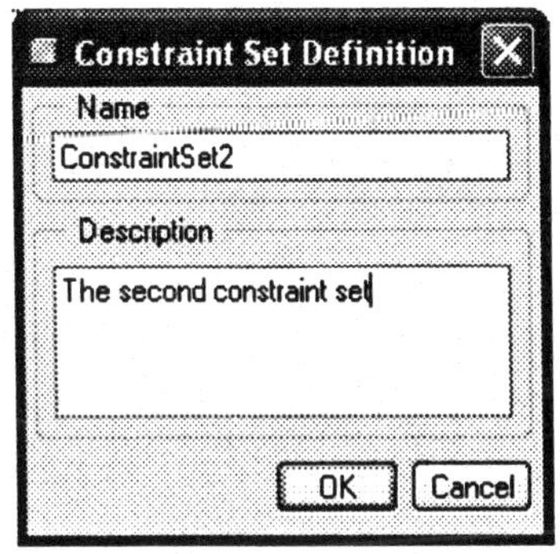

Activate the selection of surface > pick the surface region, as shown > OK.

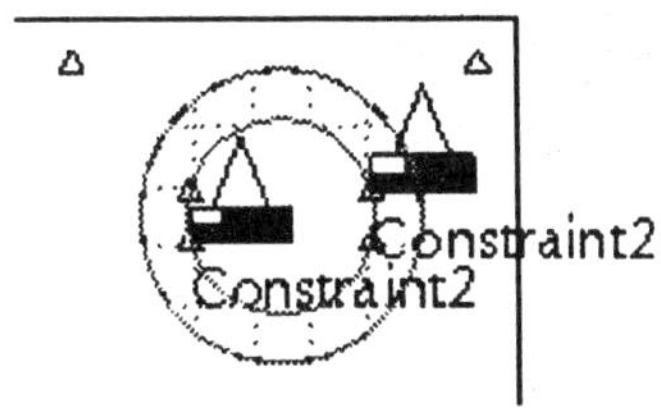

Activate the selection of coordinate system > pick CS0, as shown > **OK.**

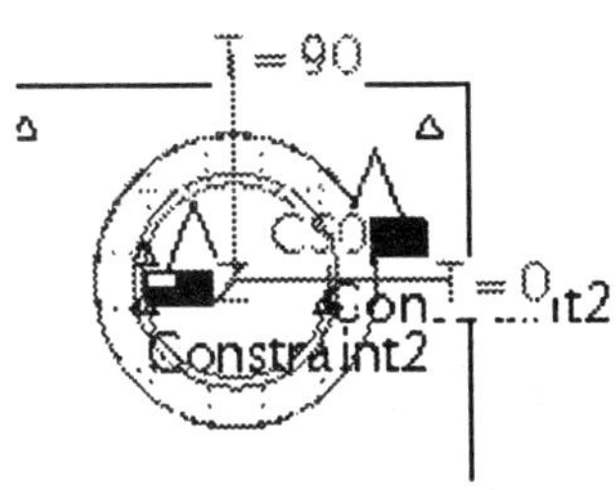

Set Translation R and Translation Z to free and fix the other 4 degrees of freedom, as shown > **OK** to complete Constraint6 in ConstraintSet2.

Repeat the above procedure to define Constraint7, Constraint8, and Constraint9, using the 3 other surface regions.

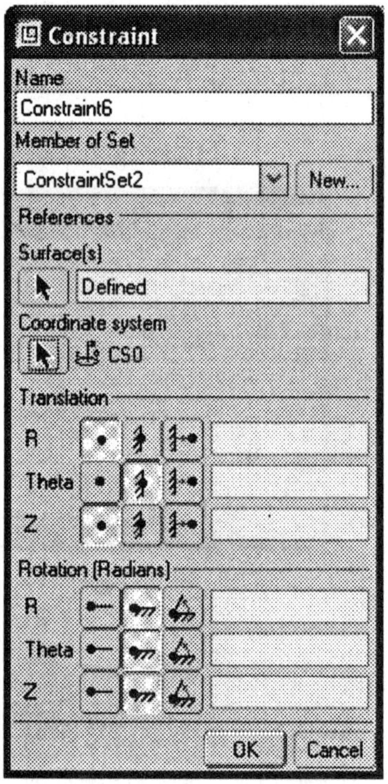
Constraint
Name
Constraint6
Member of Set
ConstraintSet2
New...
References
Surface(s)
Defined
Coordinate system
CS0
Translation
R
Theta
Z
Rotation (Radians)
R
Theta
Z
OK
Cancel

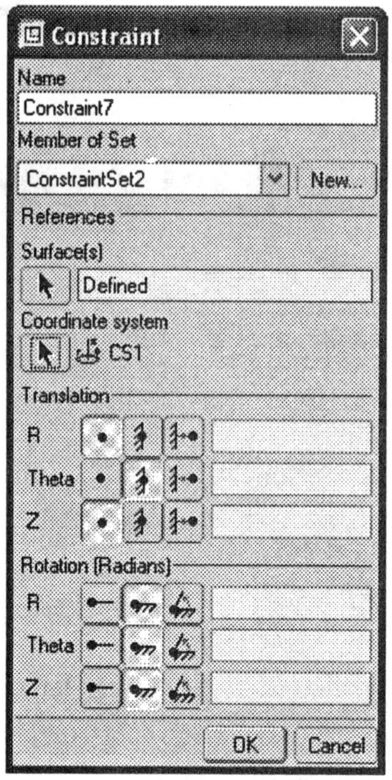
Constraint
Name
Constraint7
Member of Set
ConstraintSet2
New...
References
Surface(s)
Defined
Coordinate system
CS1
Translation
R
Theta
Z
Rotation (Radians)
R
Theta
Z
OK
Cancel

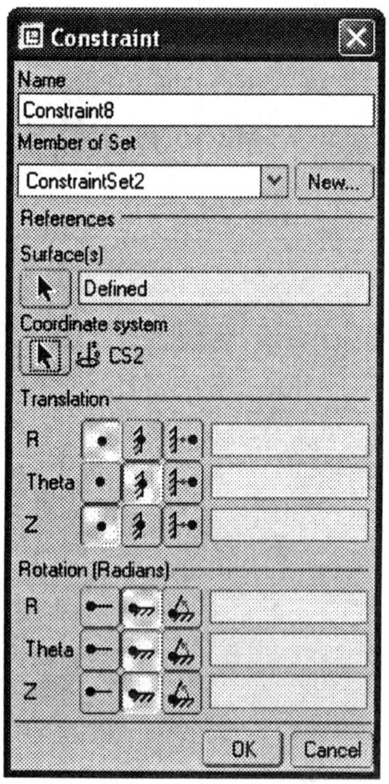
Constraint
Name
Constraint8
Member of Set
ConstraintSet2
New...
References
Surface(s)
Defined
Coordinate system
CS2
Translation
R
Theta
Z
Rotation (Radians)
R
Theta
Z
OK
Cancel

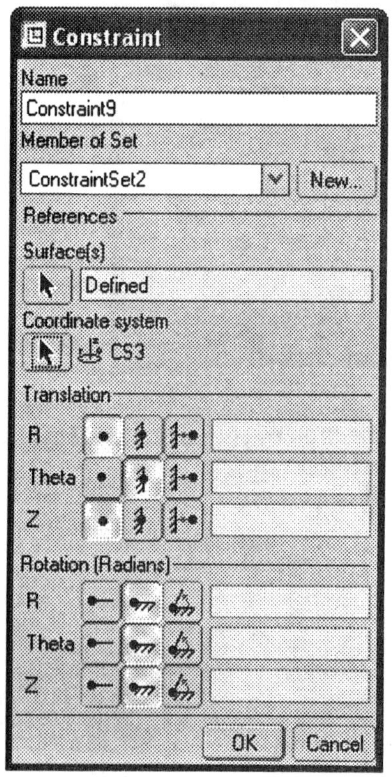
Constraint
Name
Constraint9
Member of Set
ConstraintSet2
New...
References
Surface(s)
Defined
Coordinate system
CS3
Translation
R
Theta
Z
Rotation (Radians)
R
Theta
Z
OK
Cancel

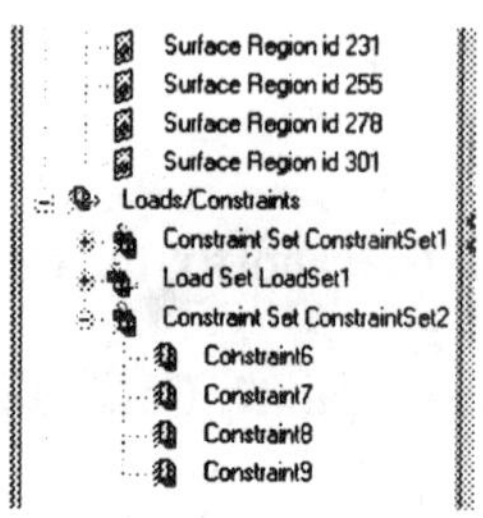
Surface Region id 231
Surface Region id 255
Surface Region id 278
Surface Region id 301
Loads/Constraints
Constraint Set ConstraintSet1
Load Set LoadSet1
Constraint Set ConstraintSet2
Constraint6
Constraint7
Constraint8
Constraint9

At this stage, it is important to define a new constraint within ConstraintSet2. The new constraint is to model the boundary condition on the bottom surface of the plate. Select Constraint > New > Surface > make sure that ConstraintSet2 is shown > pick the bottom surface of the plate. Set Transition X and Transition Z to free and fix the other 4 degrees of freedom.

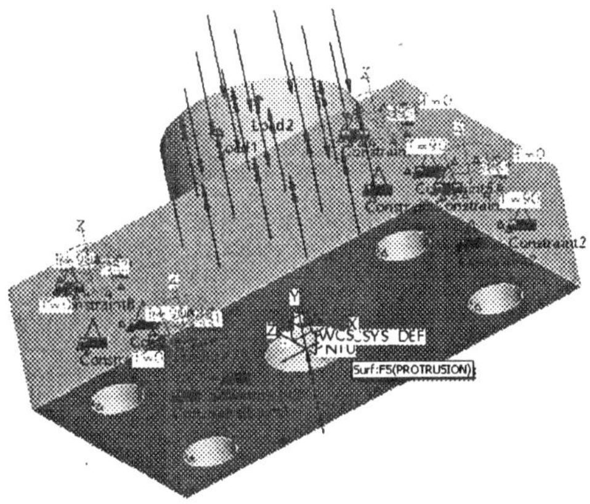

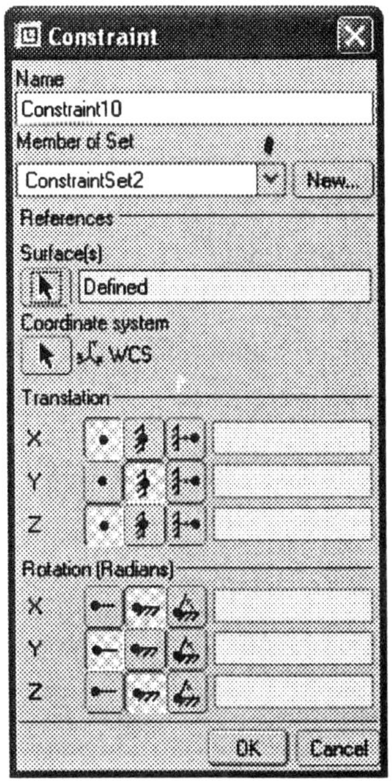

Perform a static analysis

Click the icon of performing a design study from the main toolbar > **File** > **Select New Static Analysis.**

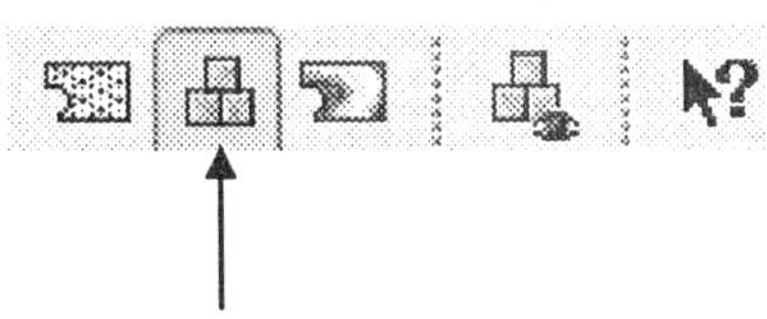

Icon of performing a design study

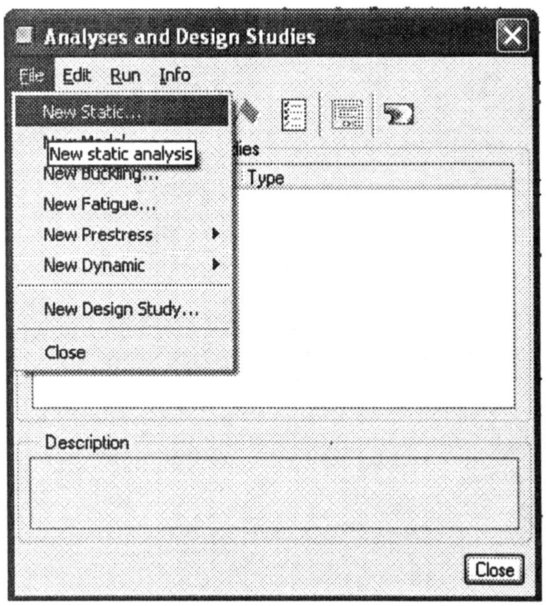

Type in *support_structure_2*

Make sure that ConstraintSet2 and LoadSet1 are highlighted > **Single-Pass Adaptive** > **OK** > **Run**

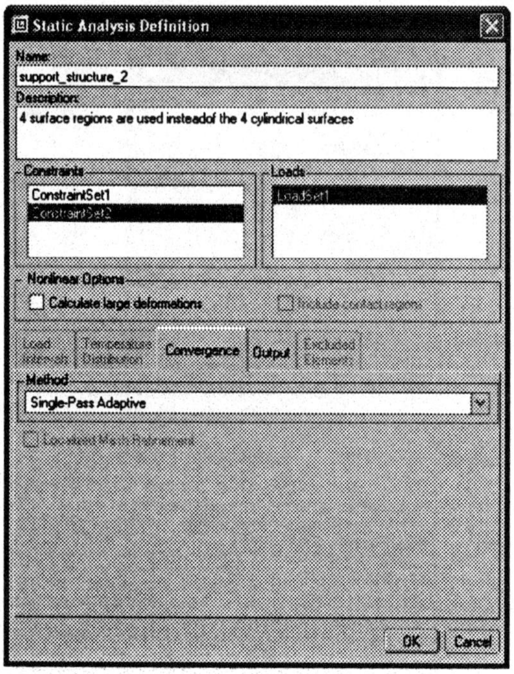

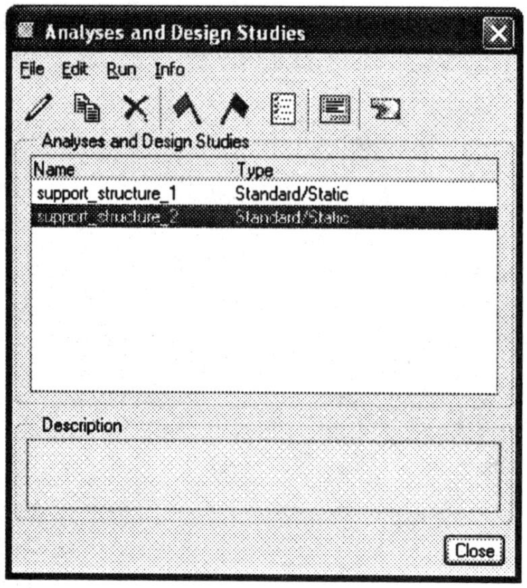

Structure > **Results** > **Insert** > **Result Window** > type in deformation distribution (surface regions) as the title > select the folder called support_structure_2 > **Fringe** > **Displacement** > **Magnitude** > in the Display Option, select Deformed and set the scale to *10* > **OK and Show.**

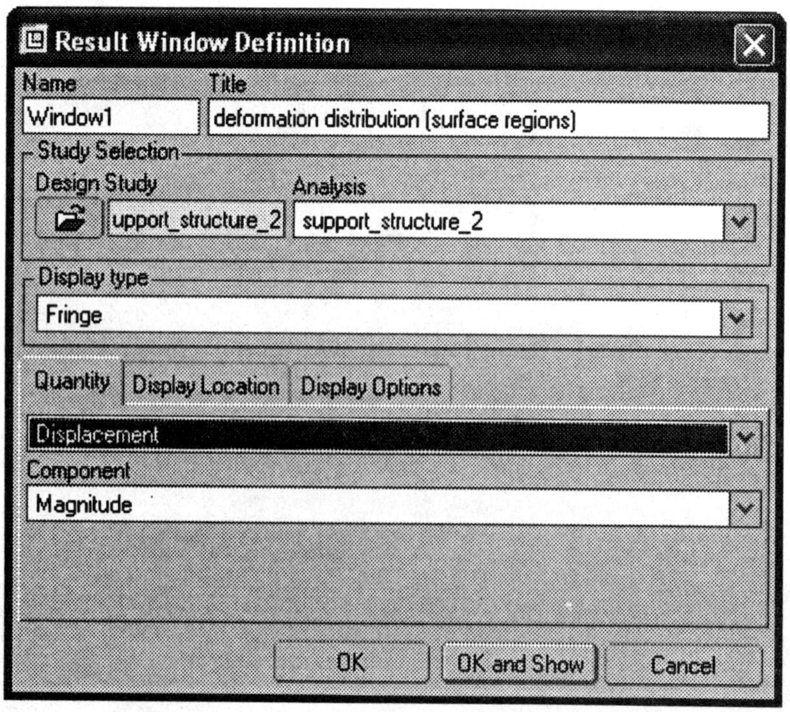

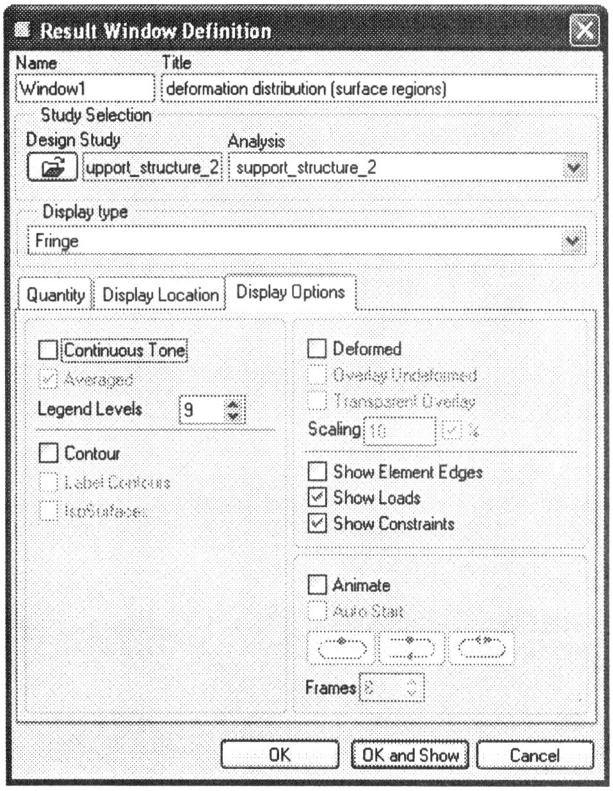
Result Window Definition
Name
Title
Window1
deformation distribution (surface regions)
Study Selection
Design Study
Analysis
upport_structure_2
support_structure_2
Display type
Fringe
Quantity
Display Location
Display Options
Continuous Tone
Averaged
Legend Levels
9
Contour
Label Contours
IsoSurfaces
Deformed
Overlay Undeformed
Transparent Overlay
Scaling
Show Element Edges
Show Loads
Show Constraints
Animate
Auto Start
Frames
OK
OK and Show
Cancel

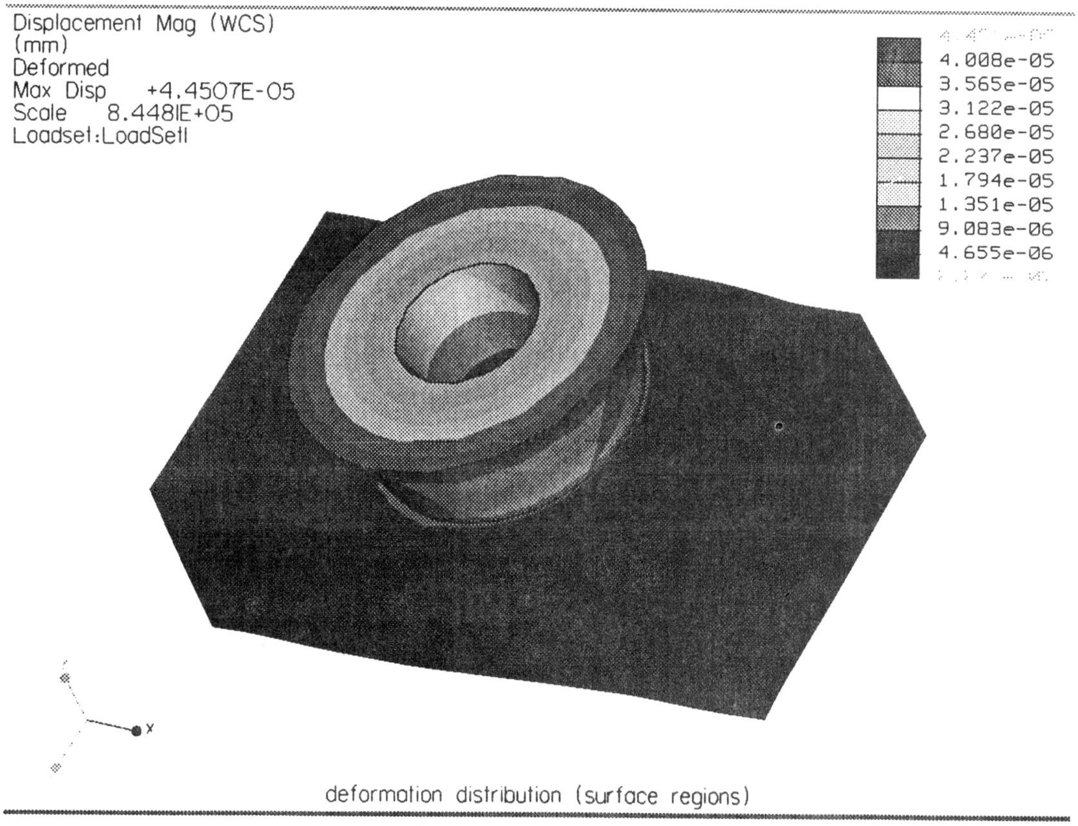
Displacement Mag (WCS)
(mm)
Deformed
Max Disp +4.4507E-05
Scale 8.4481E+05
Loadset:LoadSet1
4.008e-05
3.565e-05
3.122e-05
2.680e-05
2.237e-05
1.794e-05
1.351e-05
9.083e-06
4.655e-06
X
deformation distribution (surface regions)

Distribution of the Maximum Principal Strain

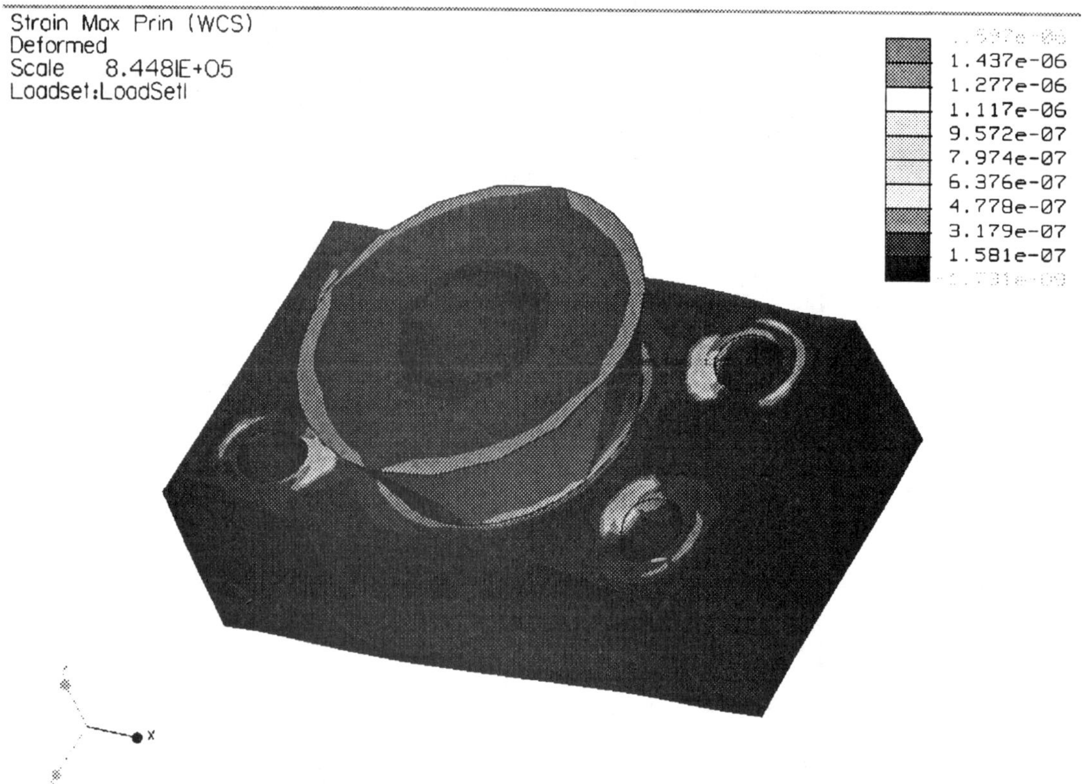

deformation distribution (surface regions)

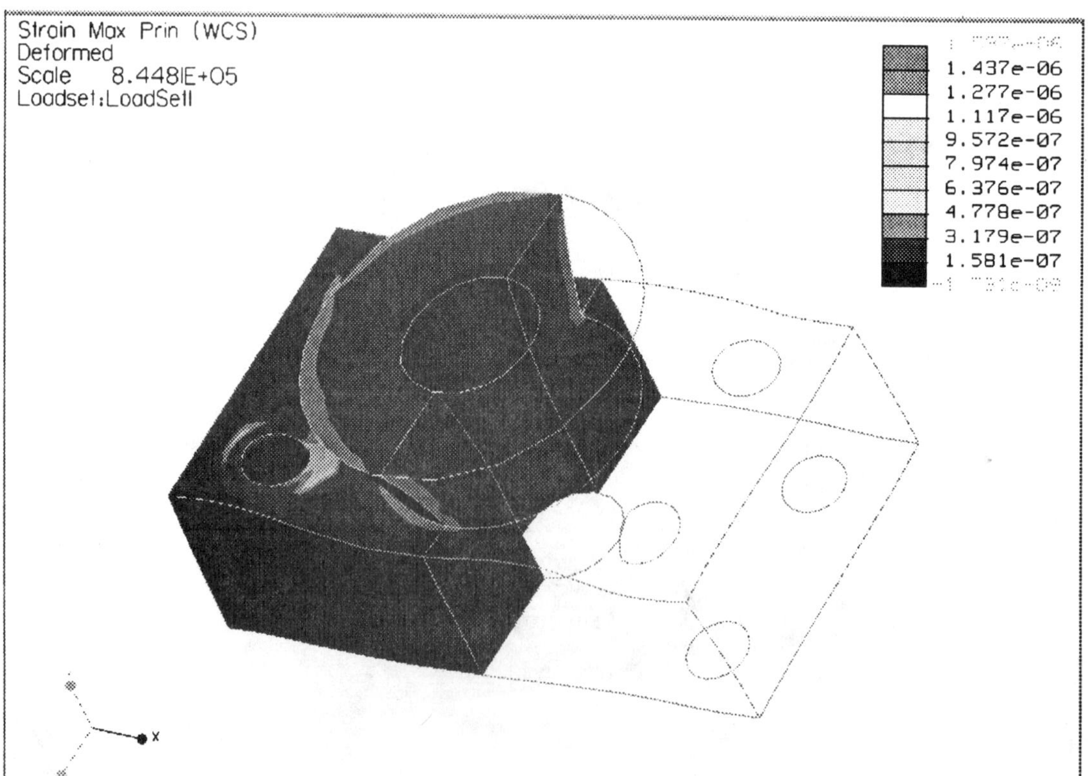

deformation distribution (surface regions)

Distribution of the Maximum Principal Strain (cutting surface)

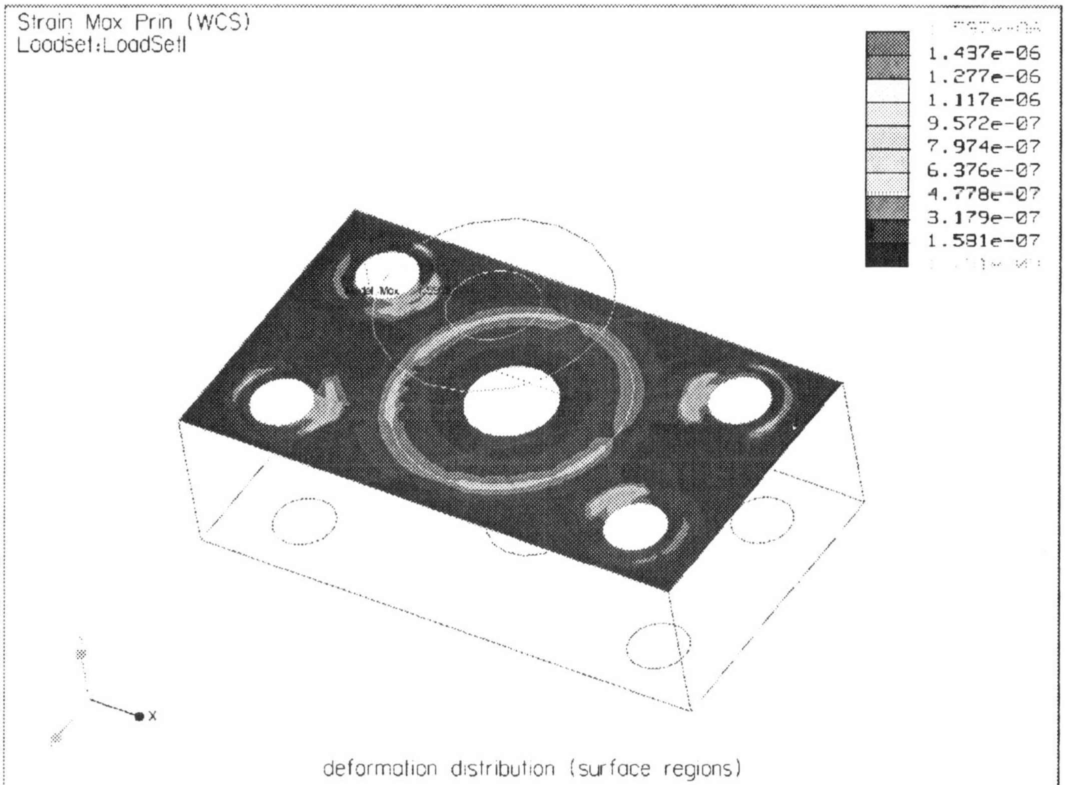

Distribution of the Von Mises Stress

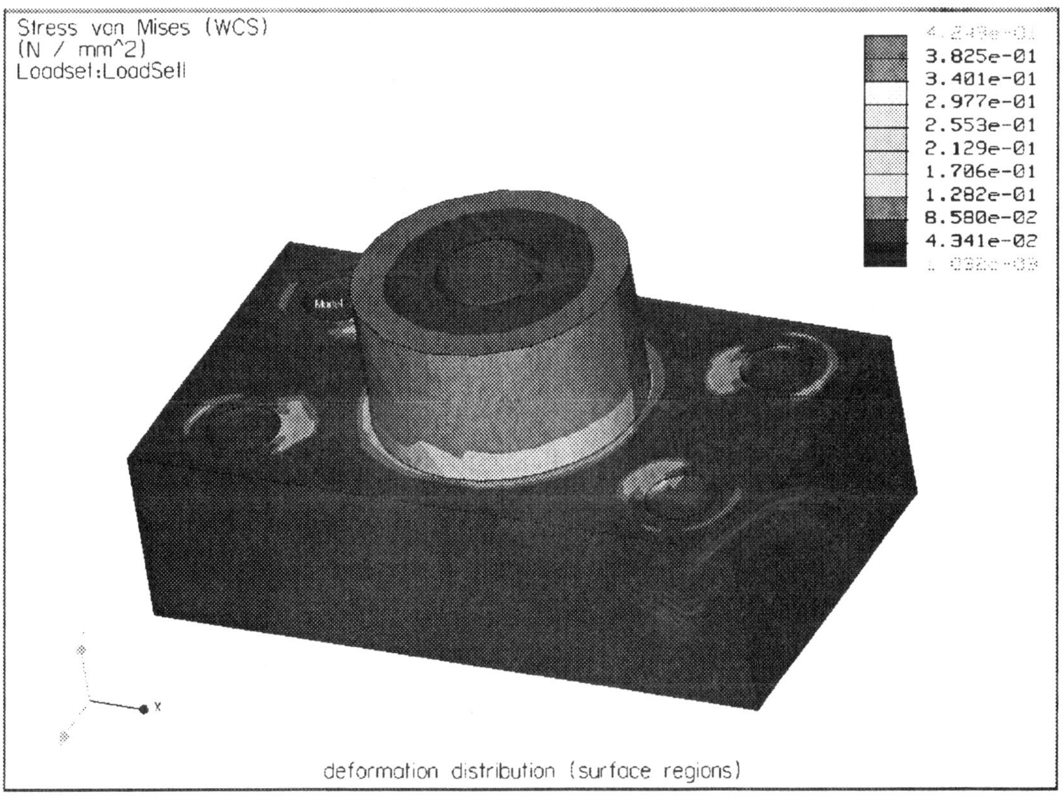

Now let us list the results obtained from the FEA evaluation in section 4.2 and the FEA evaluation in this section. In terms of the numerical values, there is little difference in the maximum displacement. However, there are certain significant differences in the maximum principal stress and the maximum von Mises stress developed under the 2 loading conditions between the 2 evaluations.

Table 4-1 The results Obtained from the 2 FEA studies.

	max principal stress (Mpa)	max von Mises (Mpa)	max displacement (μm)
Case Study 1	0.07778	0.2133	0.04232
Case Study 2	0.25999	0.4246	0.04487

4.4 FEA with an Assembly of Gear and Shaft

The following figure illustrates a gear assembly. The assembly consists of a helical gear and a shaft. The fitting between the gear component and the shaft component is an interference fit so that the force acting on a tooth or teeth of the gear is capable of driving the shaft to rotate. The material of the gear and the material of the shaft are steel material. Let us assume that only one pair of teeth is in contact, and the force acting on the tooth of the gear is 150 Newton.

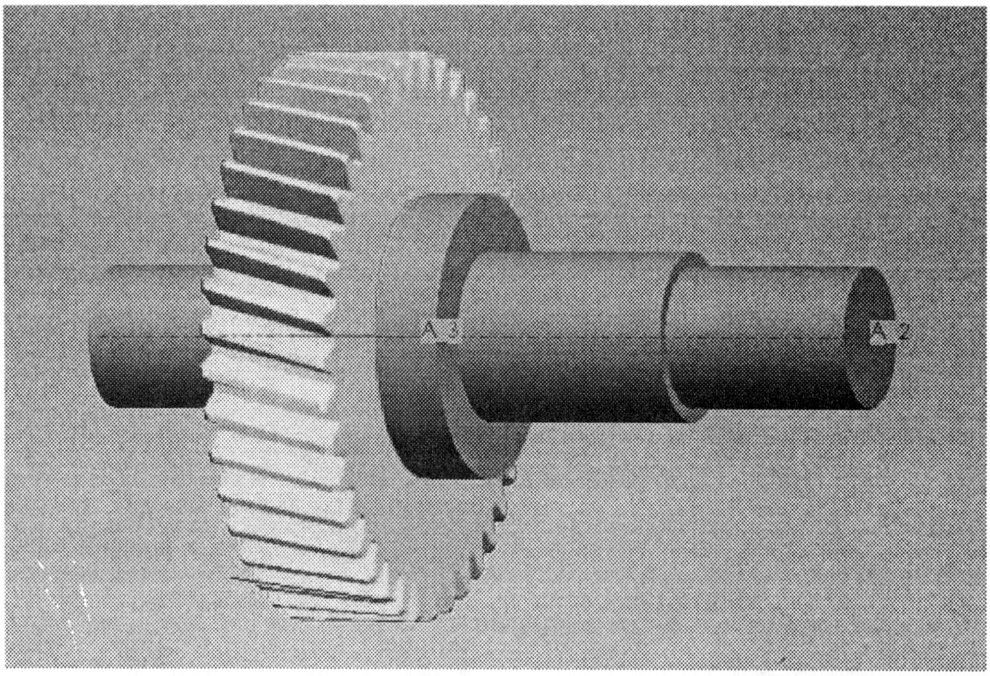

During operations of gear transmissoin, there are two failure modes, which are commonly observed. The first failure mode is the breakage of the tooth of the helical gear during operation. Such breakages occur when the force acting on the tooth induces excessively large stresses. The second failure mode is the twist of the shaft when subjected to torque due to the force acting on the tooth of the gear. In this example, we start with the study of analyzing the first failure mode. In this study, only the gear component is under consideration. To study the second failure mode, we take the entire assembly under consideration.

An engineering drawing of the gear component is shown below. We will use the shown dimensions to construct a 3D solid model of the gear component.

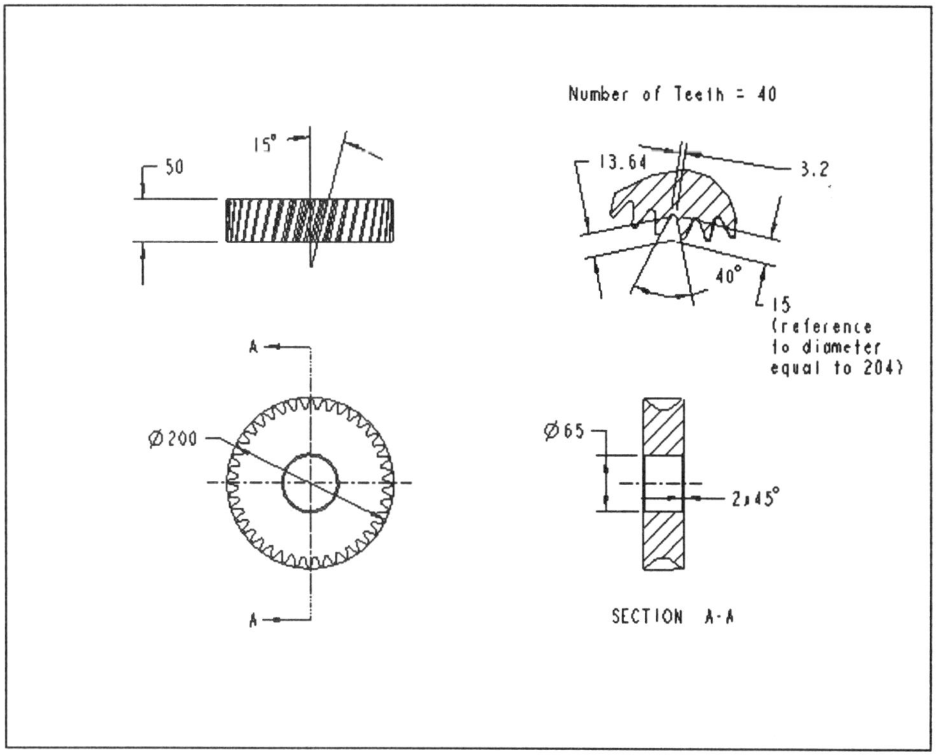

An engineering drawing of the shaft component is shown below. We will use the shown dimensions to construct a 3D solid model of the shaft component.

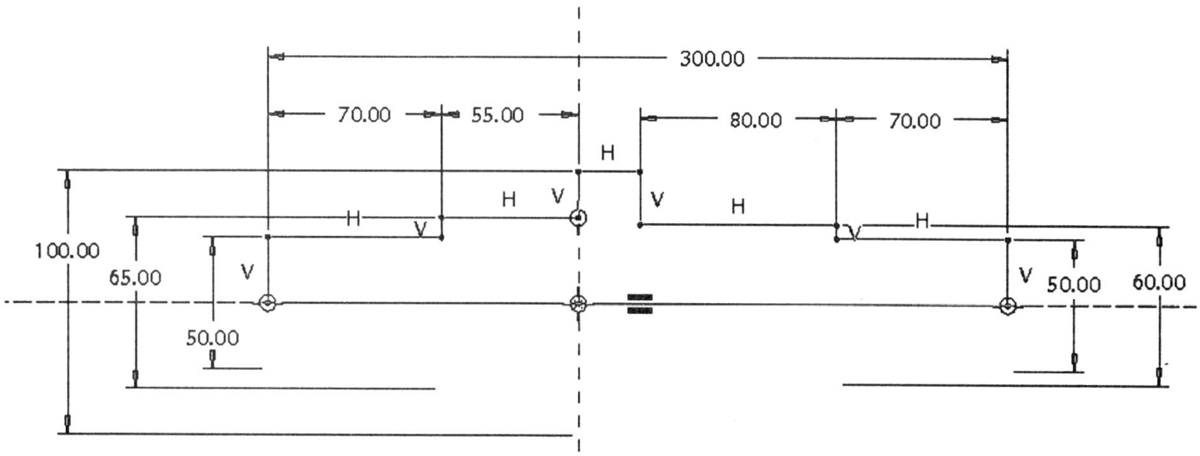

We will use an assemble process to assemble the gear component and the shaft component together before getting on performing FEA.

Step 1: Under the Pro/ENGINEER design environment, create a 3D solid model of the helical gear. The number of teeth is 40. **File > New > Part** > type *helical_gear* as the file name and clear the icon of **Use default template > OK.** Select **mmns_part_solid** (units: Millimeter, Newton, Second) and type *helical gear* in **DESCRIPTION**, and student in **MODELED_BY**, then **OK.**

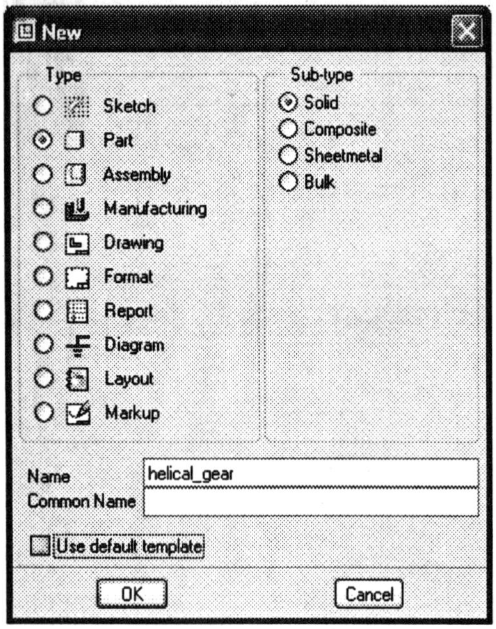

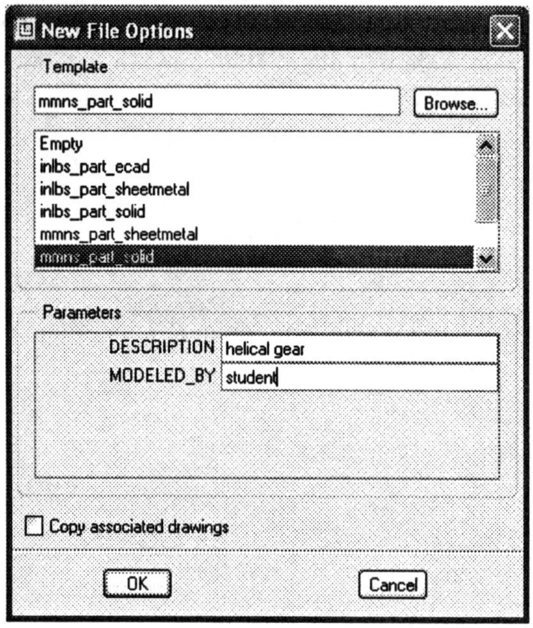

Let us create a solid cylinder: ϕ200x50 with a hole in the central location. The diameter of the hole is 65 mm. Directly select the icon of **Extrude** from the toolbar of feature creation. Make sure the icon of **Solid** is selected. Set the symmetric requirement and the depth value to 50. Click the box of **Placement > Define**.

Icon of **Extrude**

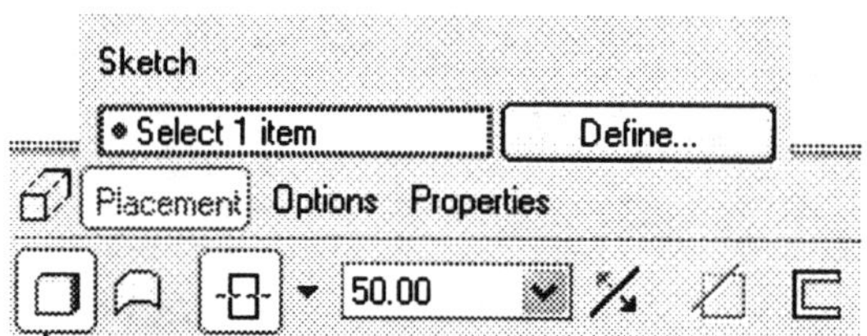

Icon of **Solid**

Select **FRONT** as the sketch plane and accept **RIGHT** to orient the sketch plane.

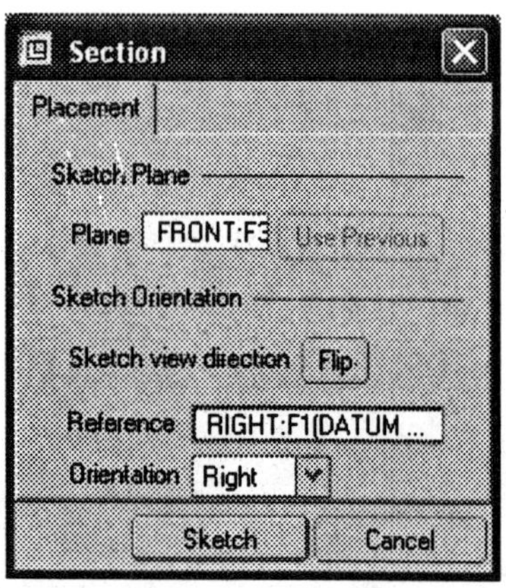

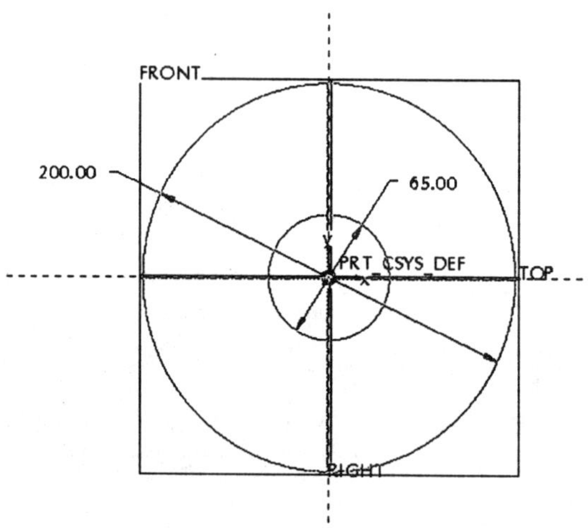

Sketch two circles. Their diameters are 200 and 65, respectively. Upon completion, select the icon of **Done** from the toolbar of sketcher, and select the icon of **Complete** from the feature control panel.

Select the icon of **Complete** to complete the feature creation

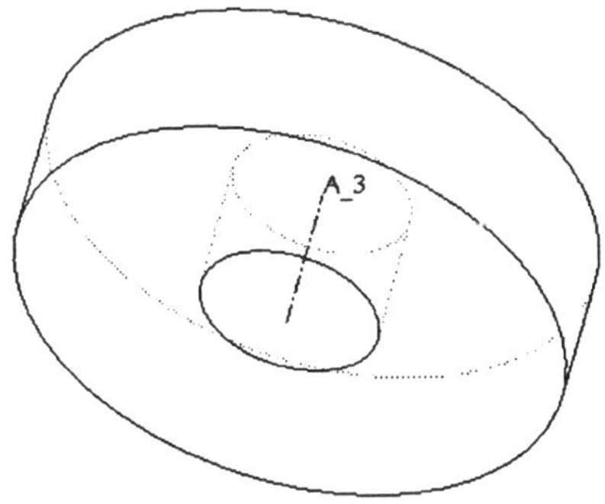

Let us add chamfers at the sharp corner with dimension equal to 2x45°. Click the icon of **Chamfer** from the toolbar of feature creation > select 45xD and type *2* as the dimension value > pick one of the 2 edges on the outer circumference first, afterwards, holding down the Ctrl key to pick the second edge > select the icon of Complete from the feature control panel.

Icon of Chamfer

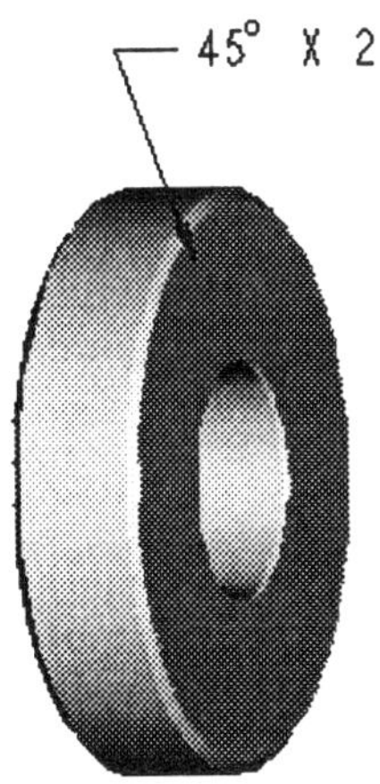

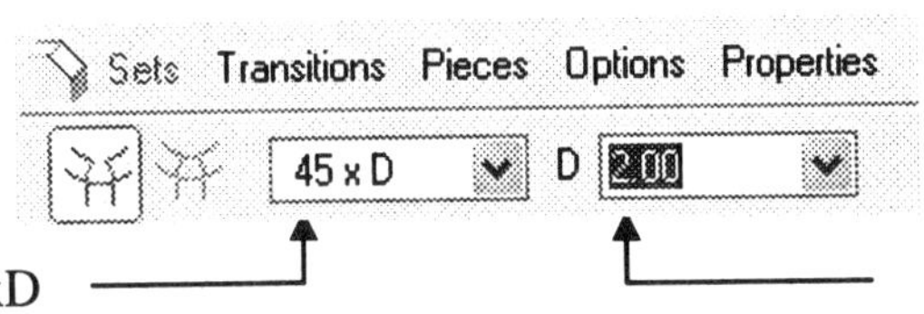

Select 45xD

Type in 2

To create teeth with a helical angle, we need a reference surface, which just goes over the cylindrical plate created. Let us assume the diameter is 204 and the width is 60.

Select the icon of **Extrude** from the toolbar of feature creation > select the icon of **Surface** > set the depth choice to Symmetry and type in 60 as the length of extrusion.

Icon of **Extrude**

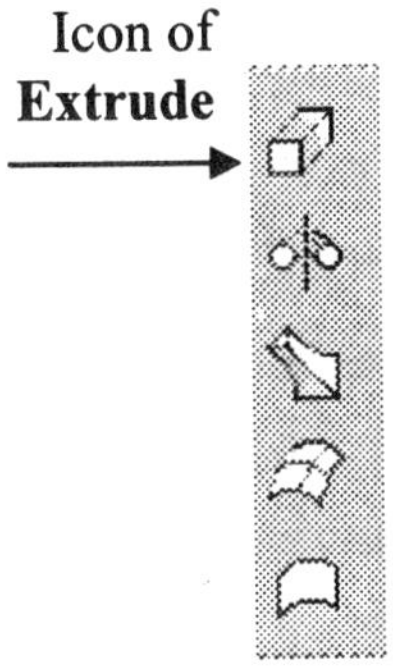

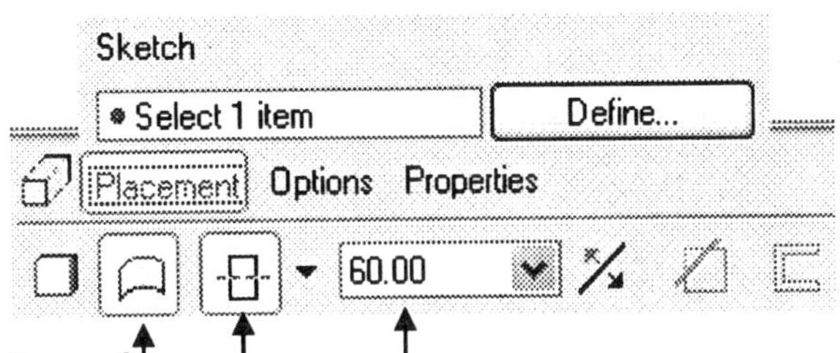

Icon of **Surface**

Symmetry

Depth setting

Select the **FRONT** datum plane as the sketch plane and accept the default setting to orient the sketch plane. Sketch a circle and the diameter dimension is 204.

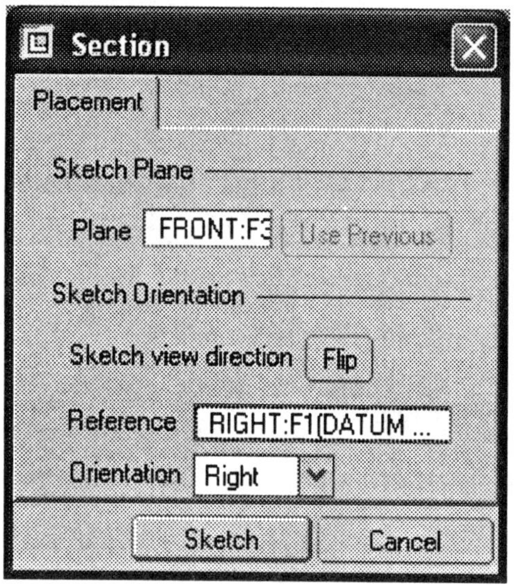

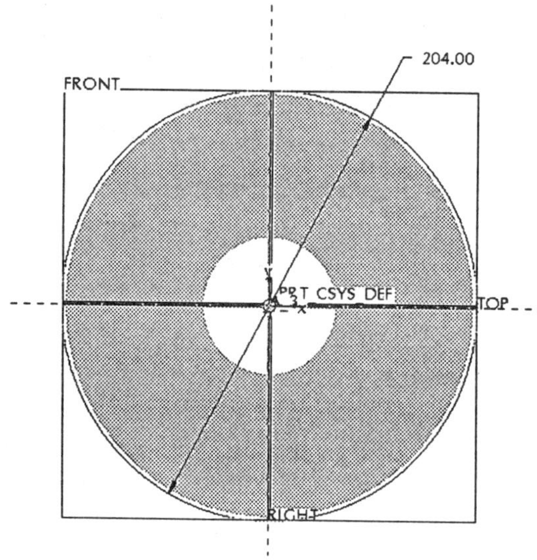

Upon completion, select the icon of **Done** from the toolbar of sketcher and select the icon of **Complete** from the feature control panel.

Select the icon of **Complete** to complete the feature creation

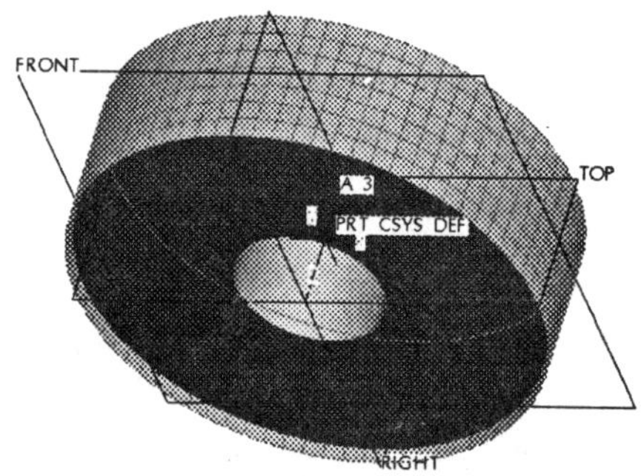

Let us sketch a datum curve (a straight line) to define the helical angle. From the toolbar of datum feature creation, select the icon of Sketch Tool. Select the **TOP** datum plane as the sketch plane. Select the RIGHT datum plane to orient the sketch plane.

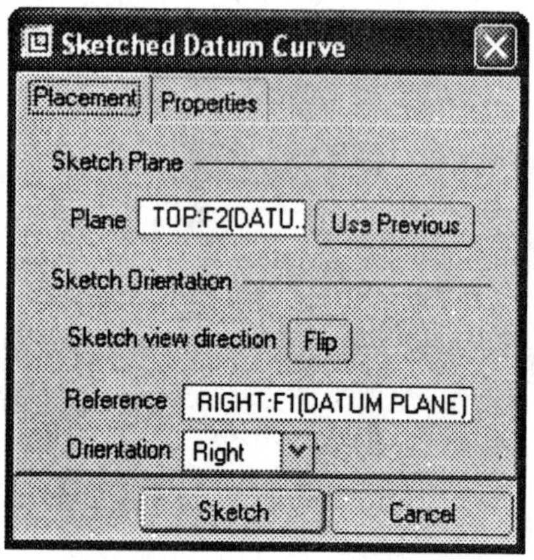

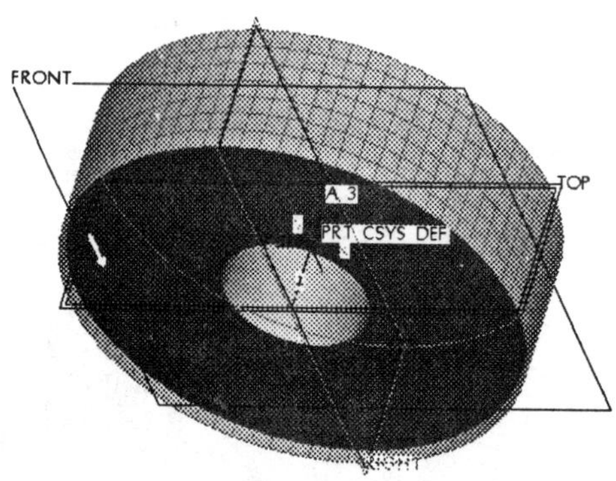

Before sketch a straight line, define a coordinate system first at a distance of 60, as shown. Afterwards, select the icon of **Line** from the toolbar of sketcher, and sketch a straight line at an angle with respect to the vertical reference or the y-axis. The angle dimension of the straight line is 15 degrees, and the length is set by specifying a dimension of 30, as shown. Upon completion, select the icon of **Done** from the toolbar of sketcher.

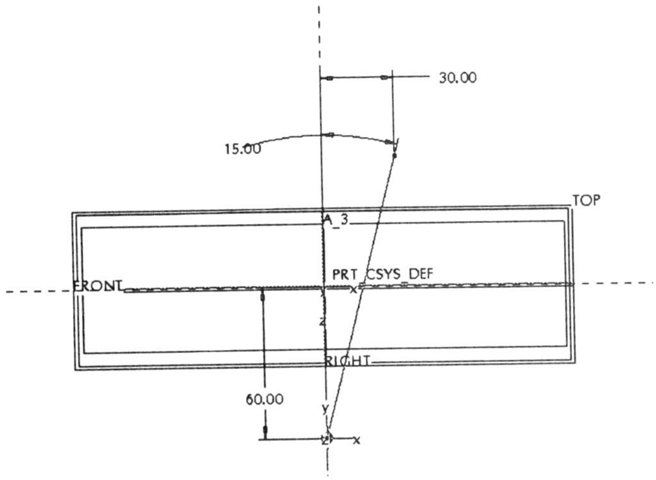

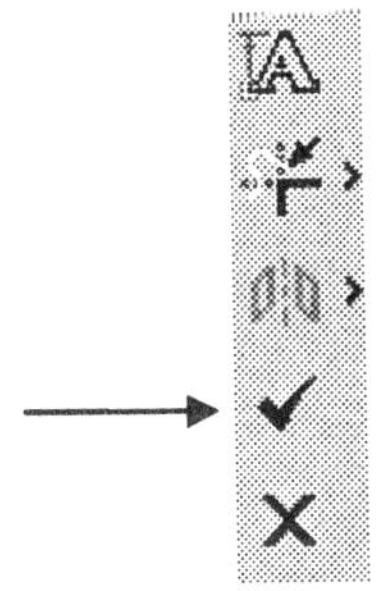

Now we need to project this straight line to the cylindrical surface. Highlight the datum curve just created. From the main toolbar, select **Edit** > **Project** > pick the upper portion of the cylindrical surface (do not pick the upper surface of the solid cylinder) so that a projected datum curve shown on the upper portion of the cylindrical surface > select the icon of **Complete** from the feature control panel.

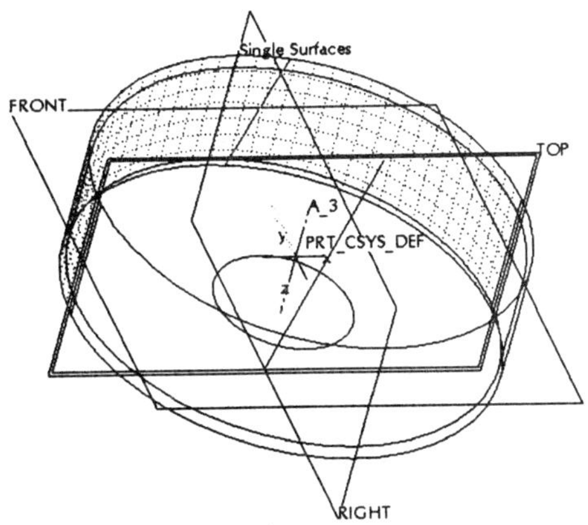

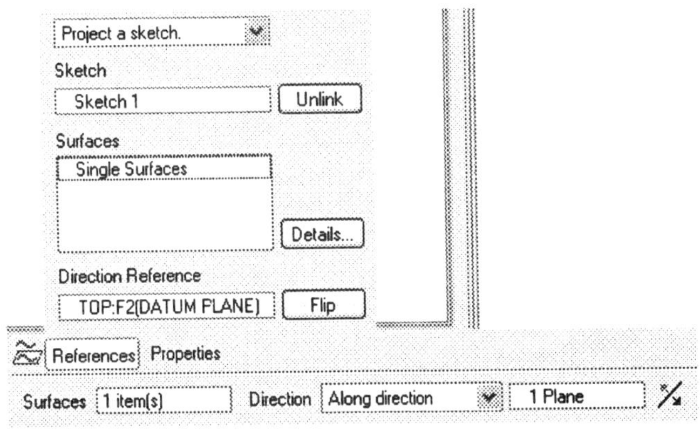

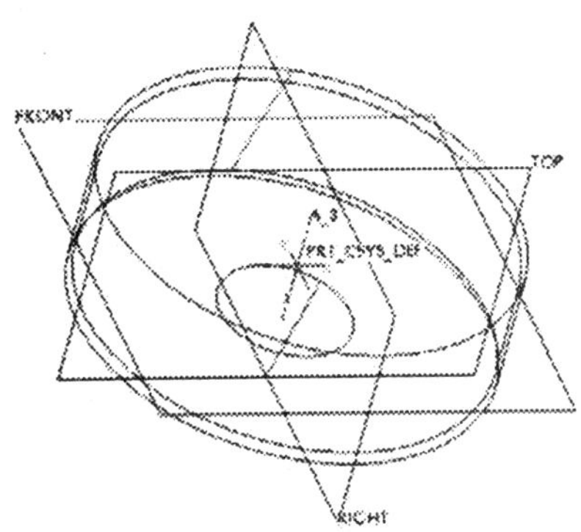

Now let us use the function called **Sweep** to cut the first tooth. From the main toolbar, select **Insert** > **Sweep** > **Cut** > **Select Traj** > **One by One** > Select pick the projected datum curve > **Done** > **OK**. Sketch a vertical centerline first.

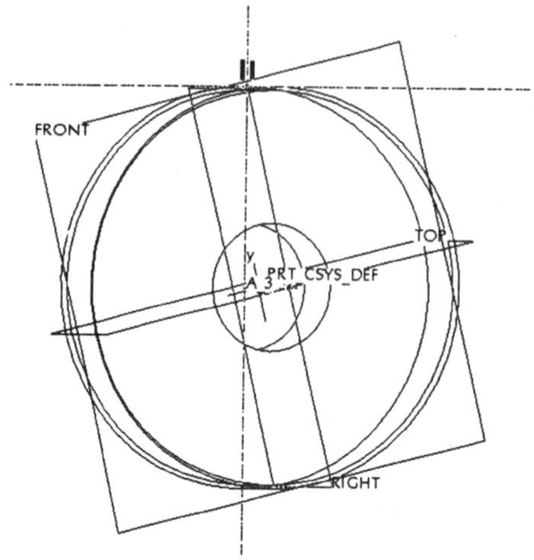

Sketch the profile of the tooth, paying attention to the symmetric issue so that the number of dimensions needed is minimized > select the icon of **Done** > **Okay** to accept the cut direction for material side > **OK**.

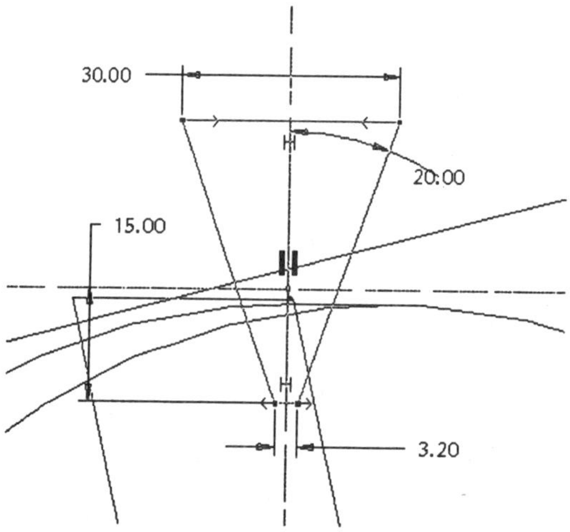

Upon completion, select the icon of **Done** from the toolbar of sketcher, and select the icon of **Complete** from the feature control panel.

In order to show the first cut tooth, we need to hide the cylindrical surface feature. From the model tree, highlight the cylindrical surface feature > right-click and hold, select **Hide.** The first cut tooth is clearly depicted.

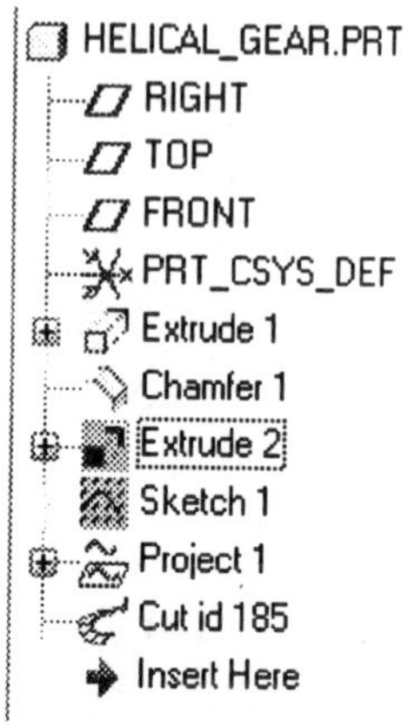

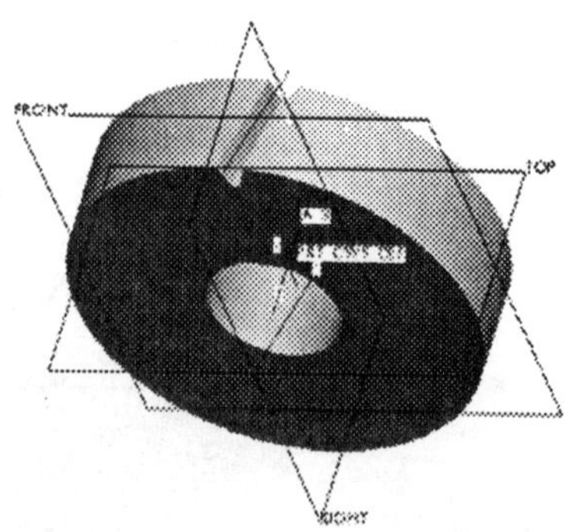

We plan to use Pattern to create the other teeth. Before doing so, we need a parameter to be used by Pattern. Use the function called **Copy** to create the second cut tooth. As a result, we have the angular parameter.. From the main toolbar, select **Edit** > **Feature Operations** > **Copy** > **Move** > **Select** > **Dependent** > **Done**

Select > Pick the first cut tooth > **Done** > **Rotate** > **Crv/Edge/Axis** > pick the axis of the hole > **Okay** to accept the direction of angle manuplation > type in 360/40 or 9 > **Done Move** > **Done** > **OK**.

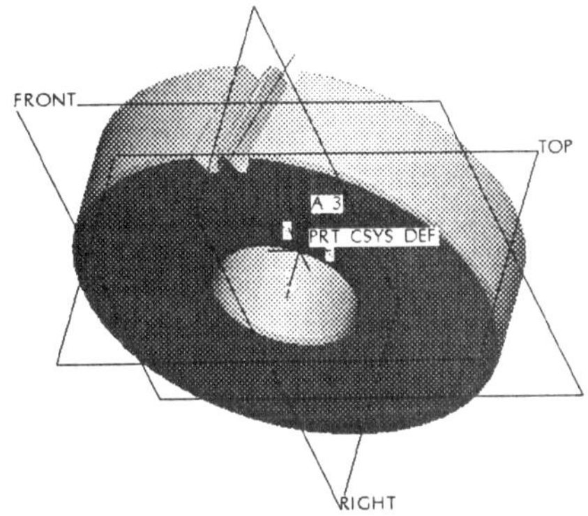

Use **Pattern** to Obtain the other 38 teeth.

From the model tree, highlight the group feature > righ-click and hold, select **Pattern** > on the display, select the angle of 9 degrees > type in 39 as the number of teeth, which includes the current one > select the icon of **Complete**.

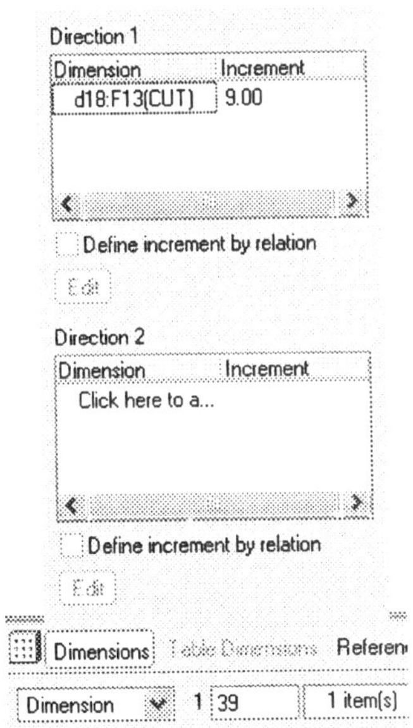

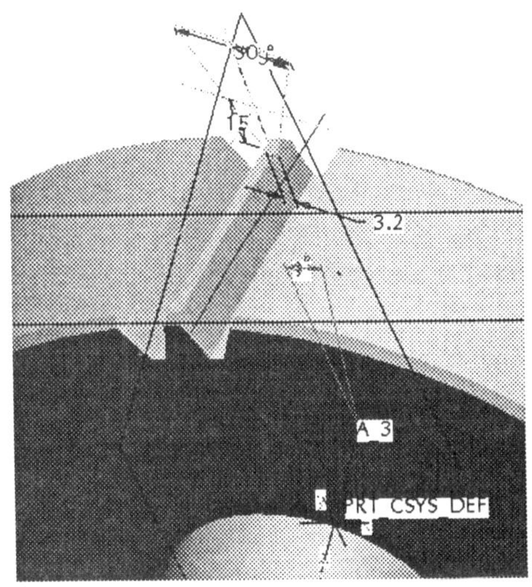

In order to define a load acting on the surface of a tooth, which has a helical angle, we need to define a loca coordinate system so that the load condition can be added with ease.

First define a datum point serving as the origin of the local coordinate system. Select the icon of creating datum point from the toolbar of datum feature creation > select the projected curve and, while holding down the Ctrl key, pick the **FRONT** datum plane > **Close**.

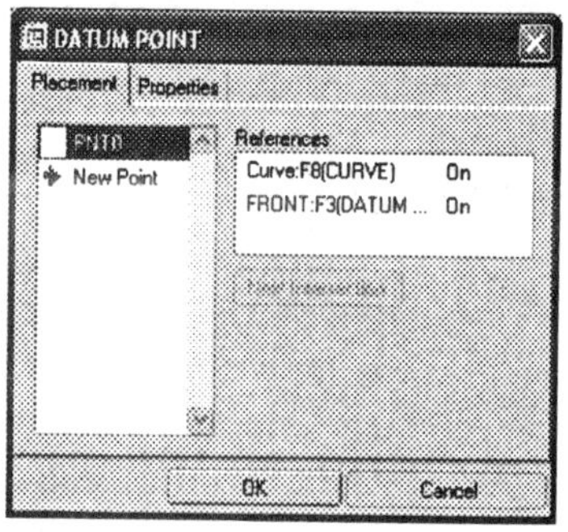

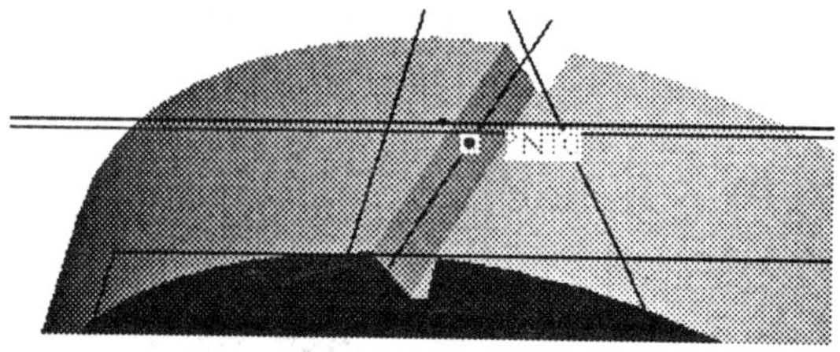

Create three datum planes. DTM1 is a plane normal to the projected curve.

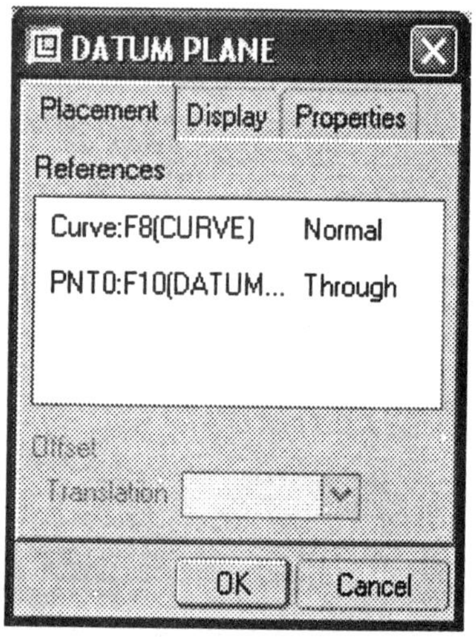

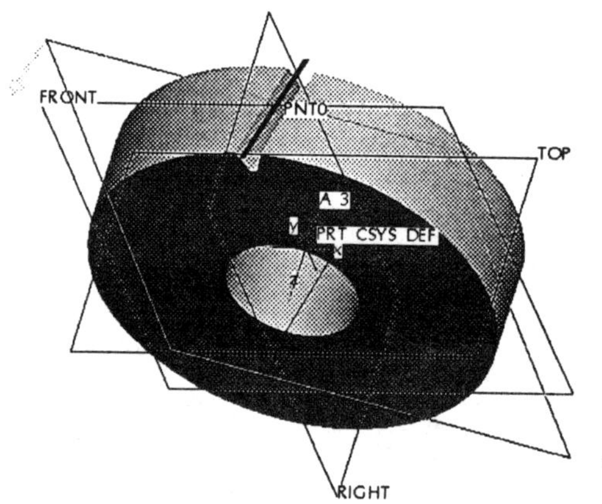

DTM2 is a plane normal to DTM1.

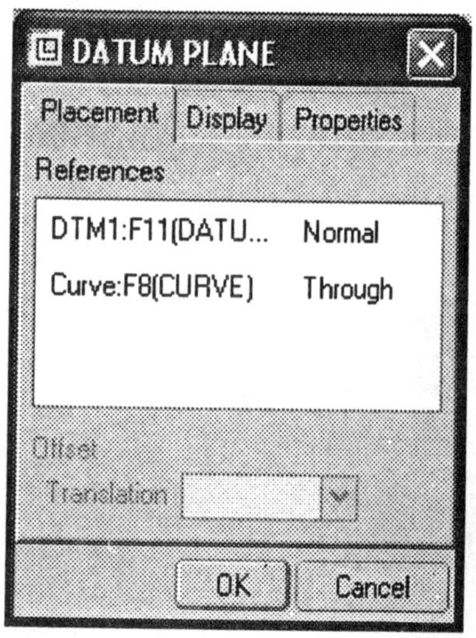

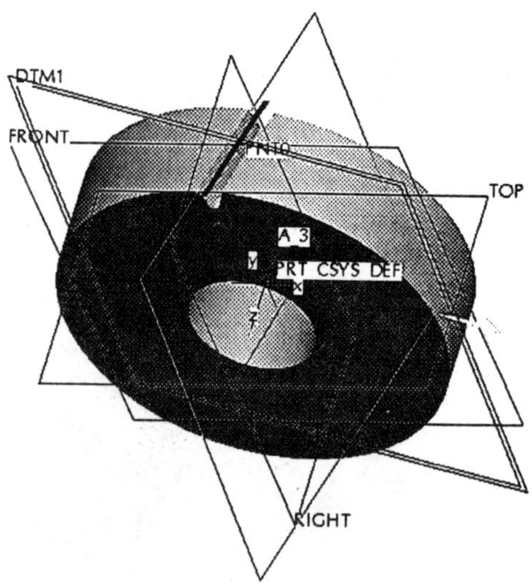

DTM3 is a plane normal to both DTM! And DTM2.

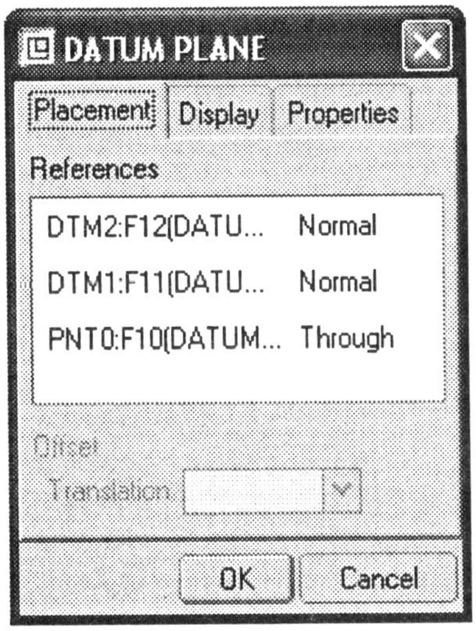

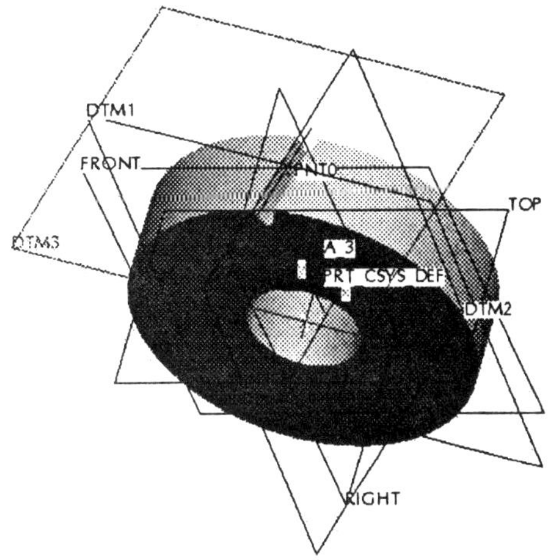

To create a local coordinate system using DTM1, DTM2 and DTM3, select the icon of creating coordinate system from the toolbar of datum feature creation > pick the three datum planes (holding down the Ctrl key) > orient the direction of the local coordinate system, as shown. Note that the direction of the x-axis will be the direction of the load to be applied.

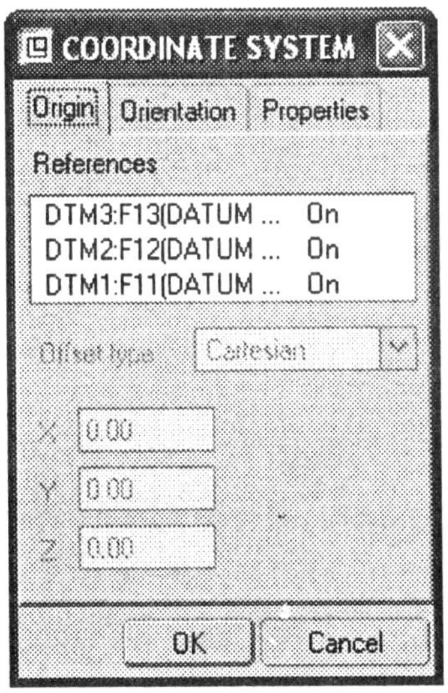

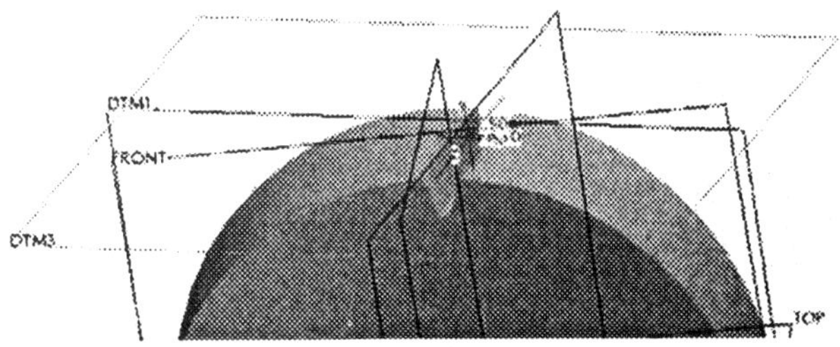

Step 2: Under the Pro/MECHANICA environment, define the material type.
Applications > **Mechanica** > **Continue** > **Structure** > **OK**.

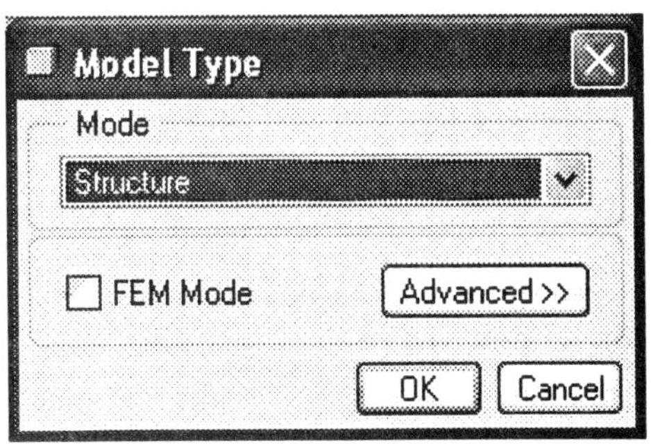

From the toolbar of functions, select the icon of **Materials** > select **STEEL**

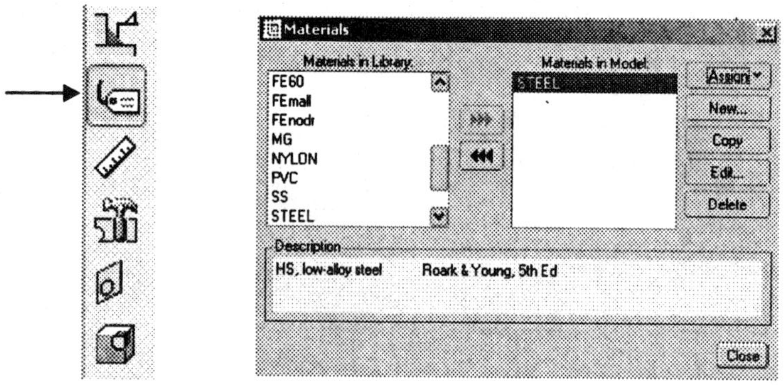

Assign > **Part** > select the beam structure from the screen display > **OK** > **Close.**

Step3: Define the fixed end constraint condition and load condition.

From the toolbar of functions, select the icon of **Constraints** > select the cylindrical surface of the hole. Set the 6 degrees of freedom to be fixed.

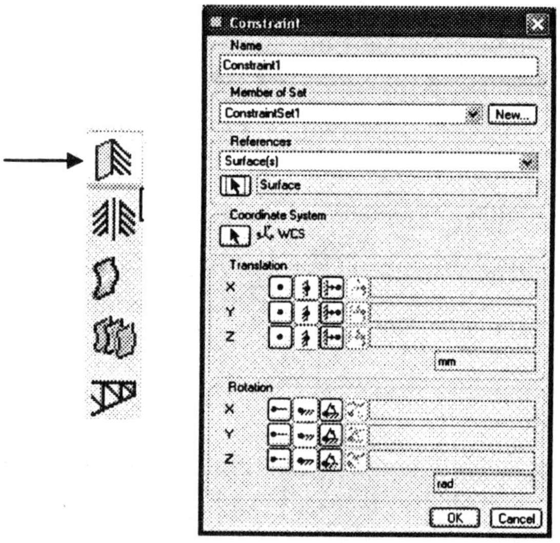

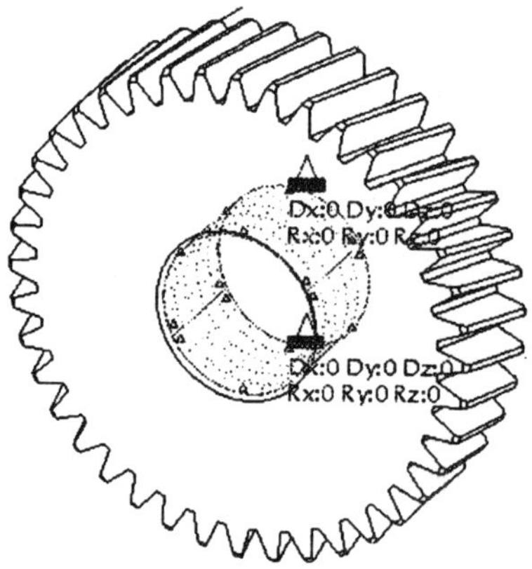

To define the load condition, from the toolbar of functions, select the icon of **Force/Moment** Loads > Surface(s) > pick the surface of tooth > select **CSO** > specify 200 in the X Force Component > **OK**.

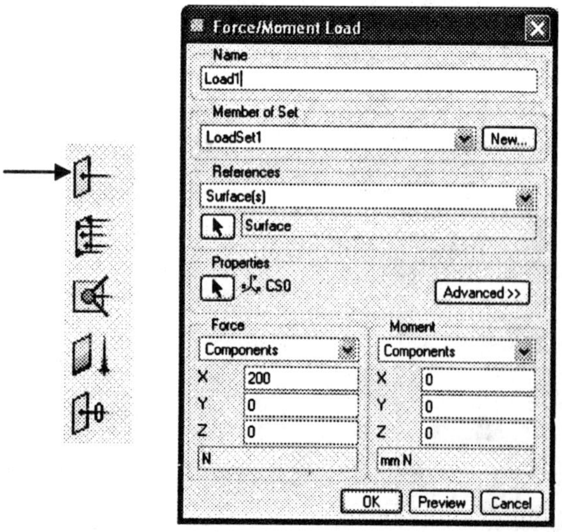

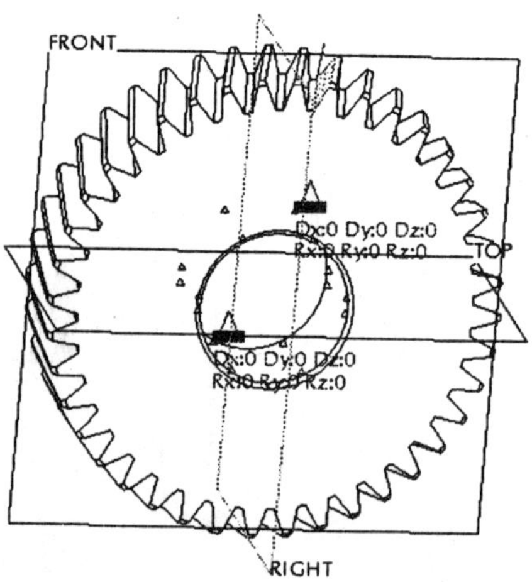

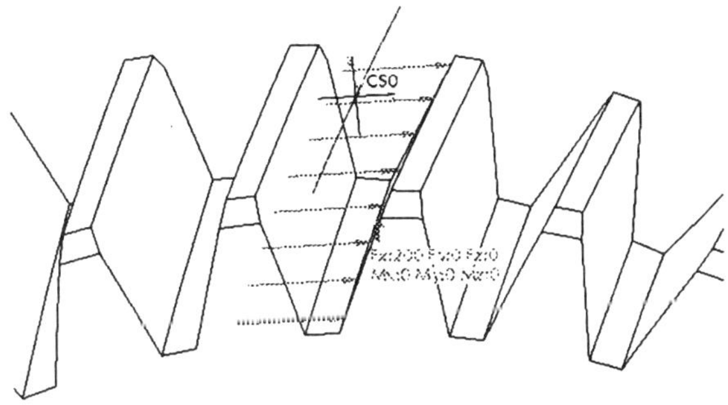

Step 4: Under the Pro/MECHANIC, set up **Analyses** and **Run.**
Select the icon of **Run an Analysis** > **File** > **New Static** > type in *helical_gear* > **OK** > **Run.**

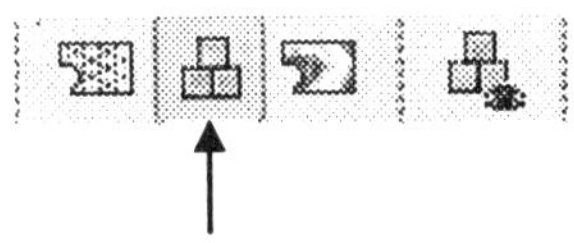

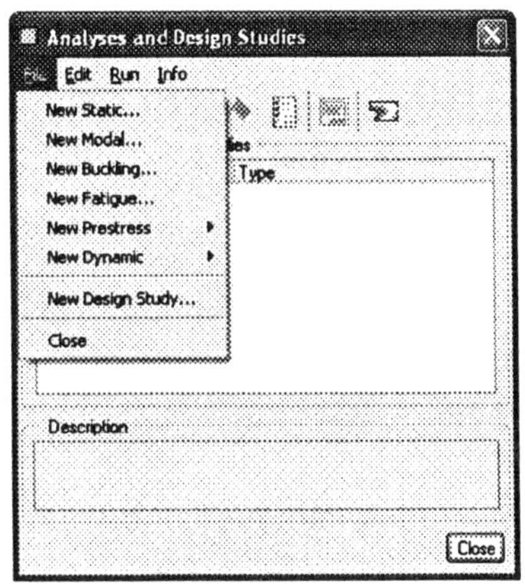

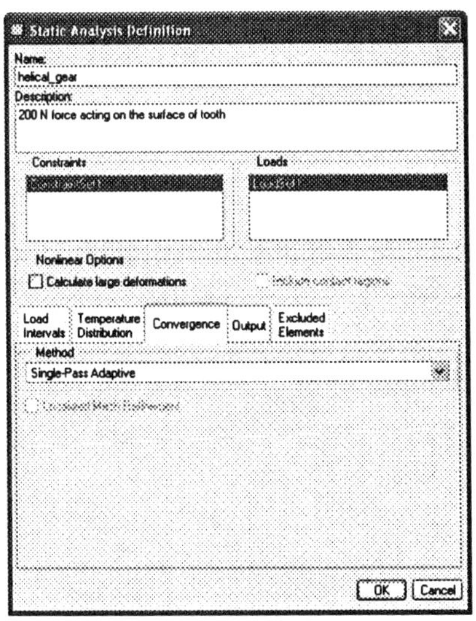

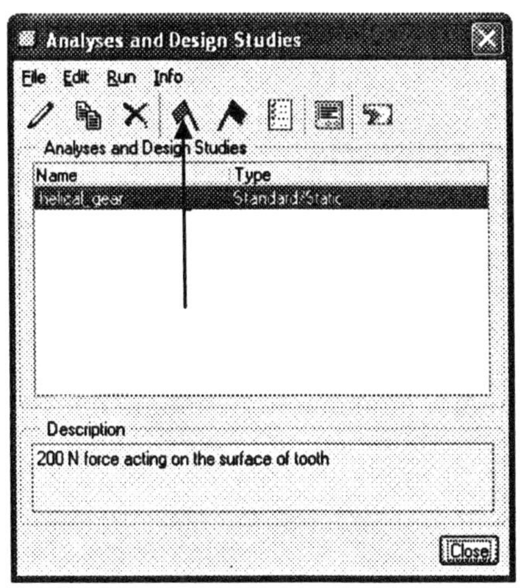

Step 5: Under the Pro/MECHANICA, study the **Results**
Select the icon of **Results.**
Type *Tooth Deformation* as the title > locate the Design Study folder called helical_gear.
Select **Fringe** > choose **Displacement** and **Magnitude** > **OK and Show**.

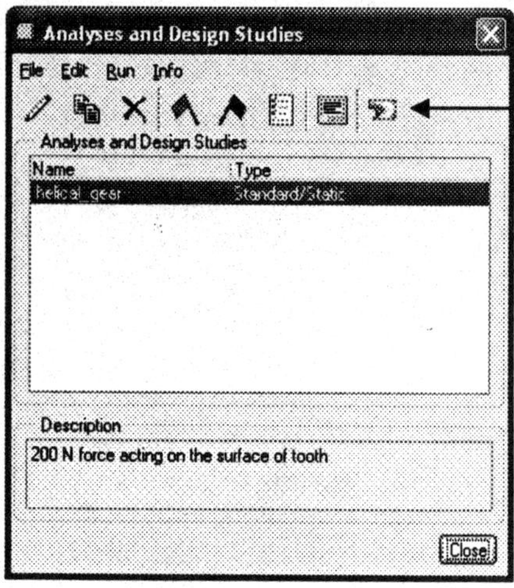

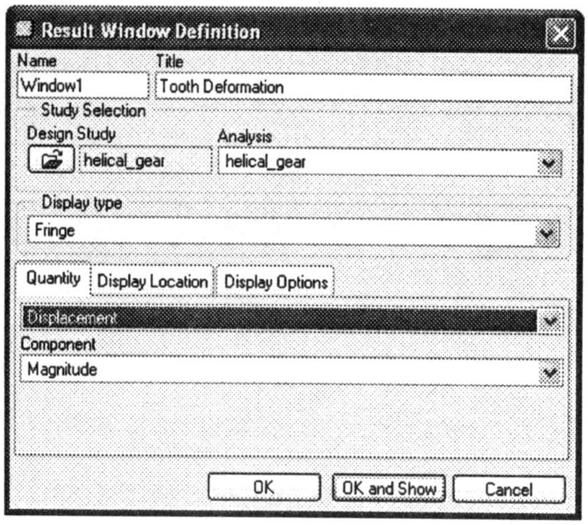

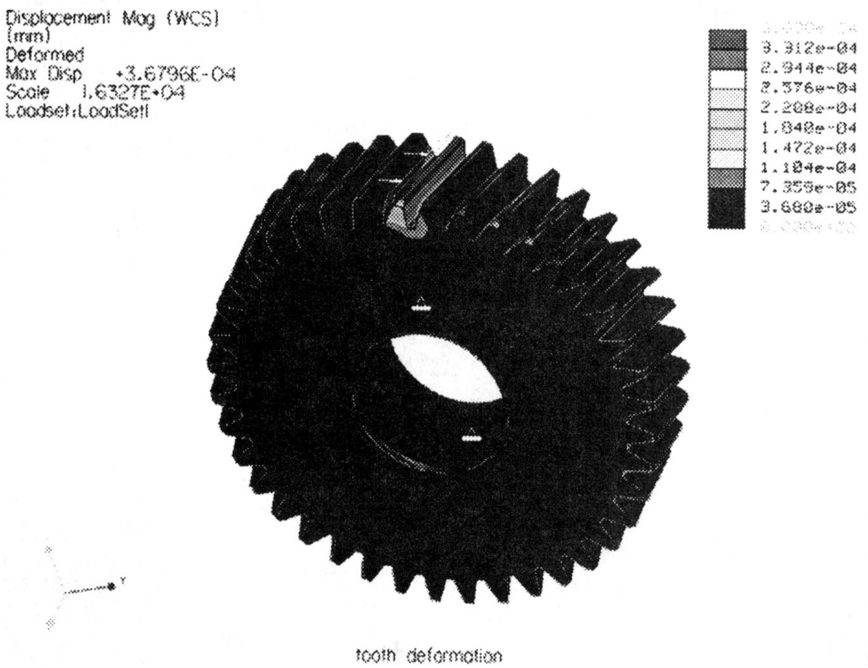

To plot a distribution of stress, from the main toolbar, select **Edit** > **Copy**. Select **Stress** > **Max Principal** > **OK and Show.** Users should be able to obtain the plot of the maximum principal stress distribution.

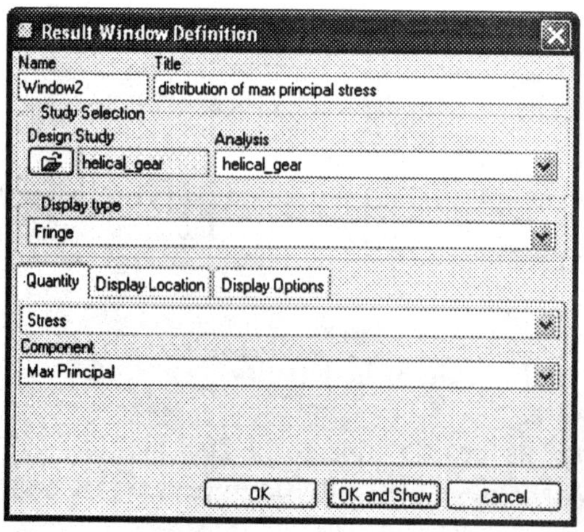

Step 6: Under the Pro/ENGINEER design environment, create a 3D solid model of the shaft component. **File > New > Part** > type *shaft_for_helical_gear* as the file name and clear the icon of **Use default template > OK.** Select **mmns_part_solid** (units: Millimeter, Newton, Second) and type *Shaft* in **DESCRIPTION**, and *student* in **MODELED_BY**, then **OK.**

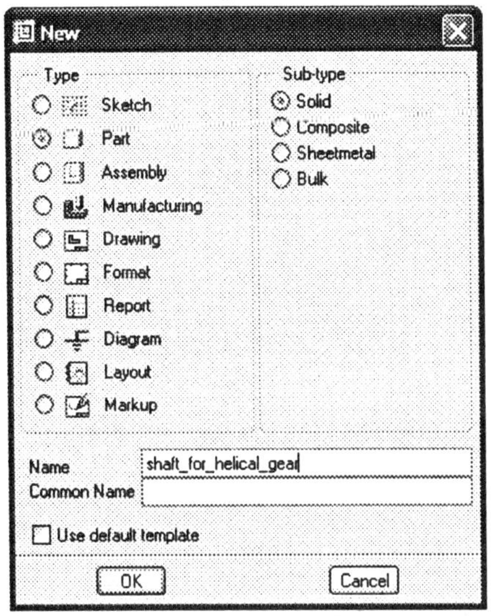

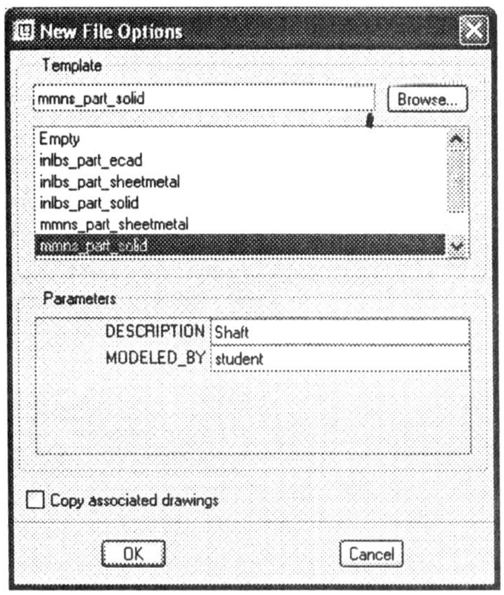

We use Revolve to create the solid model. Select the icon of **Revolve** from the toolbar of feature creation. Make sure the icon of **Solid** is selected. Click the box of **Placement** > **Define**, and select **FRONT** as the sketch plane, accept **RIHGT** to orient the sketch plane > click the box of **Sketch**.

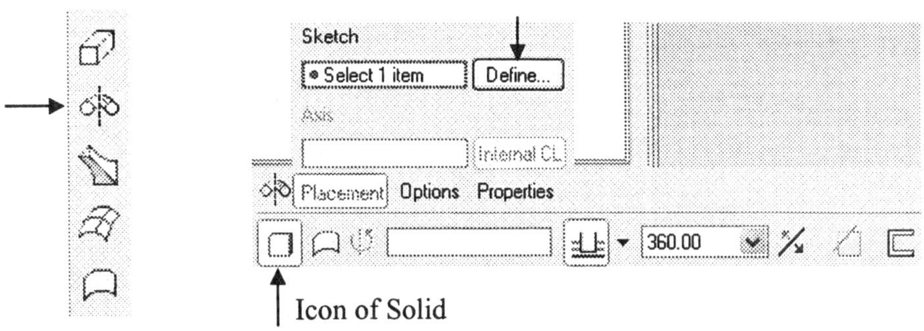

Sketch a centerline in the horizontal direction first. This centerline will be served as the centerline for revolving. Select the icon of line and sketch the following shape. Pay attention to dimensioning the diameters.

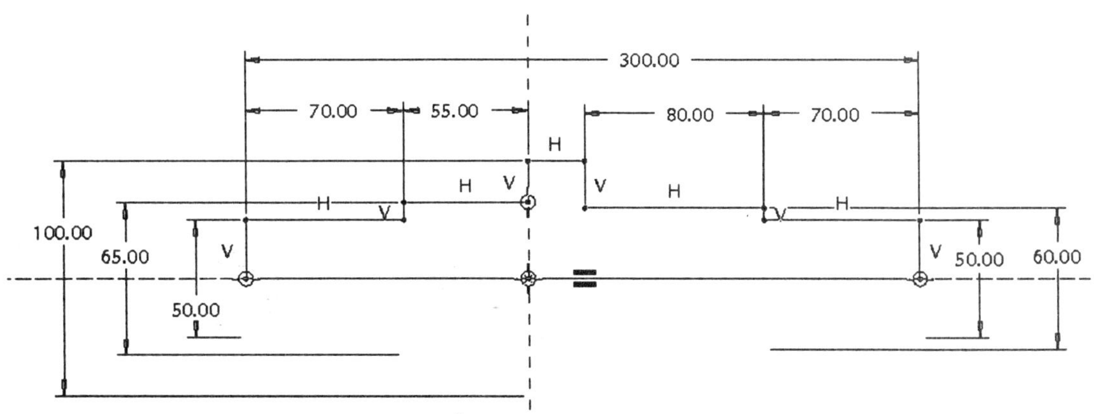

Upon completion, select the icon of **Done** from the toolbar of sketcher. On the dashboard, select the icon of **Complete** from the feature control panel

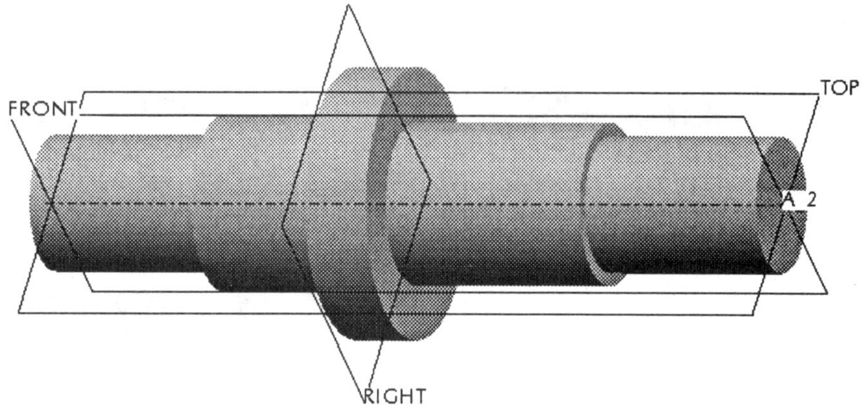

Step 7: Under the Pro/ENGINEER design environment, we assemble the gear and shaft components.

File > New > Assembly > type *gear_shaft_assembly* as the file name and clear the box of **Use default template > OK.**

Select **mmNs_asm_design** (units: mm, Newton, and second) and type in gear shaft assembly in **DESCRIPTION**, and student in **MODELED_BY**, then **OK.**

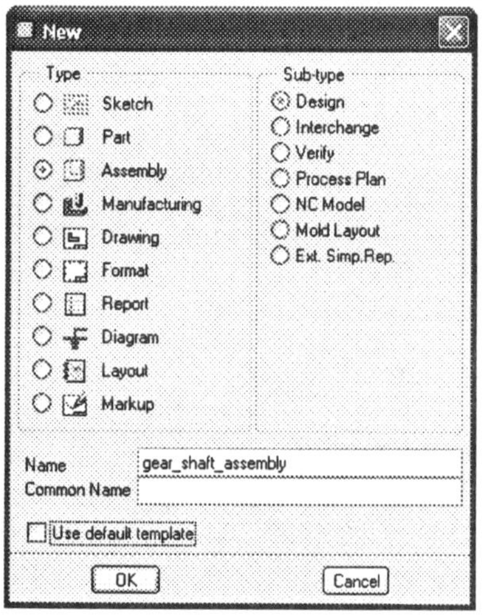

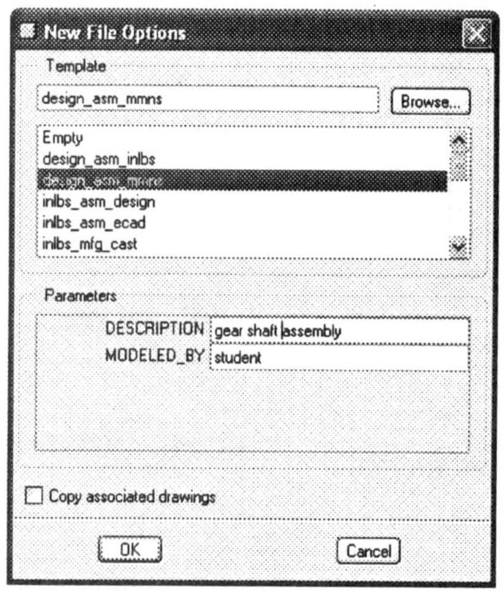

Assemble the shaft component with the default coordinate system of the assembly model.

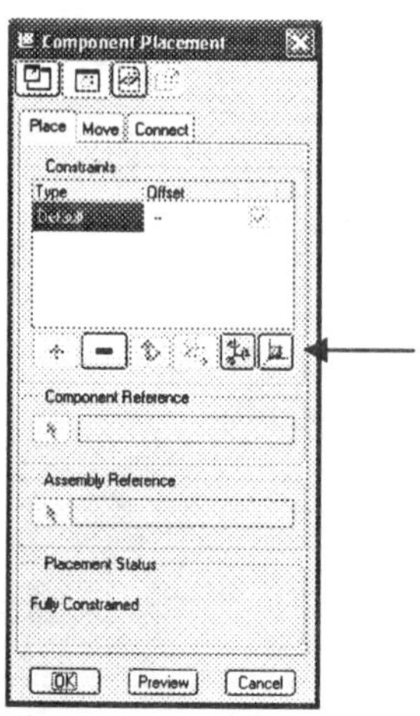

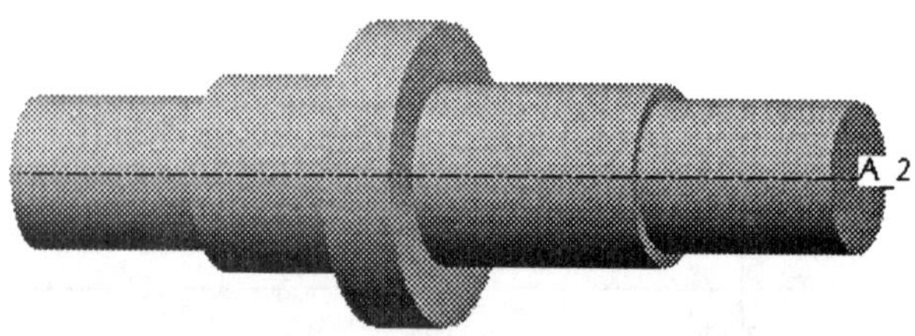

Assemble the gear component with the assembly. Pay attention to the helical angle orientation with respect to the shaft position.

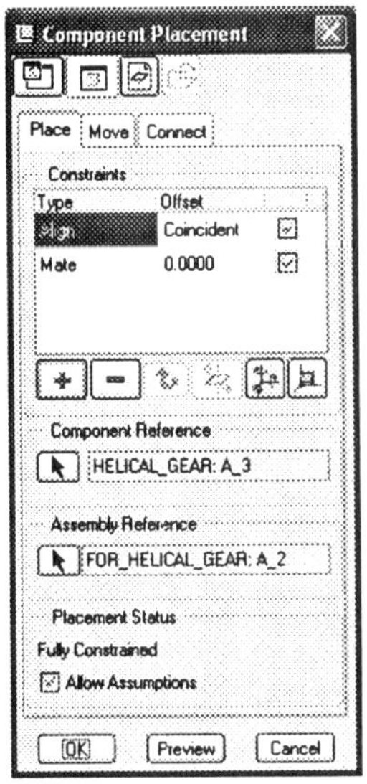

Step 8: Under the Pro/MECHANICA environment, define the 2 material types for the shaft and gear components. Both material types are steel.

Applications > **Mechanica** > **Continue** > **Structure** > **OK**.

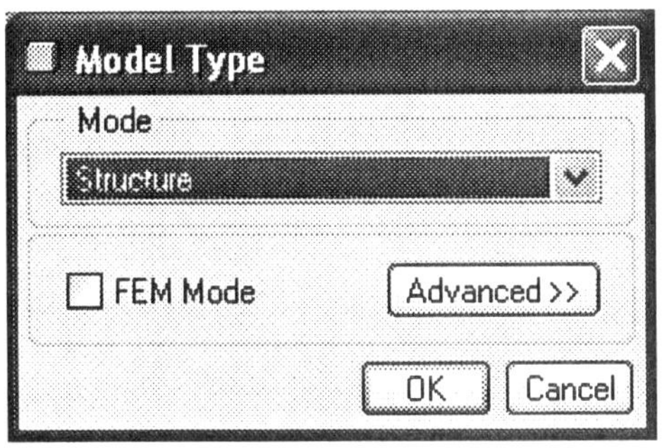

From the toolbar of functions, select the icon of **Materials** > select **STEEL**

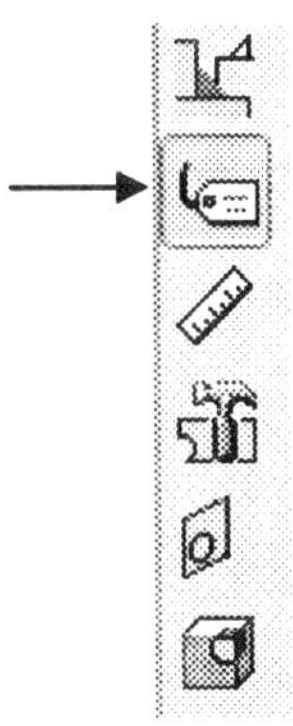

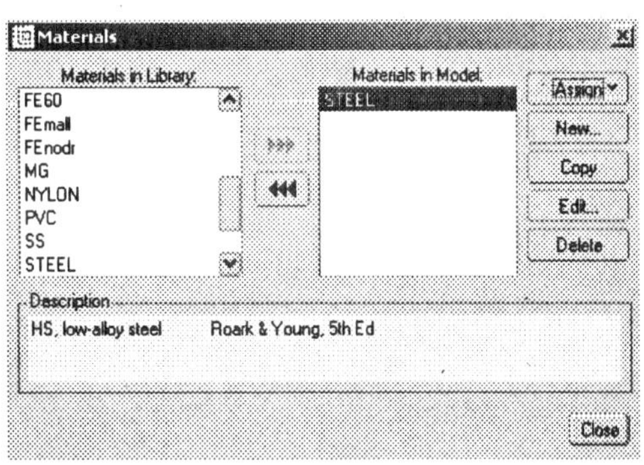

Assign > Part > select the shaft and the gear from the screen display (holding down the Ctrl key) > **OK** > **Close.**

Step 9: Under the Pro/MECHANICA environment, define the fixed end constraint condition, (assume that both ends of the shaft are fixed)

From the toolbar of functions, select the icon of **Constraints** > select the two end surfaces of the shaft > **OK**. Set the 6 degrees of freedom to be fixed > **OK**.

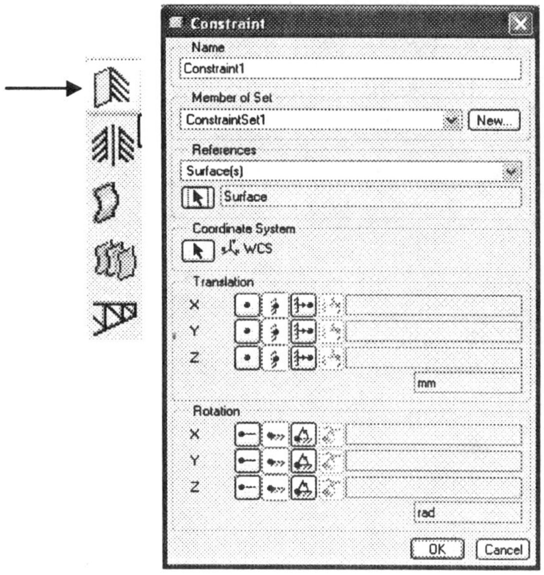

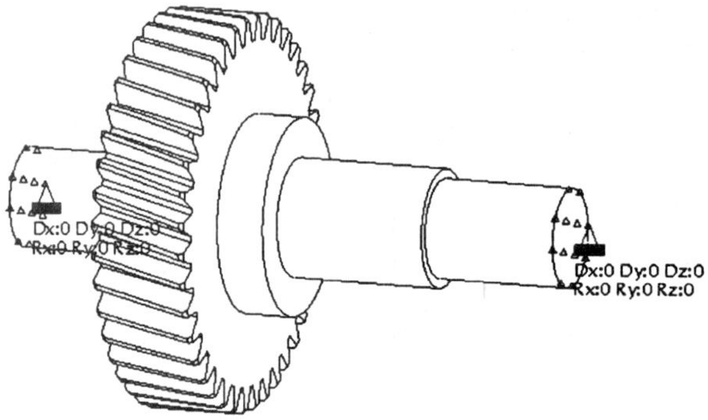

Step 10: Under the Pro/MECHANICA environment, define the load condition.

To define a torque (200 x cos15° x 100 = 193000 N-mm) as the load acting on the gear and be transmitted to the shaft component. We will define 2 loads. The first load is the torque acting on the flat surface on the left (front) side of the gear component, and the magnitude is 193000 (N-mm). The second load is the force acting on the flat surface on the left side of the gear component, and the magnitude is 200 x sin15° = 51.8 (N-mm).

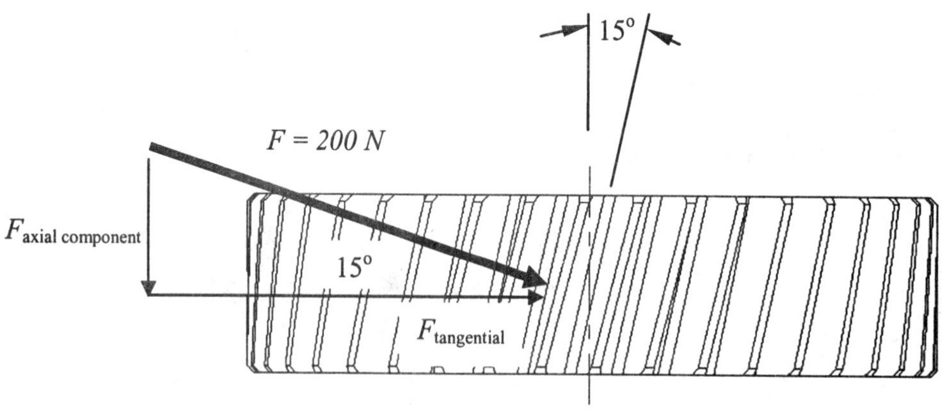

$$F_{\tan gential} = F\cos 15° = 200 * \cos 15° = 193$$

$$F_{axial} = F\sin 15° = 200 * \sin 15° = 51.8$$

Before defining the 2 loads, let us define a datum point associated with the helical gear component for defining the torque load. Open the file of helical_gear. We go back to the Part Mode.

Select the icon of creating datum point from the toolbar of datum feature creation. Pick the axis of the hole and the front surface to PNT1, as shown.

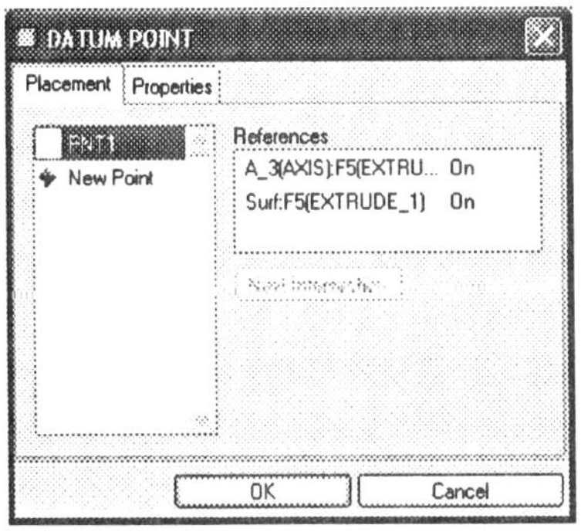

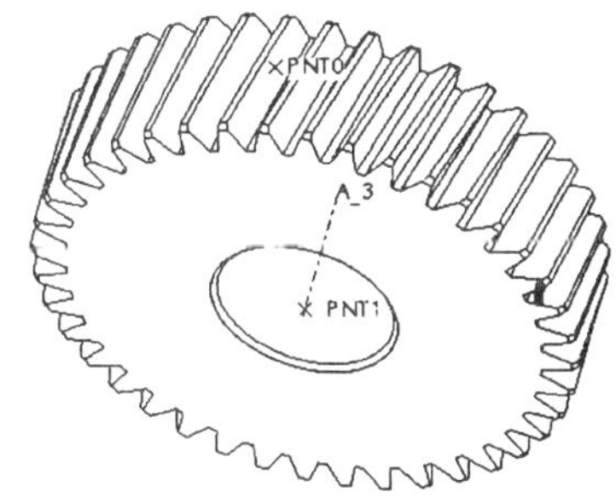

Now let us go back to the Assembly Mode. Select the icon of **Force and Moment Load** from the toolbar of functions > **Surfaces** > to define the surface, select the flat surface on the left side of the gear as shown > click the box of **Advanced** and make sure that **Total Load at Point** is selected, and PNT1 is selected > type in the magnitude of 193000 in the x component box of **Moment** > **OK**.

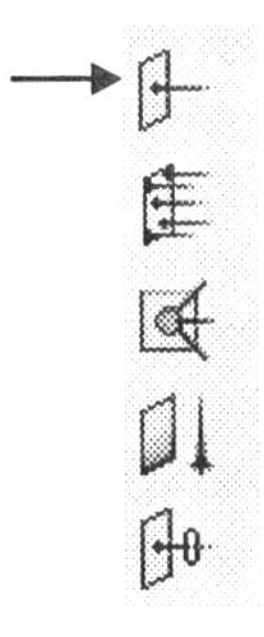

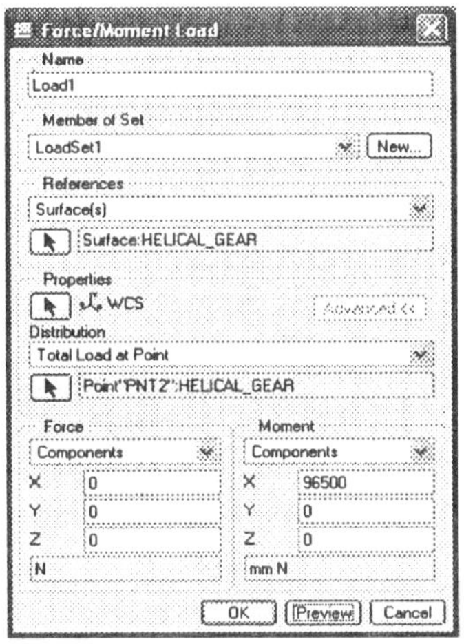

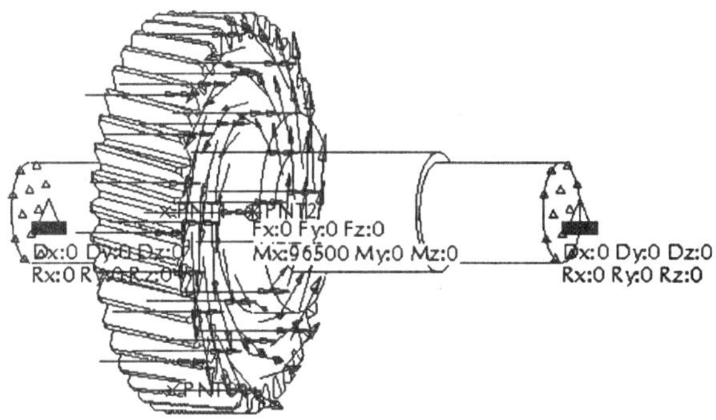

To define load 2, this is the force acting on the flat surface on the left side of the helical gear component, select the icon of Force/Moment Load from the toolbar of functions > Surface(s) > pick the flat surface on the left side of the gear component > type in 51.8 as the X-component > OK.

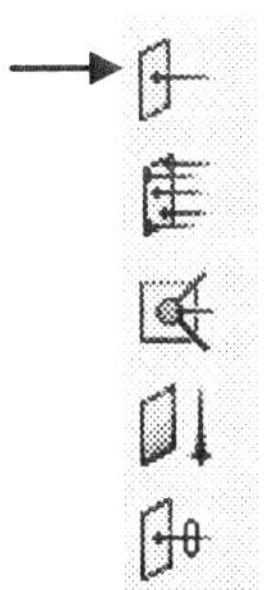

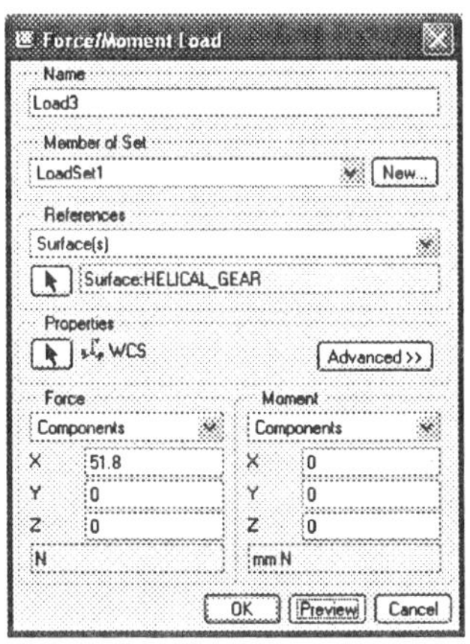

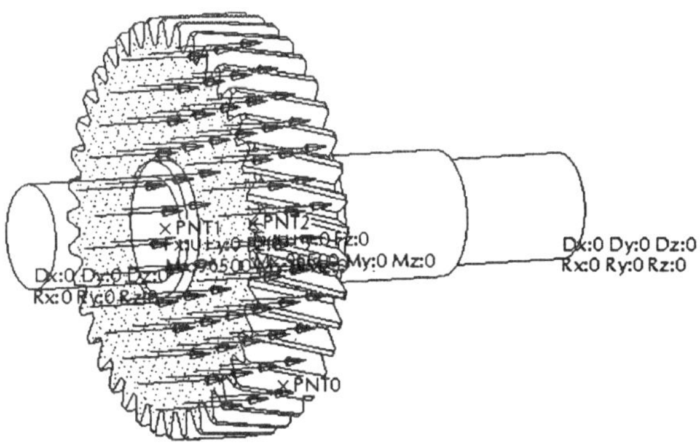

Step 11: Under the Pro/MECHANICA environment, select the icon of **Run an Analysis** > **File** > **New Static** > *type gear_shaft_assembly* as the name of the design study folder > **OK** > **Run.**

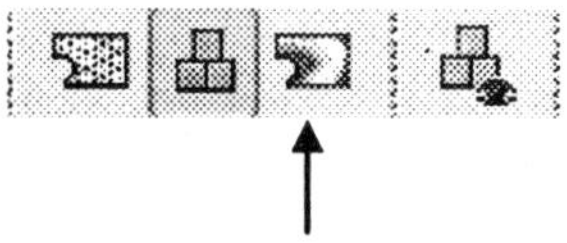

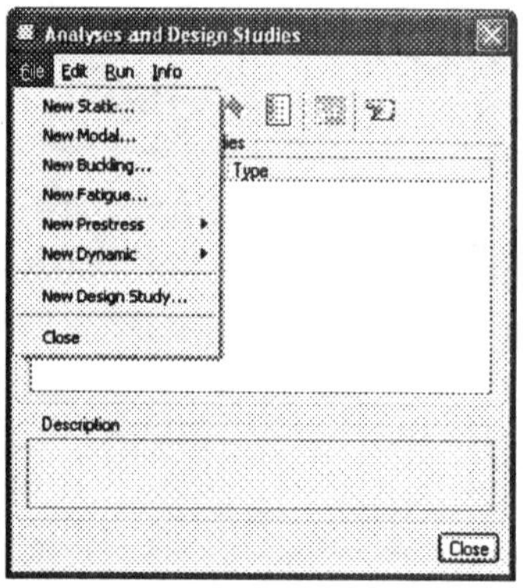

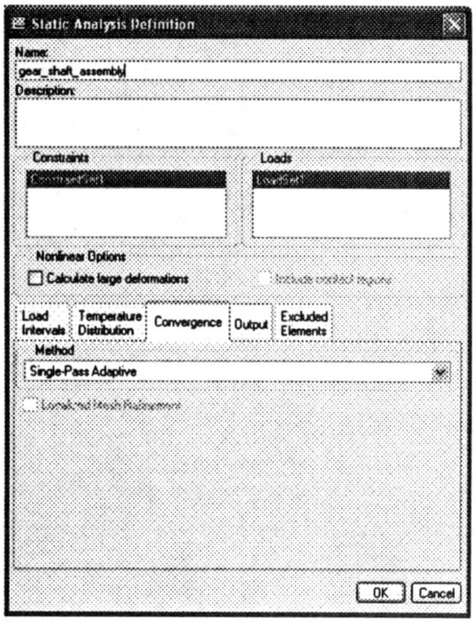

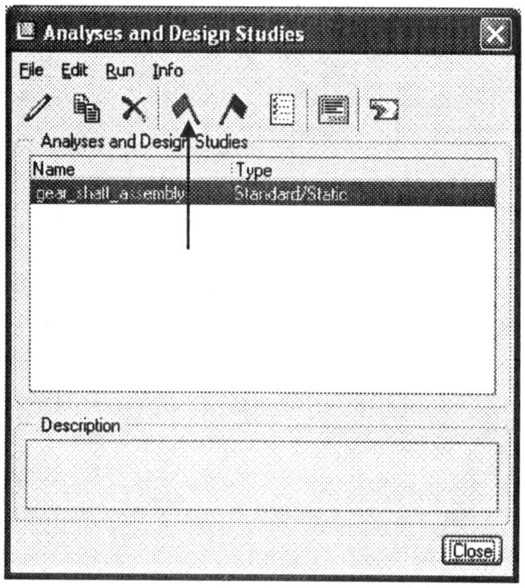

Select the icon of **Results.**

Type *gear_shaft_eformation* as the title > locate the Design Study folder called *gear_shaft_assembly*. Select **Fringe** > choose **Displacement** > **OK and Show**.

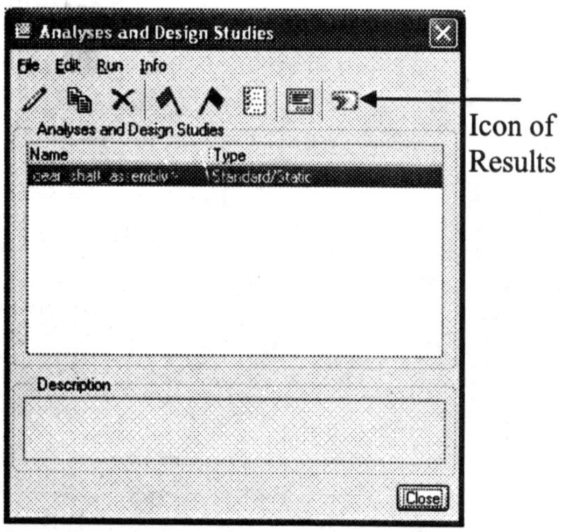

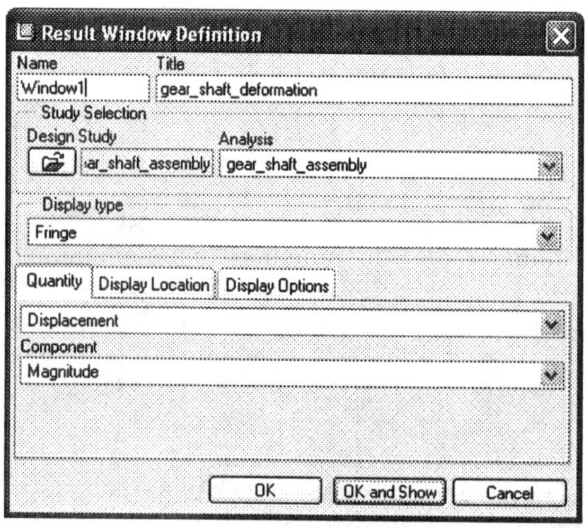

Plot of the deformation pattern of the assembly

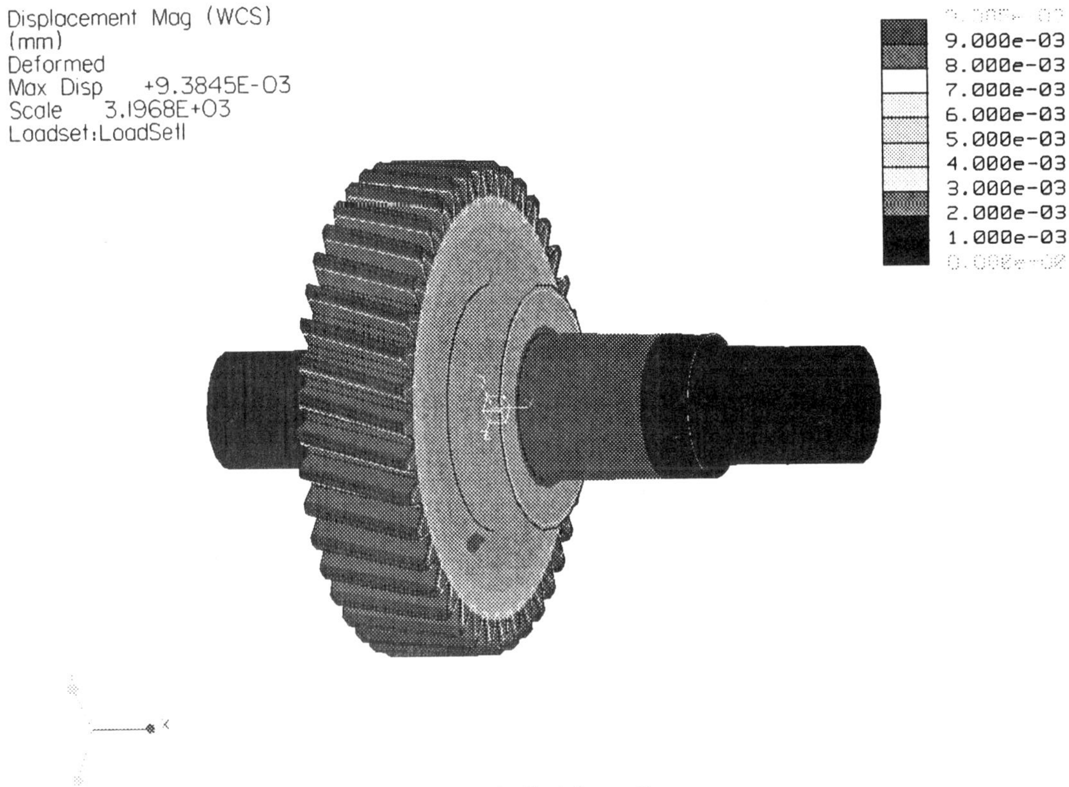

gear_shaft_deformation

Plot of the von Mises stress distribution of the assembly

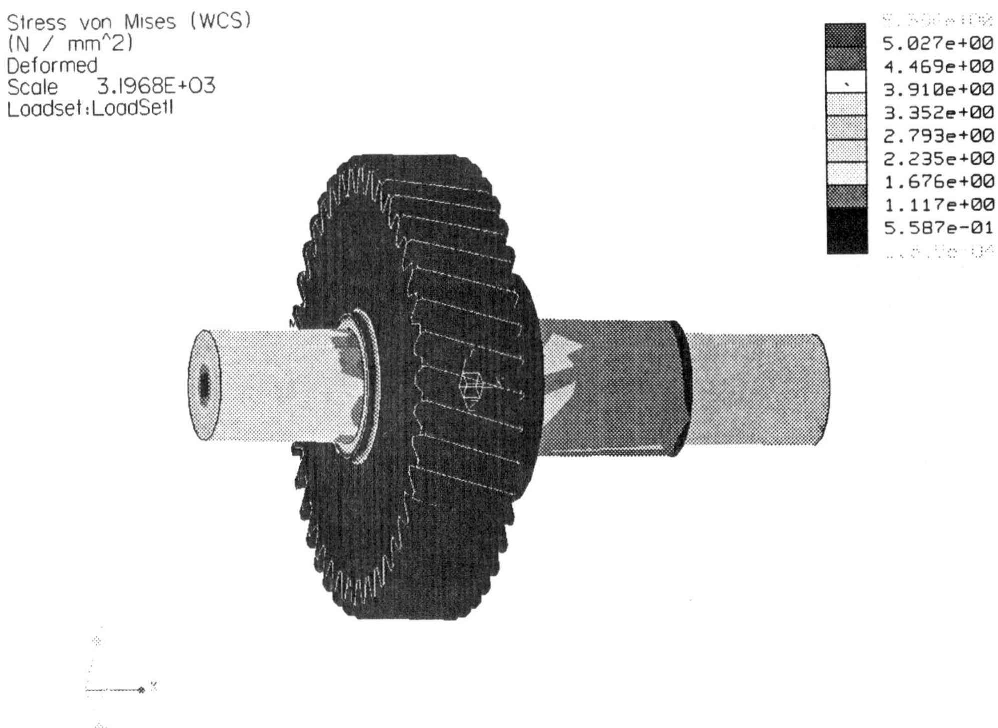

distribution of the von Mises stress

Users are also able to plot the displacement and/or stress distribution at the component level after running the analysis with the assembly structure. Let us create a new window and have a new definition for the gear component. From the main toolbar, select **Edit** > **Copy** > type max principal stress of gear > click Display Location > change **All** to **Volume**.

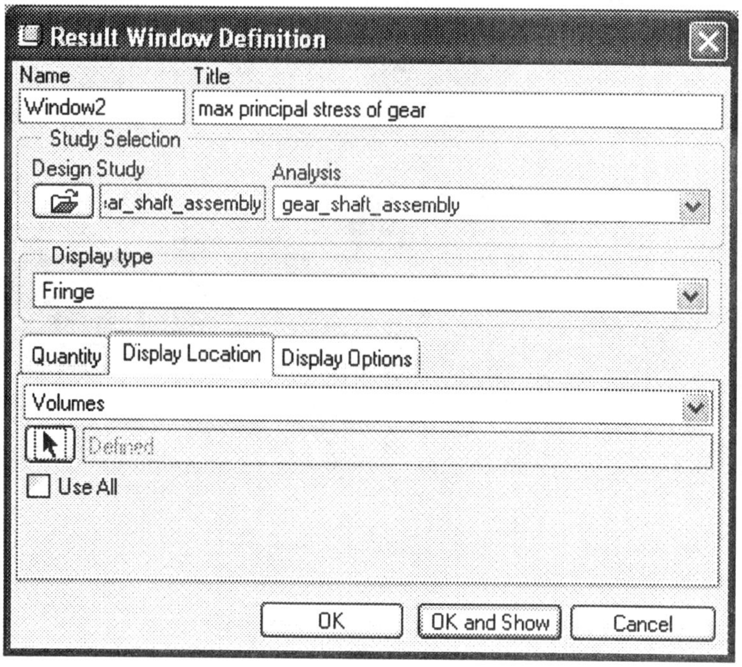

Pick the icon of arrow to select the component from the assembly > pick the gear component, which is highlighted after picking > middle button of mouse to accept the selection > **OK and Show**.

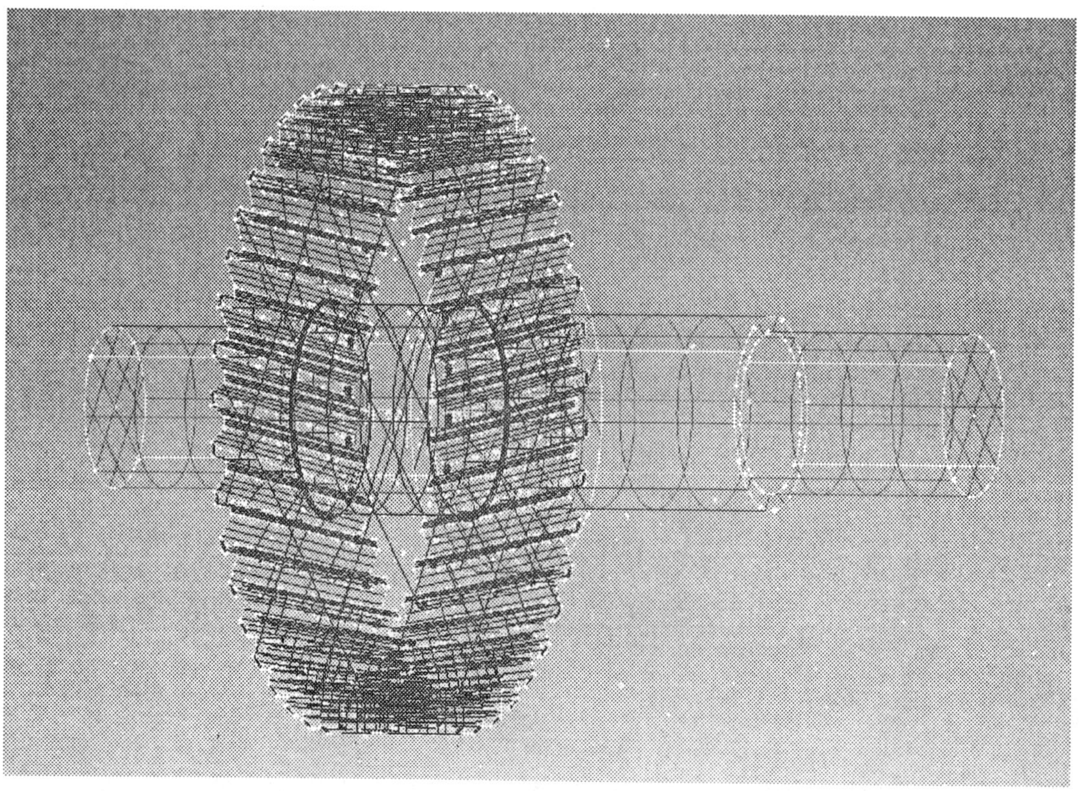

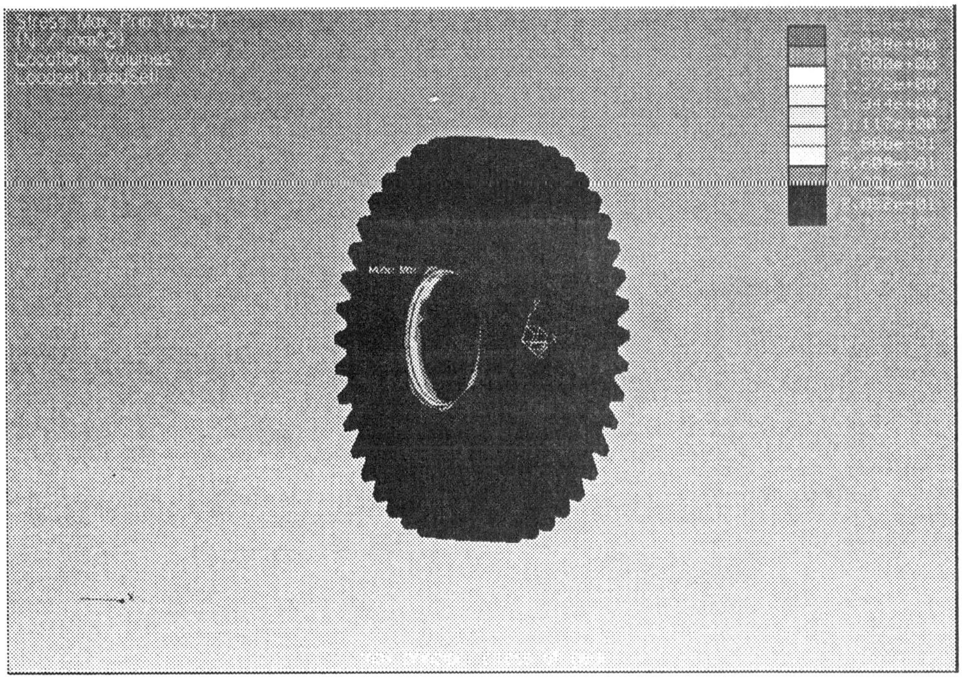

Let us create a new window and have a new definition for the shaft component. From the main toolbar, select **Edit > Copy >** type max principal stress of shaft > click Display Location > change **All** to **Volume**.

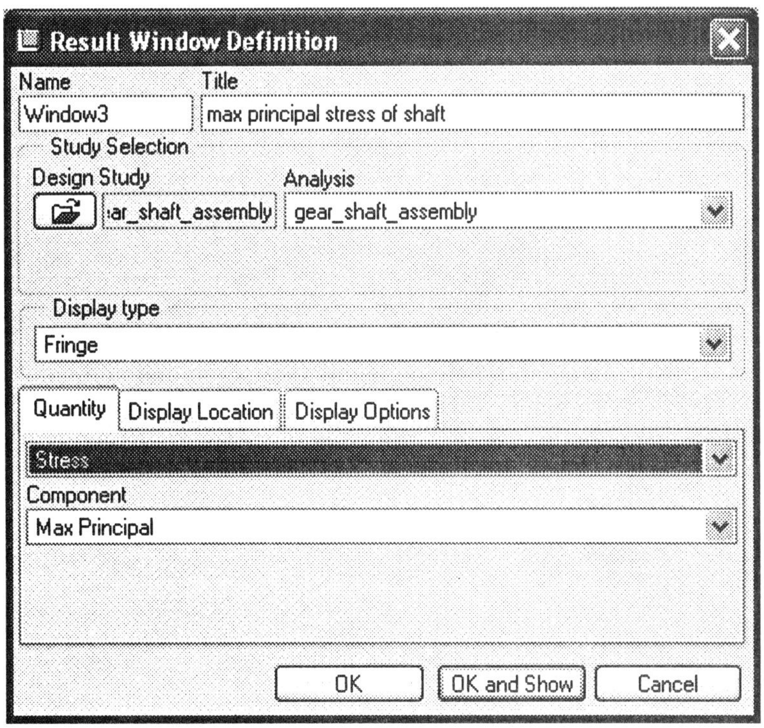

Pick the icon of arrow to select the component from the assembly > click the gear component first as the de-selection step > pick the shaft component, which is highlighted after picking > middle button of mouse to accept the selection > **OK and Show**.

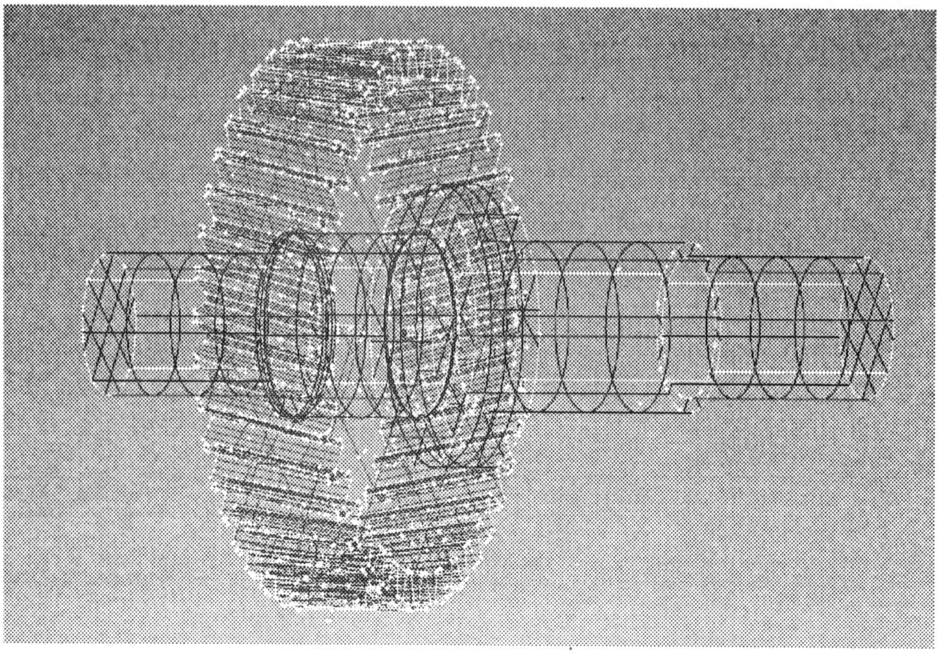

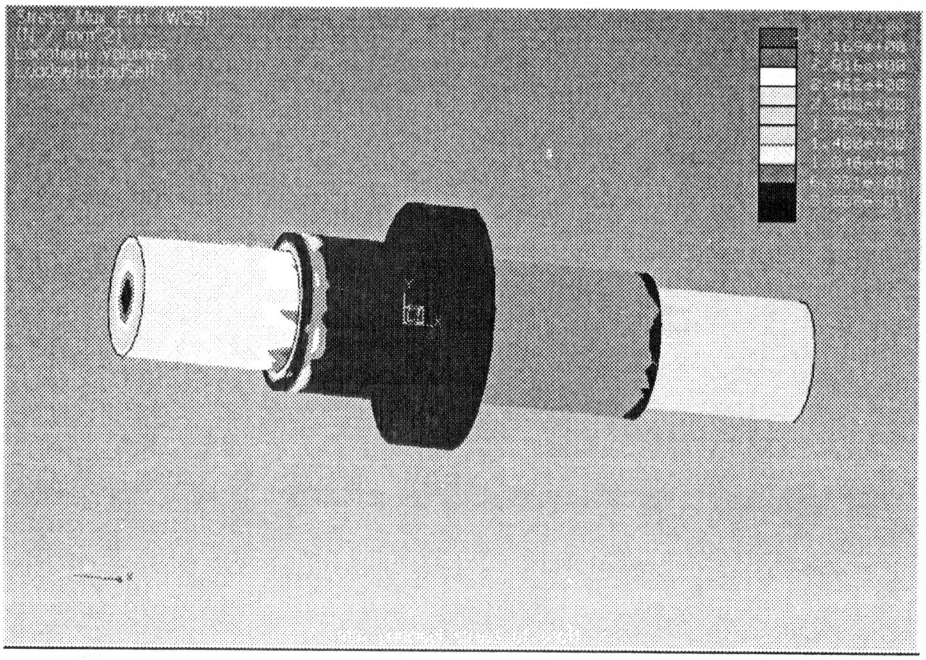

4.5 Modeling of a Load Condition due to Temperature Variation

As illustrated in section 4.4, an assembly consists of several components. The material type of each of the components in the assembly can vary. Some components are made of steel. Some components are made of copper. A few components could be made of plastic material. Differenct types of material demonstrate different characteristics in terms of material properties, such as Young's modulus, Poisson's ratio, coefficient of thermal expansion, etc. The mismatch of material properties in an assembly often leads to strain and stress developments.

For example, a change in the body temperature of a component leads to a change in its dimensions, such as the length and/or width. Such changes in dimension will induce strain and stress. To evaluate these effects quantitatively, the coefficient of thermal expansion plays a critical role. In the following, we present a case study, which deals with an assembly consisting of 2 layers and base substrate. The two layers are the cement layer and the glass layer. The base substrate represents a dentin structure, which is replaced by a type of composite material in research. Therefore, this layer structure resembles a crown-cement-dentin system used by the scientists and dentists in dentistry in the research of dental materials.

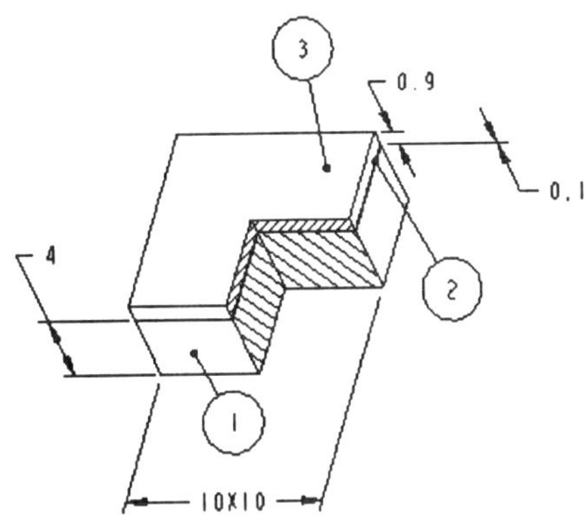

INDEX	PART NAME	MATERIAL
3	TOP_LAYER	glass
2	MIDDLE_LAYER	cement
1	BASE	composite

Let us examine the geometrical characteristics. The base structure is a rectangular block. Its size is 10 x 10 x 4 mm, as illustrated. The thickness of the cement layer is 0.1 mm. The thickness of the glass layer is 0.9 mm.

The properties of the three types of material are shown below:

	glass	cement	composite
mass density (g/cm^3)	5.0	1.3	1.7
coef. of thermal expansion (/C)	5.00E-06	6.00E-05	3.00E-05
Young's modulus (N/mm^2)	8000	6000	15000
Poisson's ratio	0.27	0.33	0.33

Investigate the stress development and the pattern of deformation when subjected temperature variation

The temperature variation: reference temperature: 20°C
model temperature: 40°C.

For the boundary condition, we assume that the dentin structure is fixed to the ground through a local contact.. The surface region is a circle. The diameter is 3 mm. Plot the deformation pattern and distribution of the maximum principal stress when subjected to the temperature variation.

Step 1: Under the Pro/ENGINEER design environment, create each of the three components, namely, cement_layer, glass_top and root.

Cement_layer.prt

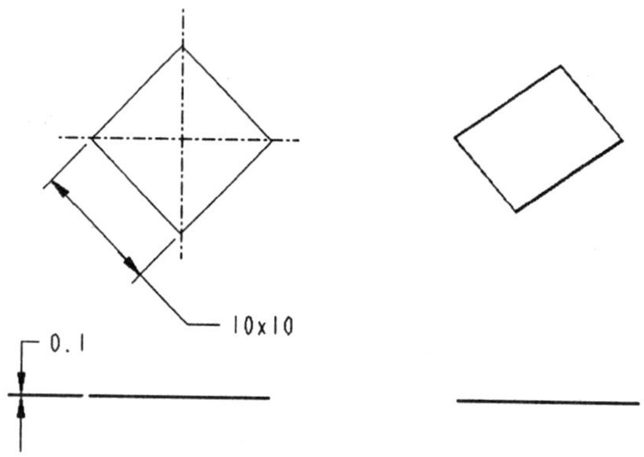

Glass_top.prt

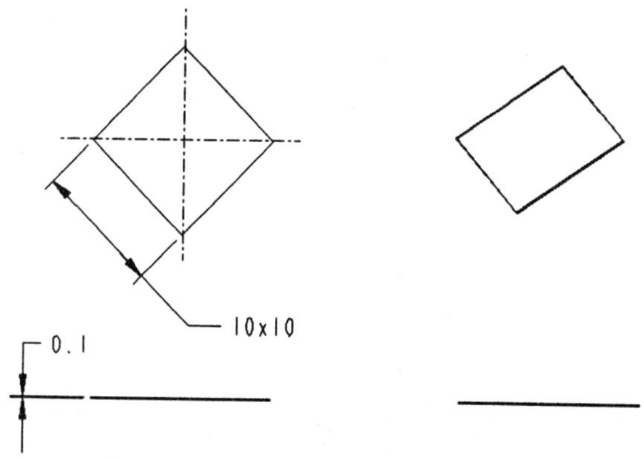

Root.prt

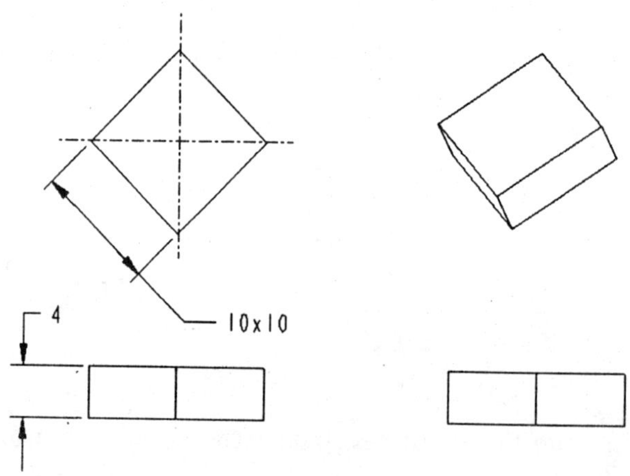

Let us create a 3D solid model for the root component first. From the menu tool bar, select **File > New > Part** > type in *root* as the file name and clear the icon of **Use default template > OK.**

Select **mmns_part_solid** (units: Millimeter, Newton, Second) and type *root* in **DESCRIPTION,** and *student* in **MODELED_BY**, then **OK.**

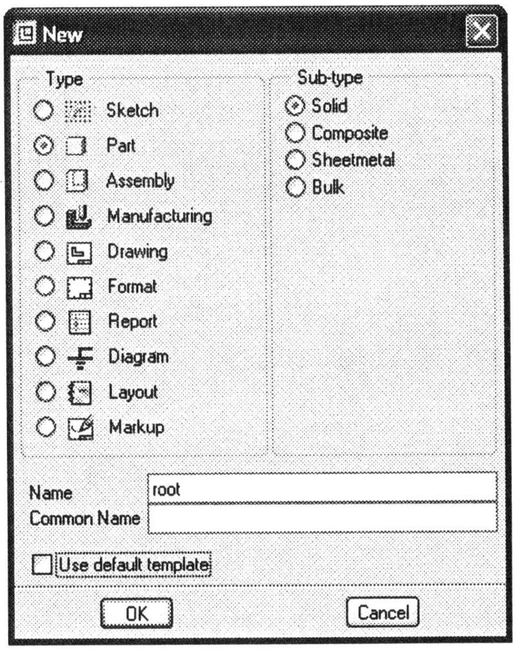

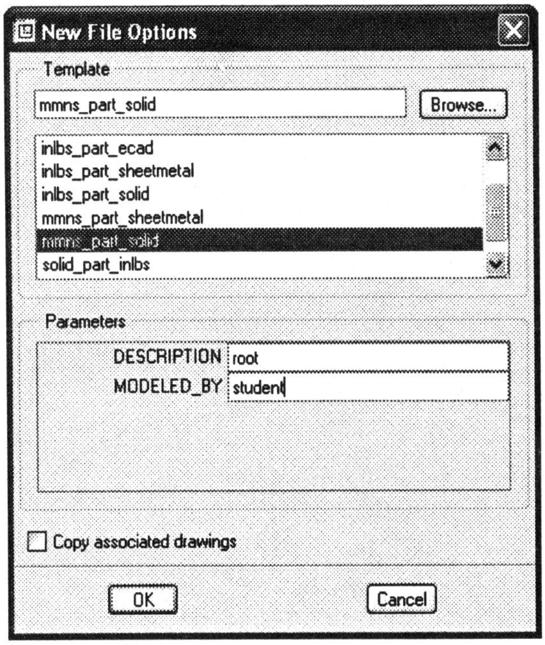

From the toolbar of feature creation, select the icon of **Extrude.**

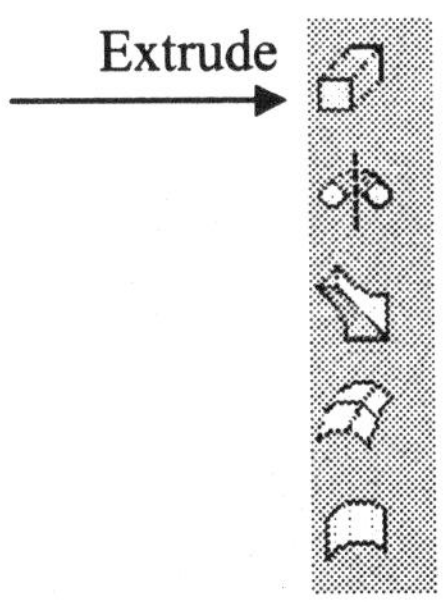

The dashboard appears. Set the depth value to 4 and click the box of **Placement** > **Define**.

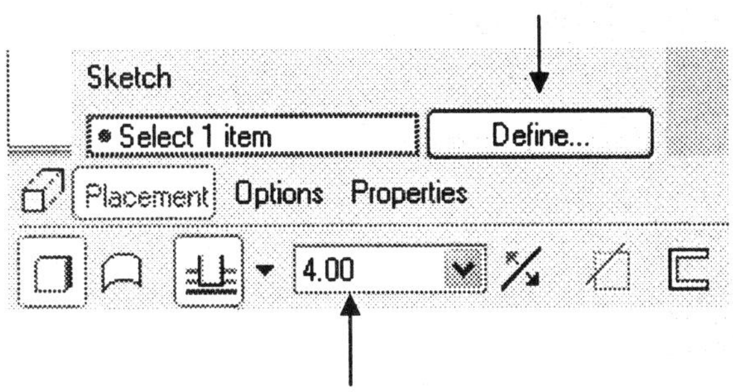

Pick the **TOP** datum plane as the sketch plane, and accept the default orientation setting

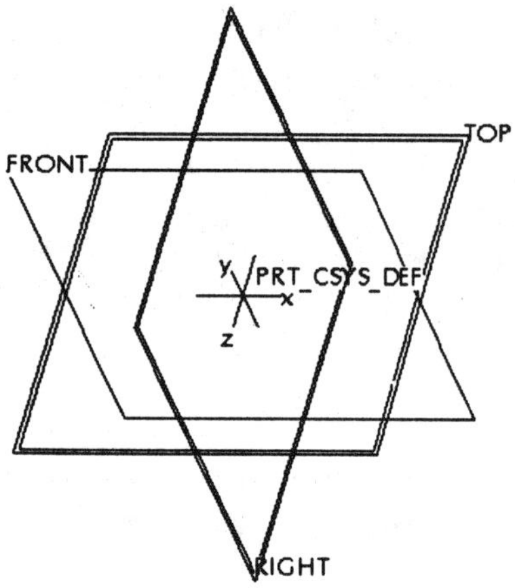

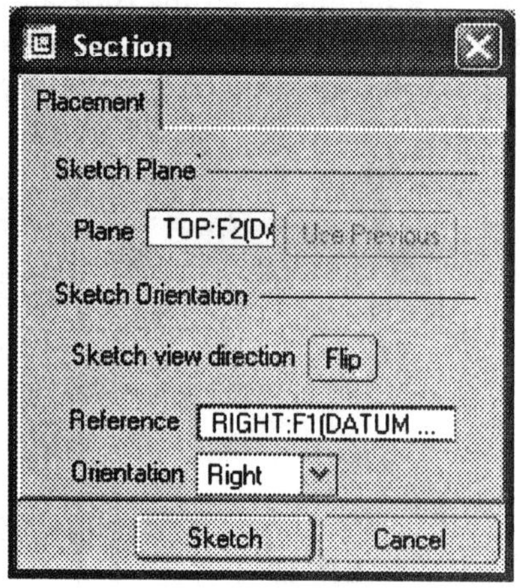

Sketch two centerlines along the x-axis and z-axis, respectively. Pick the icon of **line** and sketch a square, as shown. The dimension of each side is 10.

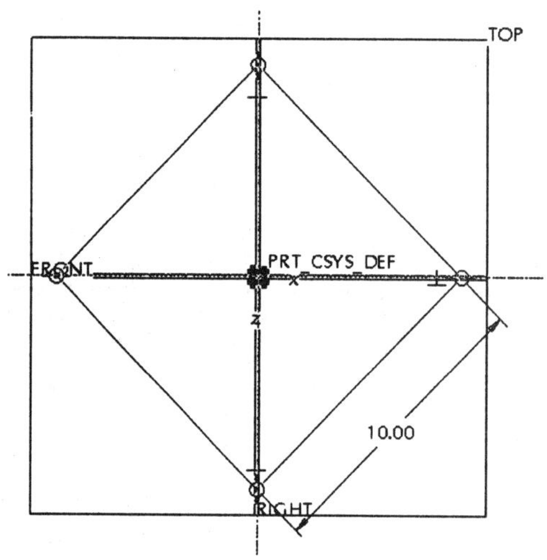

Upon completing the sketch, select the icon of **Done** from the toolbar of sketcher. Select the icon of **Complete** from the feature control panel.

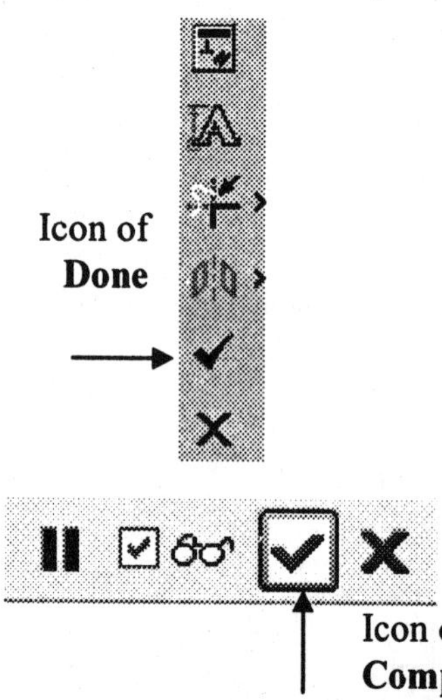

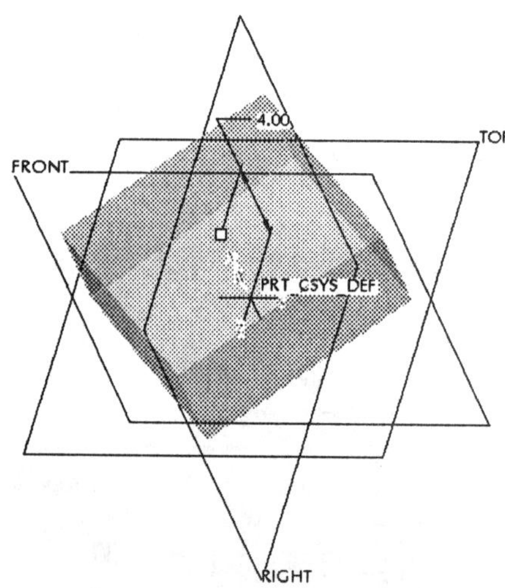

To create a 3D solid model for the cement_layer component, we make use of the created 3D solid model for the root component. From the main toolbar, select File > Make a Copy > type in cement_layer > OK.

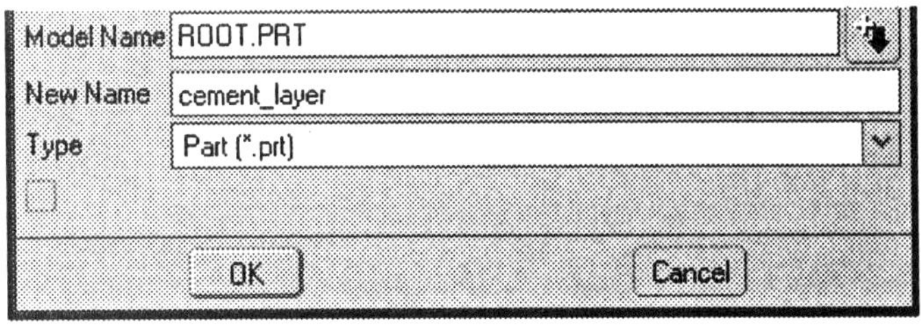

When this process is completed, we open the saved file, called cement_layer.prt. The file should be in the working directory and open it after locating it.

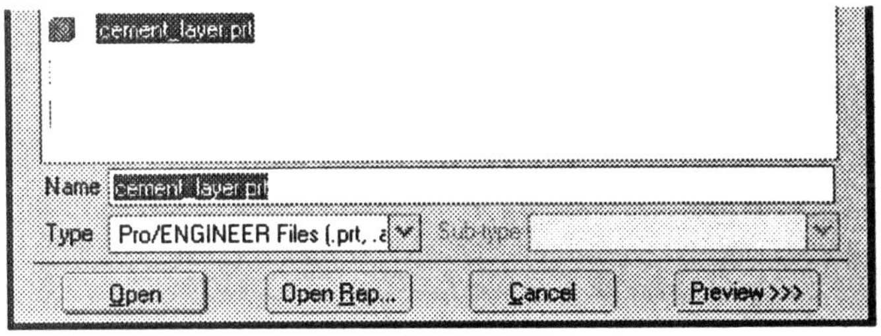

From the model tree, highlight the extrude feature – Extrude 1, right click and hold, pick **Edit Definition** > change the thickness value from 4 to 0.1.

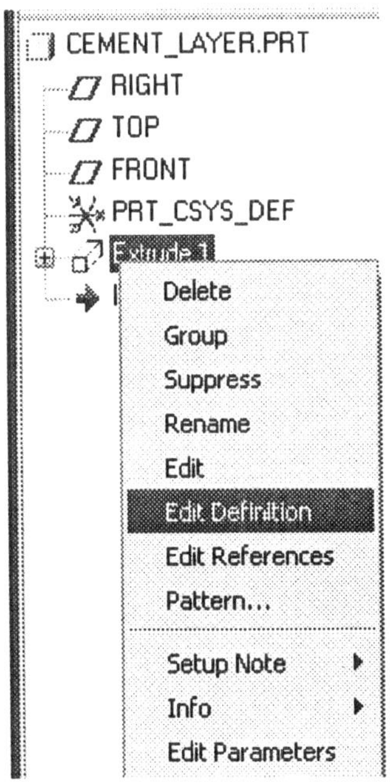

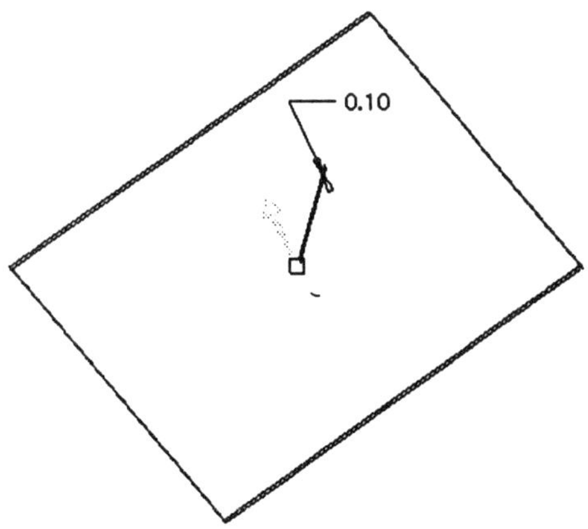

Select the icon of **Complete** from the feature control panel, completing the creation of a 3D solid model for the cement_layer component.

Icon of **Complete**

Following the same procedure, we utilize the 3D solid model for the cement_layer component to create a 3D solid model for the glass_top component. From the menu toolbar, select File > Make a Copy > type in glass_top > OK.

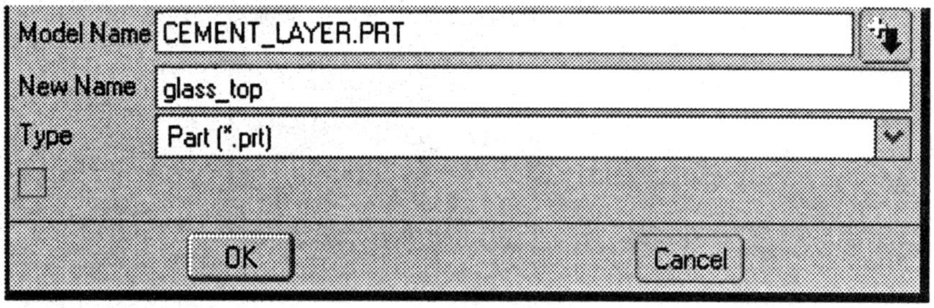

When this process is completed, we open the saved file, called cement_layer.prt. The file should be in the working directory and open it after locating it.

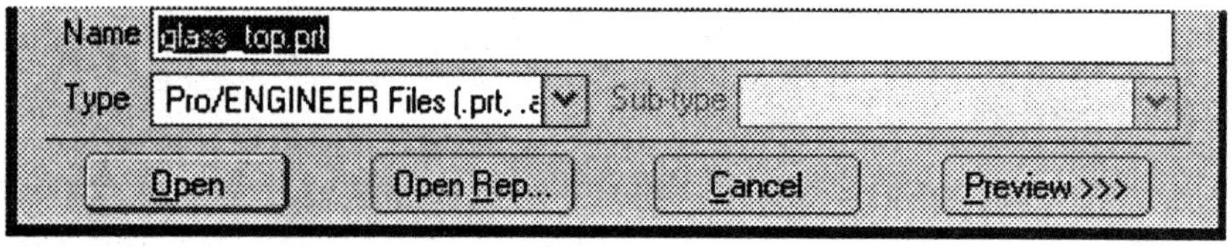

From the model tree, highlight the protrusion feature, right click and pick **Edit Definition** > change the thickness value from 0.1 to 0.9.

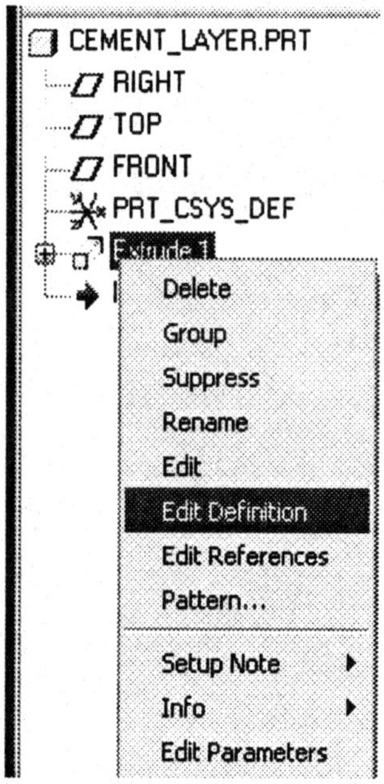

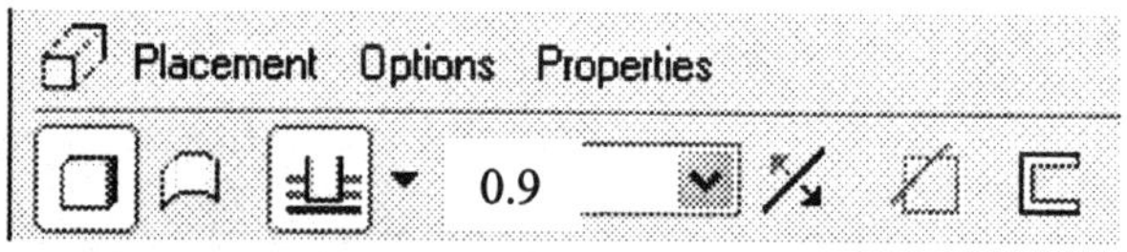

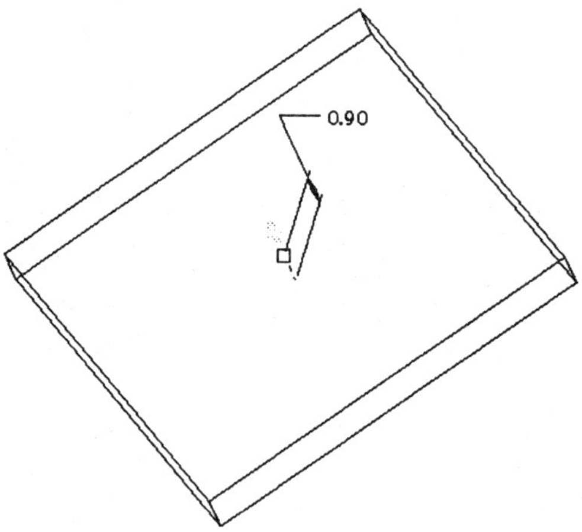

Select the icon of **Complete** from the feature control panel, completing the creation of a 3D solid model for the glass_top component.

Icon of **Complete**

Step 2: Under the Pro/ENGINEER design environment, open the root.prt and define a datum curve at the bottom surface. Afterwards, under the Pro/Mechanica, define a surface region using the defined datum curve.

Directly select the icon of sketch tool > pick the bottom surface as the sketch plane and accept the default setting to orient the sketch plane. The default setting is using the RIGHT datum plane and positioning it at the right side > **Sketch**.

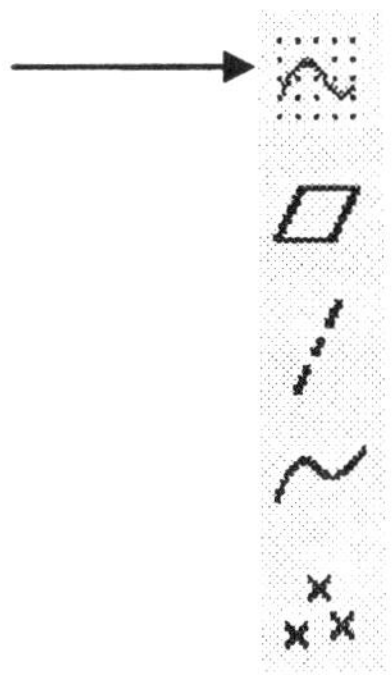

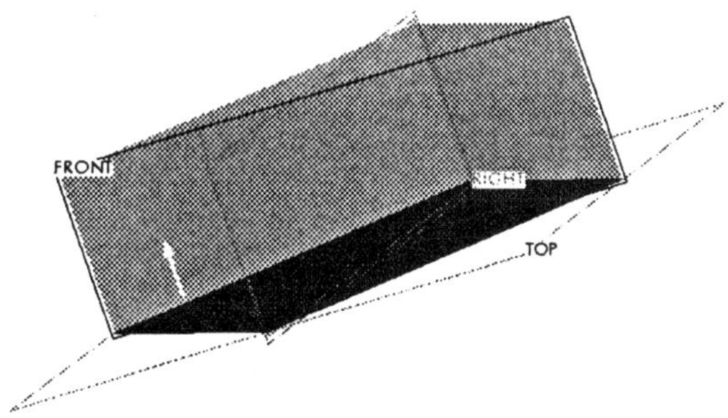

Select the icon of circle and sketch a circle. The diameter dimension is 5, as shown.

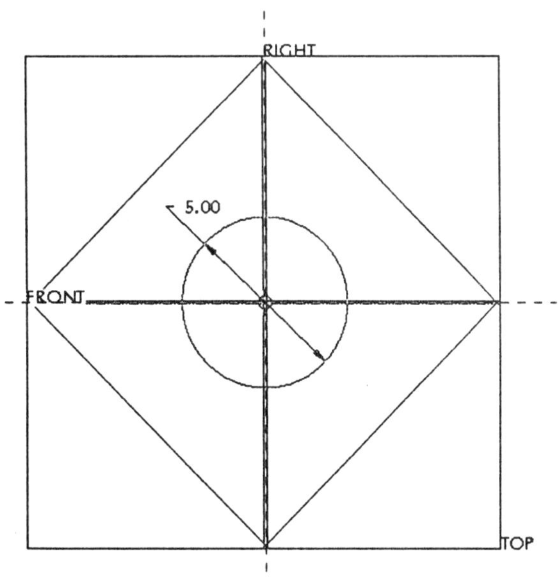

Upon completing the sketch, select the icon of **Complete** from the feature control panel.

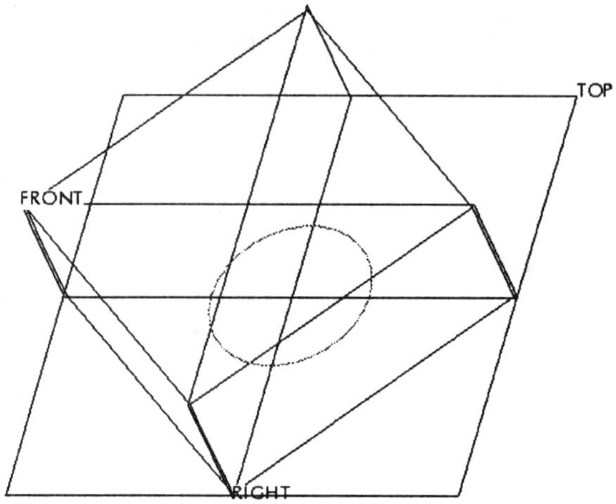

Under the **Pro/MECHANICA** environment, define a surface region using the defined datum curve.

From the main toolbar, select **Applications** > **Mechanica**

Click **Continue** after checking the unit system > **Structure > OK**.

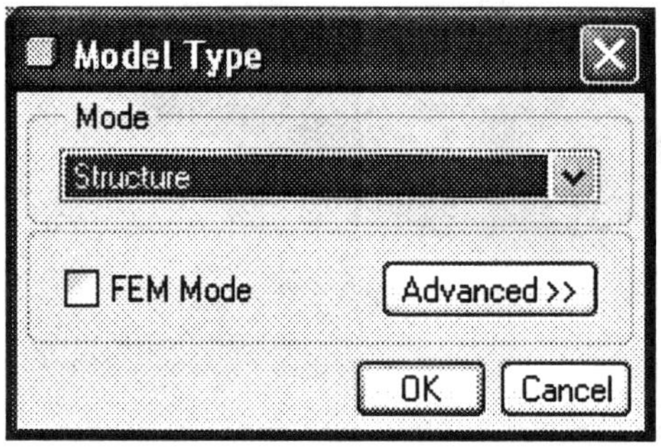

To define the first surface region, select the icon of **surface region** > **Create > Select > Done.** Pick the datum curve first, as shown > pick the surface to be split afterwards > **OK > Done > OK**.

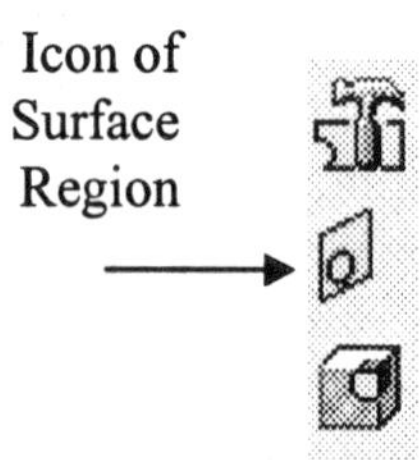

Step 3: Assemble the three components to define a three-layer structure. From the menu tool bar, select **File > New > Assembly** > type in *flat_layers* as the file name and clear the icon of **Use default template > OK.**

Select **mmns_asm_design** (units: Millimeter, Newton, Second) and type *flat layer structure* in **DESCRIPTION**, and *student* in **MODELED_BY**, then **OK.**

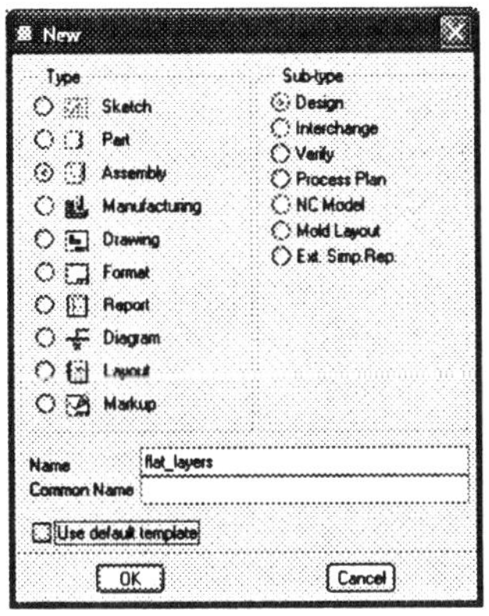

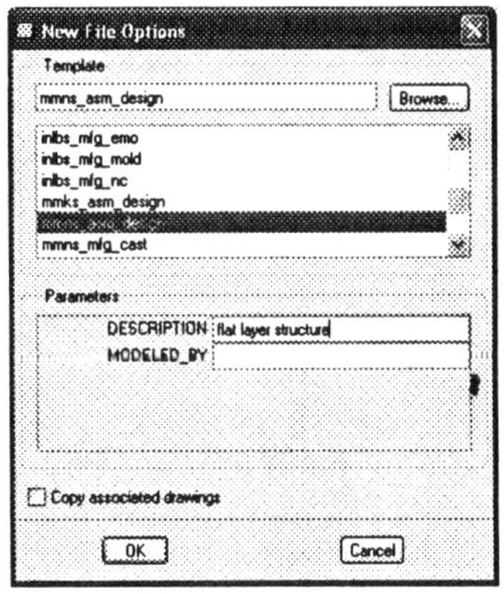

Select the icon of **adding component** > from In Session, select *rcot.prt* > **Open**

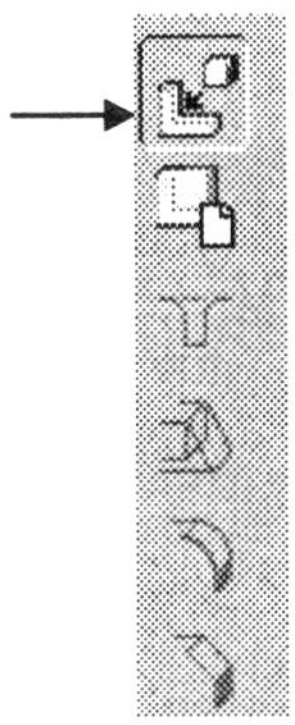

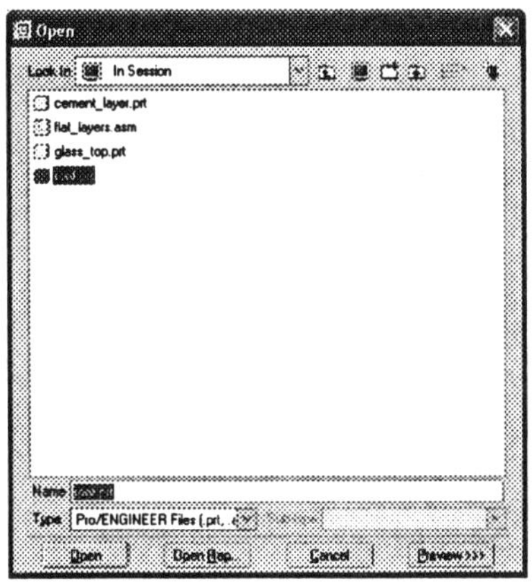

On the window of Component Placement, click the icon of assembling the component at the default location > the placement status is fully constrained, OK.

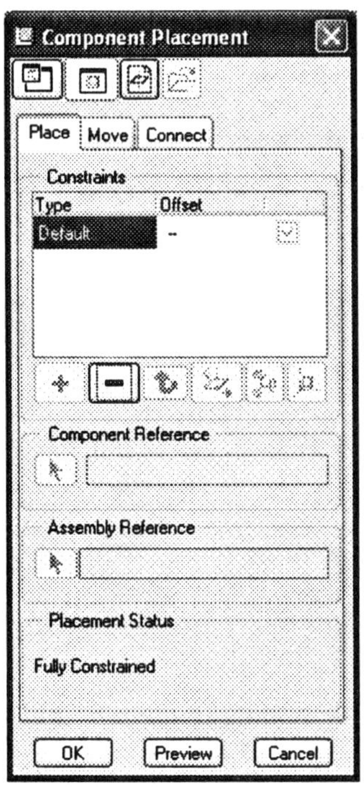

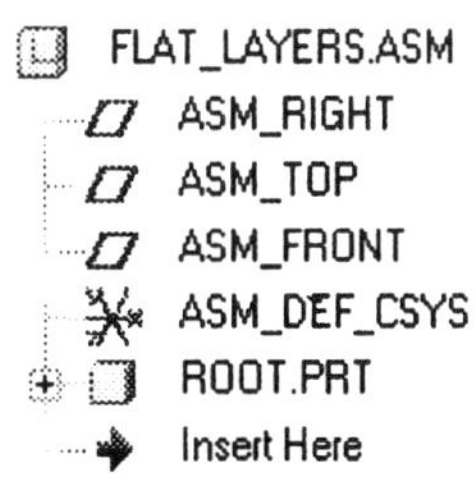

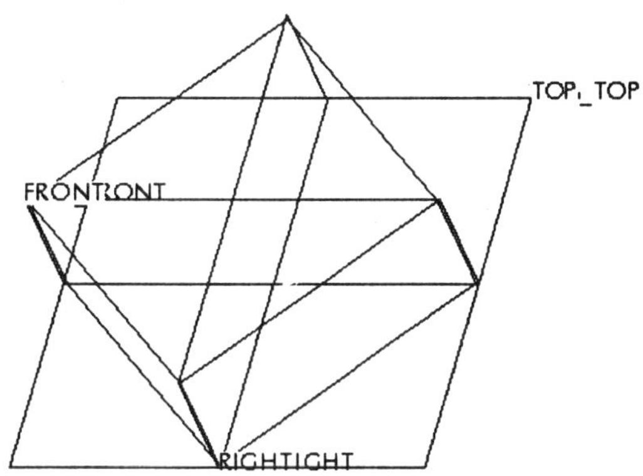

To assemble the cement_layer component, select the icon of **adding component** > from In Session, select *cement_layer.prt* > **Open**.

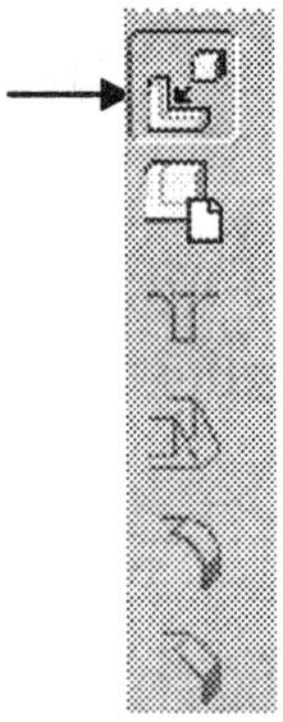

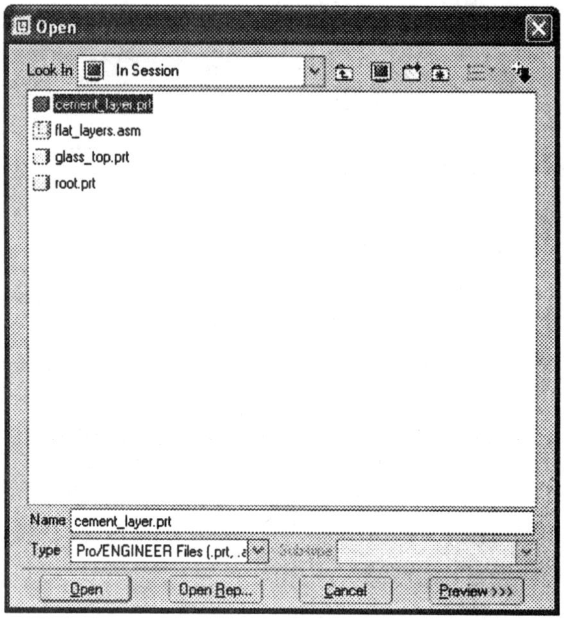

Use the **Mate** function to allow the top surface of the root component to be in contact with the bottom surface of the cement_layer component.
Use the **Mate** function to allow the top surface of the root component to be in contact with the bottom surface of the cement_layer component.

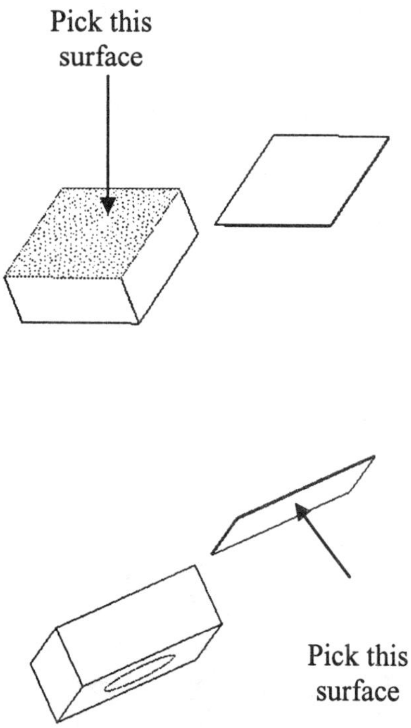

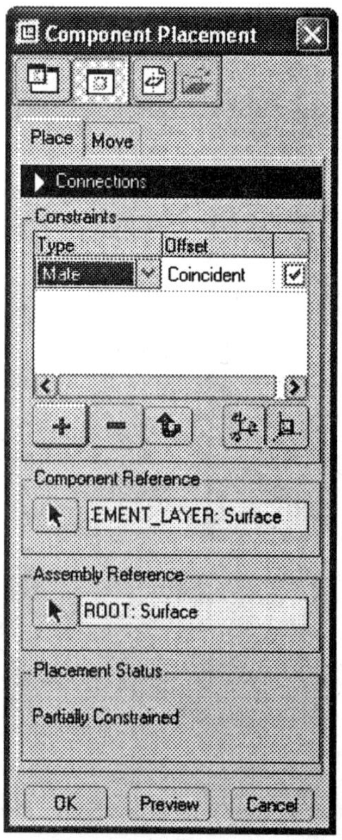

Add the second constraint, which is the alignments of the 2 sided surfaces, as shown.

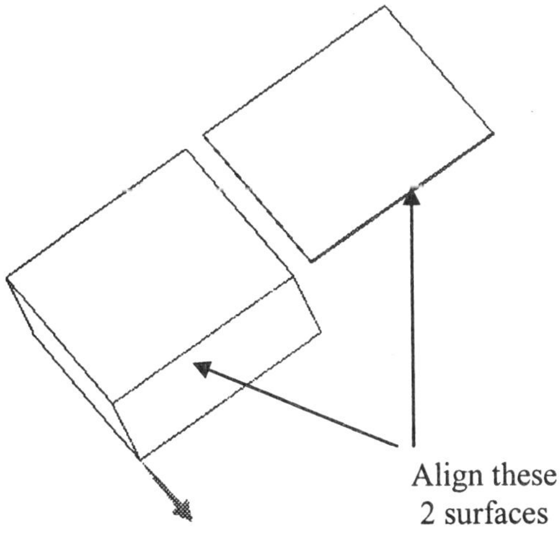

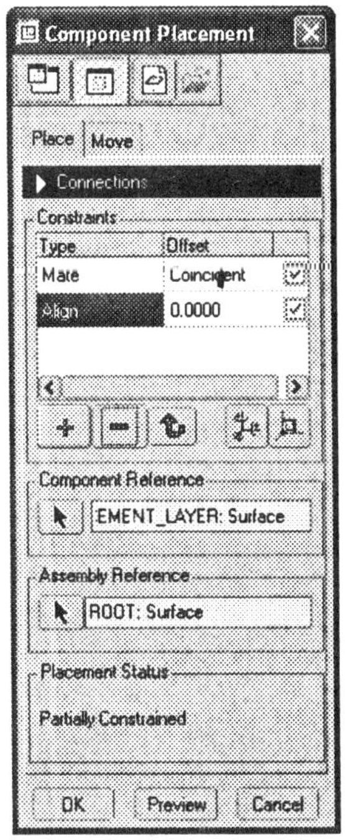

Add the third constraint, which is the alignments of the other 2 sided surfaces, as shown.

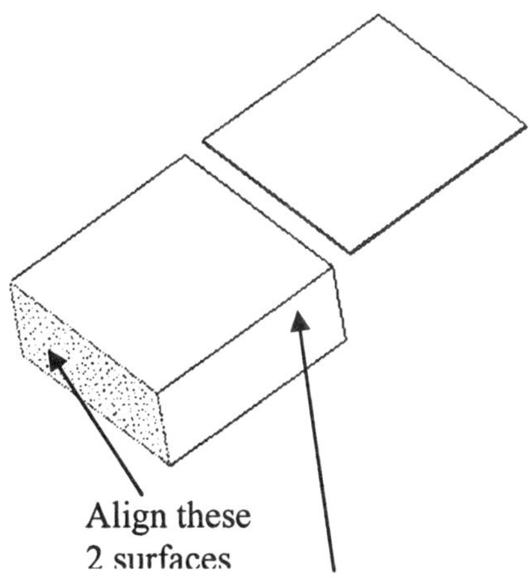

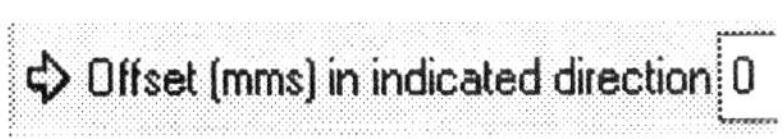

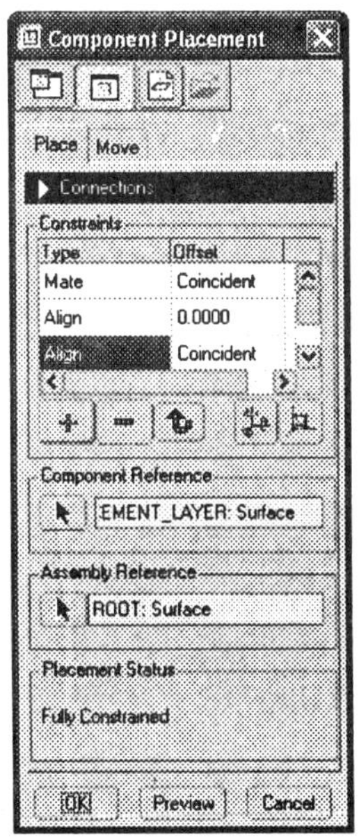

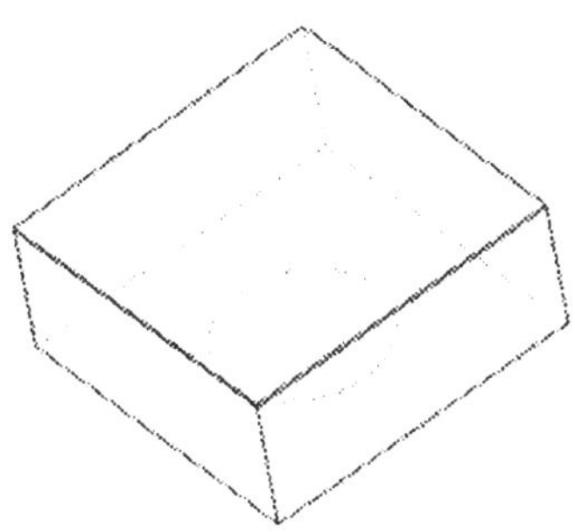

To assemble the glass_top component, select the icon of adding component > from In Session, select *glass_top.prt* > Open.

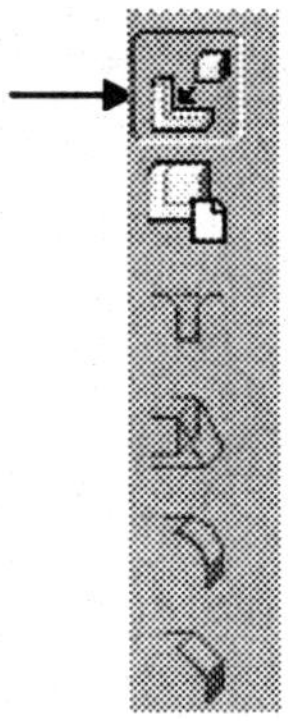

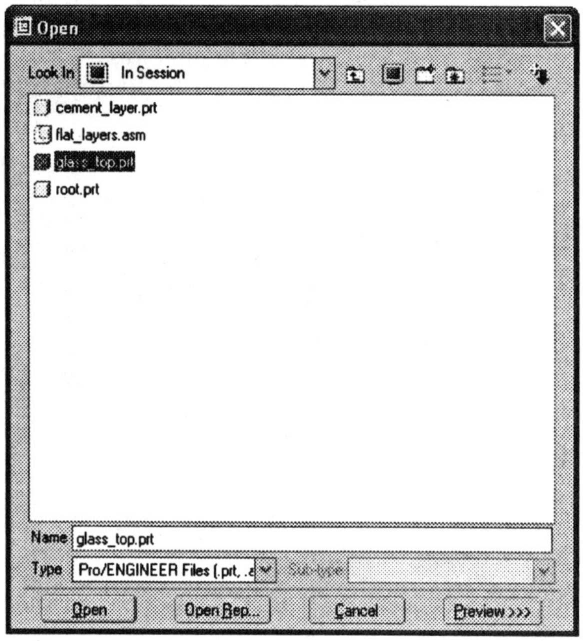

Use the Mate function to allow the top surface of the cement_layer component to be in contact with the bottom surface of the glass_top component.

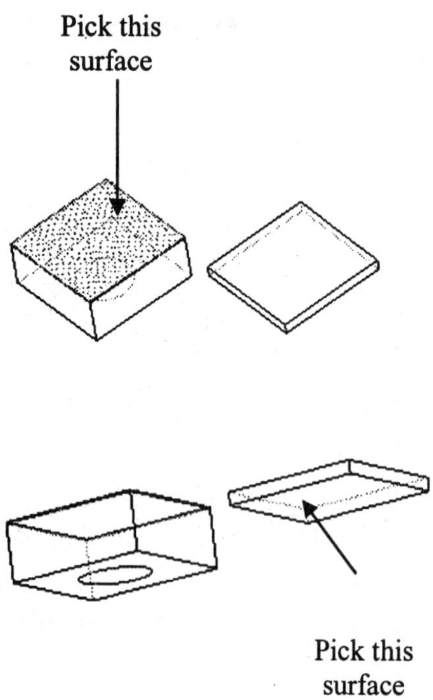

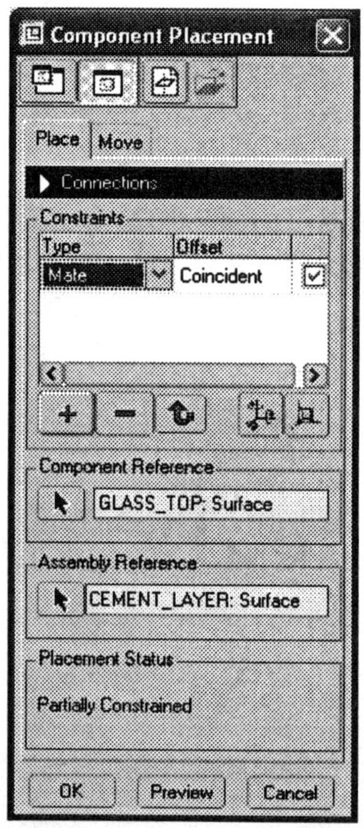

Add the second constraint, which is the alignments of the 2 sided surfaces, as shown.

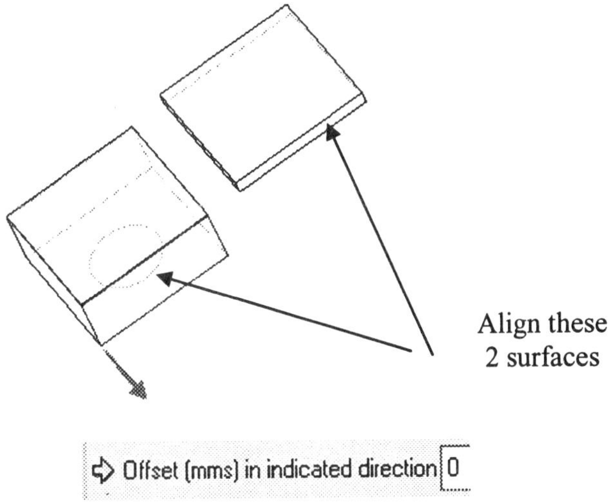

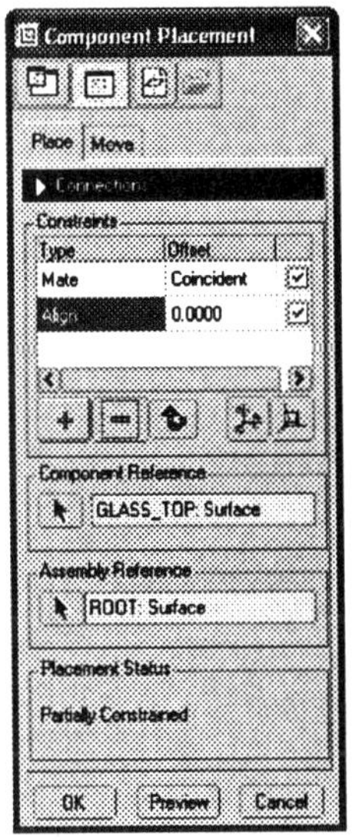

Add the third constraint, which is the alignments of the other 2 sided surfaces, as shown.

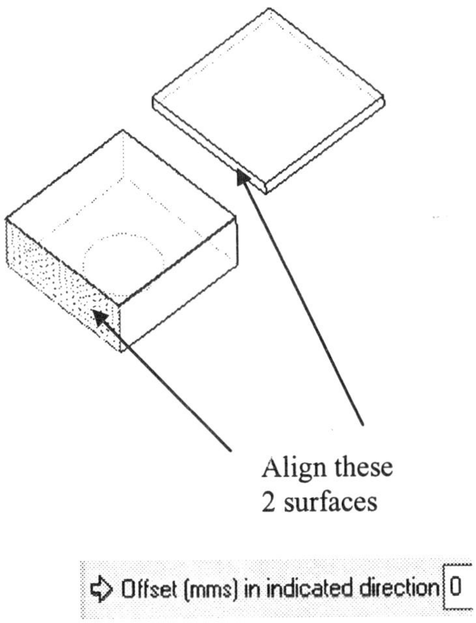

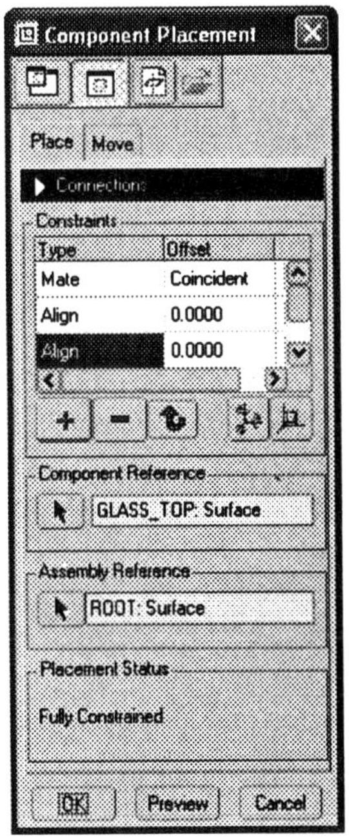

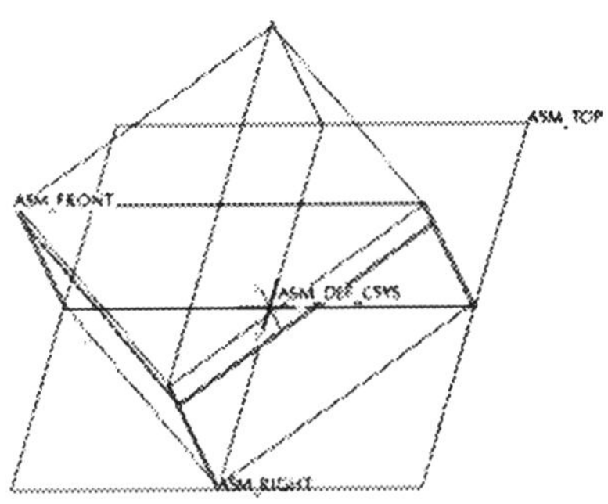

Step 4: Under the **Pro/MECHANICA** environment, specify the three different types of material, and then assign each of the material type to its corresponding component.

From the main toolbar, select **Applications** > **Mechanica**

Click **Continue** after checking the unit system > **Structure > OK**.

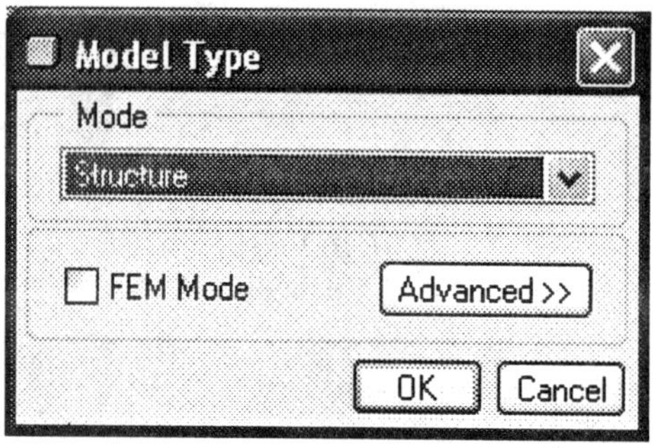

From the toolbar of functions, select the icon of **Materials > New.**

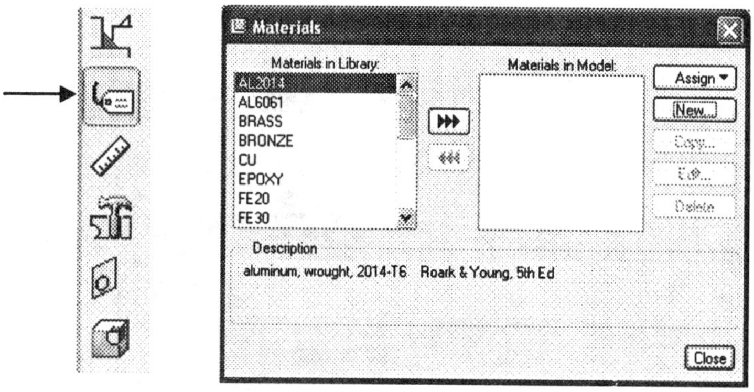

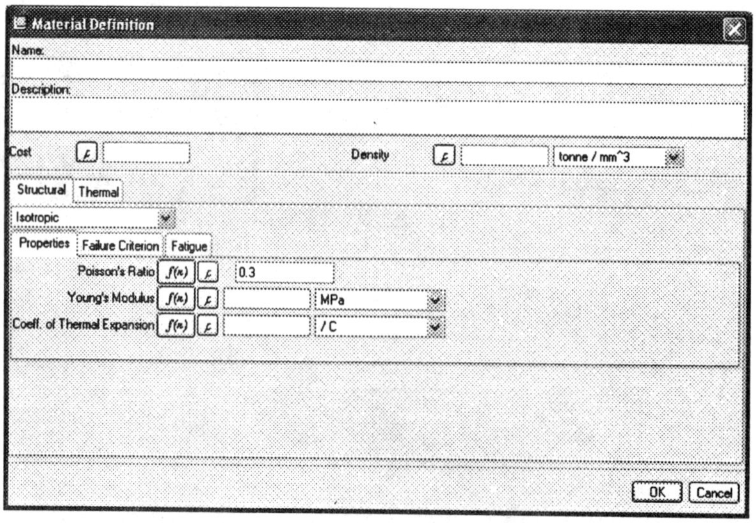

For the glass_top component, fill in the table of material properties as follows > OK.

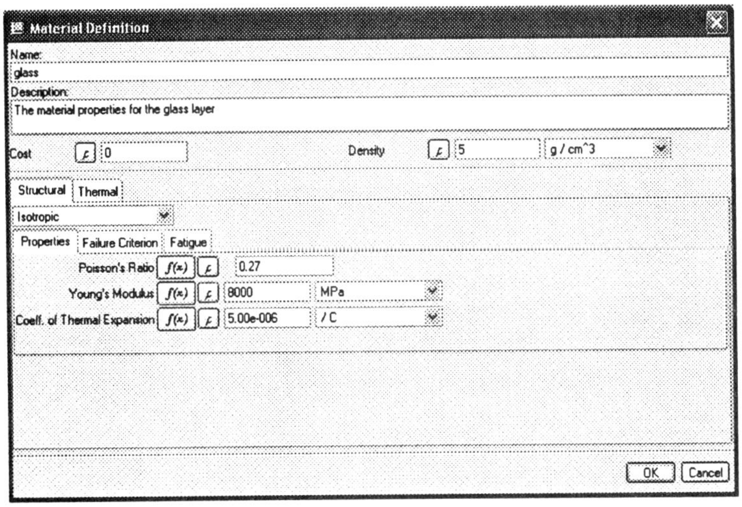

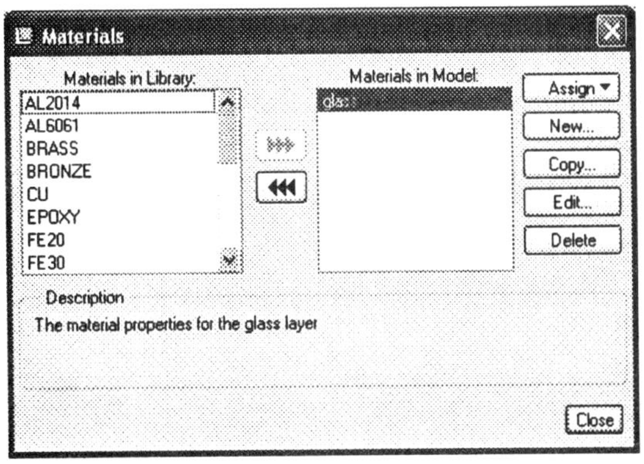

To save the defined material properties in the library, click the arrow toward the Materials in Library > Yes.

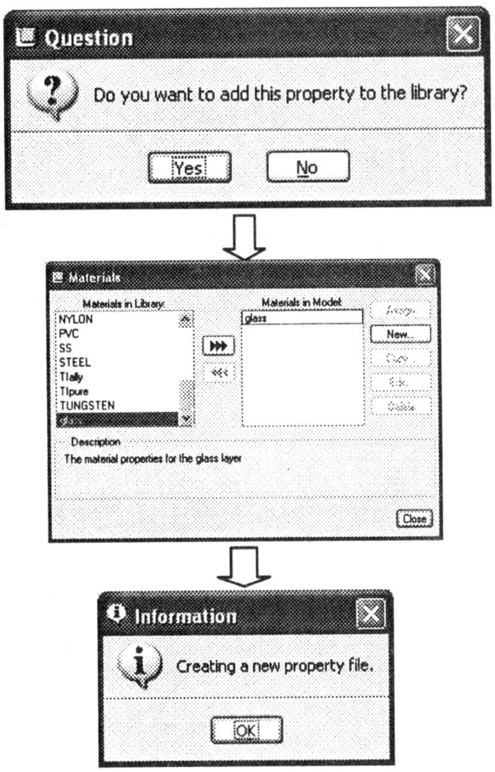

To add the thermal properties of the glass material, click Edit from the window of Materials.

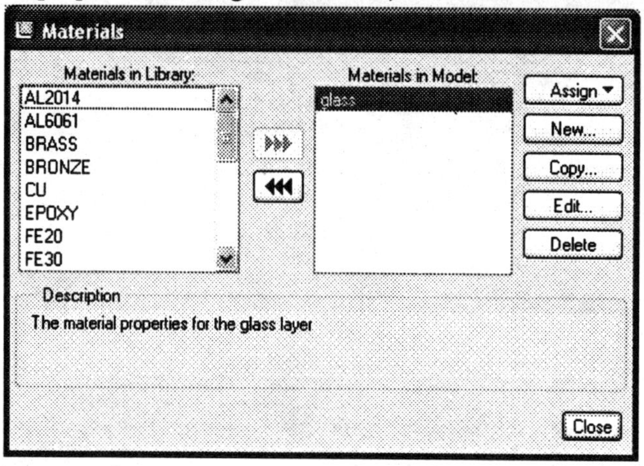

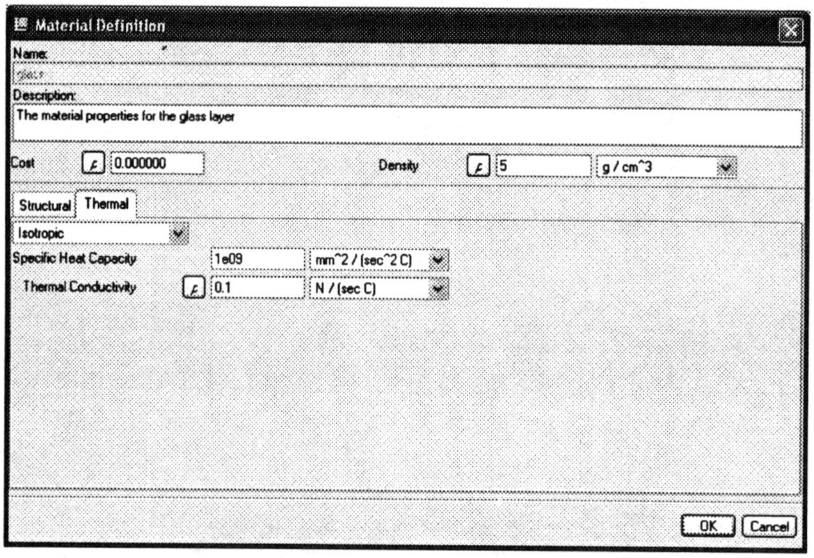

Repeat the above steps to define the material properties for the cement and the composite. For the cement layer, fill in the table of material properties as follows:

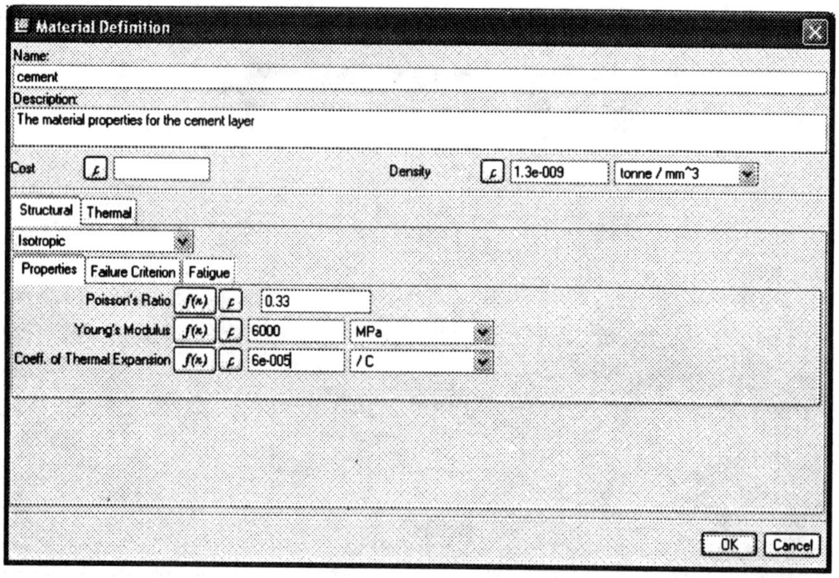

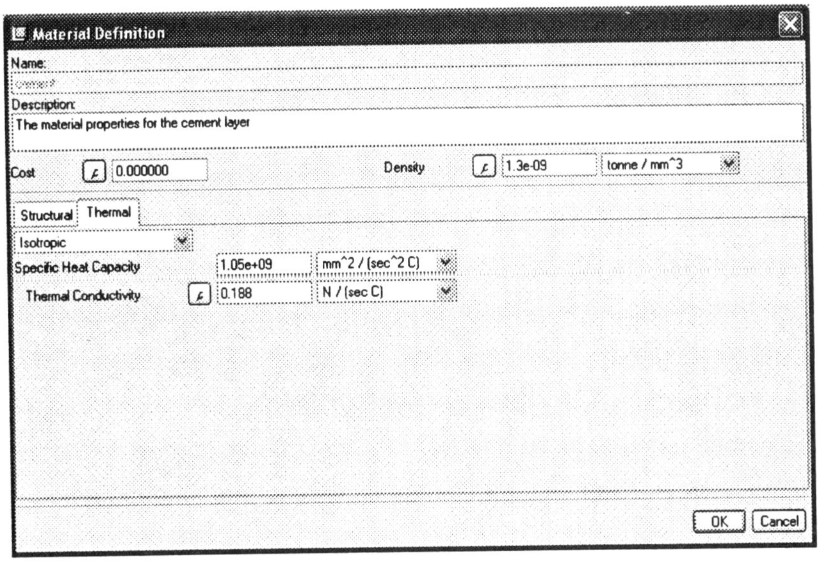

For the root substrate, fill in the table of material properties as follows:

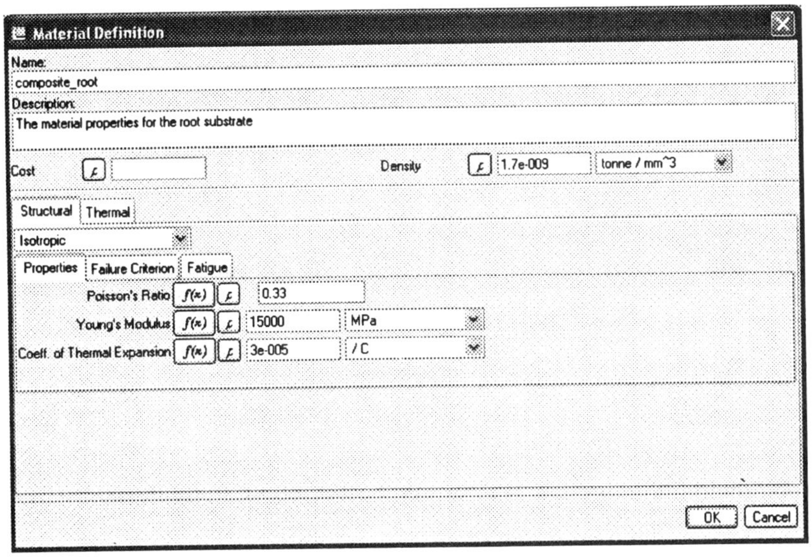

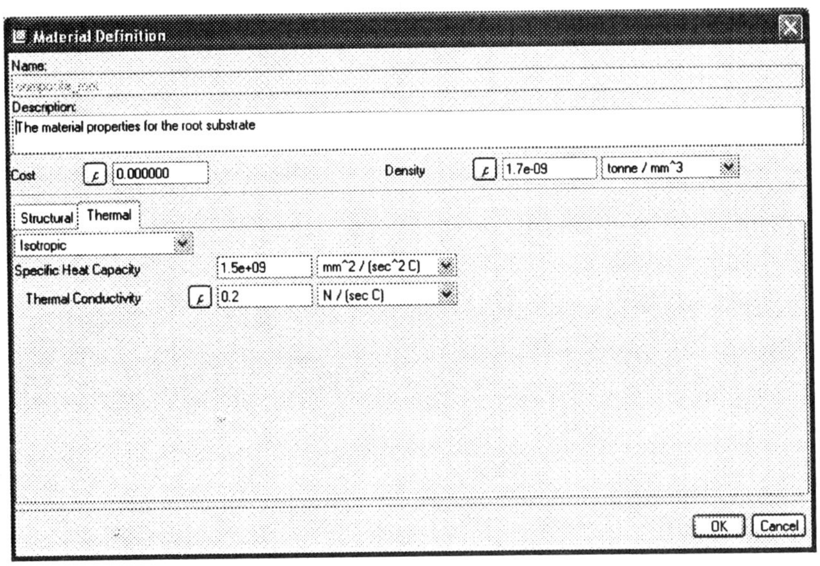

Step 5: Under the **Pro/MECHANICA** environment, define the constraint condition.

From the toolbar of functions, select the icon of **Constraints >** select the surface region defined at the bottom surface > **OK** > accept the default settings, namely, fixing all 6 degrees of freedom > **OK.**

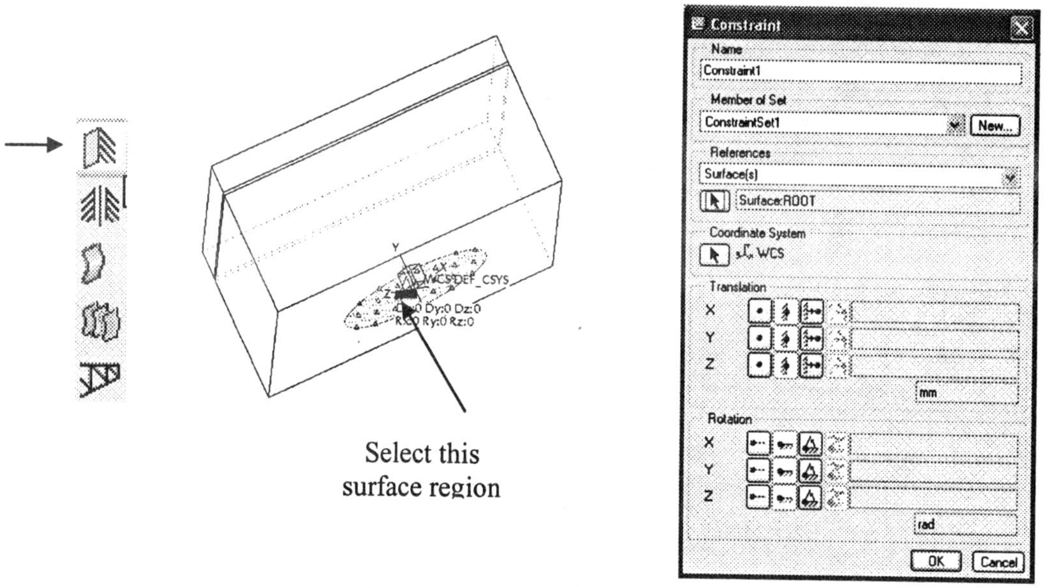

Step 6: Under the Pro/MECHANICA environment, define the load condition.
For both cases,

The temperature variation:

reference temperature:	20°C
model temperature:	40°C

From the toolbar of functions, select the icon of **Global Temp Load >** type 40 in the box of Model Temperature and 20 in the Reference Temperature > **OK.**

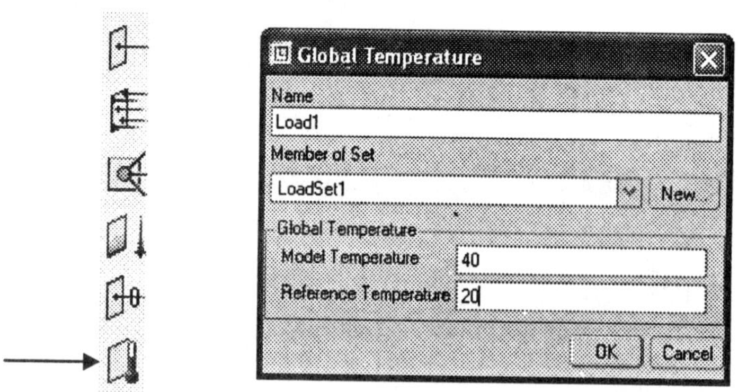

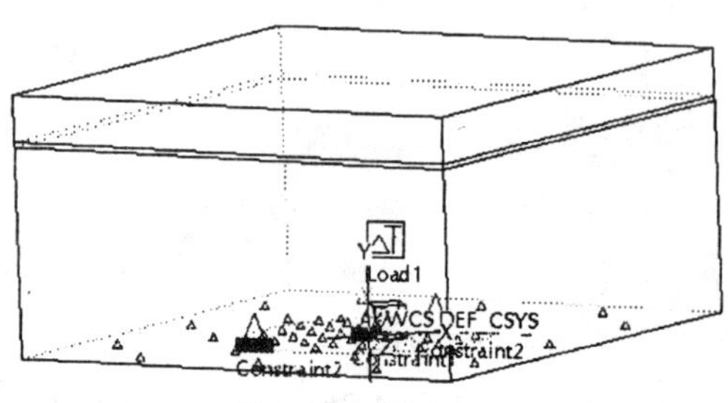

Step 7: Perform a new static analysis called flat_layers_1.

Analyses > File > New Static > type *flat_layers_1* and make a note that the surface region on the bottom surface of the root component is used.

Make sure that ConstraintSet1 and LoadSet1 are highlighted > **Single-Pass Adaptive > OK > Run.**

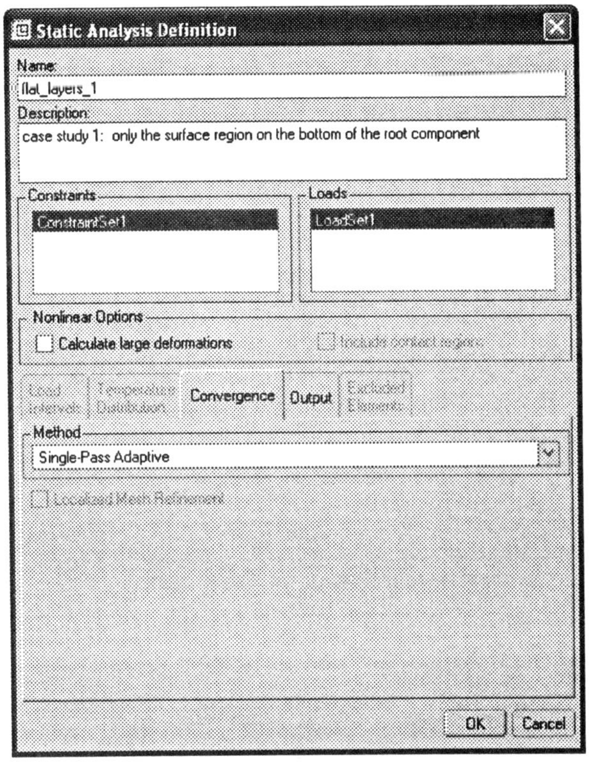

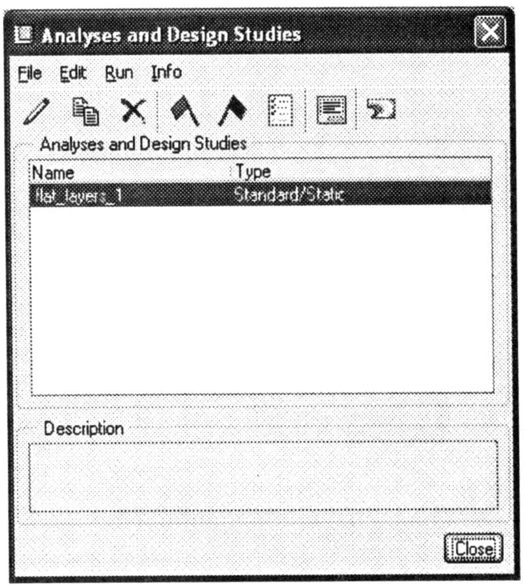

Structure > Results > Insert > Result Window > type in deflection distribution (assembly) as the title > select the folder called flat_layers_1 > Fringe > Displacement > Magnitude > in the Display Option, select Deformed and set the scale to 10 > **OK and Show.**

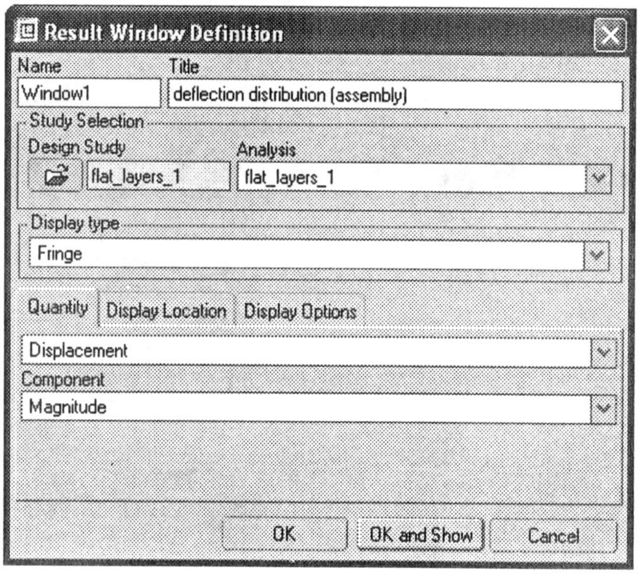

To show the deflection distribution for each of the three components, from the menu bar, select Edit > Copy > type in deflection distribution (glass) > Fringe > Displacement > Magnitude > activate Display Location > select Volume > activate the selection and pick the glass_top and middle button > OK and Show.

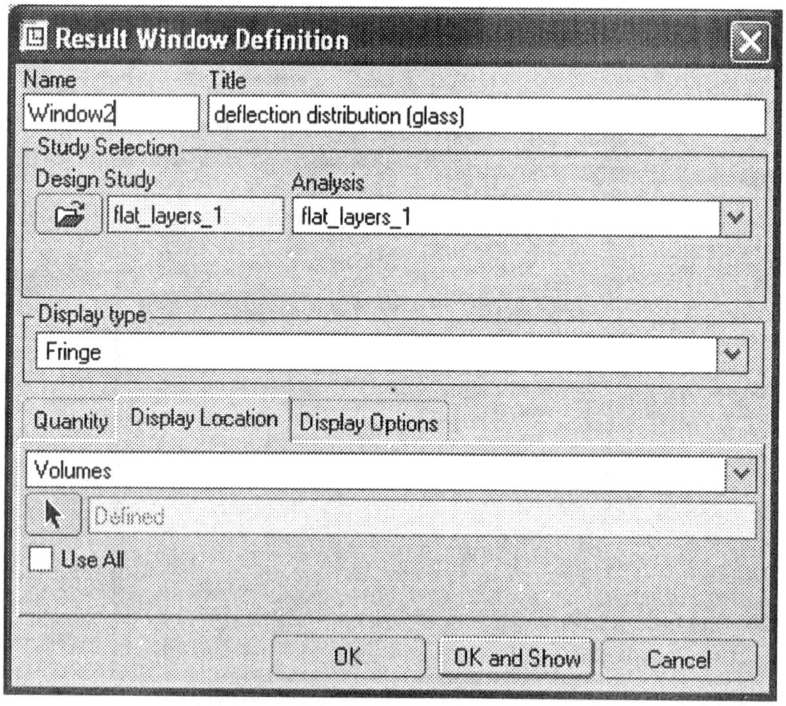

Repeat the above procedure to show the deflection distribution for the cement layer and the root component.

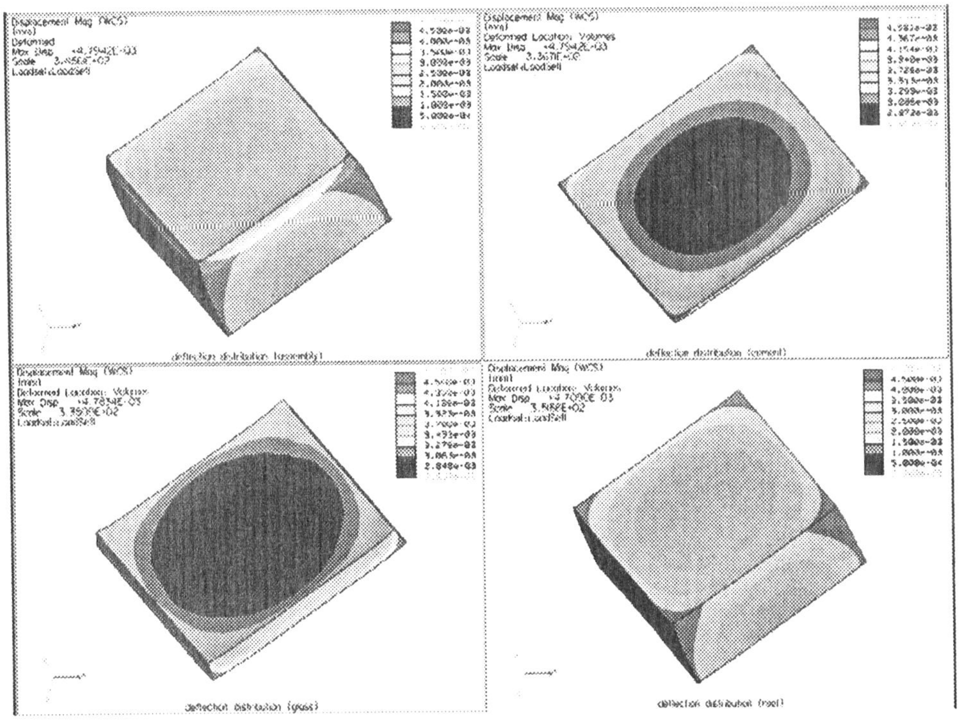

Repeat the above procedure to obtain 4 plots, which show the 4 distributions of the maximum principal stress.

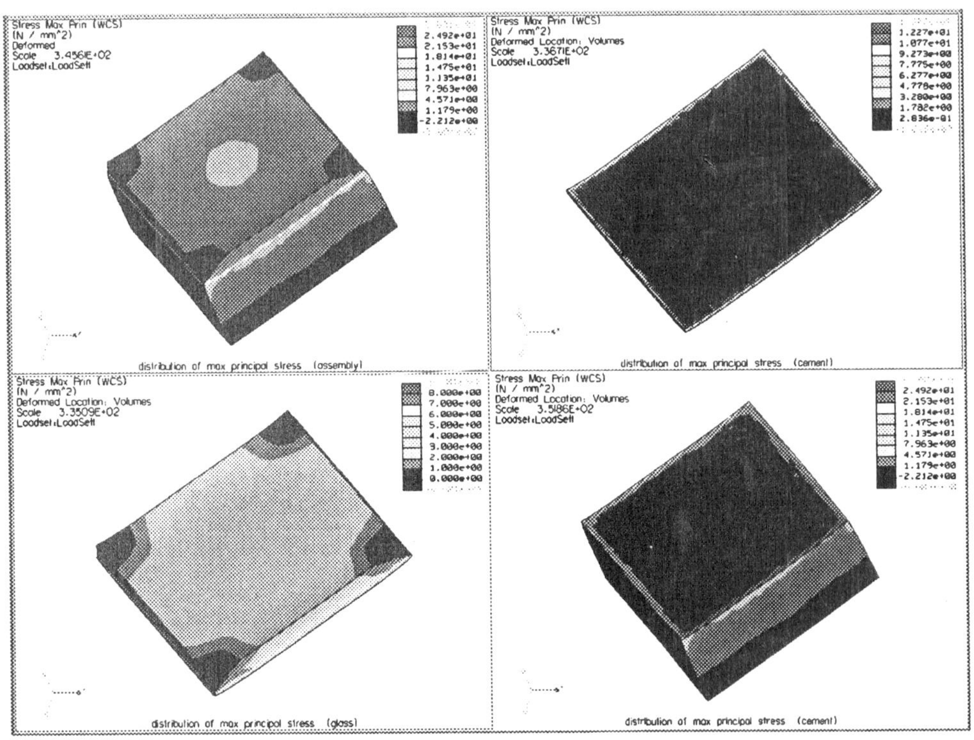

4.6 Use of Symmetry in Performing FEA

In this section, we will perform FEA with a 3D solid model, which is a half of the entire 3D model. We will also perform FEA with a 3D solid model, which is quarter of the entire 3D model. We will continue working with the 3D solid model of the assembly constructed in section 4.5 If we examine that 3D solid model closely, The geometry of the 3D solid model of the assembly is symmetric about the two planes as shown.

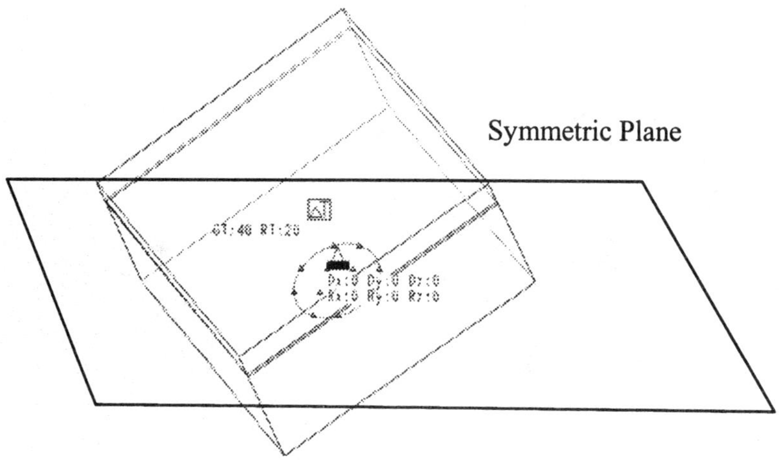

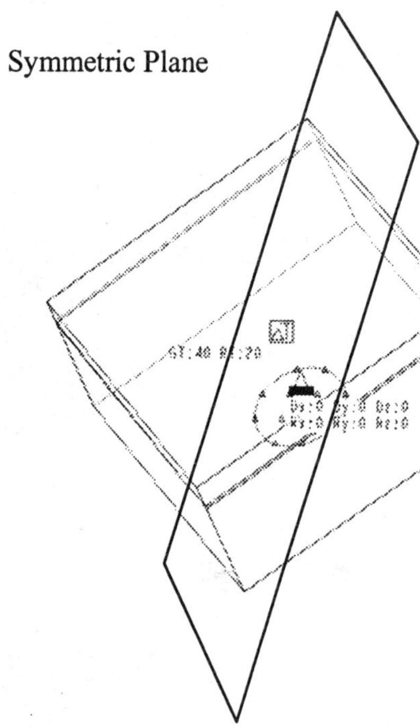

It is also very important that the constraint condition is also symmetric about those two symmetrical planes. The load condition is a temperature variation, which is symmetric about those two symmetrical planes as well.

To obtain a half model, under the Pro/ENGINEER design environment, let us open the assembly file called flat_layers. Select the icon of Extrude Tool.

Select the icon of Cut and set the depth choice to Through All. Activate Placement > Define.

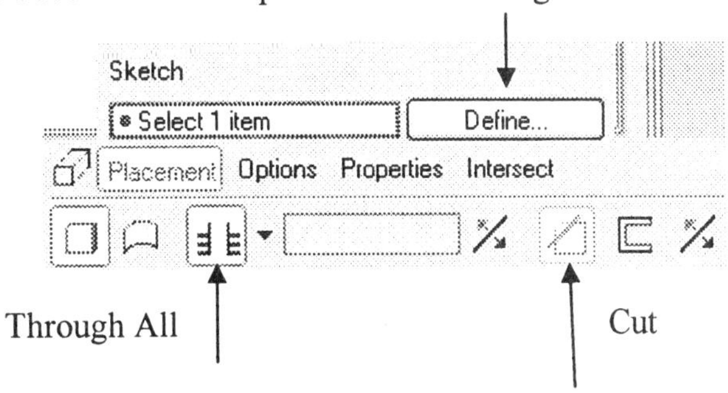

Pick the **ASM** datum plane (in this case, ASM_TOP) as the sketch plane. Accept the default setting to orient the sketch plane > **Sketch**.

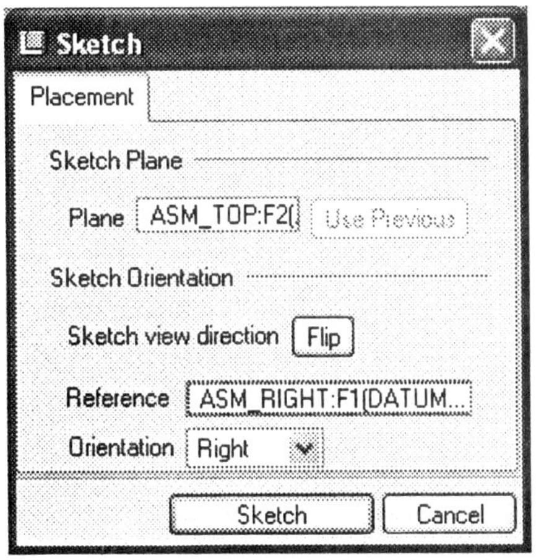

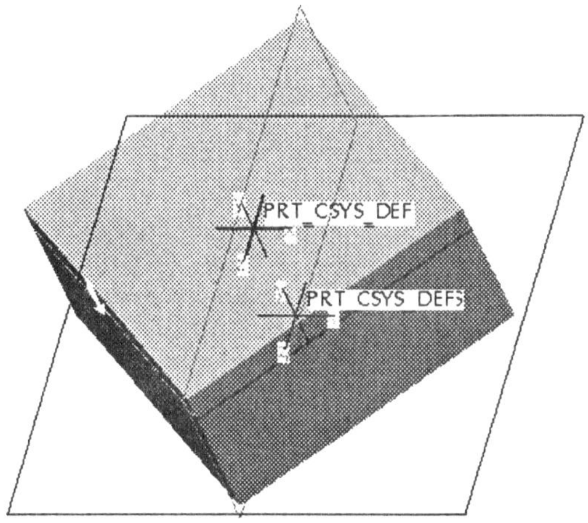

Select the icon of rectangle from the toolbar of sketcher. Sketch a rectangle, whose dimensions should be sufficient large so that half of the 3D solid model of the assembly can be removed.

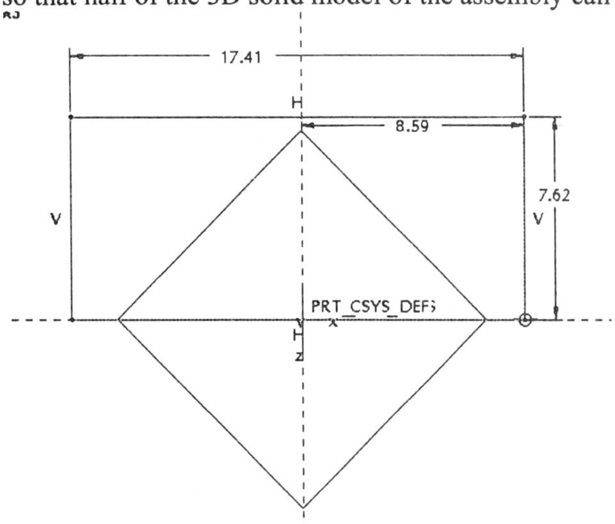

Upon completing the sketch, select the icon of **Done** from the toolbar of sketcher. Select the icon of **Complete** from the feature control panel.

Icon of **Done**

Icon of **Complete**

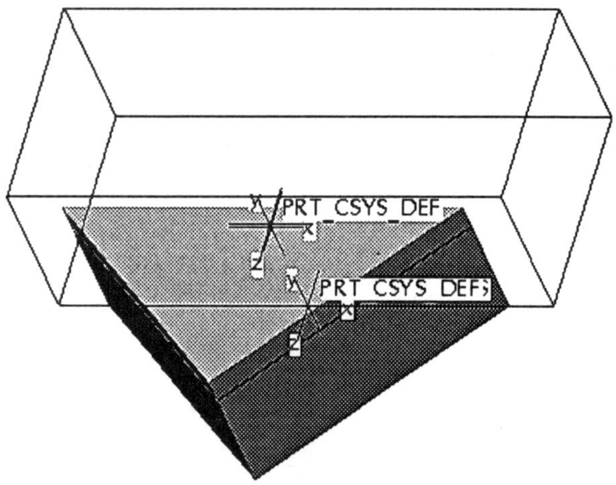

Now let us add the constraint conditions to this half model. First, we set the bottom surface fixed to the ground.

From the toolbar of functions, select the icon of **Constraints** > select the surface region defined at the bottom surface > **OK** > accept the default settings, namely, fixing all 6 degrees of freedom > **OK**.

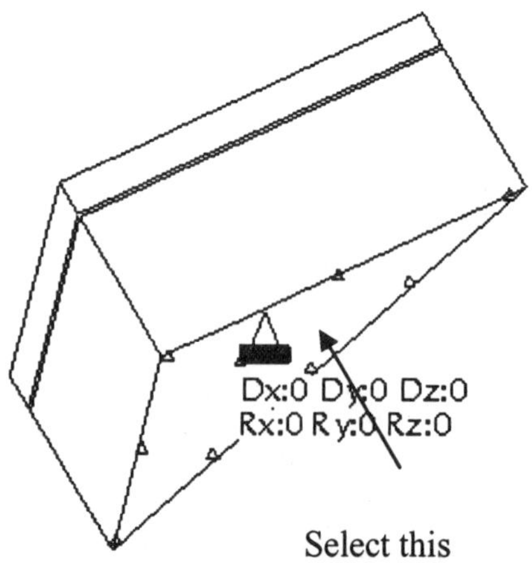

Select this bottom surface

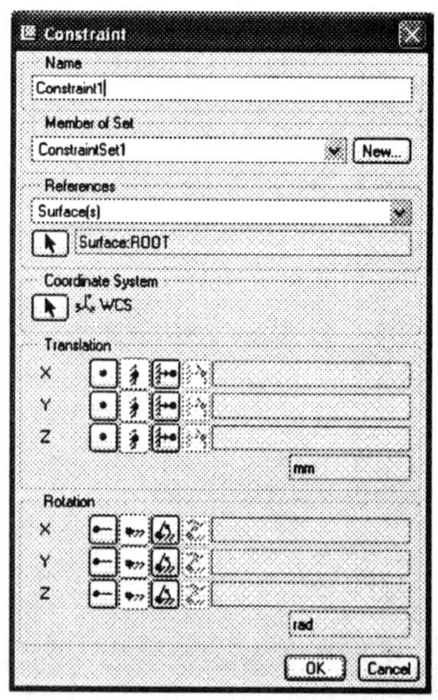

It is important to add a constraint condition on the cutting plane. For the half model of the layer structure, the additional constraint condition is to select the 3 surfaces from the 3 components, and set Translation Z to zero, Rotation X to zero and Rotation Y to zero. On the other hand, set Translation X to be free, Translation Y to be free and Rotation Z to be free, as.shown below.

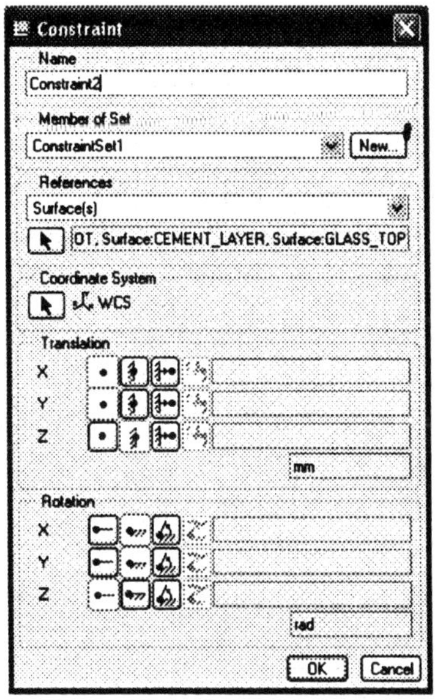

Select the 3 surfaces from the 3 components

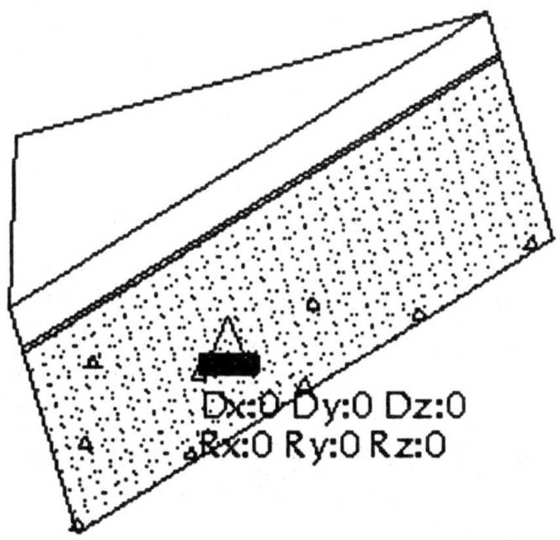

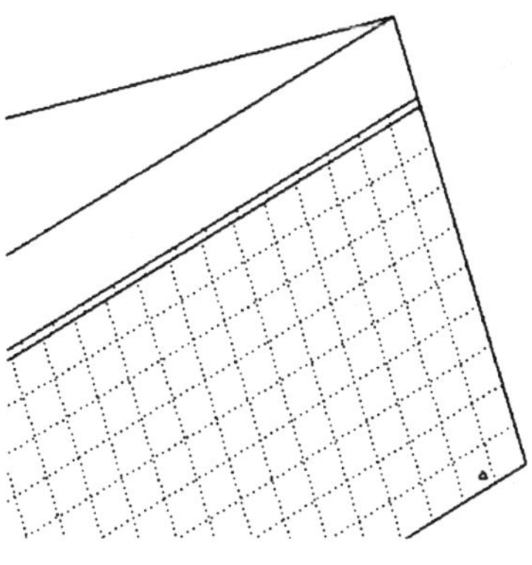

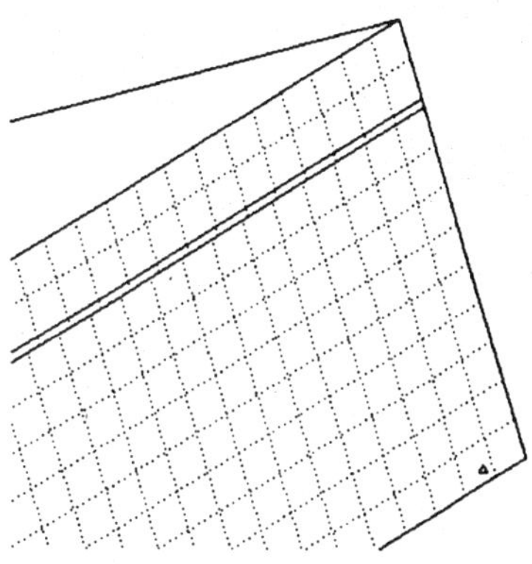

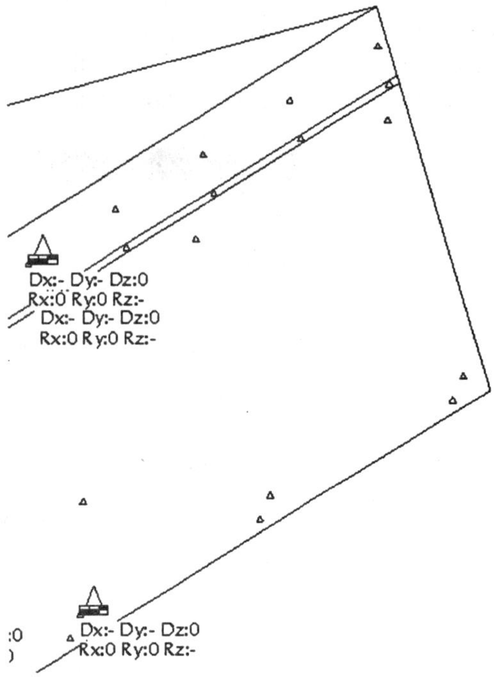

If we are dealing with a quarter model, then we need to add another constraint condition on the second cutting plane. Readers are asked to create a quarter model when doing the exercise on your own.

To set the load condition, there is no difference as compared the work we did when dealing with the entire 3D solid model of the assembly.

The temperature variation:

reference temperature:	20°C
model temperature:	40°C

From the toolbar of functions, select the icon of **Global Temp Load** > type 40 in the box of Model Temperature and 20 in the Reference Temperature > **OK**.

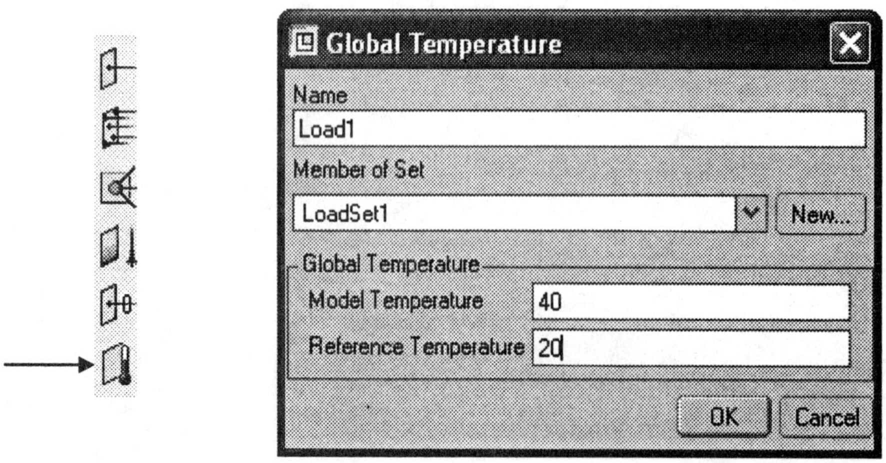

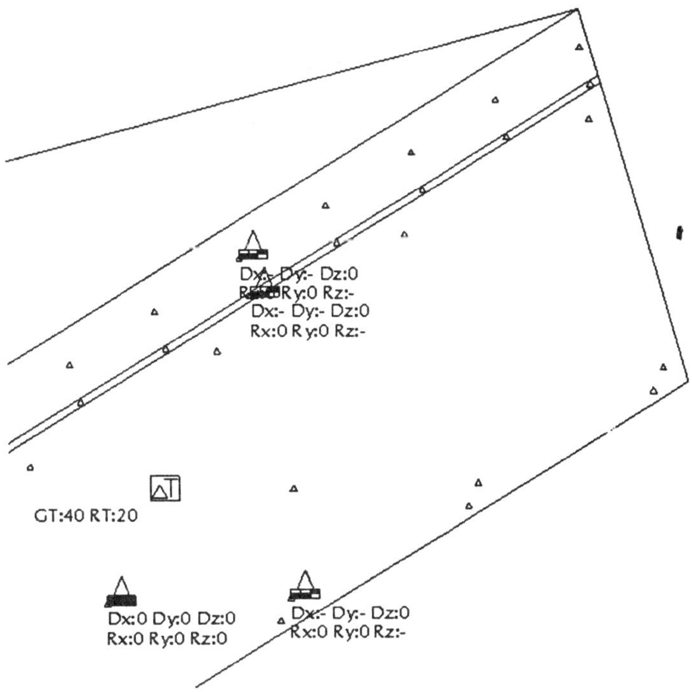

We use the same information on the material types for the 3 components as we used when dealing with the entire 3D solid model of the assembly.

In the following, we tabulate the results obtained from running an entire 3D solid model of the assembly, running a half model and running a quarter model in terms of the number of elements created, number of equations required and the CPU time for a given PC used to run the 3 programs.

	Number of Elements	Number of Equations	CPU Time (sec)
Full Model	2409	70443	640
Half Model	1197	37654	265
Quarter Model	637	21165	99

When examining these numbers, it is evident that the computing effort has been greatly reduced when we use the half model or the quarter model.

A natural question to ask is "Do you obtain the same results after running the three programs?" In the following we list values obtained from running these 3 programs in terms of the maximum displacement, the maximum principal stress and the maximum von Mises stress.

	Max. Displacement	Max Principal Stress	Max VM Stress
Full Model	0.005186 mm	39.10 Mpa	40.87 Mpa
Half Model	0.005216 mm	36.98 Mpa	48.73 Mpa
Quarter Model	0.004919 mm	25.30 Mpa	26.29 Mpa

When examining these numbers, it is evident that there exist differences among the results obtained from running the 3 programs. Which one is best? Users should consider these variations when they decide to use a half model or a quarter model.

In the following, we present the 3 plots, namely, the displacement distribution, the maximum principal stress distribution and the von Mises stress distribution. Each plot consists of 3 figures, which correspond to the 3 cases, namely, the full model, the half model and the quarter model.

Displacement Distributions

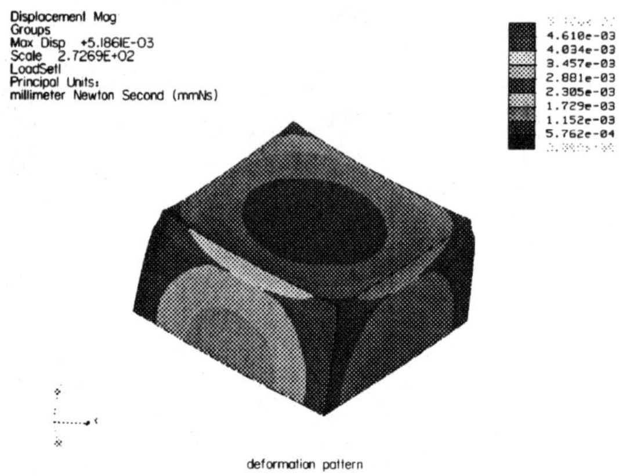

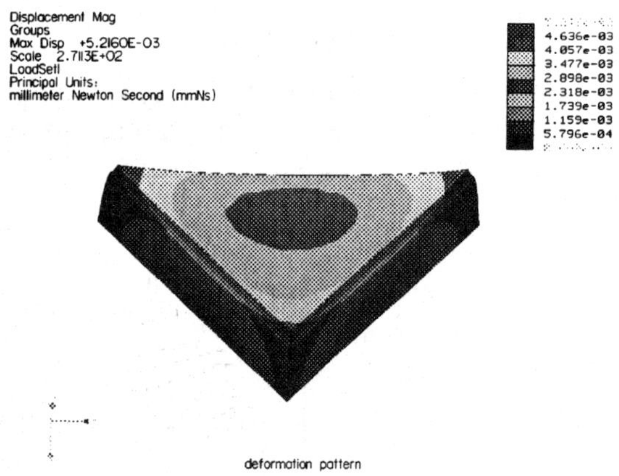

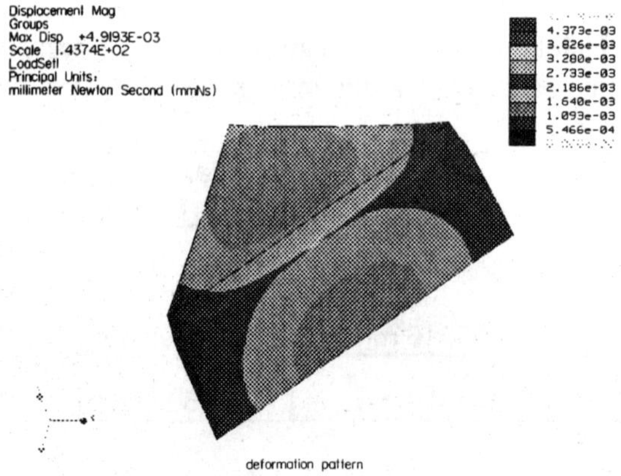

Distributions of the Maximum Principal Stress

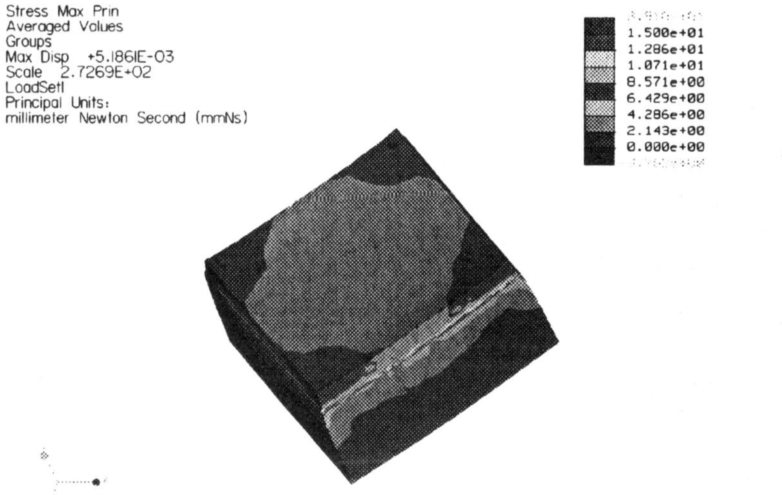

distribution of max principal stress

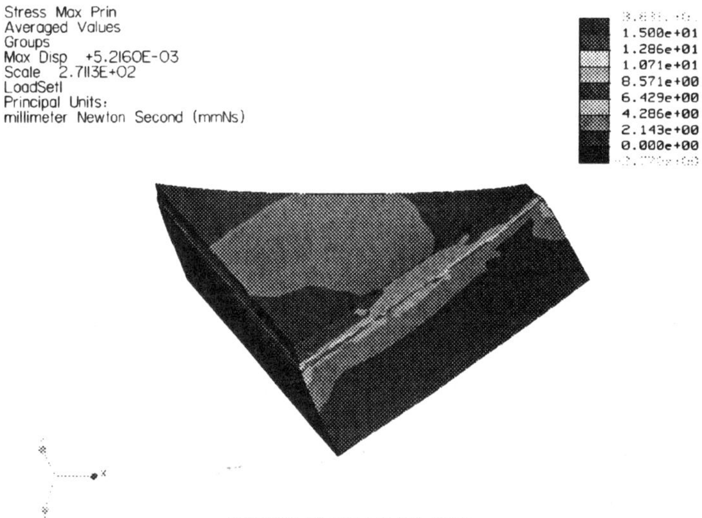

distribution of max principal stress

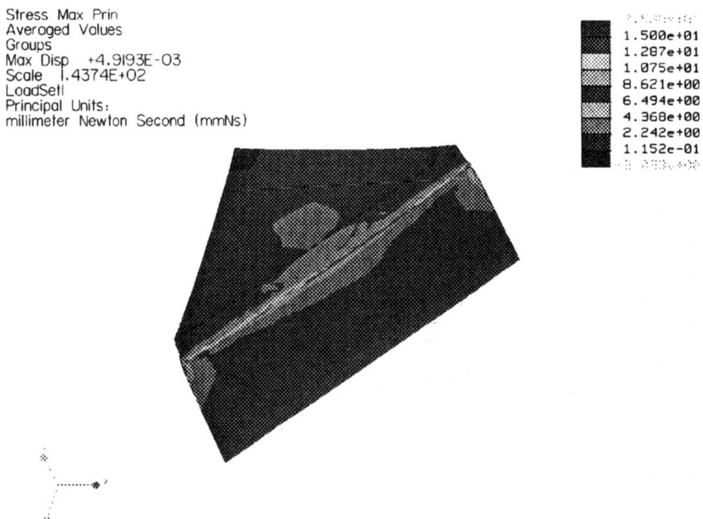

distribution of max principal stress

Distributions of the von Mises Stress

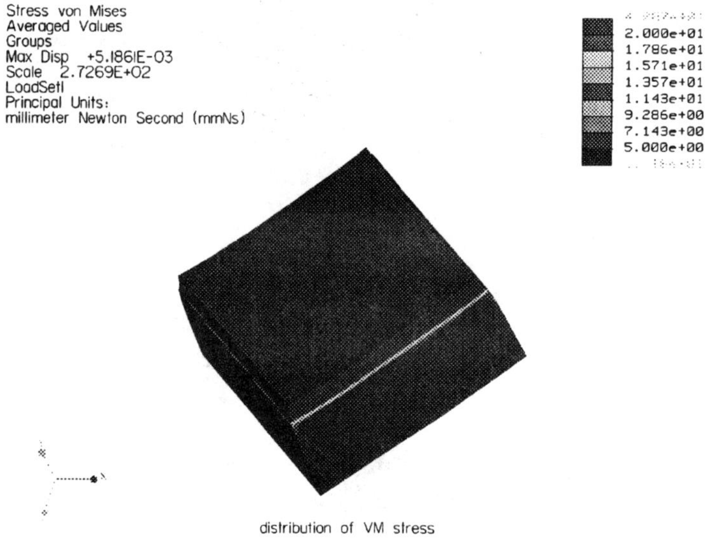

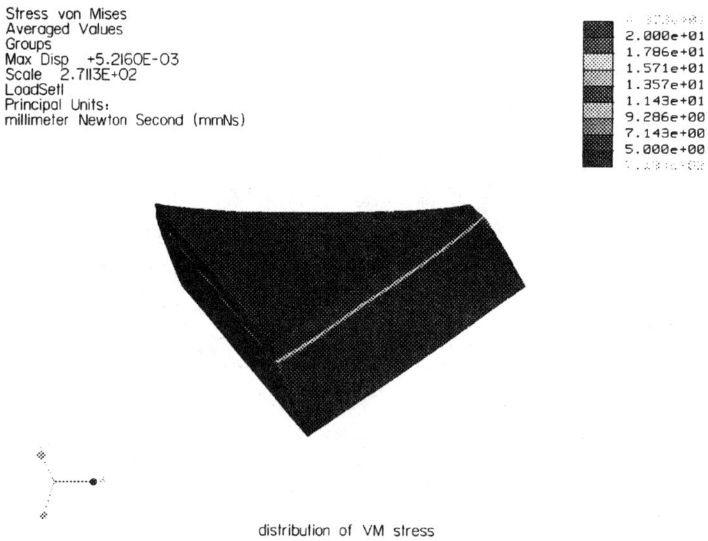

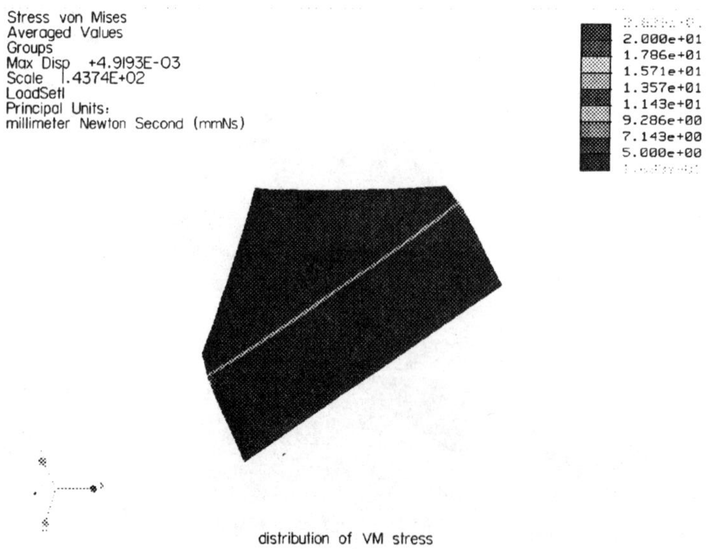

4.7 References

1. Algor Processor Reference Manual, Algor Interactive Systems, 150 Alpha Drive, Pittsburgh, PA 15238
2. O. Axelsson, Iterative Solution Methods, Cambridge University Press, New York, 1994.
3. K. J. Bathe, Finite Element Procedure, Prentice Hall, Upper Saddle River, New Jersey, 1996.
4. K. Bell, A Refined Triangular Plate Bending Finite Element, Int. J. Numer. Methods Eng., Vol. 1, No. 1 January 1969, pp. 101-122.
5. B. Carnahan, H. A. Luther, and J. O. Wikes, Applied Numerical Methods, Wiley New York, NY, 1969.
6. R. W. Clough, The Finite Element in Plane Stress Analysis, Proceedings of 2nd ASCE Conference on Electronic Computation, Pittsburgh, PA. September 1960.
7. R. D. Cook, D. S. Maikus and M. E. Plesha, Concepts and Applications of Finite Element Analysis, 3rd edition, Wiley, New York, NY, 1989.
8. J. W. Dally and W. F. Riley, Experimental Stress Analysis, McGraw Hill, 1965.
9. J. W. Dally, Design Analysis of Structural Elements, Goodington, New York, NY, 2003.
10. C. S. Desal and J. F. Abel, Introduction to the Finite Element Method, Van Nostrand Reinhold, New York, NY, 1972.
11. J. H. Earle, Graphics for Engineers, AutoCAD Release 13, Addision-Wesley, Reading, Massachusetts, 1996.
12. R. H. Gallagher, Finite Element Analysis Fundamentals, Prentice-Hall, Englewoood Cliffs, NJ, 1975.
13. J. M. Gere and S. P. Timoshenko, Mechanics of Materials, 3rd edition, PWS-KENT Publishing Company, Boston, 1990.
14. W. W. Hager, Applied Numerical Linear Algebra, Prentice-Hall, Englewood Cliffs, NJ, 1988.
15. Holland, The Finite Element Method in Plane Stress Analysis, in the Finite Element Method in Stress Analysis, I. Holland and K. Bell (eds), Tapir Press, Trondheim, Norway, 1969, Chapter 2.
16. Y. Y. Hsieh, Elementary Theory of Structure, 2nd edition, Prentice-Hall, Englewood Cliffs, NJ, 1982.
17. M. A. N. Hendriks, H Jongedijk, J. G. Rots and W. J. E. van Spange (eds), Finite Elements in Engineering and Science, DIANA Computational Mechanics, 1997.
18. K. H. Huebner, D. L. Dewhirst, D. E. Smith, and T. G. Byrom, The Finite Element Method for Engineers, 4th edition, John Wiley & Sons, Inc., 2001.
19. C.S. Krishnamoorthy, Finite Element Analysis, Theory and Programming, 2nd Ed., 1995.
20. D. L. Logan, A First Course in the Finite Element Method Using ALGOR, PWS Publishing, Boston, 1997.
21. H. C. Martin, Introduction to Matrix Methods of Structural Analysis, McGraw-Hill, New York, NY, 1966.
22. R. L. Norton, Machine Design: An Integrated Approach, Prentice Hall, Upper Saddle River, New Jersey, 1996.
23. M. Ortiz, Y. Leroy and A. Needleman, A Finite Element Method for Localized Failure Analysis, Comp. Meth. Applied Mech. Enging, 61, pp. 189-214, 1987.
24. M. Petyl, Introduction to Finite Element Vibration Analysis, Cambridge University Press, 1990.
25. T. H. H. Pian and P. Tong, Basis of Finite Element Methods for Solid Continua, Int. J. Numer. Methods Eng., Vol. 2, No. 7, April 1964, pp. 1333-1336.
26. P. G. Plockner, Symmetry in Structural Mechanics, J. of the Structural Division, American Society of Civil Engineers, Vol. 99, No. ST1, pp. 71-89, 1973.
27. J. N. Reddy, An Introduction to the Finite Element Method, McGraw-Hill Publishing Company, New York, NY, 1984.
28. T. Ross, T. Johns and Emile, Dynamic Buckling of Thin Walled Domes under External Water Pressure, Applied Solid Mechanics 2nd Conference, Ed. Tooth & Spence, Elsevier, pp. 211-224, 1987.
29. T. Ross, Finite Element Program in Structural Engineering and Continuum Mechanics, Horwood Publishing Limited, Chichester, England, 1996.

30. T. Ross, Advanced Applied Finite Element Methods, Horwood Publishing Limited, Chichester, England, 1998.
31. L. Stasa, Applied Finite Element Analysis for Engineers, Holt, Rinehart and Winston, New York, NY, 1985, pp. 101-157.
32. Y. Saad, Iterative Methods for Sparse Linear Systems, PWS Publishing Company, Boston, 1996.
33. P. Silvester, Higher-Order Polynomial Triangular Finite Element for Potential Problems, Int. J. Eng. Soc., Vol. 7, No. 8, pp. 849-861.
34. Y. Tada and G. C. Lee, Finite Element Solution to an Elastic Problem of Beams, Int. J. Numer. Methods Eng., Vol. 2, No. 2, April 1970, pp. 229-241.
35. R. L. Taylor, On Completeness of Shape Functions for Finite Element Analysis, Int. J. Numer., Methods Eng., Vol. 4., No. 1, 1972, pp. 17-22.
36. S. P. Timoshenko and D. H. Young, Theory of Structure, McGraw-Hill, Kogaksha Limited, 1965.
37. M. J. Turner, R. W. Clough, H. C. Martin, and L. J. Topp, Stiffness and Deflection Analysis of Complex Structures, J. of the Aeronautical Sciences, Vol. 23, No. 9, pp. 805-824, Sept. 1956.
38. O. C. Zienkiewicz, The Finite Element Method, third edition, McGraw-Hill, New York, NY, 1977.
39. O. C. Zienkiewicz and R. L. Taylor, The Finite Element Method, 4th edition, Vol. 1, McGraw-Hill (UK), London, 1989.

4.8 Exercises

1. The following engineering drawing illustrates an assembly. The assembly consists of two components. One is the support and the other is a shaft.

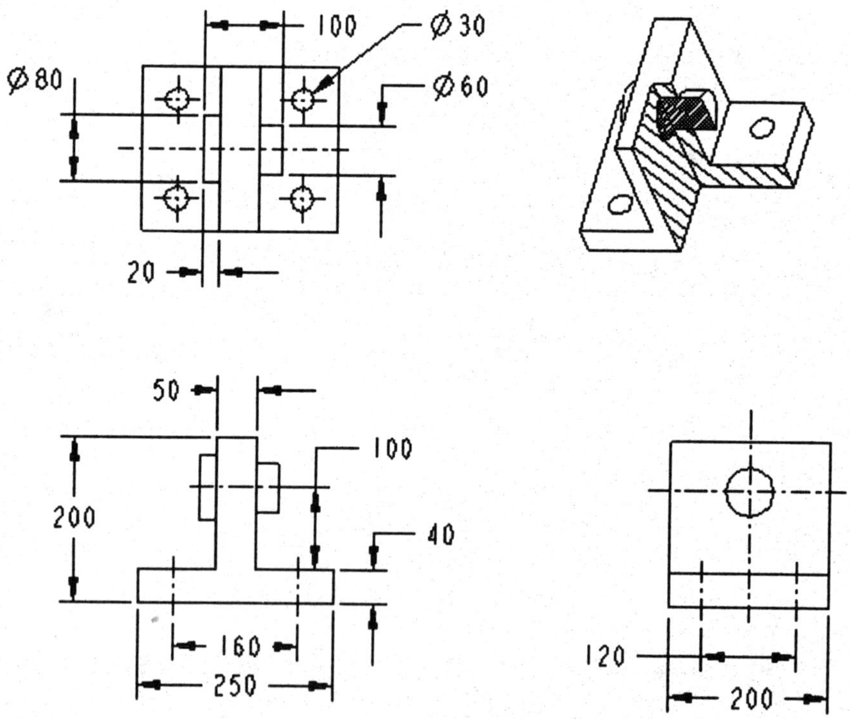

The dimensions of the support component are shown below. Pay attention to the four (4) holes, which are prepared for 4 bolts. The diameter is 30 mm. There are clearances between the bolts and the holes. The entire assembly is fixed to the ground through the 4 bolts, which are not shown here.

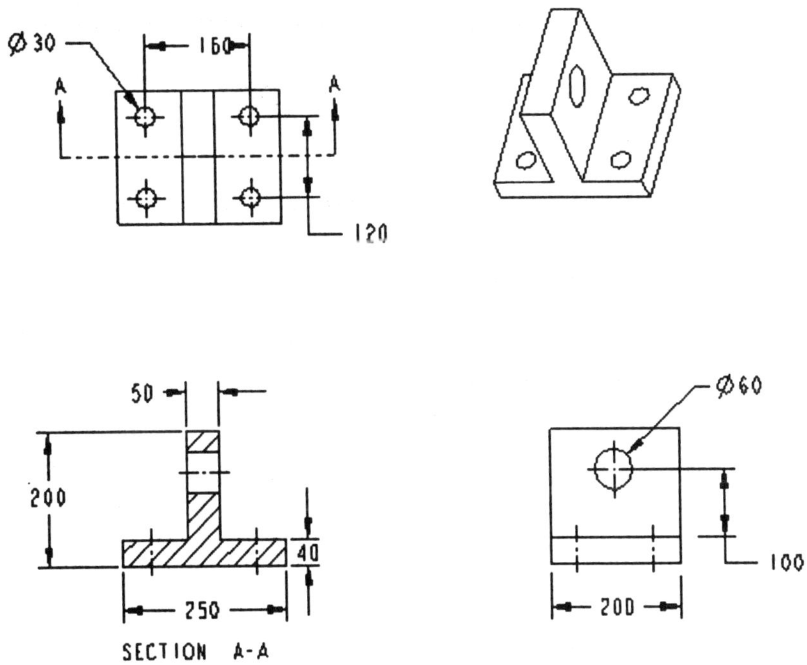

The dimensions of the shaft component are shown below. Pay attention to the end on the right side. A load is acting on the surface uniformly. The magnitude is 300 Newton.

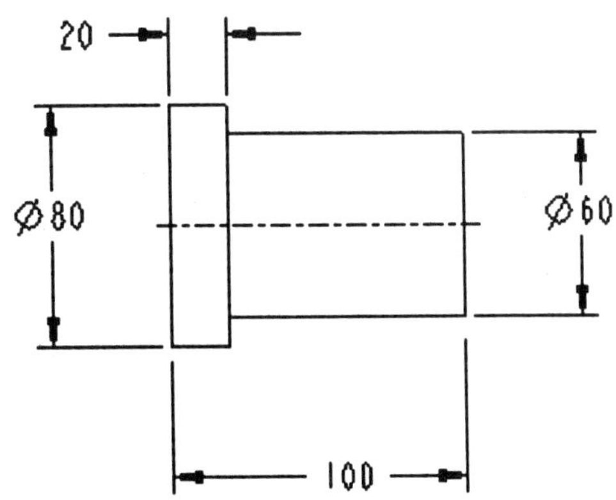

The 3D model of the assembly and the two components are available for performing an engineering analysis. The materials of both components are steel.

support structure	maximum displacement in magnitude	max principal stress	max Von mises stress
Location (x, y, z)			

shaft structure

	maximum displacement in magnitude	max principal stress	max Von mises stress
Location (x, y, z)			

2. A structure consisting of three components has been designed. The structure is shown below. This structure will be put on a tensile-test machine. The lower part (steel) is fixed on the table or the ground. The upper part (aluminum) is subjected to a load, which goes up so that the entire structure is under a tensile stress condition. The middle part is a cement layer, which glues the upper and lower parts together. The purpose of this test is to understand the role played by the middle part in terms of its thickness value, material type and the value of the inclined angle as shown.

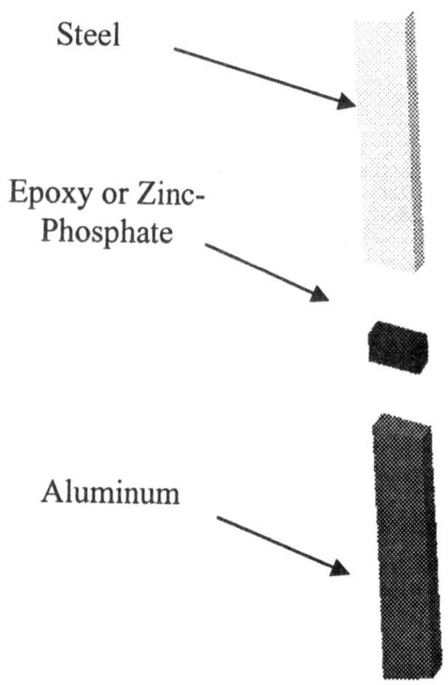

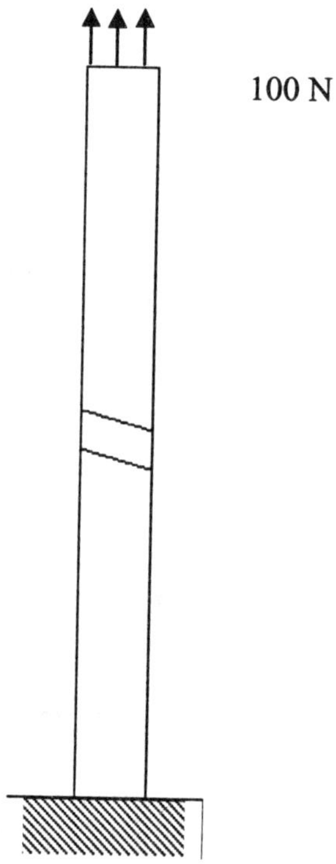

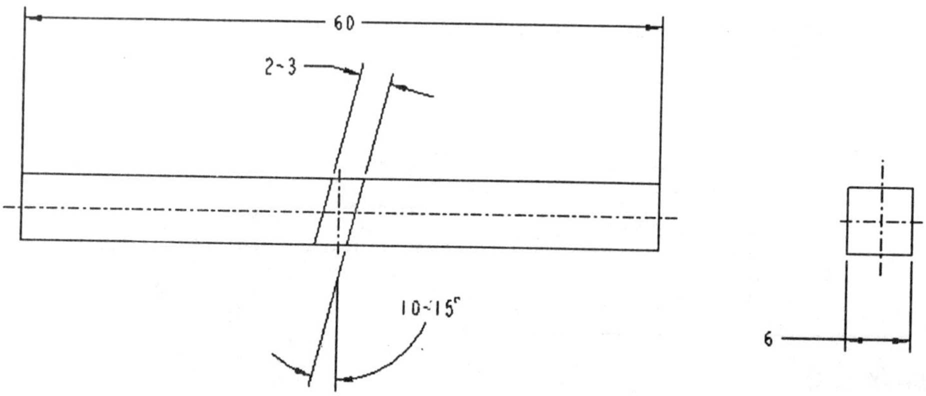

1) Run the system under the following conditions: the thickness of the middle part is 2 mm, the cement material is epoxy and the inclination angle is 10 degrees. Plot the distribution of the maximum principal stress, identify the location of the maximum principal stress, and identify the magnitude of the maximum principal stress.
2) Run the system under the following conditions: the thickness of the middle part is 2 mm, the cement material is zinc-phosphate and the inclination angle is 10 degrees. Plot the distribution of the maximum principal stress, identify the location of the maximum principal stress, and identify the magnitude of the maximum principal stress. Compare the two locations. Is there a shift in the location of the maximum principal stress? Compare the 2 magnitudes of the maximum principal stress. Is the value increased or reduced as a result of the material type change?
3) Run the system under the following conditions: the thickness of the middle part is 2 mm, the cement material is zinc-phosphate and the inclination angle is 15 degrees. Plot the distribution of the maximum principal stress, identify the location of the maximum principal stress, and identify the magnitude of the maximum principal stress. Compare the two locations obtained in (2) and (3). Is there a shift in the location of the maximum principal stress? Compare the 2 magnitudes of the maximum principal stress obtained in (2) and (3). Is the value increased or reduced as a result of the material type change?
4) Run the system under the following conditions: the thickness of the middle part is 3 mm, the cement material is zinc-phosphate and the inclination angle is 15 degrees. Plot the distribution of the maximum principal stress, identify the location of the maximum principal stress, and identify the magnitude of the maximum principal stress. Compare the two locations obtained in (3) and (4). Is there a shift in the location of the maximum principal stress? Compare the 2 magnitudes of the maximum principal stress obtained in (3) and (4). Is the value increased or reduced as a result of the material type change?
5) Carry out a systematic study and obtain a quantitative relation which characterizes the value of the maximum principal stress as a function of the three design parameters associated with the middle part component.

3. A crown-cement-root system is shown below. To reduce the computation effort, a quarter of the entire model is used. The geometry and dimensions of each of three components are shown below.

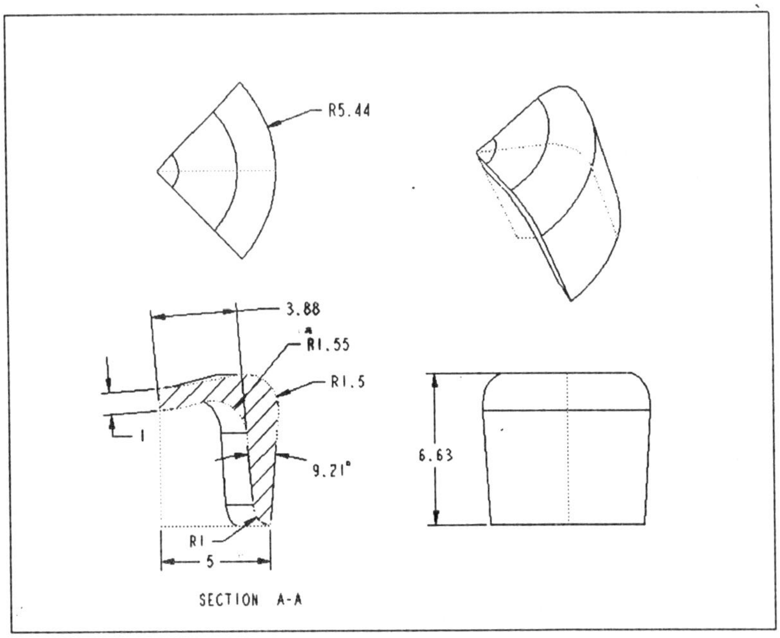

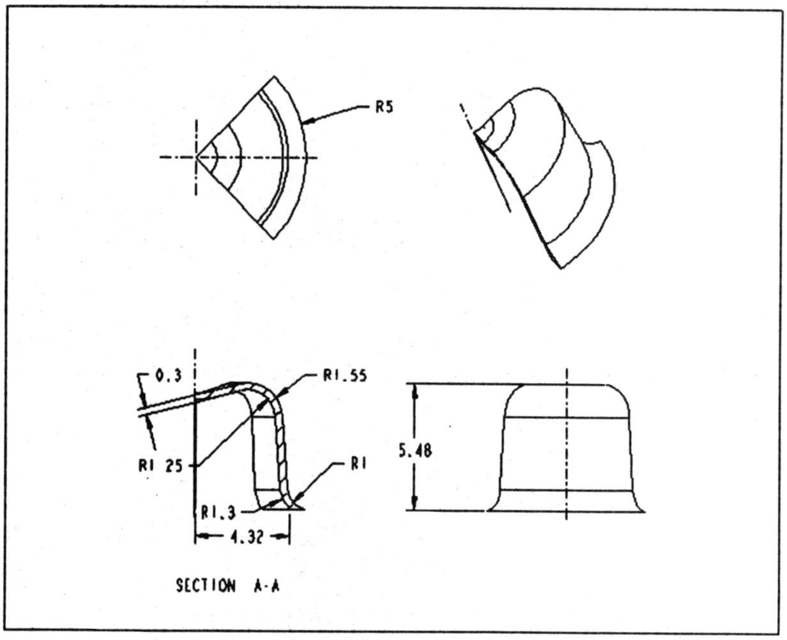

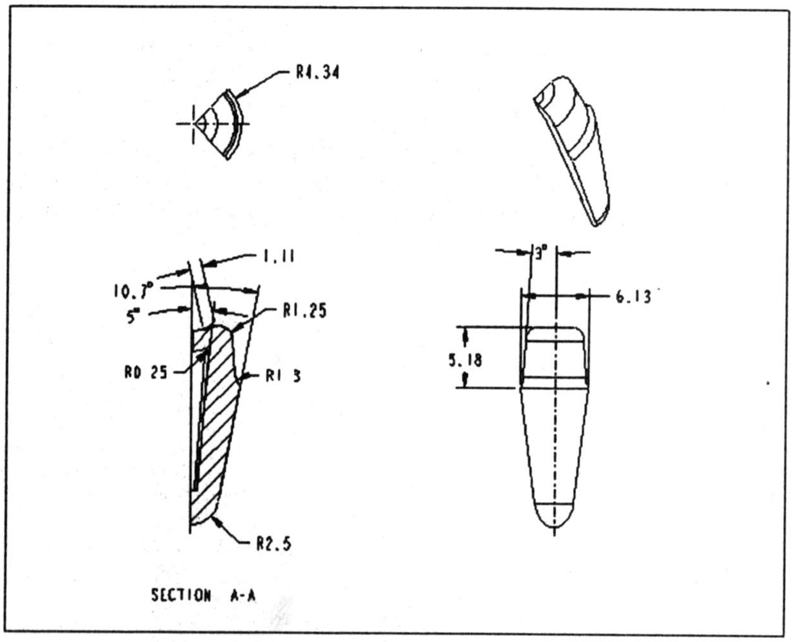

Regarding the material properties, we will use the glass, cement and composite defined in Problem 1.

Regarding the temperature variation, let us have two cases. In case 1, we assume that the patient is drinking hot coffee. In case 2, we assume that the patient is having ice cream.

Case 1:

The temperature variation:	reference temperature:	20°C
	model temperature:	75°C

Case 2:

The temperature variation:	reference temperature:	20°C
	model temperature:	8°C

For the boundary condition, we assume that the lower portion of the root is fixed. Investigate the stress development and the pattern of deformation when subjected temperature variation.

4. A geometrical model for a crown-cement-dentin-core system is shown below. The load magnitude is 200 N, acting on the curved surface with a circular area (diameter is 1 mm).

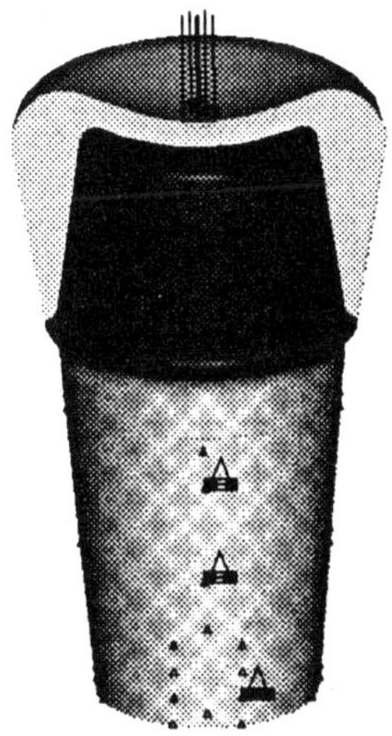

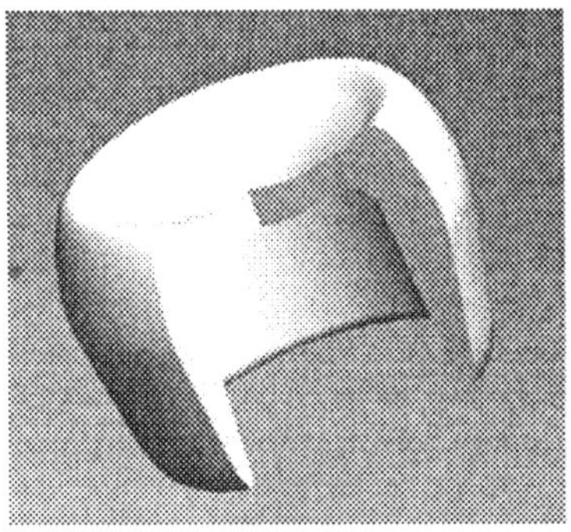

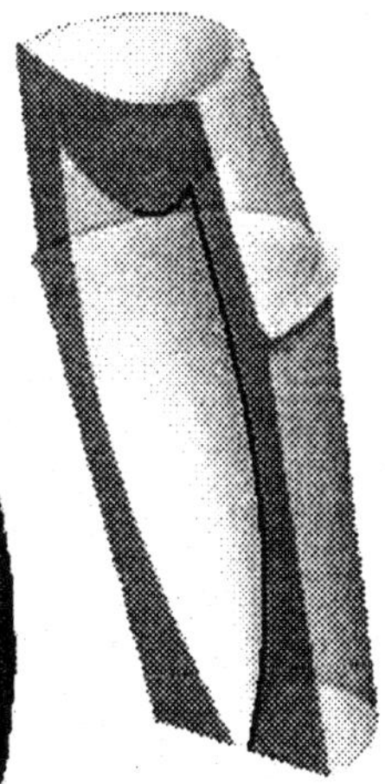

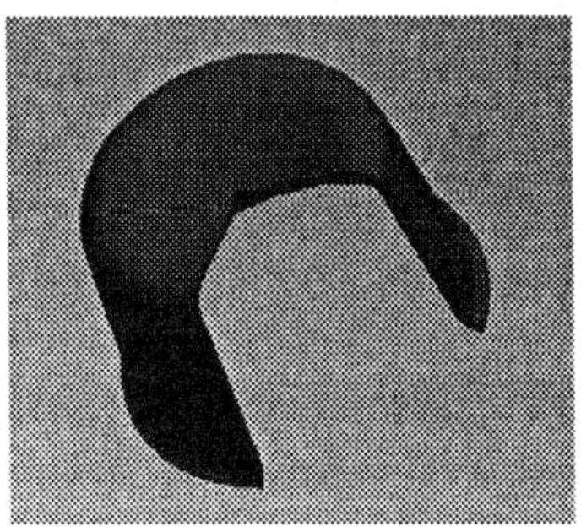

(1) Create these 4 components using the geometry shown in Problem 3 where a quarter model is created. In Problem 3, users used 90 degrees for the revolving operation. In this exercise, users should use 360 degrees to create the full-size models.

(2) Illustrate the maximum principal stress distribution in each of the four components.

(3) Assume that the load direction is at an inclined angle equal to 20 degrees, as shown. Illustrate the maximum principal stress distribution in each of the four components. For setting the material type for the crown, cement and dentin components, refer to the 3 material types used in section 4.5. The numerical values for Young's modulus and Poisson's ratios are listed on pp. 206-207. For the material type of the core component, let us choose STEEL to simulate the filling material after the root canal operation.

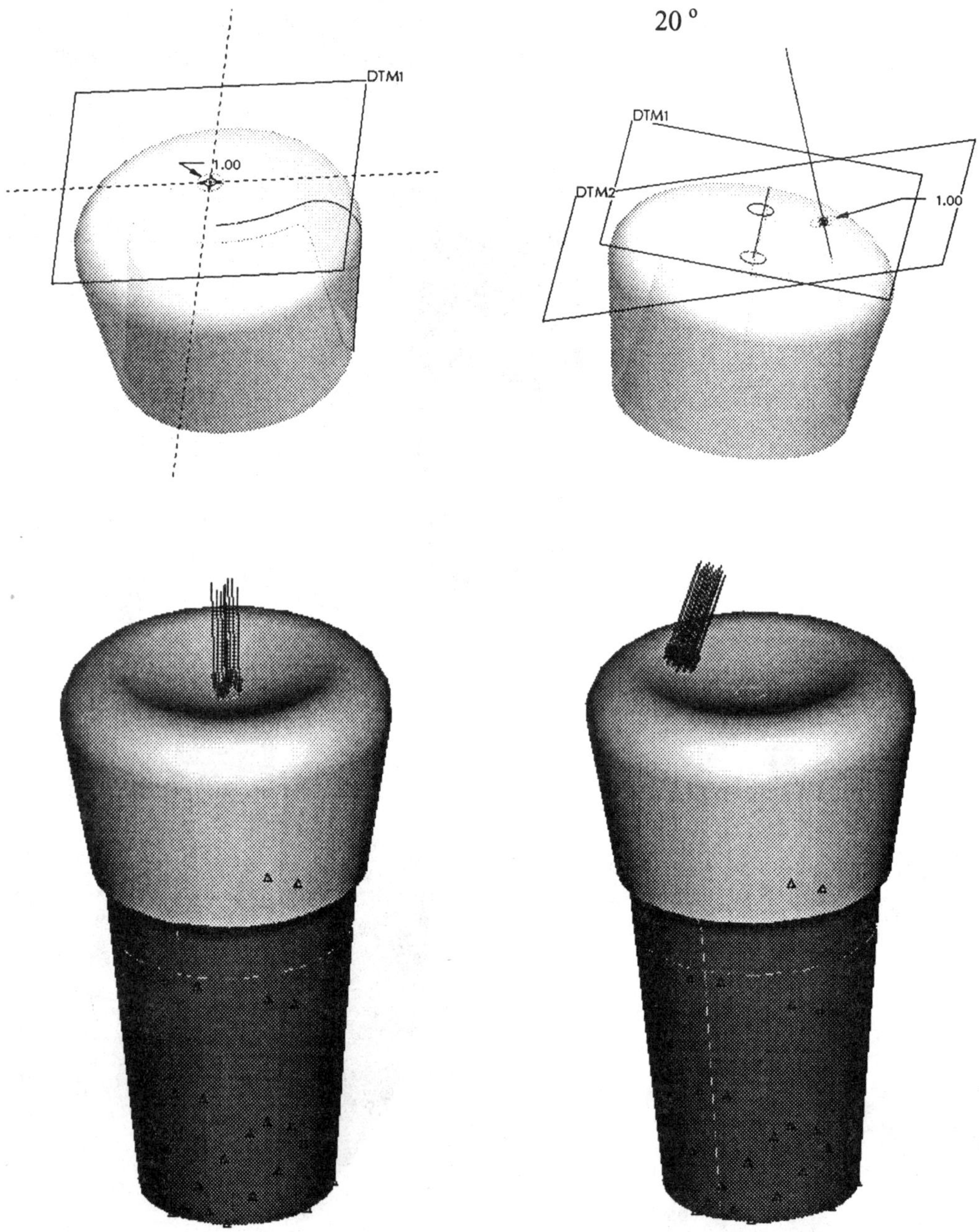

CHAPTER 5

CONCEPTS OF PRINCIPAL STRESSES AND VON MISES STRESS

5.1 Introduction

In the previous chapters, we evaluate the deflection and stress distributions, using FEA, when a mechanical structure is subjected to a loading condition and/or loading conditions. In those stress evaluations, we focused our attention to the maximum principal stress and von Mises stress. In the subject of engineering mechanics, principal stresses are referred to as normal stresses. A large value of the normal stress induced by the loading condition could be a leading cause to a system failure, such as fracture, breakage, etc. In engineering mechanics, von Mises stress is an indication of the applied stress, which is related to yielding, and an extra amount of such plastic deformation is also a leading cause to a system failure. In the chapter, we demonstrate the procedure to calculate the principal stresses and von Mises stress in the process of carrying out FEA.

5.2 Evaluation of the Stress Condition at a Given Location

In this section, we provide users with a method to evaluate the stress condition at a given location. As illustrated below, a beam has a rectangular cross section with its left end fixed and the right end set to free. The two dimensions of the rectangular section are 30 mm and 10 mm. The overhanging length of the beam is 180 mm.

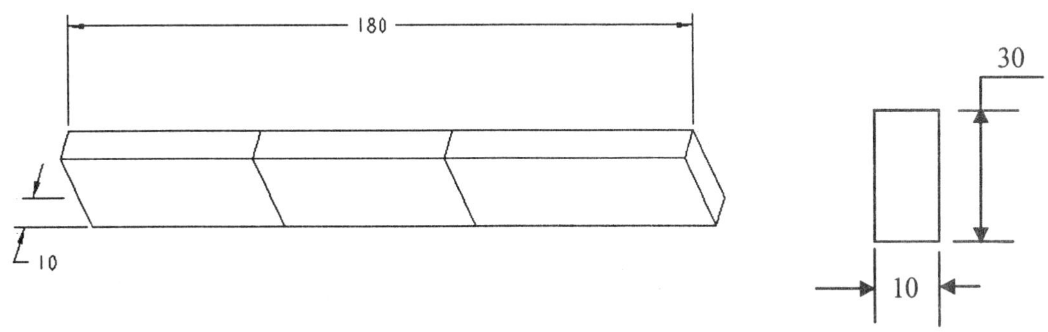

In the process of carrying out a structural analysis, such as the study of a stress distribution or the study of a strain distribution, people may be interested in knowing the value of stress or the value of strain at a given location or at a set of locations. In order to obtain the numerical values, those locations must be nodes in performing FEA so that performance indices, such as maximum principal stress, von Mises stress, etc., can be established. To do so, datum points must be defined at those locations before carrying out FEA.

Generally speaking, the process of creating datum points that are on surfaces is not too complicated as we experienced when working with the examples in the previous chapters. However, for those locations which are not on surfaces, instead of, which are inside the solid model, special caution has to be taken to make sure that node assignments are assured at those locations.

As illustrate in the following, there are 6 locations where the evaluation of stress and/or strain is required. Four (4) of those 6 locations are inside the solid model. They are PNT0, PNT1, PNT2 and PNT3. Both PNT4 and PNT5 are located on the boundary surface.

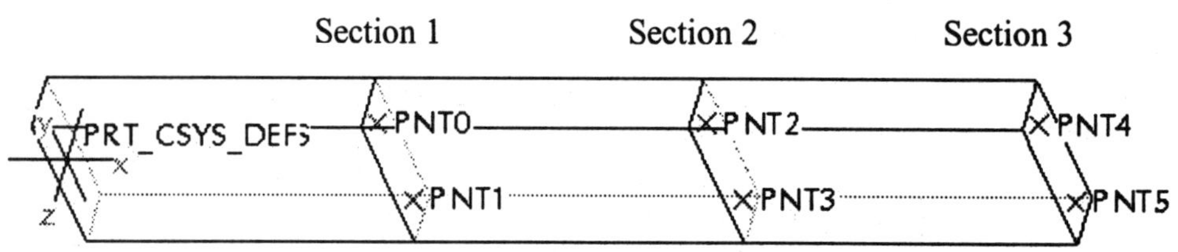

In this example, we will use an approach, which divides the entire beam into three portions. They are independent components. We call them beam_left, beam_middle and beam_right. Their material assignments are identical among the three components. We assemble them together for the evaluation. In this way, the node assignment to each location can be assured.

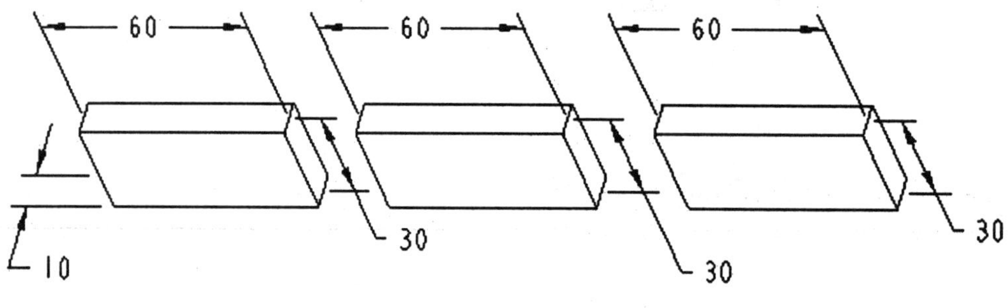

Step 1: Create a 3D solid model for beam_left.

From the main toolbar, select the icon of **Create a new object**.

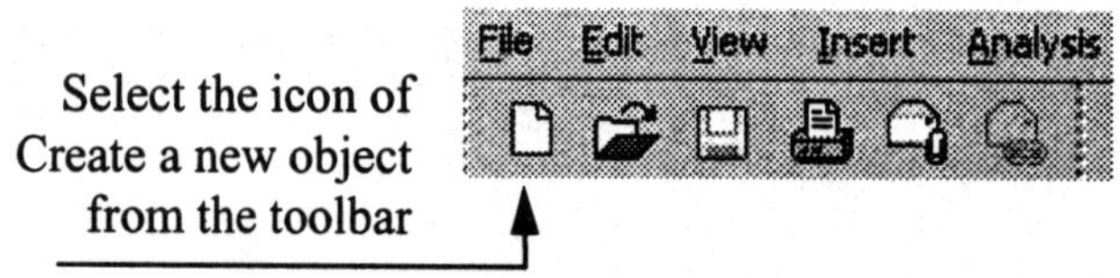

Make sure that the Part mode is selected, and type *beam _left* as the file name. Clear the box called **Use default template > OK**, and select the **mmns_part_solid**. Fill in **DESCRIPTION** by typing *fixed end part.* Fill in **MODELED_BY** by typing *student* > **OK.**

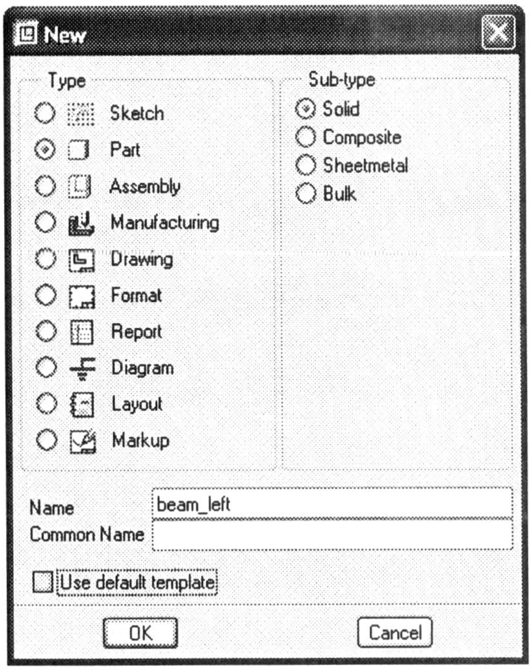

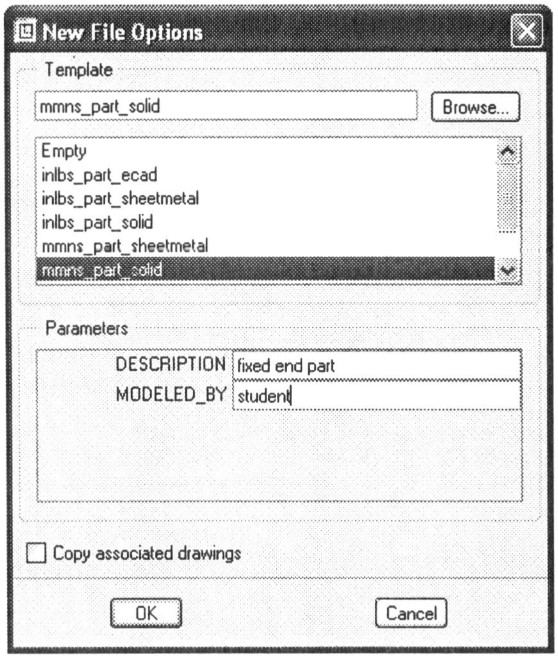

Create a 3D solid model and start with a block. The 3 dimensions are 180 x 100 x 50 mm. From the main toolbar, select **Insert** > **Extrude.**

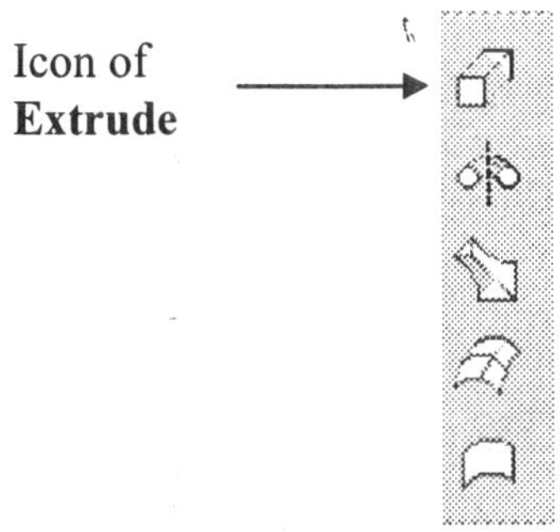

The dashboard appears. Make sure that the icon of **Solid** on the dashboard is selected. Set the depth choice to **Symmetry** and type 10 as the thickness value. Activate **Placement** to define a sketch plane > **Define.**

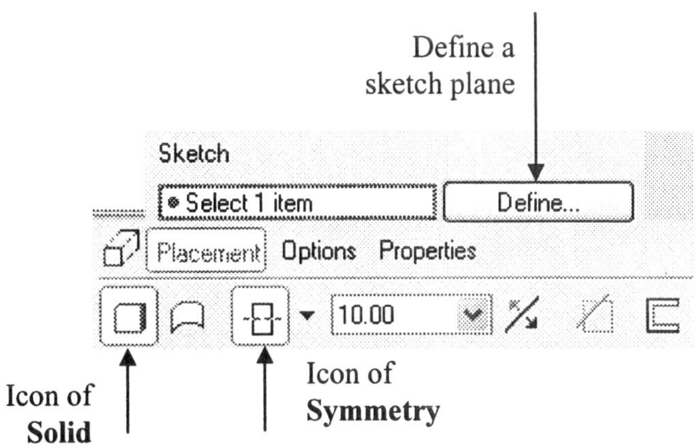

Select the **FRONT** datum plane as the sketch plane, and click the box of **Sketch** to accept the **RIGHT** datum plane as the default reference to orient the sketch plane, as illustrated below.

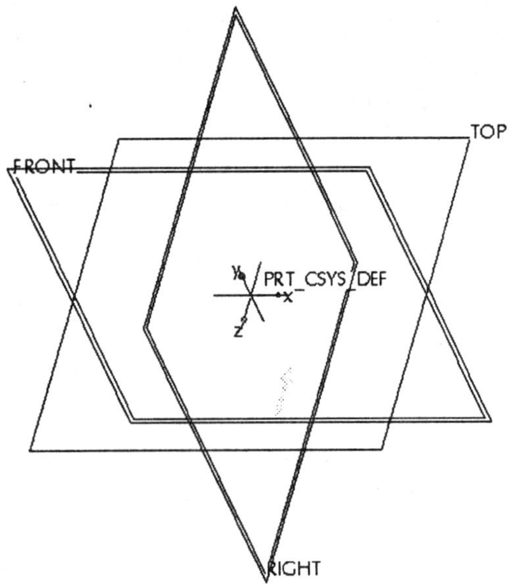

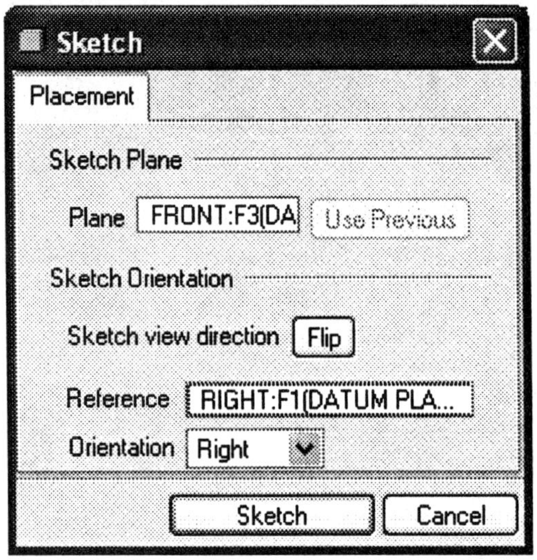

Select the icon of **Centerline** to draw a horizontal centerline or a centerline along the x-axis.

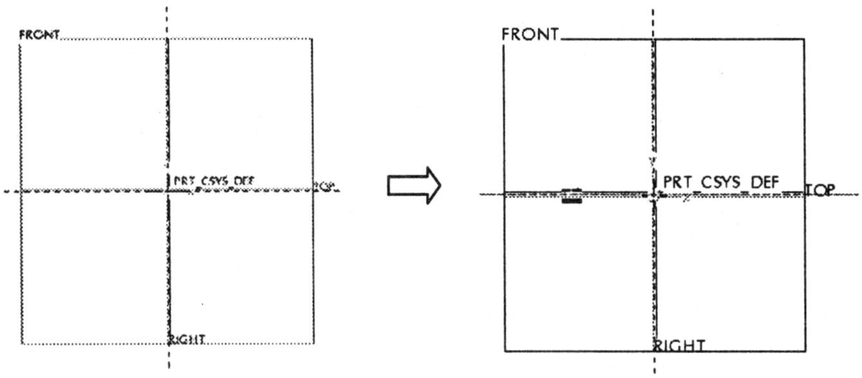

Select the icon of **Rectangle** to make a 2D sketch as shown below. The 2 dimensions are 60 and 30, respectively. Pay attention to the symmetric requirement about the horizontal centerline.

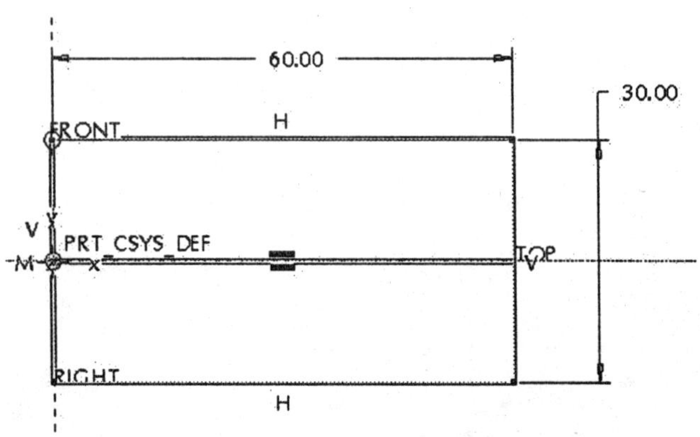

Upon completing the sketch, select the icon of **Done** from the toolbar of sketcher. On the dashboard, check the symmetry.

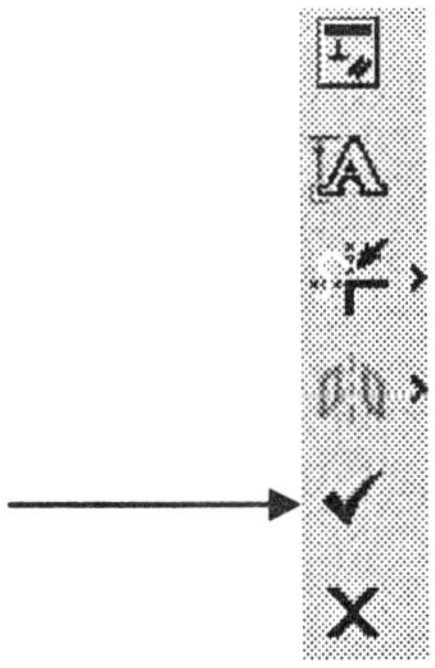

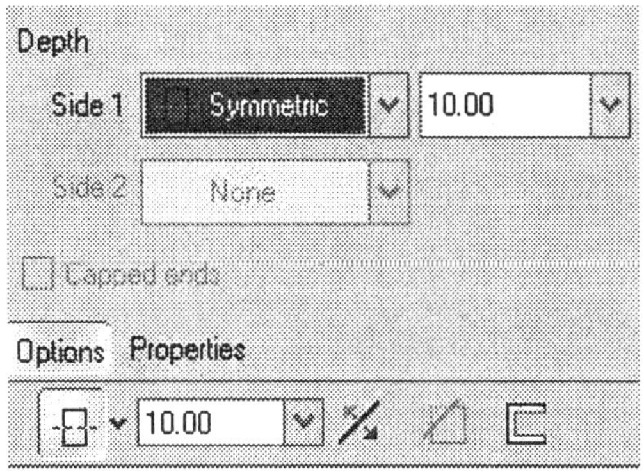

On the Feature Control Panel, pick the icon of **Complete** to complete the creation of the cylindrical feature, as shown below:

Pick the icon of **Complete** from the Feature Control Panel

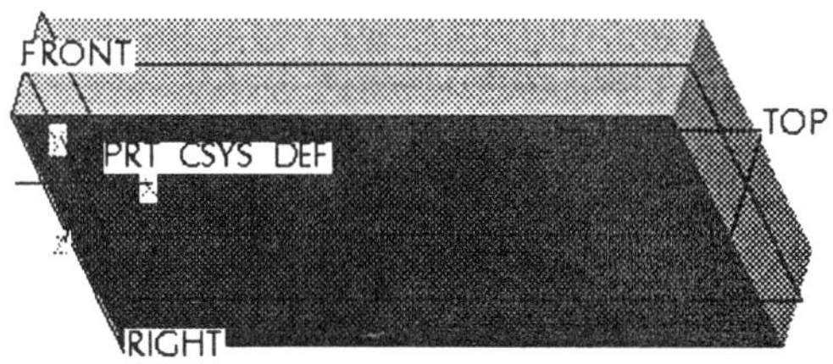

Step 2: Create an assembly

Make sure that the Assembly mode is selected, and type *beam_3_parts* as the file name. Clear the box called **Use default template**, and select the **mmns_asm_design**. Fill in **DESCRIPTION** by typing in *beam_assembly*. Fill in **MODELED_BY** by typing *student* > **OK**

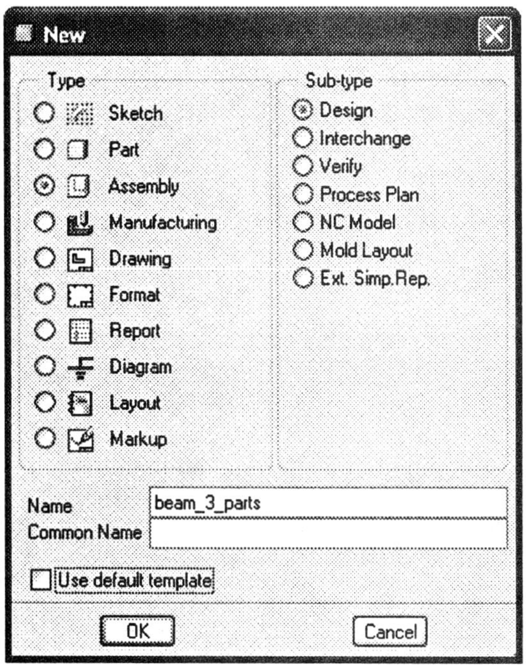

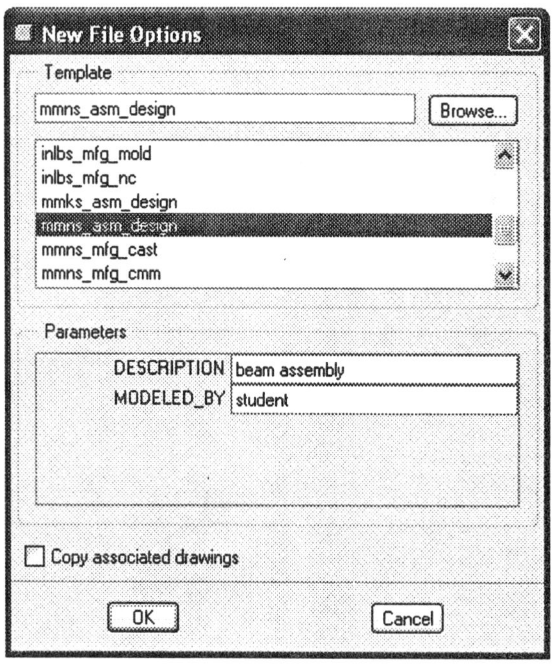

Select the icon of **Add** component to the assembly from the toolbar of assembly, and select the file of beam_left.prt > Open.

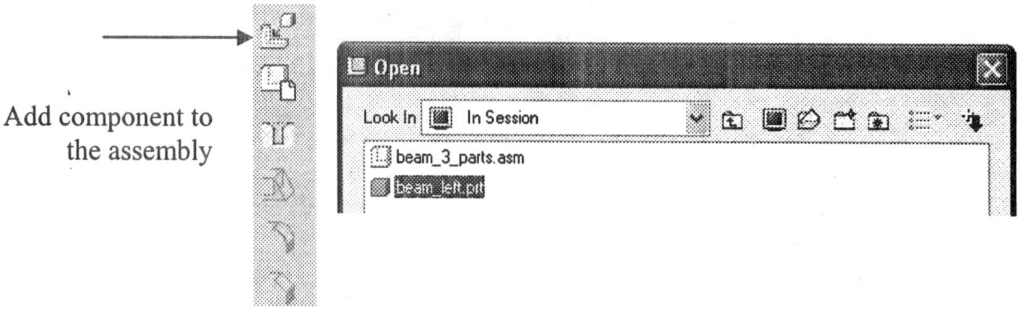

In the **Component Placement** window, select the icon of Assemble component at default location > **OK** to complete the assembly process. This is equivalent to assemble the part_default_coordinate system at the location of the assembly_default_coordinate system.

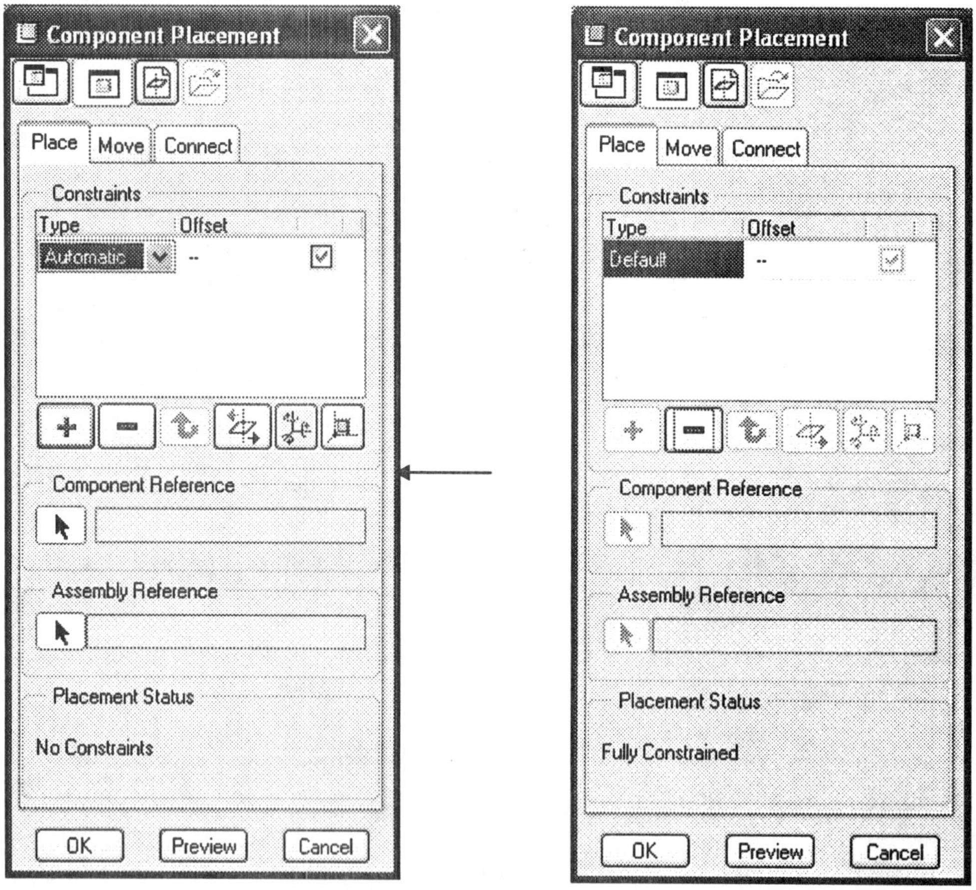

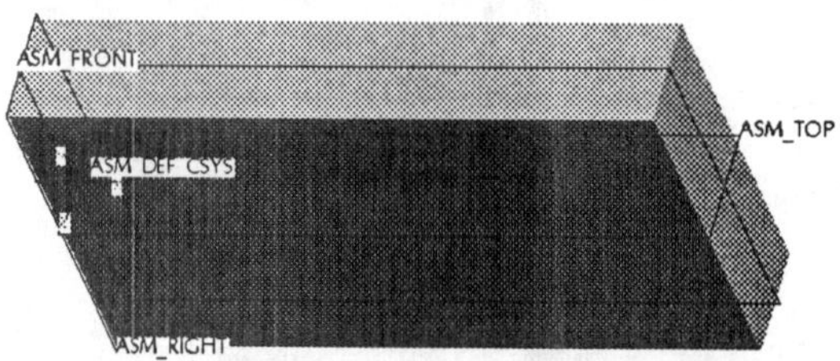

Select the icon of **Create** a component in the assembly mode from the toolbar of assembly.

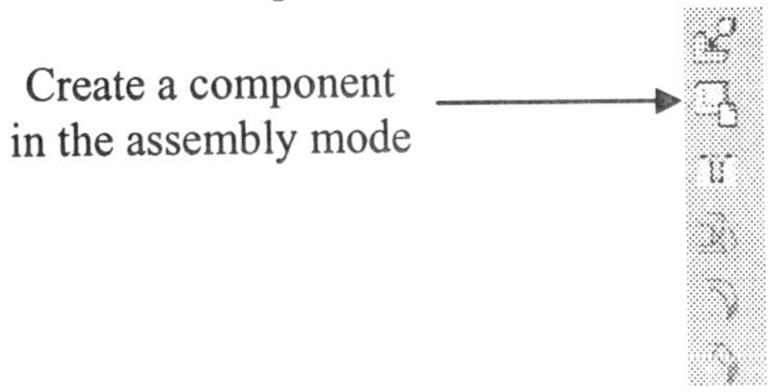

Type *beam_middle* as the name of the new component, and select the mmns_part_solid.prt as the startup file for the new component > **OK**.

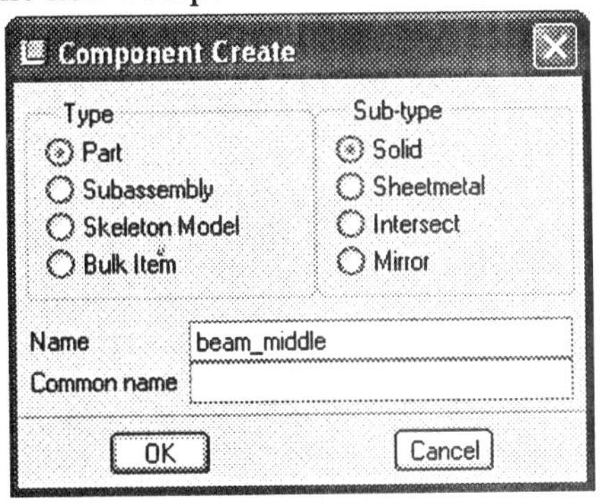

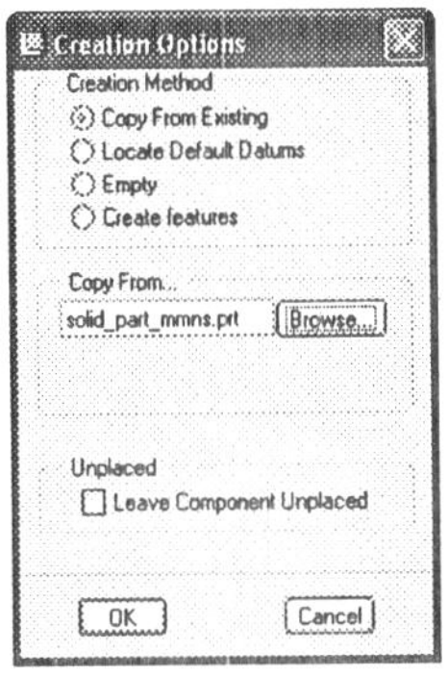

Select the icon of **Assemble component at default location** > **OK**.

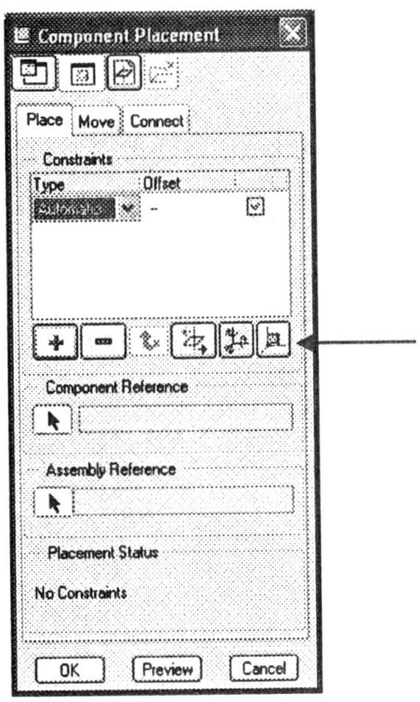

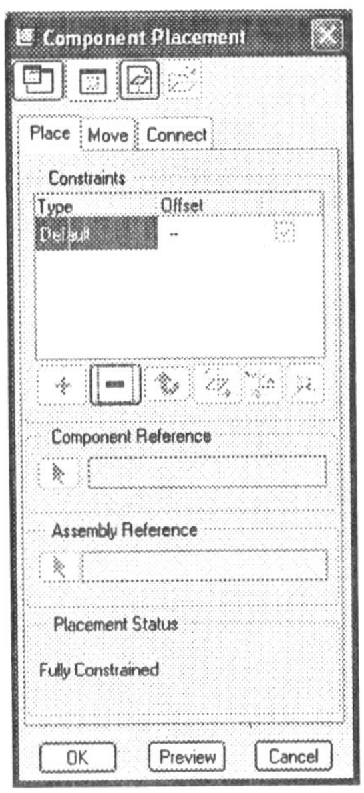

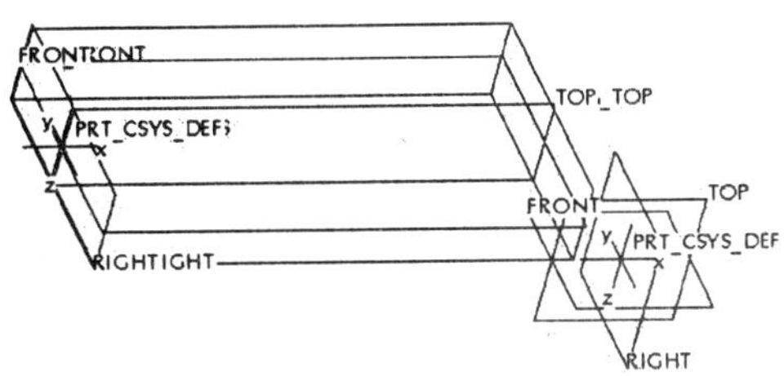

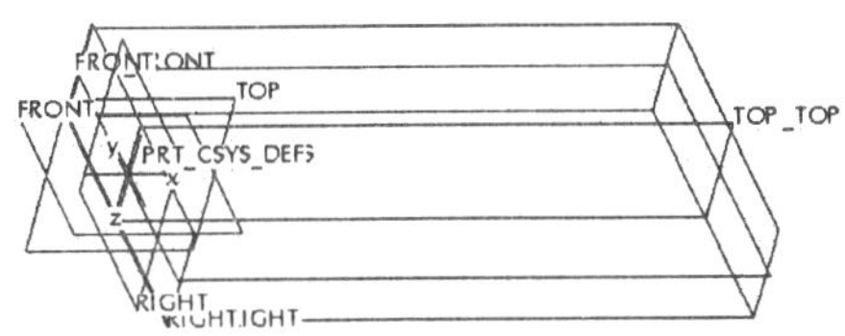

To activate beam_middle.prt in the assembly mode, pick beam_middle from the model tree, right-click and hold, select **Activate**.

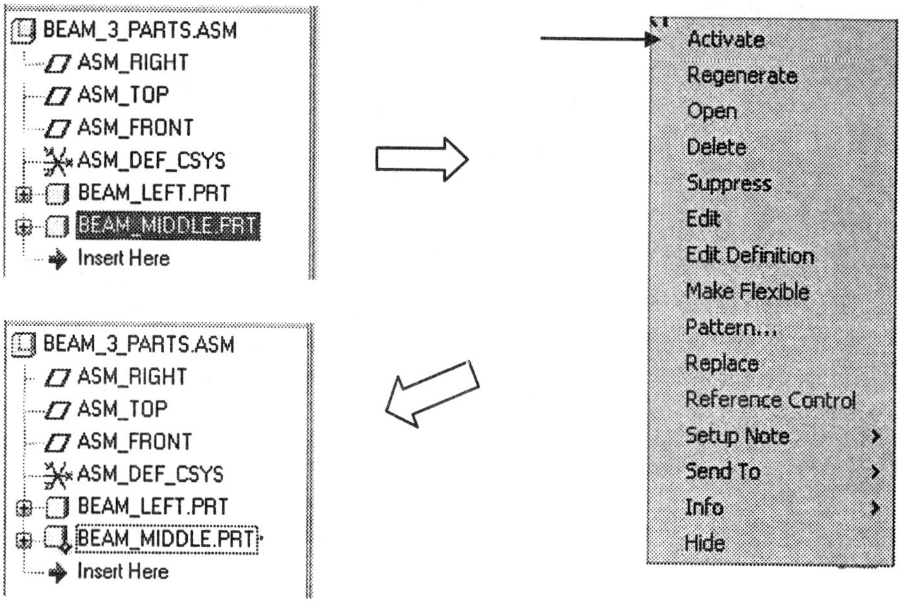

From the toolbar of datum feature, select the icon of datum plane, select the surface on the right side of beam_left and specify the offset value equal to zero > **OK**. This means that a datum plane called DTM1 has been created within beam_middle.prt.

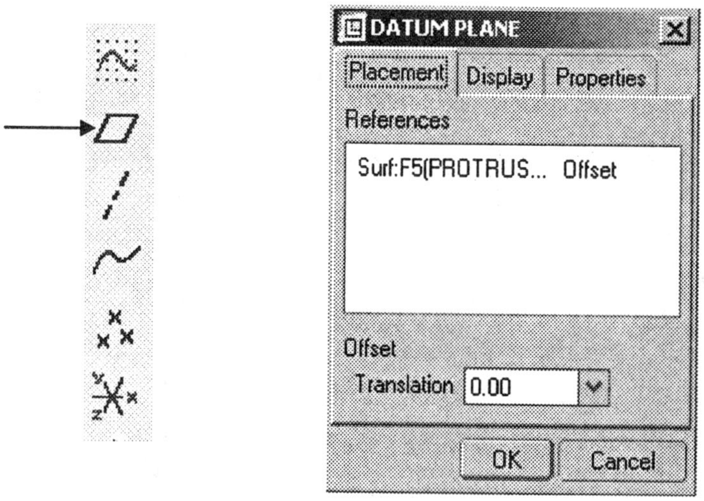

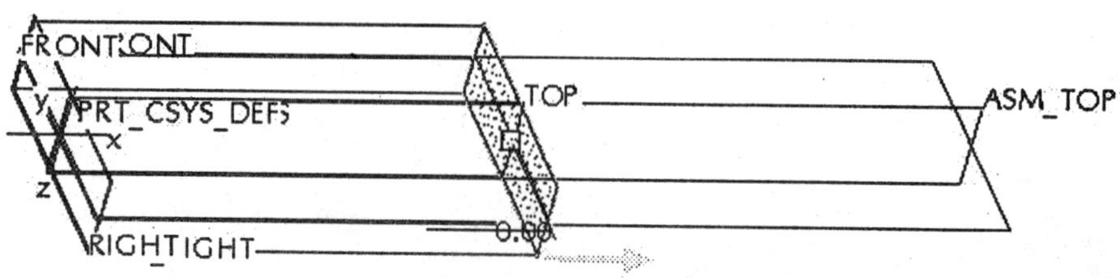

Now let us activate the assembly by right-click on beam_3_parts. Afterwards, pick beam_middle.prt > right-click and hold > select **Open**.

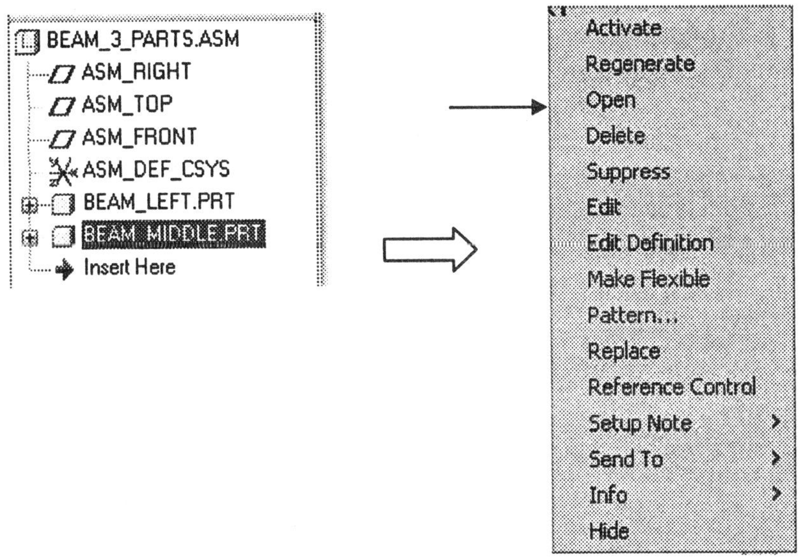

To create a 3D solid model, select the icon of **Extrude** from the toolbar of feature creation. The dashboard appears. Make sure that the icon of **Solid** on the dashboard is selected. Set the depth choice to Symmetry and type 10 as the thickness value. Activate **Placement** to define a sketch plane > **Define.**

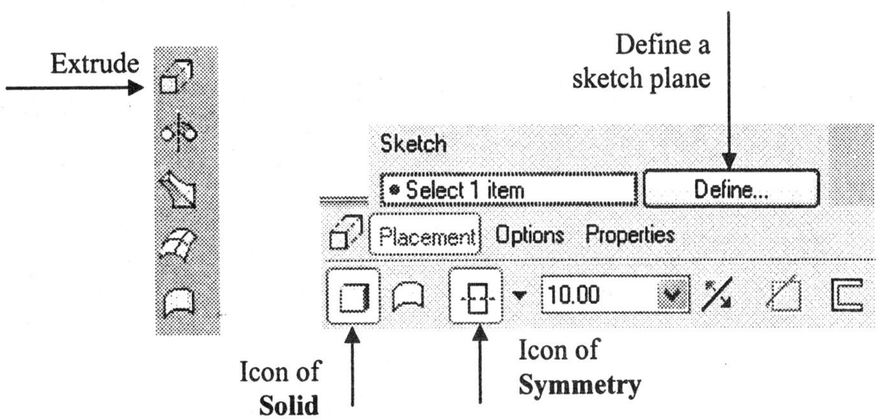

Select the **FRONT** datum plane as the sketch plane, and click the box of **Sketch** to accept the **RIGHT** datum plane as the default reference to orient the sketch plane, as illustrated below.

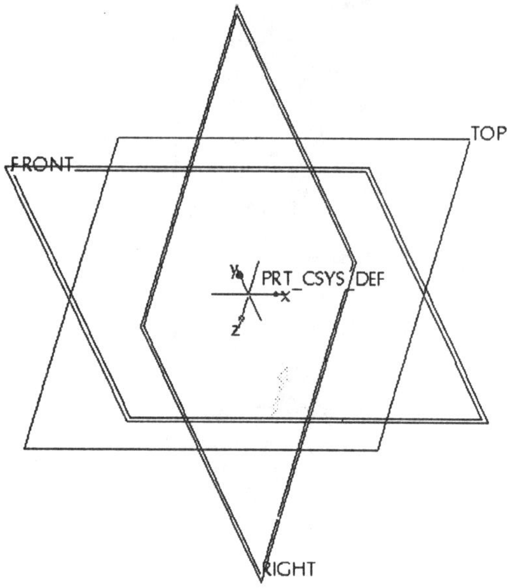

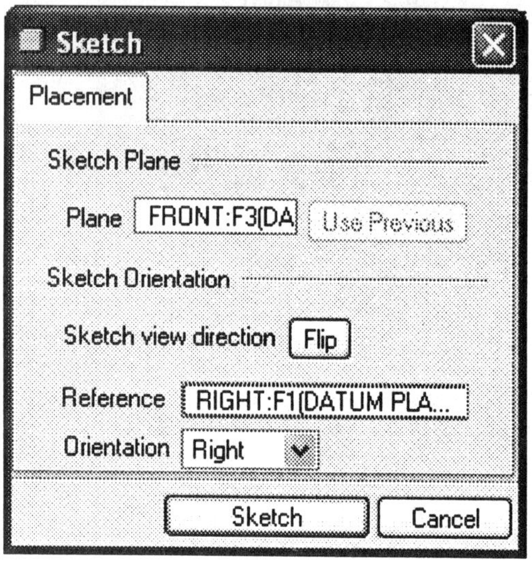

Select **DTM1** as a reference, sketch a horizontal centerline, and sketch a rectangle (60 x 30), as shown below:

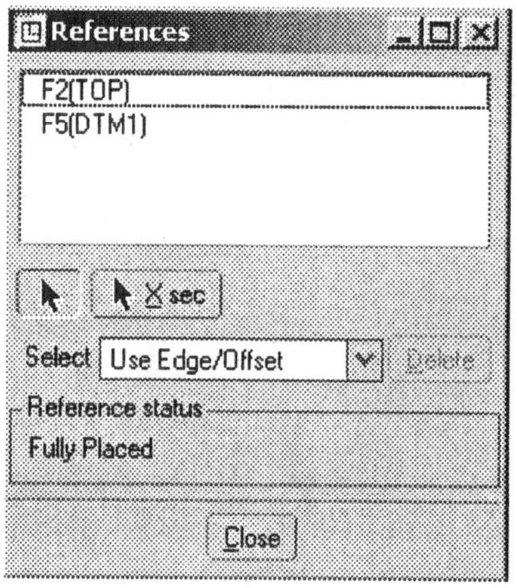

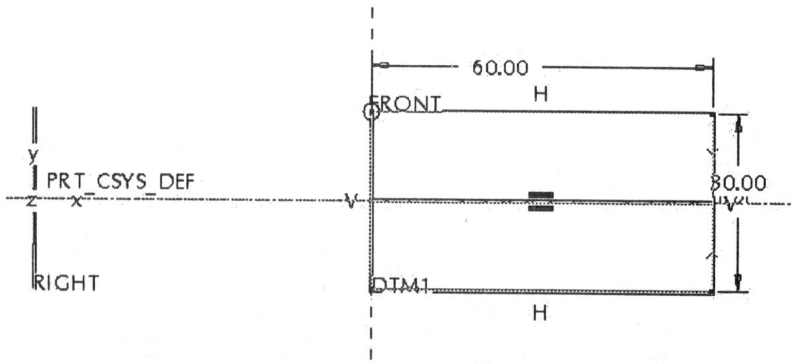

Upon completing the sketch, select the icon of **Done** from the toolbar of sketching. On the dashboard, type *10* as the depth value, as shown.

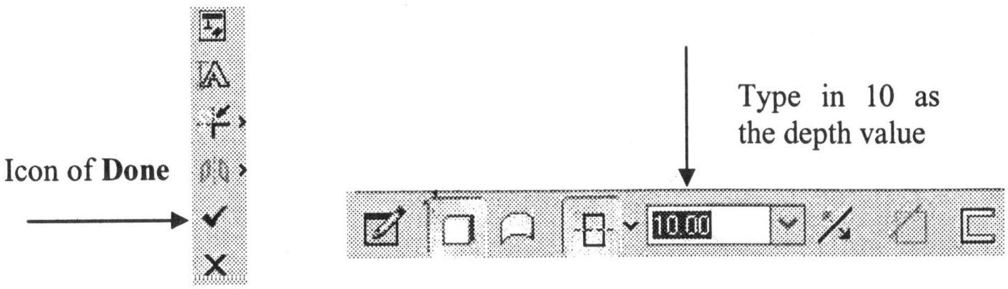

(1) Plot the distribution of the maximum principal stress.

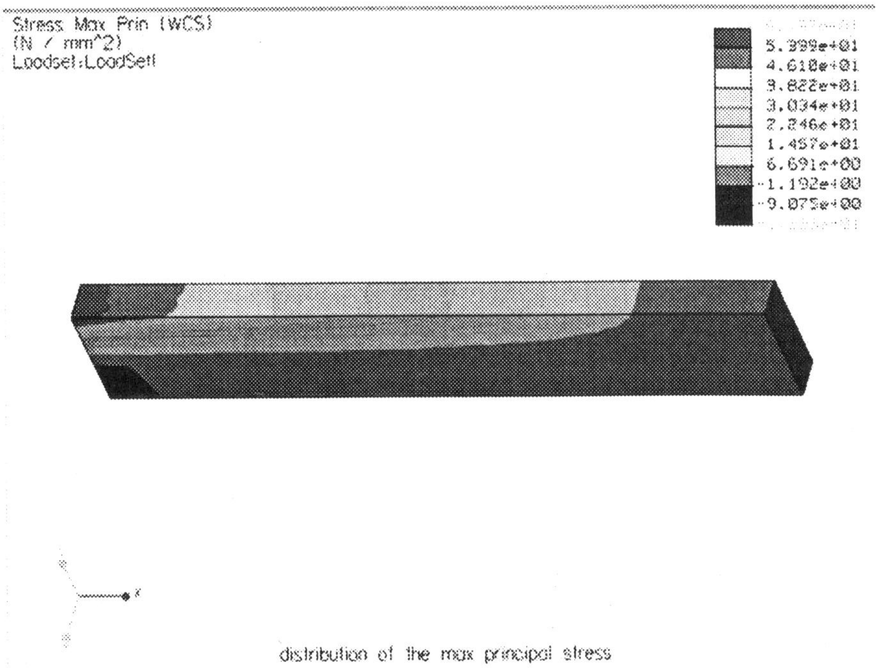

(2) Plot the three distributions of the normal stresses in the X, Y and Z directions.

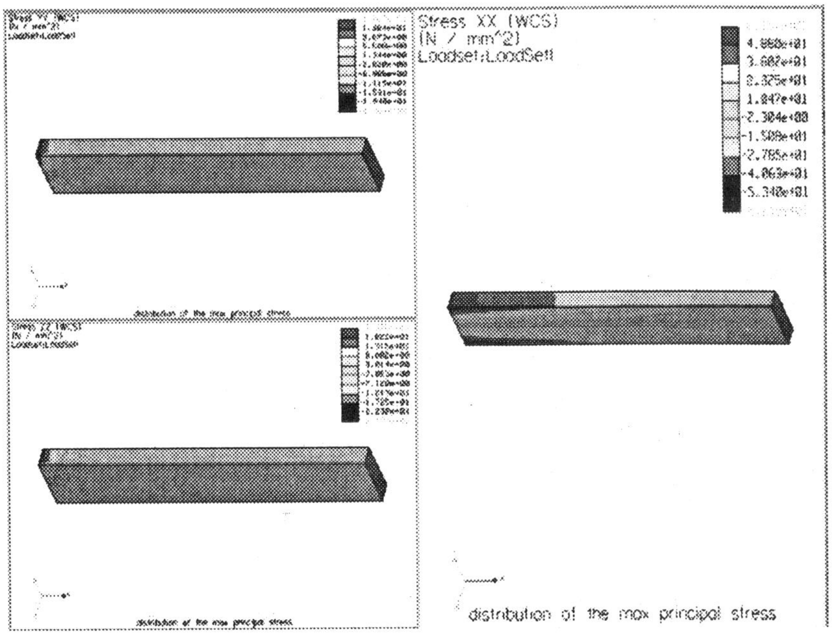

(3) Plot the three distributions of the shear stresses, namely, τ_{xy} ,τ_{xz} and τ_{yz}, respectively.

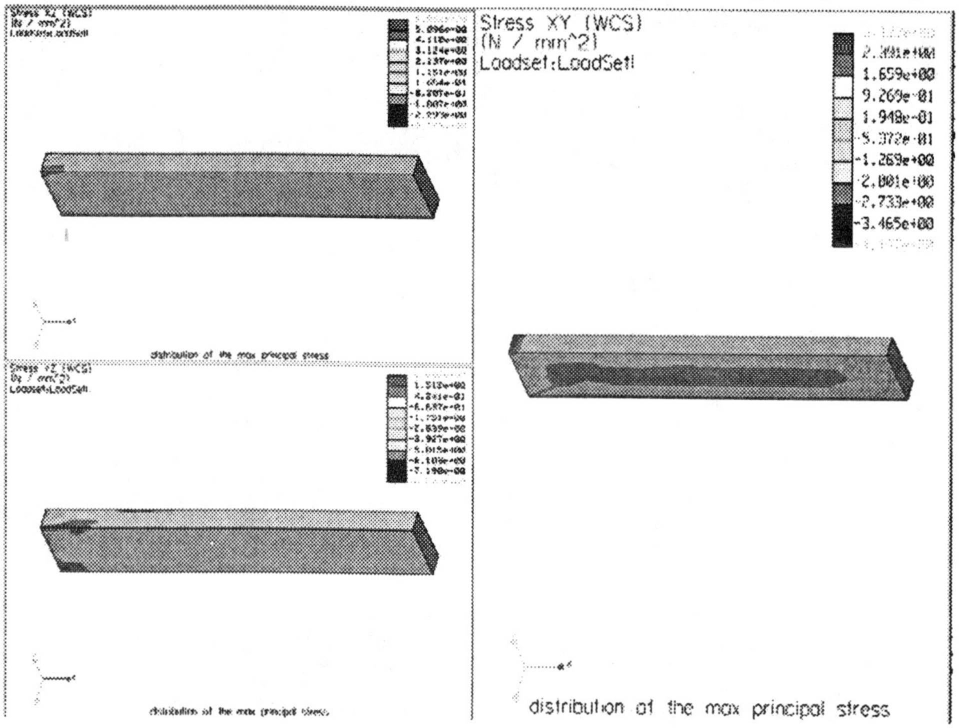

At this moment, we have known the values of the three normal stresses and three shear stresses. We are able to calculate the 3 principal stress values and the 3 associated shear stress values.

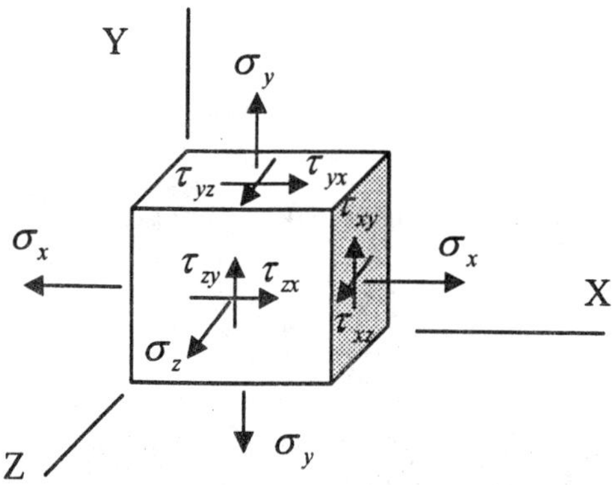

(1) Determine the principal stresses and the associated shear stresses at the location marked as PNT0. For determining the principal stresses and the associated shear stresses at the other 5 locations, readers should practice as an exercise.

Dimensions: 180 mm x 30 mm x 10 mm

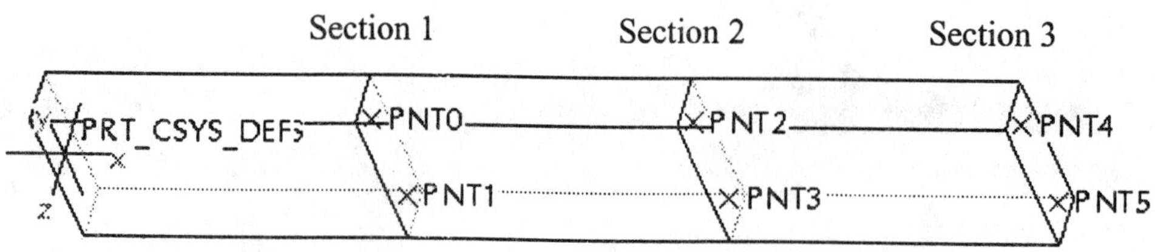

Conditions of constraints and load

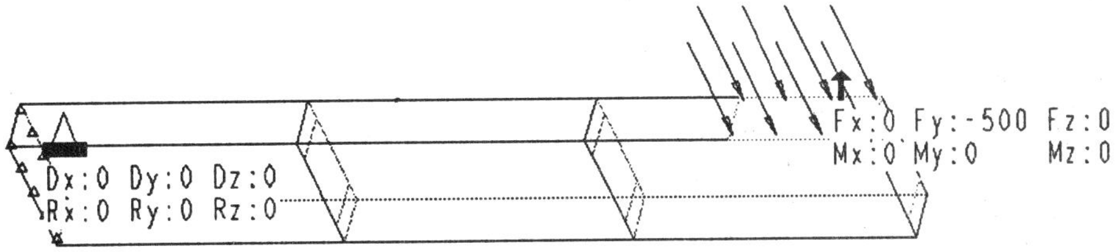

location	sigma x	sigma y	sigma z	shear xy	shear xz	shear yz
PNTO	23.49	0.1055	0.1104	-1.445	-0.1355	-0.2796

Generally speaking, a stress element is subjected to normal and shear stresses on its six faces. Common sense dictates that inclined planes cut through such an element, or transformation of the reference coordinate system, are also subjected to normal and shear stresses. The relations between the normal and shear stresses acting on the inclined surfaces can be determined based on the principle of static equilibrium after the transformation.

Of particular importance are those inclined surfaces, on which only normal stresses are acting on them with no shear stresses acting on them. Those normal stresses are called principal stresses. Generally speaking, there are three principal stresses because a stress element consists of three (3) pairs of surfaces, which are parallel to each other. The magnitudes of these three principal stresses can be obtained as the three real roots of the following cubic equation:

$$\sigma^3 - A\sigma^2 + B\sigma - C = 0$$

where

$$A = \sigma_x + \sigma_y + \sigma_z$$

$$B = \sigma_x\sigma_y + \sigma_x\sigma_z + \sigma_y\sigma_z - \tau_{xy}^2 - \tau_{xz}^2 - \tau_{yz}^2$$

$$C = \sigma_x\sigma_y\sigma_z + 2\tau_{xy}\tau_{xz}\tau_{yz} - \sigma_x\tau_{yz}^2 - \sigma_y\tau_{xz}^2 - \sigma_z\tau_{xy}^2$$

We call these 3 quantities, namely, A, B, and C, stress invariants because they do not change in value in the process of rotating the reference coordinate system from one orientation to a new orientation. In mathematics, a vector possessing this characteristic is termed as a *tensor*. Therefore, stress is a tensor. Because the relation between stress and strain through the parameter of elasticity, which is a constant, strain is also a tensor.

We use the function called Roots in MATLAB to determine σ_1, $\sigma 2$, and σ_3. The obtained results are shown below:

location	A	B	C	sigma 1	sigma 2	sigma 3
PNTO	23.4	-4.96	-4.29	23.6	-0.54	0.34

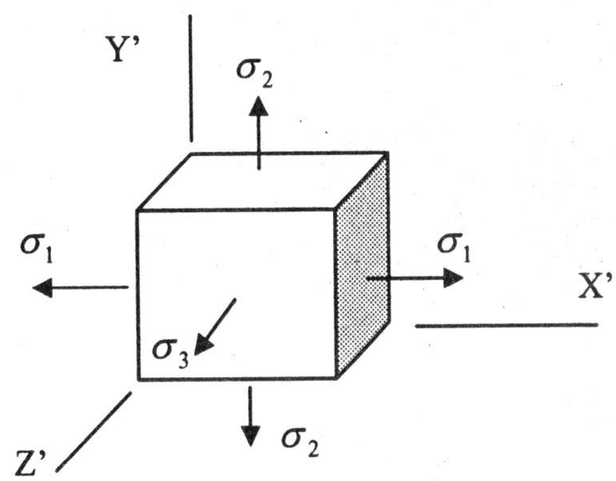

(2) Determine the maximum shear stress for each of the six locations.

Following the guideline of Mohr's circle for plane stress, we have the relations between the set of principal stresses and the set of three maximum shear stresses as follows:

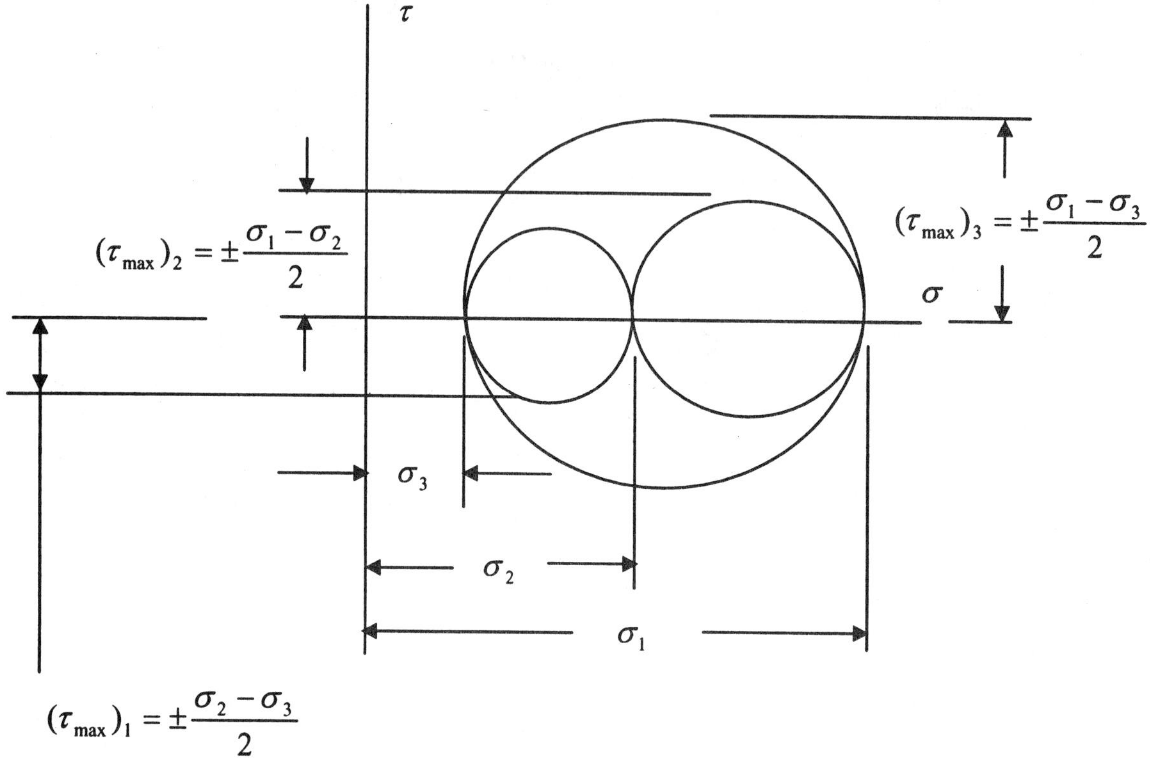

It is important to note that the absolute maximum shear stress is the numerically largest of the stresses determined from these three equations shown in the above figure.

location	sigma 1	sigma 2	sigma 3	max shear
PNTO	23.6	-0.54	0.34	12.1

(3) Determine the von Mises stress for each of the six locations.

There are several criteria to determine the onset of inelastic behavior or yield for ductile materials under the combined stresses described above. In Pro/MECHANICA, von Mises yield condition is implemented as default. Note that the von Mises yield condition is based on the principle that the strain energy of stressed component can be divided into volumetric strain energy and shear strain energy components. The volumetric strain energy is related to the mean "hydrostatic stress", leading to volume change and no distortion. The shear strain energy is related to the deviatory or distortional stress, leading to no volumetric change, but instead distorting or deviating the element from its initial cubic shape. The definition of the von Mises stress is given by

$$von\_Mises\_stress = \frac{1}{\sqrt{2}}\sqrt{(\sigma_1-\sigma_2)^2+(\sigma_2-\sigma_3)^2+(\sigma_1-\sigma_3)^2}$$

location	sigma 1	sigma 2	sigma 3	von Mises
PNTO	23.6	-0.54	0.34	23.7

5.3 References

1. Algor Processor Reference Manual, Algor Interactive Systems, 150 Alpha Drive, Pittsburgh, PA 15238
2. O. Axelsson, Iterative Solution Methods, Cambridge University Press, New York, 1994.
3. K. J. Bathe, Finite Element Procedure, Prentice Hall, Upper Saddle River, New Jersey, 1996.
4. K. Bell, A Refined Triangular Plate Bending Finite Element, Int. J. Numer. Methods Eng., Vol. 1, No. 1 January 1969, pp. 101-122.
5. B. Carnahan, H. A. Luther, and J. O. Wikes, Applied Numerical Methods, Wiley New York, NY, 1969.
6. R. W. Clough, The Finite Element in Plane Stress Analysis, Proceedings of 2nd ASCE Conference on Electronic Computation, Pittsburgh, PA. September 1960.
7. R. D. Cook, D. S. Maikus and M. E. Plesha, Concepts and Applications of Finite Element Analysis, 3rd edition, Wiley, New York, NY, 1989.
8. J. W. Dally and W. F. Riley, Experimental Stress Analysis, McGraw Hill, 1965.
9. J. W. Dally, Design Analysis of Structural Elements, Goodington, New York, NY, 2003.
10. C. S. Desal and J. F. Abel, Introduction to the Finite Element Method, Van Nostrand Reinhold, New York, NY, 1972.
11. J. H. Earle, Graphics for Engineers, AutoCAD Release 13, Addision-Wesley, Reading, Massachusetts, 1996.
12. R. H. Gallagher, Finite Element Analysis Fundamentals, Prentice-Hall, Englewoood Cliffs, NJ, 1975.
13. J. M. Gere and S. P. Timoshenko, Mechanics of Materials, 3rd edition, PWS-KENT Publishing Company, Boston, 1990.
14. W. W. Hager, Applied Numerical Linear Algebra, Prentice-Hall, Englewood Cliffs, NJ, 1988.
15. Holland, The Finite Element Method in Plane Stress Analysis, in the Finite Element Method in Stress Analysis, I. Holland and K. Bell (eds), Tapir Press, Trondheim, Norway, 1969, Chapter 2.
16. Y. Y. Hsieh, Elementary Theory of Structure, 2nd edition, Prentice-Hall, Englewood Cliffs, NJ, 1982.
17. M. A. N. Hendriks, H Jongedijk, J. G. Rots and W. J. E. van Spange (eds), Finite Elements in Engineering and Science, DIANA Computational Mechanics, 1997.
18. K. H. Huebner, D. L. Dewhirst, D. E. Smith, and T. G. Byrom, The Finite Element Method for Engineers, 4th edition, John Wiley & Sons, Inc., 2001.
19. C.S. Krishnamoorthy, Finite Element Analysis, Theory and Programming, 2nd Ed., 1995.
20. D. L. Logan, A First Course in the Finite Element Method Using ALGOR, PWS Publishing, Boston, 1997.
21. H. C. Martin, Introduction to Matrix Methods of Structural Analysis, McGraw-Hill, New York, NY, 1966.
22. R. L. Norton, Machine Design: An Integrated Approach, Prentice Hall, Upper Saddle River, New Jersey, 1996.
23. M. Ortiz, Y. Leroy and A. Needleman, A Finite Element Method for Localized Failure Analysis, Comp. Meth. Applied Mech. Enging, 61, pp. 189-214, 1987.
24. M. Petyl, Introduction to Finite Element Vibration Analysis, Cambridge University Press, 1990.
25. T. H. H. Pian and P. Tong, Basis of Finite Element Methods for Solid Continua, Int. J. Numer. Methods Eng., Vol. 2, No. 7, April 1964, pp. 1333-1336.
26. P. G. Plockner, Symmetry in Structural Mechanics, J. of the Structural Division, American Society of Civil Engineers, Vol. 99, No. ST1, pp. 71-89, 1973.
27. J. N. Reddy, An Introduction to the Finite Element Method, McGraw-Hill Publishing Company, New York, NY, 1984.
28. T. Ross, T. Johns and Emile, Dynamic Buckling of Thin Walled Domes under External Water Pressure, Applied Solid Mechanics 2nd Conference, Ed. Tooth & Spence, Elsevier, pp. 211-224, 1987.
29. T. Ross, Finite Element Program in Structural Engineering and Continuum Mechanics, Horwood Publishing Limited, Chichester, England, 1996.

30. T. Ross, Advanced Applied Finite Element Methods, Horwood Publishing Limited, Chichester, England, 1998.
31. L. Stasa, Applied Finite Element Analysis for Engineers, Holt, Rinehart and Winston, New York, NY, 1985, pp. 101-157.
32. Y. Saad, Iterative Methods for Sparse Linear Systems, PWS Publishing Company, Boston, 1996.
33. P. Silvester, Higher-Order Polynomial Triangular Finite Element for Potential Problems, Int. J. Eng. Soc., Vol. 7, No. 8, pp. 849-861.
34. Y. Tada and G. C. Lee, Finite Element Solution to an Elastic Problem of Beams, Int. J. Numer. Methods Eng., Vol. 2, No. 2, April 1970, pp. 229-241.
35. R. L. Taylor, On Completeness of Shape Functions for Finite Element Analysis, Int. J. Numer., Methods Eng., Vol. 4., No. 1, 1972, pp. 17-22.
36. S. P. Timoshenko and D. H. Young, Theory of Structure, McGraw-Hill, Kogaksha Limited, 1965.
37. M. J. Turner, R. W. Clough, H. C. Martin, and L. J. Topp, Stiffness and Deflection Analysis of Complex Structures, J. of the Aeronautical Sciences, Vol. 23, No. 9, pp. 805-824, Sept. 1956.
38. O. C. Zienkiewicz, The Finite Element Method, third edition, McGraw-Hill, New York, NY, 1977.
39. O. C. Zienkiewicz and R. L. Taylor, The Finite Element Method, 4th edition, Vol. 1, McGraw-Hill (UK),London, 1989.

5.4 Exercises

1. A cantilever beam is shown below (the geometry and dimensions are identical to the cantilever beam studied in section 5.2). The material of the beam is steel. The beam is fixed at its left end and the right end is set to free. The load condition is shown below.

Dimensions: 180 mm x 30 mm x 10 mm

Conditions of constraints and load

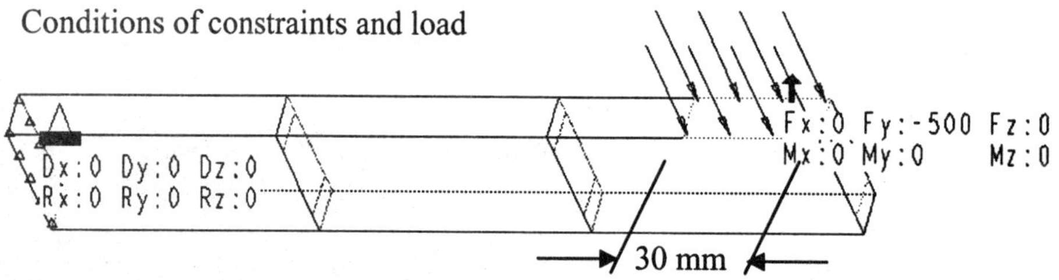

Three sections are selected from the beam as shown. On each of the three sections, two locations are chosen. As a result, a total of six locations are chosen. The stress condition for PNT0 was studied in section 5.2. In this exercise, you are asked to study the stress condition for each of the other 5 locations, marked as PNT1, PNT2, PNT3, PNT4 and PNT5.

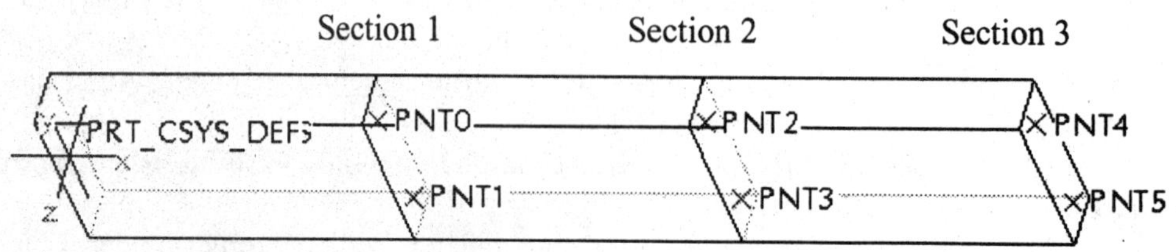

Specifically speaking, the following requirements should be met.

1) Determine the maximum principal stress for each of the six locations.
2) Determine the maximum shear stress for each of the six locations.
3) Determine the von Mises stress for each of the six locations.
4) Can we neglect the stresses acting on the two Z faces so that a plane stress model can be used?

Make sure that you use the function called Measure to define the following stress items, tabulate them and finally come up with the solutions to the above four questions.

section	location	sigma x	sigma y	sigma z	shear xy	shear xz	shear yz
1	PNT1						
2	PNT2						
2	PNT3						
3	PNT4						
3	PNT5						

Follow the procedure as indicated below to calculate the 3 principal stresses.

$$\sigma^3 - A\sigma^2 + B\sigma - C = 0$$

where

$$A = \sigma_x + \sigma_y + \sigma_z$$

$$B = \sigma_x\sigma_y + \sigma_x\sigma_z + \sigma_y\sigma_z - \tau_{xy}^2 - \tau_{xz}^2 - \tau_{yz}^2$$

$$C = \sigma_x\sigma_y\sigma_z + 2\tau_{xy}\tau_{xz}\tau_{yz} - \sigma_x\tau_{yz}^2 - \sigma_y\tau_{xz}^2 - \sigma_z\tau_{xy}^2$$

section	location	A	B	C	sigma 1	sigma 2	sigma 3
1	PNT1						
2	PNT2						
2	PNT3						
3	PNT4						
3	PNT5						

To determine the maximum shear stress, you should following the guideline of Mohr's circle for plane stress. The maximum shear stresses can be determined by

$$(\tau_{\max})_1 = \pm\frac{\sigma_2 - \sigma_3}{2}$$

section	location	sigma 1	sigma 2	sigma 3	max shear
1	PNT1				
2	PNT2				
2	PNT3				
3	PNT4				
3	PNT5				

To determine the von Mises stress at each location, you should use the following formula:

$$von\_Mises\_stress = \frac{1}{\sqrt{2}}\sqrt{(\sigma_1 - \sigma_2)^2 + (\sigma_2 - \sigma_3)^2 + (\sigma_1 - \sigma_3)^2}$$

section	location	sigma 1	sigma 2	sigma 3	von Mises
1	PNT1				
2	PNT2				
2	PNT3				
3	PNT4				
3	PNT5				

2. We performed FEA with this object in section 2.2.1 and section 2.2.2. Let us define 4 datum points on the top surface, and define another set of 4 datum points on the bottom surface, as illustrated. Use the function called Measure to calculate the maximum deflection and maximum principal stress at each of these 8 locations. Examine the numerical values and tabulate them to observe those values, which are very close to each other..

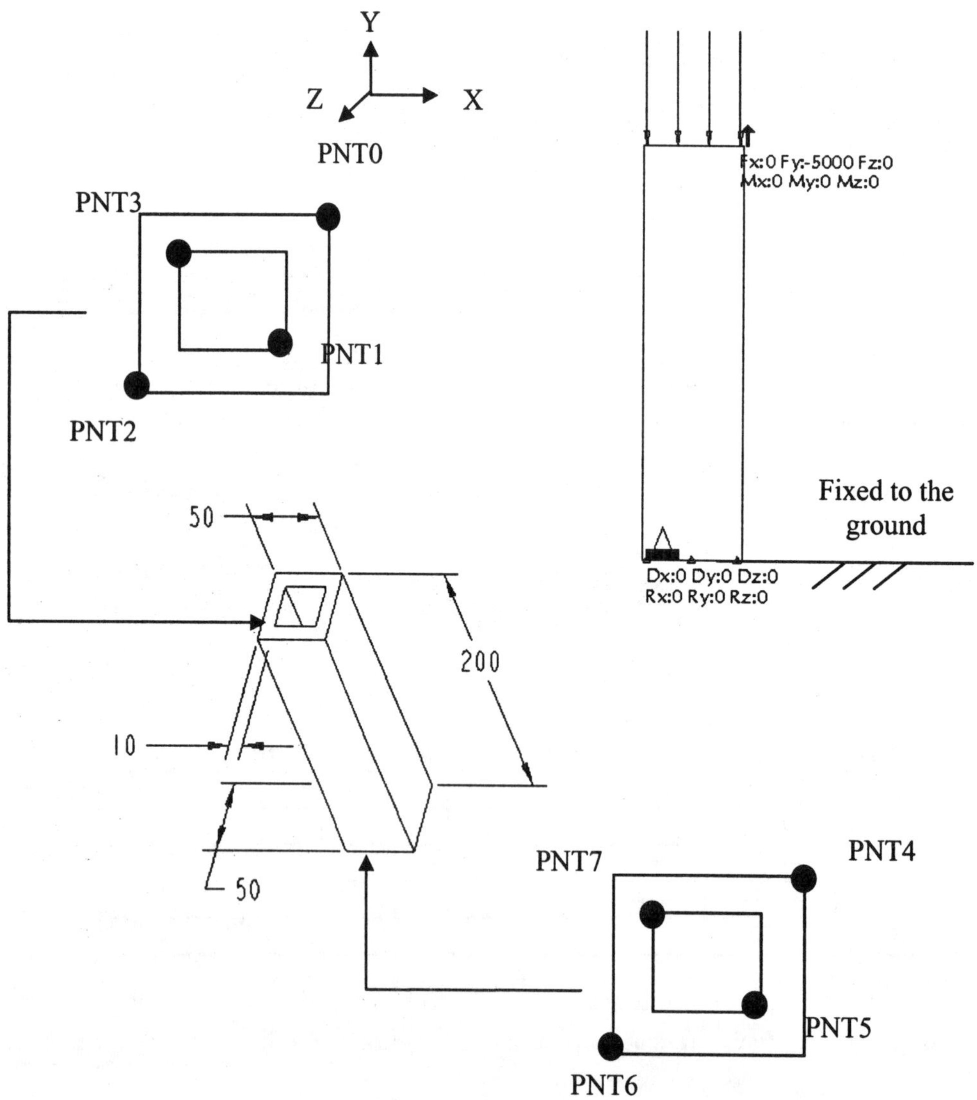

3. The following object is taken from the example studied in section 3.4. The magnitude of the bearing load was 500 Newton. Define a datum point and use the function of Measure to evaluate the maximum principal stress at the location you have chosen. Compare the evaluated value of the maximum principal stress with the Model Max Value. Can we read the maximum principal stress value at the chosen location in the computer report?

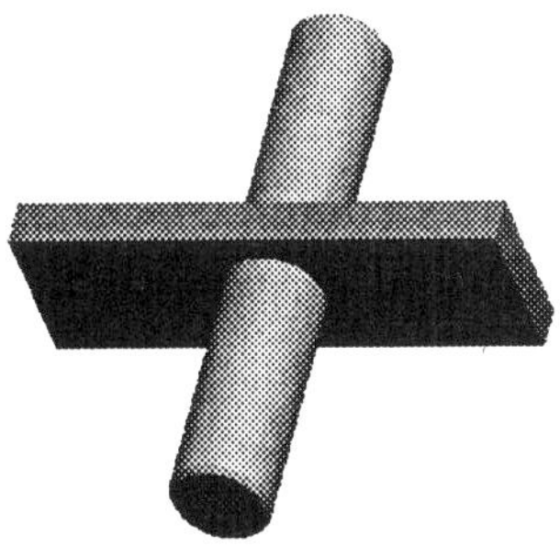

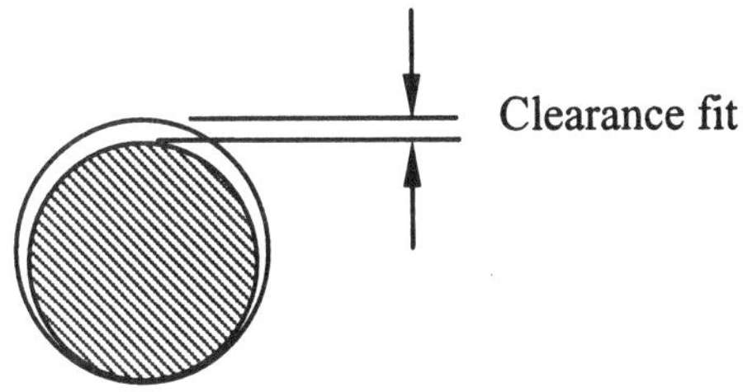

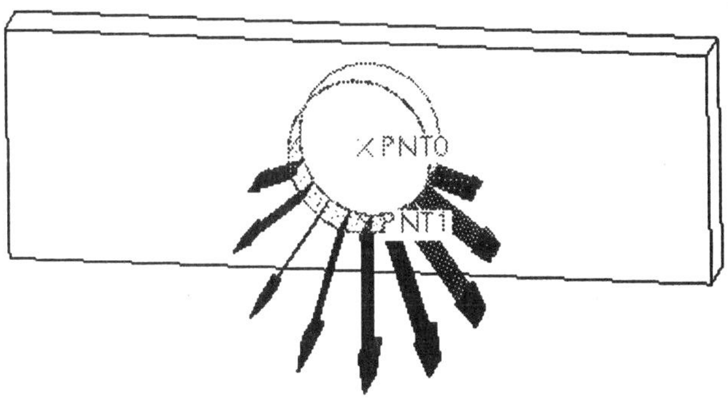

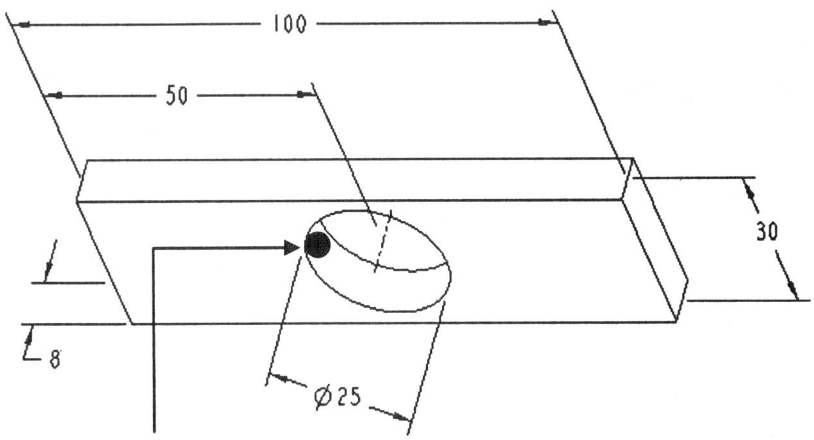

Define a datum point on the circumference of the inner circle

CHAPTER 6

IDEALIZATION: MODELING WITH SHELLS

6.1 Introduction

As discussed in Chapter 1, in order to perform a finite element analysis, we need a geometrical model which is a representation of the physical object under study. When the geometrical model is available, the finite element analysis can be carried out. In this chapter, section 2.1 is an brief interlocution to the Pro/ENGINEER Wildfire design system. An example of creating a block component is used to demonstrate the essential steps needed to create a geometrical model. Section 2.2 is an interlocution to the Pro/Mechanica system. In section 2.2, an example of a column structure subjected to a compressive load is used to demonstrate the essential steps needed for performing a finite element analysis, such as specify the material type, define a set of constraint conditions, and define a set of loading conditions so that the maximum displacement of the column structure can be numerically evaluated and located, and the occurred under the constraint and load conditions.

6.2 Concept of Surface Pairs

Before performing an analysis, we need a 3D solid model in general. In this book, we focus on using the Pro/ENGINEER Wildfire design system to create 3D solid models. The command to start

Figure 6-1 illustrates a plate used to support three plants, which are on display. The three dimensions of the plate and the three surface regions where the three plants are placed are shown below:

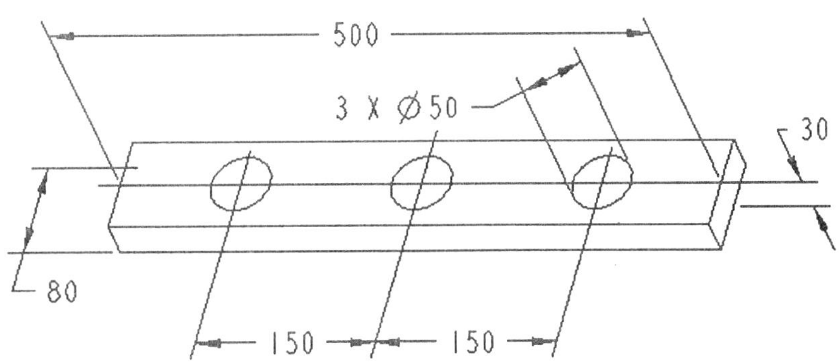

Figure 6-1 Geometrical Shape and Dimensions of the Plate

Assume that the weight of each plant is 200 N. The plate material is Nylon. Use a shell modeling approach to evaluate the deflection distribution and the distribution of the von Mises stress.

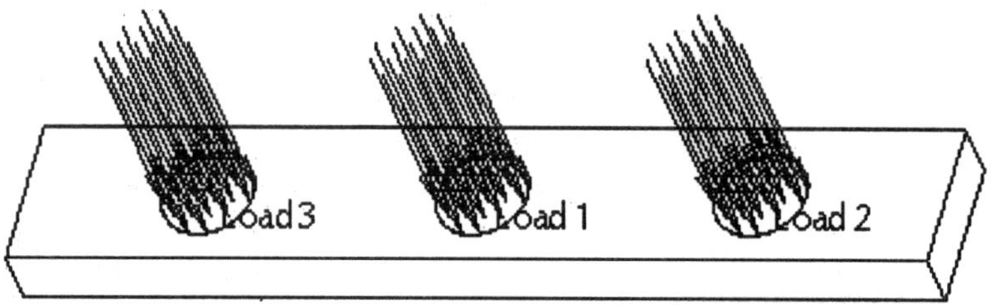

To model the constraint condition, assume the plate is fixed to the ground through two supports. The dimensions of the two supports are shown below.

Step 1: Create a 3D solid model

File > New > Part > type *plate_for_display* as the file name and clear the icon of **Use default template.** Select **mmns_part_solid** (units: Millimeter, Newton, Second) and type *plate_shell* in **DESCRIPTION**, and *student* in **MODELED_BY**, then **OK**

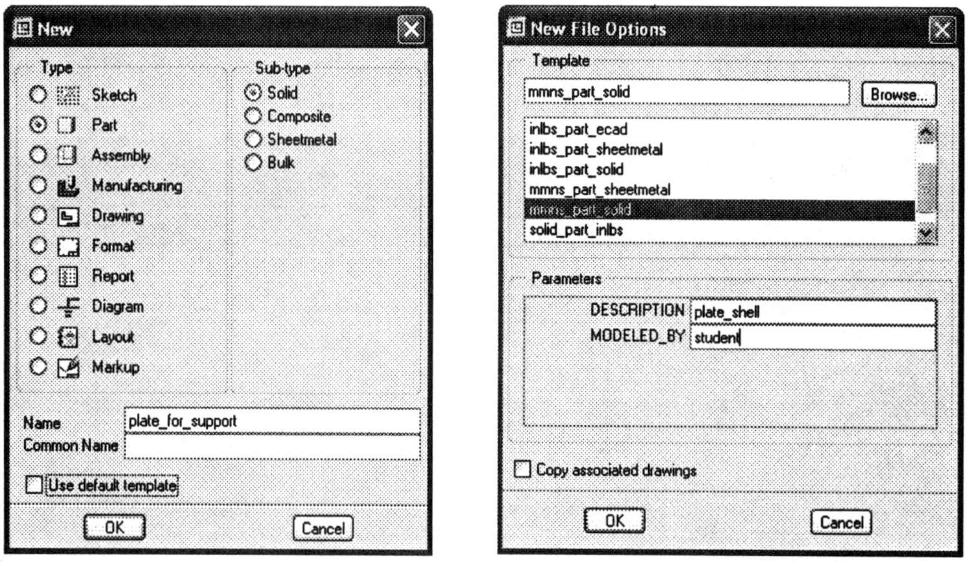

Pick up the icon of **Extrude** displayed on the toolbar of feature creation. Make sure that the icon of **Solid** is selected. Set the thickness value to 30. To initiate the process of sketching, activate **Placement > Define.**

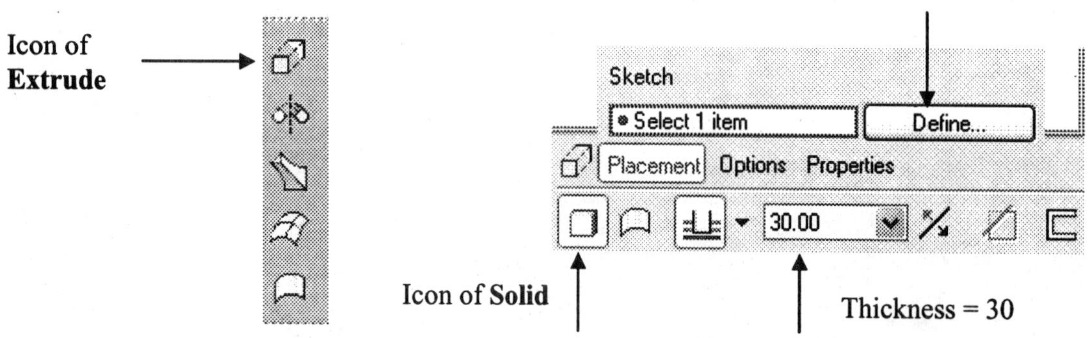

Select the **TOP** datum plane displayed on screen as the sketch plane. Pick the box of **Sketch** to accept the **RIGHT** datum plane as the default reference to orient the sketch plane, as illustrated below:

Pick **FRONT** as the sketch plane

Accept the default setting of orientation

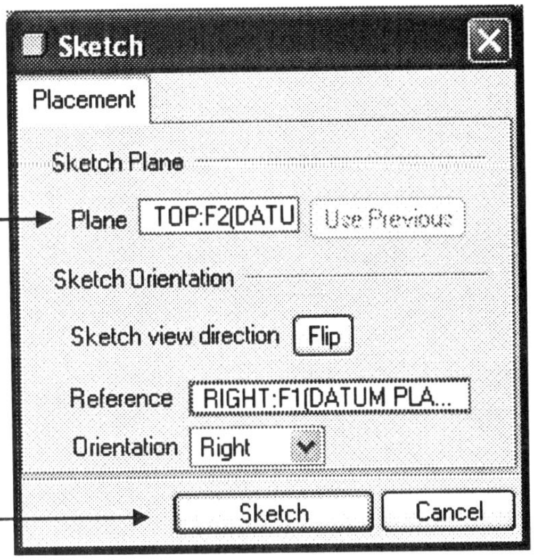

Both the sketch window and the references window appear. Let us first close the reference window. Select the icon of **rectangle** from the toolbar of sketcher and sketch a rectangle. The 2 dimensions are 200 and 50, respectively.

Both the sketch window and the references window appear. Let us first close the reference window. Select the icon of **centerline** from the toolbar of sketcher, and draw a horizontal centerline and a vertical centerline, as shown.

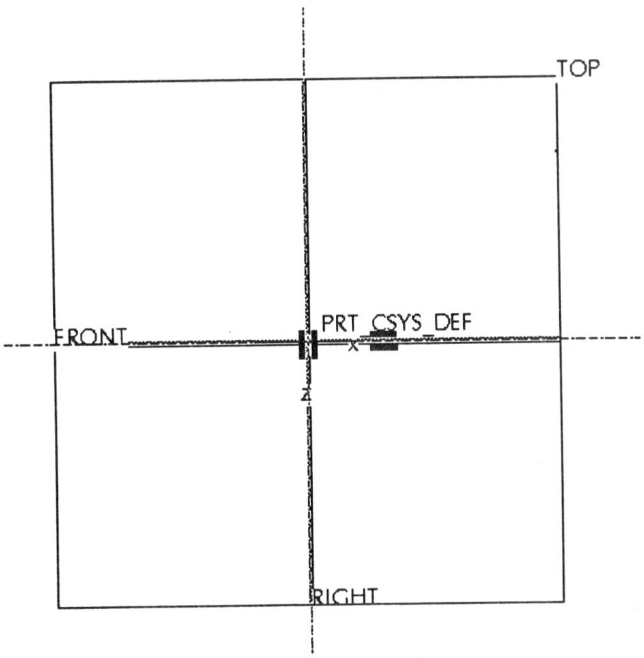

Centerline

Pick the icon of rectangle to sketch a rectangle. The two dimensions are 500 and 80, respectively. Pay attention to the symmetric requirement.

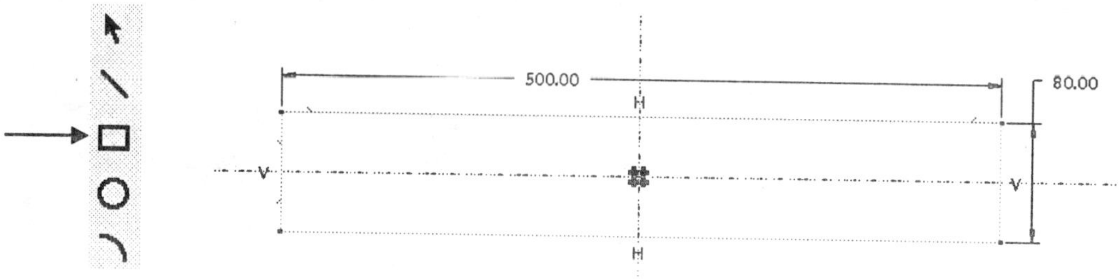

Upon completing the sketch, select the icon of **Done** from the toolbar of sketcher. On the dashboard, check the thickness value, which is equal to *30*.

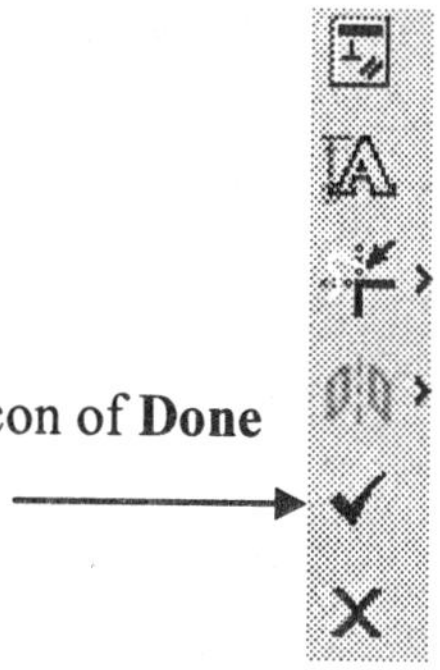

Icon of **Done**

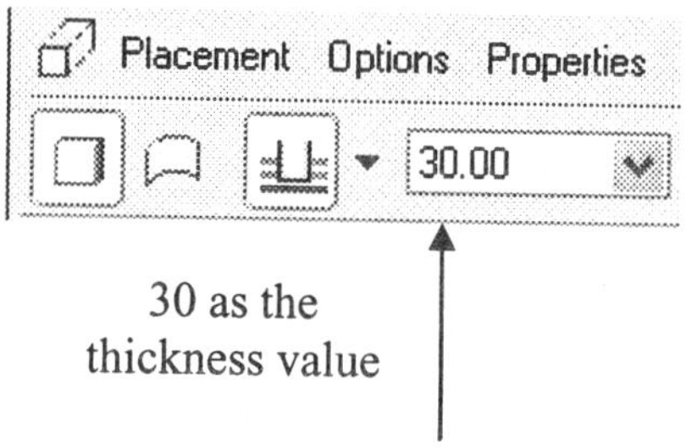

30 as the thickness value

Select the icon of **Complete** from the feature control panel, completing the creation of a 3D solid model for the column structure.

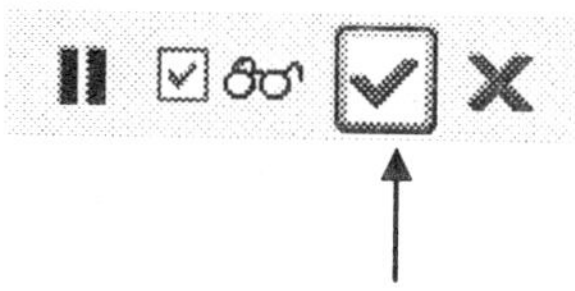

Pick the icon of **Complete** from the **Feature Control Panel**

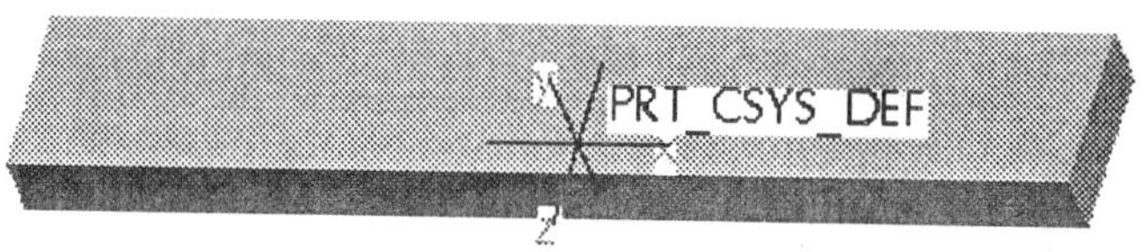

Step 2: Create five (5) datum curves.

Select the icon of **Sketch Tool** from the toolbar of datum feature creation.

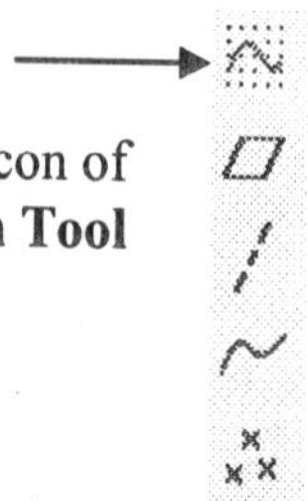

Icon of **Sketch Tool**

Pick the top surface of the plate (do not pick the **TOP** datum plane) as the sketch plane, and accept the default orientation setting, which uses the **RIGHT** datum plane and position it at the right side > **Sketch**.

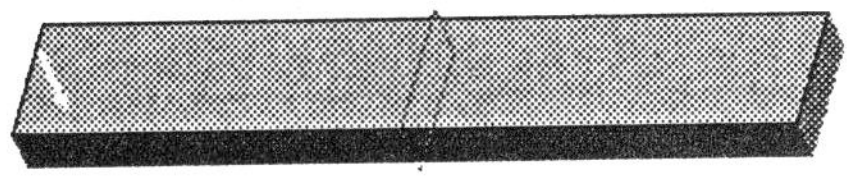

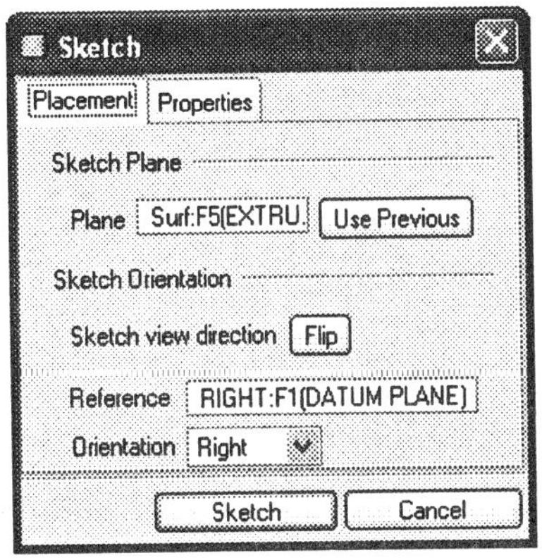

Pick the icon of circle to sketch a circle. The diameter dimension is 50. Upon completion, select the icon of **Done.**

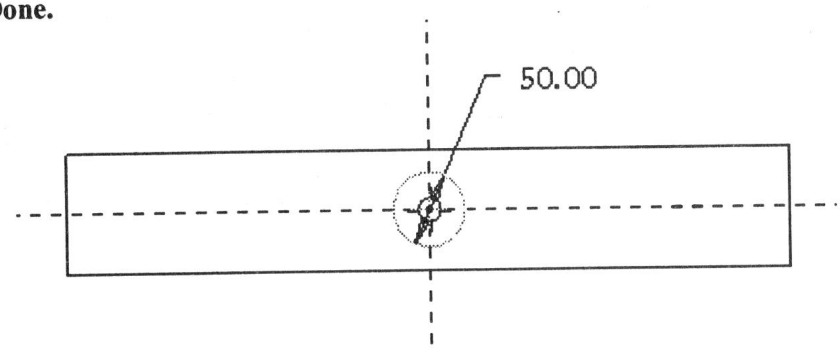

Repeat the above procedure to create the second datum curve, as shown.

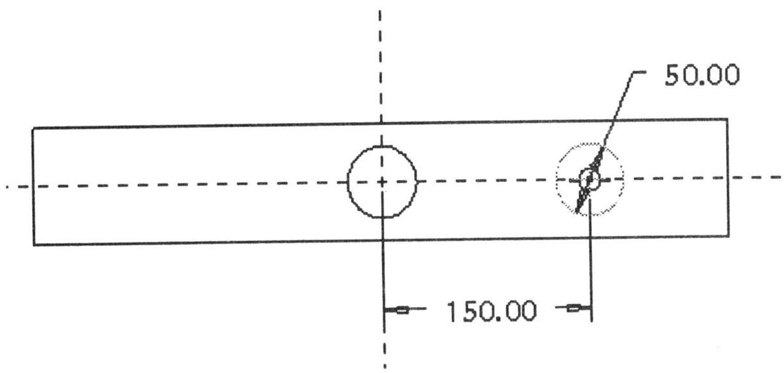

Repeat the above procedure to create the third datum curve, as shown.

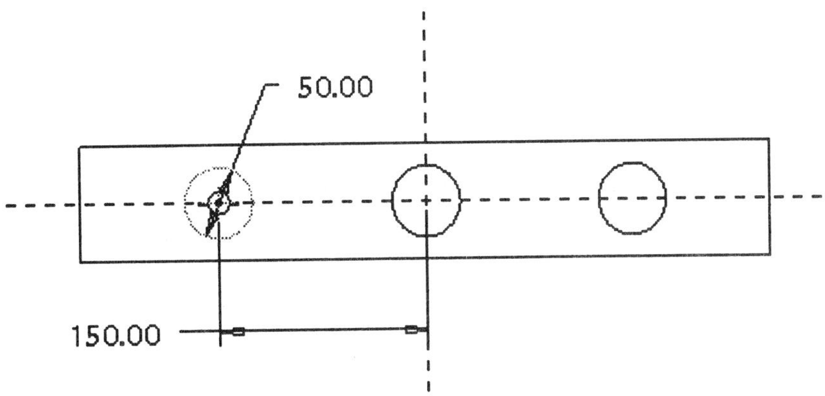

To create the fourth datum curve, we need to select the bottom surface of the plate as the sketch plane, as shown.

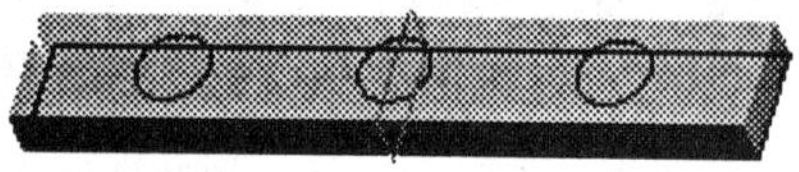

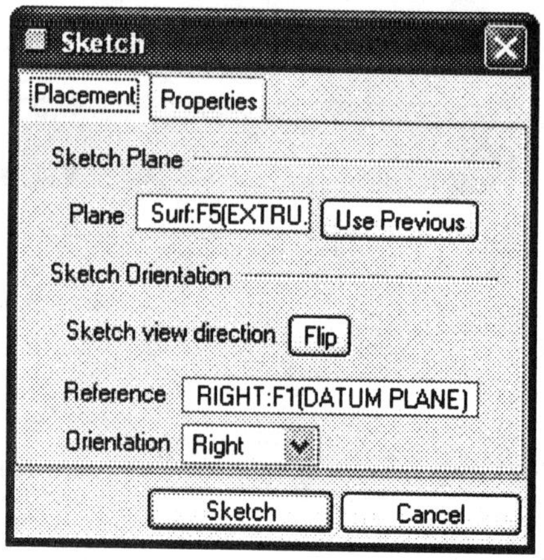

Before making the sketch, add 3 additional references, which are shown below:

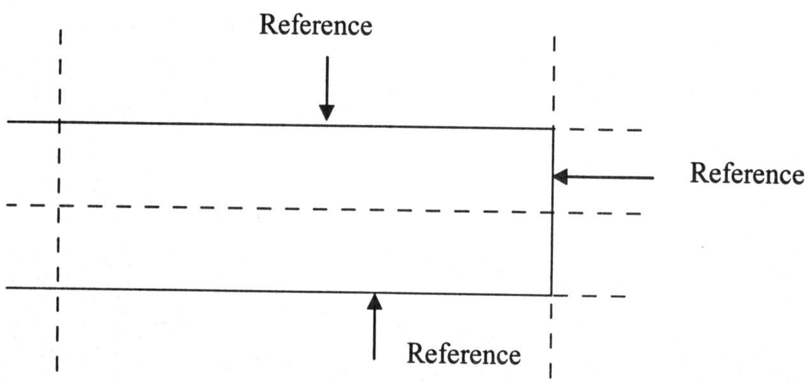

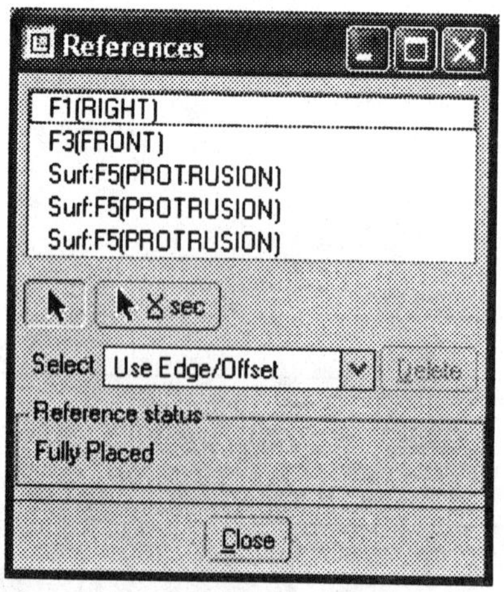

Pick the icon of **rectangle** to sketch a rectangle. Only one dimension is needed because there are 3 sides already selected as the references. The needed dimension is the width of the rectangle, which is equal to 40. Upon completion, select the icon of **Done**.

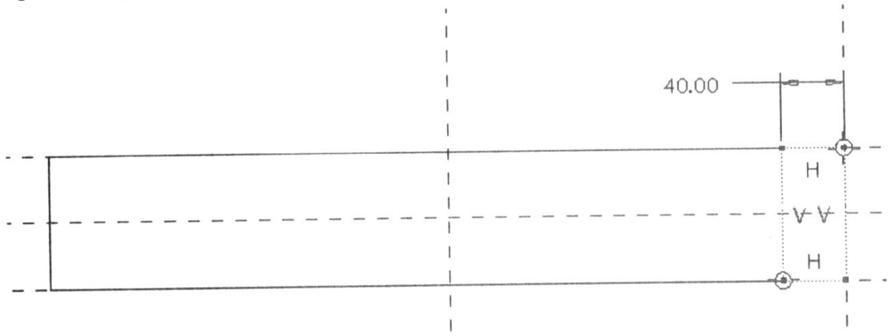

Repeat the above procedure to create the 5th datum curve, as shown.

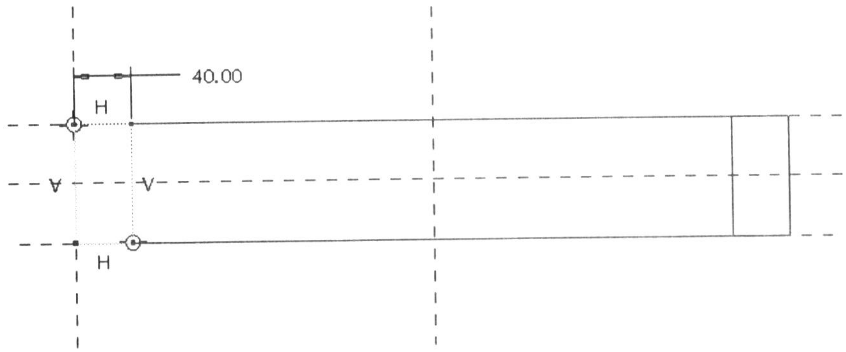

Step 3: Perform FEA under **Pro/Mechanica**

From the menu toolbar, select **Applications** > **Mechanica**. Click **Continue** after checking the unit system.

Select **Structure** from the Model Type window > **OK**.

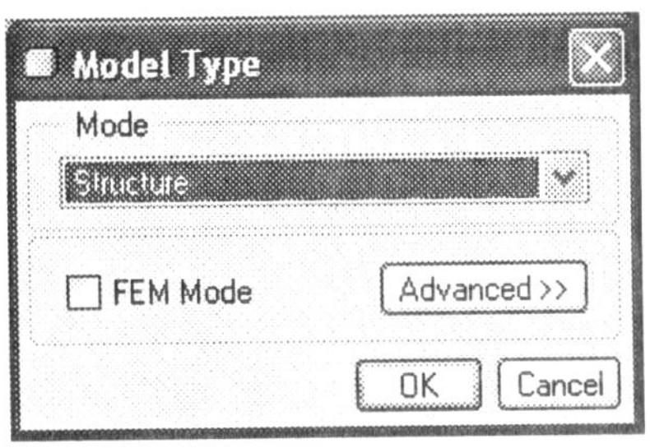

From the toolbar of functions, select the icon of **Materials** > select **NYLON.** Click **Assign** > **Part** > pick the solid model of plate on display > **OK** > **Close.**

Icon of Materials

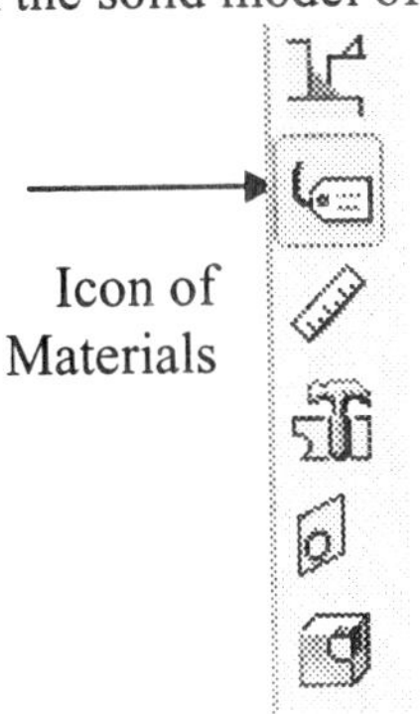

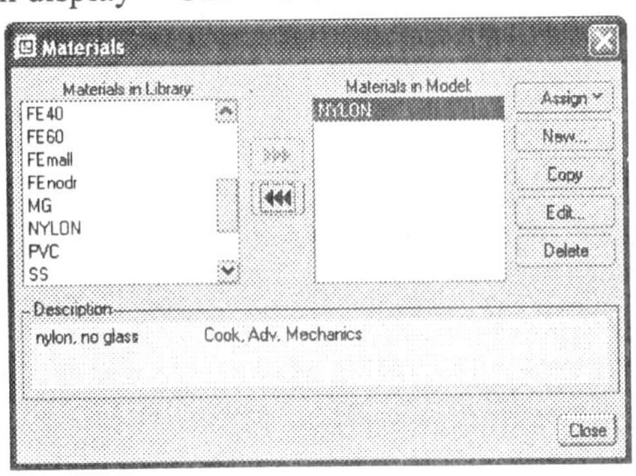

Now let us define the 5 surface regions. From the toolbar of Functions, select the icon of **Surface Region** > **Select** > **Done.**

Select the datum curve created previously, and select the top surface of the plate as the surface to be split, from which the surface region is separated > **OK** > **Done** > **OK**.

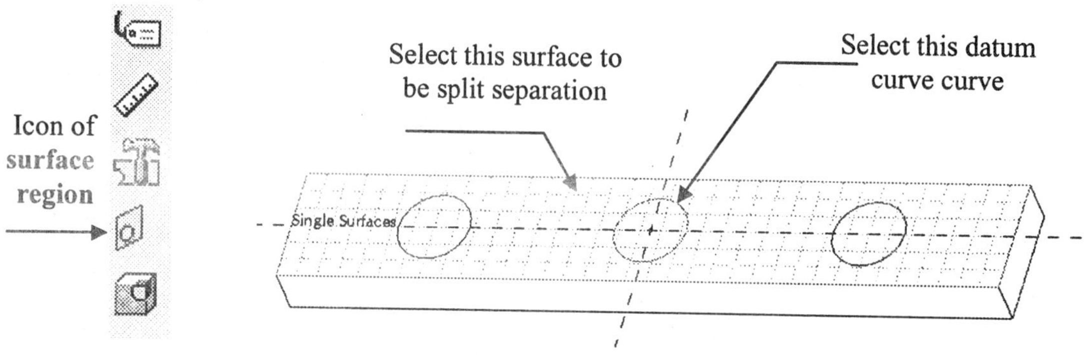

Repeat the above process 2 more times so that the 3 surface regions on the top of the plate are defined.

To define the two surface regions on the bottom of the plate, we follow the above procedure as well.

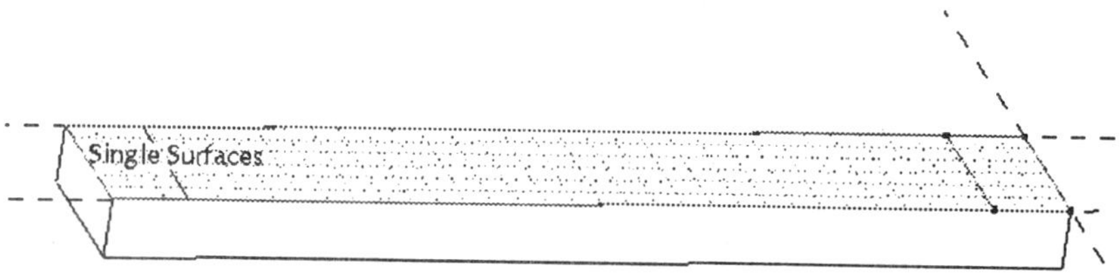

From the model tree, the 5 created surface regions are listed under Simulation Feature.

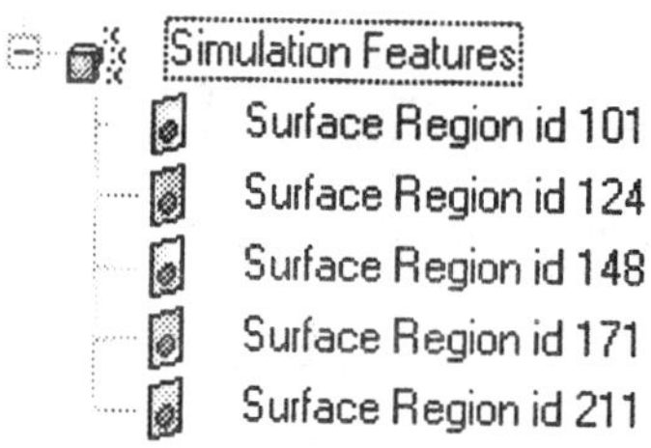

To define the shell mode, from the toolbar of **Functions**, select the icon of **New Shell Pair.**

New > Constant > pick the top surface of the plate first, while holding down the **Ctrl** key, pick the three surface regions located on the top surface.

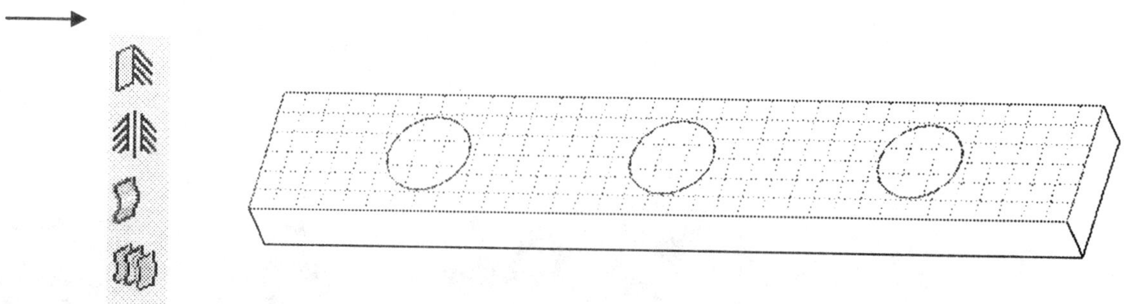

Release the **Ctrl** key and rotate the 3D solid model so that the bottom surface can be easily picked up.

Hold down the Ctrl key, and pick the bottom surface and the 2 surface regions located on the bottom surface > OK > OK.

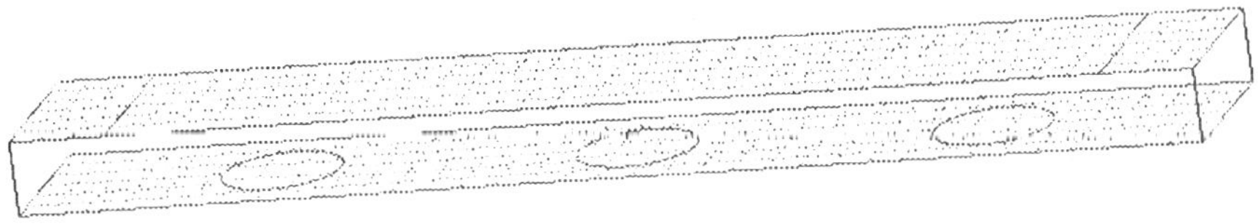

To verify the shell model, **Compress** > **Shells only**

Show Compress

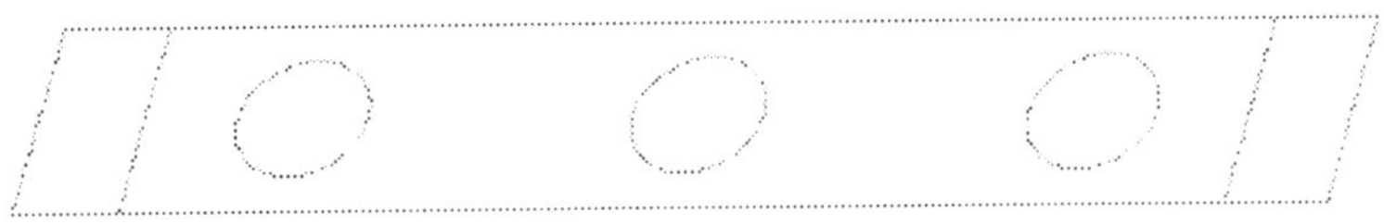

Show Original

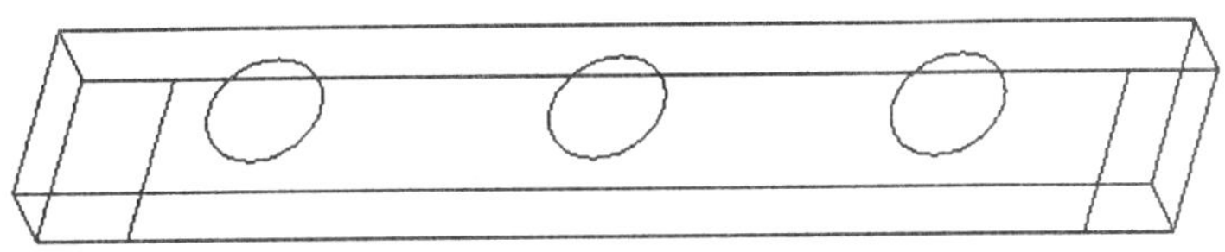

Show Both

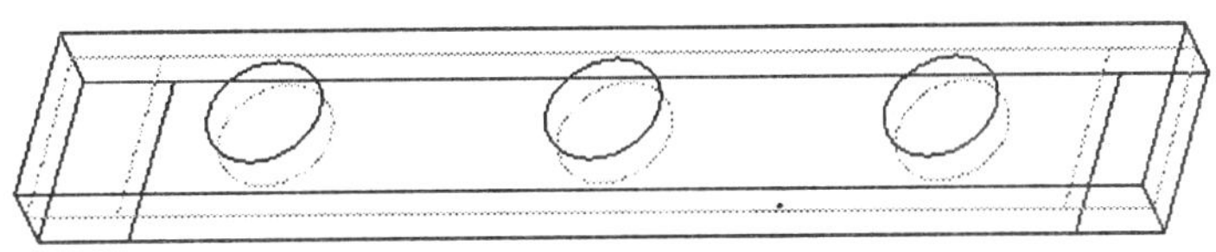

To define the constraint condition, select the icon of **New Displacement Constraint** from the toolbar of functions > **Surface(s)** > pick the surface region at the right side on the bottom surface of the plate. Because the constraint condition is "fixed to the ground", we set all 6 degrees of freedom equal to zero, which is the default setting > **OK**.

To define the second constraint, select the icon of **New Displacement Constraint** from the toolbar of functions > **Surface(s)** > pick the surface region at the left side on the bottom surface of the plate. Because the constraint condition is "fixed to the ground", we set all 6 degrees of freedom equal to zero, which is the default setting > **OK**.

Icon of New Displacement Constraint

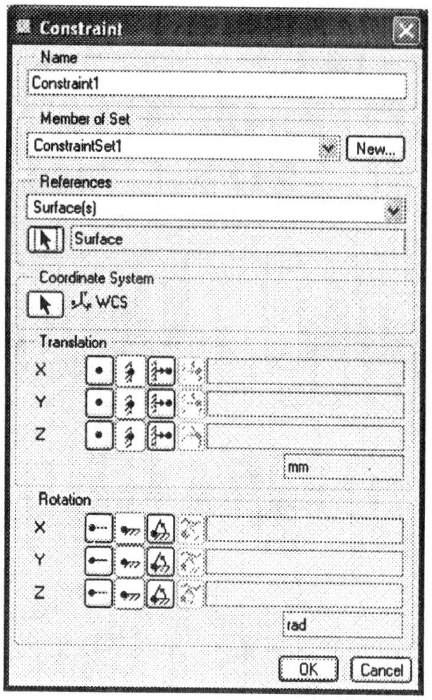

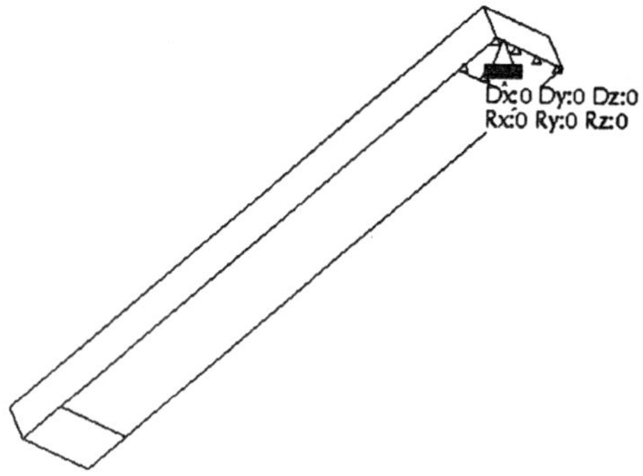

To define the second constraint, select the icon of **New Displacement Constraint** from the toolbar of functions > **Surface(s)** > pick the surface region at the left side on the bottom surface of the plate. Because the constraint condition is "fixed to the ground", we set all 6 degrees of freedom equal to zero, which is the default setting > **OK**.

Icon of New Displacement Constraint

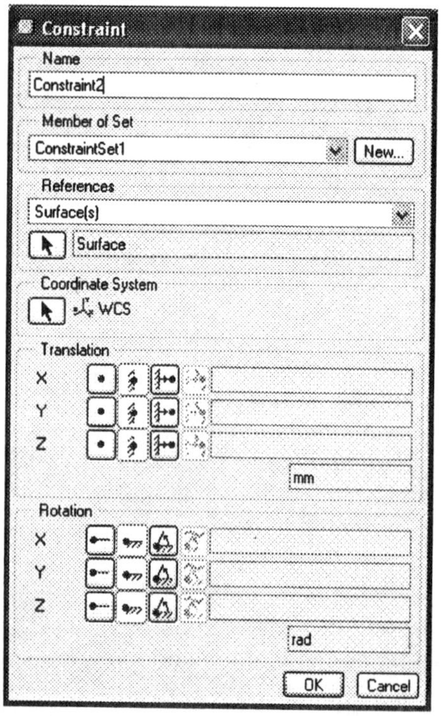

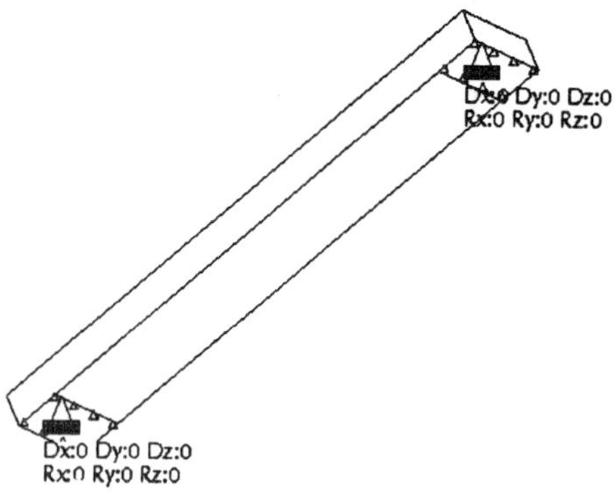

To define the load condition on the top surface of the plate, select the icon of **New Force/Moment Load** from the toolbar of functions > **Surface(s)** > pick the surface region at the right side > **OK** to accept the surface selection.

Because the load is acting along the negative Y direction, we type -200 in the Y force component box > **OK**.

Icon of New Force/Moment Load

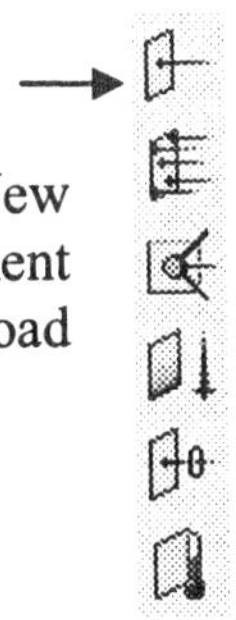

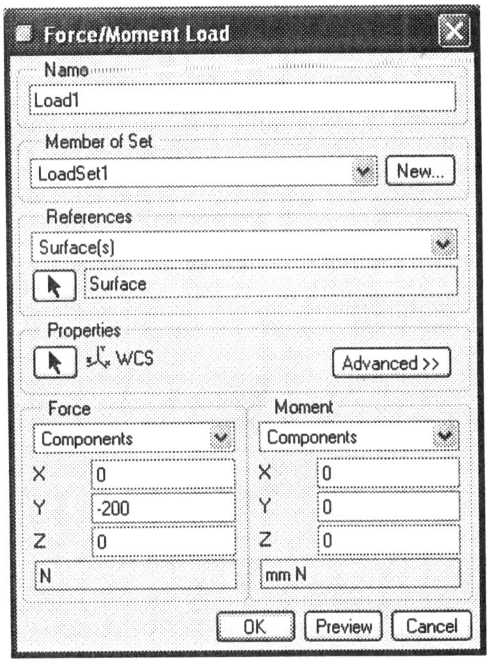

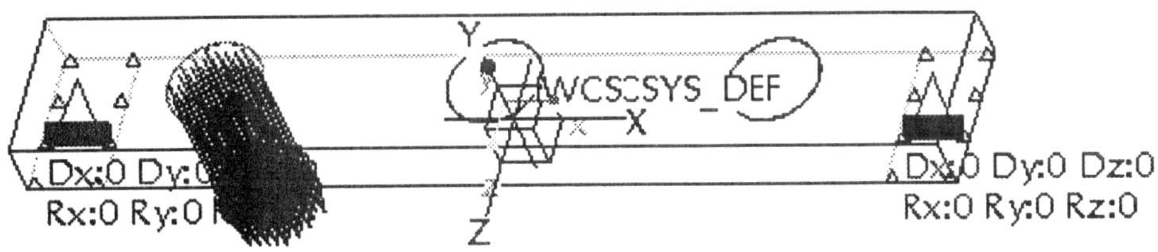

Repeat this procedure, to define Load2 and Load3 at the other 2 surface regions with the magnitude equal to -200 N.

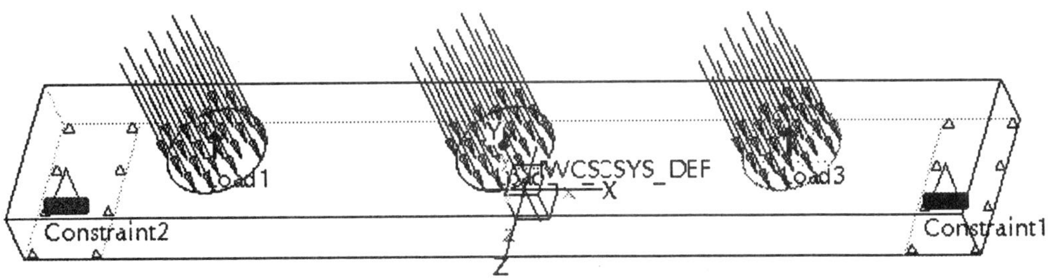

Step 4: Run the FEA program

Under the Pro/MECHANICA, set up **Analyses** and **Run** it

Select the icon of **Run an Analysis** from the main toolbar > **File** > **New Static** > type *plate* as the name of the analysis folder > **OK.** Click the box of **Run.**

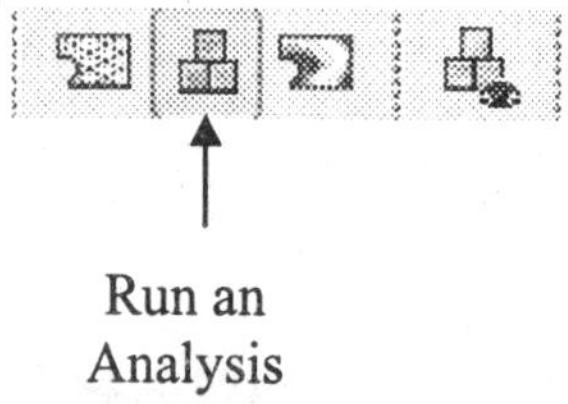

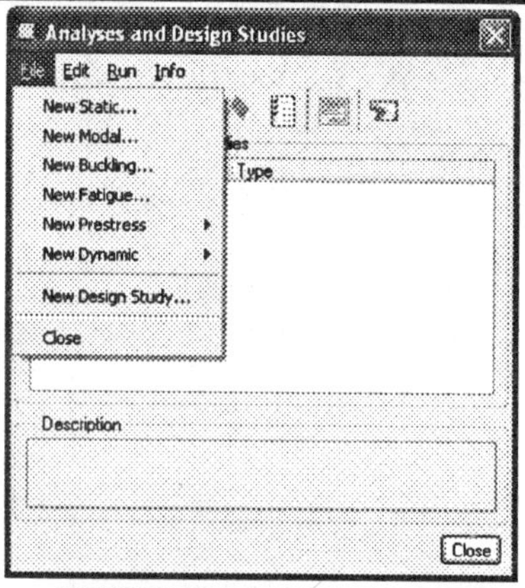

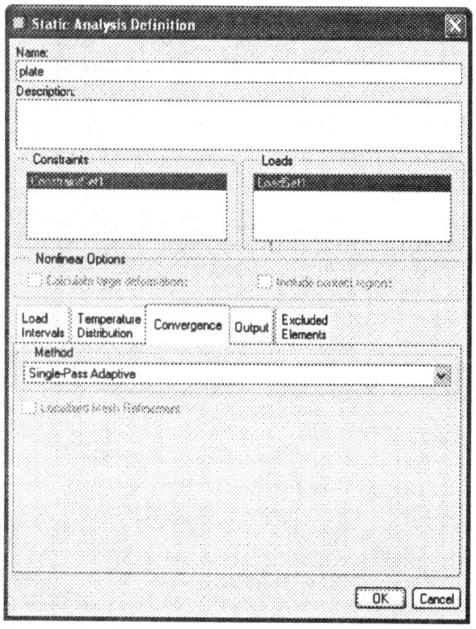

There is always a message displayed on screen. The message is asking the user "Do you want error detection?" Click **Yes**. In this way a warning message will appear if an abnormal operation occurs.

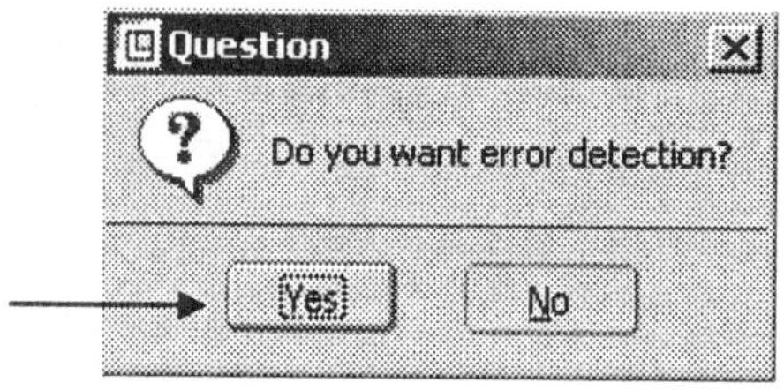

To monitor the computing process, users may click the icon of **Display study status**.

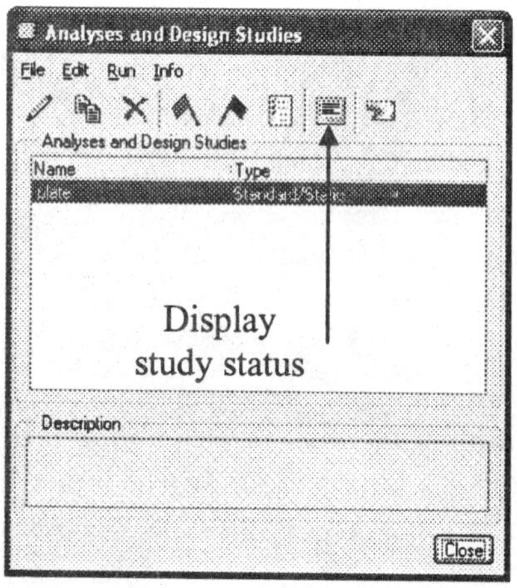

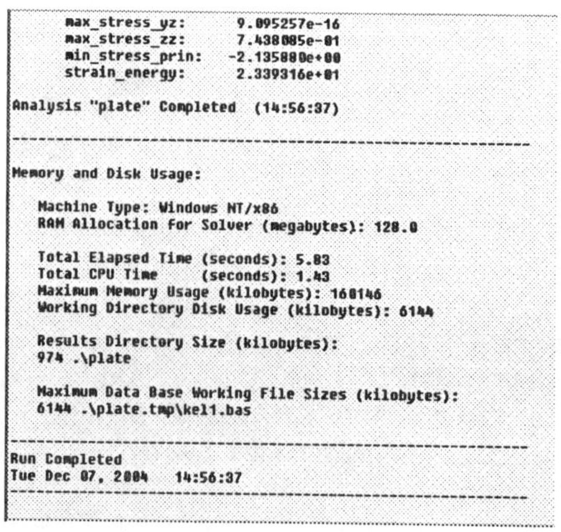

```
        max_stress_yz:        9.095257e-16
        max_stress_zz:        7.438085e-01
        min_stress_prin:     -2.135880e+00
        strain_energy:        2.339316e+01

Analysis "plate" Completed  (14:56:37)

------------------------------------------------------------

Memory and Disk Usage:

    Machine Type: Windows NT/x86
    RAM Allocation for Solver (megabytes): 128.0

    Total Elapsed Time (seconds): 5.83
    Total CPU Time     (seconds): 1.43
    Maximum Memory Usage (kilobytes): 168146
    Working Directory Disk Usage (kilobytes): 6144

    Results Directory Size (kilobytes):
    974 .\plate

    Maximum Data Base Working File Sizes (kilobytes):
    6144 .\plate.tmp\kel1.bas

------------------------------------------------------------
Run Completed
Tue Dec 07, 2004   14:56:37
------------------------------------------------------------
```

Step 5: Examine the results obtained from running the FEA program

Click the icon of **Review results** > **Displacement** > **Magnitude** > **OK and Show.** Note that if the folder called *plate* does not appear in the Result Window Definition, click the icon of folder under Design Study, locate it and open it.

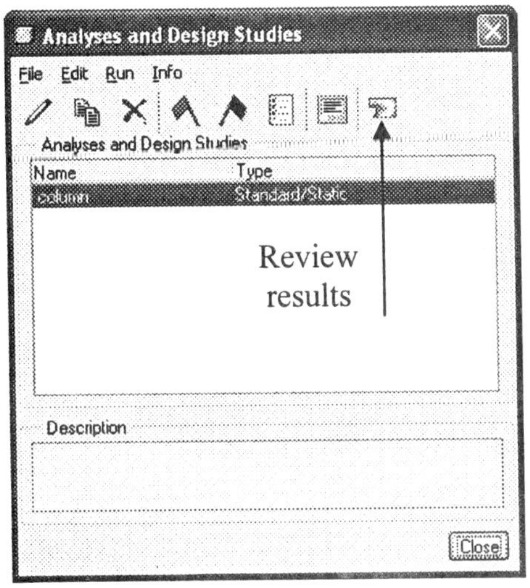

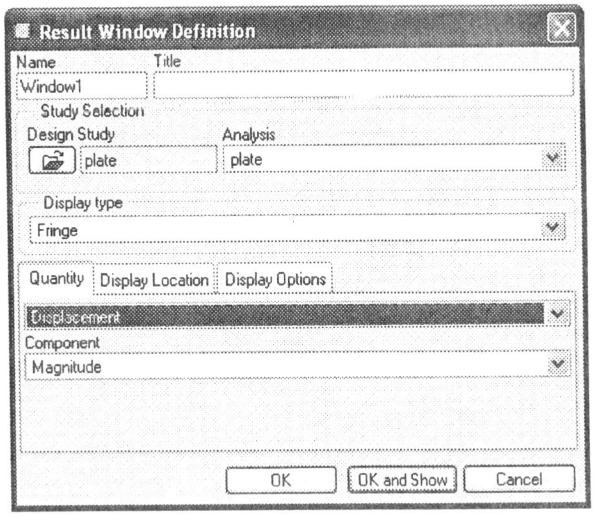

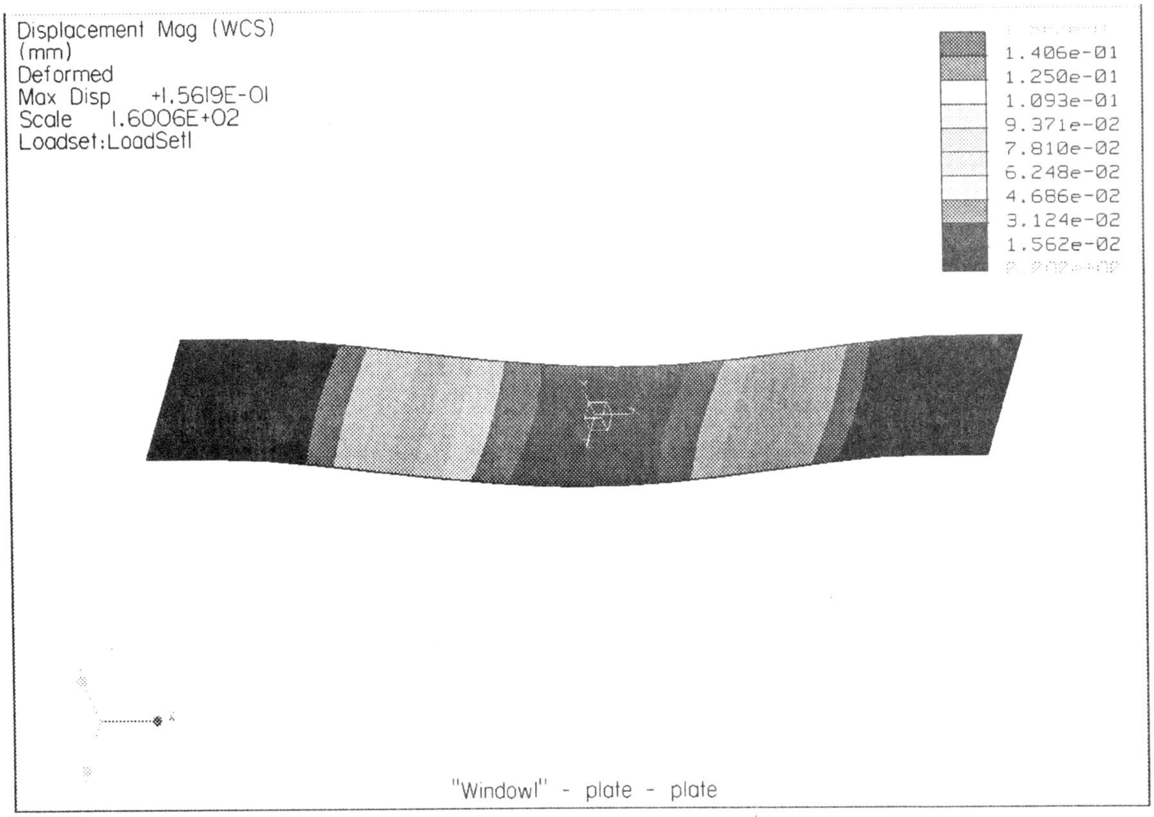

To study the distribution of von Mises stress, select **Edit** from the main toolbar > **Copy** and a new Result Window appears > **Stress** > **von Mises** > type Distribution of von MIses as the title > **OK and Show.**

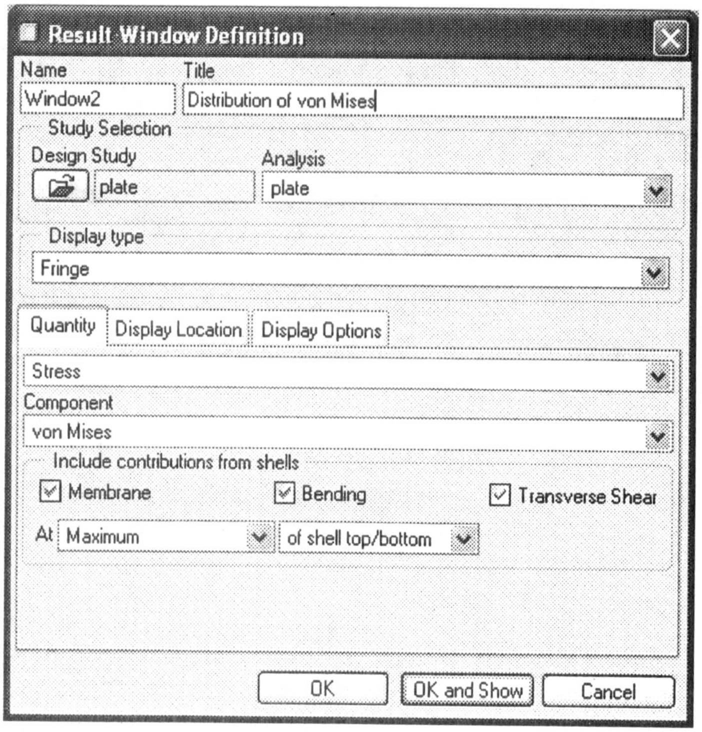
Result Window Definition
Name
Title
Window2
Distribution of von Mises
Study Selection
Design Study
Analysis
plate
plate
Display type
Fringe
Quantity
Display Location
Display Options
Stress
Component
von Mises
Include contributions from shells
Membrane
Bending
Transverse Shear
At
Maximum
of shell top/bottom
OK
OK and Show
Cancel

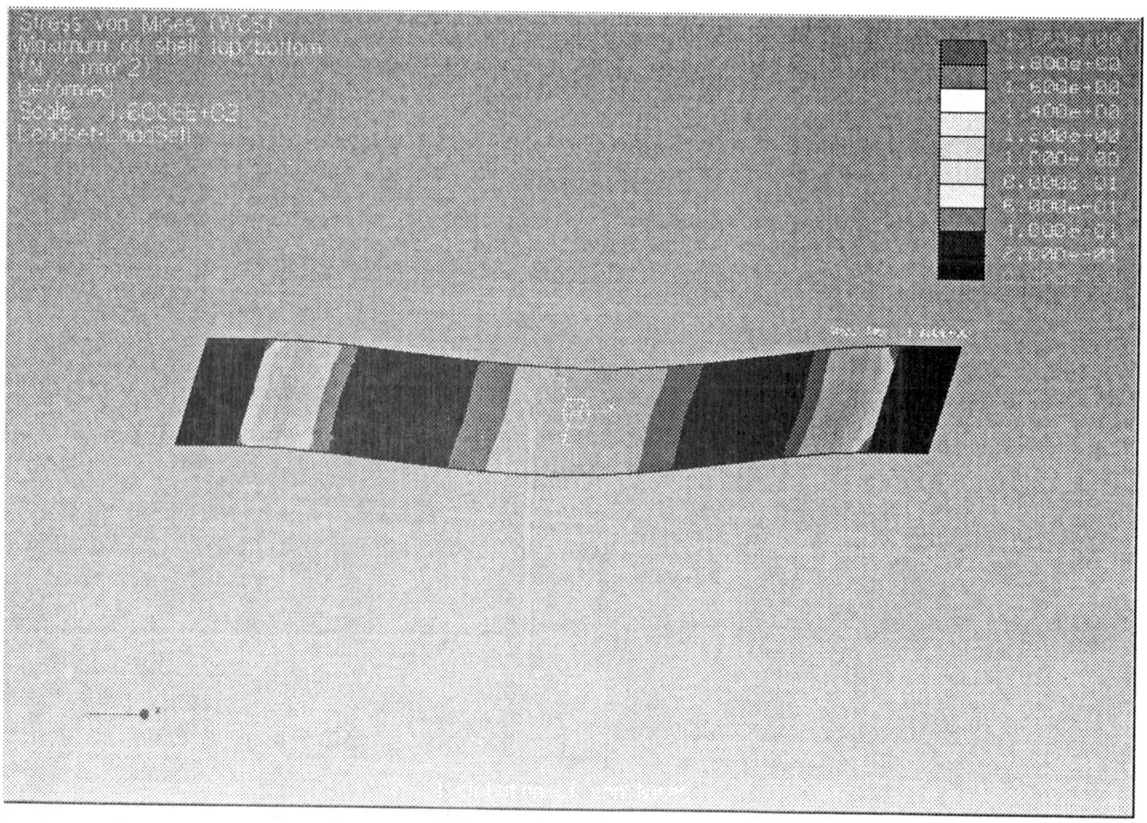
Stress von Mises (WCS)
Maximum of shell top/bottom
(N / mm^2)
Deformed
Scale 1.6000E+02
Loadset:LoadSet1

6.3 Shell Modeling Using Auto Detect

In section 6.2, we discussed the creation of a shell model, or the creation of a middle surface, by picking a pair of surfaces, which are parallel to each other. Under certain circumstances, if the geometrical model is created such a way that the software system may be capable of automatically detecting pairs of surfaces and generate the middle surfaces. The following example is used to demonstrate this unique characteristic in shell modeling.

A t-shaped bracket is shown below. This object consists of two plates. The thickness of both plates is 25 mm. The dimensions of the plate in the horizontal position are 250 x 250 mm. The dimensions of the plate in the vertical position are 250 x 250 mm.

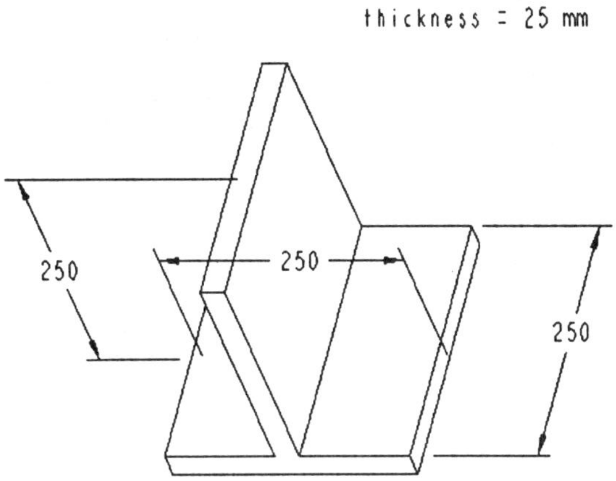

Assume that the load acts on the top surface of the support structure. The load is a uniformly distributed load with a total value equal to 500 Newton. The entire support structure is held to the ground by the two lateral surfaces, as shown. Assume that the material type of the two plates is steel.

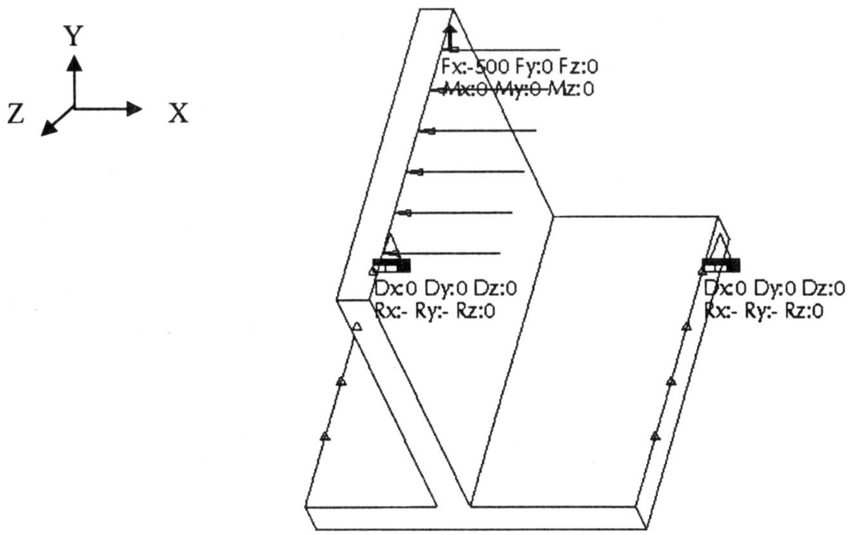

There are 2 objectives for performing FEA:

(1) Find the location of the maximum displacement on the beam when subjected to this uniformly distributed load and specify its magnitude.
(2) Find the location of the maximum principal stress and specify this magnitude.

Step 1: Create a 3D solid model

File > New > Part > type in *support* as the file name and clear the icon of **Use default template > OK.** Select **mmns_part_solid** (units: Millimeter, Newton, Second) and type in *shell model* in **DESCRIPTION**, and *student* in **MODELED_BY**, then **OK**

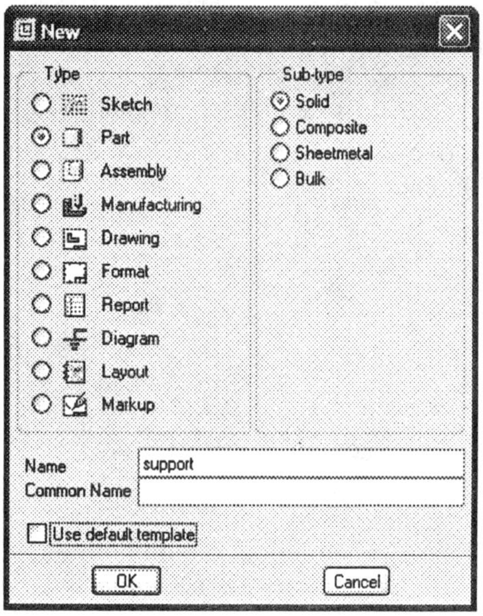

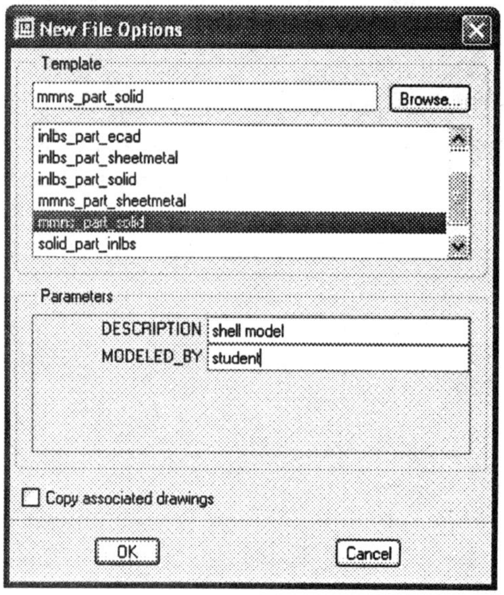

To create the plate in the horizontal position, from the main toolbar, select **Insert > Extrude.** From the dashboard, select **Thin Protrusion** > set the thickness value to 25 > select the symmetric depth setting > the depth value equal 250. Activate **Placement** to define a sketch plane > **Define.**

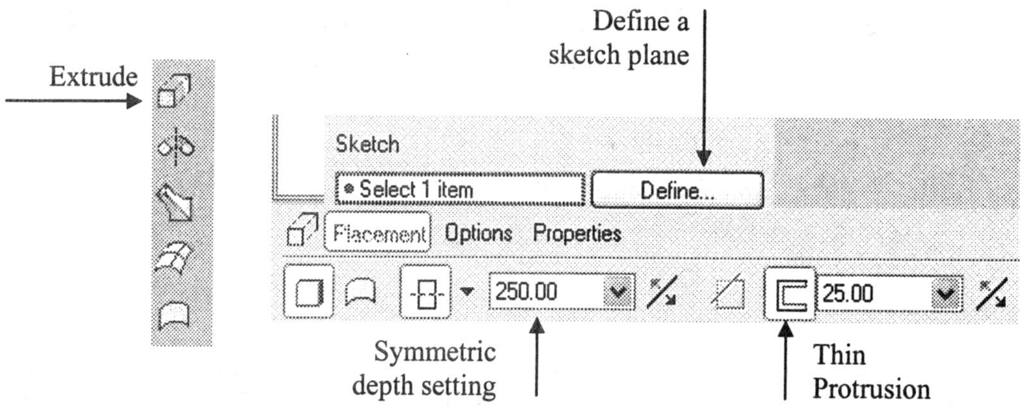

Select the **FRONT** datum plane as the sketch plane, and click the box of **Sketch** to accept the **RIGHT** datum plane as the default reference to orient the sketch plane, as illustrated below.

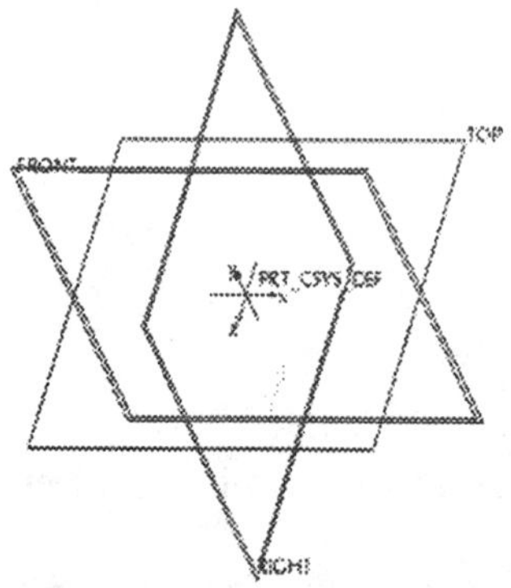

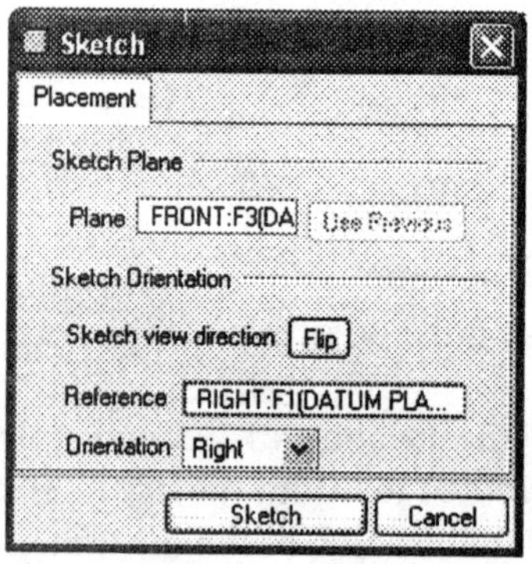

Select the icon of **Centerline** from the toolbar of sketcher to draw a centerline along the y-axis. Select the icon of **Line** to sketch a line along the x-axis. This line is symmetric about the y-axis. The length is 250.

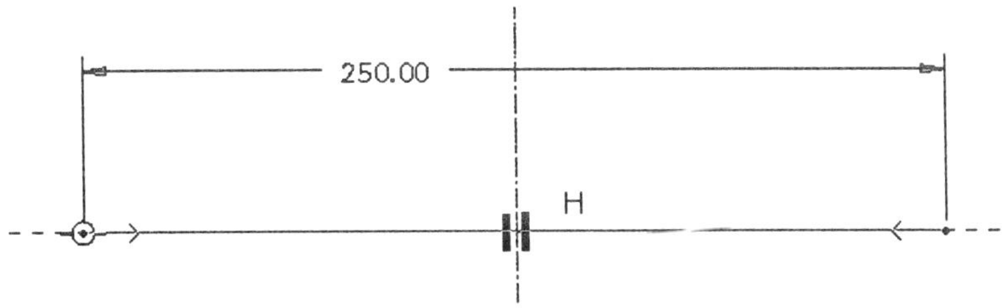

Upon completion, pick the icon of **Done** from the toolbar of sketcher. From the dashboard, select the icon of **Complete** from the feature control panel, thus completing the creation of the first feature.

Pick the icon of **Complete** from the Feature Control Panel

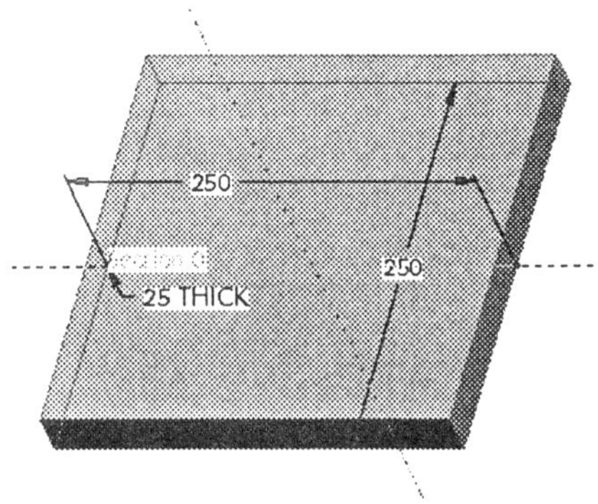

To create the vertical plate, from the main toolbar, select **Insert > Extrude.** From the dashboard, select **Thin Protrusion** > set the thickness value to 25 > select the symmetric depth setting > the depth value equal 250. Activate **Placement** to define a sketch plane > **Define.**

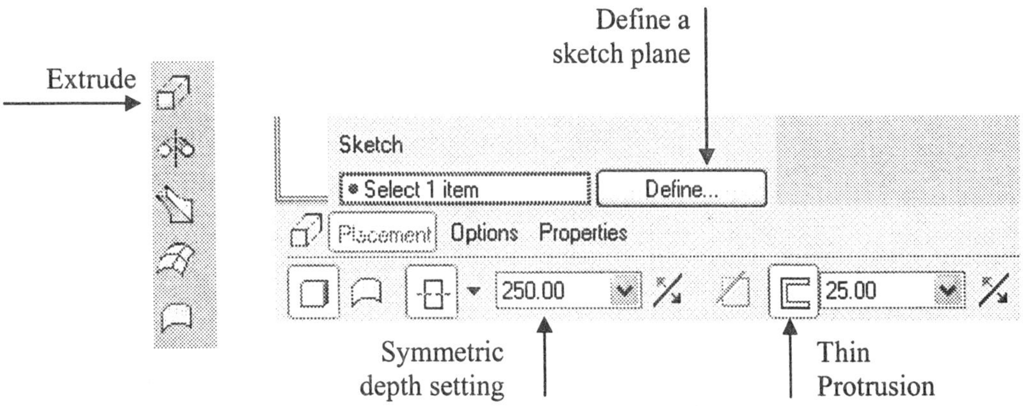

Select the **FRONT** datum plane as the sketch plane, and click the box of **Sketch** to accept the **RIGHT** datum plane as the default reference to orient the sketch plane, as illustrated below.

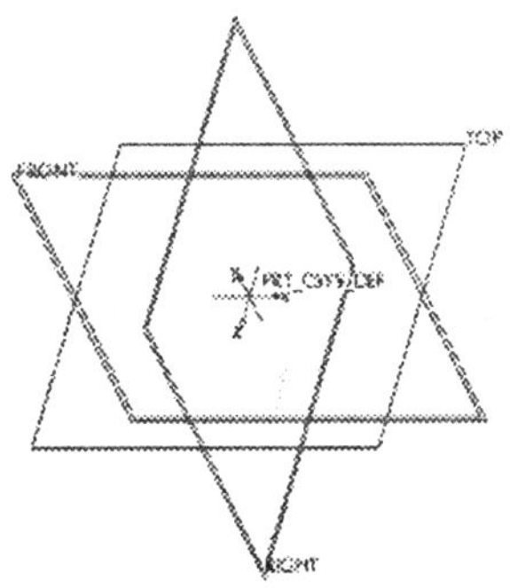

Select the icon of **Line** to sketch a line along the y-axis. The length is 250.

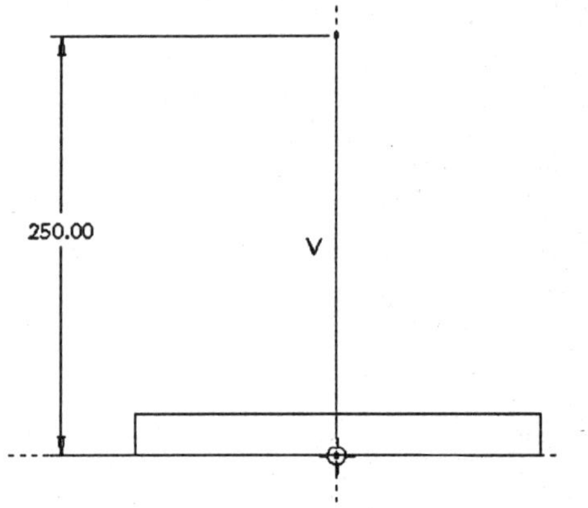

Upon completion, pick the icon of **Done** from the toolbar of sketcher. From the dashboard, select the icon of **Complete** from the feature control panel, thus completing the creation of the second feature.

Pick the icon of **Complete** from the Feature Control Panel

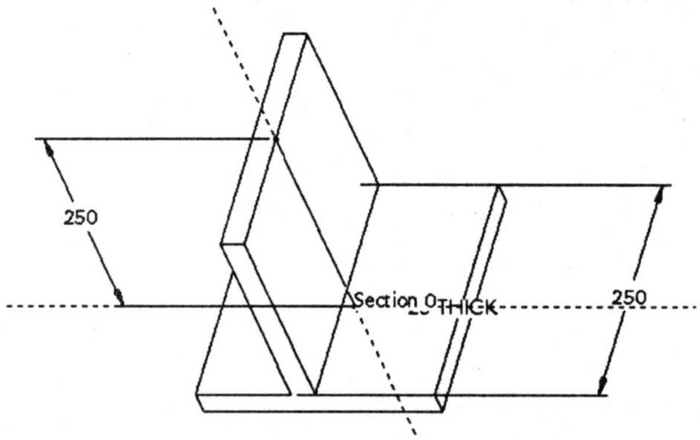

Step 2: FEA under Pro/MECHANICA.

Applications > **Mechanica** > **Continue** > **Structure** > **OK**.

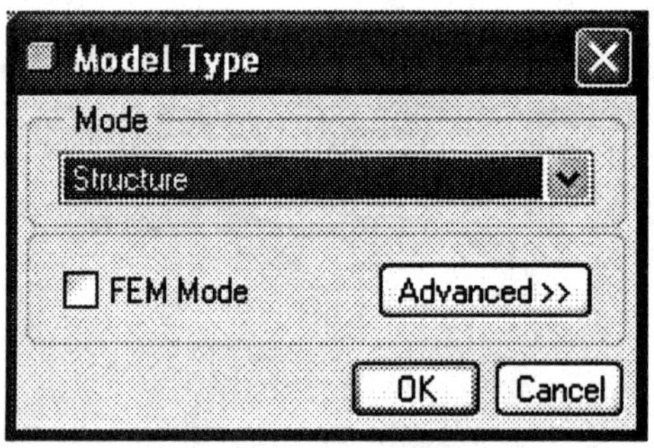

From the toolbar of functions, select the icon of **Materials >** select **STEEL**

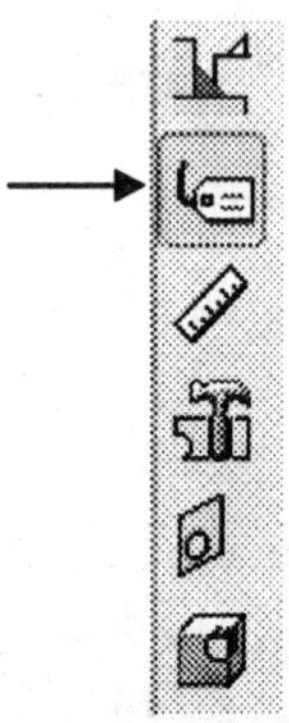

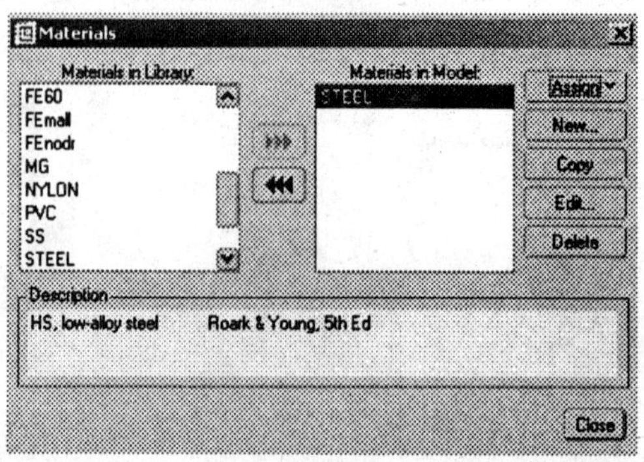

Assign > **Part** > select the support structure from the screen display > **OK** > **Close**

To define the shell model, from the toolbar of functions, select the icon of **New Shell Pair** > **Auto Detect** > **Compress** > **Shells Only**.

Show Compress

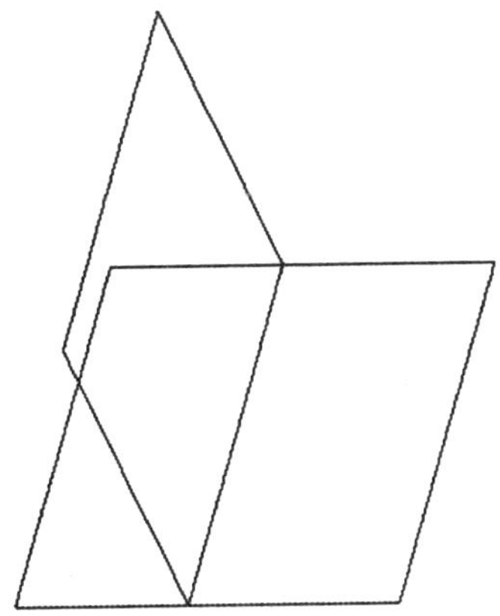

Show Original

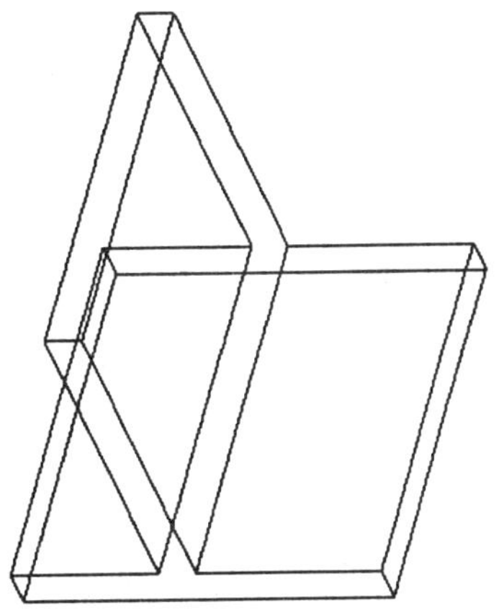

Show Both

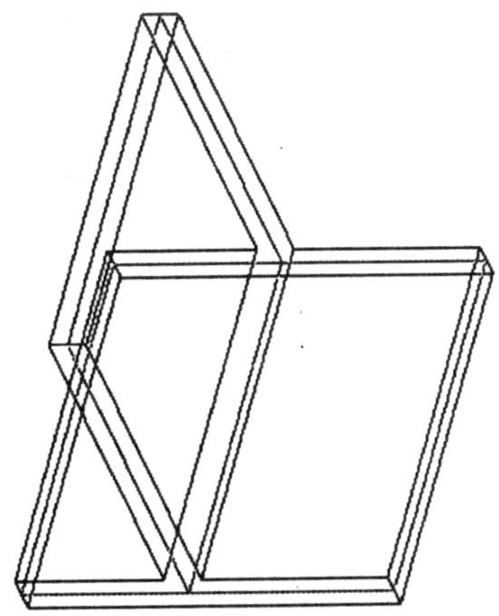

From the toolbar of functions, select the icon of **Constraints** > select **Edge(s)/Curve(s)** > pick the edge on the left side > **OK**.

Note the constraint conditions for fixing a straight line are, in this case, fixing Translation X, Translation Y, Translation Z and Rotation Z.

Repeat this procedure to define the second constraint for fixing the edge on the right side.

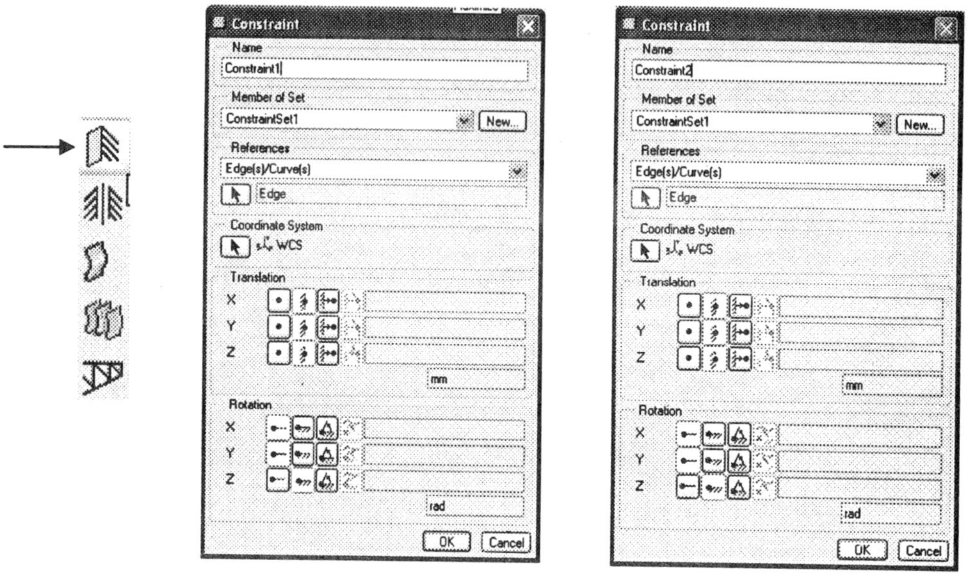

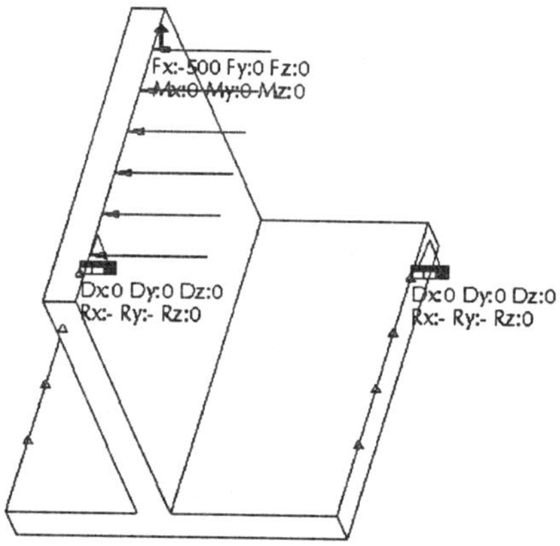

To define a uniform load acting on the top edge in the negative x direction, select the icon of **force and moment** from the toolbar > **Edge(s)/Curve(s)** > pick the edge on the top surface > **OK** > click **Advanced** > change the distribution to **Total Load** and **Uniform** > type in *-500* in X of the Force box > **OK**.

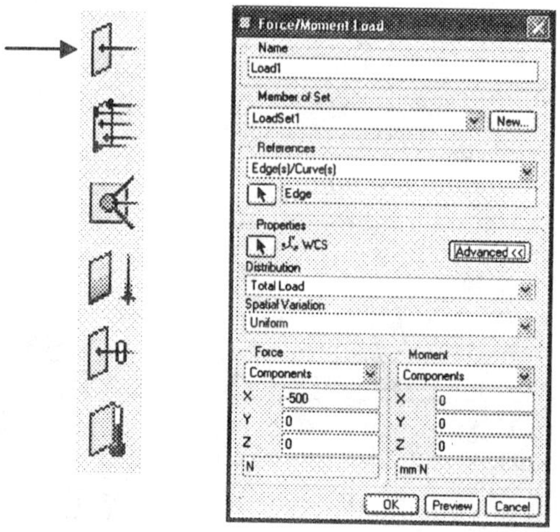

Perform a static analysis

Click the icon of performing a design study from the main toolbar > **File** > Select **New static analysis**.

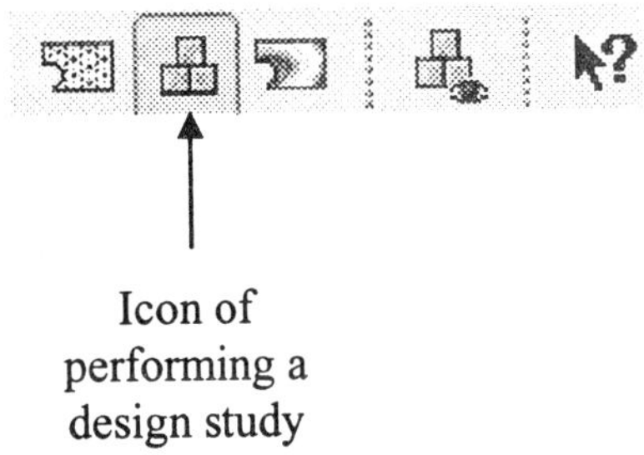

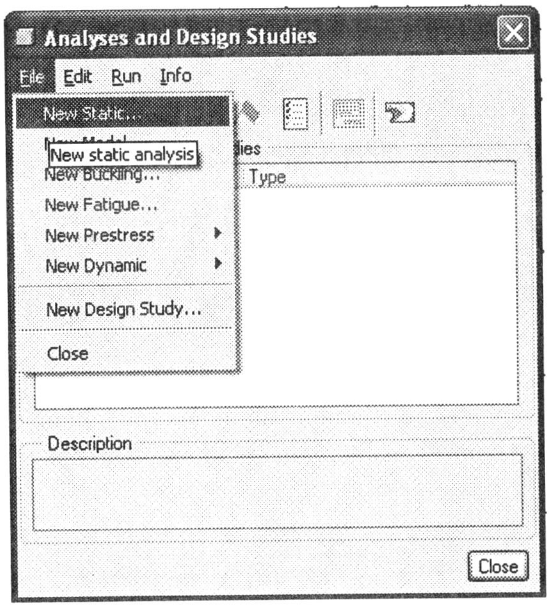

Type t_bracket as the name of design study.

Make sure that ConstraintSet1 and LoadSet1 are highlighted > Single-Pass Adaptive > OK > Run

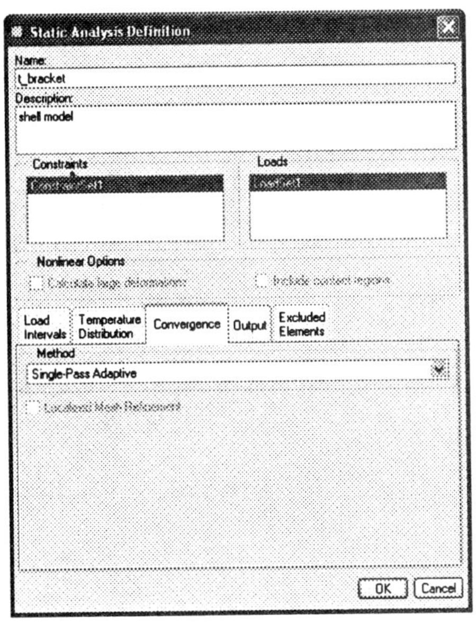

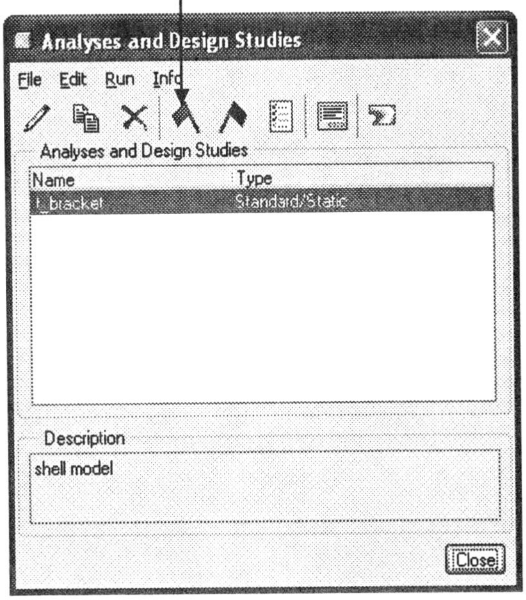

To review the results, click the icon of review results > type in deformation distribution (cylindrical surfaces) as the title of the result window > select the folder called support_structure_1 > Fringe > Displacement > Magnitude.

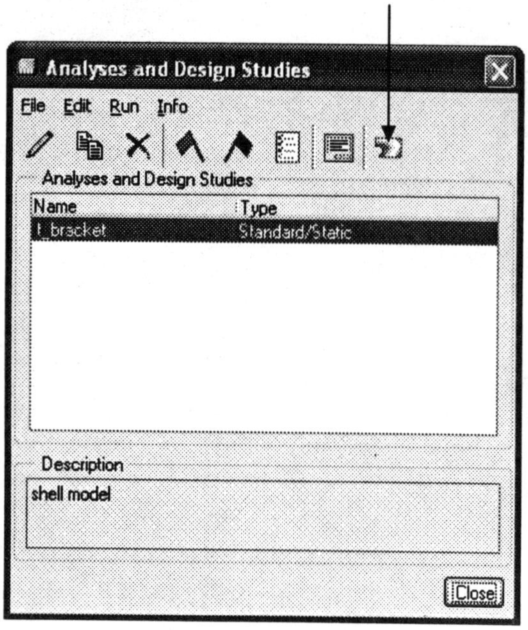

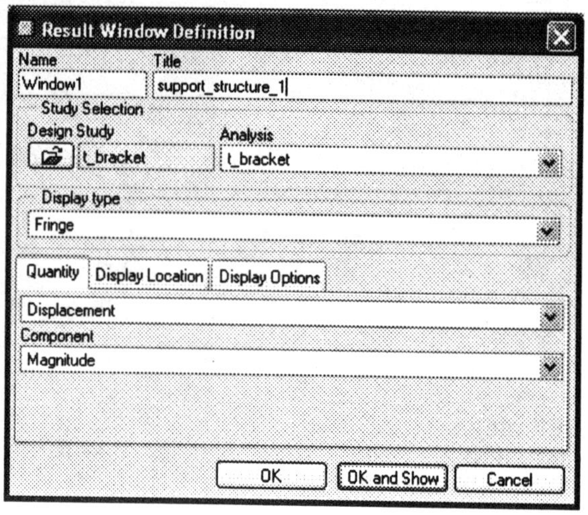

The deformation distribution

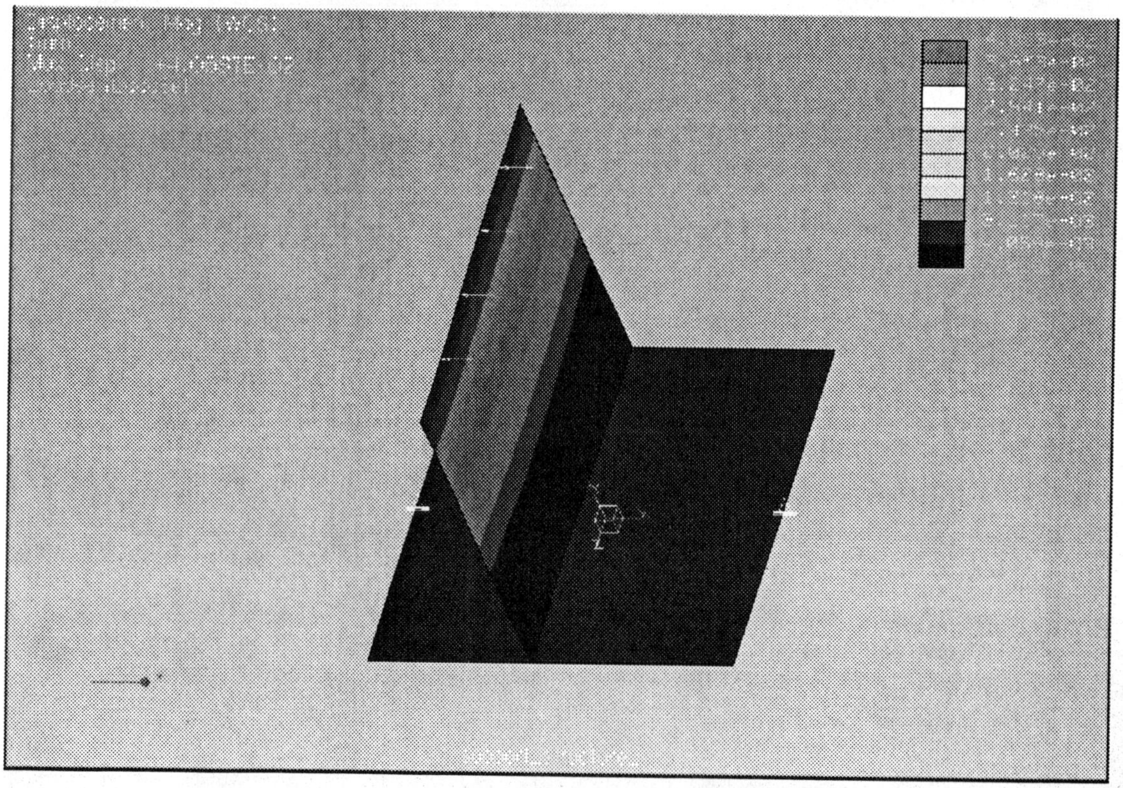

To study the distribution of the von Mises stress distribution, select **Edit** from the main toolbar > **Copy** and a new Result Window appears > **Stress** > **von Mises** > type von Mises stress distribution > **OK and Show**.

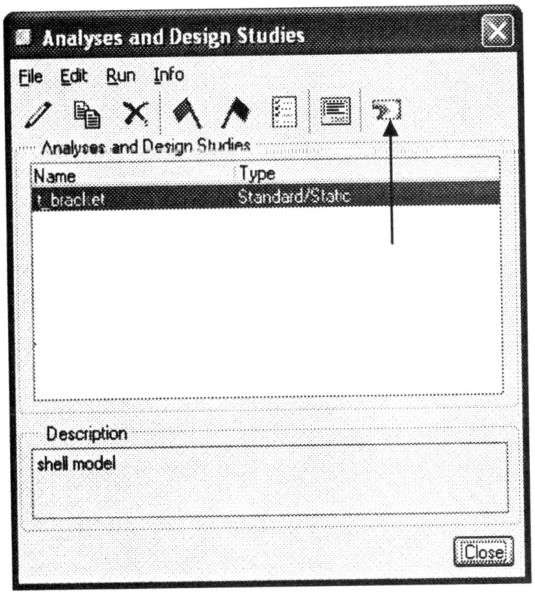

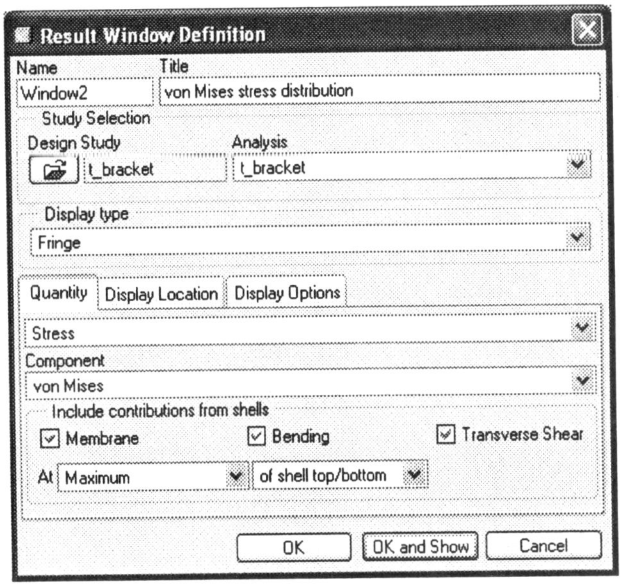

The von Mises stress distribution

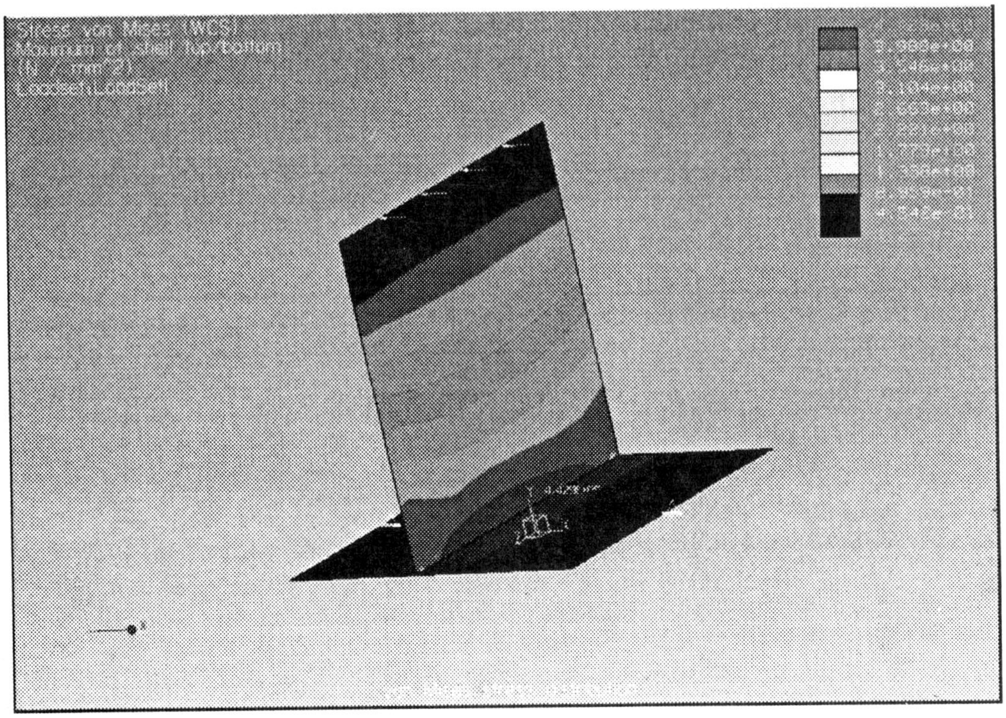

6.4 Case Study 1: Shell Modeling for Bath Tub

In section 6.3, we used **Auto Detect** to create the middle surfaces of a shell model for a t-bracket component. In the process of creating the geometrical model of the t-bracket, we used **thin wall structure** to create the 2 plate features. In this section, we present a case study. The component under study is a bath tub, as shown below.

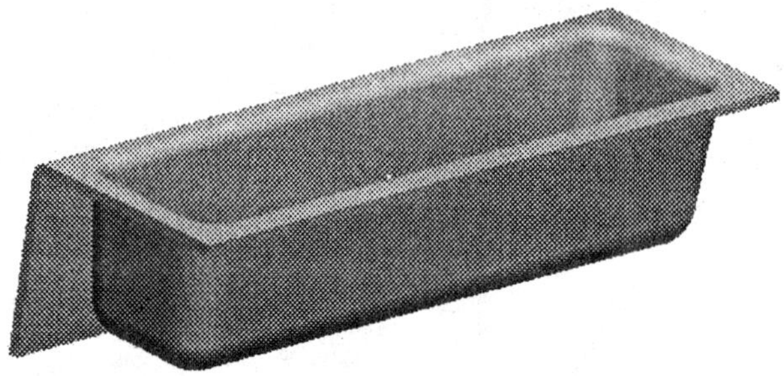

The geometrical shape of the shown bath tub is rather complicated. To illustrate the dimensions involved in creating a 3D solid model of the bath tub, an engineering drawing is shown in Figure 6-2.

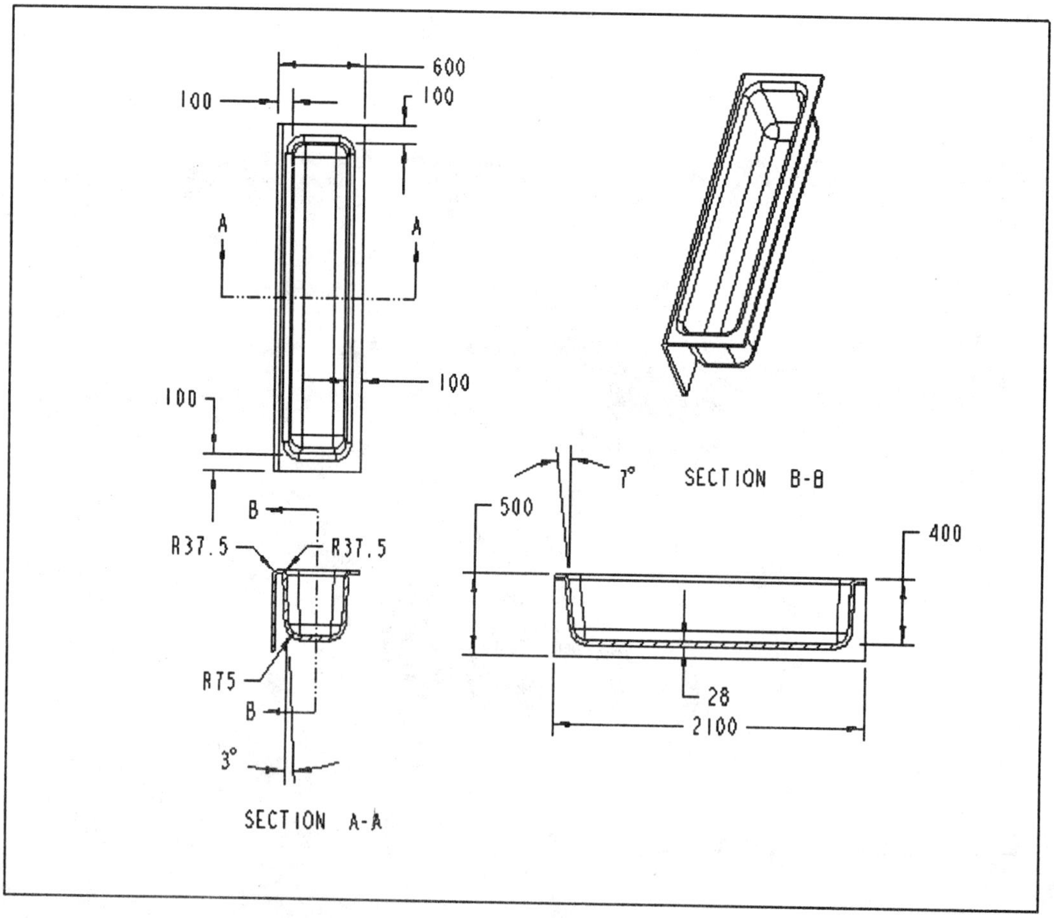

Figure 6-2 Geometrical Shape and Dimensions of a Bath Tub

Step 1: Create a 3D solid model

File > **New** > **Part** > type *bath_tub* as the file name and clear the icon of **Use default template.** Select mmns_part_solid (units: Millimeter, Newton, Second) and type in *bath tub structure* in **DESCRIPTION**, and *student* in **MODELED_BY**, then **OK.**

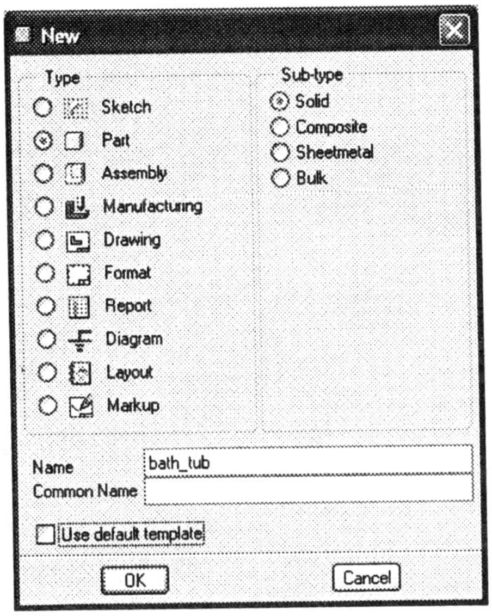

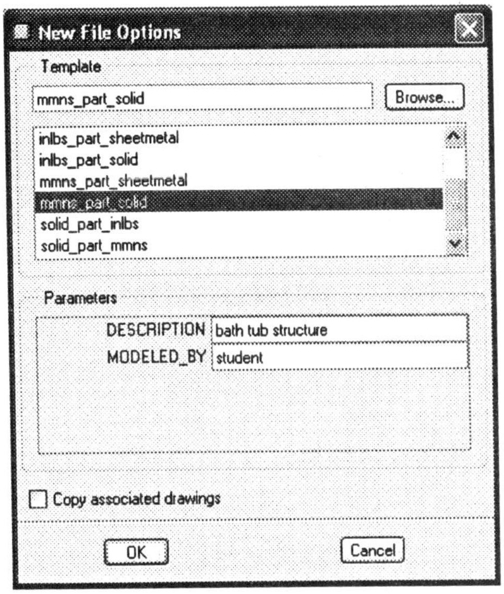

To create the horizontal plate, from the main toolbar, select **Insert > Extrude.** From the dashboard, set the depth value to 2100. Activate **Placement** to define a sketch plane > **Define.**

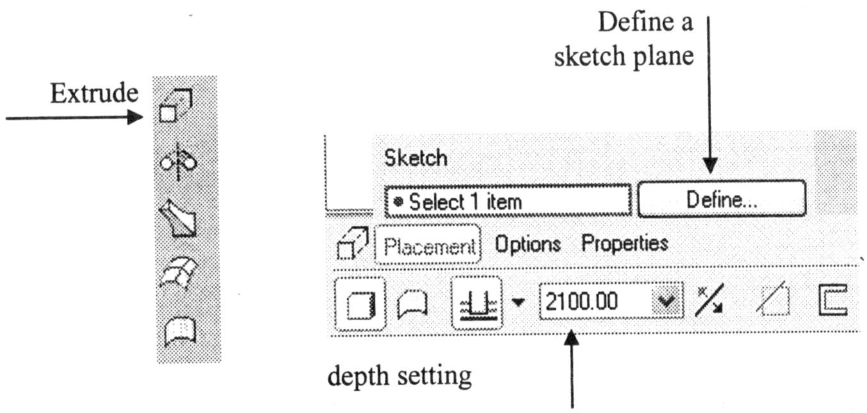

Select the **FRONT** datum plane as the sketch plane, and click the box of **Sketch** to accept the **RIGHT** datum plane as the default reference to orient the sketch plane, as illustrated below.

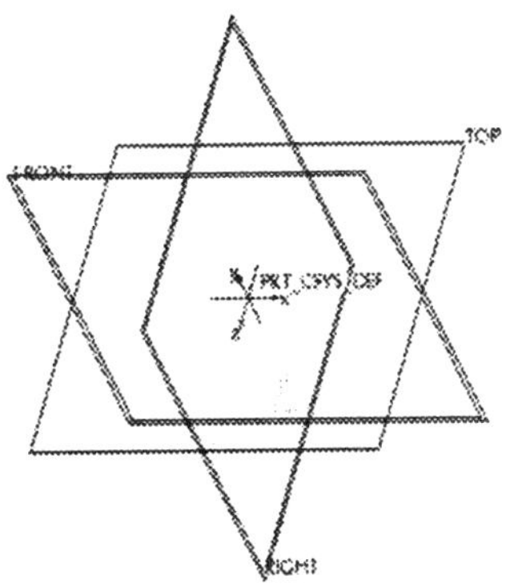

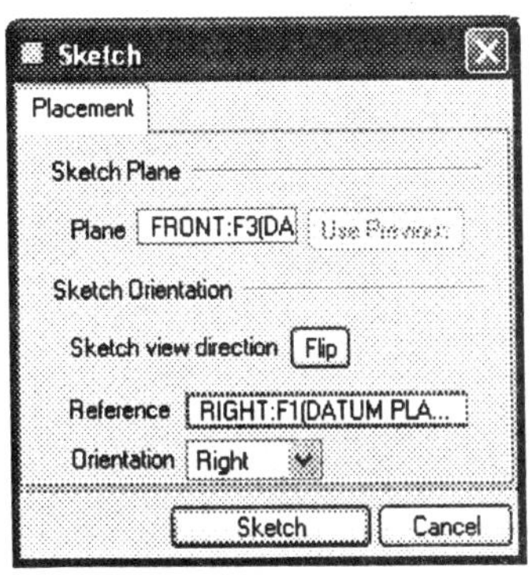

Both the sketch window and the references window appear. Let us first close the reference window. Select the icon of **rectangle** and sketch a rectangle. The 2 dimensions are 600 and 500, as shown.

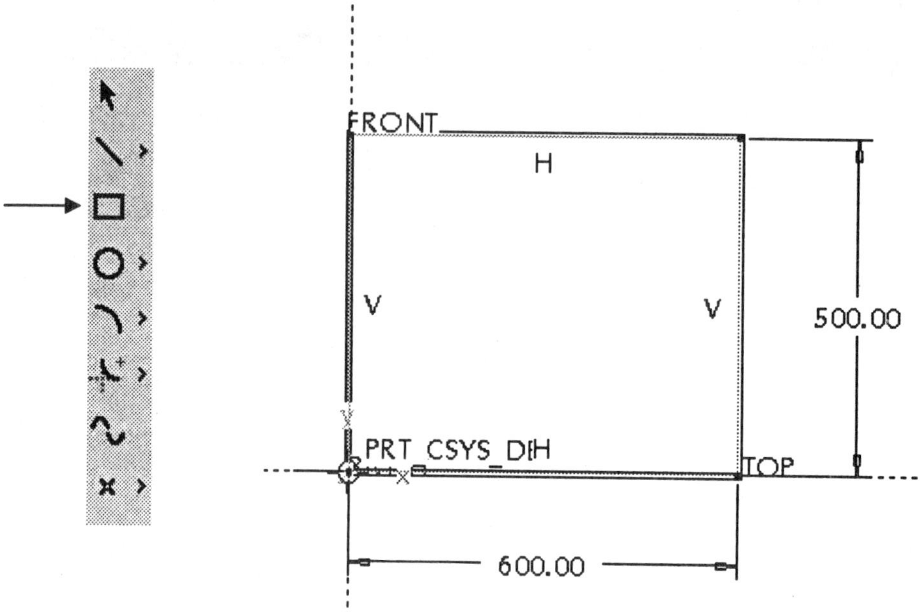

Upon completion, pick the icon of **Done** from the toolbar of sketcher. From the dashboard, select the icon of **Complete** from the feature control panel, thus completing the creation of the first feature.

Pick the icon of **Complete** from the Feature Control Panel

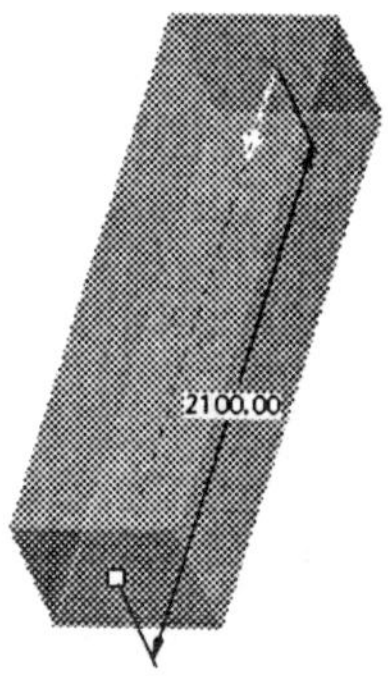

Step 2: Create a cut feature using **Extrude**. The cut feature is a rectangular block so that the thickness value of the wall is equal to 100 mm. The depth of the cut feature is 400 mm.

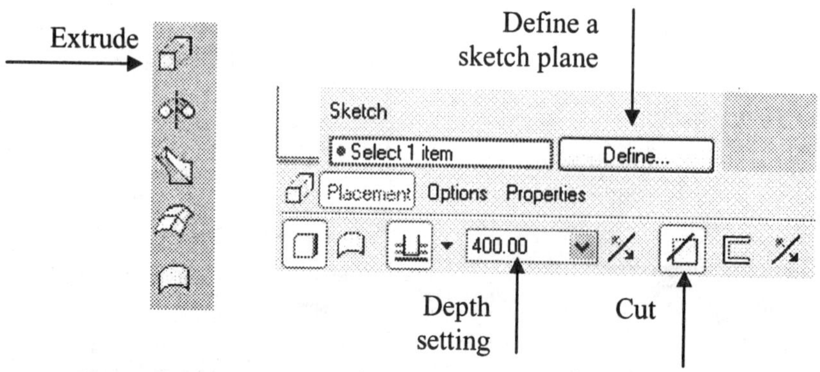

Select the top surface of the block as the sketch plane, and accept the default orientation setting.

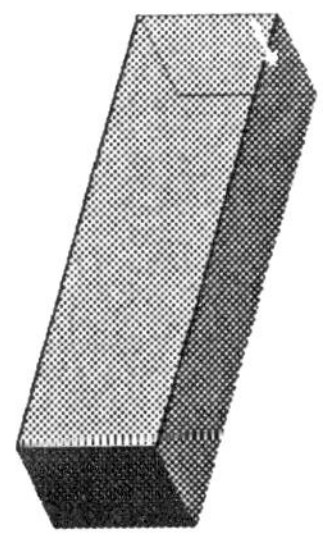

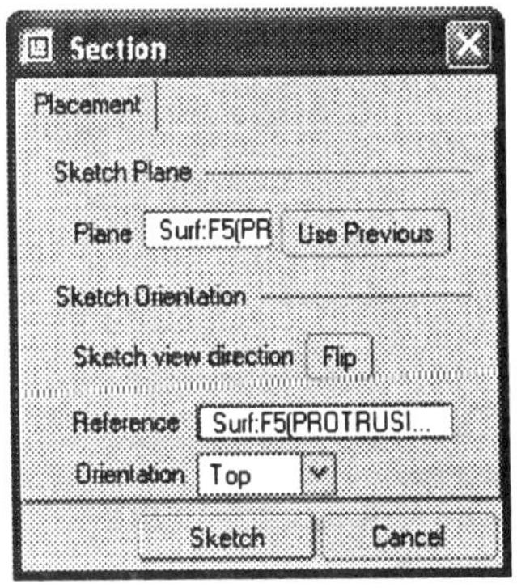

Add 2 new references before sketching the rectangle.

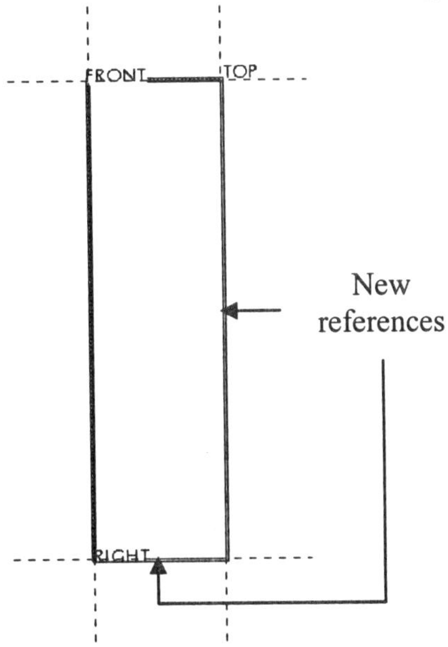

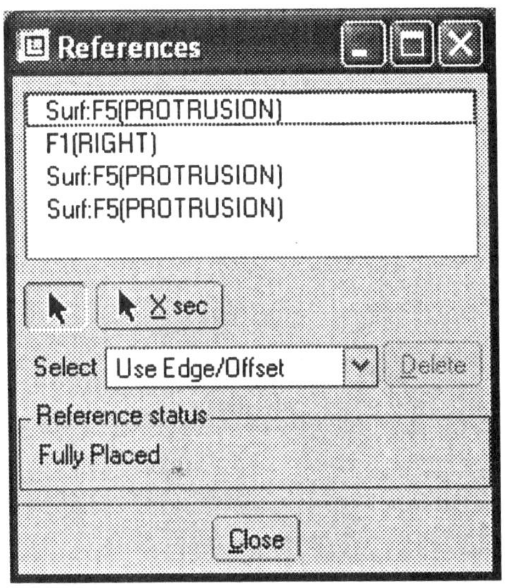

Pick the icon of **rectangle** and sketch a rectangle so that the thickness value at each side is 100 mm, as shown.

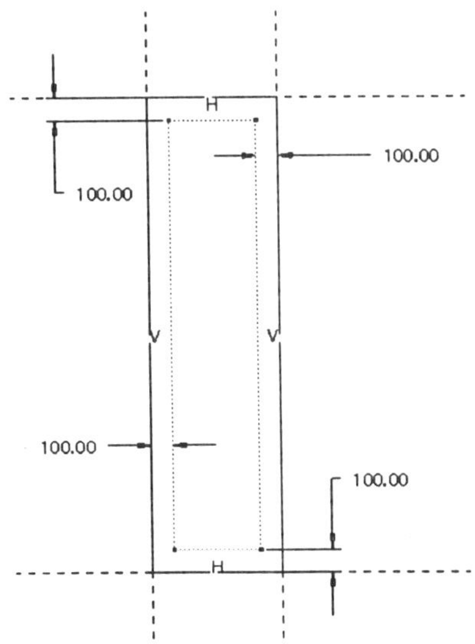

Upon completing the sketch, select the icon of **Done** from the toolbar of sketcher. On the dashboard, select the icon of **Complete**.

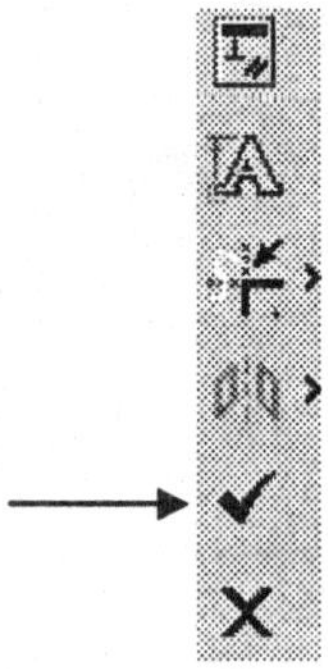

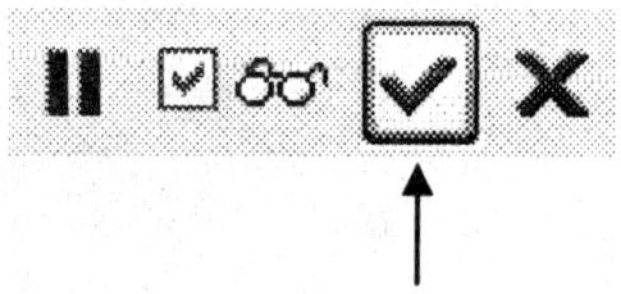

Pick the icon of **Complete** from the Feature Control Panel

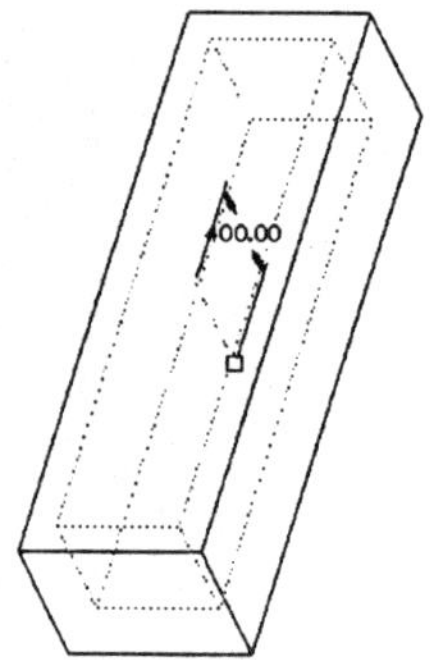

Step 3: Create drafts on three surfaces so that water can flow downward.

Pick the icon of **Draft** from the toolbar of pick and place. Activate **References** > select the 3 surfaces as the draft surfaces. For the draft hinges, select the top surface of the block, as shown below, and specify the draft angle equal to 3 degrees.

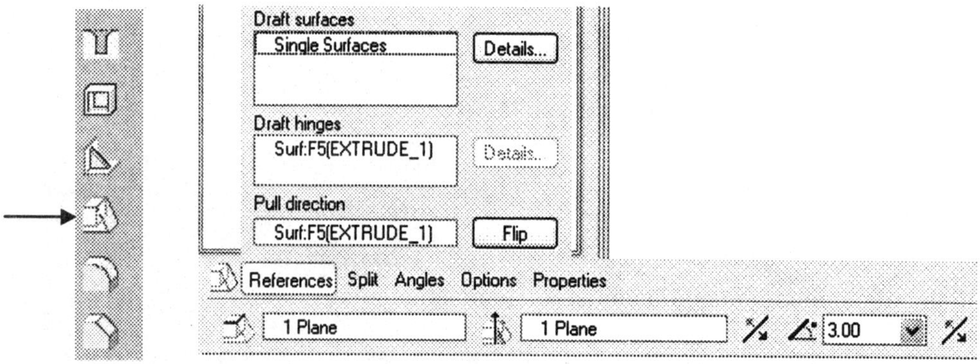

For the draft hinges, select the top surface of the block, as shown below, and specify the draft angle equal to 3 degrees.

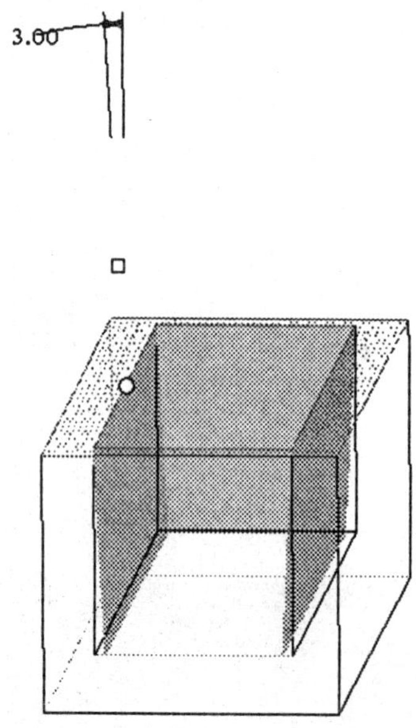

Select the icon of **Complete** from the feature control panel, completing the creation of the third feature (draft) of a 3D solid model for the bath tub.

Pick the icon of **Complete** from the Feature Control Panel

Step 4: Create a draft on the fourth surface so that water can flow downward.

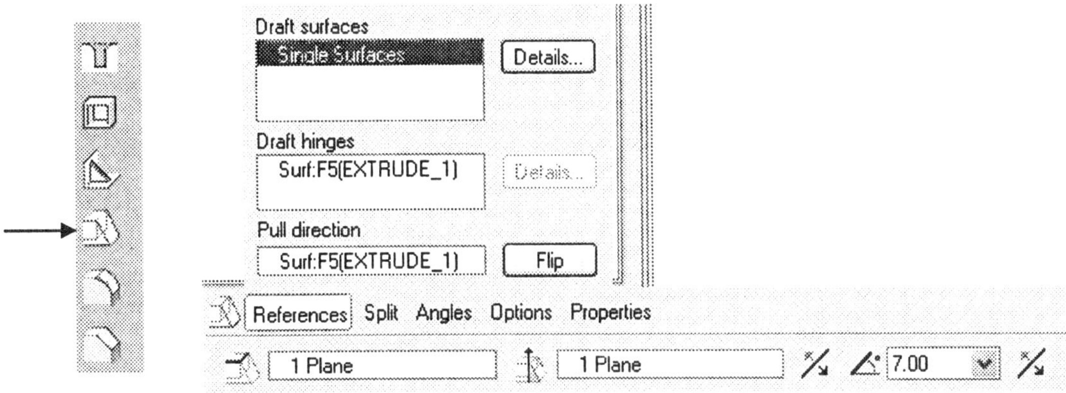

Activate References > select the surface as the draft surface. For the draft hinges, select the top surface of the block, as shown below, and specify the draft angle equal to 7 degrees.

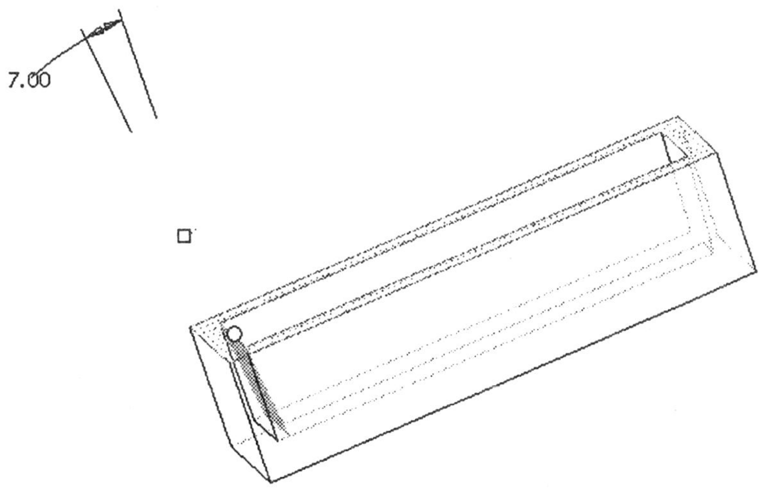

Select the icon of **Complete** from the feature control panel, completing the creation of the fourth feature (draft) of a 3D solid model for the bath tub.

Pick the icon of **Complete** from the Feature Control Panel

Step 5: Create rounds on the corners of connecting the inside surfaces.
Pick the 8 internal edges, as shown, and specify 75 as the radius dimension.

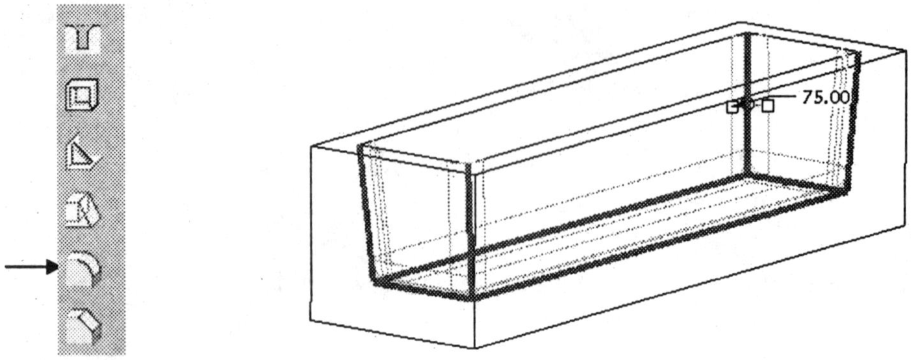

Step 6: Create rounds on the 4 edges on the top surface of the bath tub.
Pick the 4 edges around the cut feature on the top surface, as shown, and specify 37.5 as the radius dimension.

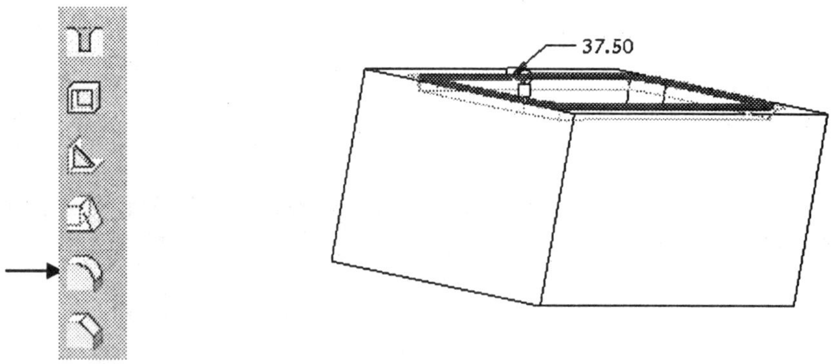

Step 7: Create a round along the edge on the top surface of the bath tub.
Pick the edge, as shown, and specify 37.5 as the radius dimension.

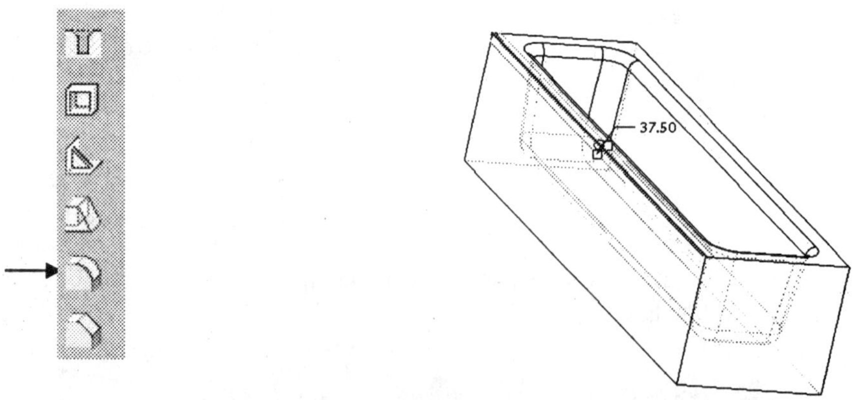

Step 8: Create a thin wall structure using **Shell**.
Pick the 4 surfaces shown in the following figure to remove in the operation of creating a thin wall structure. Specify the thickness value to 12.5.

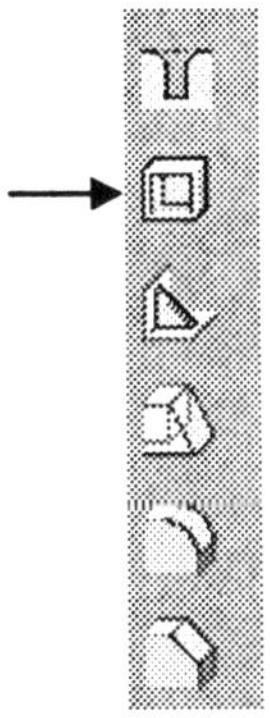

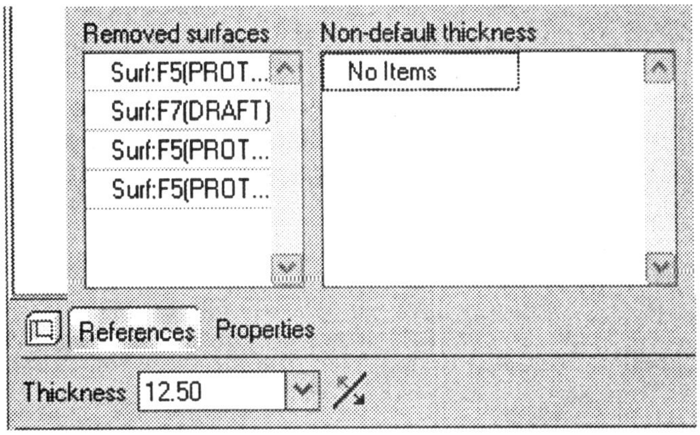

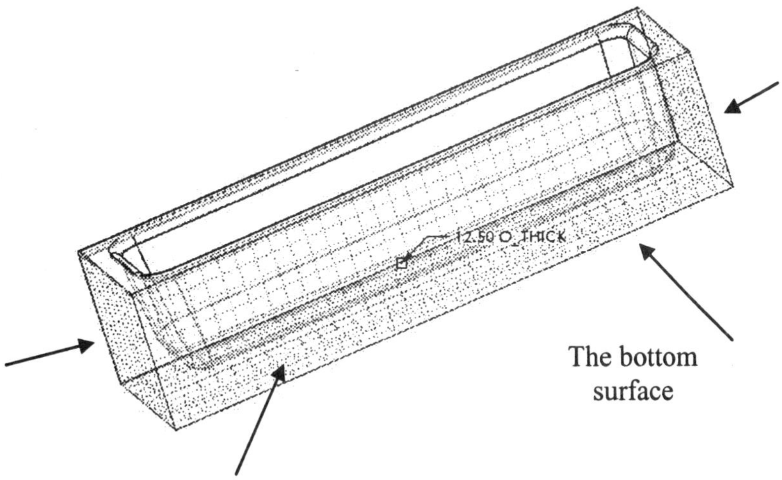

Select the icon of **Complete** from the feature control panel, completing the creation of the thin wall structure as a 3D solid model for the bath tub.

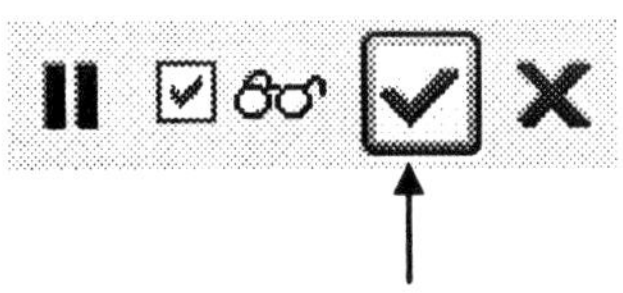

Pick the icon of **Complete** from the Feature Control Panel

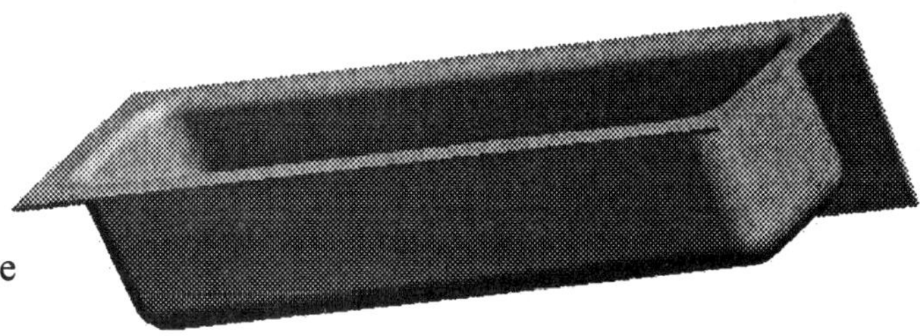

Step 9: FEA under Pro/MECHANICA.
Applications > Mechanica > Continue > Structure > OK.

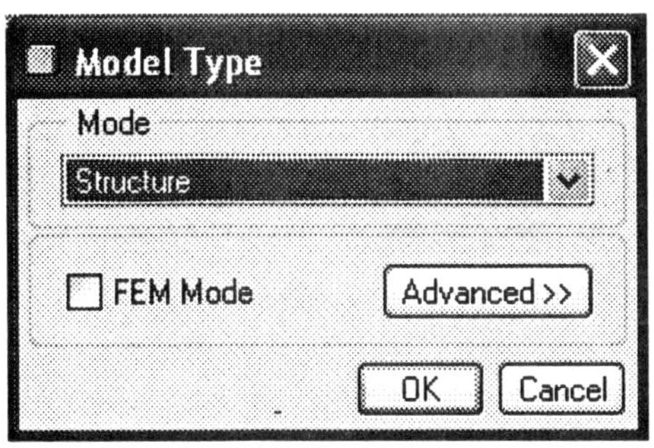

From the toolbar of functions, select the icon of **Materials** > select **STEEL.**

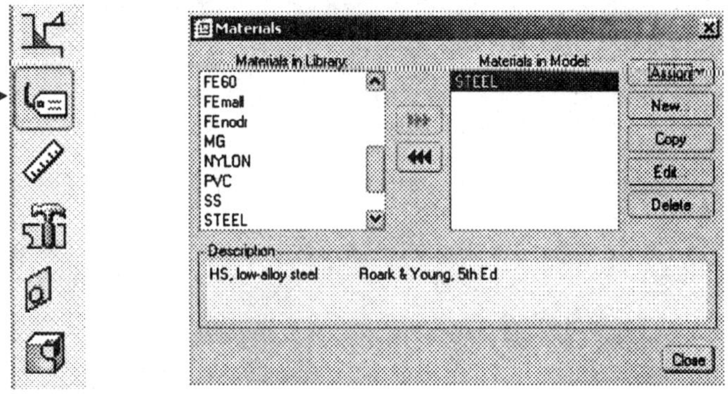

Assign > **Part** > select the bath tub model > **OK** > **Close**

To define a shell model, from the toolbar of functions, select the icon of **New Shell Pair** > **Auto Detect** > **Compress** > **Shells Only**.

Show Compress

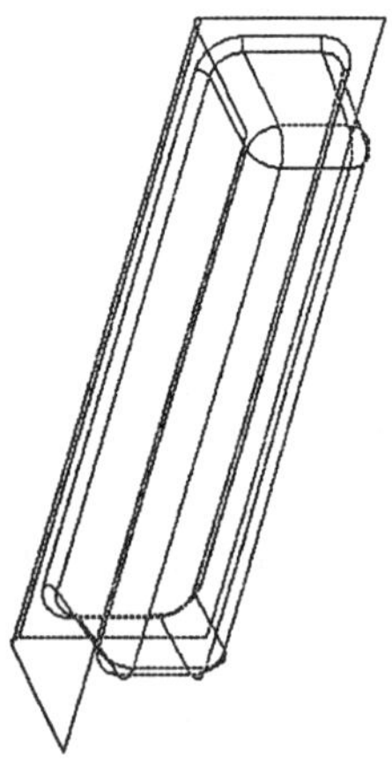

Show Original

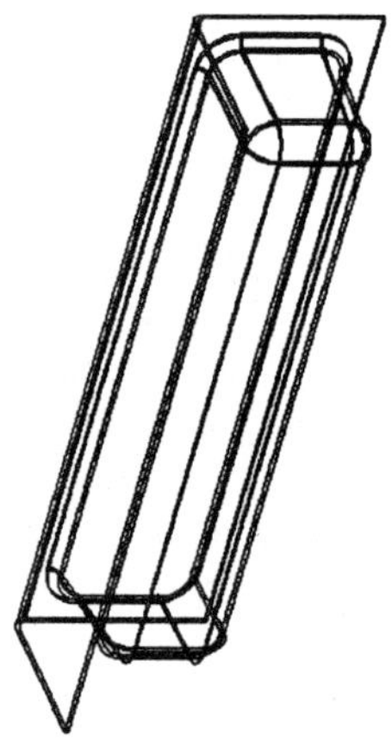

Show Both

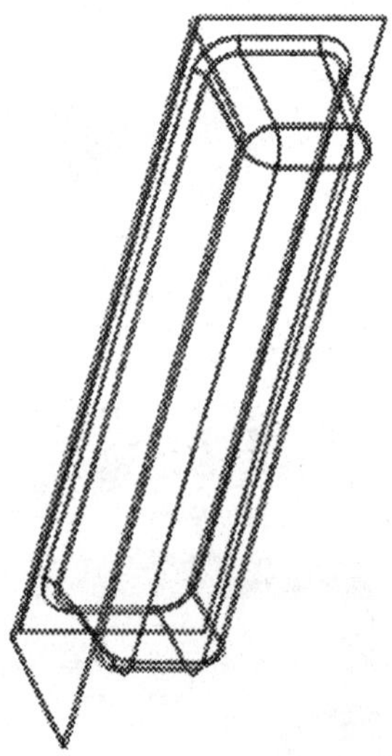

To define the constraint conditions on the bottom surface of the bath tub, we need surface regions. Let us create three (3) datum curves on the bottom of the bath tub.

Select the icon of **sketch tool** from the toolbar of datum creation. Pick the bottom surface of the bath tub as the sketch plane, and accept the default orientation setting.

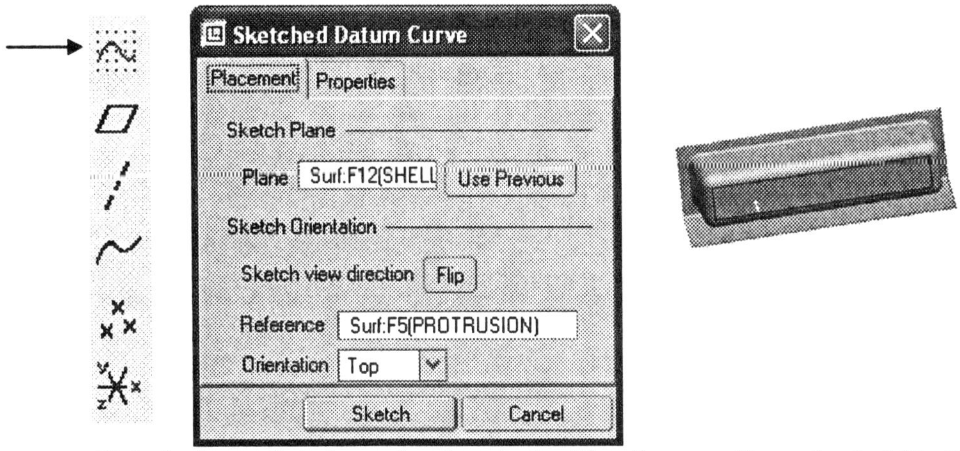

Pick the icon of circle to sketch a circle. The diameter dimension is 150. The 2 position dimensions are 300 and 1700, as shown. Upon completion, select the icon of **Done**. Repeat the above procedure to create the other 2 datum curves. To define the sketch plane, use **Use Previous**. The 2 position dimensions are 370 and 1050, respectively.

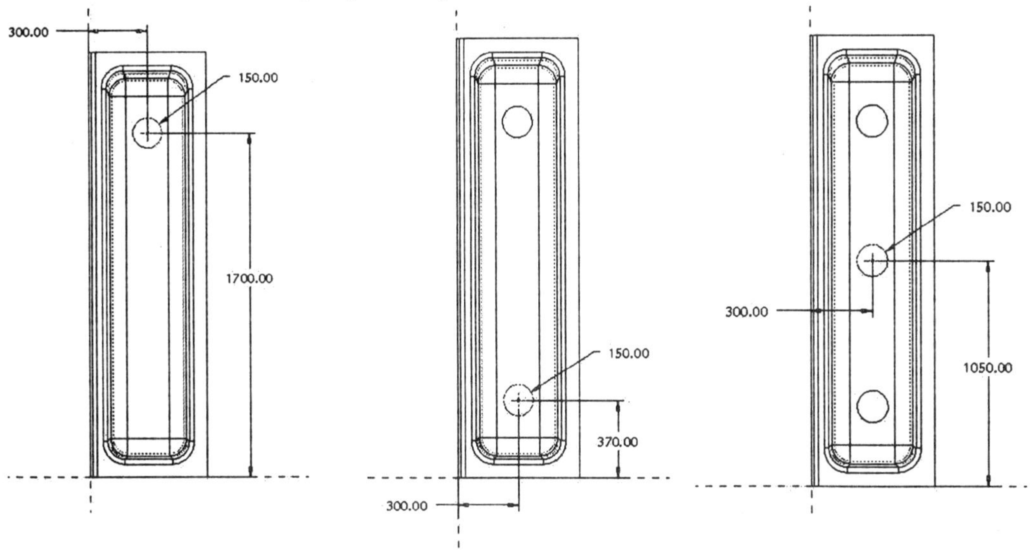

In order to define the load condition, we also need a surface region. Select the icon of sketch tool from the toolbar of datum creation, and pick the "top" surface as the sketch plane and accept the default orientation setting.

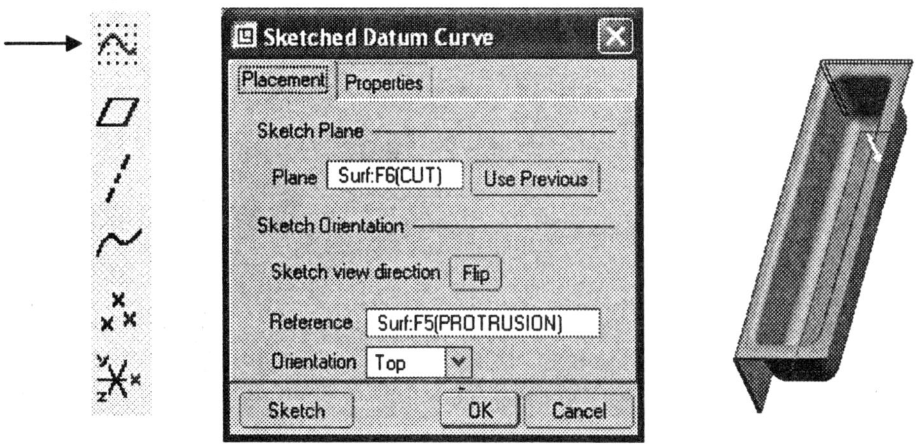

Pick the icon of **rectangle** and sketch a rectangle. The 2 size dimensions are 200 and 100, respectively. The 2 position dimensions are 1350 and 250, as shown.

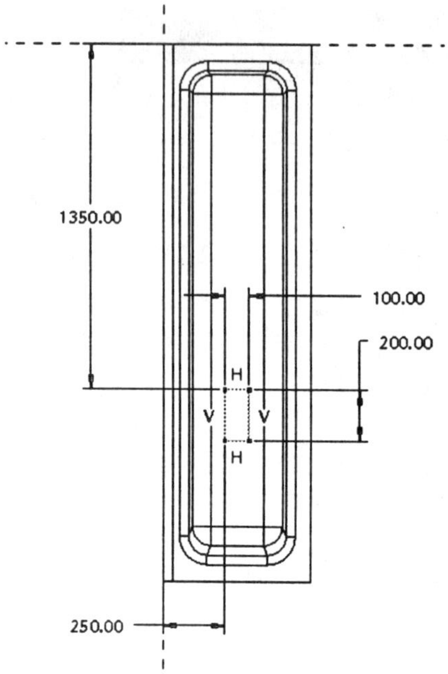

Now let us define the 4 surface regions. From the toolbar of functions, select the icon of **surface region** > from the pop up window, select **Select** > **Done.**

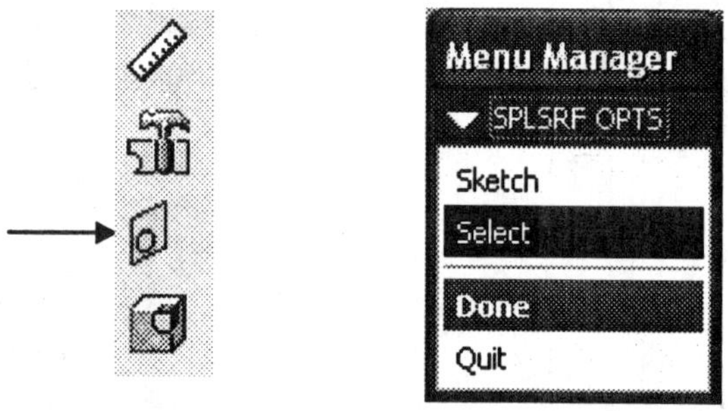

Select the circular datum curve created previously, and select the bottom surface of the bath tub as the surface to be split, from which the surface region is separated > **OK > Done > OK**. Repeat the above procedure to define the other 2 circular surface regions.

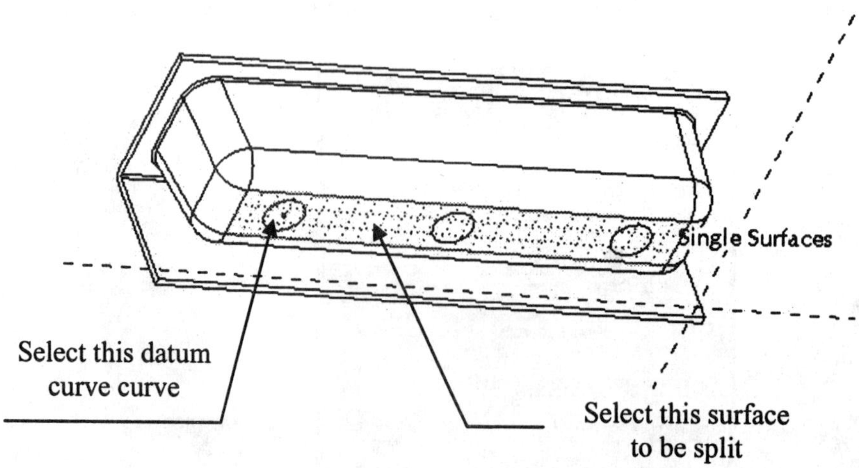

Following the same procedure to define the surface region for the loading area, from the toolbar of functions, select the icon of **surface region** > from the pop up window, select **Select** > **Done.**

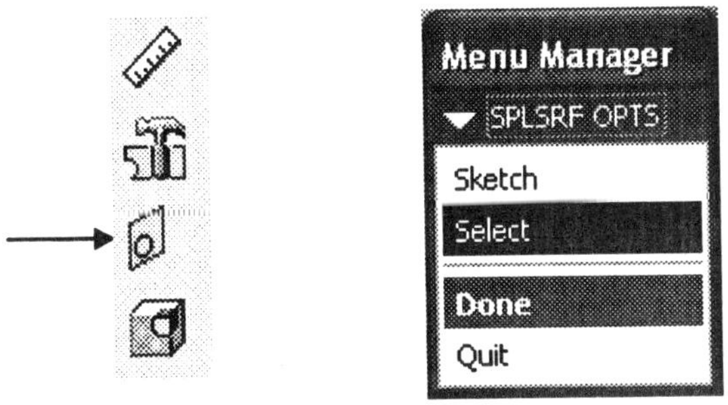

Select the rectangular datum curve created previously, and select the "top" surface of the bath tub as the surface to be split, from which the surface region is separated > **OK** > **Done** > **OK**.

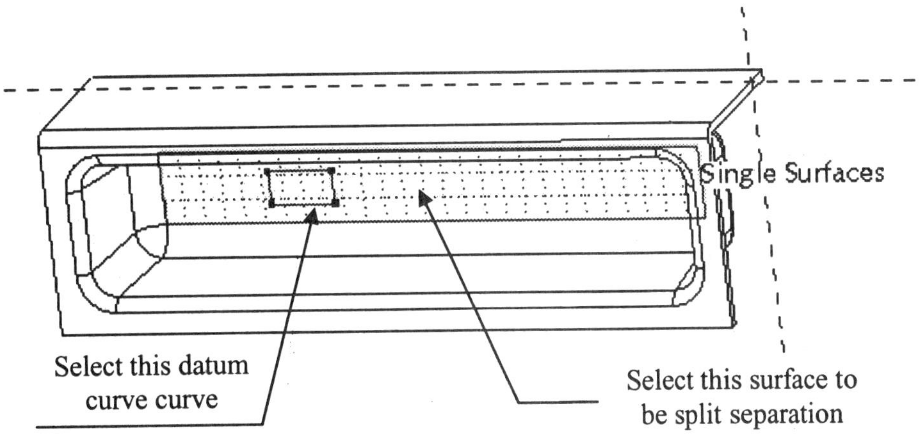

Now let us check the defined shell mode, from the toolbar of functions, select the icon of **New Shell Pair** > **Compress** > **Shells Only** > **ShowCompress,** all the 4 surface regions have been mapped on the middle-surface of the shell model > **Done** > **Done/Return**.

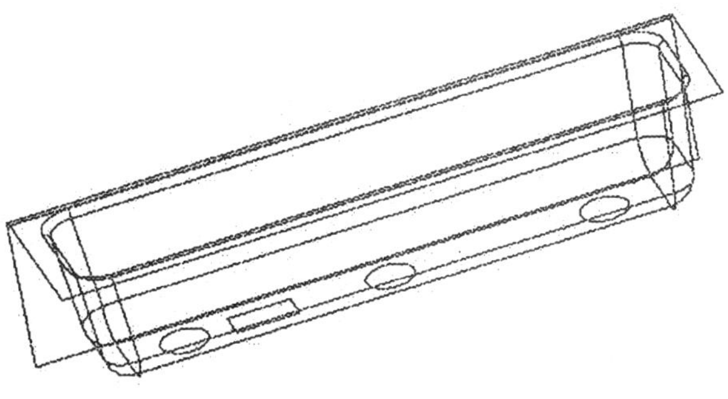

To define the constraint condition, from the toolbar of functions, select the icon of **Constraints** > select **Surface(s)** > pick one of the three surface regions defined on the bottom surface of the bath tub > fix all 6 degrees of freedom > **OK**.

Repeat this procedure to define the second and the third constraints.

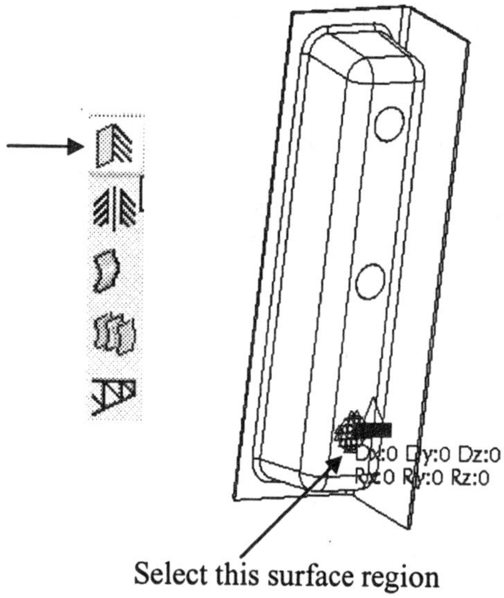

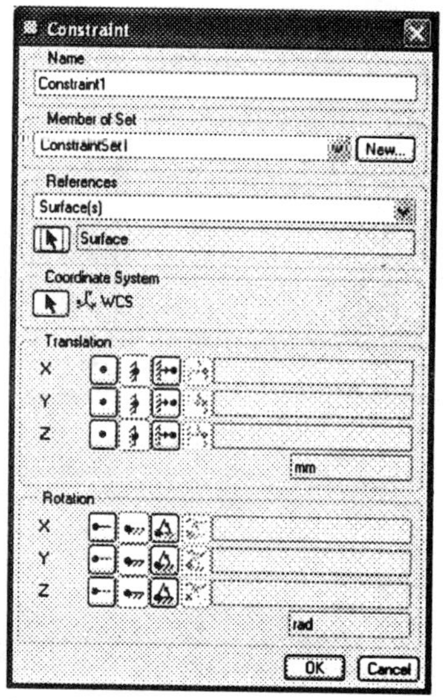

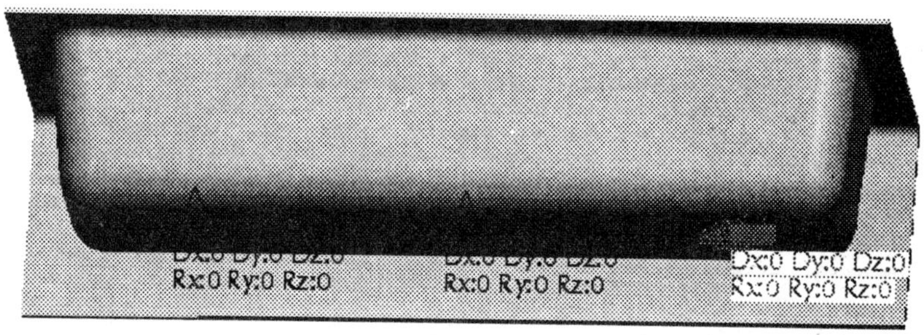

To define a uniform load acting on the rectangular area in the negative y direction, select the icon of **force and moment** from the toolbar of functions > **Surface** > pick the rectangular surface region > **OK** > click **Advanced** > change the distribution to **Total Load** and **Uniform** > type *-5000* in Y of the Force box > **OK**.

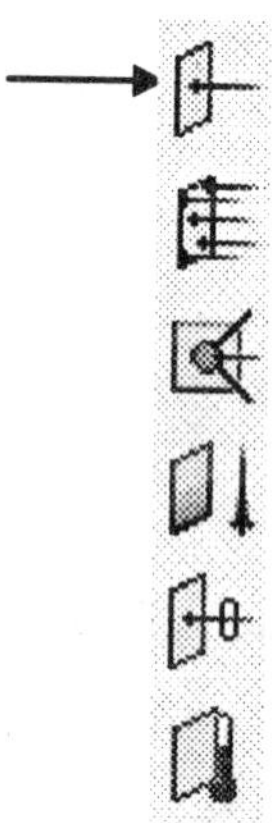

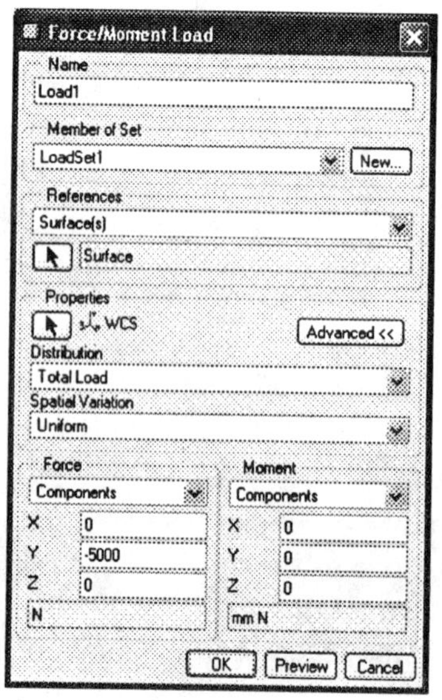

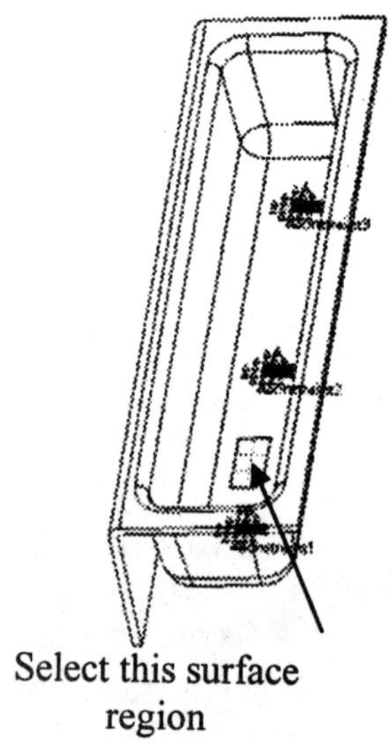

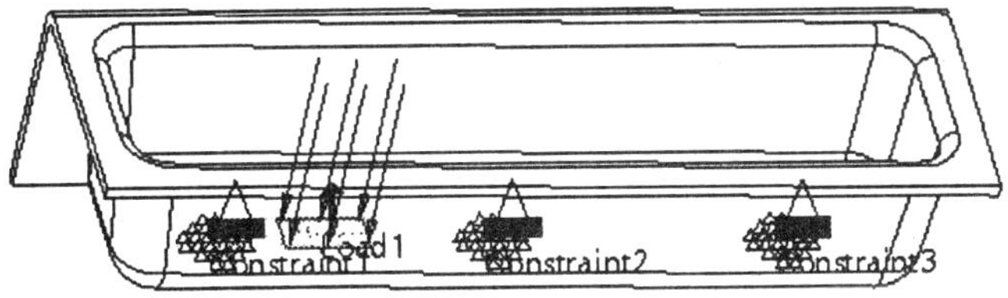

Perform a static analysis

Click the icon of **performing a design study** from the main toolbar > **File** > Select **New static analysis**.

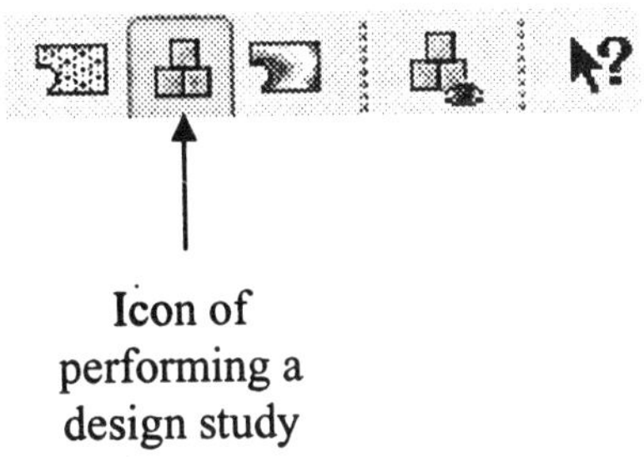

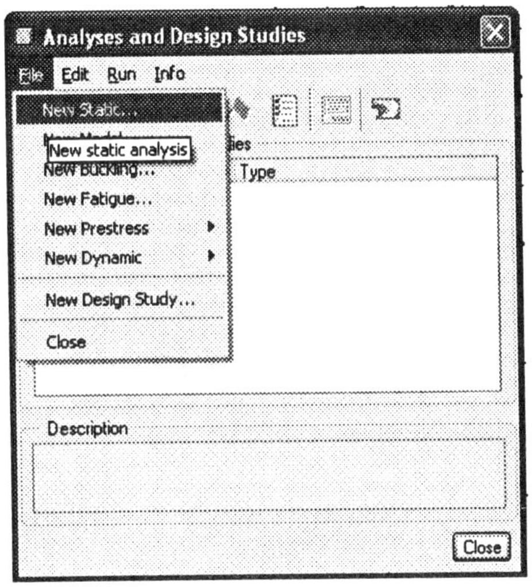

Type *bathtub_1* as the name of design study.

Make sure that ConstraintSet1 and LoadSet1 are highlighted > Single-Pass Adaptive > OK > Run.

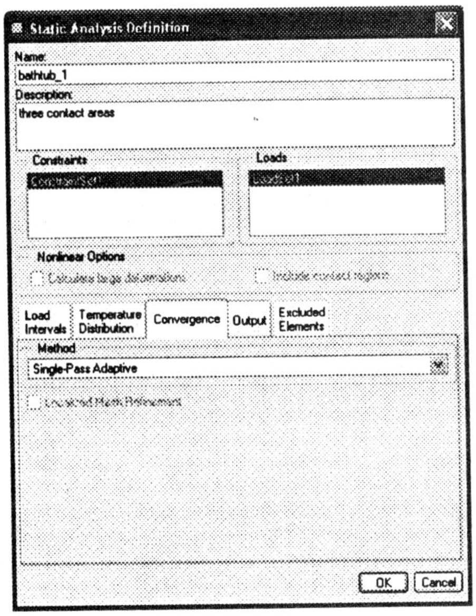

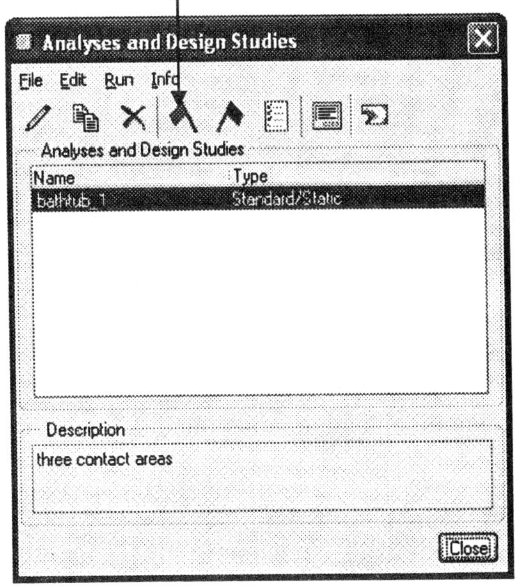

To review the results, click the icon of review results > type *deflection distribution* as the title of the result window > select the folder called bathtub_1 > **Fringe** > **Displacement** > **Magnitude** > **OK and Show**.

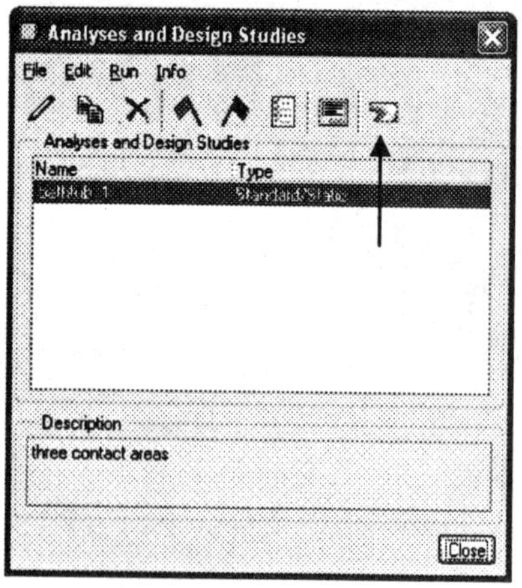

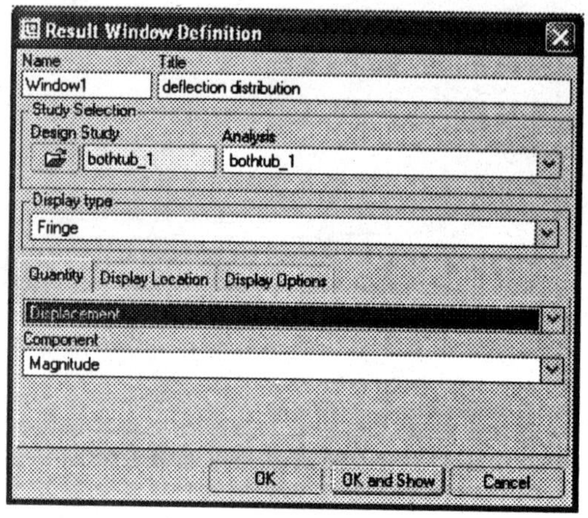

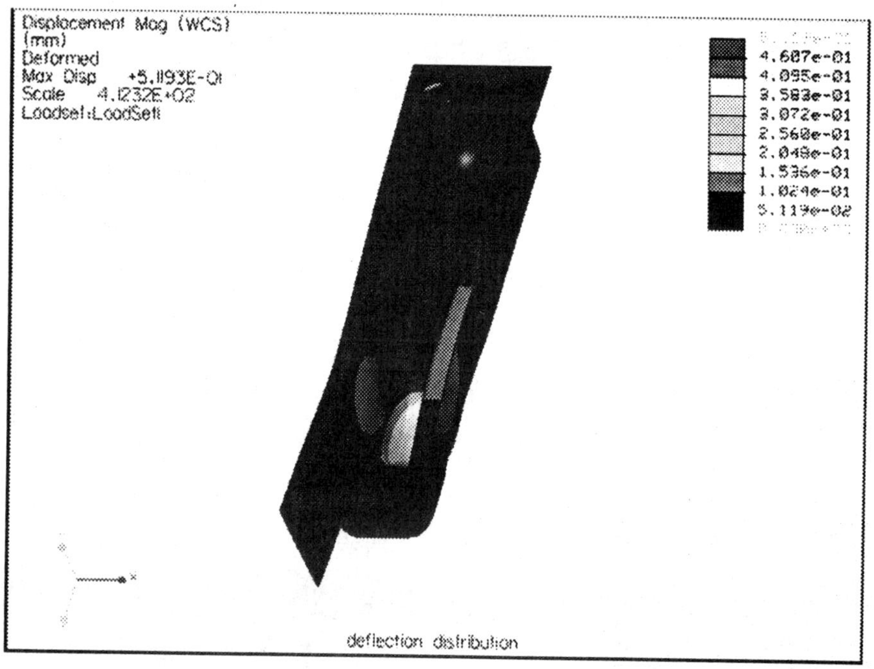

To examine the distribution of von Mises stress, **Edit > Copy** > type *distribution of von Mises stress* > select **Stress** and **von Mises > OK and Show**

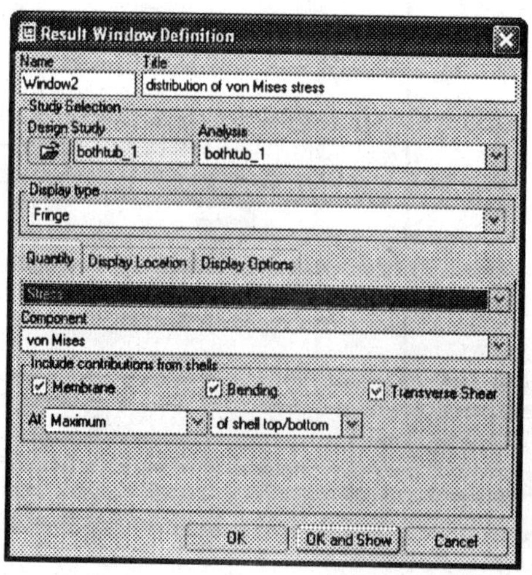

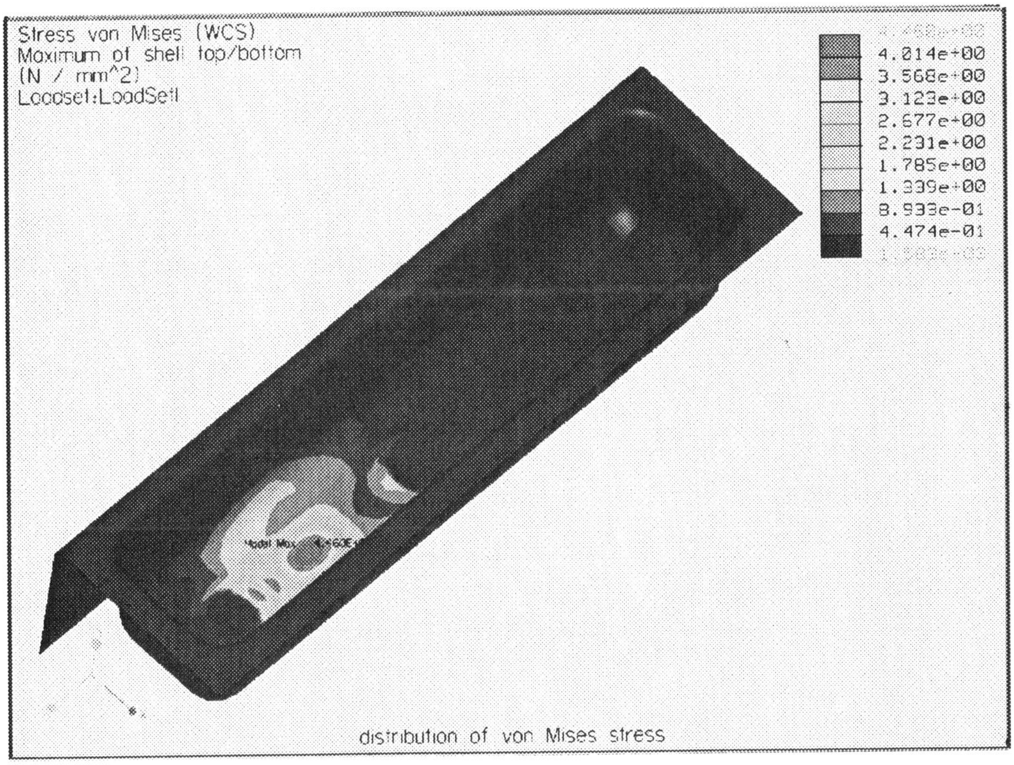

Maximum von Mises stress is equal to 4.46 MPa

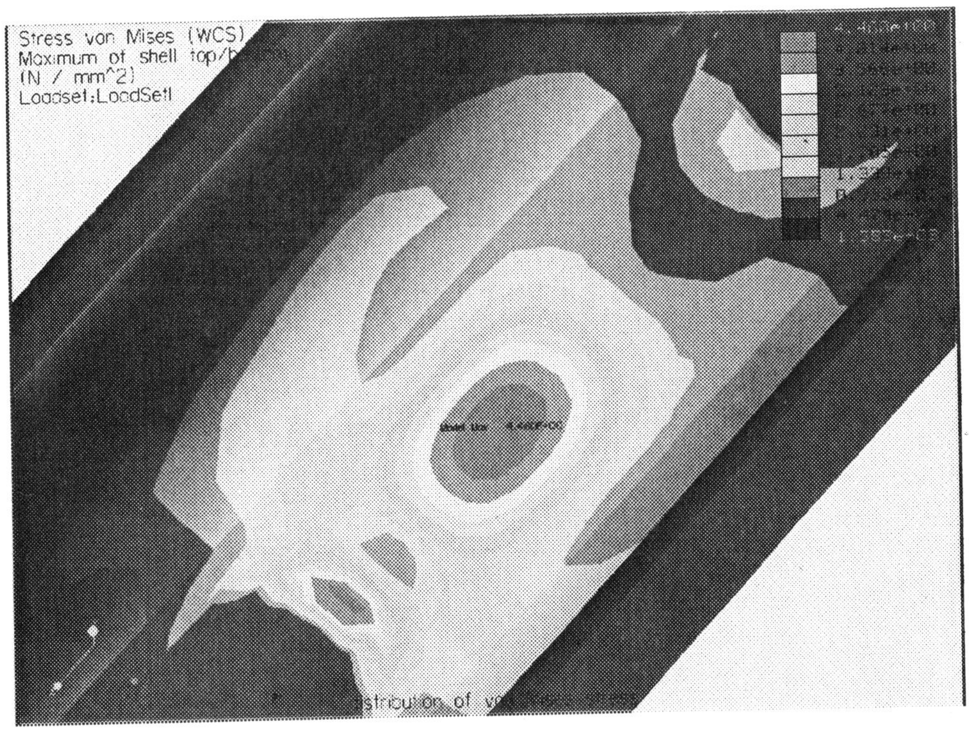

To enhance the constraint condition by adding two more surface regions at the bottom surface of the bath tub. Select the icon of sketch tool from the toolbar of datum creation, and pick the bottom surface as the sketch plane and accept the default orientation setting.

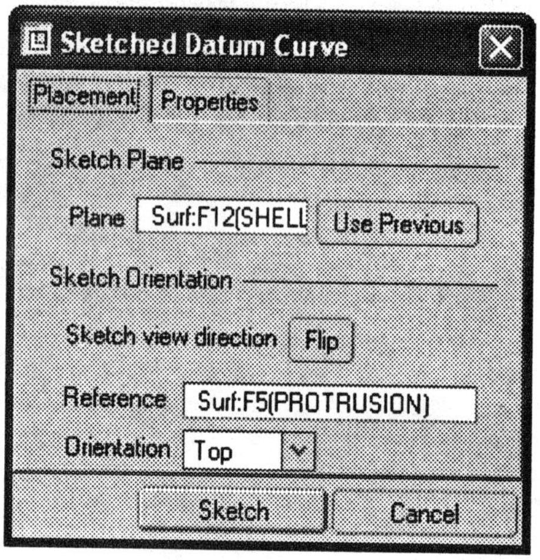

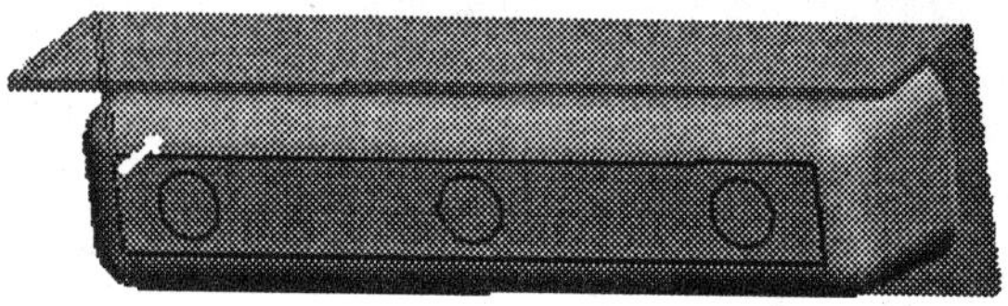

Pick the icon of circle and sketch a circle. The diameter dimension is 150. The 2 position dimensions are 1400 and 300, as shown.

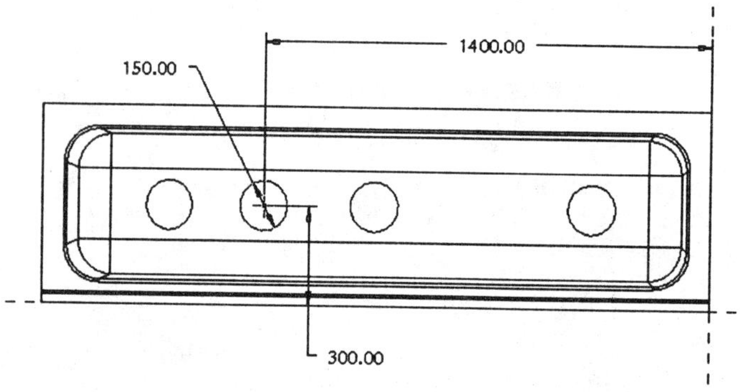

Repeat the above procedure to add another datum curve.

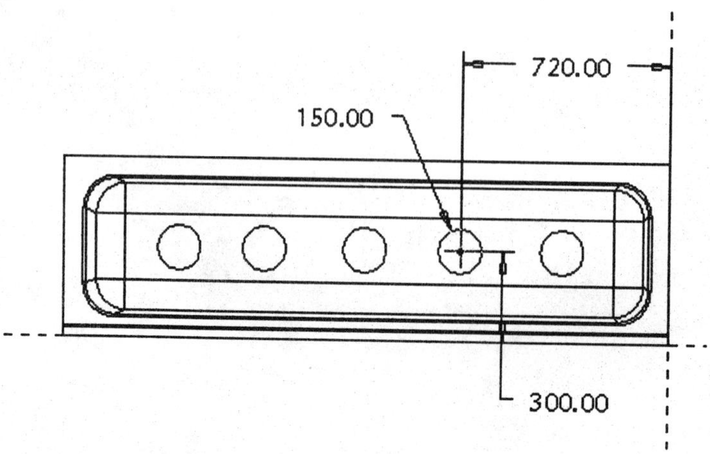

Now let us define the 2 newly added surface regions. From the toolbar of functions, select the icon of **surface region** > from the pop up window, select **Select** > **Done.**

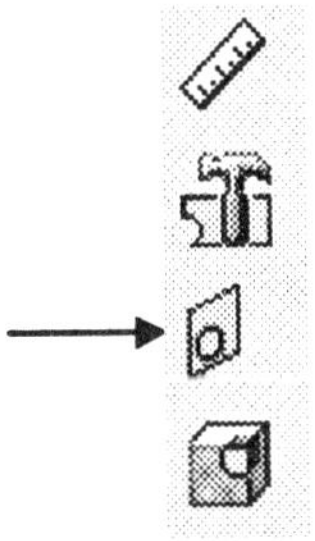

Select the first added circular datum curve, and select the bottom surface of the bath tub as the surface to be split > **OK** > **Done** > **OK**.

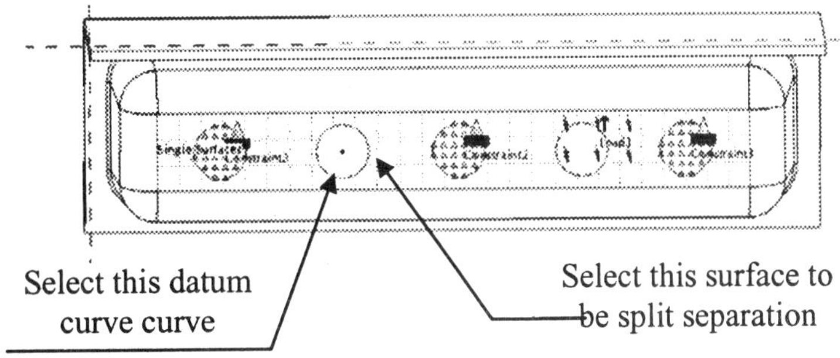

Repeat the above procedure to create another surface region.

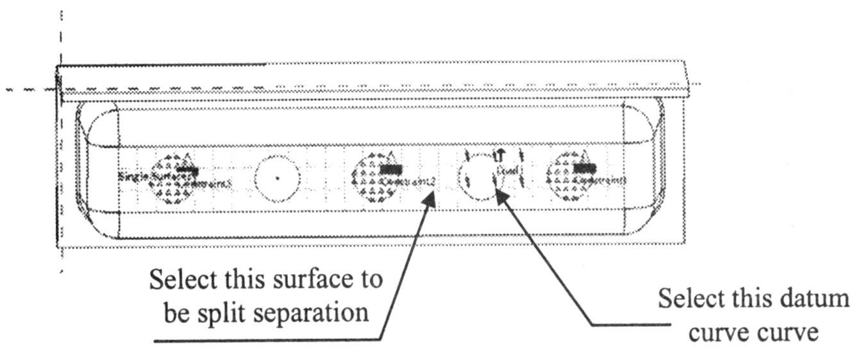

To verify the shell mode, from the toolbar of functions, select the icon of **New Shell Pair** > **Compress** > **Shells Only** > **ShowCompress,** all the 6 surface regions have been mapped on the middle-surface of the shell model > **Done** > **Done/Return**.

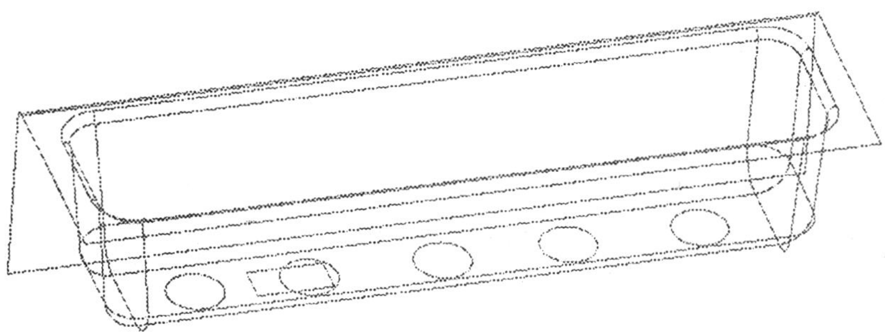

To add 2 new constraints, from the toolbar of functions, select the icon of **Constraints** > select **Surface(s)** > pick one of the two surface regions just defined on the bottom surface of the bath tub > fix all 6 degrees of freedom > **OK**.

Repeat this prócedure to define the second added constraint.

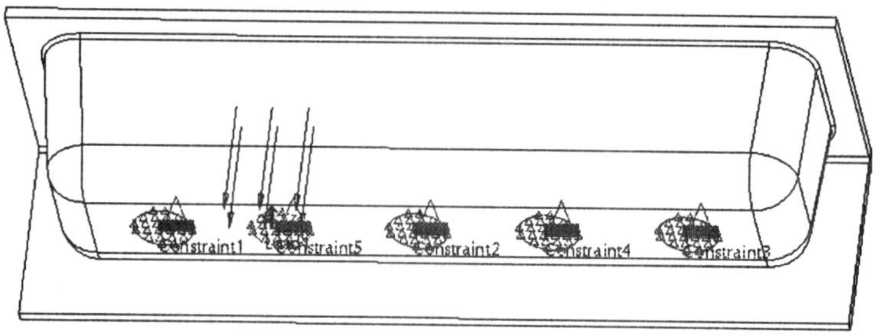

Perform a new static analysis

Click the icon of performing a design study from the main toolbar > File > Select New static analysis.

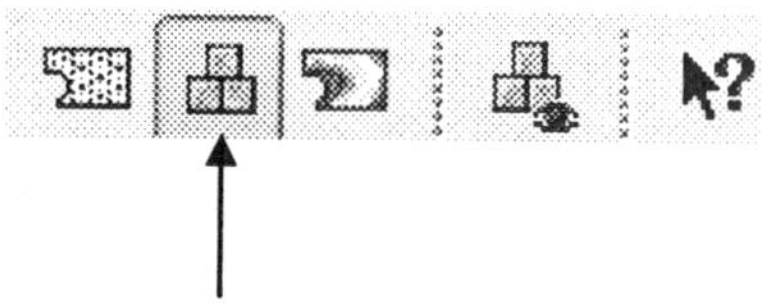

Icon of performing a design study

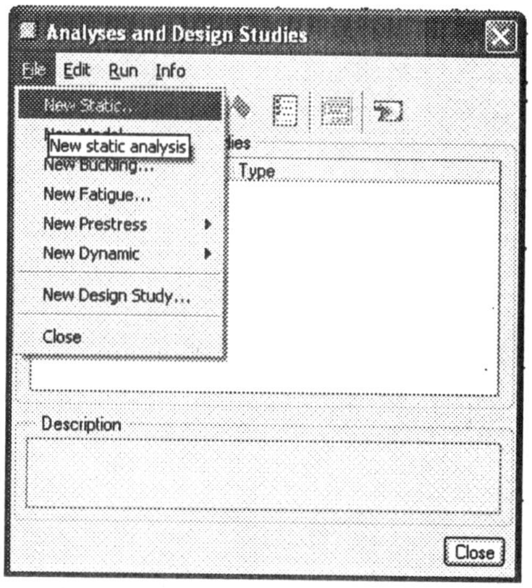

Type *bathtub_2* as the name of design study.

Make sure that ConstraintSet1 and LoadSet1 are highlighted > Single-Pass Adaptive > **OK** > **Run**.

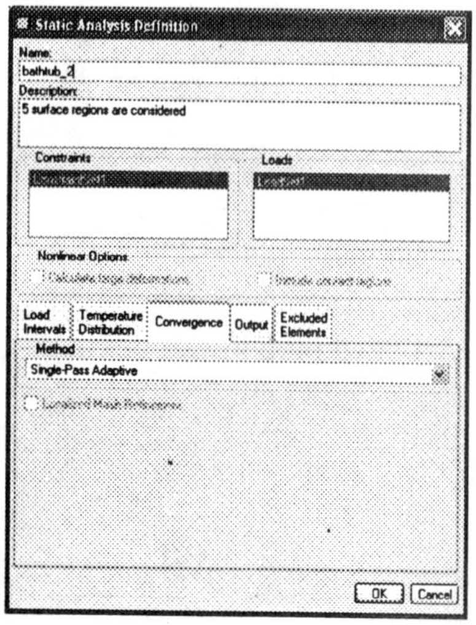

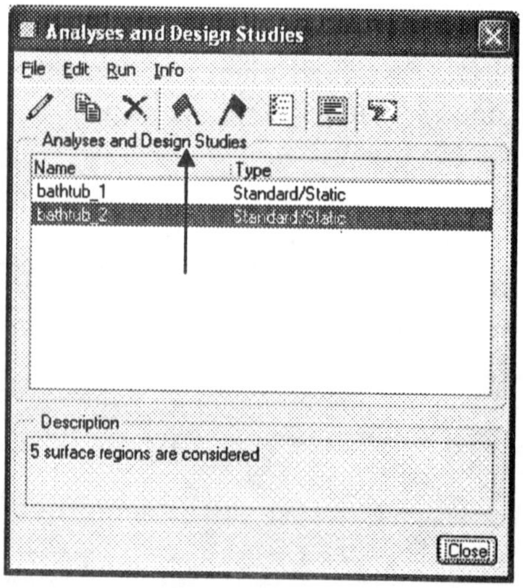

To review the results, click the icon of review results > type *deflection distribution (new constraint condition)* as the title of the result window > select the folder called bathtub_2 > **Fringe** > **Displacement** > **Magnitude** > **OK and Show**.

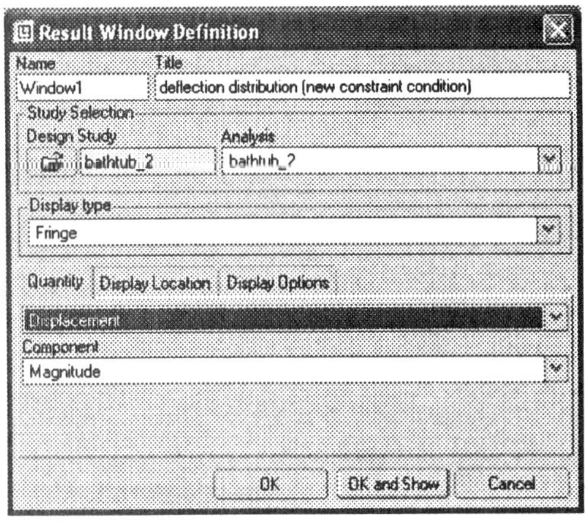

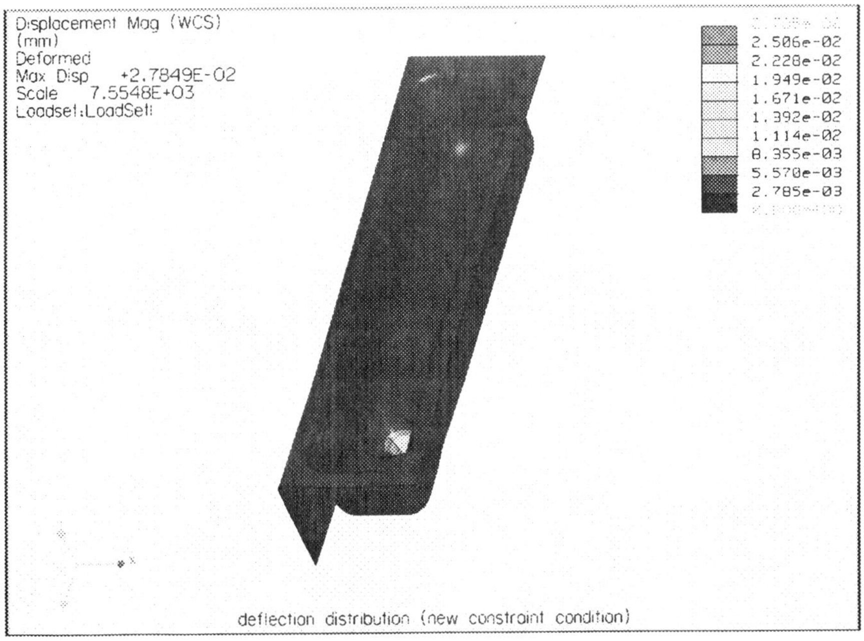

To examine the distribution of von Mises stress, **Edit > Copy** > type *distribution of von Mises stress* > select **Stress** and **von Mises > OK and Show.**

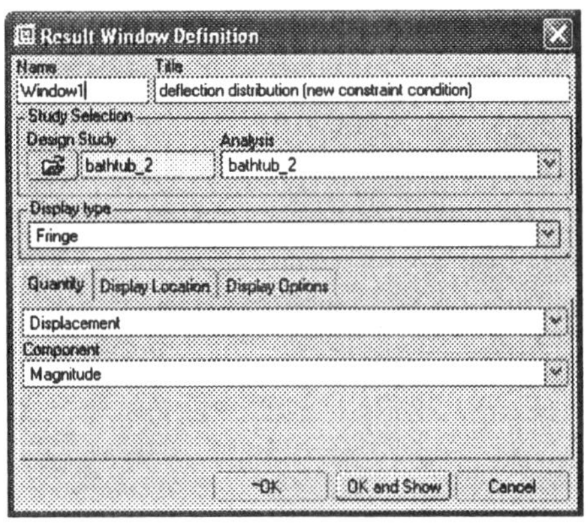

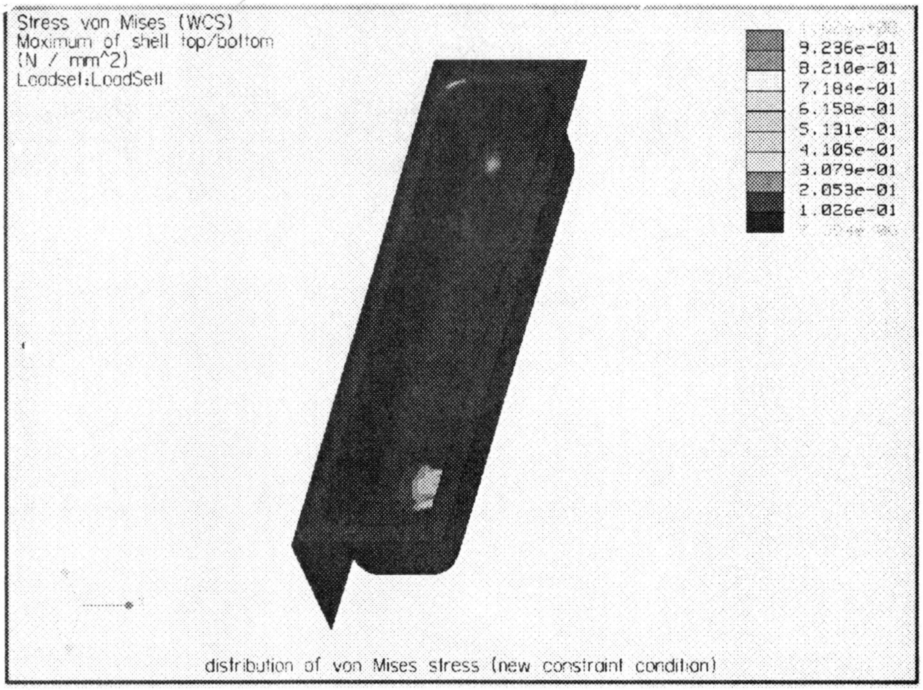

Maximum von Mises stress is equal to 1.03 MPa.

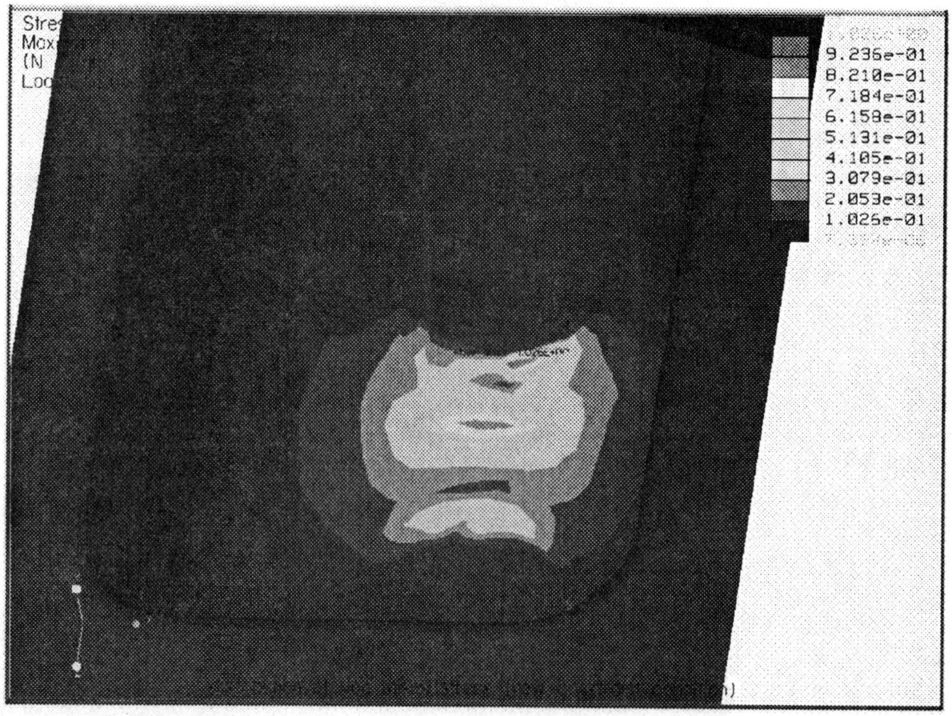

6.5 Case Study 2: Shell Modeling for a Welding Structure

In this section, we will work with an assembly. The assembly is a welding strucgture. As shown below, there are two components. One is called Base, and the other U_channel. The base component is fixed to a base through a set of 6 holes using bolts. The U_channel component is designed to hold an object, whose weight is 1000 Newton. The thickness of the two components is 8 mm. There will be 8 spot welds to connect the two components. The size of each spot weld is 20 mm. The material for the two components and the welds is assumed to be STEEL.

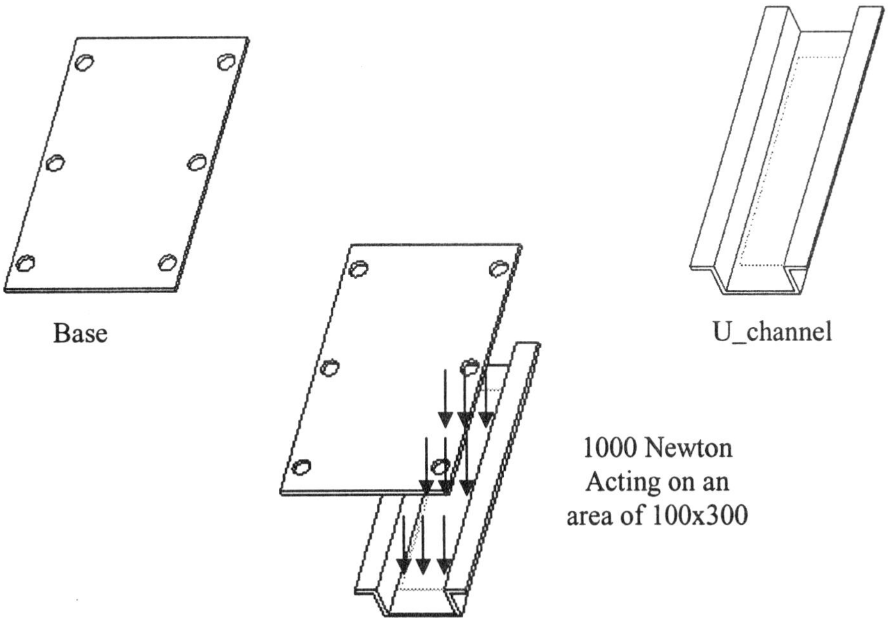

The geometrical shape and key dimensions of the welding structure are shown below. To illustrate the dimensions involved in creating two (2) 3D solid models for the 2 components and a 3D solid model of the welding assembly, an engineering drawing is shown below.

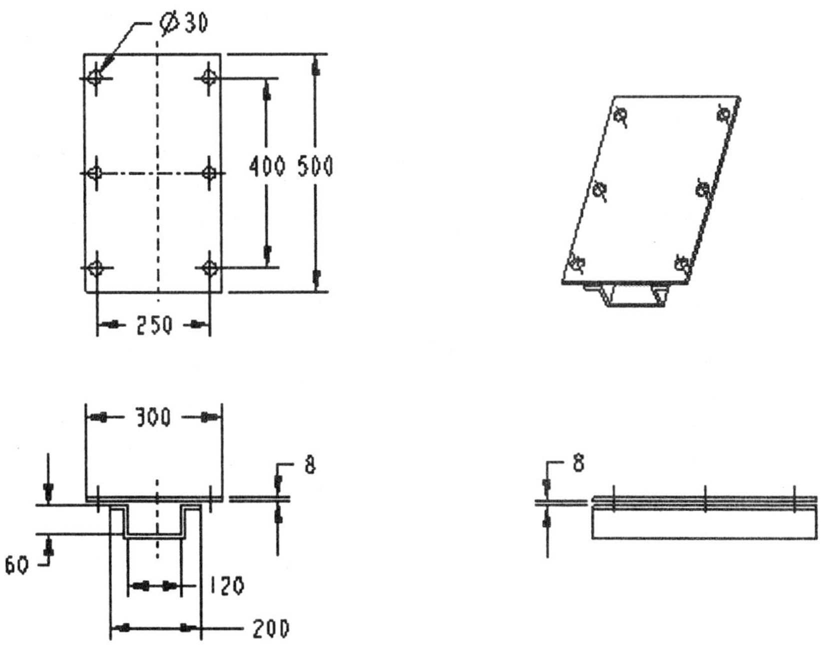

Step 1: Create a 3D solid model for the base component

File > New > Part > type *base* as the file name and clear the icon of **Use default template > OK.**

Select mmns_part_solid (units: Millimeter, Newton, Second) and type *base* in DESCRIPTION, and *student* in MODELED_BY, then **OK**.

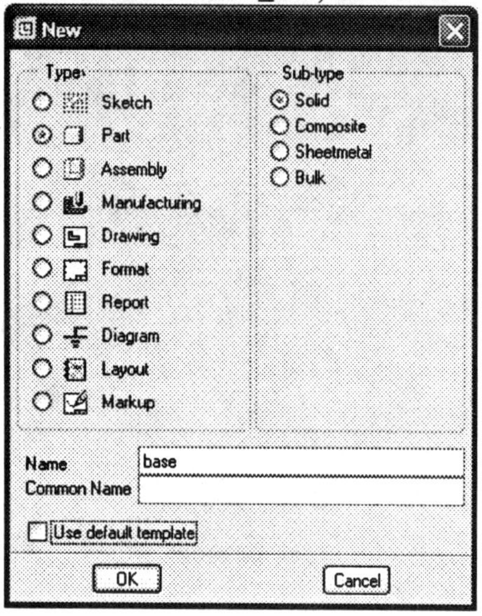

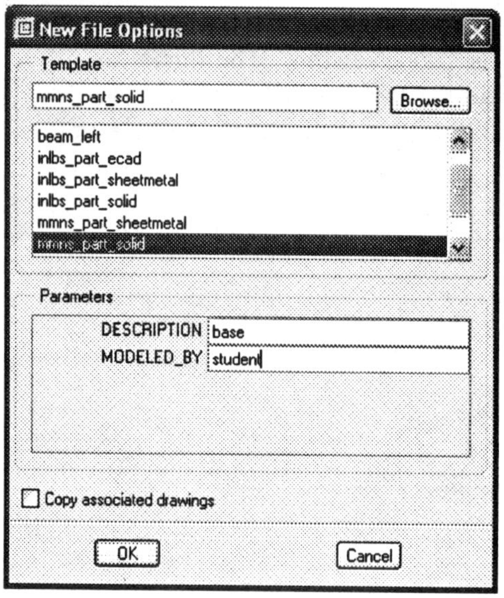

Feature 1: a plate with the dimensions: 300 x 500 x 8 mm.

Directly pick up the icon of **Extrude** from the toolbar of feature creation.

To create the horizontal plate, from the main toolbar, select **Insert > Extrude.** From the dashboard, select **Thin Protrusion** > set the thickness value to *8* > select the symmetric depth setting > set the depth value equal *500*. Activate **Placement** to define a sketch plane > **Define.**

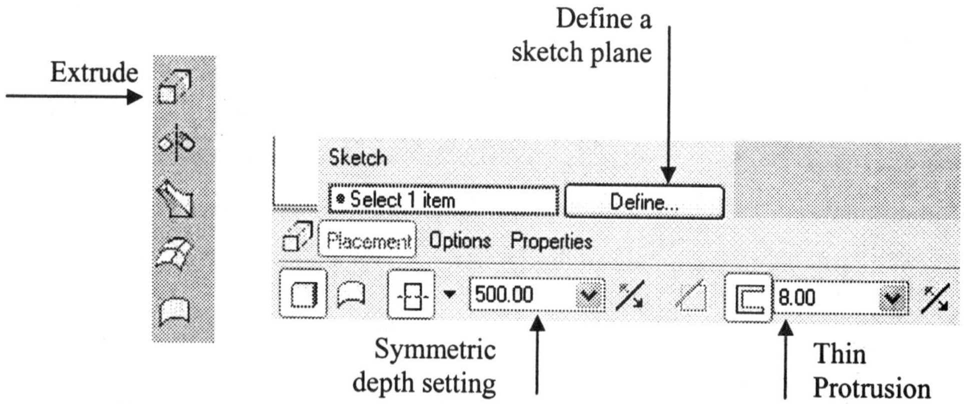

Select the **FRONT** datum plane as the sketch plane, and click the box of **Sketch** to accept the **RIGHT** datum plane as the default reference to orient the sketch plane, as illustrated below.

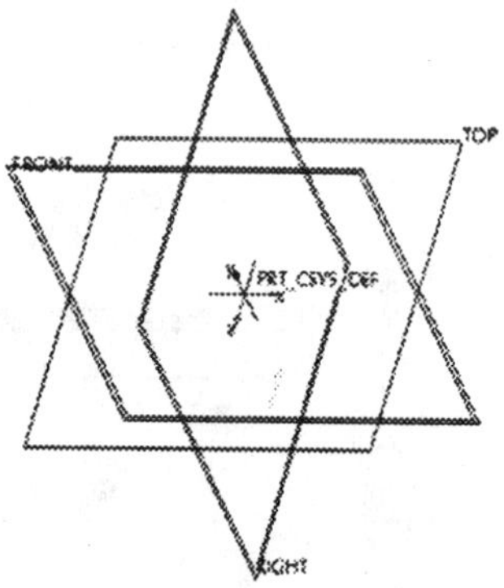

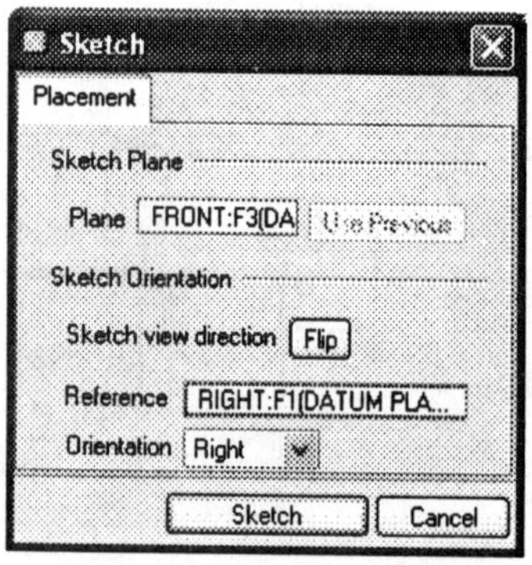

Select the icon of **Centerline** from the toolbar of sketcher to draw a centerline along the y-axis. Select the icon of Line to sketch a line along the x-axis. The line is symmetric about the y-axis. The length is 300.

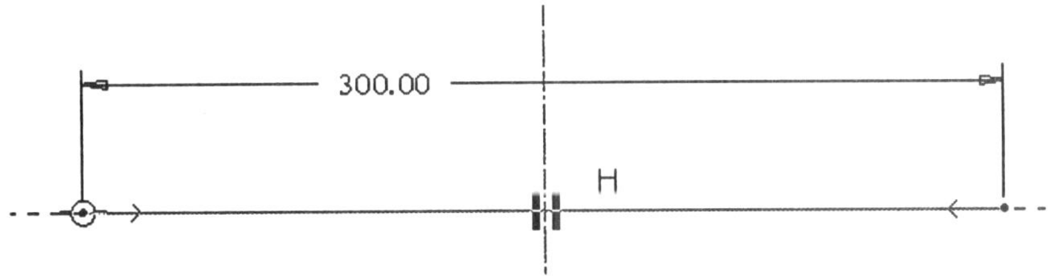

Upon completion, pick the icon of **Done** from the toolbar of sketcher. Note that the side to which the material is added with respect to the sketched line. The position of the TOP datum plane is the bottom of the horizontal plate.

Select the icon of **Done** to complete the 2D sketch

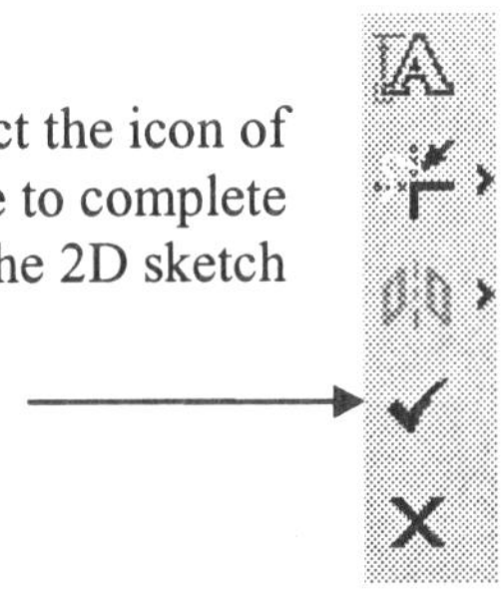

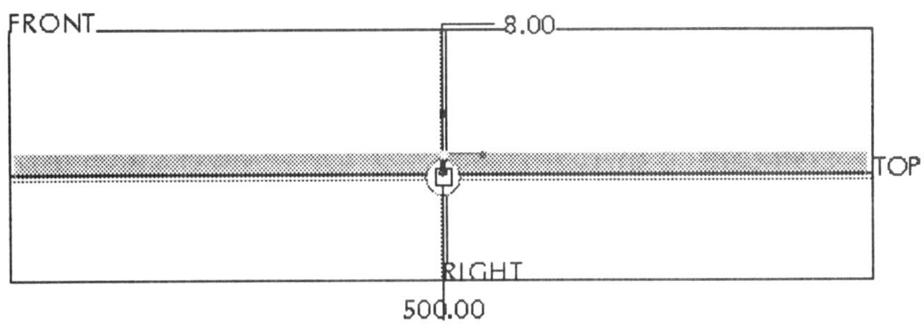

From the dashboard, select the icon of **Complete** from the feature control panel, thus completing the creation of the first feature.

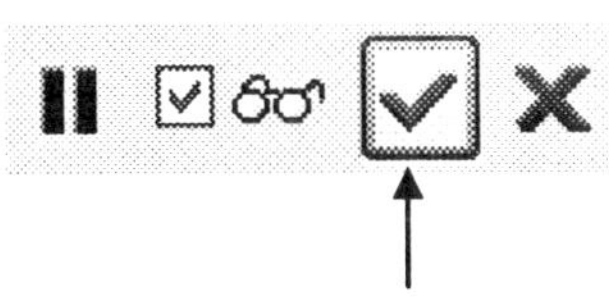

Pick the icon of **Complete** from the Feature Control Panel

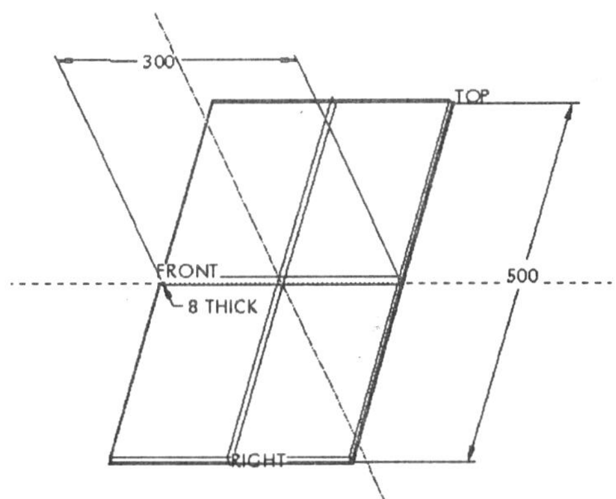

Create 6 through-holes on the thin plate.

Directly pick up the icon of Extrude from the toolbar of feature creation. From the dashboard, select the icon of **Cut** and set the depth choice to **Through All**. Activate **Placement** to define a sketch plane > **Define.**

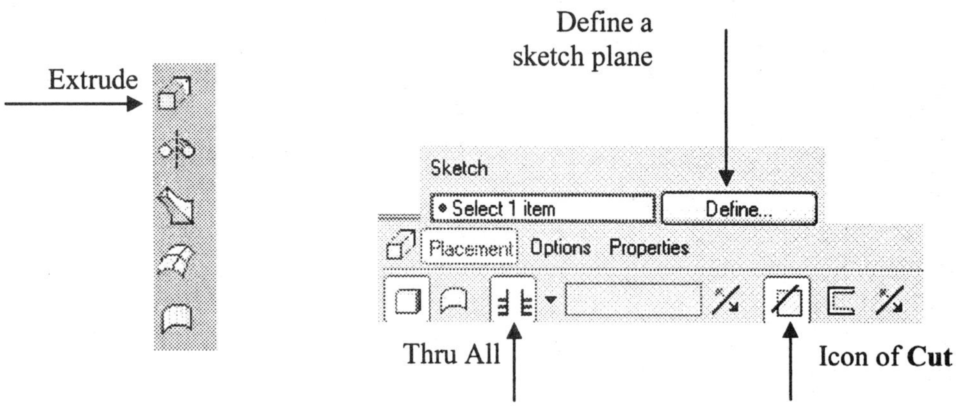

Select the top surface of the plate as the sketch plane and accept the **RIGHT** datum plane being positioned on the right side to orient the sketch plane. Click the box of **Sketch.**

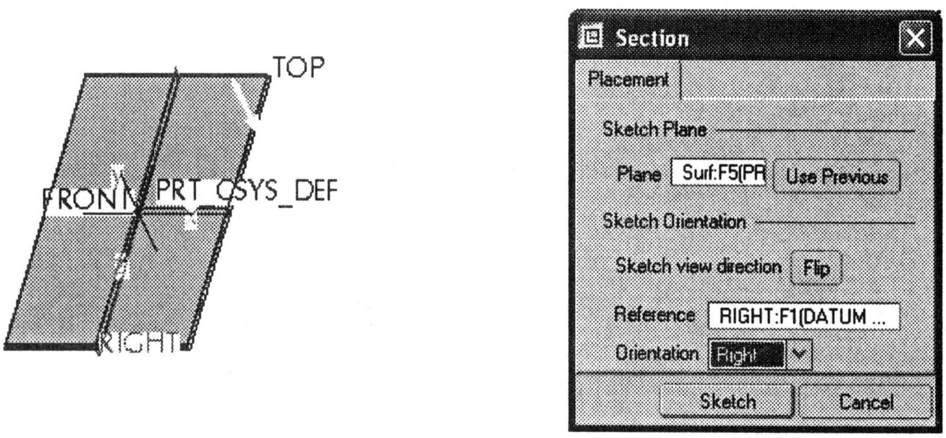

Make sure that the **RIGHT** datum plane and the **FRONT** datum plane are selected as the references.

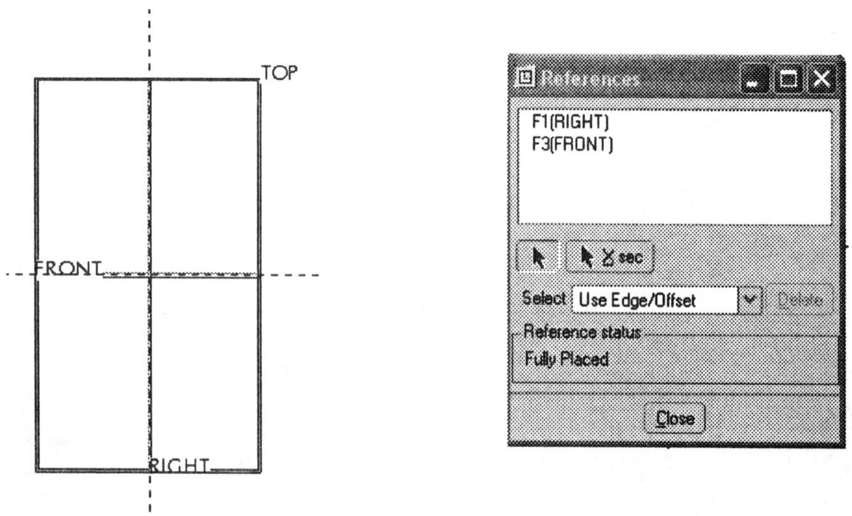

Sketch the 2 centerlines along the x-axis and z-axis, respectively. Sketch 6 holes equal in diameter. The diameter is 30. The 6 holes are symmetric about the two centerlines, as shown.

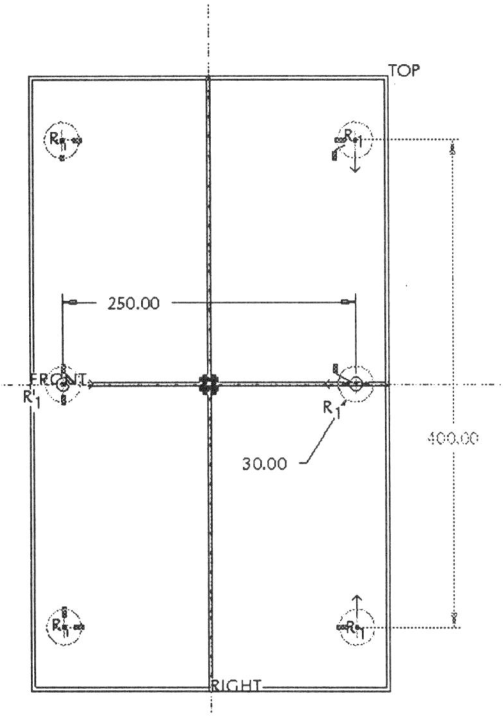

Select the icon of **Done** to complete the 2D sketch. Select the icon of **Complete** from the feature control panel, thus completing the 3D modeling of the hole feature.

Select the icon of **Done** to complete the 2D sketch

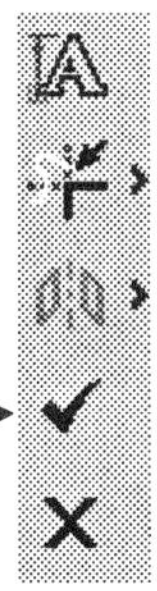

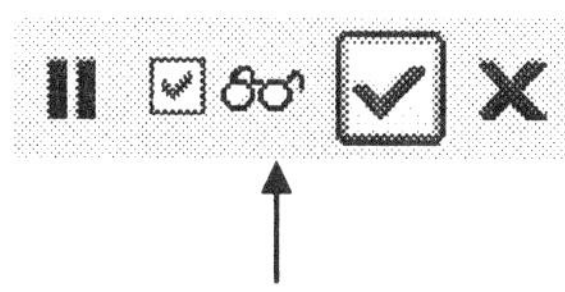

Select the icon of **Complete** to complete the creation of the thin plate feature

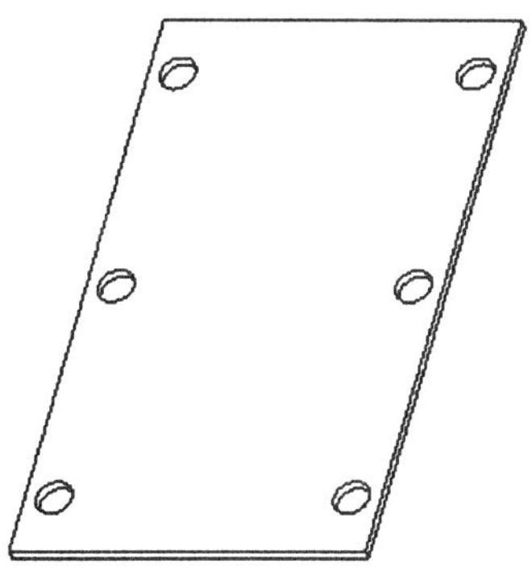

Step 2: Create a 3D solid model for the U-shape component

File > New > Part > type *u_shape* as the file name and clear the icon of **Use default template > OK.** Select **mmns_part_solid** (units: Millimeter, Newton, Second) and type *u-shape* in **DESCRIPTION**, and *student* in **MODELED_BY**, then **OK.**

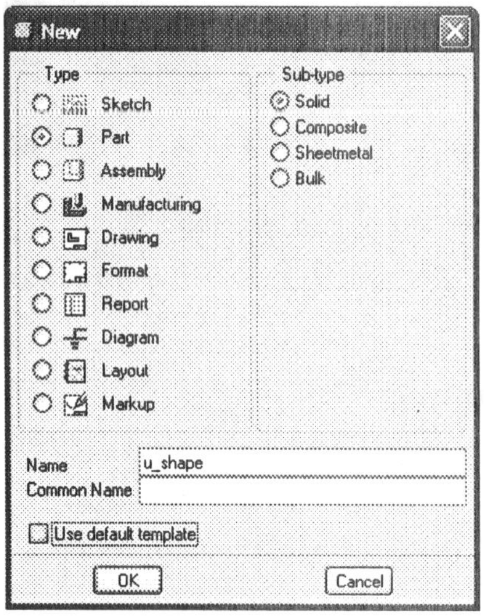

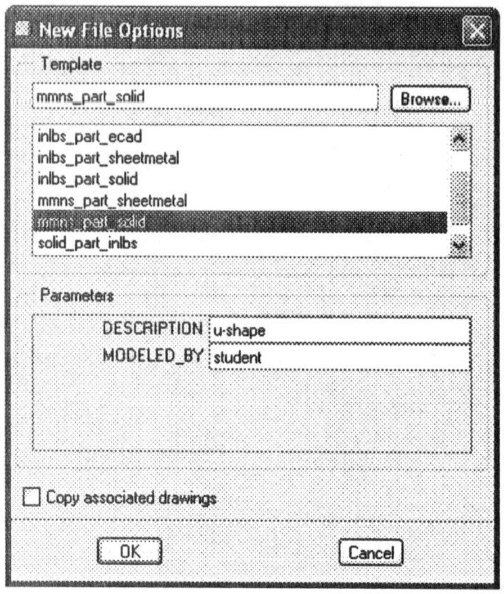

Feature 1: a bent U-shaped plate.

Directly pick up the icon of **Extrude** from the toolbar of feature creation.

From the dashboard, select **Thin Protrusion** > set the thickness value to *8* > select the symmetric depth setting> depth equal *500.* Activate **Placement** to define a sketch plane > **Define.**

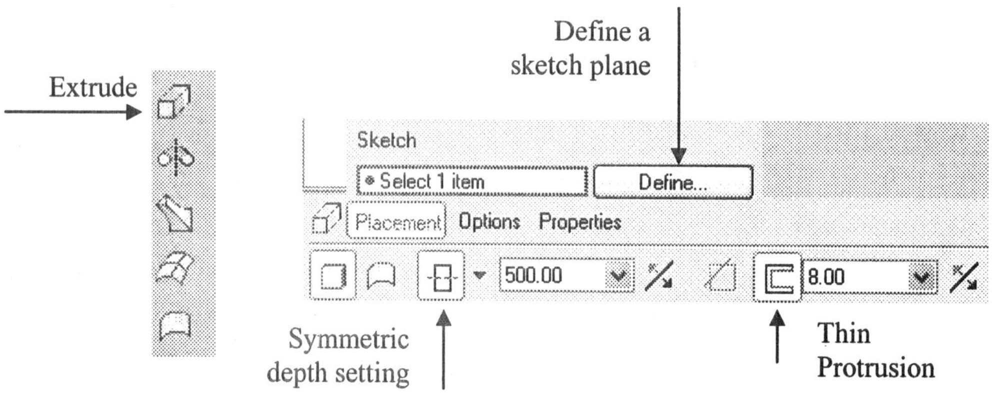

Select the **FRONT** datum plane as the sketch plane and accept the **RIGHT** datum plane being positioned on the right side to orient the sketch plane. Click the box of **Sketch.**

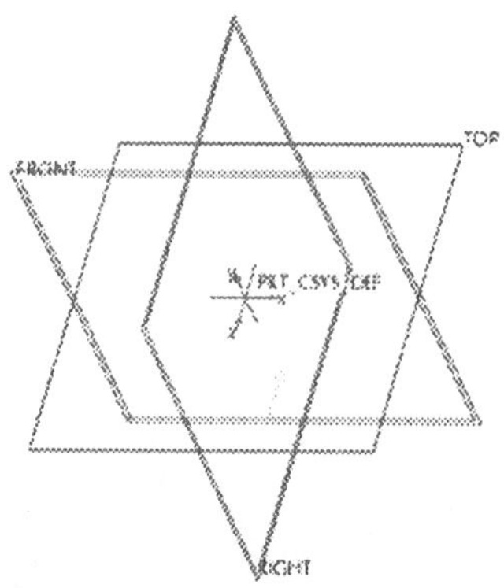

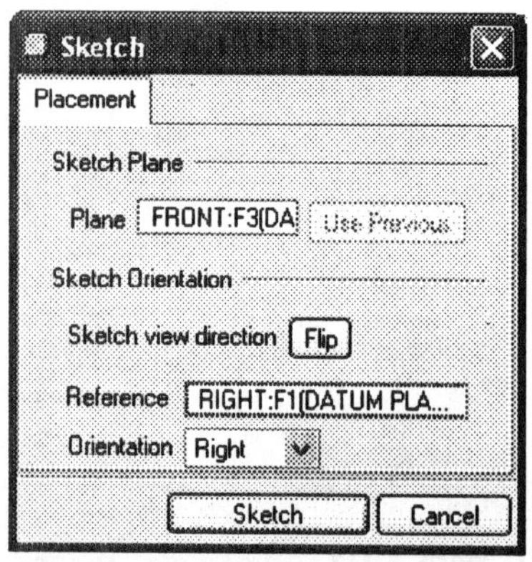

Accept the two references selected by the system default. The selected references are the projection of the **RIGHT** datum plane on the **FRONT** datum plane, and the projection of the **TOP** datum plane on the **FRONT** datum plane. These two projections are coincident with the y-axis and x-axis, respectively.

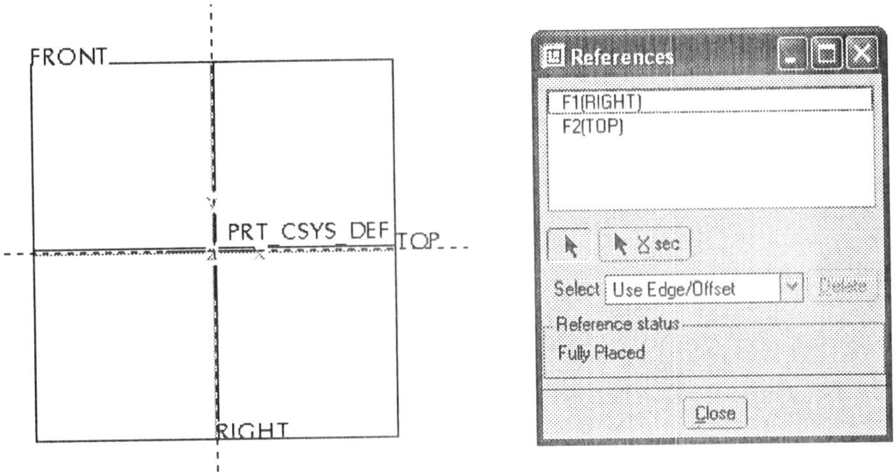

Sketch a centerline along the y-axis. Afterwards, prepare the following sketch, which is symmetric about the y-axis, as shown.

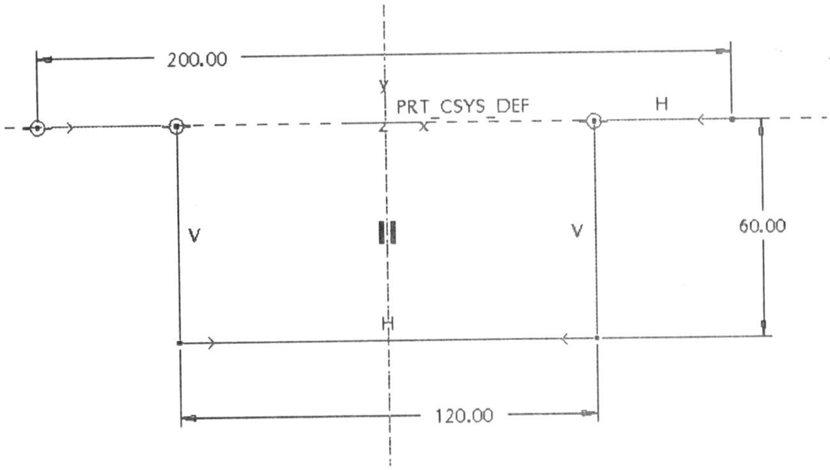

Select the icon of **Done** to complete the 2D sketch. Select the icon of **Complete** from the feature control panel, thus completing the 3D modeling of the hole feature.

Select the icon of **Done** to complete the 2D sketch

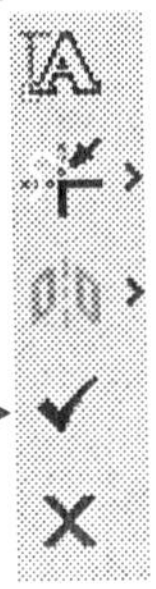

Select the icon of **Complete** to complete the creation of the thin plate feature

Pay special attention to the material side

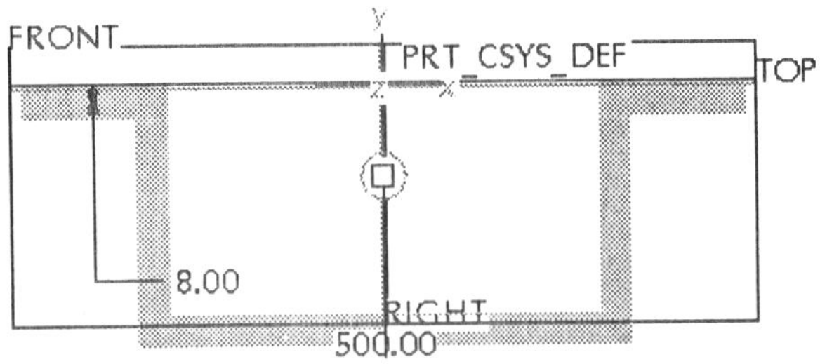

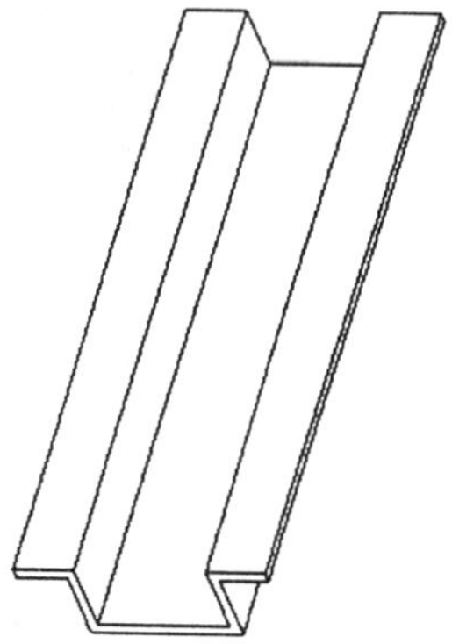

Step 3: Create a 3D solid model for the assembly, which consists of the base and the U-shape components.

File > **New** > **Part** > type *spot_welding* as the file name and clear the icon of **Use default template** > **OK.** Select **mmns_part_solid** (units: Millimeter, Newton, Second) and type in *spot welding* in **DESCRIPTION**, and *student* in **MODELED_BY**, then **OK.**

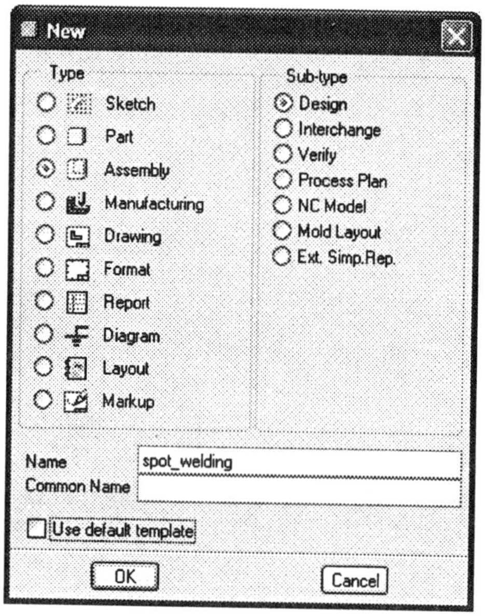

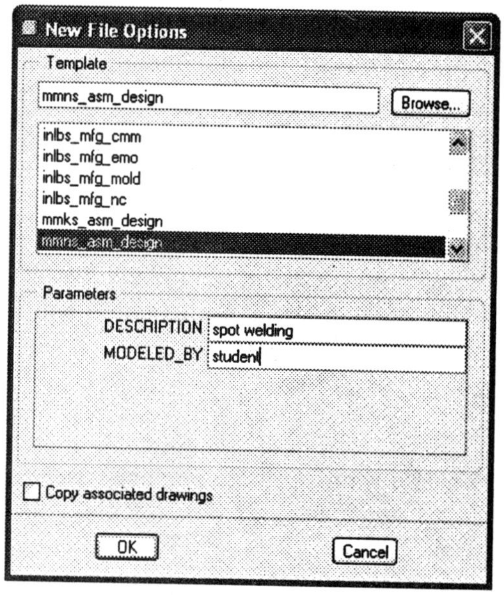

Directly pick up the icon of Add component from the toolbar of feature creation.

Icon of **Add component**

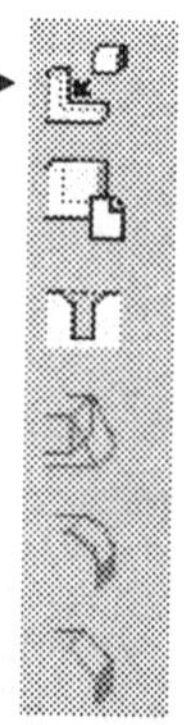

Select the file called base.prt from the **In Session** window > **Open**.

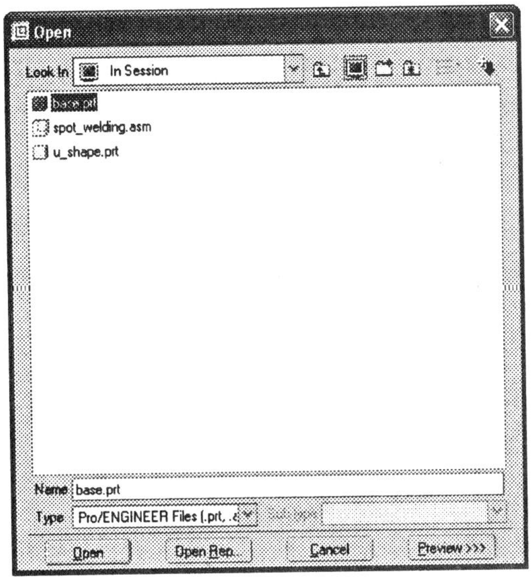

Select the icon of **Default** > **OK** to complete assembling the base component.

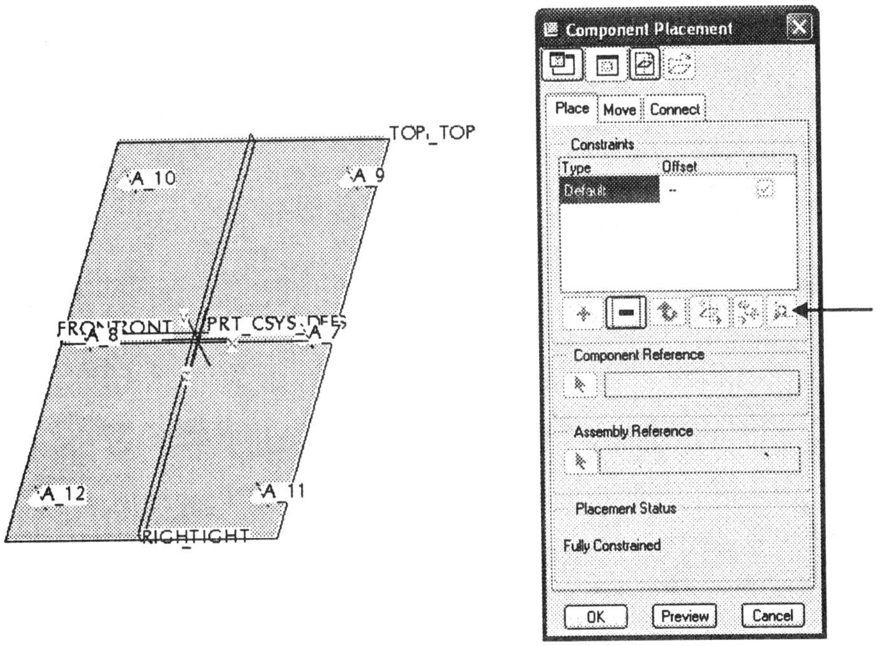

Pick up the icon of Add component from the toolbar of feature creation, again.

Icon of **Add component**

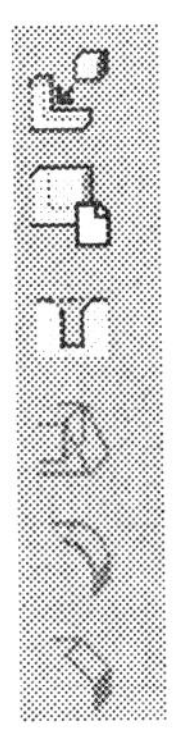

Select the file called u_shape.prt from the In Session window > **Open**.

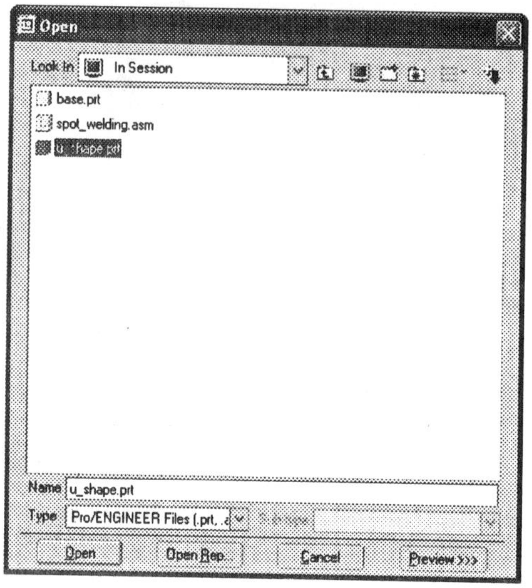

Select the icon of default > **OK** to complete assembling the u_shape component.

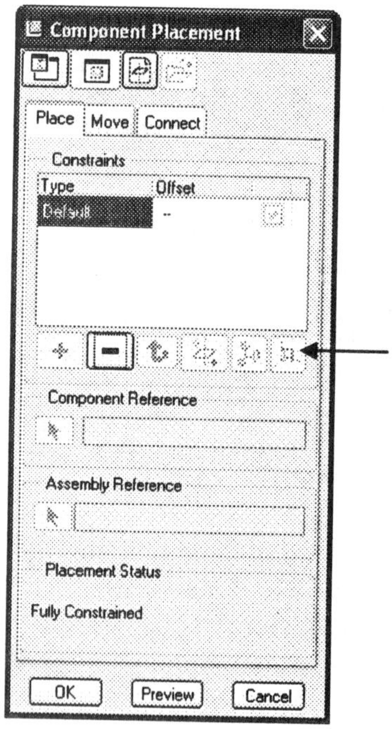

Step 4: Open the file called base.prt to define 8 datum points for spot welds.
From the model tree of the assembly window, highlight U_SHAPE > right click and pick Open.
Click the icon of **datum point tool** and expand it, and select Offset coordinate system.

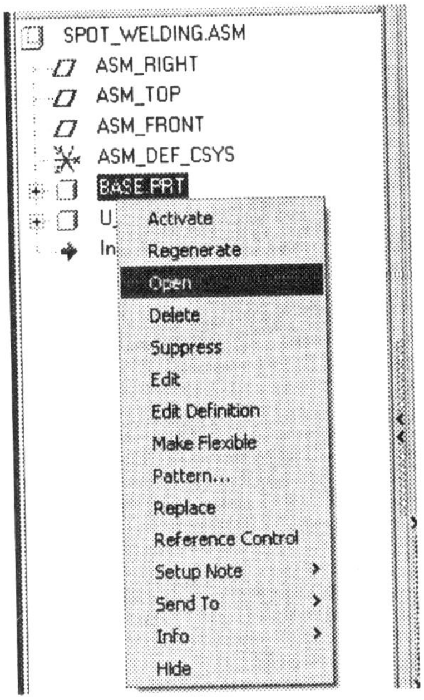

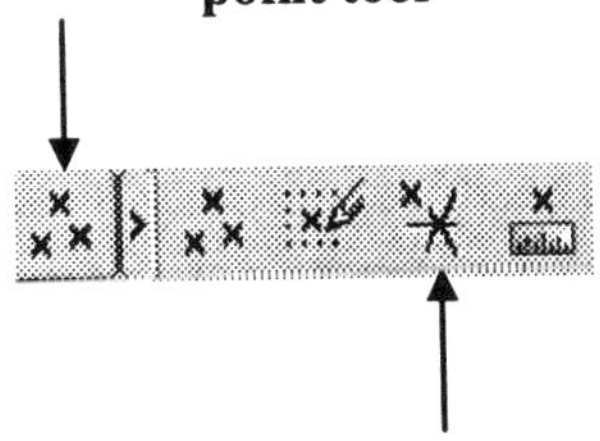

Type in the coordinates of the 8 datum points > **OK**.

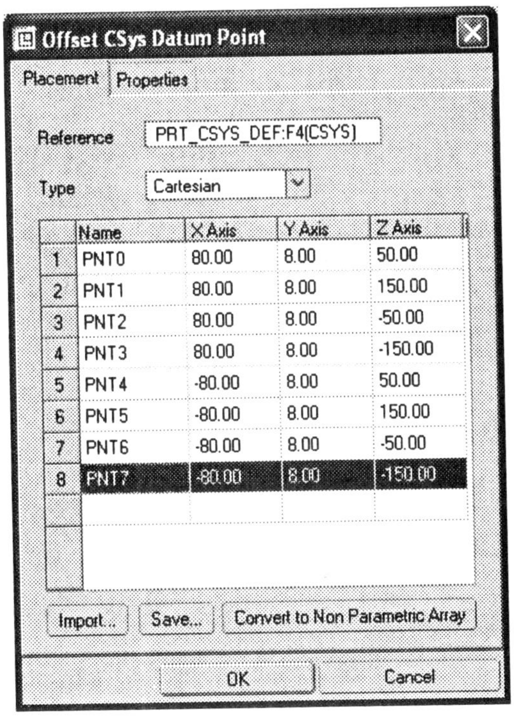

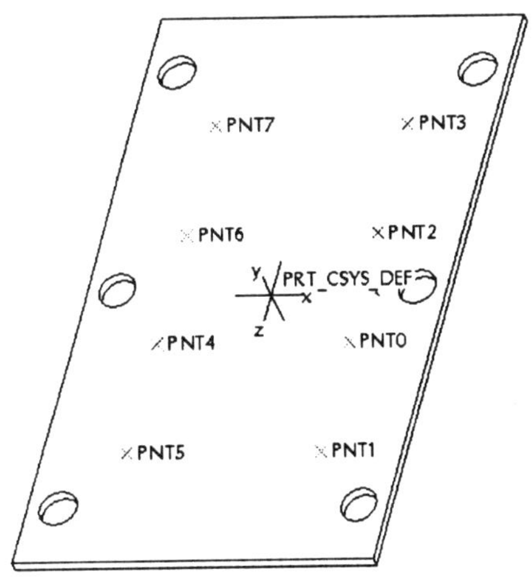

Step 5: Define the shell model and the cylindrical coordinate system for the base component under **Pro/Mechanica**

Applications > Mechanica > Continue > Structure > OK.

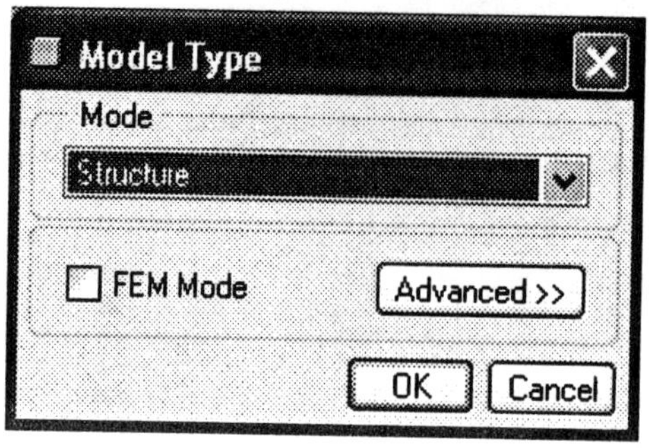

To define the shell model, from the toolbar of functions, select the icon of **New Shell Pair** > **Auto Detect** > **Compress** > **Shells Only**.

ShowCompress

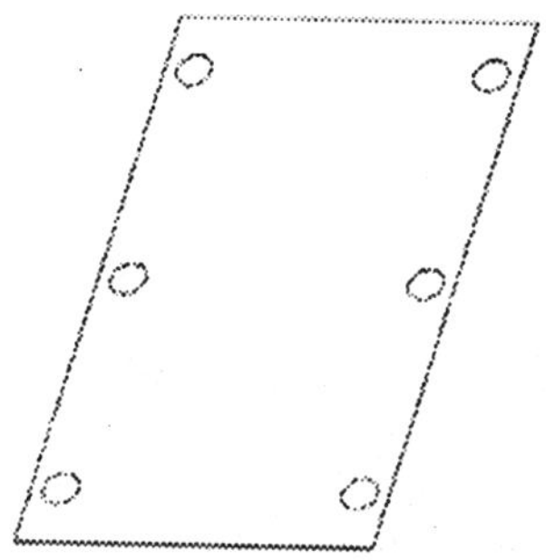

ShowOriginal

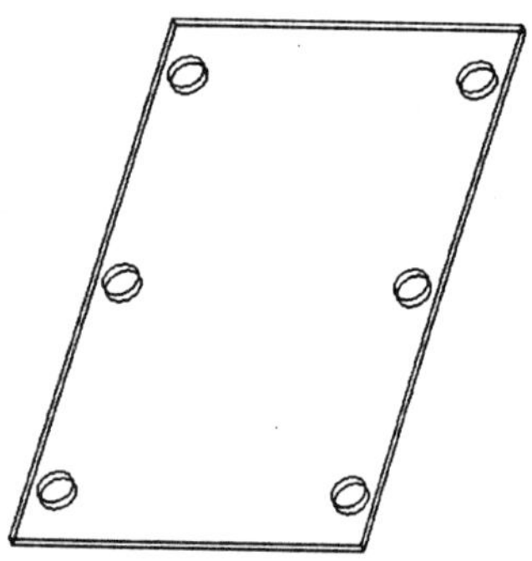

Show Both

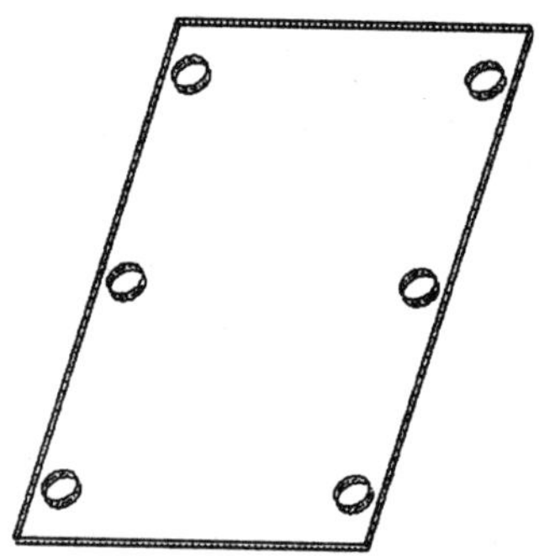

However, to model the welding structure, we need to move the compression surface to the top surface of the base component. To do so, **Edit > Pair Place** > move the compression surface to the top surface of the base component.

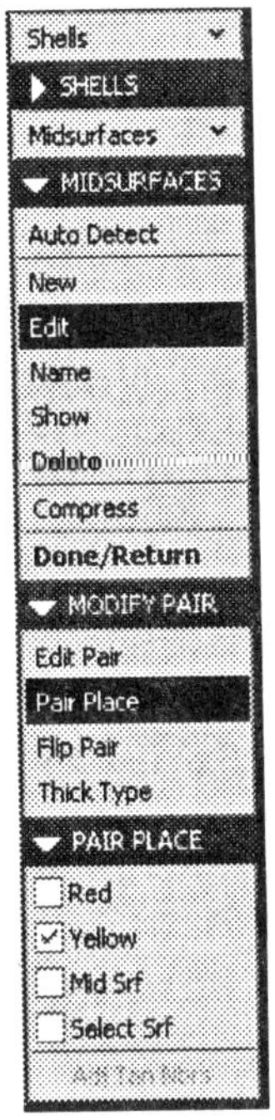

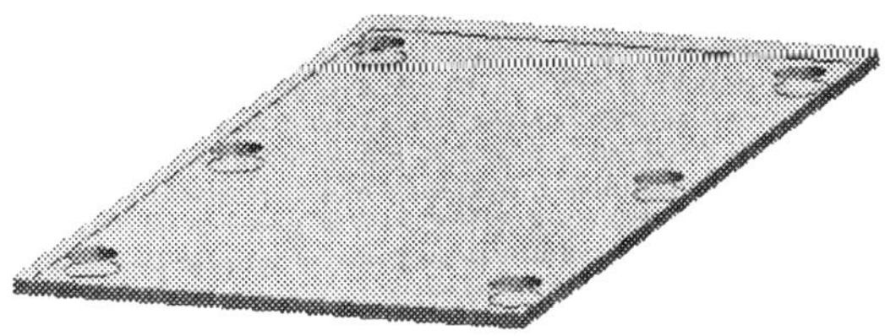

To define the constraint condition, we need to have six cylindrical coordinate systems at the central locations of the 6 holes.

Select the icon of creating a coordinate system from the toolbar of datum feature creation > pick an axis and an edge to create a Cartesian coordinate system.

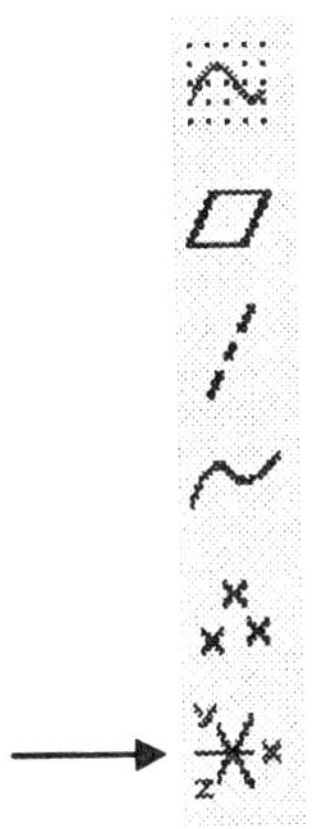

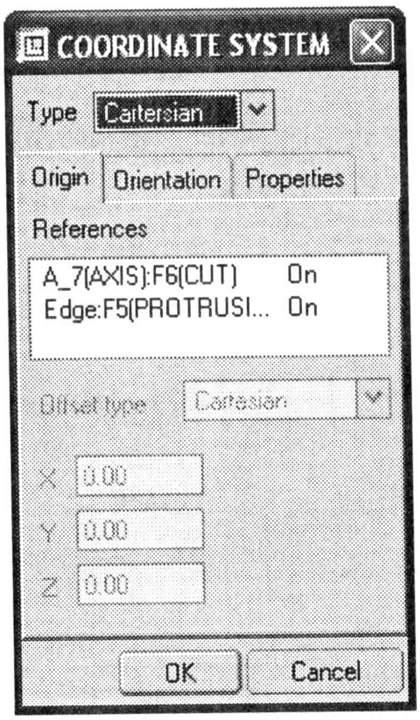

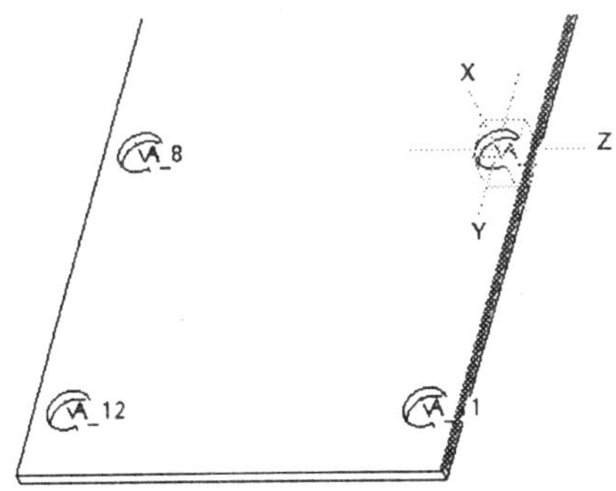

Note that the vertical axis is now marked as the x-axis. In order to convert to a cylindrical coordinate system, the vertical axis has to be set as the z-axis. Before converting the Cartesian coordinate system to a cylindrical coordinate system, we need to re-orient the Cartesian coordinate system so that the vertical axis is the z-axis. For changing the orientation, click Orientation > set the vertical axis to be Z > change Cartesian to Cylindrical in the box called Type > **OK**.

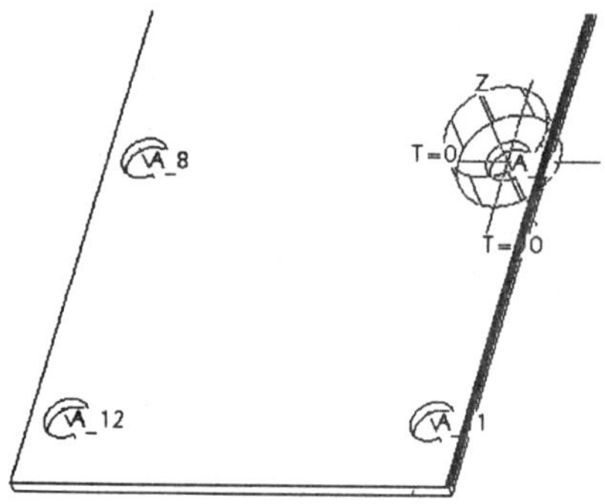

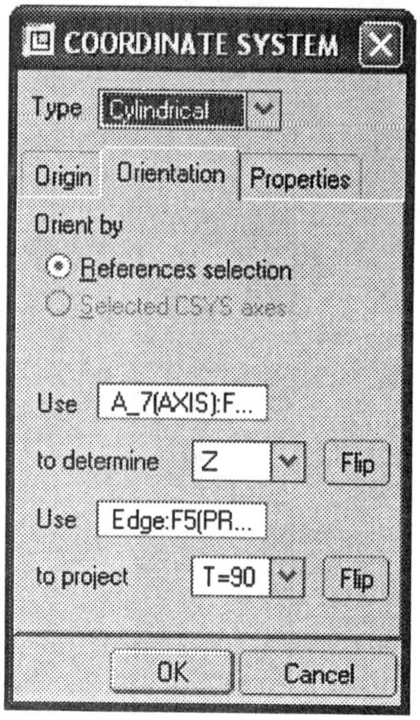

Repeat the above procedure to create the other 5 cylindrical coordinate system, as shown.

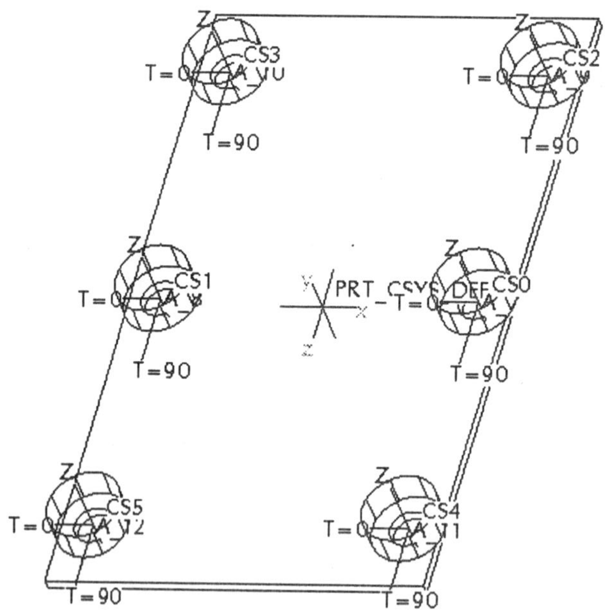

Step 5: Define the shell model and a surface region for defining the load for the u-shape component under **Pro/Mechanica**

Applications > Mechanica > Continue > Structure > OK.

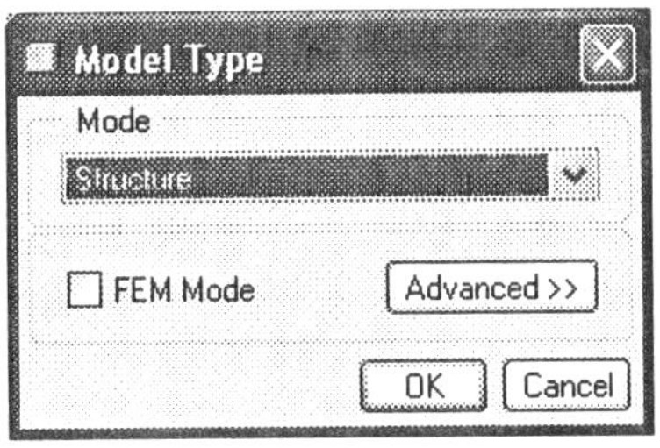

To define the shell model, from the toolbar of functions, select the icon of **New Shell Pair** > **Auto Detect** > **Compress** > **Shells Only**.

ShowCompress

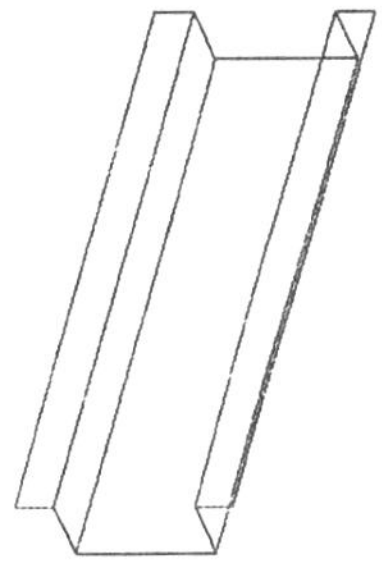

ShowOriginal

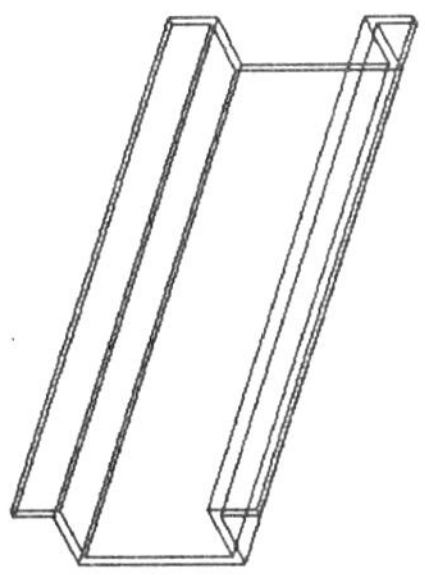

Show Both

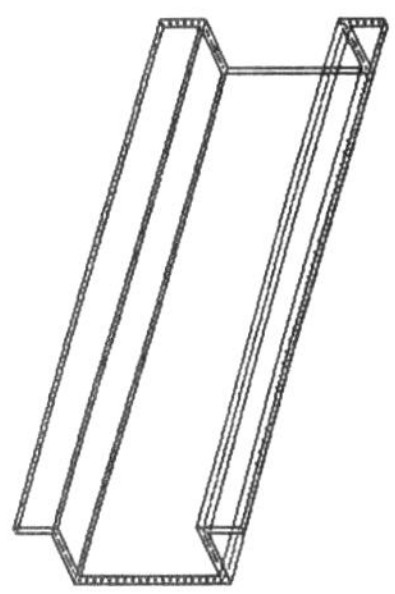

However, to model the welding structure, we need to move the compression surface to the bottom surface of the u-shape component. To do so, **Edit** > **Pair Place** > move the compression surface to the "bottom" surface of the base component.

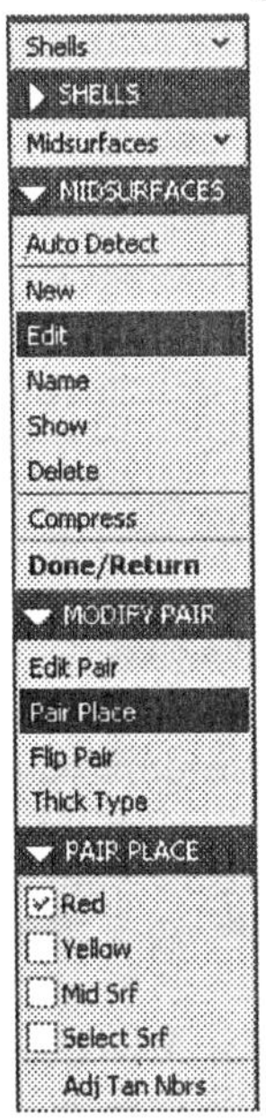

You need to repeat the above procedure 3 more times because there are 4 pairs of surfaces, which form the shell model.

Define a surface region (100x300) on the u-shape component. First go back to the Pro/ENGINEER design environment. Select the icon of datum curve from the toolbar of datum feature creation. Select the surface, as shown as the sketch plane and accept the default orientation.

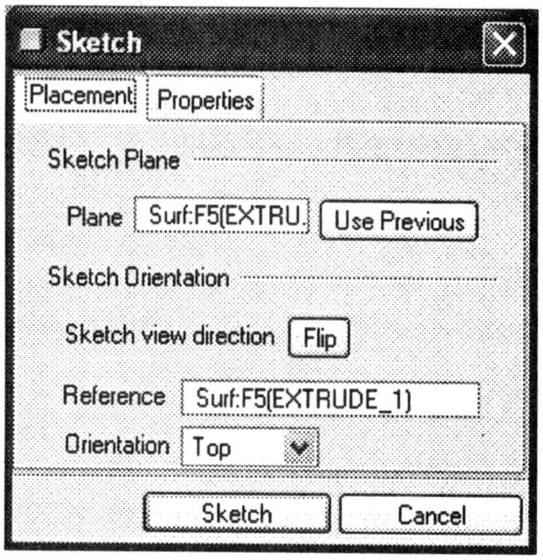

First make sure that the FRONT datum plane and the RIGHT datum plane are selected as references. Sketch 2 centerlines along the x-axis and z-axis. Afterwards, select the icon of rectangle and sketch a rectangle, which is symmetric about the 2 centerlines. The 2 dimensions are 100 and 300, respectively.

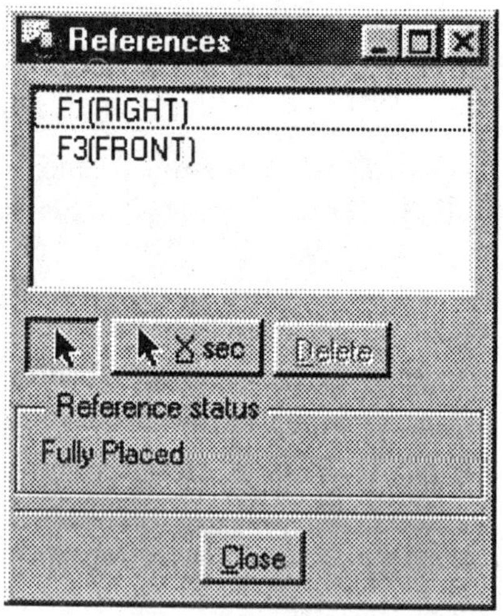

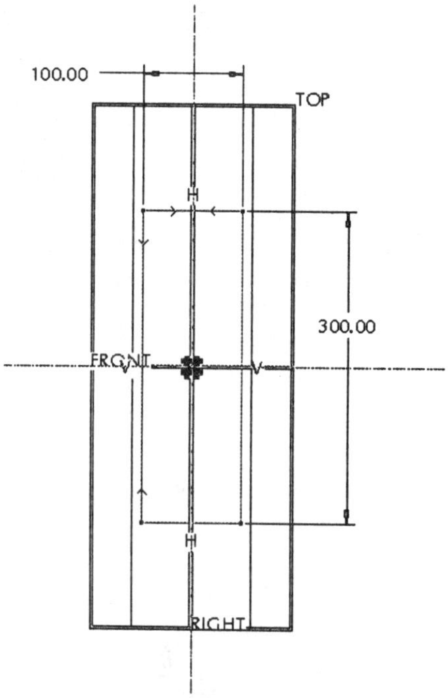

Go back to Pro/Mechanica to define a surface region using the sketched datum curve. Select the icon of surface region > Select > Done > pick the datum curve > pick the surface to be split > **OK > Done > OK**.

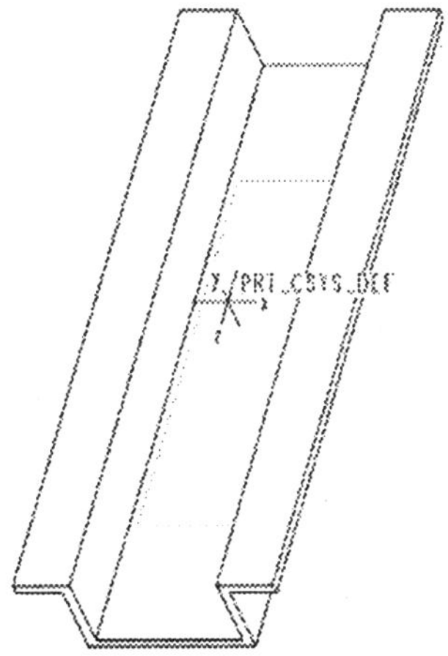

Step 6: Perform FEA with the assembly model.

Go back to the assembly window.

From the menu bar, select **Applications** > **Mechanica**

Click **Continue** after checking the unit system > **Structure** > **OK**.

First let us check the shell model for the assembly. From the main toolbar, select Insert > Midsurface > Compress > Shells only.

ShowCompress

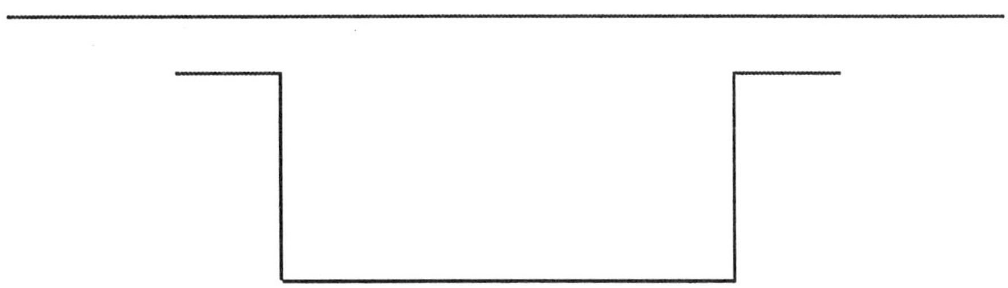

ShowBoth

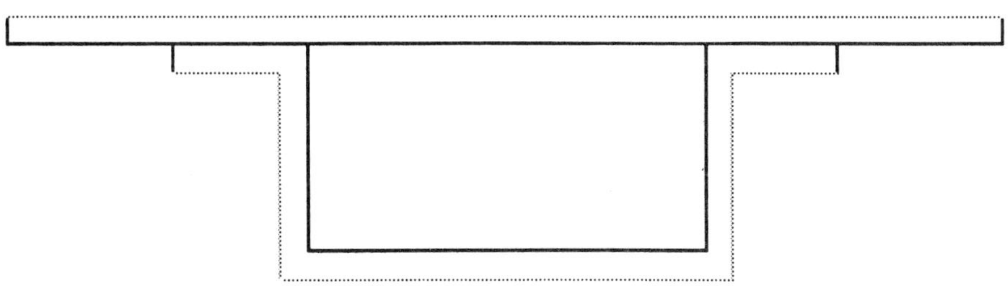

From the toolbar of functions, select the icon of **Materials >** select **STEEL.**

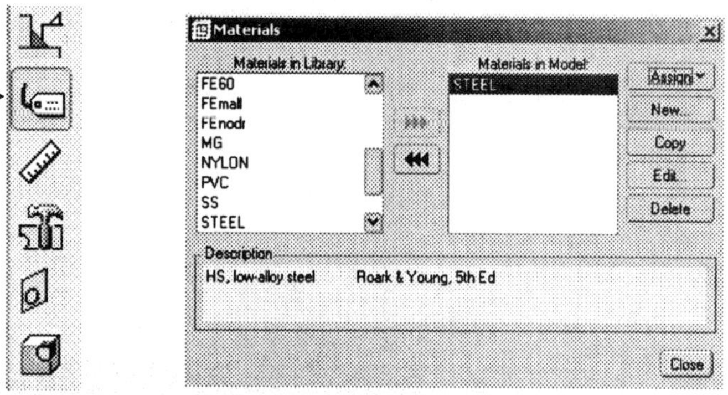

Assign > **Part** > pick the two components (holding down the Ctrl key for picking the second component) > **OK** > **Close.**

Define the bolt constraint conditions.

Select the icon of **Constraint** > > select **Edge(s)/Curve(s)** > pick the two edges to define the first hole > **OK**. Make sure that CS0 is selected as the reference coordinate system. Make sure that Trans R is left as free to characterize the presence of clearance between the hole and the bolt during the assembly.

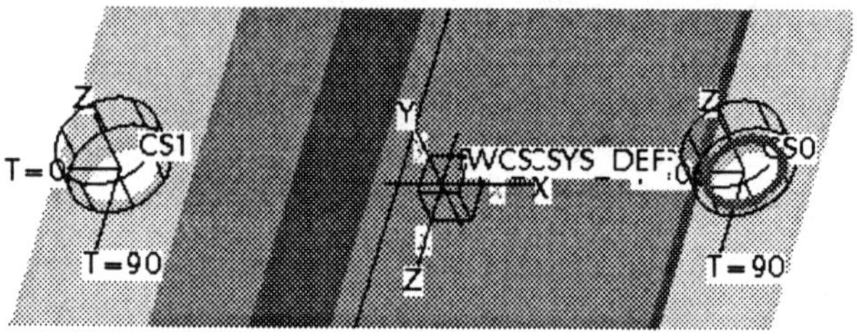

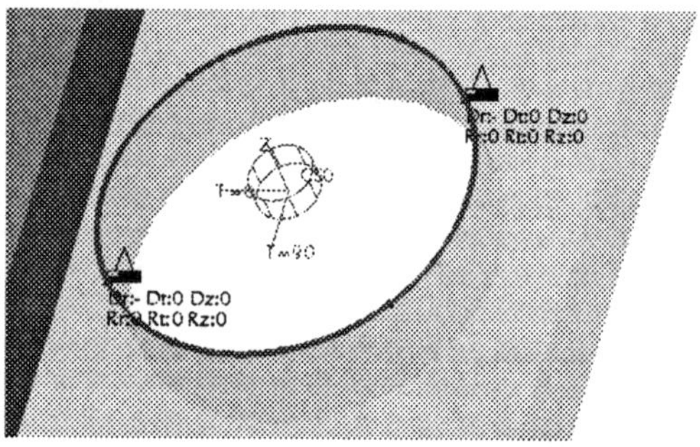

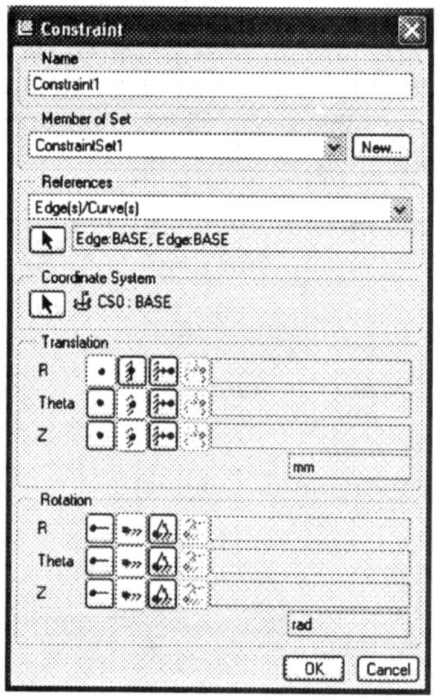

Repeat this procedure for the other five holes to complete the definition of the constraint set.

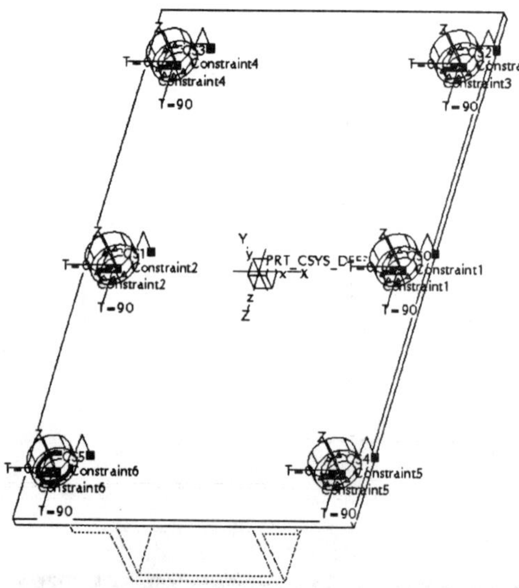

Define the 8 spot welds.

A spot weld is considered as a beam element. This beam element, when it is defined, is fused to the surface over an area defined by Diameter of the weld and material type specified. The length of this beam element is determined by the location of the fusion points on each surface.

From the main toolbar, select Insert > Connection > Spot Weld > Create > select the top surface of the base component as the first surface of the contact, and select the bottom surface of the u-shape component on the right side as the second surface of the contact.

Select > pick PNT0 first, holding down the Ctrl key, pick PNT1, PNT2 and PNT3 > **OK.** Specify the diameter of the spot weld: 20 mm.

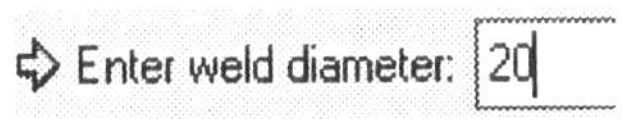

Select STEEL as the material type > **OK**.

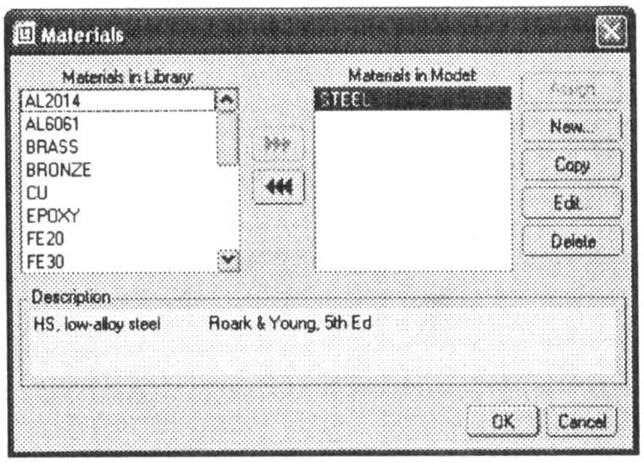

Repeat the procedure for the other side of the model in terms of selecting the two contact surfaces, picking up PNT4, PNT5, PNT6, and PNT7, specifying the diameter of the spot weld and the material type > OK > Done/Return.

Define the load condition.

The load is a uniform load of 1000 Newton is applied on the surface region created with the u-shape component. Select the icon of **force and moment** from the toolbar of functions > **Surface** > pick the rectangular surface region > **OK** > click **Advanced** > change the distribution to **Total Load** and **Uniform** > type *-1000* in Y of the Force box > **OK**.

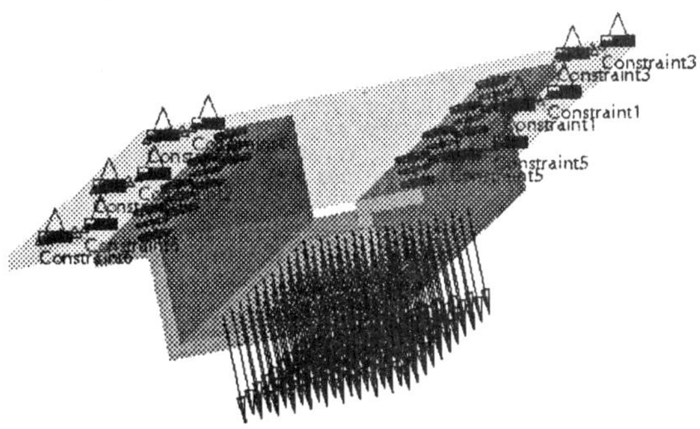

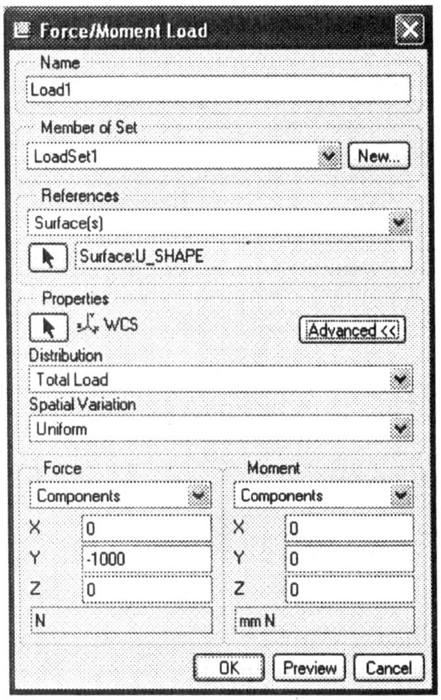

Perform a static analysis

Click the icon of performing a design study from the main toolbar > **File** > Select **New static analysis**.

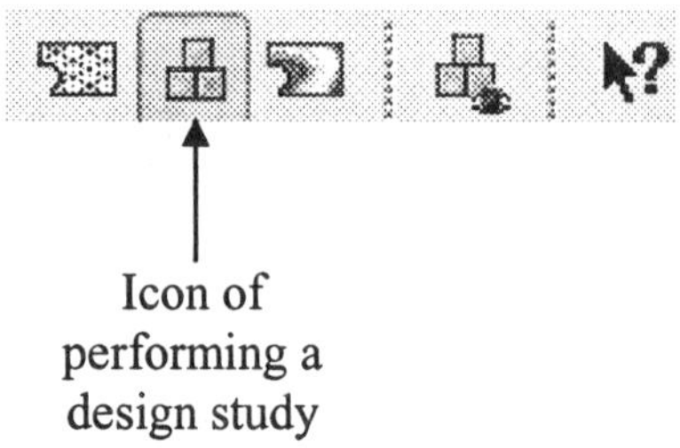

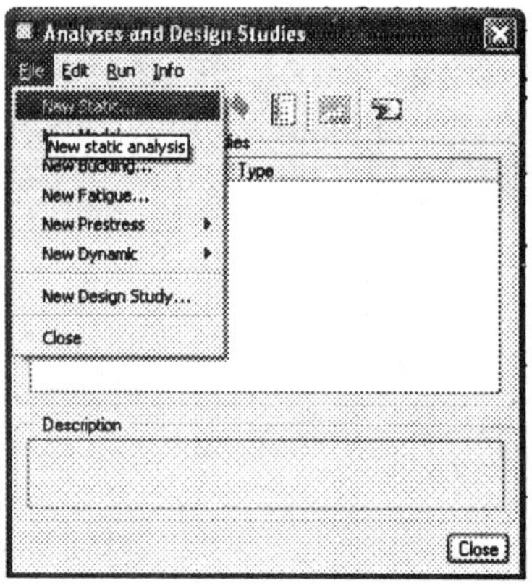

Type *spot_weld* as the name of design study.

Make sure that ConstraintSet1 and LoadSet1 are highlighted > Single-Pass Adaptive > **OK** > **Run**.

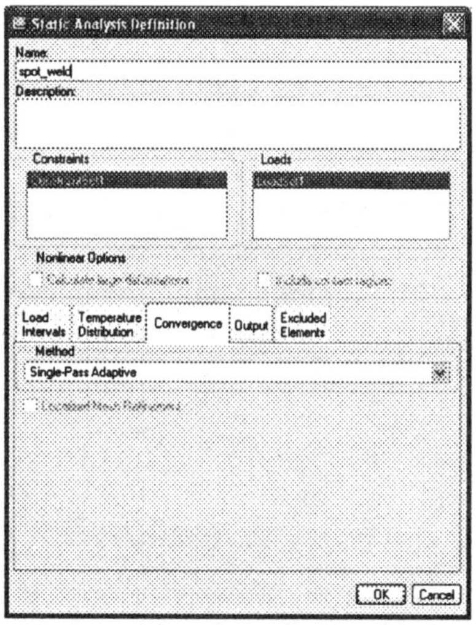

To review the results, click the icon of review results > type *deflection distribution* as the title of the result window > select the folder called bathtub_1 > **Fringe** > **Displacement** > **Magnitude** > **OK and Show**.

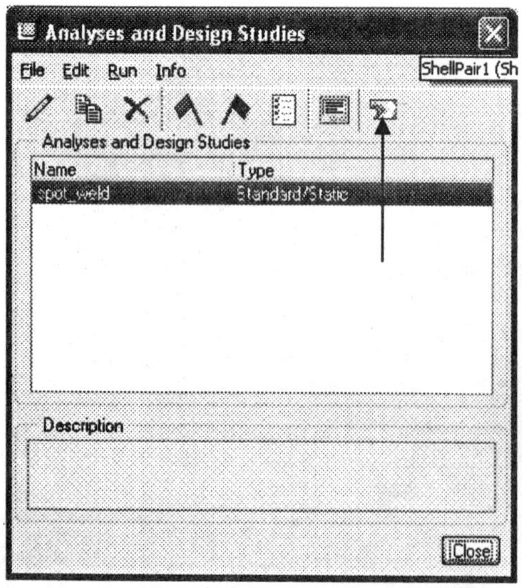

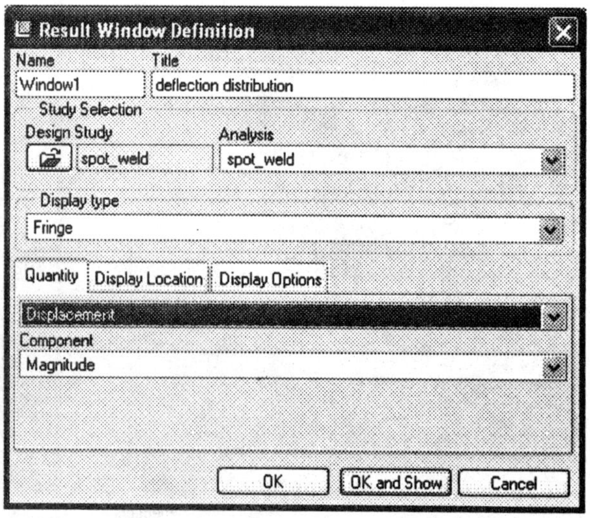

Displacement Distribution

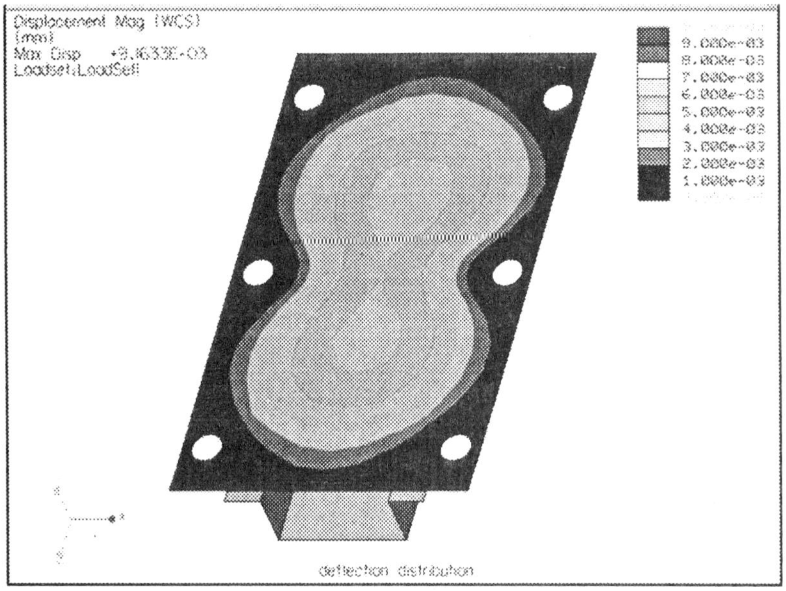

Activate **Display Options** > select **Deformed** > **OK and Show**

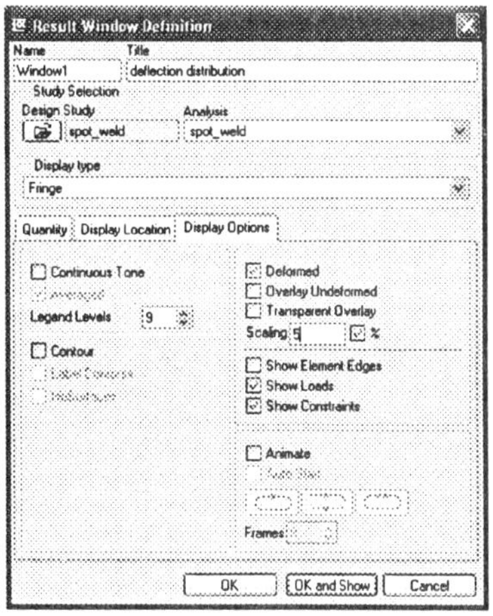

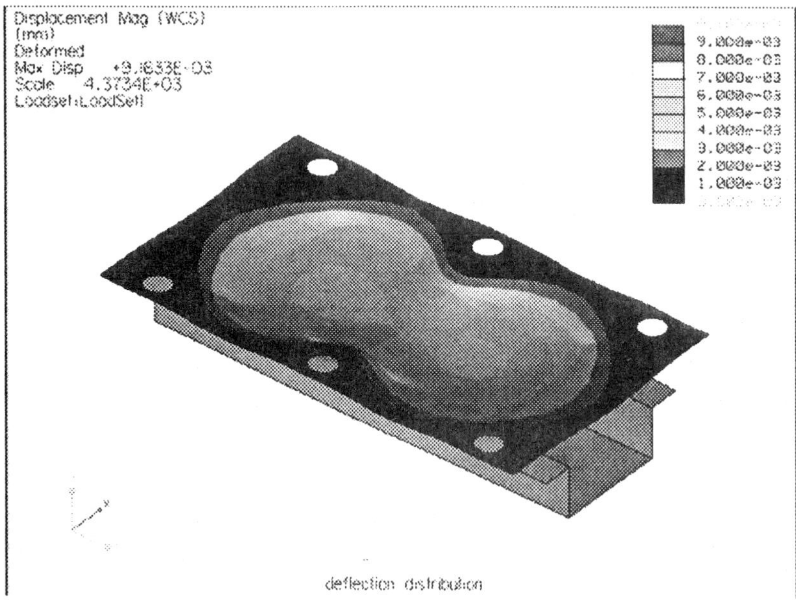

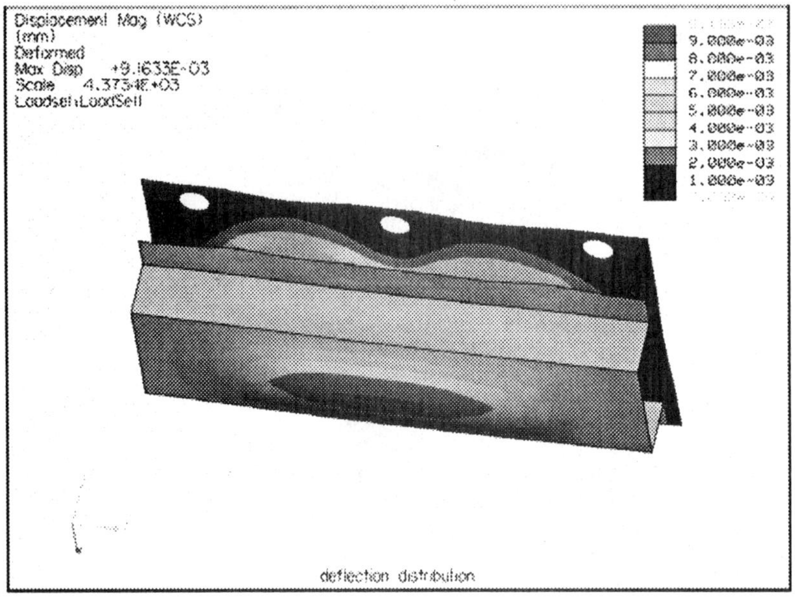

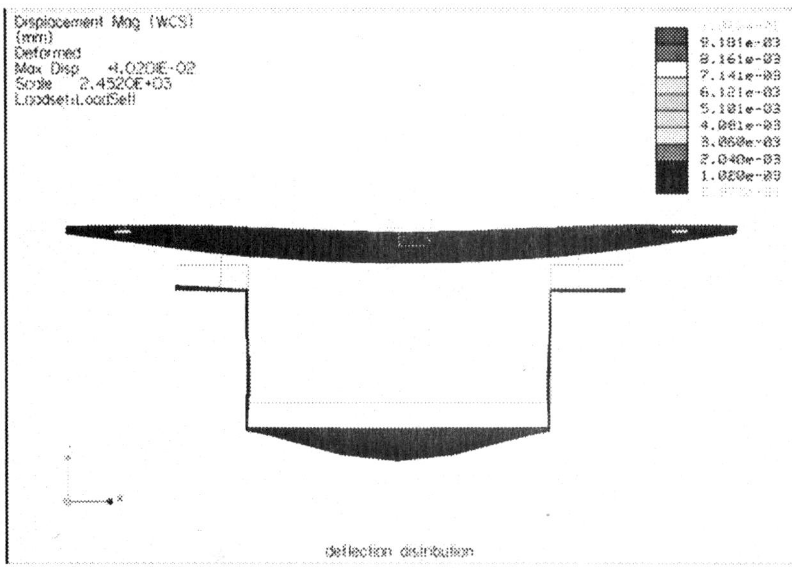

To examine the distribution of maximum principal stress, **Edit** > **Copy** > type in distribution of max principal stress > select **Stress** and **Max principal** > **OK and Show.**

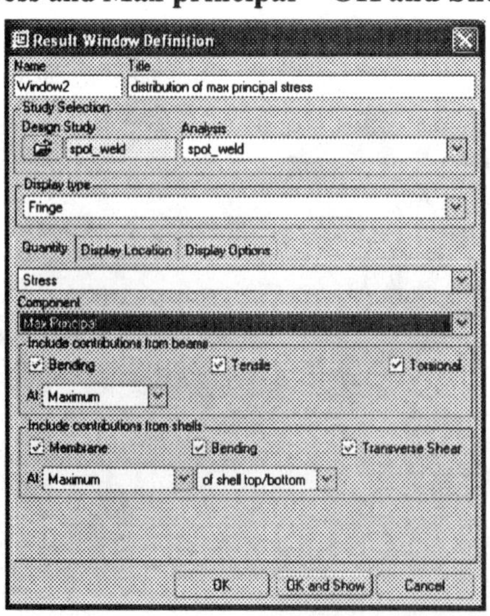

Distribution of the Maximum Principal Stress

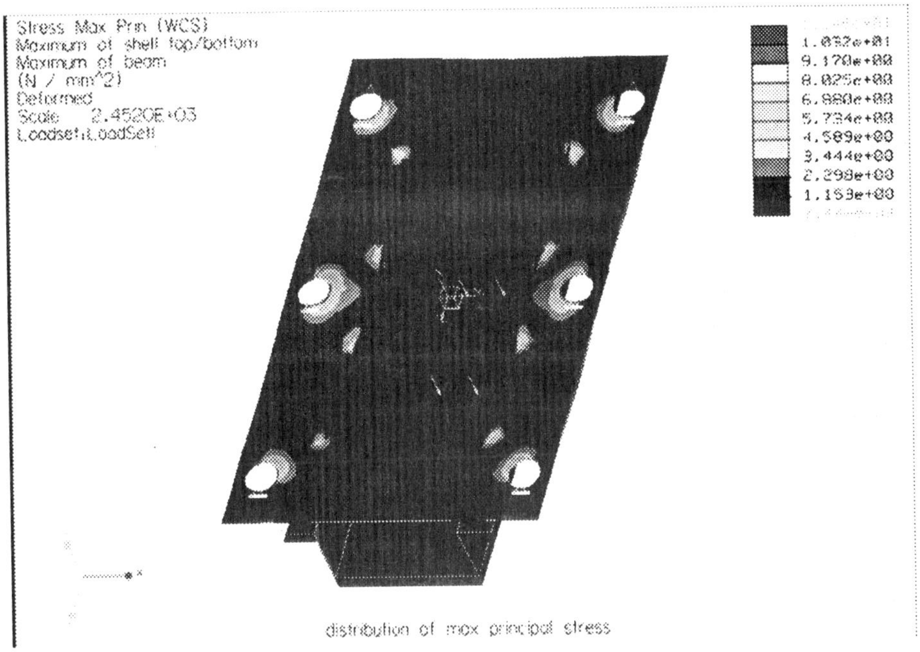

6.6 References

1. Algor Processor Reference Manual, Algor Interactive Systems, 150 Alpha Drive, Pittsburgh, PA 15238
2. O. Axelsson, Iterative Solution Methods, Cambridge University Press, New York, 1994.
3. K. J. Bathe, Finite Element Procedure, Prentice Hall, Upper Saddle River, New Jersey, 1996.
4. K. Bell, A Refined Triangular Plate Bending Finite Element, Int. J. Numer. Methods Eng., Vol. 1, No. 1 January 1969, pp. 101-122.
5. B. Carnahan, H. A. Luther, and J. O. Wikes, Applied Numerical Methods, Wiley New York, NY, 1969.
6. R. W. Clough, The Finite Element in Plane Stress Analysis, Proceedings of 2nd ASCE Conference on Electronic Computation, Pittsburgh, PA. September 1960.
7. R. D. Cook, D. S. Maikus and M. E. Plesha, Concepts and Applications of Finite Element Analysis, 3rd edition, Wiley, New York, NY, 1989.
8. J. W. Dally and W. F. Riley, Experimental Stress Analysis, McGraw Hill, 1965.
9. J. W. Dally, Design Analysis of Structural Elements, Goodington, New York, NY, 2003.
10. C. S. Desal and J. F. Abel, Introduction to the Finite Element Method, Van Nostrand Reinhold, New York, NY, 1972.
11. J. H. Earle, Graphics for Engineers, AutoCAD Release 13, Addision-Wesley, Reading, Massachusetts, 1996.
12. R. H. Gallagher, Finite Element Analysis Fundamentals, Prentice-Hall, Englewoood Cliffs, NJ, 1975.
13. J. M. Gere and S. P. Timoshenko, Mechanics of Materials, 3rd edition, PWS-KENT Publishing Company, Boston, 1990.
14. W. W. Hager, Applied Numerical Linear Algebra, Prentice-Hall, Englewood Cliffs, NJ, 1988.
15. Holland, The Finite Element Method in Plane Stress Analysis, in the Finite Element Method in Stress Analysis, I. Holland and K. Bell (eds), Tapir Press, Trondheim, Norway, 1969, Chapter 2.
16. Y. Y. Hsieh, Elementary Theory of Structure, 2nd edition, Prentice-Hall, Englewood Cliffs, NJ, 1982.
17. M. A. N. Hendriks, H Jongedijk, J. G. Rots and W. J. E. van Spange (eds), Finite Elements in Engineering and Science, DIANA Computational Mechanics, 1997.

18. K. H. Huebner, D. L. Dewhirst, D. E. Smith, and T. G. Byrom, The Finite Element Method for Engineers, 4th edition, John Wiley & Sons, Inc., 2001.
19. C.S. Krishnamoorthy, Finite Element Analysis, Theory and Programming, 2nd Ed., 1995.
20. D. L. Logan, A First Course in the Finite Element Method Using ALGOR, PWS Publishing, Boston, 1997.
21. H. C. Martin, Introduction to Matrix Methods of Structural Analysis, McGraw-Hill, New York, NY, 1966.
22. R. L. Norton, Machine Design: An Integrated Approach, Prentice Hall, Upper Saddle River, New Jersey, 1996.
23. M. Ortiz, Y. Leroy and A. Needleman, A Finite Element Method for Localized Failure Analysis, Comp. Meth. Applied Mech. Enging, 61, pp. 189-214, 1987.
24. M. Petyl, Introduction to Finite Element Vibration Analysis, Cambridge University Press, 1990.
25. T. H. H. Pian and P. Tong, Basis of Finite Element Methods for Solid Continua, Int. J. Numer. Methods Eng., Vol. 2, No. 7, April 1964, pp. 1333-1336.
26. P. G. Plockner, Symmetry in Structural Mechanics, J. of the Structural Division, American Society of Civil Engineers, Vol. 99, No. ST1, pp. 71-89, 1973.
27. J. N. Reddy, An Introduction to the Finite Element Method, McGraw-Hill Publishing Company, New York, NY, 1984.
28. T. Ross, T. Johns and Emile, Dynamic Buckling of Thin Walled Domes under External Water Pressure, Applied Solid Mechanics 2nd Conference, Ed. Tooth & Spence, Elsevier, pp. 211-224, 1987.
29. T. Ross, Finite Element Program in Structural Engineering and Continuum Mechanics, Horwood Publishing Limited, Chichester, England, 1996.
30. T. Ross, Advanced Applied Finite Element Methods, Horwood Publishing Limited, Chichester, England, 1998.
31. L. Stasa, Applied Finite Element Analysis for Engineers, Holt, Rinehart and Winston, New York, NY, 1985, pp. 101-157.
32. Y. Saad, Iterative Methods for Sparse Linear Systems, PWS Publishing Company, Boston, 1996.
33. Y. Tada and G. C. Lee, Finite Element Solution to an Elastic Problem of Beams, Int. J. Numer. Methods Eng., Vol. 2, No. 2, April 1970, pp. 229-241.
34. M. J. Turner, R. W. Clough, H. C. Martin, and L. J. Topp, Stiffness and Deflection Analysis of Complex Structures, J. of the Aeronautical Sciences, Vol. 23, No. 9, pp. 805-824, Sept. 1956.
35. O. C. Zienkiewicz, The Finite Element Method, third edition, McGraw-Hill, New York, NY, 1977.
36. O. C. Zienkiewicz and R. L. Taylor, The Finite Element Method, 4th edition, Vol. 1, McGraw-Hill (UK),London, 1989.

6.7 Exercises

1. A plastic plate is used to support three plants, which are on display. The three dimensions of the plate and the three surface regions where the three plants are placed are shown below.

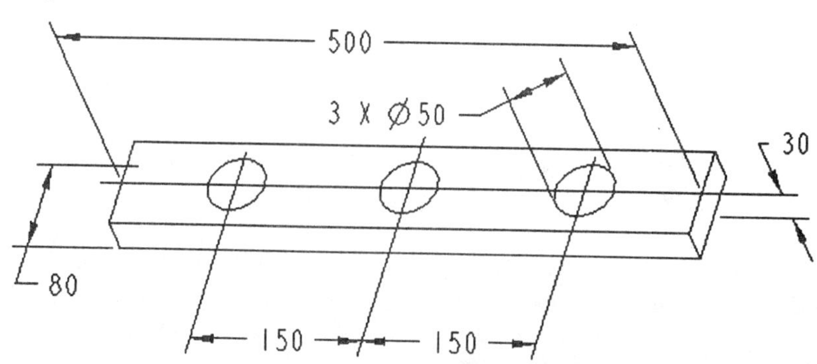

Assume that the weight of each plant is 200 N. The plate material is Nylon. Use a shell modeling approach to evaluate the deflection distribution and the distribution of the von Mises stress.

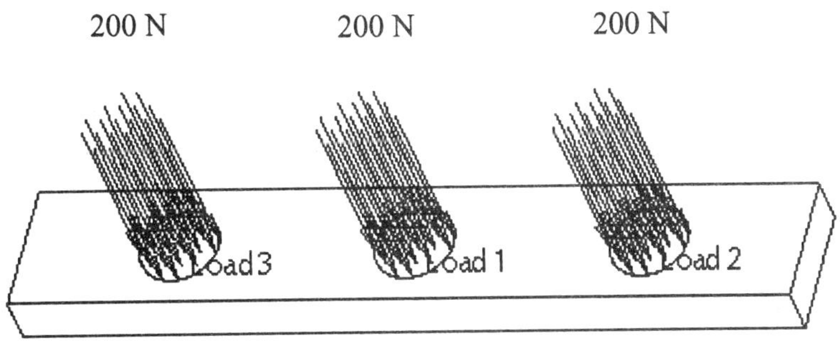

Case 1: the plate is fixed to the two supports. The dimensions of the two supports are shown in the figure.

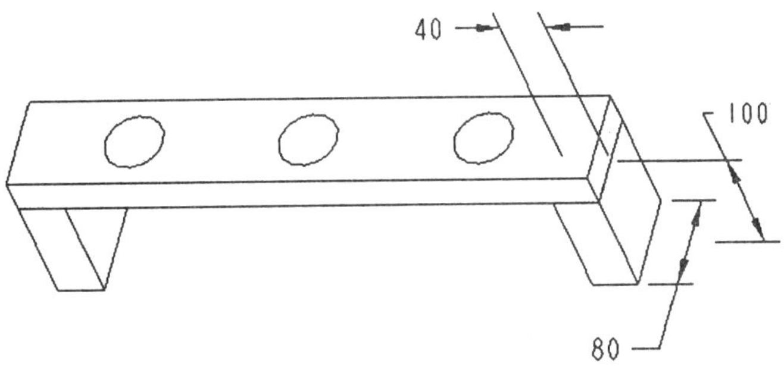

Case 2: the plate is fixed to the 2 supports through the use of 4 bolts and 4 washers.

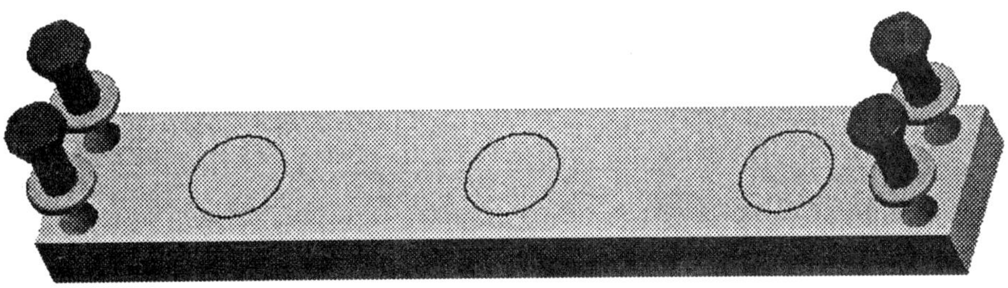

2. Use a shell modeling approach. A welding structure is shown below. It consists of 2 components, namely, a base plate and a top plate, as shown below:

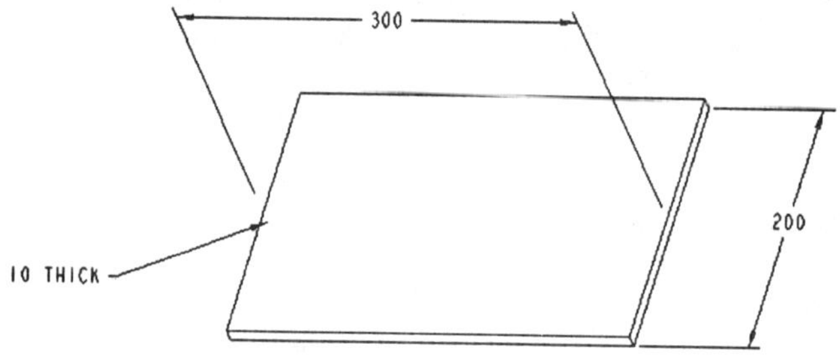

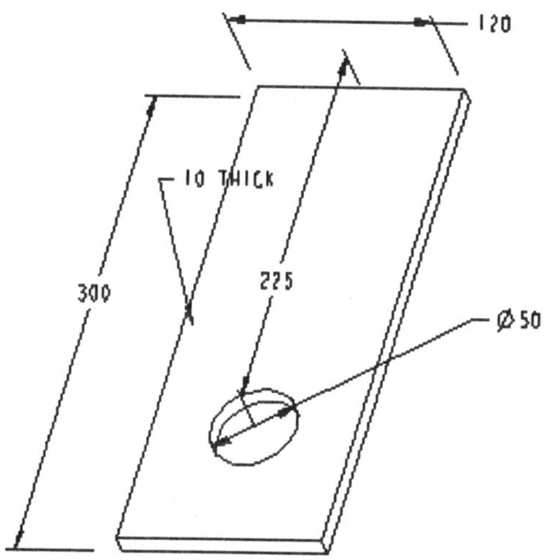

In your previous homework, you dealt with a spot welding structure. This structure is not a spot welding structure. The welding part is a rectangular shaped string. As a result, you need to assemble it with the base plate and the top plate. The welding part is shown below:

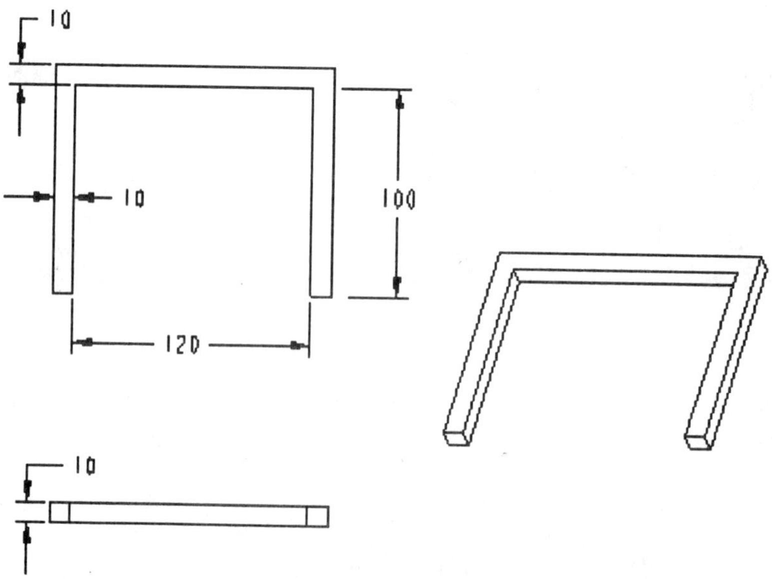

The following figure shows the entire welding structure after the assembly.

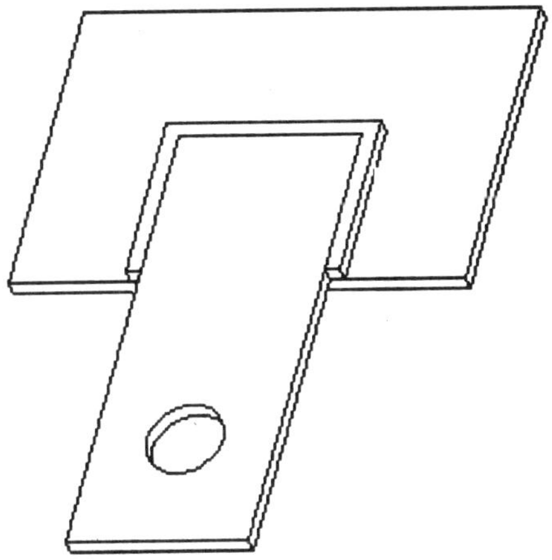

(1) You are asked to create a 3D solid model for the base plate component.
(2) You are asked to create a 3D solid model for the top plate component.
(3) You are asked to create a 3D solid model for the welding part.
(4) You are asked to create an assembly, which consists of the base plate, the top plate and the welding part.
(5) You are asked to create a shell model for each of the three components.
(6) You are asked to verify your shell model at the assembly level.
(7) You are asked to use Edit > Pair Place to adjust the locations of the compressed planes so that the entire assembly is in the following setting.

Top View

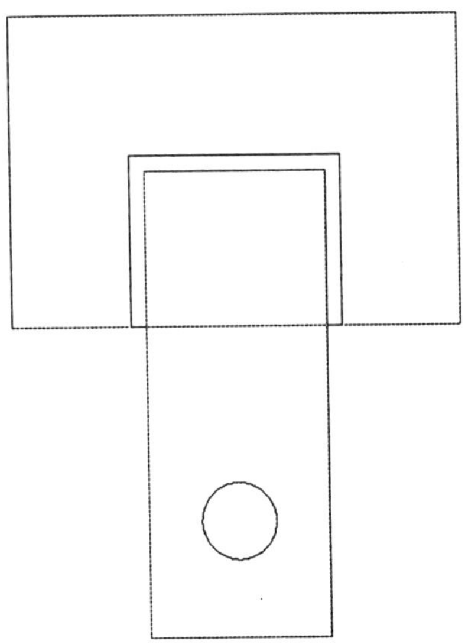

Right Side View

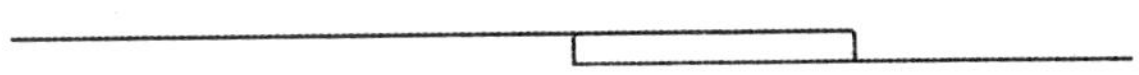

(8) Assume that the material for the base plate and the top plate components is STEEL, as listed in the Pro/E material library. Assume that the material for the weld part is TUNGSTEN, as listed in the Pro/E material library.

(9) Perform two (2) design studies, assuming that the bottom surface of the base plate is fixed to the ground:

Design study 1: Bearing load with magnitude equal to 1000 N.
Design study 2: Uniform load with magnitude equal to 1000 N.

In each design study, prepare 2 plots of the displacement distribution and the max principal stress distribution. Identify the locations of the max displacements and the max principal stress of the entire assembly structure.

CHAPTER 7

APPLICATIONS ON MODAL AND BUCKLING ANALYSES

7.1 Introduction

Finite element methods have been successfully used in structural analyses in engineering. They have also been widely used in other fields in engineering. In the chapter, we present 2 applications. The first application is in the field of structural vibration. The second application is in the field of structural stability. Two examples will be used in each application to demonstrate the procedure of application.

7.2 Applications on Modal Analyses

As we discussed in Chapter 1, finite element methods treat a continuous system as an assembly of elements through discretization. The global stiffness matrix obtained through an assembly of individual stiffness matrices characterizes the system response when subjected to a load condition or load conditions. The global stiffness matrix has complex eigenvalues. Transformation of this global stiffness matrix into a non-diagonal matrix which is called the modal form is viewed as modal analysis, which focuses on the mode shapes of the vibrating system.

The following figure illustrates a driving board used in sport competition. As illustrated, a person is jumping from the board and diving to the swimming pool.

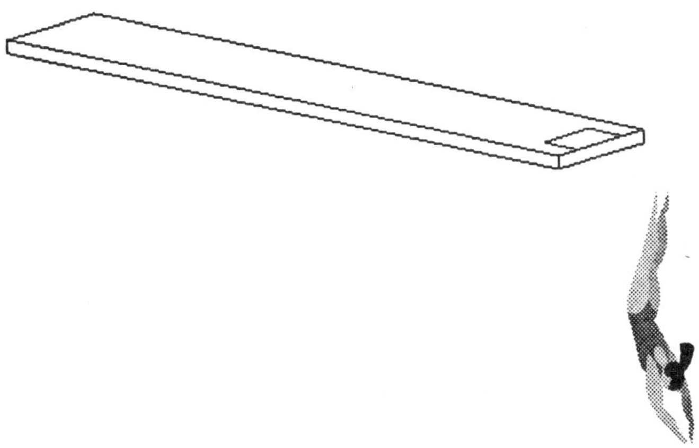

Assume the load condition is 600 Newton and action area is a rectangle. The two dimensions of the rectangular area are 400 x 250 mm. The geometrical shape of the diving board is a rectangular plate. The three dimensions are 4000 x 750 x 75 mm. We assume that the back end of the diving board is a fixed end.

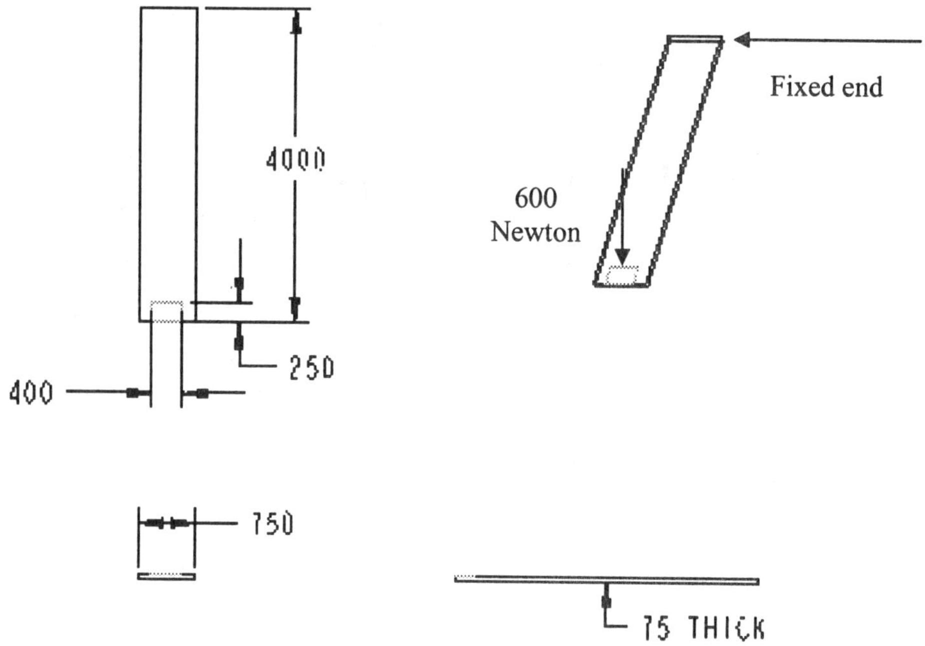

Step 1: Under the Pro/ENGINEER design environment, create a 3D solid model of the diving platform.

From the main toolbar, select **File** > **New** > **Part** > type diving_*platform* as the name of the file > clear the icon of **Use default template** > **OK**.

Select **mmns_part_solid** (unit system: millimeter, Newton and Second) and type *diving platform* in **DESCRIPTION** and *student* in **MODELED_BY** > **OK**.

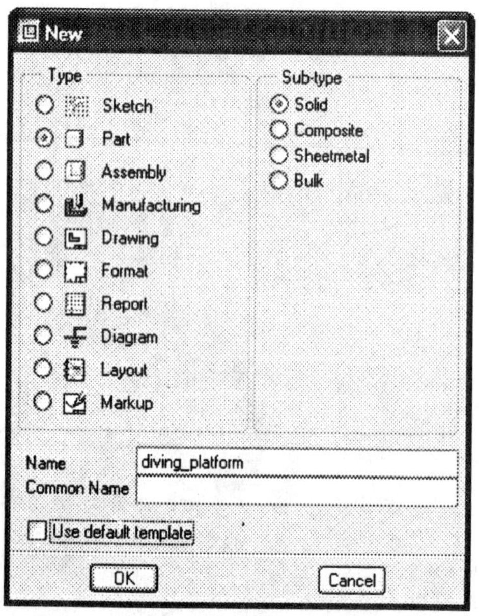

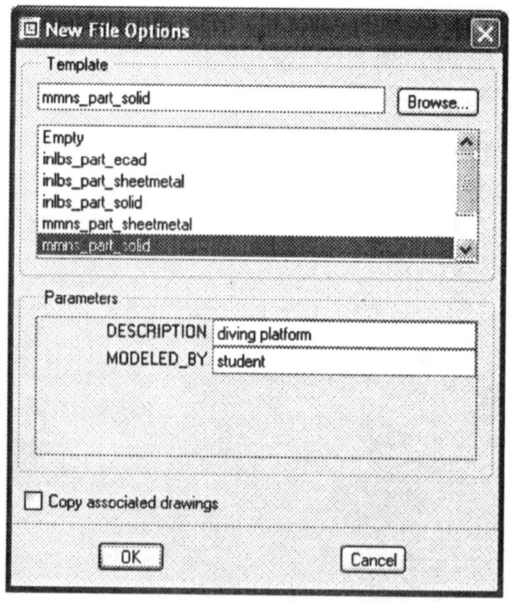

Let us use **Thin Protrusion** to create the 3D solid model so that **Auto Detect** can be used to generate a shell model later.

From the toolbar of feature creation, select the icon of **Extrude.**

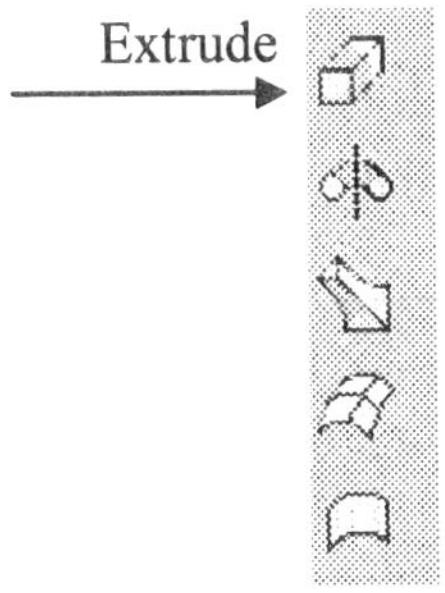

The dashboard appears. Select the icon of thicken sketch, set the thickness value to *75* and set the depth value to *4000*. Click the box of **Placement** > **Define**.

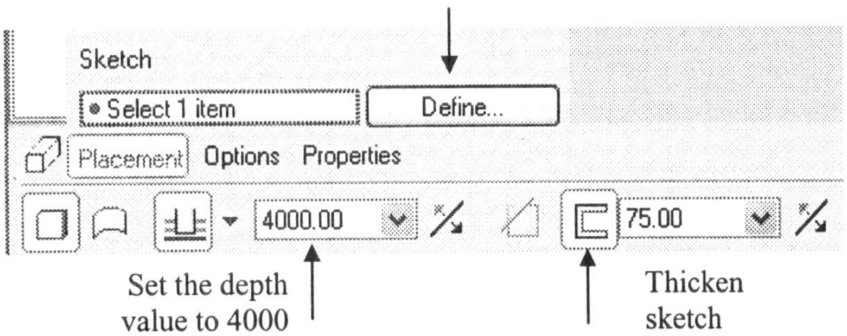

Select the **FRONT** datum plane as the sketch plane, and click the box of **Sketch** to accept the **RIGHT** datum plane as the default reference to orient the sketch plane, as illustrated below.

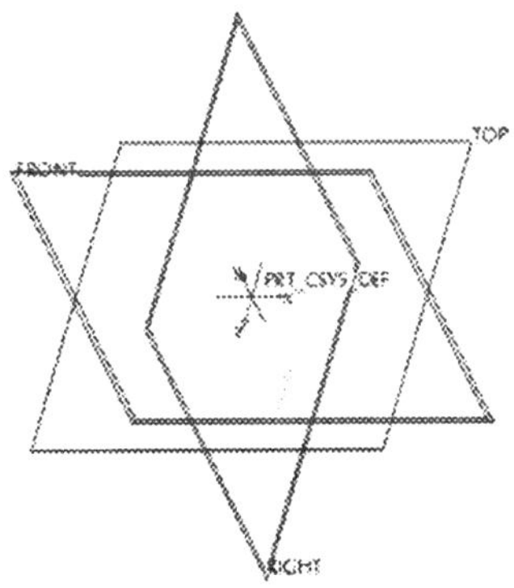

Select the icon of **Centerline** from the toolbar of sketcher to draw a centerline along the y-axis. Select the icon of Line to sketch a line along the x-axis. The line is symmetric about the y-axis. The length is 750.

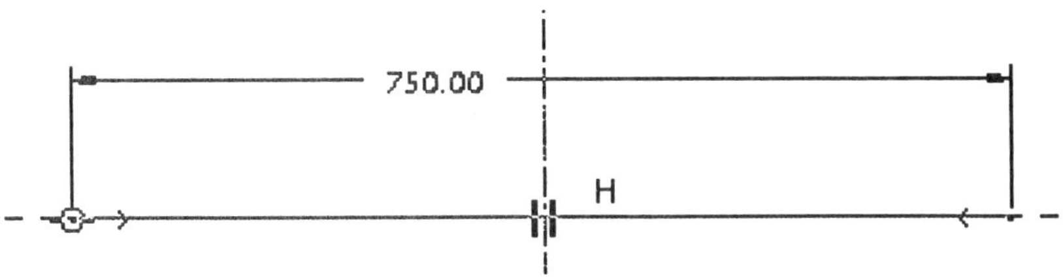

Upon completion, pick the icon of **Done** from the toolbar of sketcher. From the dashboard, select the icon of **Complete** from the feature control panel, thus completing the creation of the first feature.

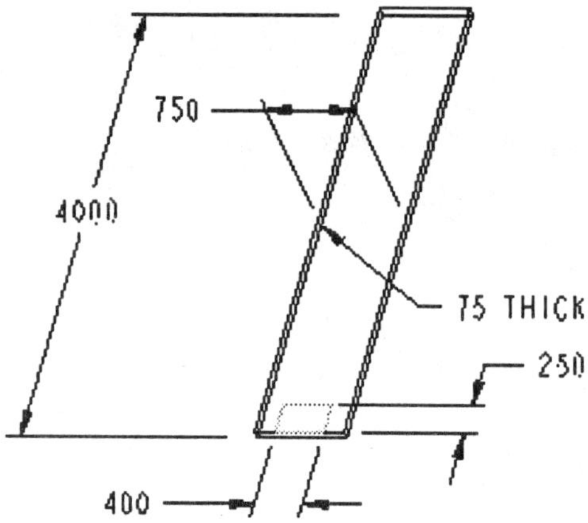

Note that a datum curve is needed for defining a surface region where a load can be applied. To sketch a datum curve, select the icon of sketch tool from the toolbar of datum feature creation > pick the top surface of the platform as the sketch plane and accept the default setting to orient the sketch plane.

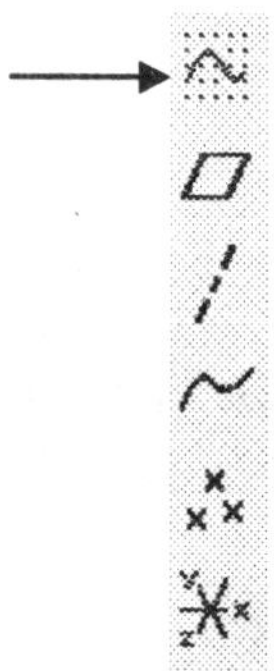

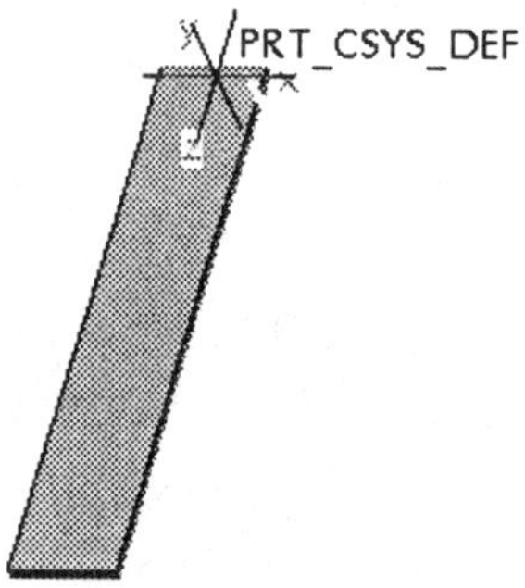

Select the RIGHT datum plane and the front surface of the platform as the 2 references, sketch a vertical centerline along the z-axis. Sketch a rectangle. The 2 dimensions are 400x250 respectively.

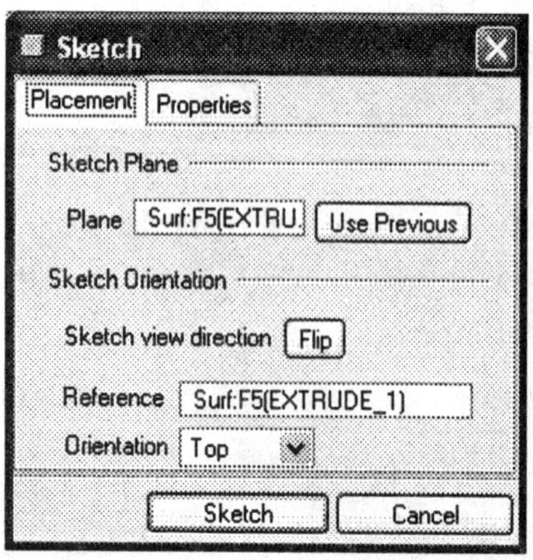

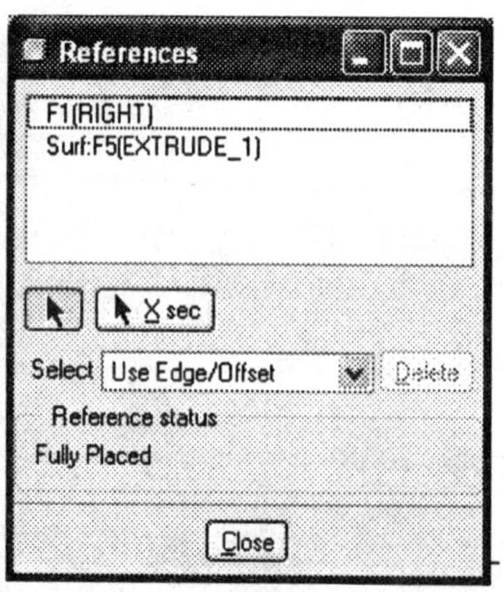

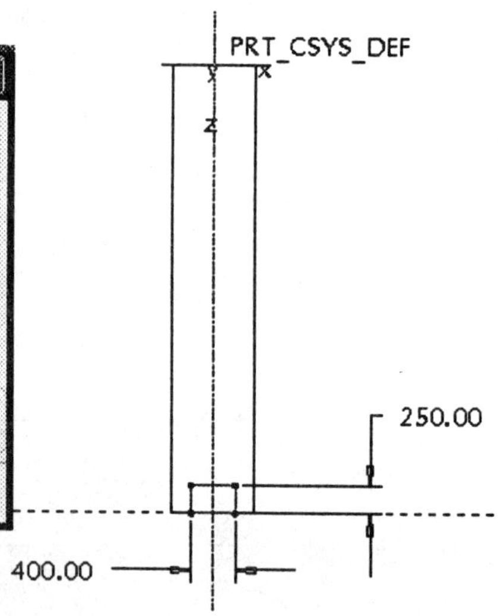

Step 2: Under the Pro/MECHANICA environment, specify the material type. Select **Nylon.**
Applications > **Mechanica** > **Continue** > **Structure** > **OK.**

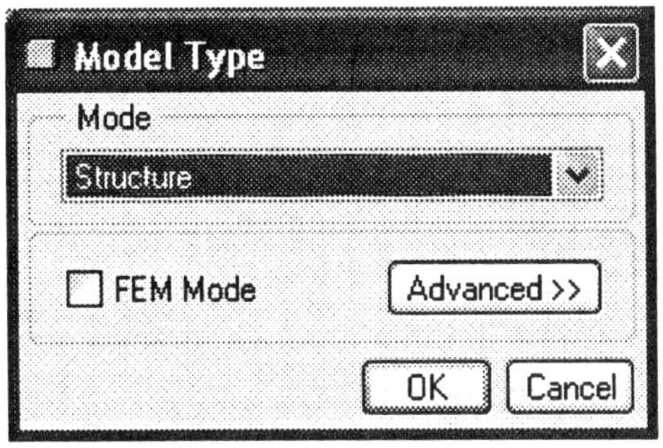

From the toolbar of functions, select the icon of **Materials** > select **NYLON.**

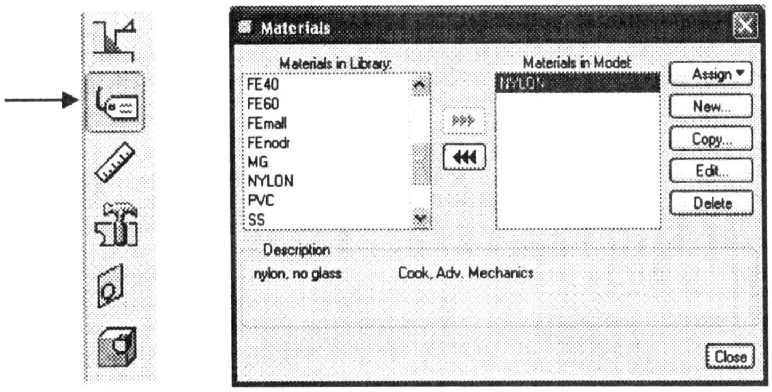

Assign > **Part** > select the platform structure from the screen display > **OK** > **Close**.

Step 3: Under the Pro/MECHANICA environment, define the surface region.

To define the surface region, select the icon of **surface region** > **Select > Done**. Pick the datum curve first, as shown > pick the top surface to be split afterwards > **OK > Done > OK.**

Icon of Surface Region

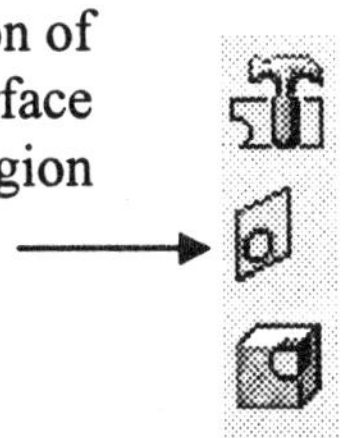

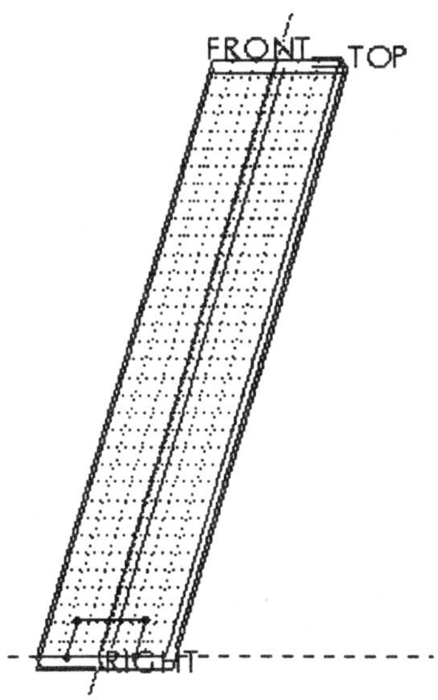

Step 4: Under the Pro/MECHANICA, define a shell model.

From the toolbar of functions, select the icon of **New Shell Pair** > **Auto Detect** > **Compress** > **Shells Only**.

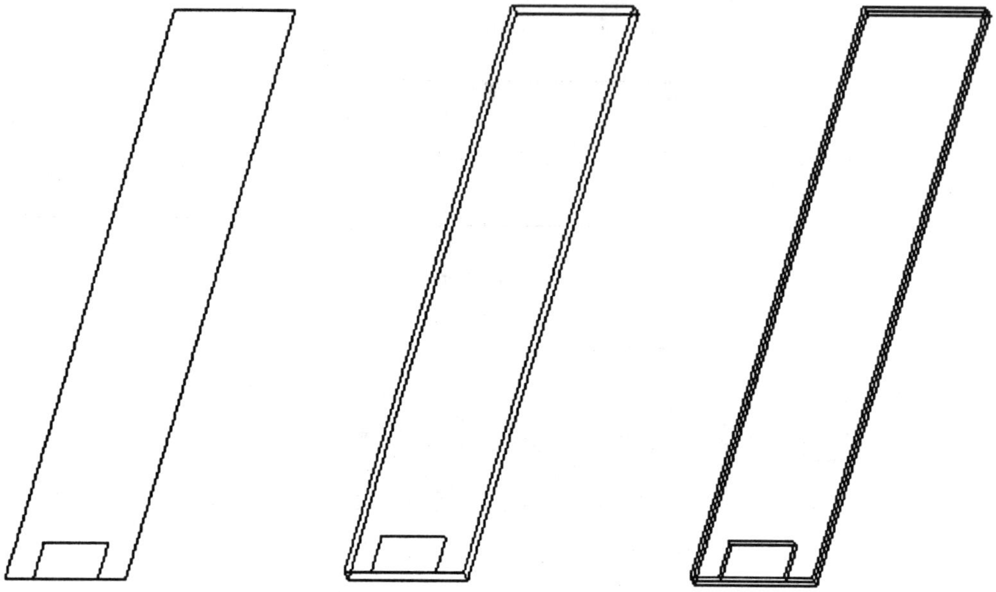

Step 5: Under the Pro/MECHANICA environment, define the constraint condition: Fixed end.

Note that we need to use Edge/Curve, and cannot use Surface.

From the toolbar of functions, select the icon of **Constraints** > select **Edge(s)/Curve(s)** > pick the edge on the back side > **OK**.

Fix all 6 degrees of freedom > **OK**.

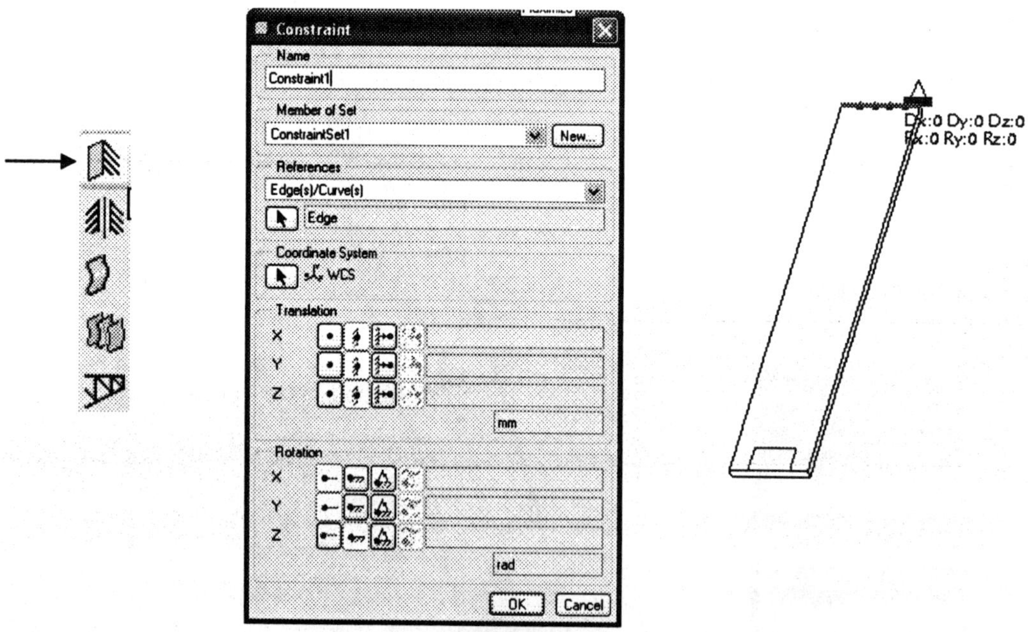

Step 6: **Under the Pro/MECHANICA environment, define the load condition.**

To define a uniform load acting on the surface region in the negative y direction, select the icon of **force and moment** from the toolbar > **Surface** > pick the surface region > **OK** > click **Advanced** > change the distribution to **Total Load** and **Uniform** > type in *-600* in Y of the Force box > **OK**.

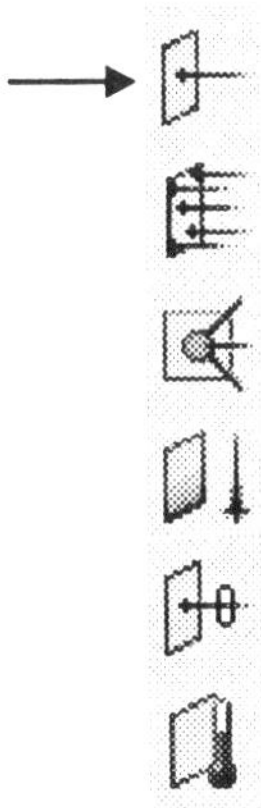

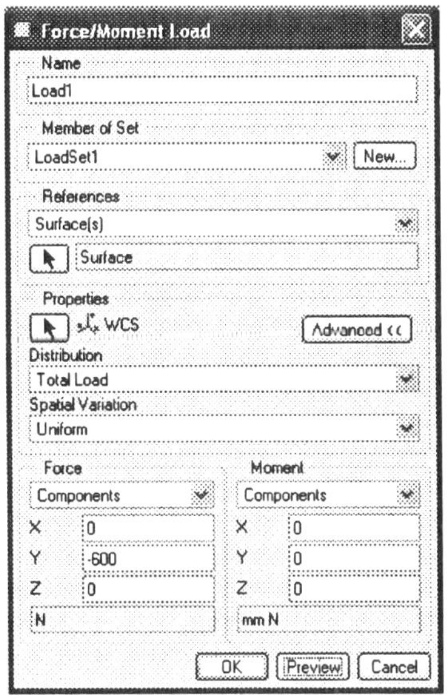

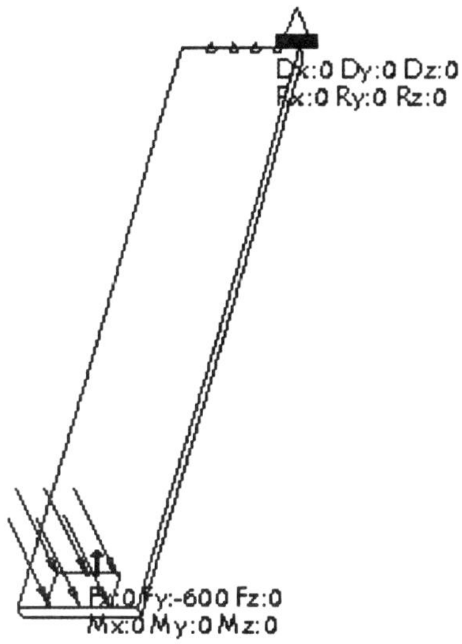

Step 7: Perform a modal analysis and run it.
Click the icon of performing a design study from the main toolbar > File > Select New Modal.

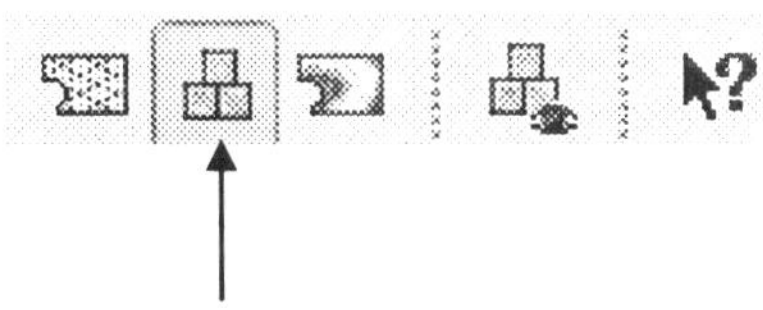

Icon of performing a design study

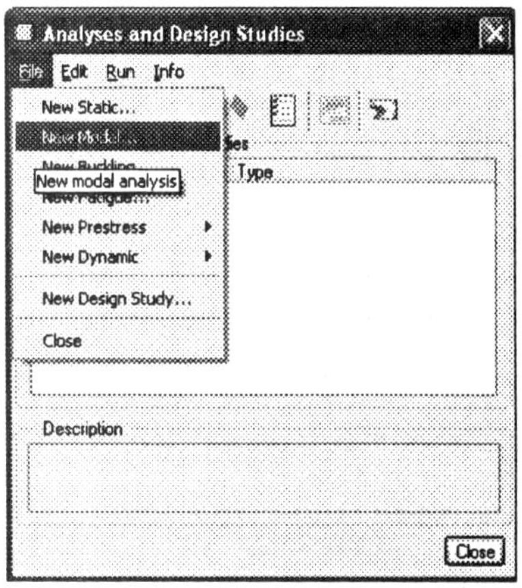

Type the file name: *modal_beam* and run the program.

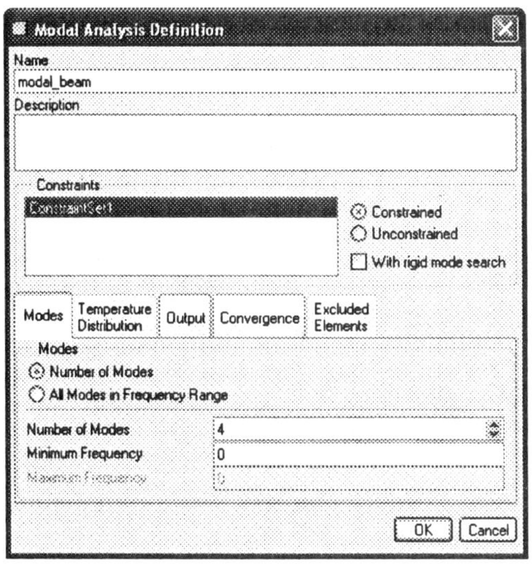

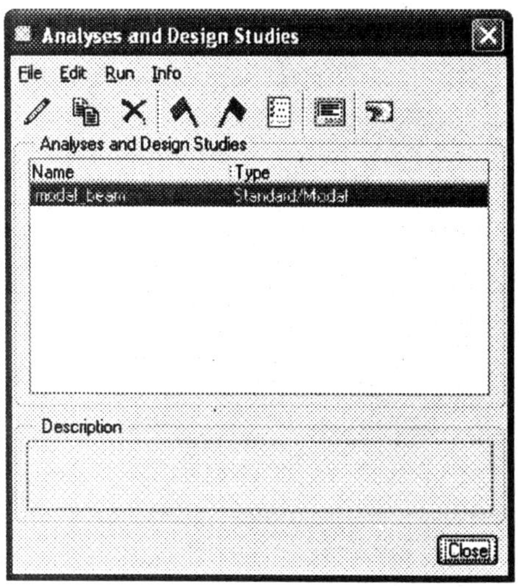

Note that the load condition defined is not shown. Let us run the program.

Step 8: Display the results by showing the shapes of the first 4 modes.

```
Number of Modes: 4

   Mode   Frequency (Hz)
   ----   --------------
      1   1.412859e+00
      2   8.819251e+00
      3   1.352389e+01
      4   1.386584e+01
```

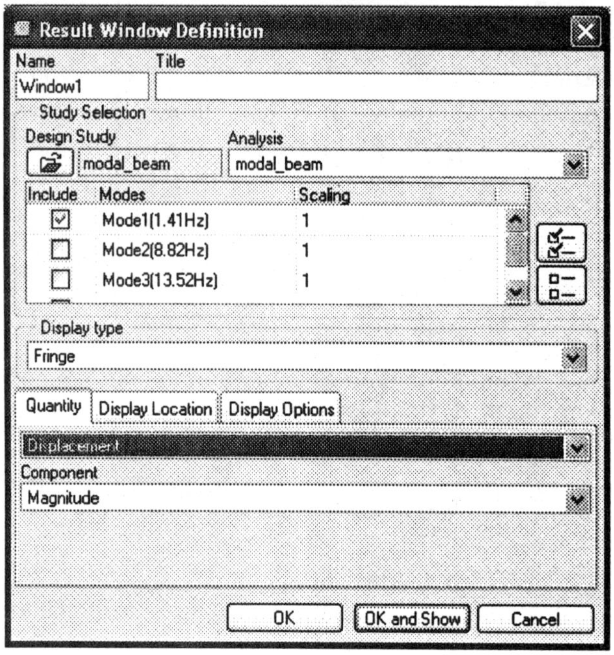

Frequency of Mode 1:	1.41 Hz
Frequency of Mode 2:	8.82 Hz
Frequency of Mode 3:	13.5 Hz
Frequency of Mode 4:	13.9 Hz

The 4 mode shapes are shown below:

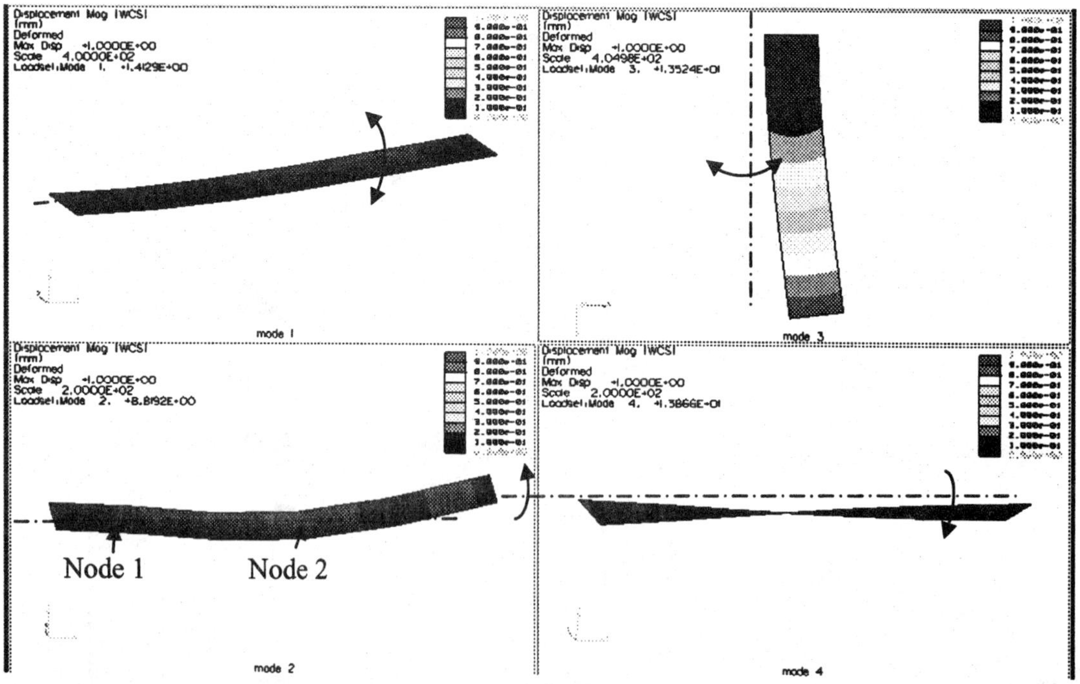

If readers are asked a question: If the load condition is changed from the current 600 Newton to 750 Newton while maintaining its direction downwards, will the first 4 mode shapes change accordingly? To verify your judgment, run the program again with the load condition equal to 750 Newton, instead of

600 Newton. Readers should recognize that the magnitude of the load has no effect on the 4 mode shapes. From the analytical viewpoint, mode shapes are independent of the input format.

If the material type is being changed, will the 4 mode shapes change accordingly? How about the numerical values of the frequencies of the 4 modes? We suggest that readers also try a new run with a material type different from Nylon as we selected. Readers should observe that the 4 mode shapes will change in terms of the frequencies of the 4 new modes.

7.3 Applications on Designing a Music Tuning Fork

The following figure illustrates a tuning fork used to tune a piano. A tuning fork is a simple two-pronged metal instrument with a handle and tines that form a U shape. Tuning forks, usually made of steel, aluminum, or magnesium-alloy, will vibrate at a set frequency to produce a musical tone when struck and therefore is used for many applications such as piano tuning as well as assessing a persons ability to hear. A tuning fork acts as an example of a simple mechanical system with normal modes of resonance. A tuning fork can also be modeled as two particles of equal mass *m* moving in the *x-y* plane under the influence of restoring forces. Tuning forks work by holding the handle and striking the tines with a force great enough for the tines to vibrate at a frequency, resulting in a sound.

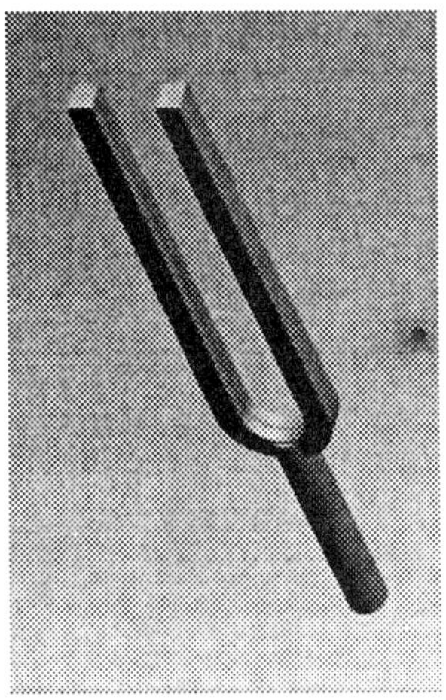

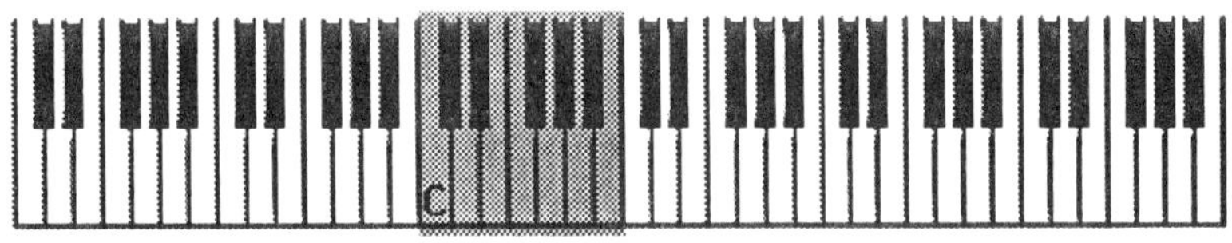

The geometry of a music fork is shown below. Assume that the material is steel. Determine the frequencies of the first 4 modes and the mode shapes of the first 4 modes.

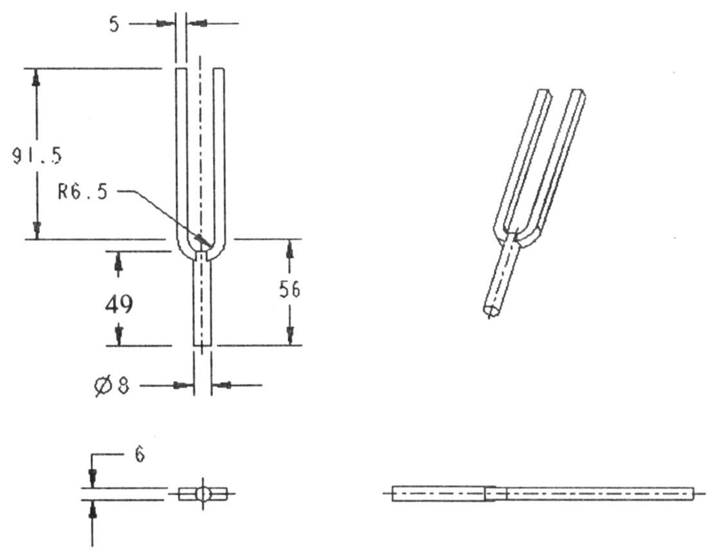

Step 1: Under Pro/ENGINEER, create a 3D solid model of the tuning fork.

Use **Sweep** to create the U-shape first (create the U-shape datum curve before sweeping).

From the main toolbar, select **File** > **New** > **Part** > type *fork_g* as the name of the file > clear the icon of **Use default template** > **OK**. Select **mmns_part_solid** (unit system: millimeter, Newton and Second) and type *fork G tune* in **DESCRIPTION** and student in **MODELED_BY** > **OK**.

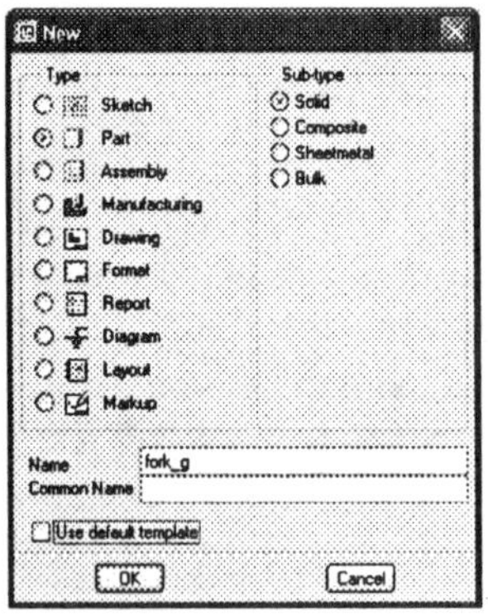

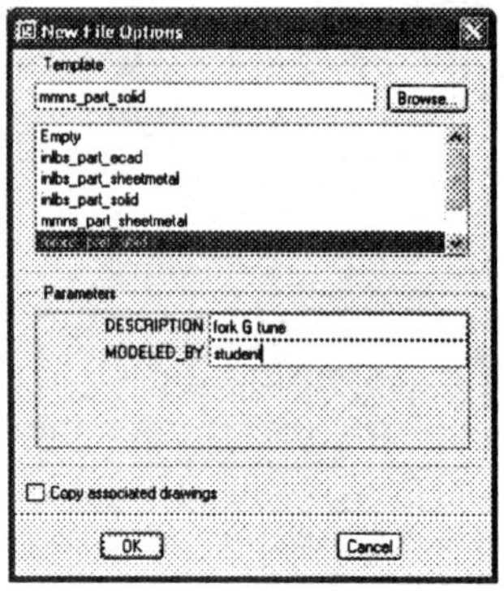

From the main toolbar, select **Insert** > **Sweep** > **Protrusion** > **Sketch Traj**.

Setup **New** > **Plane** > Pick the **FRONT** datum plane as the sketch plane > **Okay** and **Default** for orientation setting.

Sketch a circle and the diameter is 13, first.

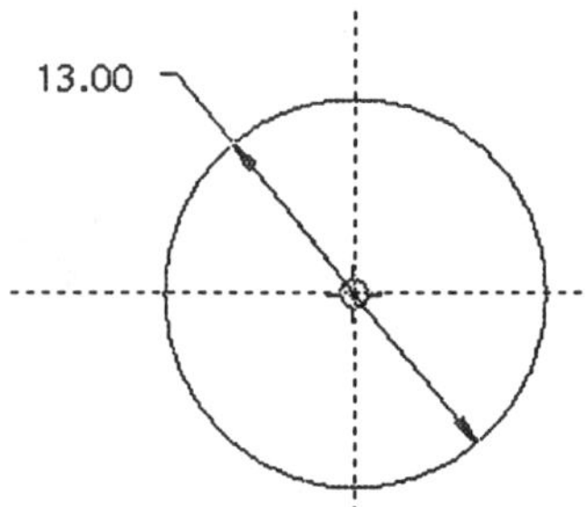

Sketch 2 lines tangential to the sketched circle. The dimension of length is 91.5, as shown. Afterwards, delete the extra 2 segments, as shown > select the icon of Done from the toolbar of sketcher.

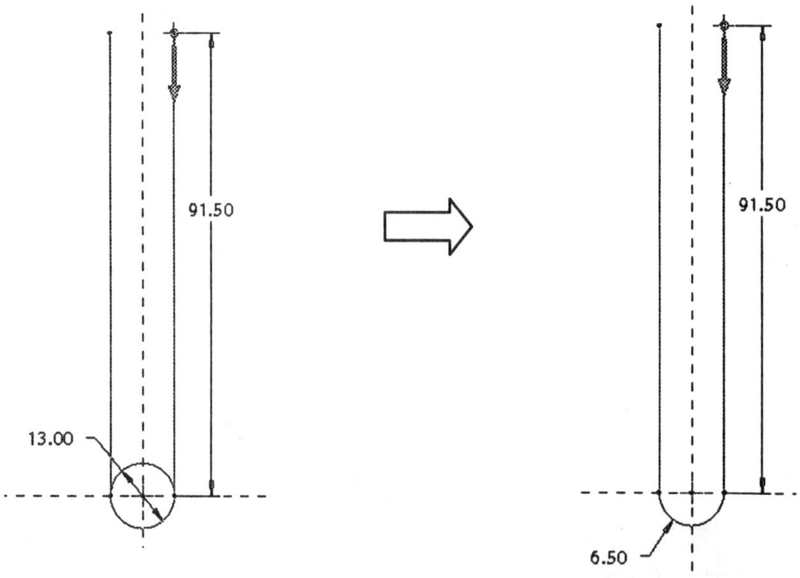

Sketch a rectangle, and the 2 dimensions are 5 and 6, respectively > select the icon of **Done** > **OK**.

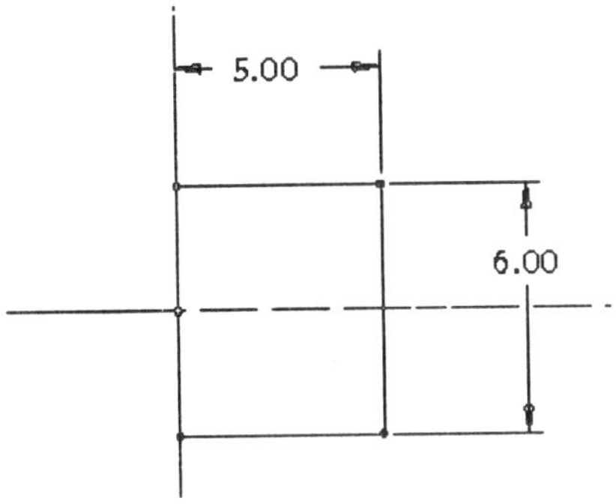

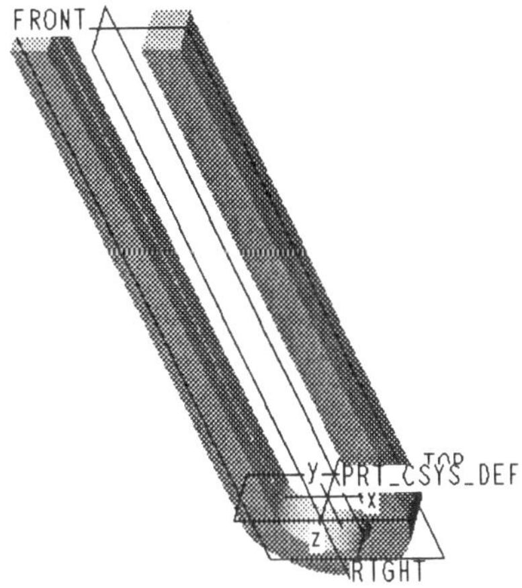

Use **Extrude** to add a cylinder of $\phi 8$ to the U-shape. The extrusion starts at a plane which is 56 away from the TOP datum plane. The extruding distance is set to 49.

To add this feature, the first step is to create a datum plane. Select the icon of creating datum plane > pick the **TOP** datum plane > type *56* as the distance of translation (you may need to type -56) > **OK**.

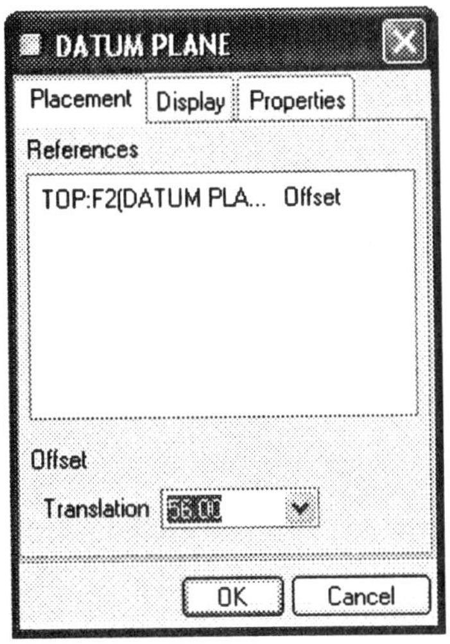

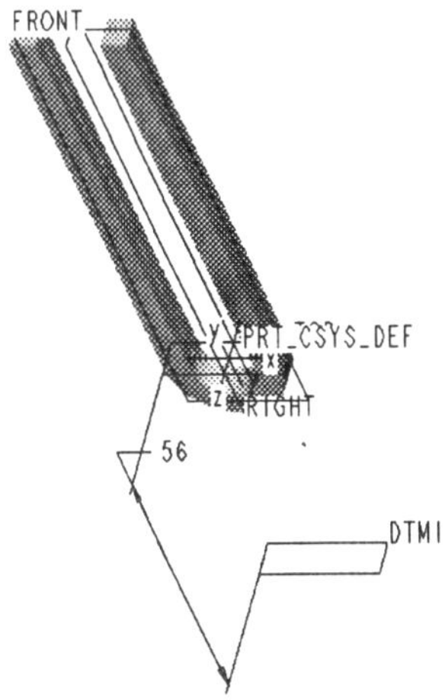

The second step is to use Extrude to create the cylindrical feature. Select the icon of **Extrude** from the toolbar of feature creation > Set the depth value to 49 > **Placement** > **Define**.

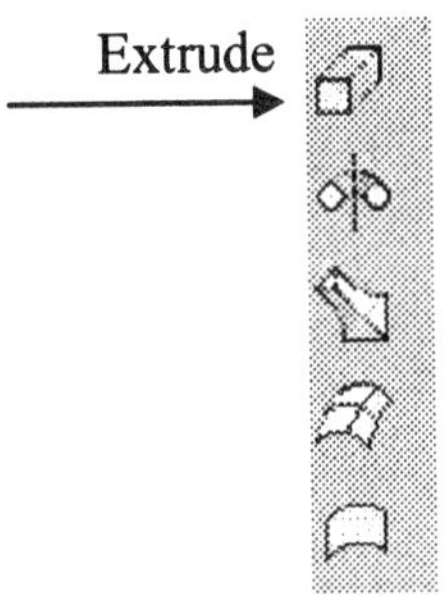

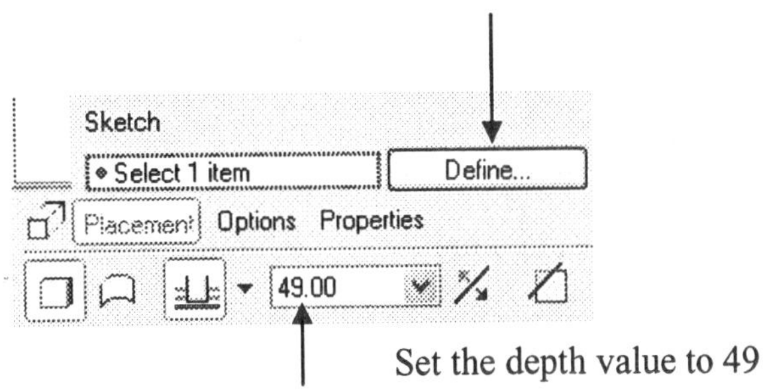

Pick **DTM1** as the sketch plane and accept the default setting to orient the sketch plane. Sketch a circle and the dimension of the diameter is 8 > select the icon of **Done** from the toolbar of sketcher > select the icon of **Complete** from the feature control panel.

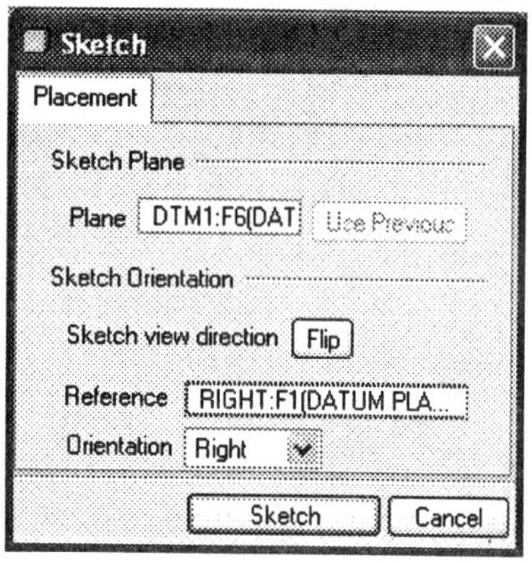

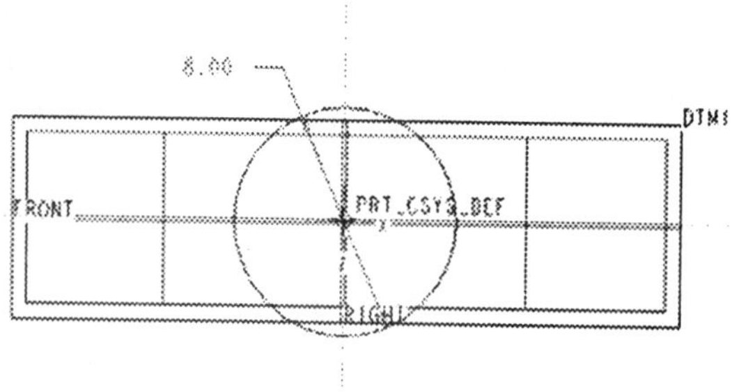

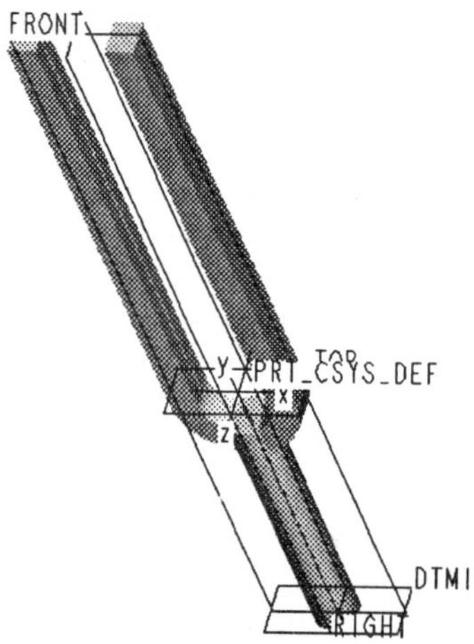

Step 2: Under the Pro/MECHANICA environment, specify the material type. Select STEEL. **Applications** > **Mechanica** > **Continue** > **Structure** > **OK.**

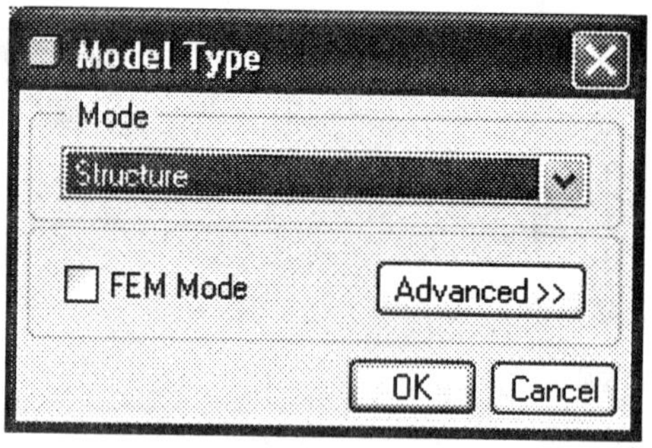

From the toolbar of functions, select the icon of **Materials** > select **STEEL.**

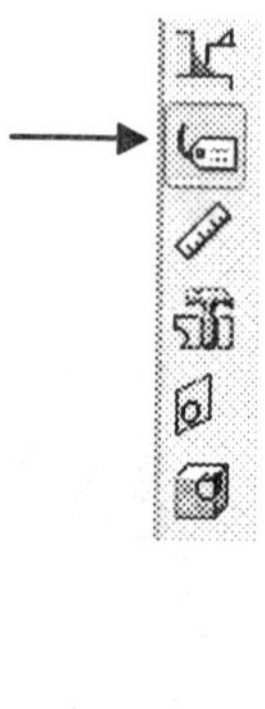

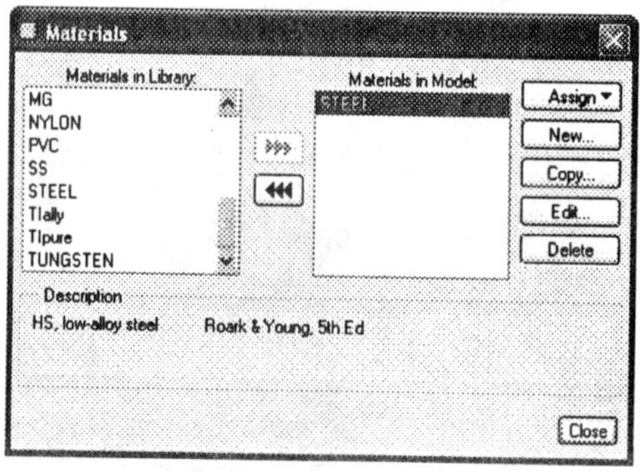

Assign > **Part** > select the platform structure from the screen display > **OK** > **Close**.

Step 3: Under the Pro/MECHANICA environment, define the constraint condition. The condition is to fully constrain the handle surface of the fork (the cylindrical protrusion).

Model > Constraints > New > Surface > **select the cylindrical surface, and fix the six degrees of freedom** > OK.

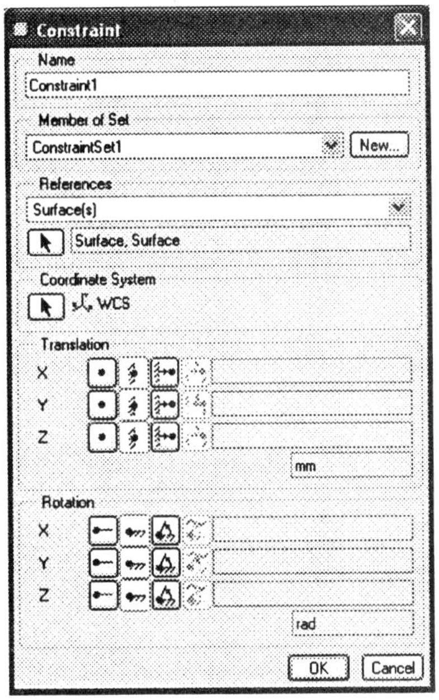

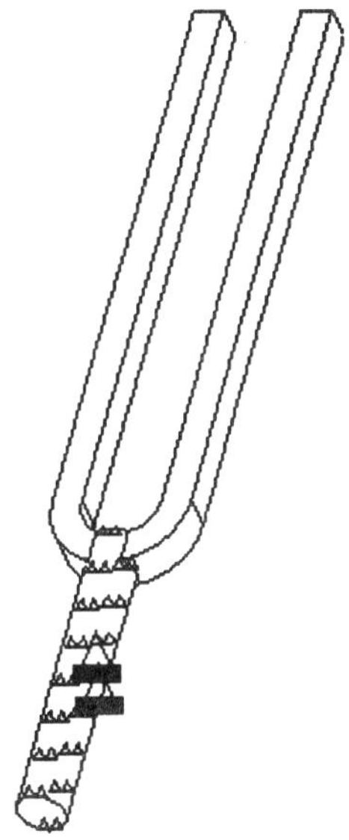

Step 4: Create an analysis and run it.

Note that there is no load defined. It is important to remember that the shapes of individual modes are independent of the direction and magnitude of the exciting force.

Click the icon of performing a design study from the main toolbar > **File** > Select **New Modal**.

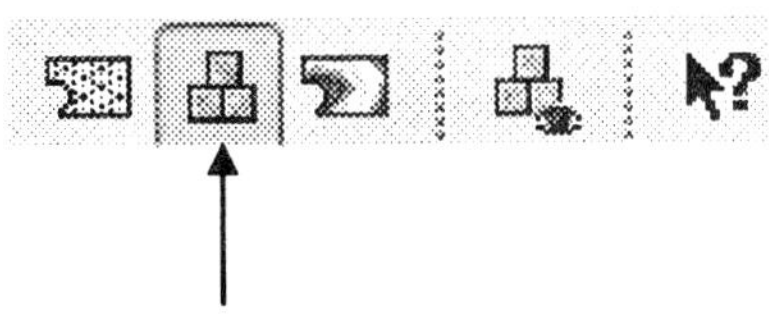

Icon of performing a design study

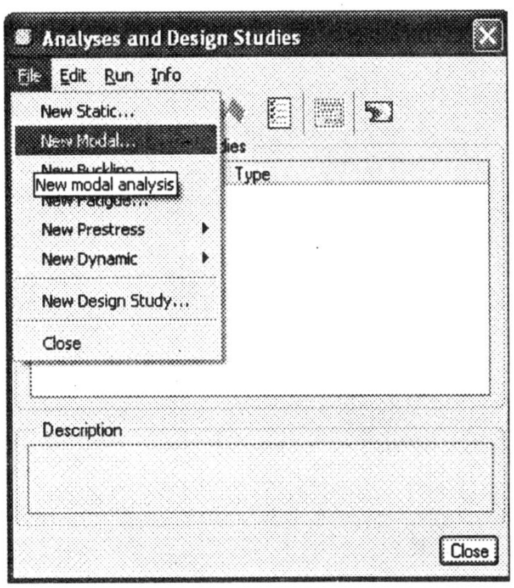

Type the file name: *modal_fork* and run the program.

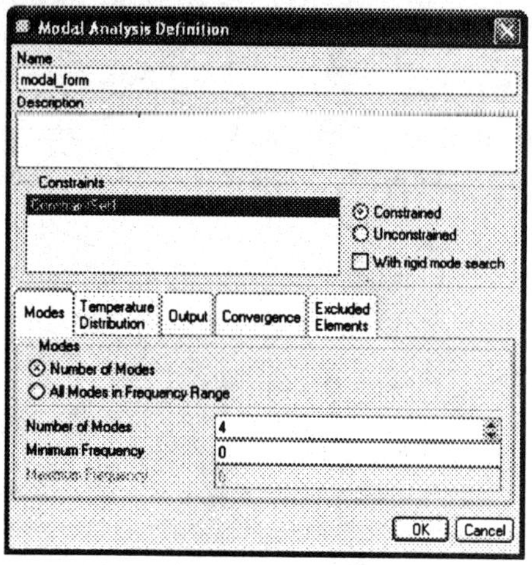

Number of Modes: 4

Mode	Frequency (Hz)
1	3.810660e+02
2	3.826385e+02
3	4.405395e+02
4	4.418484e+02

Step 5: Display the results by showing the shape of the first mode through animation.

Note that the frequency of the first mode is 381Hz, which is very close to 384 Hz, the frequency of G-sharp. The student may search this information on the web. Two (2) of the web addresses are

http://www.indigo.com/tuning/tuning.html.

http://mathforum.org/library/drmath/view/52470.html.

Let us examine the first mode shape for the G note fork, for a frequency of 381 Hz is illustrated below.

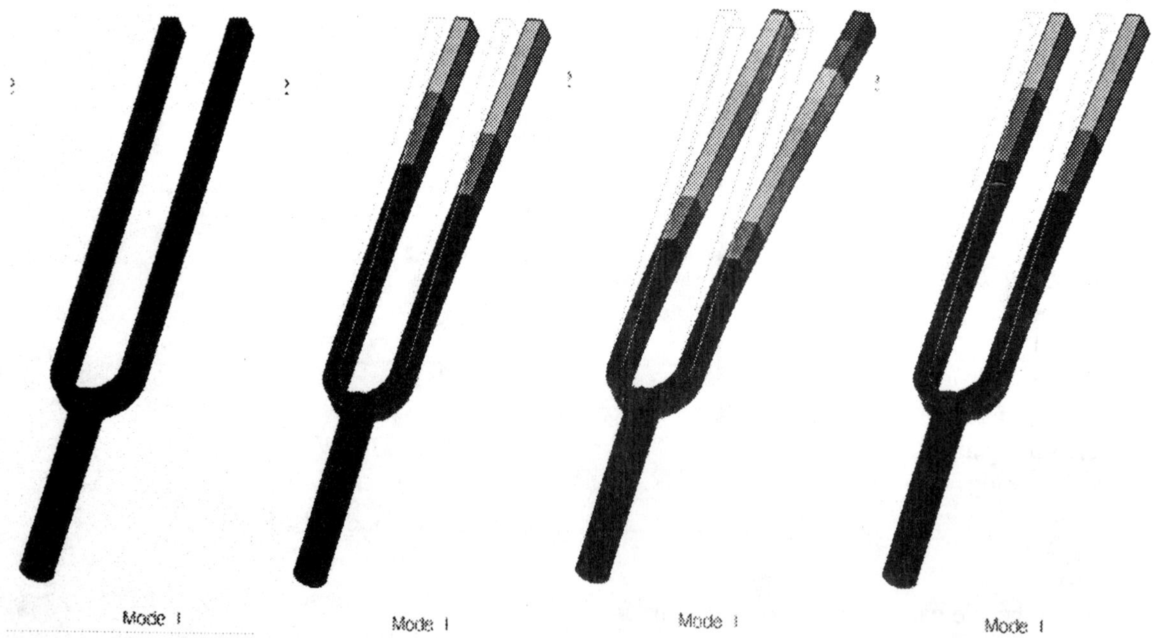

The first mode is the dominant mode of fork vibration. The frequency corresponding to this mode is the frequency of the musical note denoted G. Tuning a piano is based on this note, and thus this frequency. The way the tuning fork is excited directly corresponds to the harmonics heard. Impacting one of the tines with a hard surface excites the tuning fork. The direction of impact determines which harmonics, or modes, will be present. The excitation force and mode shape for mode 1 is shown below:

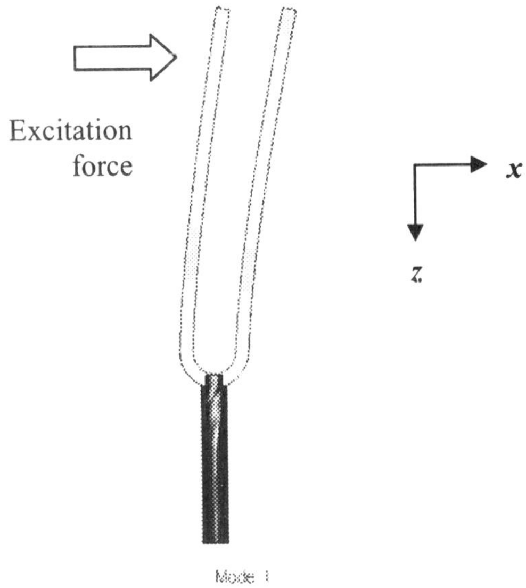

This figure shows that the excitation force must be along the same direction as the mode motion. For mode 1, the excitation force should therefore be along the x-direction.

For mode 2, the tines move in the x-z plane in directions either towards, or away from each other. Two opposite forces oriented along the x-direction excite mode 2. This phenomenon is illustrated below. So, by striking only one tine of the tuning fork along the x-direction, the first mode will be excited without presence of the second mode.

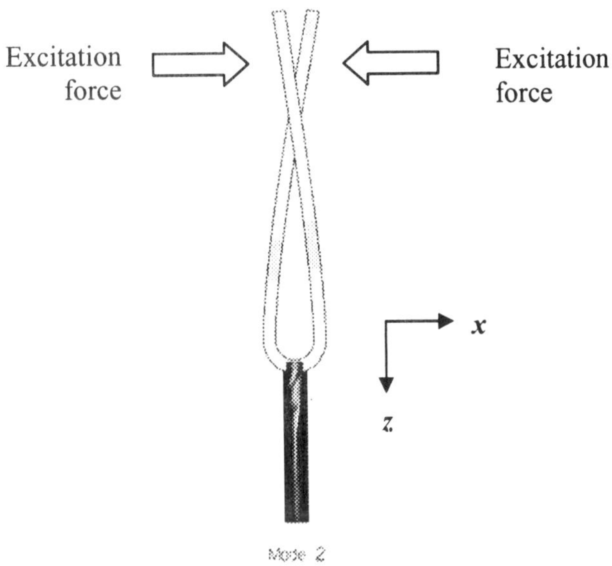

Mode 3 consists of motion in the y-z plane along the y-direction. To excite this mode, an excitation force must strike one of the tines along the y-direction, as shown. Again, by striking a tine in the x-direction, only mode 1 will be excited. So in summary, by striking one of the tuning fork tines along the x-direction, mode 1 will be the dominant mode of oscillation, which causes the mode 1 frequency to be the dominant tone. While the higher harmonics of modes 2 and 3 will still be present, by striking the fork in the above manner, their roles in the tone will be reduced.

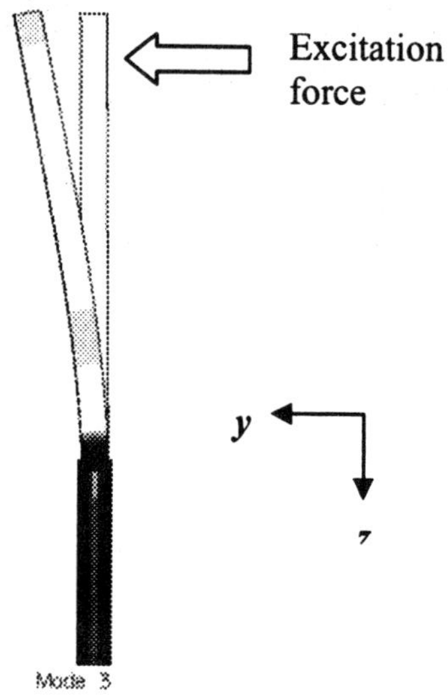

7.4 Applications on Buckling Analyses

Structures fail in a variety of ways, depending on their geometrical shapes, types of material, kinds of loads, and boundary conditions. Structures made of ductile materials, such as metals, may stretch and/or bend excessively when subjected to severe and unexpected large loads. The locations of failure could be those sections having small areas or with sharp corners where the maximum stress developed exceeds the strength limit. For brittle materials, cracking is a major concern causing structural failures. When cycling or repeated loads are present, fatigue is the form of failure.

The following figure shows a column. The lower end is a fixed end and the upper end is a free end. The dimension of its length is L, and the dimension of its section is a x a. When the ratio of L to a is larger than 10, the column is viewed as a slender column. When such a slender column, or such a slender bar, is subjected to a compressive load, such as P, the column could fail if it bends and deflects laterally.

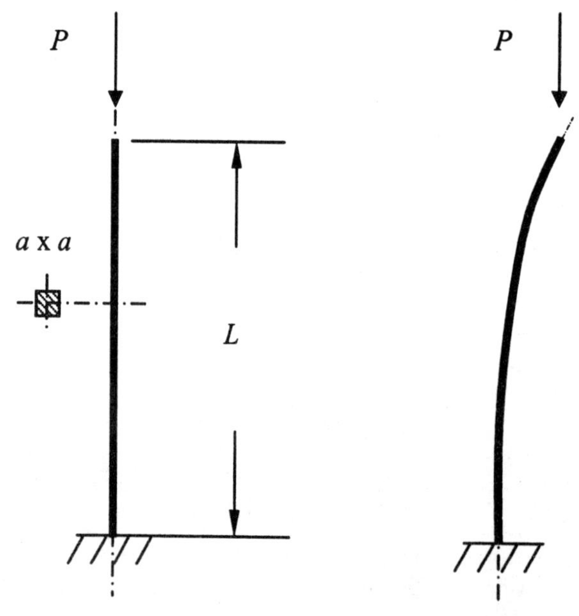

This type of failure is called buckling. Under an increasing axial load, the lateral deflection increases too, and eventually the column collapses completely. Have you seen this kind of failure in your daily life? For example, a tornado storm hits an area, hundreds of houses collapse under the heavy blow exerted by the tornado. Buckling is usually one of the major failure modes.

Evaluation of the load, which leads to a buckling failure mode, is a technical challenge because there is a significant amount of factors related to such an evaluation. For example, variation of the boundary condition, or the forms of the supports at both ends, can lead to a different mode shape (the first mode shape) when buckling occurs.

The following figure illustrates three types of the support form, resulting three different mode shapes at the onset of buckling. As a result, the critical load leading to buckling could be significantly different from one another.

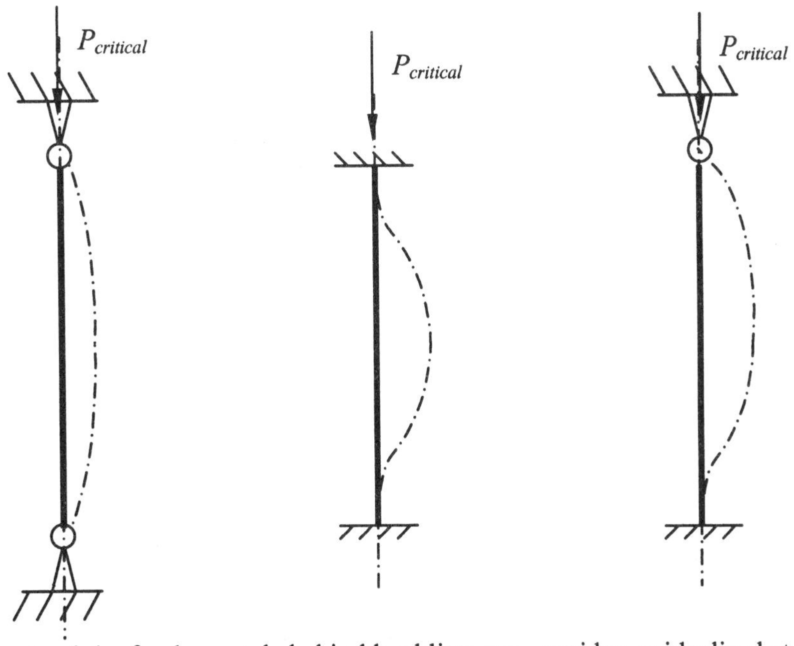

To understand the fundamentals behind buckling, we consider an idealized structure shown in the following figure. The slender bar is a rigid body that is pinned at the base and supported by an elastic spring at the top. The stiffness of the spring, or the spring constant is *K*.

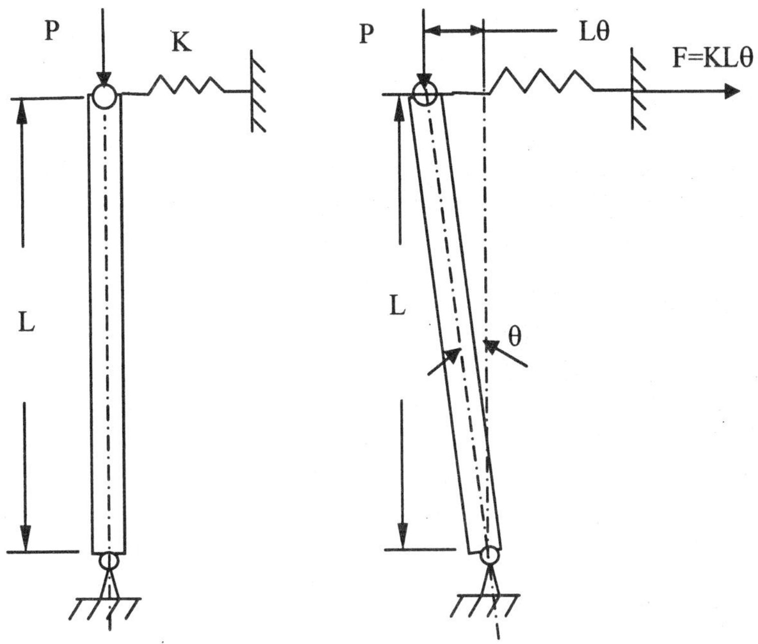

Assume that the slender bar is subjected to a load, P, which is perfectly aligned with its axis in the vertical direction. Assume that there is no stretch of the spring at its initial stage so that the slender bar is at its equilibrium position. Note that this equilibrium is unstable. This means that a small disturbance from the lateral side could lead the slender bar move away from its equilibrium position.

The analysis of this slender bar – spring system starts from giving a small rotation of the slender bar with respect to its axis, simulating a circumstance where an external force of disturbance causes such a small rotation of the slender bar. Because the small rotation gives rise to the presence of a distance, which is equal to L*θ, it induces two moments acting on the slender bar:

The first moment is due to the acting compression force, P. The moment is called overturning moment, and it is given by

$$\textit{Overturning Moment} = P*(L*\theta)$$

The second moment is due to the restoring force exerted by the spring. The moment is called restoring moment, and is given by

$$\text{Restoring Moment} = K*(L*\theta)*L$$

If Overturning Moment > Restoring Moment, the slender bar – spring system is unstable.

If Overturning Moment < Restoring Moment, the slender bar – spring system is stable.

If Overturning Moment = Restoring Moment (it is called a bifurcation point), the slender bar – spring system is in its neutral equilibrium, or at its bifurcation point. Under such a circumstance, the critical load for the buckling analysis can be determined by equating the two moment equations:

$$P_{critical}*(L*\theta) = K*(L*\theta)*L$$
$$P_{critical} = K*L$$

Therefore, $P < P_{critical} = K*L$, the system is stable, and $P > P_{critical} = K*L$, the system is unstable. To quantify the stability condition, a new parameter called Buckling Load Factor (BLF) is introduced. It is defined by a ratio of the applied force to the critical load:

$$BLF = \frac{P_{critical}}{P_{applied}}$$

Under the Pro/MECHANICA environment, a buckling load factor is calculated for each of the buckling mode shapes. Therefore, the buckling analysis is a linear eigenvalue bifurcation instability analysis. It is important to note that before defining a buckling analysis, a static analysis must be defined before hand. In the static analysis, Pro/MECHANICA calculates the stress stiffening of the 3D solid model due to the applied forces. In the buckling analysis, Pro/MECHANICA calculates the model's elastic stiffness due to geometry and material properties. Afterwards, Pro/MECHANICA uses these two solutions to calculate the BLF.

In the following figure, a slender column is shown. The length of the column is 4000 mm. The cross-section is a rectangle. The two dimensions of the rectangle are 75 and 50, respectively. Assume that one end of the column is fixed to the ground, as shown. The loading condition is uniform load acting at the other end. The magnitude of the load is 5000000 Newton.

Now let us determine the numerical value of the critical load, *Pcritical*, under which buckling failure occurs for the slender bar shown below. The left end of the slender bar is the fixed end. The right end is free to move. The compression load is acting on the center of the hole with the diameter equal to 500000 mm.. The material of the slender column is steel.

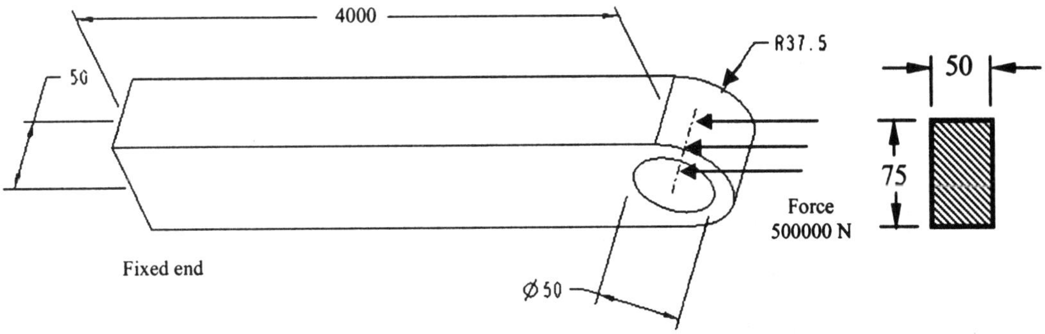

Step 1: create a 3D solid model

File > **New** > type in *slender_column* as the file name and select the mmNs as the unit system.

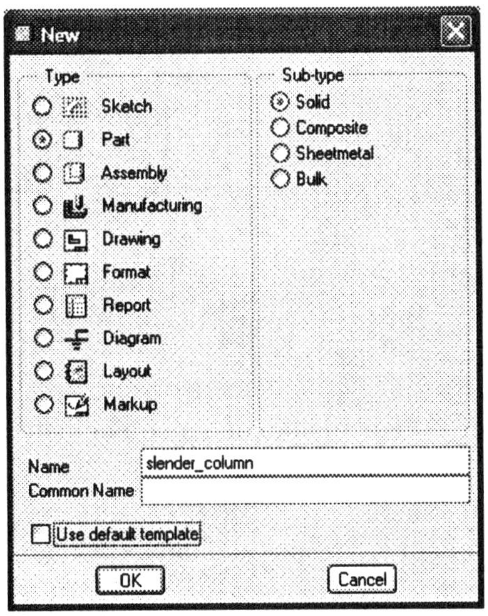

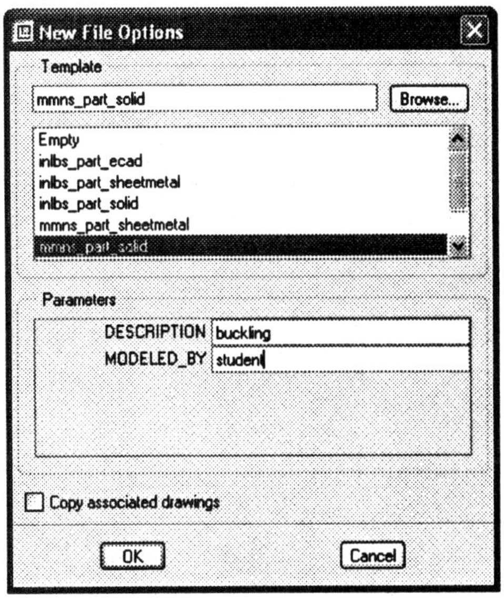

From the toolbar of feature creation, select the icon of **Extrude.**

The dashboard appears. Select the icon of symmetry, set the depth value to *50*. Click the box of **Placement** > **Define**.

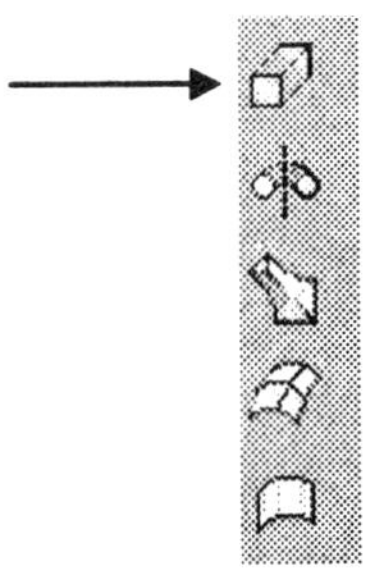

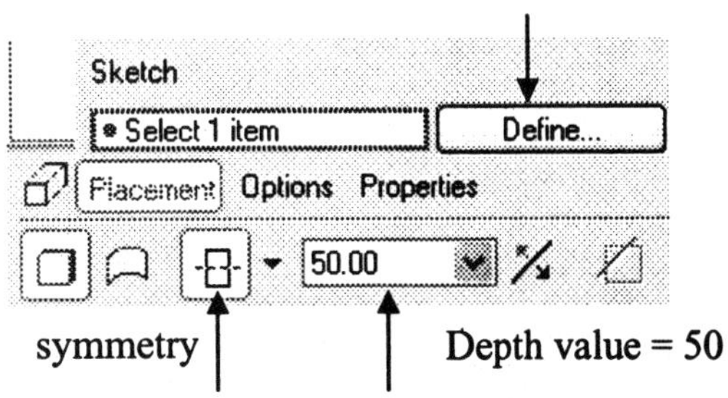

Select the **FRONT** datum plane as the sketch plane, and click the box of Sketch to accept the **RIGHT** datum plane as the default reference to orient the sketch plane > **Sketch**.

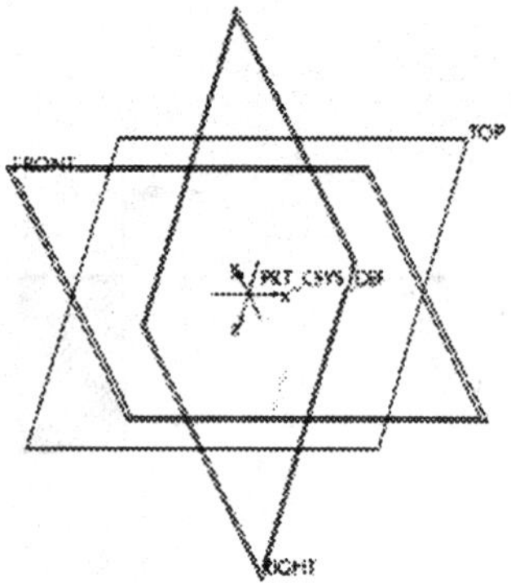

Sketch a horizontal centerline. Sketch an arc and line to form a closed sketch, as shown below:

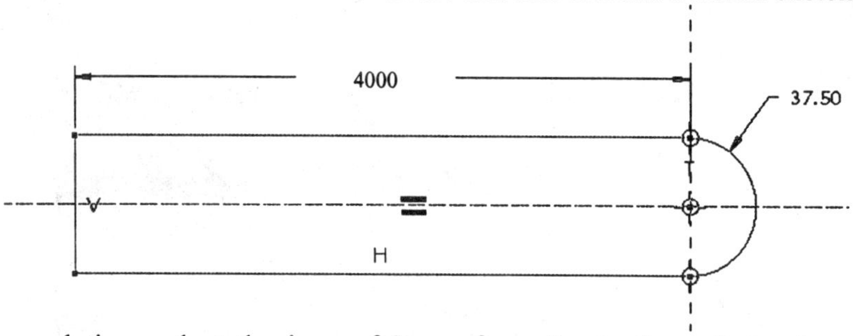

Upon completion, select the icon of **Done** from the toolbar of sketcher. Select the icon of **Complete** from the feature control panel.

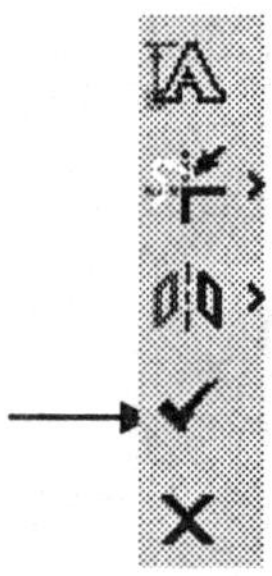

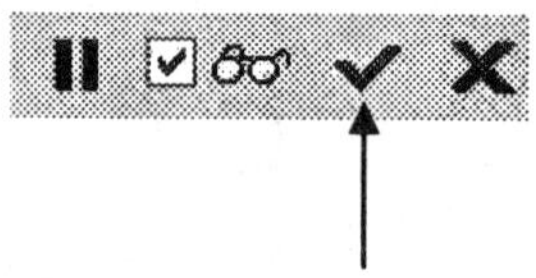

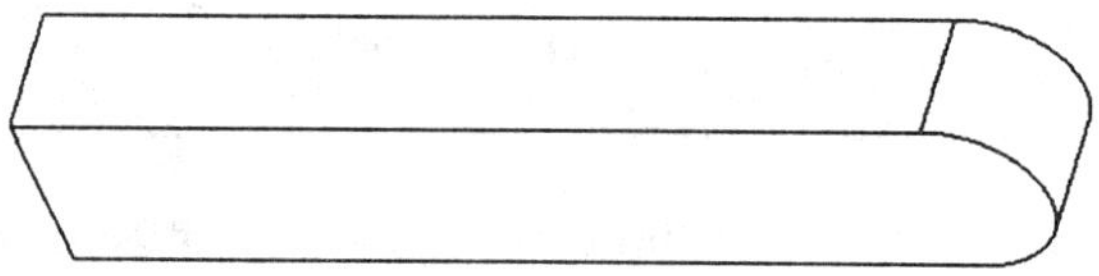

Pay special attention to the process of creating the hole at the right end of the created column for controlling the location and the direction of the load to be specified.

From the toolbar of feature creation, select the icon of **Extrude**. On the dashboard, select the icon of **Cut** and set the depth choice to **Though all** > click the box of **Placement** > **Define**.

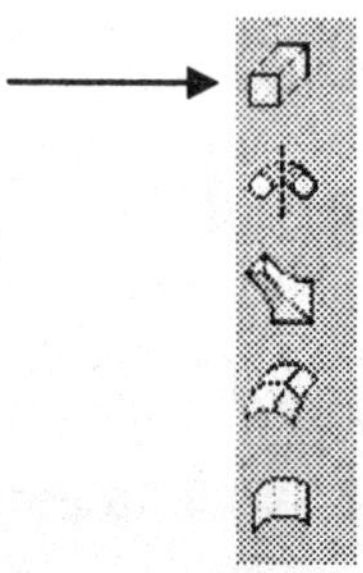

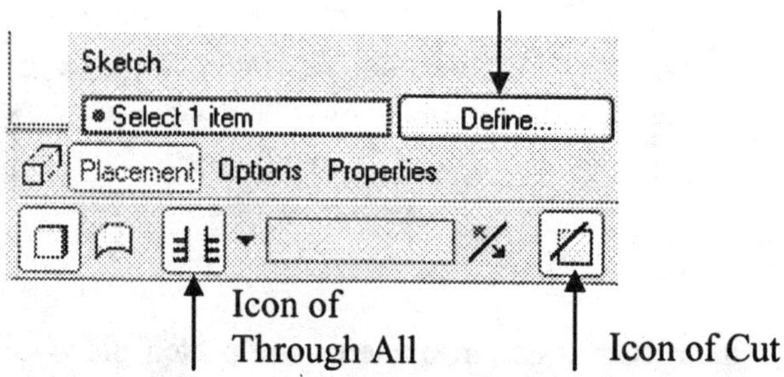

Pick the front surface of the column and accept the default form to orient the sketch plane > select the icon of **Creating arc** and sketch a half circle, as shown > specify the radius equal to 25.

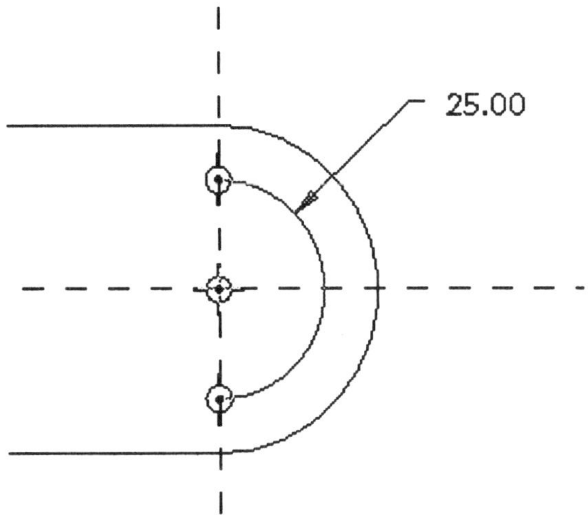

Afterwards, sketch the other half circle, as shown.

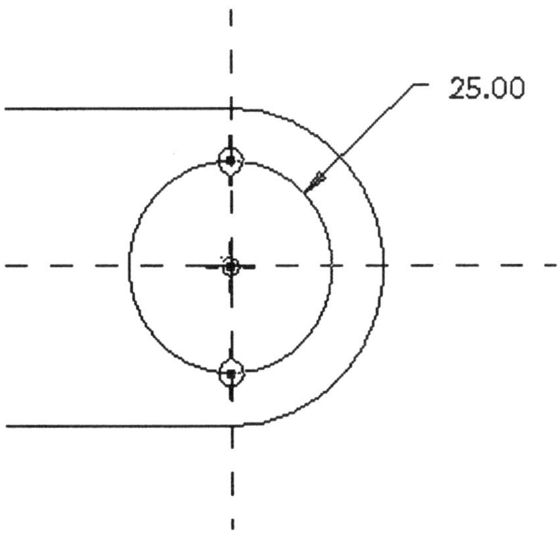

Upon completion, select the icon of **Done** from the toolbar of sketcher. Select the icon of **Complete** from the feature control panel.

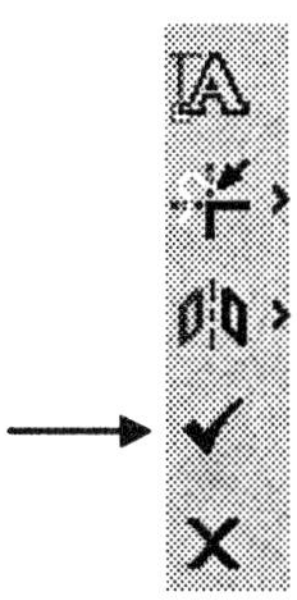

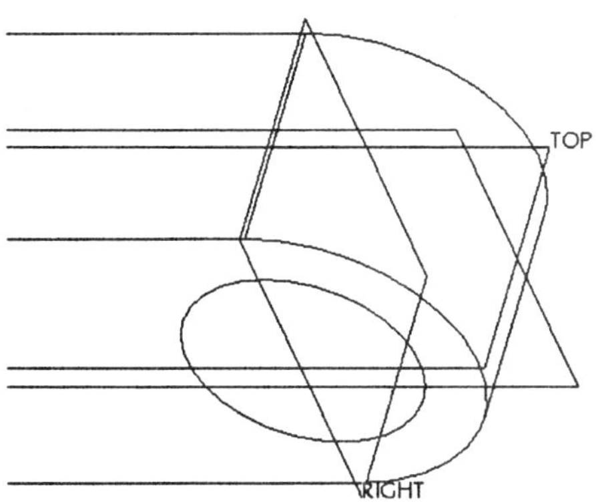

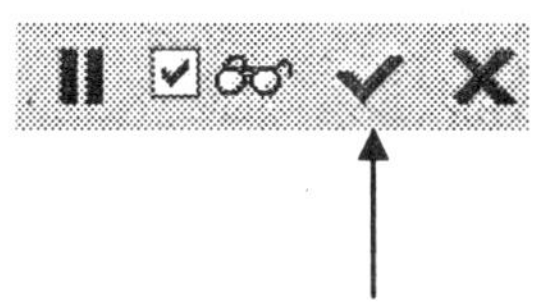

To create an axis of the hole feature, select the icon of creating axis from the toolbar of creating datum feature > let us select the **TOP** and **RIGHT** datum planes > **OK**.

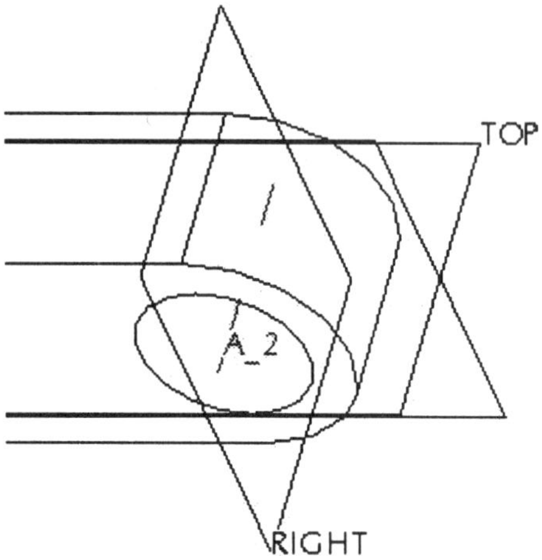

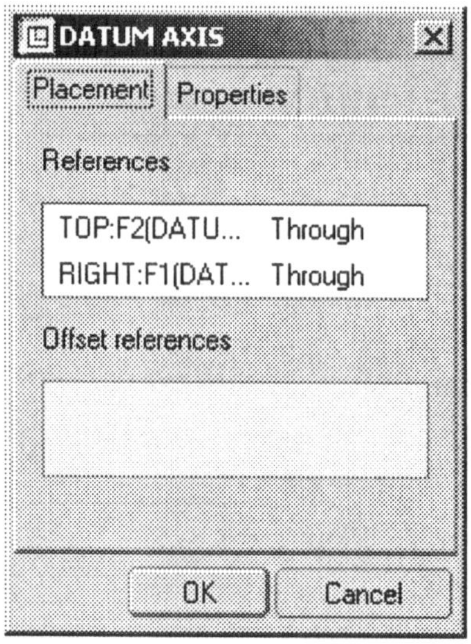

To create a datum point at the middle of the created axis, select the icon of creating point from the toolbar of creating datum feature > select both the created axis and the FRONT datum plane.

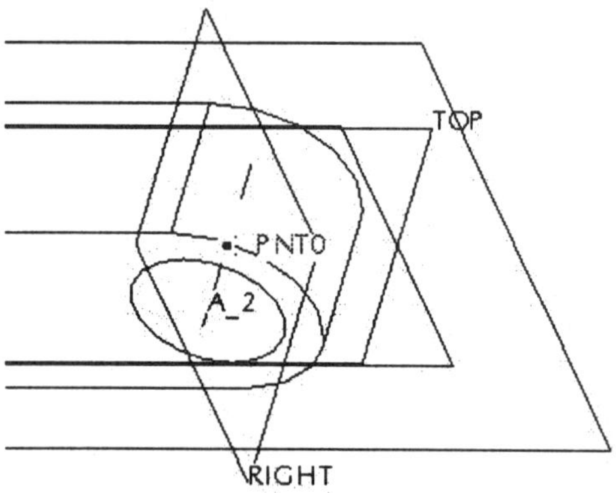

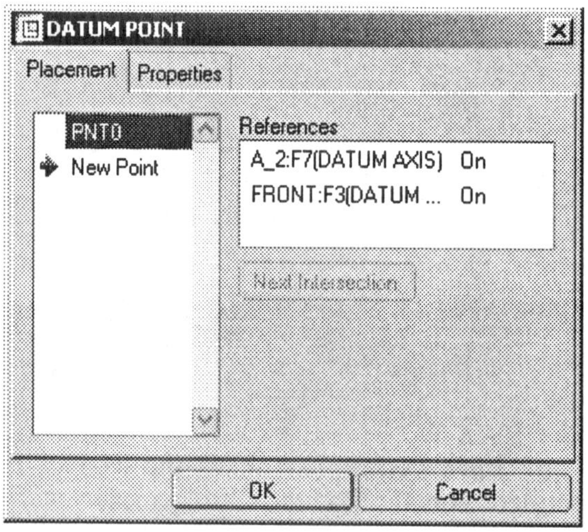

Step 2: under the Pro/Mechanica environment, specify the material type, the constraint condition and the load condition.

Applications > Mechanica > Continue > Structure > OK.

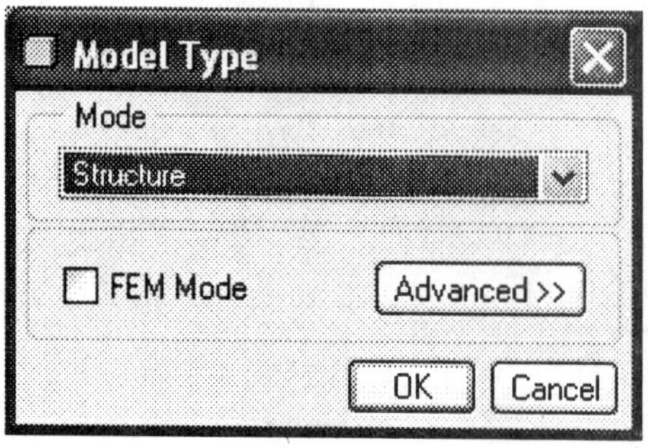

From the toolbar of functions, select the icon of **Materials > select STEEL.**

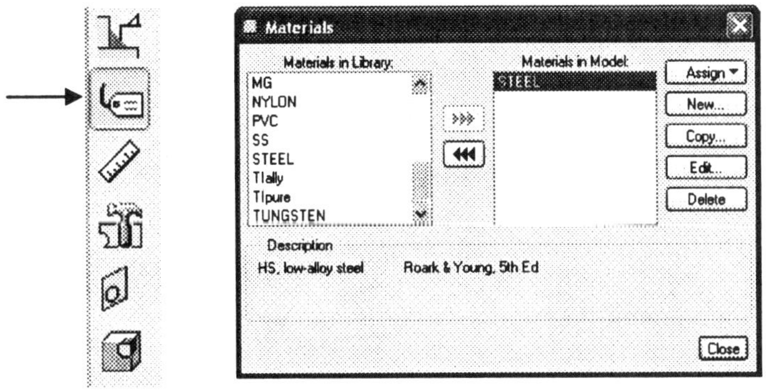

Assign > **Part** > select the column structure from the screen display > **OK** > **Close**.

To define the constraint condition, select the icon of constraint > select the surface on the left end and fix all six degrees of freedom > OK.

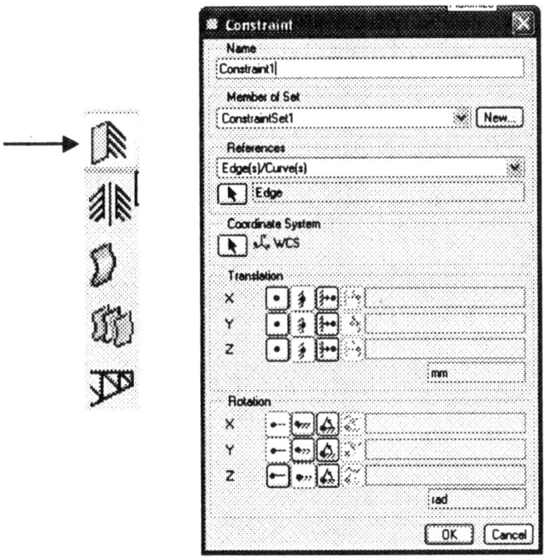

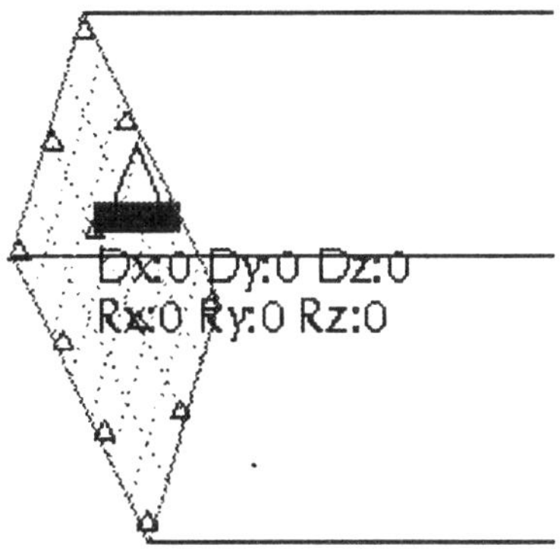

To define the load condition, select the icon of Force/Moment > pick the half cylindrical surface on the left side > OK. Click Advanced > select Total Load at Point and select PNT0, which was created in the part modeling. Specify the magnitude equal to -500000 N, along the x-direction.

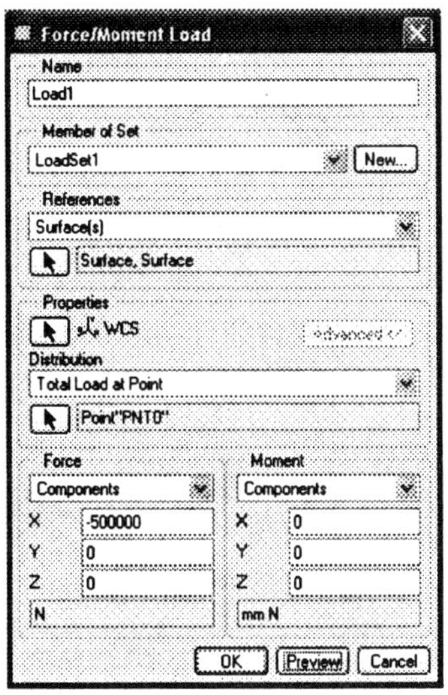

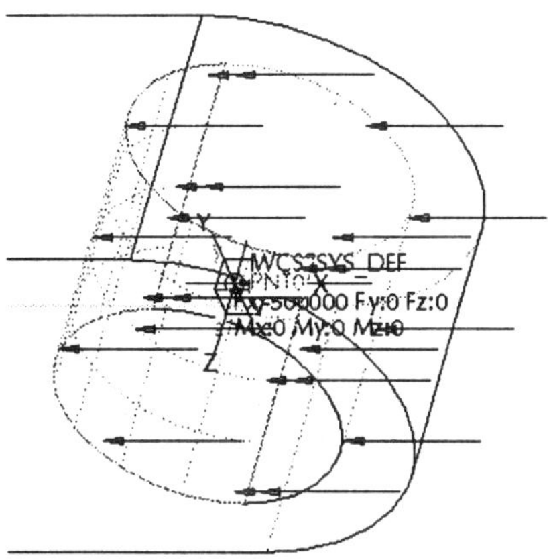

Step 3: perform buckling analysis

Click the icon of **performing a design study** from the main toolbar > **File** > Select **New Static.** Type *static_for_buckling* as the file name > **OK**. Run the static analysis.

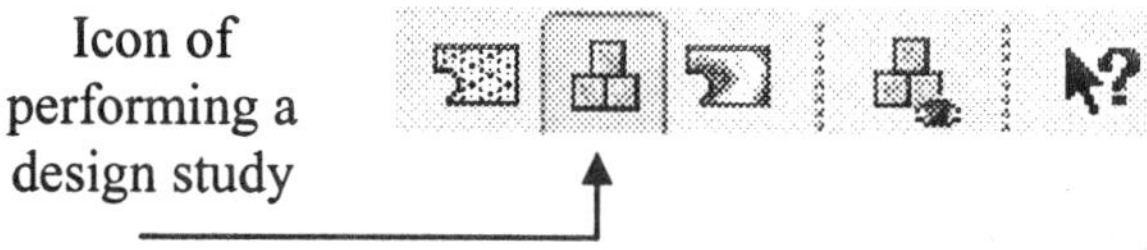

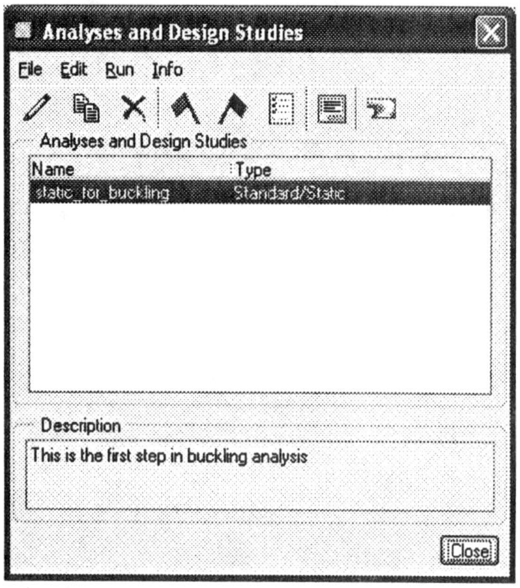

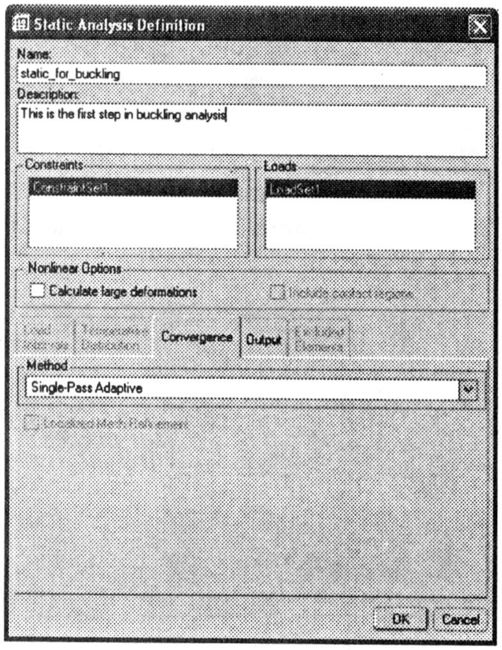

File > **New Buckling** > type *buckling_1* as the name. Specify the number of buckling modes to be 3. Therefore, we expect to have 3 **BLF** values. Each of the 3 **BLF** values is associated with its own buckling mode. Check the box called Use static analysis results from previous design study > **OK**.

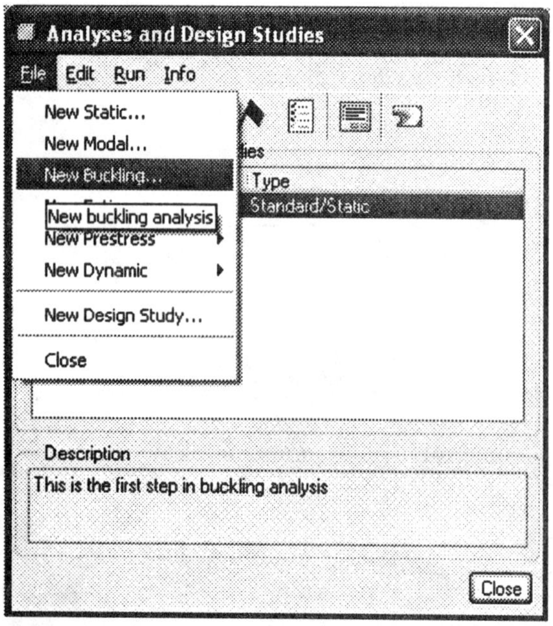

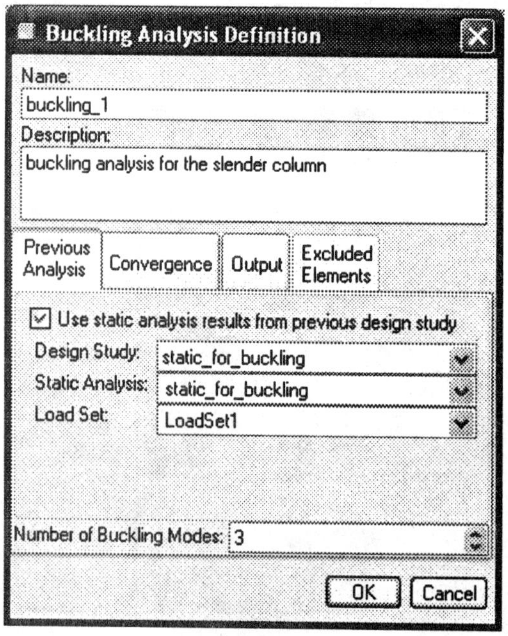

Run the buckling analysis.

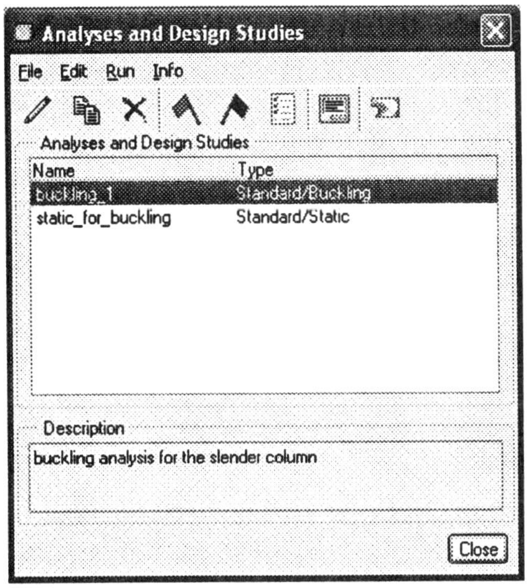

The obtained BLF values are shown below. They are 0.0483, 0.1086 and 0.4358 for the 3 modes, respectively. All these 3 values are less than 1, indicating that they are all candidates for buckling failure. Note that these BLF values are associated with the magnitude of the applied load, which is 500000 N.

```
Number of Modes: 3

    Mode     B. L. F.
    ----   ------------
       1   4.830472e-02
       2   1.086358e-01
       3   4.358160e-01
```

To plot the three mode shapes, select Results > Insert > Result Window > type in The first mode shape as the title > select the folder called buckling_1 > select Mode 1 > select Displacement > from Display Option, select Deformed > OK and Show.

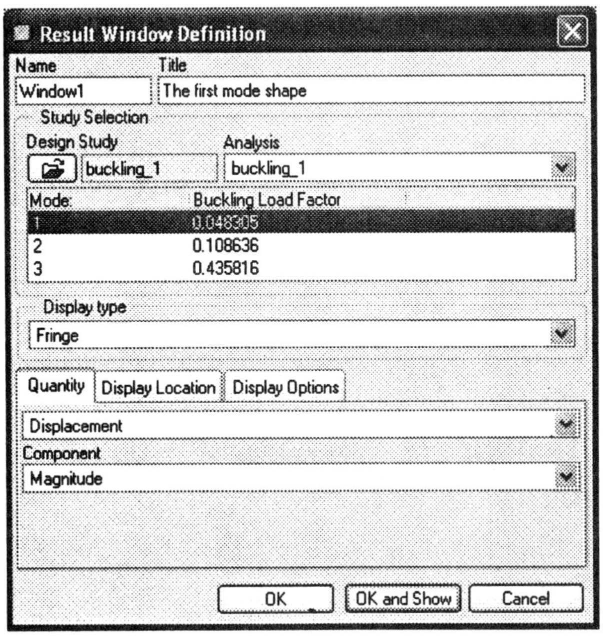

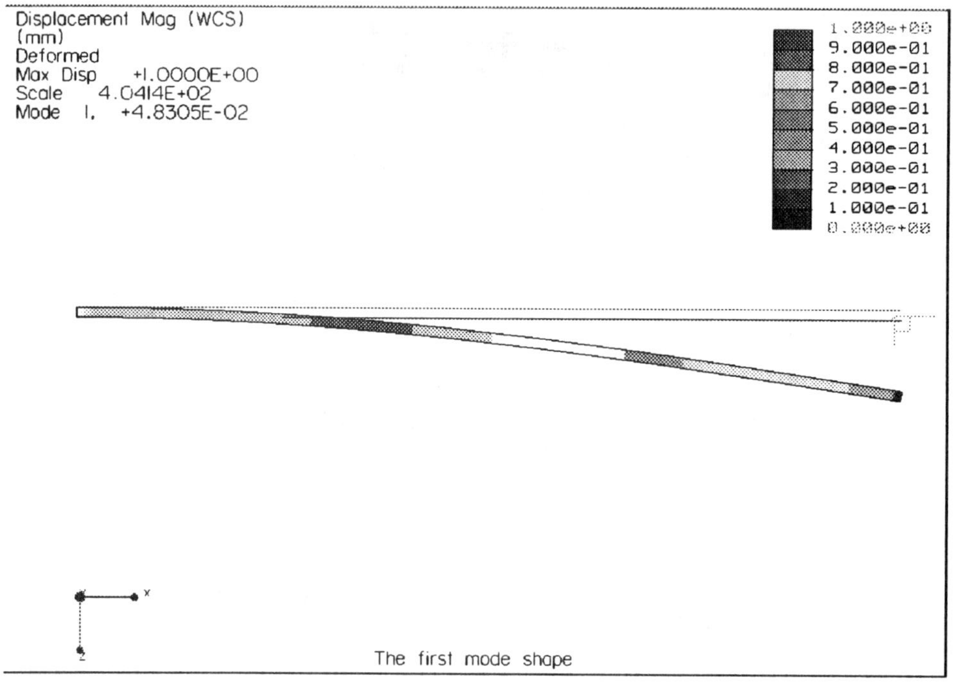

To plot the second mode shape, Edit > Copy > change the first mode to second mode > select Mode 2 > OK and Show.

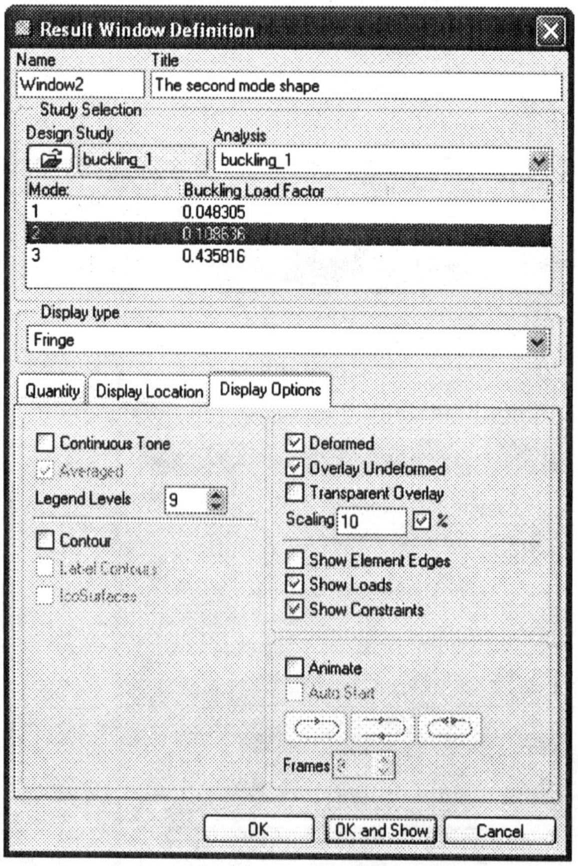

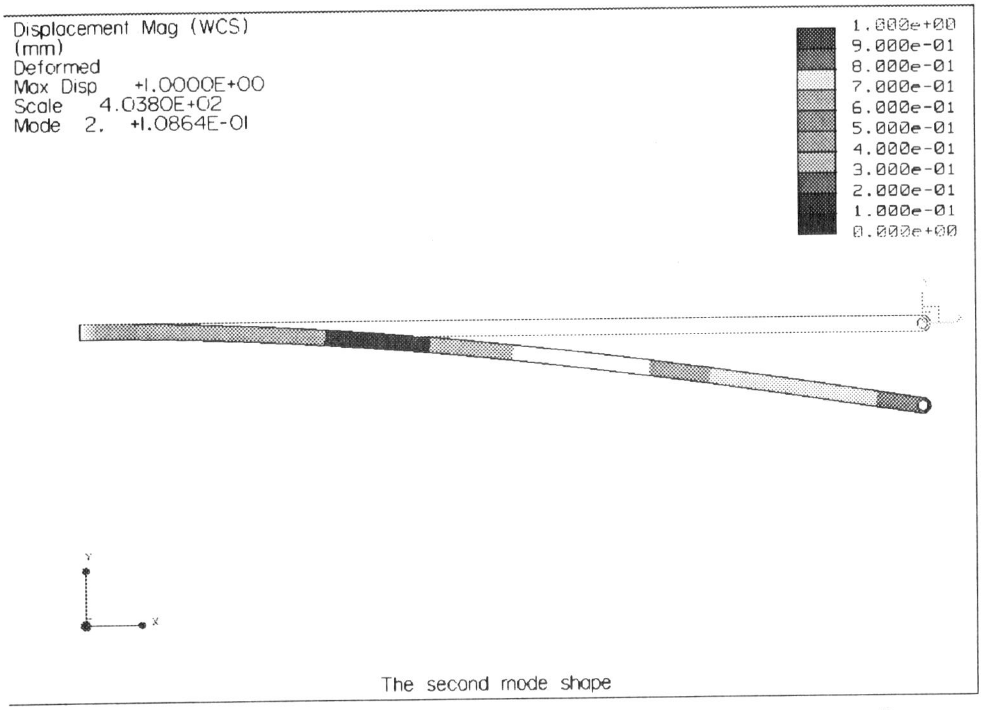

Edit > Copy > change the first mode to third mode > select Mode 3 > OK and Show.

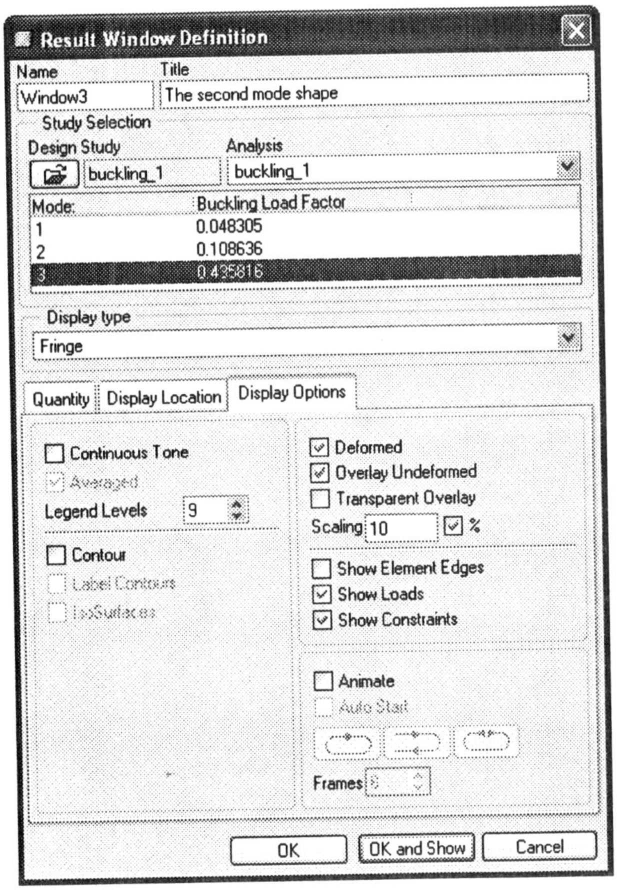

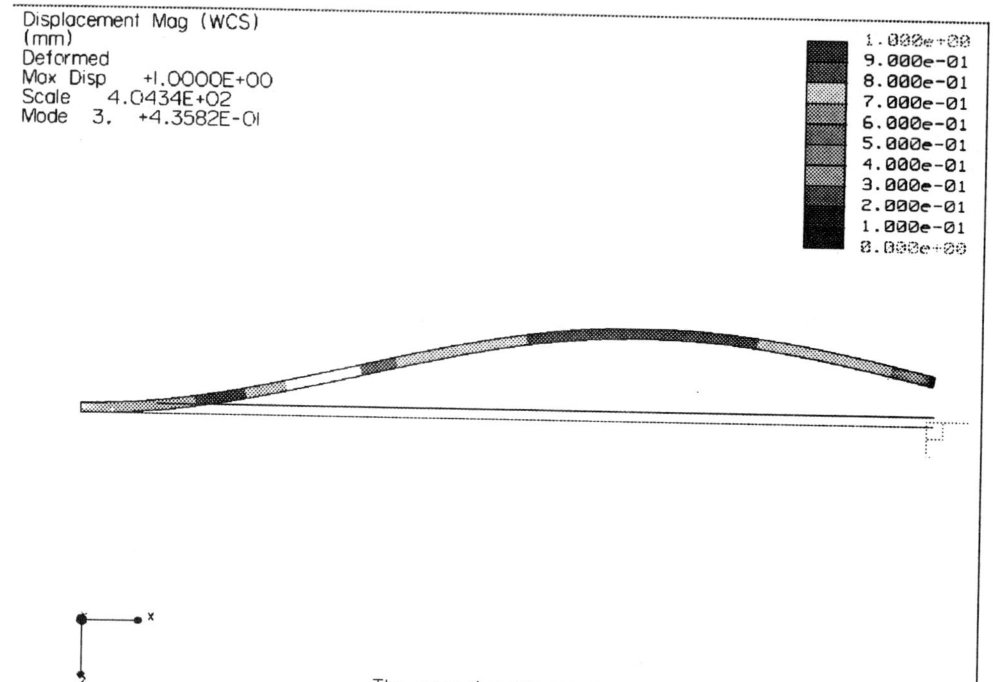

The second mode shape

Calculations of the critical loads for the three buckling modes are shown below:

$$P_{critical\_mode\_1} = 0.048305 * 500000 = 24153 \text{ N}$$
$$P_{critical\_mode\_2} = 0.108636 * 500000 = 54318 \text{ N}$$
$$P_{critical\_mode\_3} = 0.435816 * 500000 = 217908 \text{ N}$$

Because $P_{critical\_mode\_1} = 24153$ N is the smallest one among the three critical loads, the first mode is most likely to be the failure mode. However, attention should also be paid to the second and third modes because their associated critical loads are smaller than the applied load.

7.5 Buckling Analysis of a Thin Disk

Perform a buckling analysis for a thin disk. Its geometrical shape and three key dimensions are shown below. The material of the disk is steel. The inner hole is fixed to the ground and the disk is subjected to a radial (inward) load. Its magnitude is 20 N/mm^2.

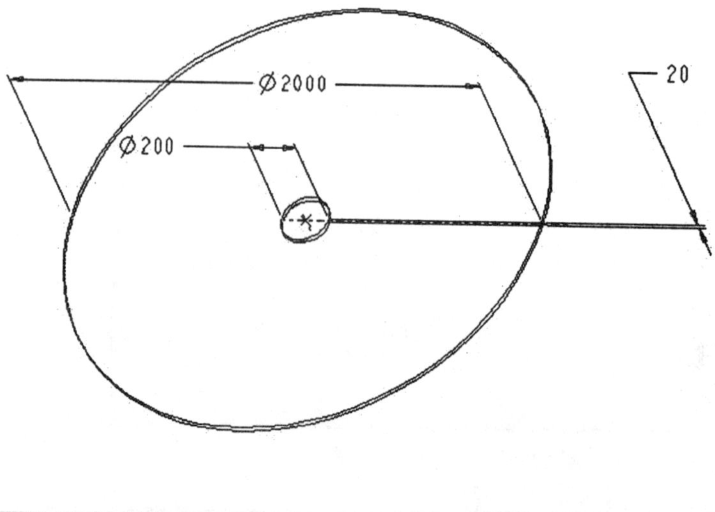

Step 1: Under the Pro/ENGINEER design system, create a 3D solid model of the disk.

File > **New** > type in *disk* the file name and select the mmNs as the unit system > **OK**. Type *buckling analysis* under **DESCRIPTION** and *student* under **MODELED_BY**.

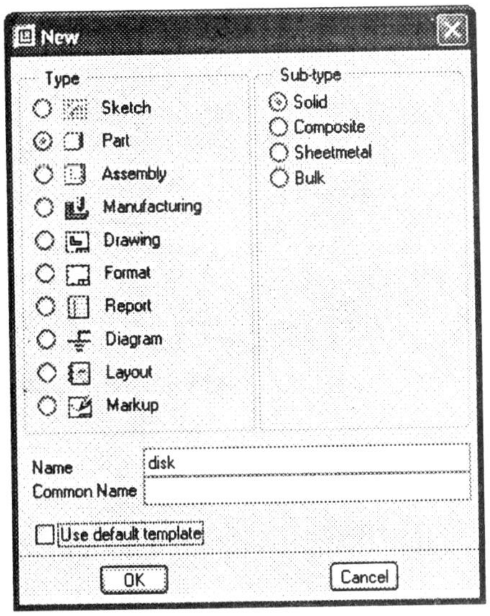

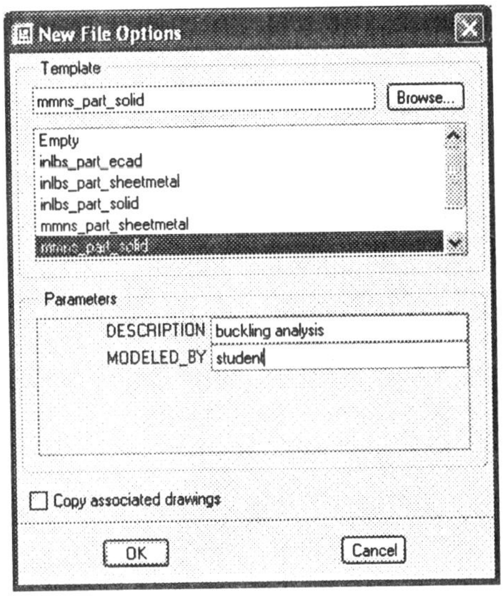

Directly select the icon of Extrude from the toolbar of feature creation. From the dashboard, set the thickness value to 20. Click the icon of Placement > Define.

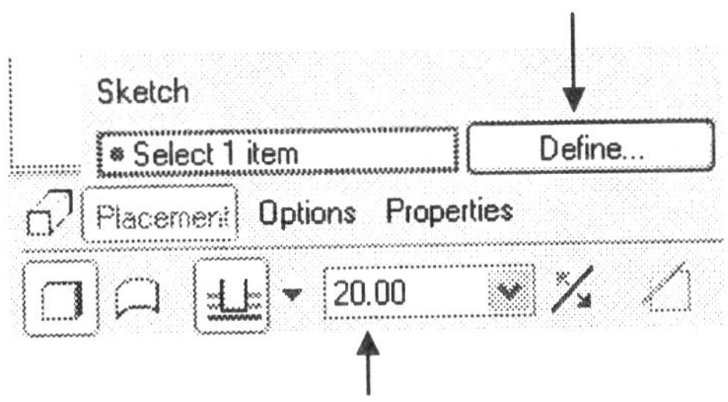

Select the TOP datum plane as the sketch plane and accept the default setting to orient the sketch plane > **Sketch**.

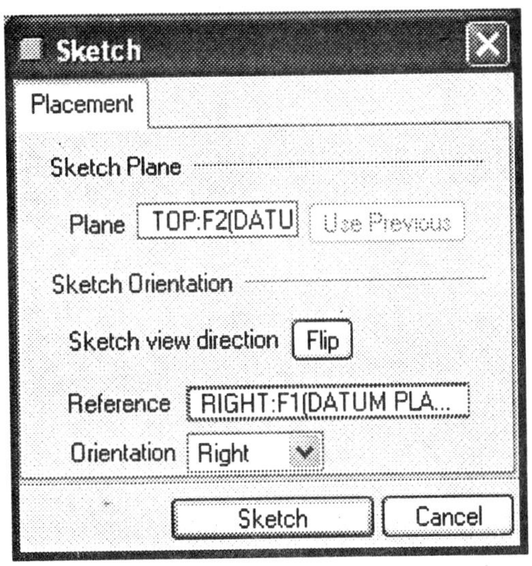

Pick the icon of circle to sketch 2 circles. The 2 diameters are 2000 and 200, respectively.

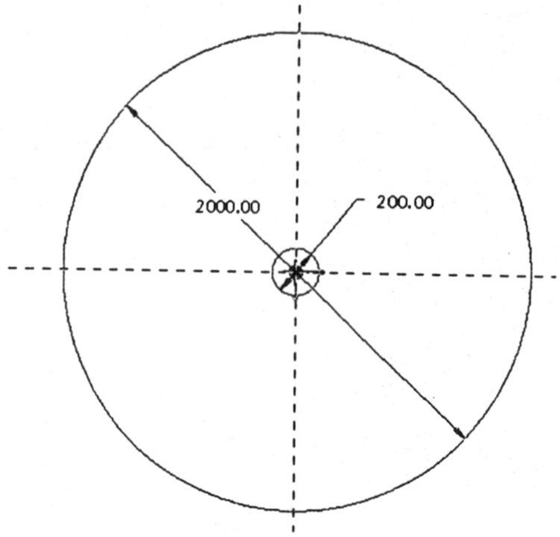

Upon completion, select the icon of **Done** from the toolbar of sketcher. Select the icon of **Complete** from the feature control panel.

Step 2: Under the Pro/MECHANICA environment, define the material type.
Applications > **Mechanica** > **Continue** > **Structure** > **OK.**

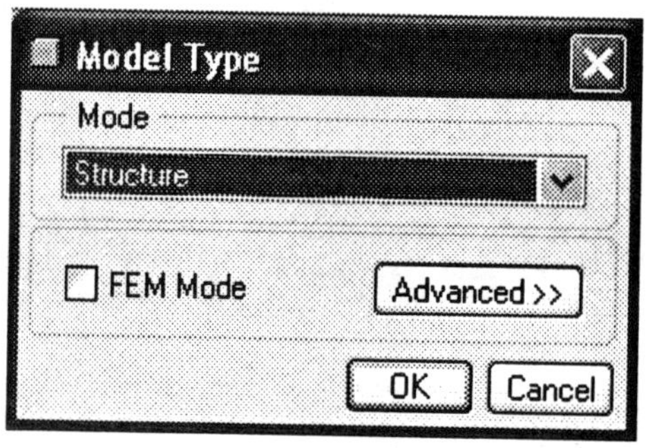

From the toolbar of functions, select the icon of **Materials** > select **STEEL.**

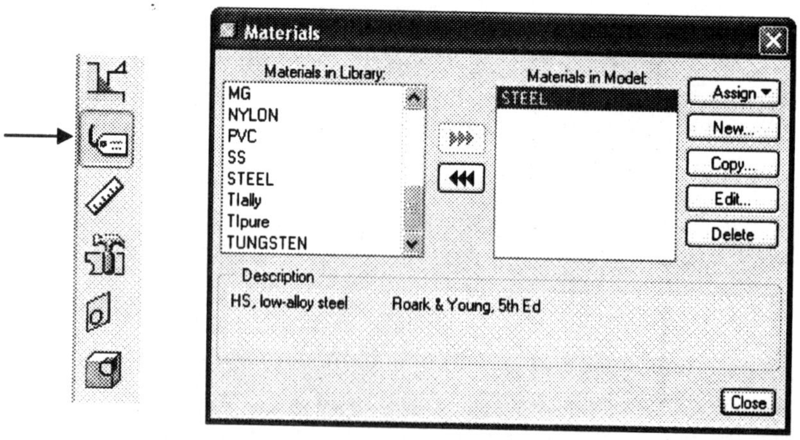

Assign > **Part** > select the column structure from the screen display > **OK** > **Close.**

Step 3: Under the Pro/MECHANICA environment, define the constraint condition and the load condition.

To define the constraint condition, select the icon of constraint > select the two halves of the cylindrical surface, and fix all six degrees of freedom.

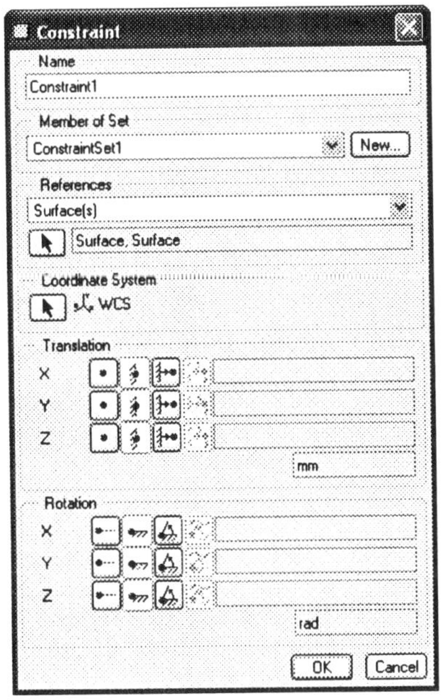

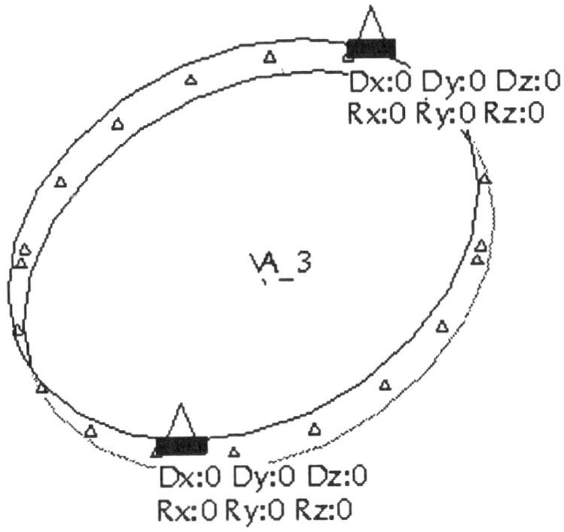

In order to specify a radial load, a cylindrical coordinate system is needed. Use the three datum planes to create the required cylindrical coordinate system, CSO.

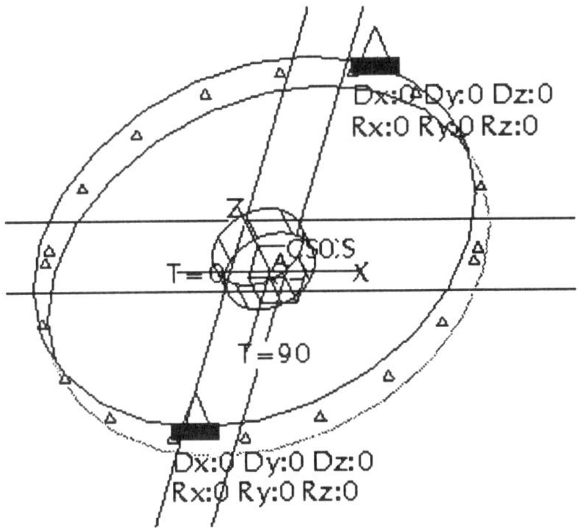

Define the radial load, it is a surface load along the outer circumference.

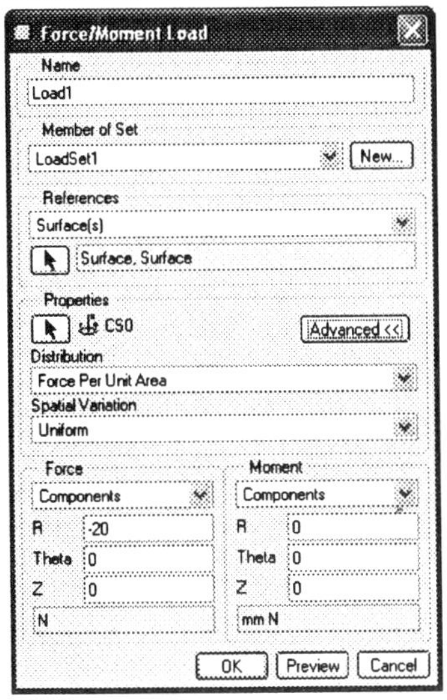

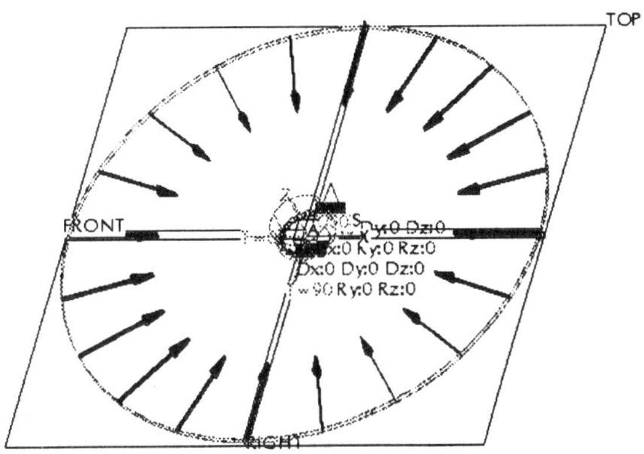

Step 4: Prepare two (2) analyses. The first one is a static analysis, and the second one is the buckling analysis. Make sure to set the number of modes to 4.

Step 5: Run the static analysis first. Then run the buckling analysis.

The obtained BLF values are shown below. They are 0.8246, 0.8408, 1.065, and 1.067 for the four modes, respectively. Among these four values, only the BLF value associated with mode 1 and mode 2 are less than 1, indicating that these 2 modes are the candidates for buckling failure. Note that these BLF values are associated with the magnitude of the load applied, which is 20 N/mm^2. Because the total area where the load is acting on is equal to 125,600 mm^2, the total load acting in the radial direction is equal to 2,512,000 N.

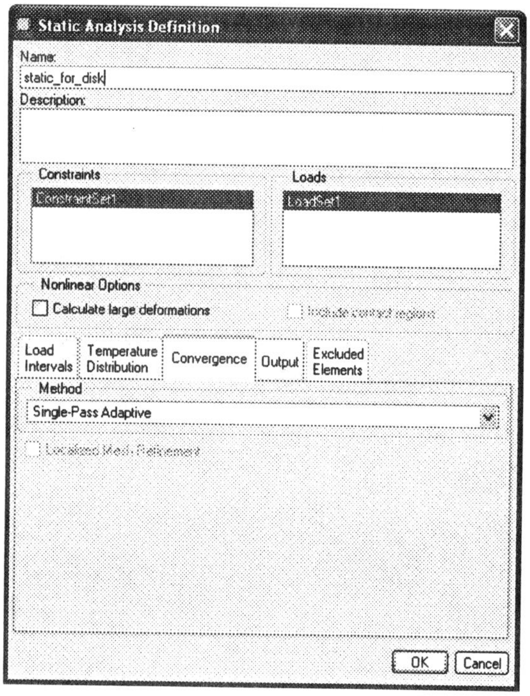

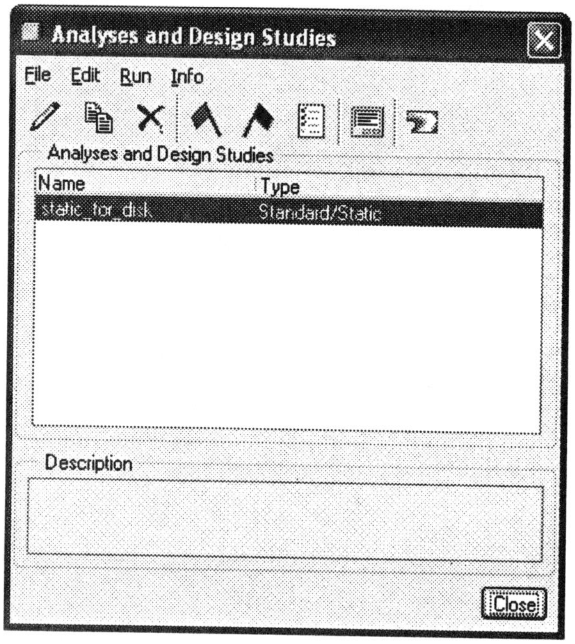

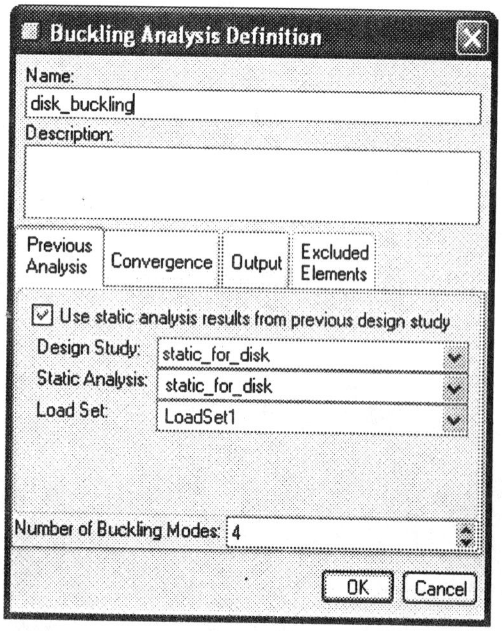

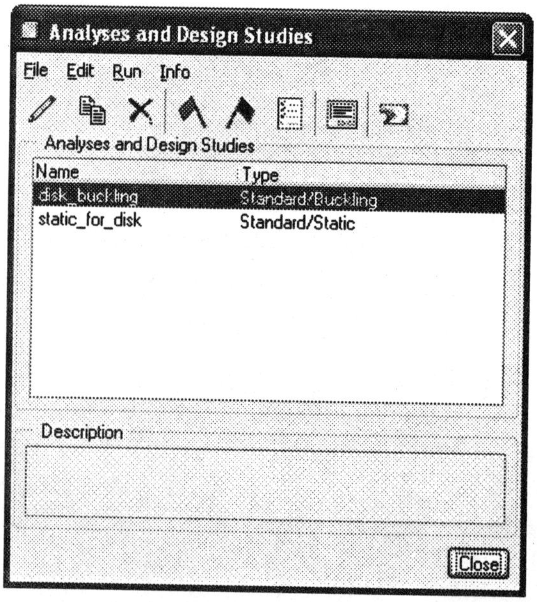

Number of Modes: 4

Mode	B. L. F.
1	8.246007e-01
2	8.408427e-01
3	1.065356e+00
4	1.067407e+00

The four mode shapes are shown below. Note that the first mode shape and the second mode shape are essentially duplicates of each other. So are the third mode shape and the fourth mode shape. This phenomenon indicates that the eigen values associated with the two duplicate modes are close or equal to each other. As a result, Pro/MECHANICA finds two different poles of the same buckling shape.

The first mode shape

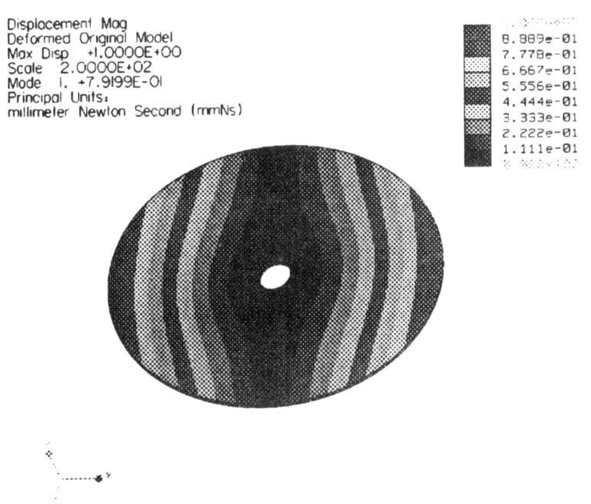

The first mode shape

The second mode shape

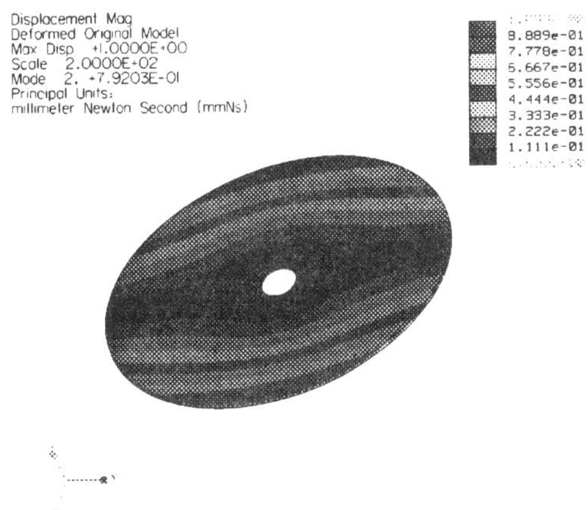

The second mode shape

The third mode shape

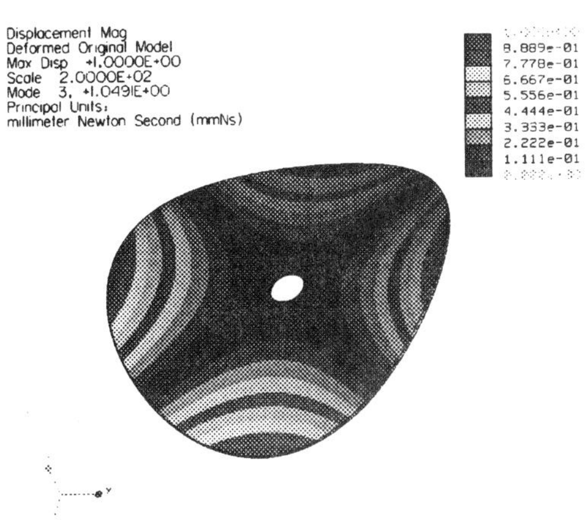

The third mode shape

The fourth mode shape

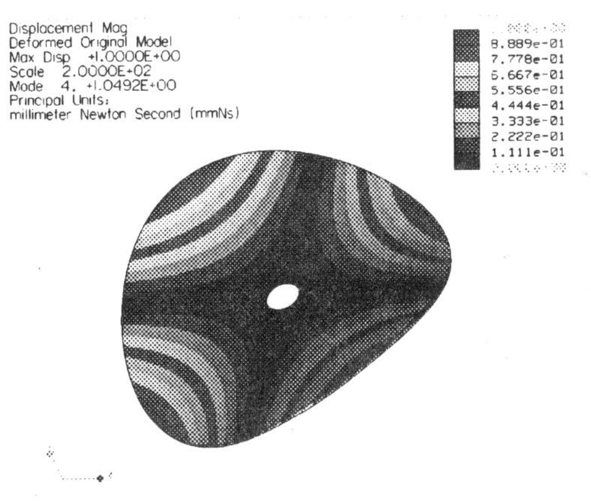

The fourth mode shape

When a BLF value is greater than 1, this means that the magnitude of the critical load is greater than the currently applied load. Therefore, buckling failure will not occur with the associated mode shape. As a result, we only need to calculate two critical loads for the first and second buckling modes.

$$P_{critical\_mode\_1} = 0.8246 * 2512000 = 2{,}071{,}395 \text{ N}$$
$$P_{critical\_mode\_3} = 0.8408 * 2512000 = 2{,}112{,}090 \text{ N}$$

When the total applied load reaches 1,990,000 N, or the unit-applied load reaches 16.5 N/mm^2, buckling failure will occur, and the mode shape will be the shape of the first mode.

7.6 References

1. Algor Processor Reference Manual, Algor Interactive Systems, 150 Alpha Drive, Pittsburgh, PA 15238
2. O. Axelsson, Iterative Solution Methods, Cambridge University Press, New York, 1994.
3. K. J. Bathe, Finite Element Procedure, Prentice Hall, Upper Saddle River, New Jersey, 1996.
4. K. Bell, A Refined Triangular Plate Bending Finite Element, Int. J. Numer. Methods Eng., Vol. 1, No. 1 January 1969, pp. 101-122.
5. B. Carnahan, H. A. Luther, and J. O. Wikes, Applied Numerical Methods, Wiley New York, NY, 1969.
6. R. W. Clough, The Finite Element in Plane Stress Analysis, Proceedings of 2nd ASCE Conference on Electronic Computation, Pittsburgh, PA. September 1960.
7. R. D. Cook, D. S. Maikus and M. E. Plesha, Concepts and Applications of Finite Element Analysis, 3rd edition, Wiley, New York, NY, 1989.
8. J. W. Dally and W. F. Riley, Experimental Stress Analysis, McGraw Hill, 1965.
9. J. W. Dally, Design Analysis of Structural Elements, Goodington, New York, NY, 2003.
10. C. S. Desal and J. F. Abel, Introduction to the Finite Element Method, Van Nostrand Reinhold, New York, NY, 1972.
11. J. H. Earle, Graphics for Engineers, AutoCAD Release 13, Addision-Wesley, Reading, Massachusetts, 1996.
12. R. H. Gallagher, Finite Element Analysis Fundamentals, Prentice-Hall, Englewoood Cliffs, NJ, 1975.
13. J. M. Gere and S. P. Timoshenko, Mechanics of Materials, 3rd edition, PWS-KENT Publishing Company, Boston, 1990.
14. W. W. Hager, Applied Numerical Linear Algebra, Prentice-Hall, Englewood Cliffs, NJ, 1988.
15. Holland, The Finite Element Method in Plane Stress Analysis, in the Finite Element Method in Stress Analysis, I. Holland and K. Bell (eds), Tapir Press, Trondheim, Norway, 1969, Chapter 2.
16. Y. Y. Hsieh, Elementary Theory of Structure, 2nd edition, Prentice-Hall, Englewood Cliffs, NJ, 1982.
17. M. A. N. Hendriks, H Jongedijk, J. G. Rots and W. J. E. van Spange (eds), Finite Elements in Engineering and Science, DIANA Computational Mechanics, 1997.
18. K. H. Huebner, D. L. Dewhirst, D. E. Smith, and T. G. Byrom, The Finite Element Method for Engineers, 4th edition, John Wiley & Sons, Inc., 2001.
19. C.S. Krishnamoorthy, Finite Element Analysis, Theory and Programming, 2nd Ed., 1995.
20. D. L. Logan, A First Course in the Finite Element Method Using ALGOR, PWS Publishing, Boston, 1997.
21. H. C. Martin, Introduction to Matrix Methods of Structural Analysis, McGraw-Hill, New York, NY, 1966.
22. R. L. Norton, Machine Design: An Integrated Approach, Prentice Hall, Upper Saddle River, New Jersey, 1996.
23. M. Ortiz, Y. Leroy and A. Needleman, A Finite Element Method for Localized Failure Analysis, Comp. Meth. Applied Mech. Enging, 61, pp. 189-214, 1987.
24. M. Petyl, Introduction to Finite Element Vibration Analysis, Cambridge University Press, 1990.

25. T. H. H. Pian and P. Tong, Basis of Finite Element Methods for Solid Continua, Int. J. Numer. Methods Eng., Vol. 2, No. 7, April 1964, pp. 1333-1336.
26. P. G. Plockner, Symmetry in Structural Mechanics, J. of the Structural Division, American Society of Civil Engineers, Vol. 99, No. ST1, pp. 71-89, 1973.
27. J. N. Reddy, An Introduction to the Finite Element Method, McGraw-Hill Publishing Company, New York, NY, 1984.
28. T. Ross, T. Johns and Emile, Dynamic Buckling of Thin Walled Domes under External Water Pressure, Applied Solid Mechanics 2nd Conference, Ed. Tooth & Spence, Elsevier, pp. 211-224, 1987.
29. T. Ross, Finite Element Program in Structural Engineering and Continuum Mechanics, Horwood Publishing Limited, Chichester, England, 1996.
30. T. Ross, Advanced Applied Finite Element Methods, Horwood Publishing Limited, Chichester, England, 1998.
31. L. Stasa, Applied Finite Element Analysis for Engineers, Holt, Rinehart and Winston, New York, NY, 1985, pp. 101-157.
32. Y. Saad, Iterative Methods for Sparse Linear Systems, PWS Publishing Company, Boston, 1996.
33. P. Silvester, Higher-Order Polynomial Triangular Finite Element for Potential Problems, Int. J. Eng. Soc., Vol. 7, No. 8, pp. 849-861.
34. Y. Tada and G. C. Lee, Finite Element Solution to an Elastic Problem of Beams, Int. J. Numer. Methods Eng., Vol. 2, No. 2, April 1970, pp. 229-241.
35. R. L. Taylor, On Completeness of Shape Functions for Finite Element Analysis, Int. J. Numer., Methods Eng., Vol. 4., No. 1, 1972, pp. 17-22.
36. S. P. Timoshenko and D. H. Young, Theory of Structure, McGraw-Hill, Kogaksha Limited, 1965.
37. M. J. Turner, R. W. Clough, H. C. Martin, and L. J. Topp, Stiffness and Deflection Analysis of Complex Structures, J. of the Aeronautical Sciences, Vol. 23, No. 9, pp. 805-824, Sept. 1956.
38. O. C. Zienkiewicz, The Finite Element Method, third edition, McGraw-Hill, New York, NY, 1977.
39. O. C. Zienkiewicz and R. L. Taylor, The Finite Element Method, 4th edition, Vol. 1, McGraw-Hill (UK),London, 1989.

7.7 Exercises

1. Design a turning fork for the C note. The frequency of the first mode for the C note is 260.97 Hz. The geometrical shape shown below is for a turning force for the G-sharp fork. There are 2 dimensions are allowed to adjust so that a turning fork for the C note can be designed. The 2 dimensions are 5 mm and 6 mm, respectively. Plot the modal shapes of Mode 1, Mode 2, Mode 3 and Mode 4. What are the frequencies associated with these 4 modes?

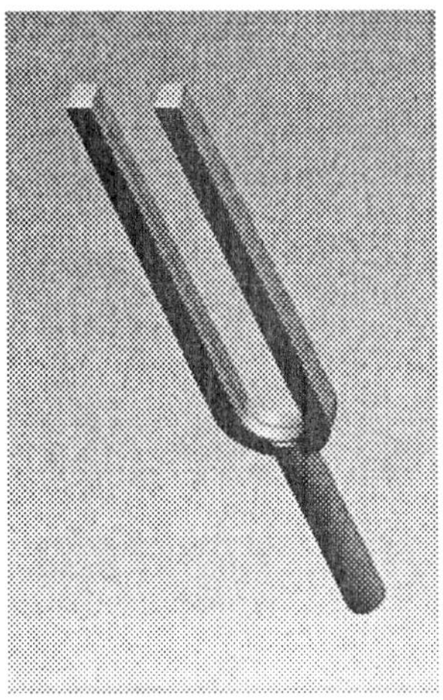

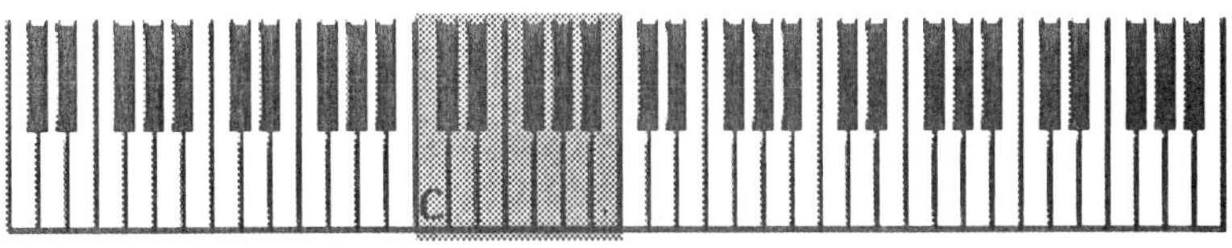

Adjustable dimension for the C note

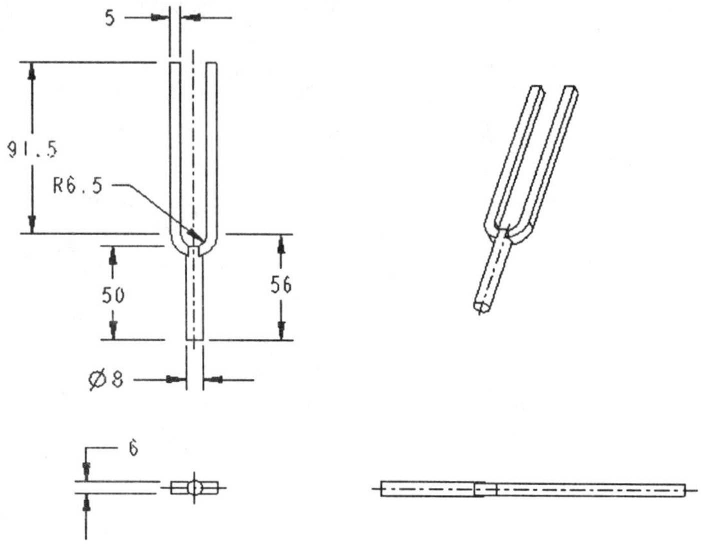

Adjustable dimension for the C note

2. Determine the first 4 modal shapes and their associated frequencies for the beam structure shown below. There are 2 cases. In case 1, the exciting force acts in the Y direction and downward. In case 2, the exciting force acts in the Z direction and towards the inside of the paper. Plot the 4 modal shapes for each of the 2 cases.

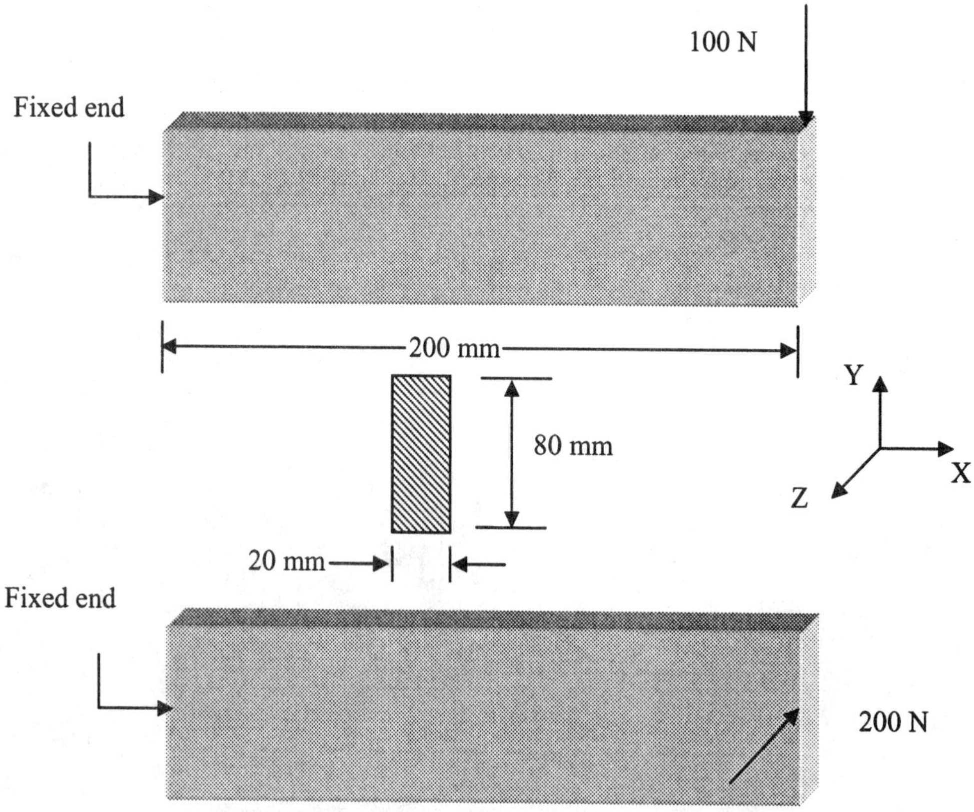

3. In section 7.4, a column structure is under investigation for buickling stability. At that time, we assumed that one end is firmly fixed to the ground. Now let us investigate the effects due to the boundary condition on the mode shape and BLF magnitude. In this exercise, let us change the constraint condition on the left side of the slender column from a fixing end to a pin joint with a small surface region to restrict the movement in the x-direction. Therefore, first cut a through hole at the left side, as shown. Afterwards, define the constraint condition which is closely related to a pin joint.

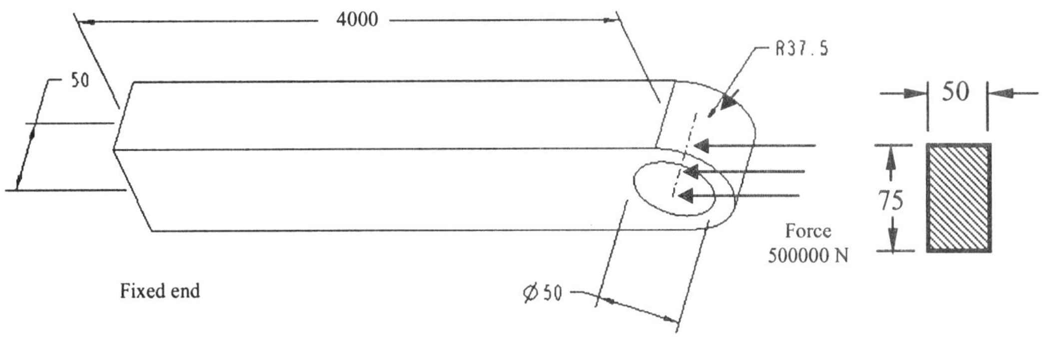

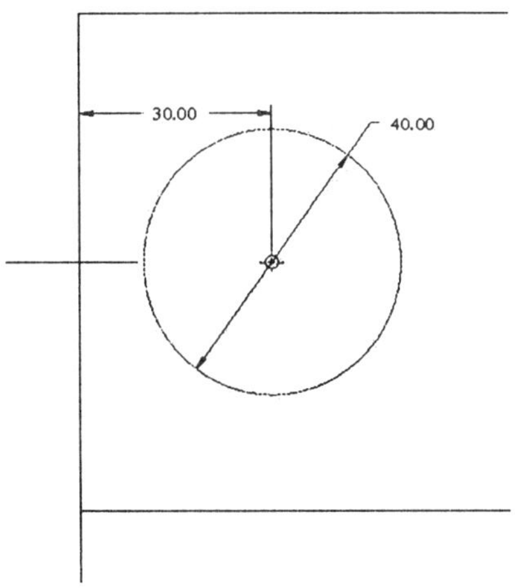

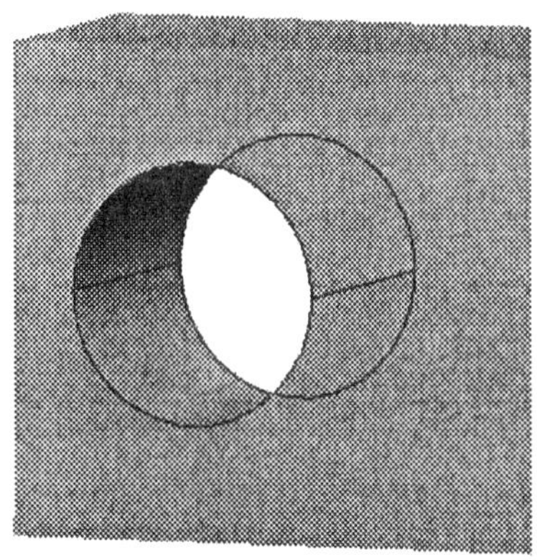

Construct a surface region at the bottom surface

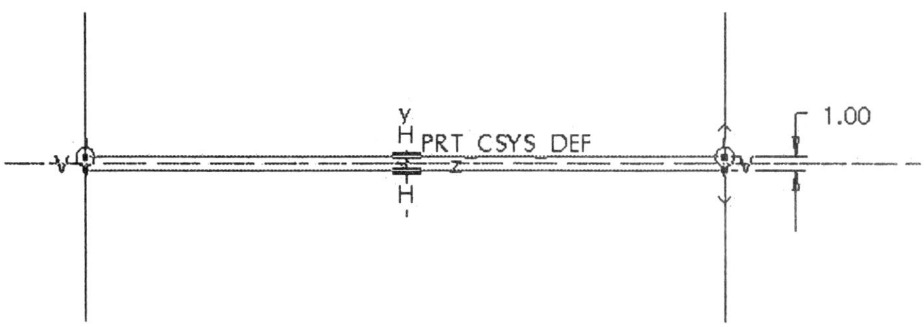

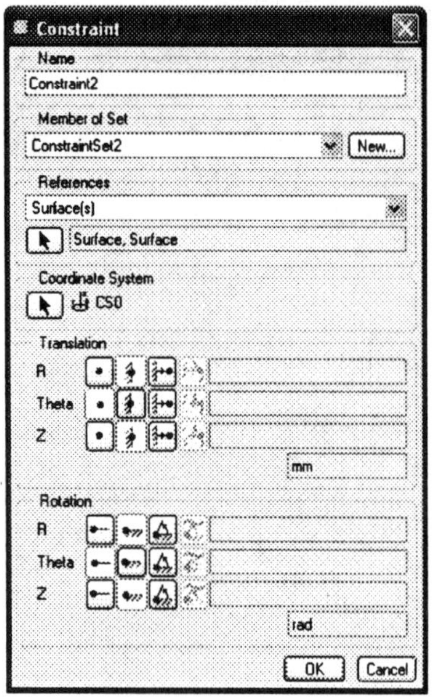

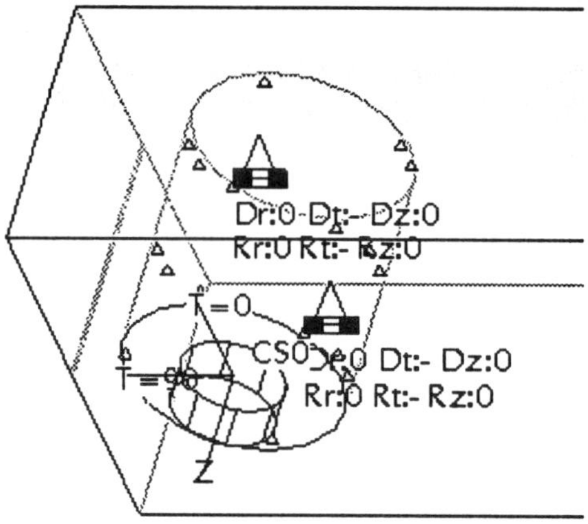

Determine the first 3 modal shapes and their associated BLF values.

CHAPTER 8

EVALUATION OF ENGINEERING DESIGN AND DESIGN OPTIMIZATION

8.1 Introduction

Design engineers design components and assemble them in their offices or under a virtual environment. A question, which is naturally to be asked, would be "Do their designed components or products meet the required specifications?" More important is such a question "What is the safety factor of the designed component or the product when considering the possible failure mode(s)?" Making design changes is a common practice for the purpose of improving the performance of the component and/or product or reducing the cost in the realization process. In this chapter, we present a case study. The component in this study is a bracket. The entire evaluation process represents an initial evaluation, a re-design process and re-evaluation so that numerical values of certain parameters of the designed bracket have been modified so that the performance of the re-design bracket meets the design specifications

8.2 Fundamentals of a Decision Making Process

The process of evaluation and redesign is a decision process. Before getting on the case study, let us go over the general guidelines and mathematical formulas used in a decision making process. For example, a company manufactures furniture. The company produces two types of furniture based on the demand in market. They are chairs and tables. The raw material used to manufacture the chairs and the tables is wood. Certainly, workers are needed to shape and assemble the components of the chairs and the tables. We will use normalized numbers to quantitatively describe this production process.

We assume that the objective of making a decision is to maximize the profit from selling the chairs and tables in market. Assume that the profit gained from selling a chair is \$4 and the profit for selling a table is \$5. We may mathematically represent this objective function as

Maximize the profit: profit is determined by $4*x_1 + 5*x_2$
where x_1 and x_2 are the number of the chairs and the number of the tables sold in market.

How many chairs and how many tables can be manufactured? There are certain constraint conditions, which limit the number of chairs and the number of table to be manufactured. Assume that total wood volume available at this moment for making furniture is 10 units. The number of workers available is 36. In order to make a chair, the consumption of wood volume is 1 unit. To make a table, the consumption of wood volume is 2 units. Therefore, the first constraint condition is

$$x_1 + 2x_2 \le 10 \quad \text{(wood)}$$

To make a chair, 6 workers are needed. To make a table, 6 workers are needed. Therefore, the second constraint condition is

$$6x_1 + 6x_2 \le 36 \quad \text{(workers)}$$

For example, we may add a new constraint condition, indicating that the maximum number of chairs needed in market is 4. Therefore, the third constraint condition is

$$x_1 \le 4 \quad \text{(limit)}$$

When combining the objective function and the constraint conditions together, we characterize such a decision process mathematically using the following format:

$$\text{Maximize } \ Z = 4x_1 + 5x_2$$

Subject to

$$x_1 + 2x_2 \le 10 \quad \text{(wood)}$$

$$6x_1 + 6x_2 \le 36 \quad \text{(workers)}$$

$$x_1 \le 4 \quad \text{(limit)}$$

$$x_1 \ge 0 \quad \text{(non - negative)}$$

$$x_2 \ge 0 \quad \text{(non - negative)}$$

The solution to this so-called optimization process is the maximum profit achievable is \$28 when $x_1 = 2$ and $x_2 = 4$. Such a combination makes the fullest use of the availability of wood and the number of workers available while it meets the limit on the number of tables set by the market demand.

In the following we use a graphical method to demonstrate the procedure of obtaining such an optimal solution. In this example, we have 2 design variables. As a result, we start the search for an optimal solution in a 2 dimensional space, as shown below. We only consider those regions where both x_1 and x_2 are positive.

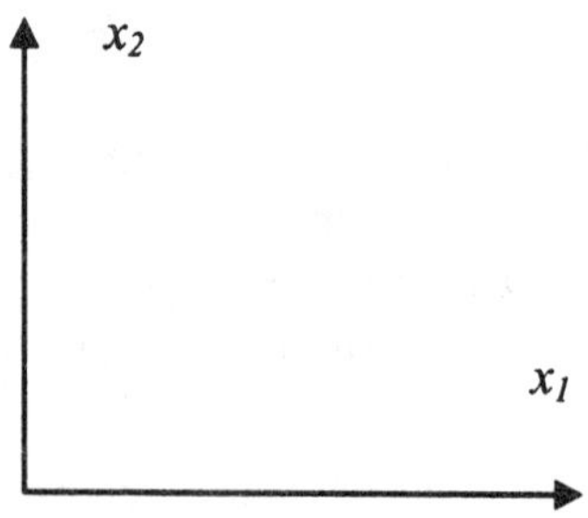

$$x_1 \le 0 \quad \text{(non - negative)}$$

The second step in the search process is to identify a feasible region to search. In this example, there are three constraint conditions. Mathematically speaking, each constraint condition can be represented by a straight line.

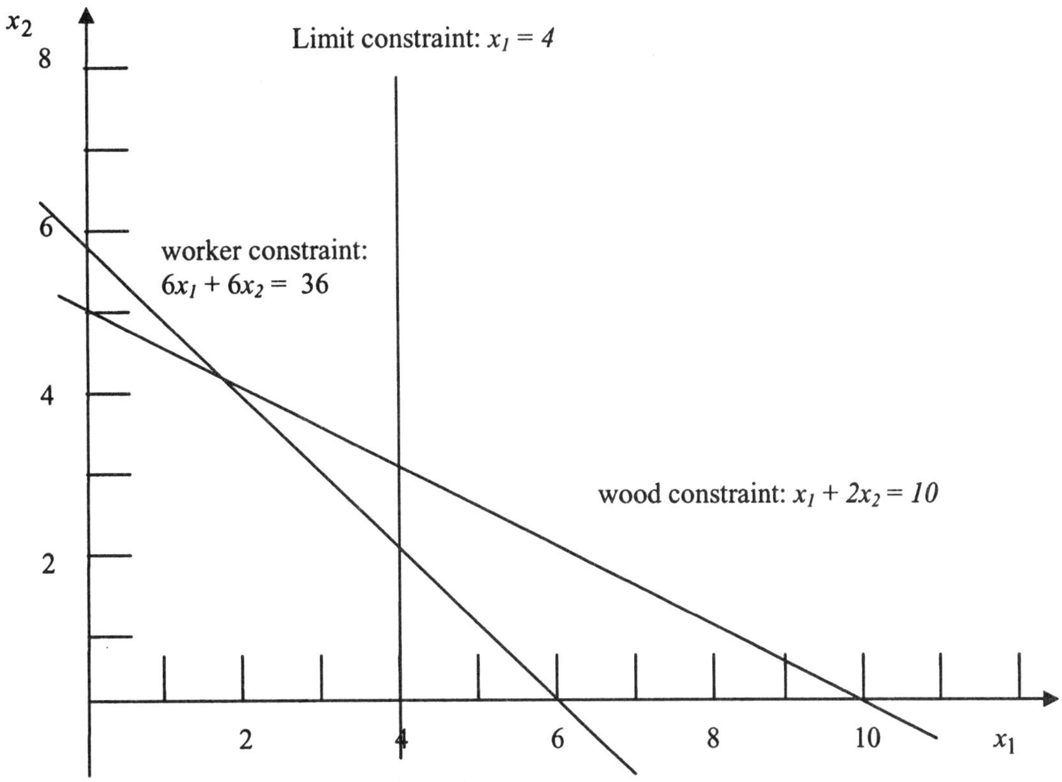

As illustrated, the three constructed lines intersect with each other. They also intersect with the x_1-axis and x_2-axis. When connecting all these segments, a closed area is formed. The shadow area shown in the following figure represents the feasible region to search for an optimal solution.

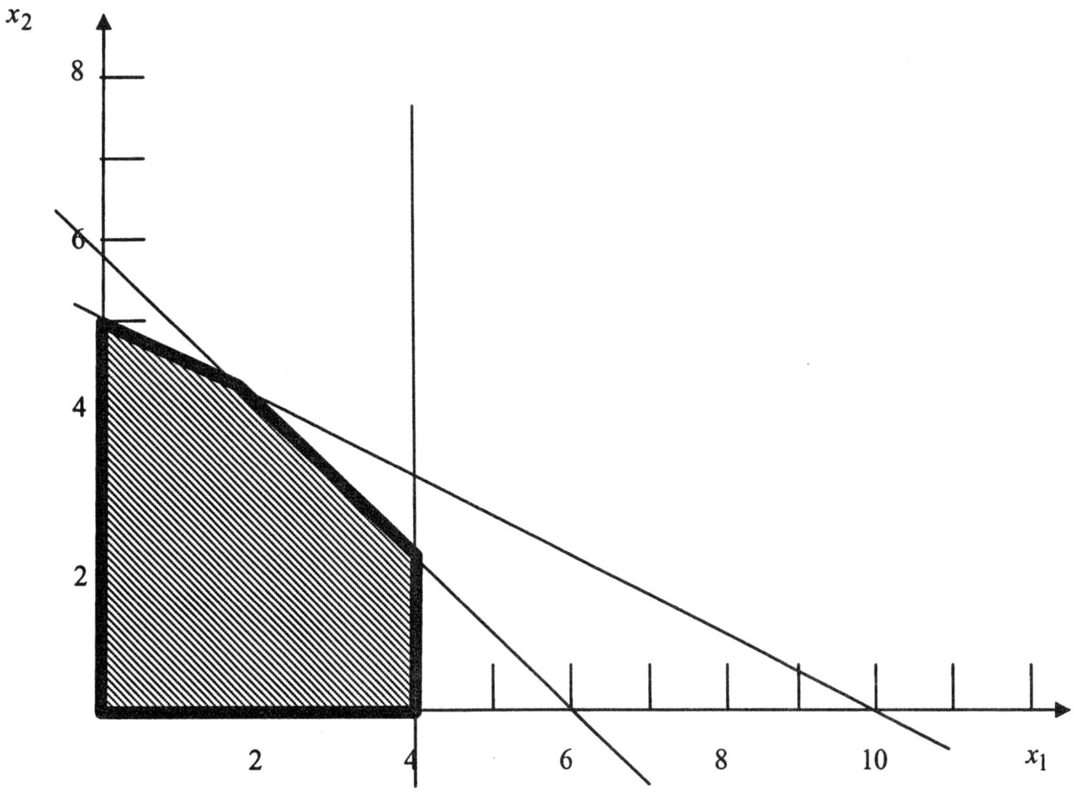

How to derive at an optimal solution? Readers may imagine moving an inclined line starting from the origin (x1 = x2 = 0) in a parallel fashion. The slope of this inclined line is determined by the 2 coefficients in the objective function. When this inclined line reaches its limiting position, the location ($x_1 = 2$ and $x_2 = 4$) is the optimal solution.

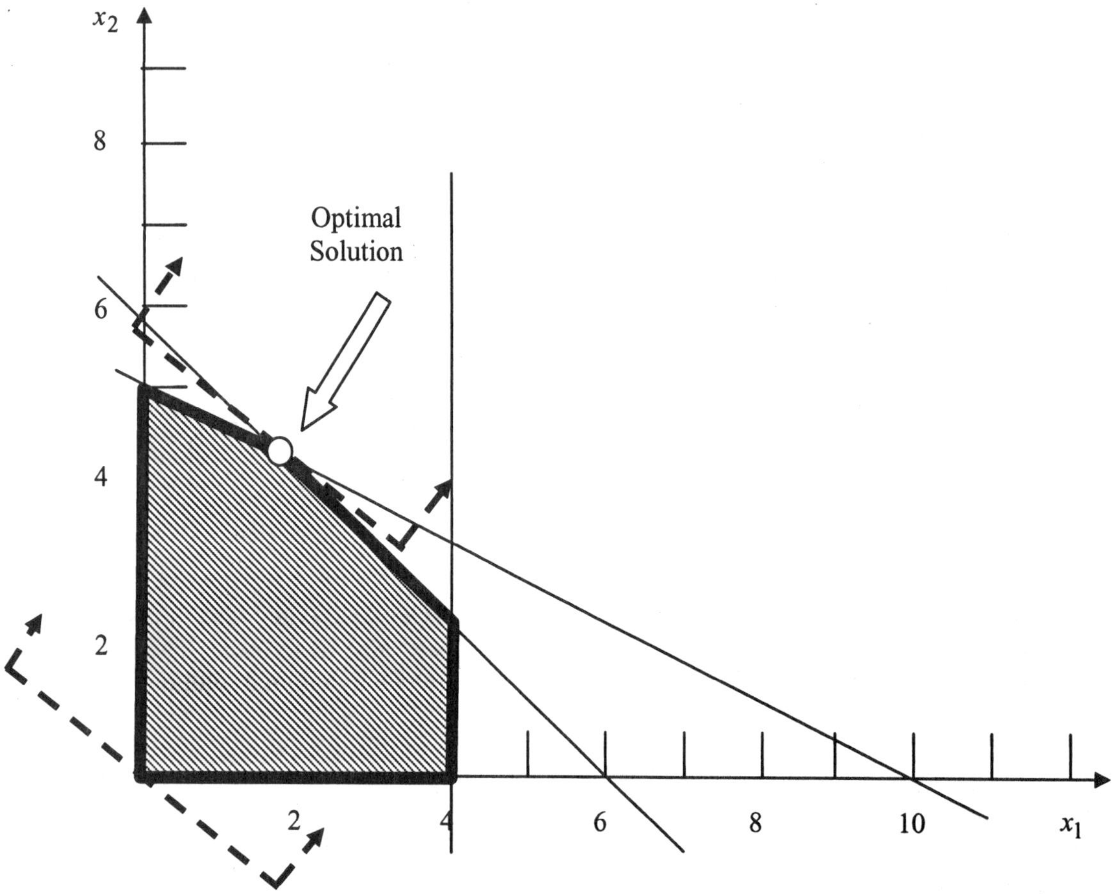

8.3 Case Study: Redesign a Bracket Component

The geometric shape of a bracket is shown below with its dimensions. The material is steel. The objective of this re-design process is to reduce the weight of the bracket mass so that the cost of the bracket can be reduced accordingly in terms of raw material utilization, packaging, and transportation.

In the re-design process, there are certain constraints. The first constraint is that the maximum von Mises stress should be kept below 85 MPa. The reason is that the yield stress of the steel material is about 225 MPa. For a safety factor of 2.5, the limit for the maximum von Mises stress should be set at 85 MPa. However, we are informed that the maximum von Mises stress at this design stage when it is subjected to loading is 90 MPa, a violation to the safe constraint condition.

The second constraint is the limit on the maximum displacement magnitude. The current value of this performance index is 0.028 mm. It has to be kept below 0.02 mm.

The third constraint is related to the product process. The bracket components will be fabricated using a punch process where a mold is used. We are limited to change a few dimensions among the many shown on the following engineering drawing. Therefore, we have to select the design parameters with care in the redesign process so that the changes of their dimensions can be accommodated in the process of modifying the mold design accompanied.

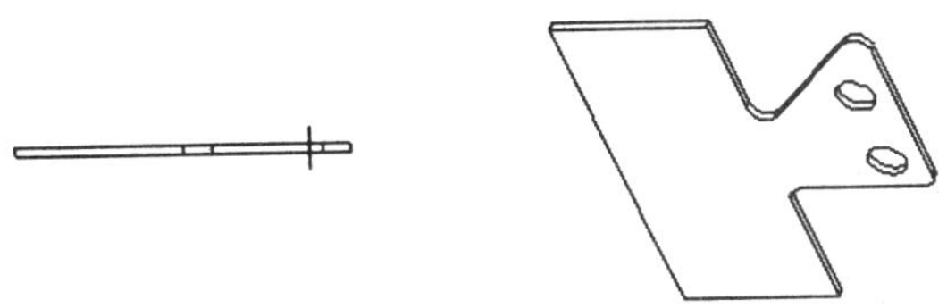

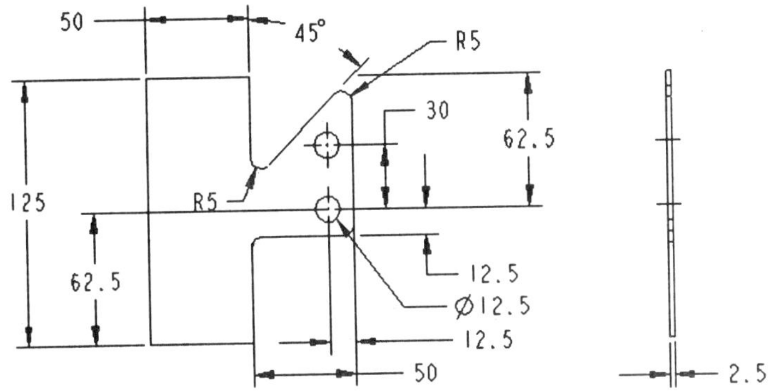

The boundary condition is that the top surface and the bottom surface on the right side of the bracket are fixed.

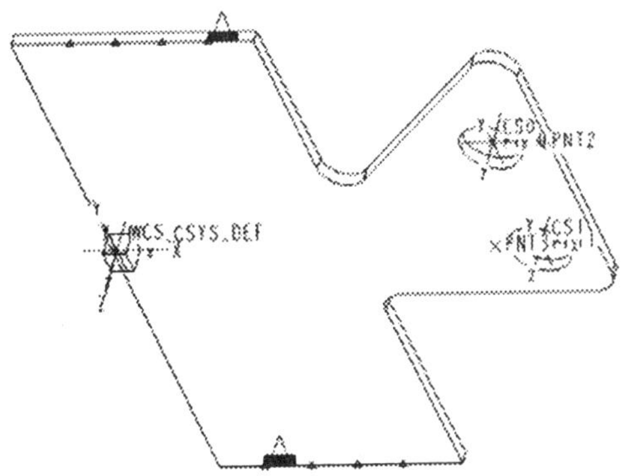

The loading condition consists of 2 bearing loads. One is acting on the upper hole and the other on the lower hole. The magnitudes of these 2 bearing load are 1001.25 Newton.

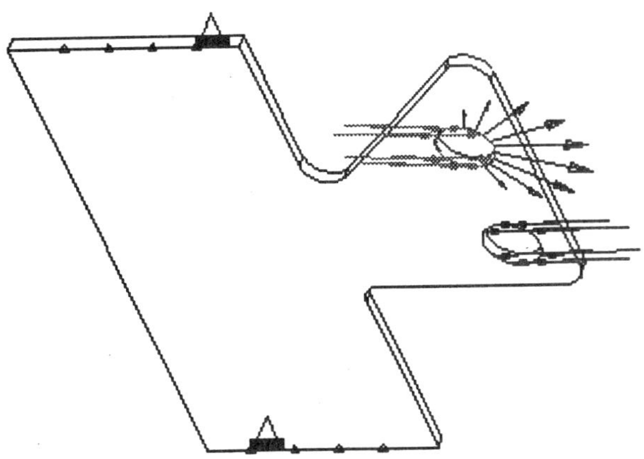

Step 1: Under the **Pro/ENGINEER** design environment, create a 3D solid model of the bracket. Assume that the file name is *bracket.prt.*

In the process of creating this bracket, pay special attention to the two inner fillets. They are created as equally to each other. The same is applied to the two outer tips. Their dimensions will be used as two (2) design parameters in the optimization process.

Step 2: Define design parameters.

Assume that the following 4 dimensions will be used as the design parameters.

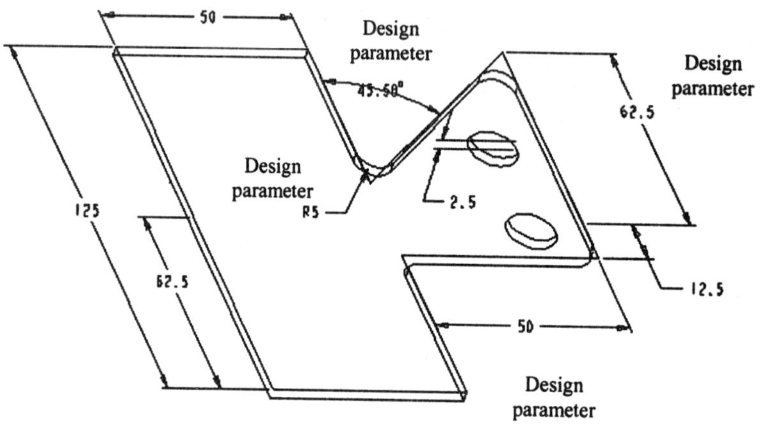

- The angle dimension: 45°
- The distance between TOP to the tip of the nose: 62.5
- The radius of the inner fillets: R5
- The over width: 50

Let us first use symbolic names for the 4 design parameters selected. From the model tree, highlight the protrusion feature > right click and pick **Edit**.

From the geometry on display, pick the dimension of 45 degrees, right click and pick **Properties**. Click the box called **Dimension Text**. In the box called **Name**, type in *ang*.

Repeat the above procedure by picking the dimension of 62.5 and type in *top,* picking the dimension of R5 and type in *inner_fillets* and picking the dimension of 50 and type over_width.

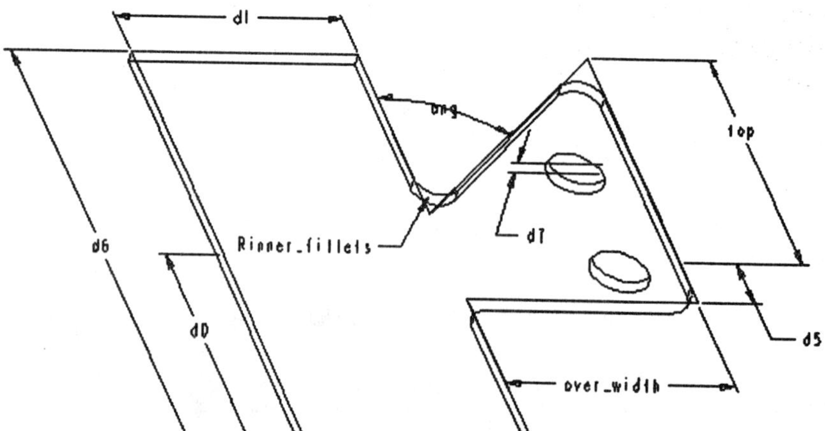

To switch between the dimensions in numerical values and the dimensions in symbolic names, from the main toolbar: **Info > Switch Dimensions**

Step 3: Use Relations to Assure the Compatibility of Geometric Shape in the Process of Optimization.

Note that the tip has to rotate as the angle dimension varies to maintain the desired part shape. The following relation is established.

From the main toolbar, **Tools** > **Relations** > in the window of **Relations**, type in the following relation, using three lines:

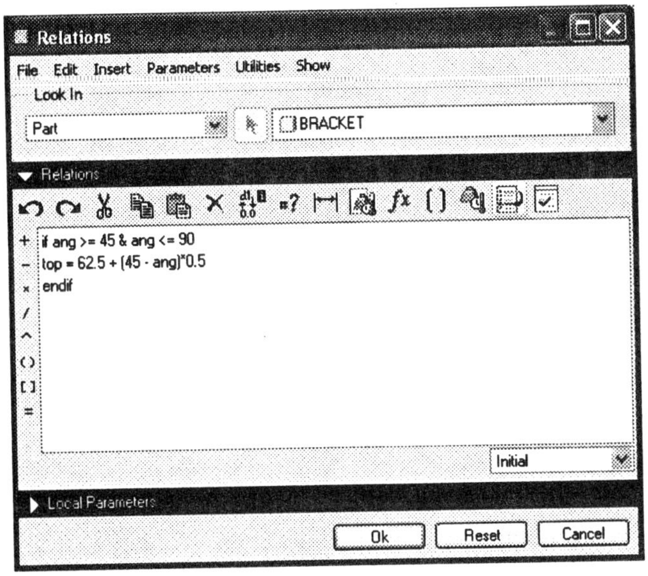


```
if ang >= 45 & ang <= 90
top = 62.5 + (45-ang)*0.5
endif
```

Upon completion, click **OK**.

To verify the compatibility and if the defined relation works, highlight the protrusion feature displayed on the model tree, right click and select Edit.

From the geometry on display, select the angle dimension and change it from 45 to 60 > click the icon of **Regenerate,** which is on the menu bar.

Select the angle dimension and change it from 60 to 90 > click the icon of **Regenerate,** which is on the main toolbar.

Select the angle dimension and change it back from 90 to 45 > **Regenerate.**

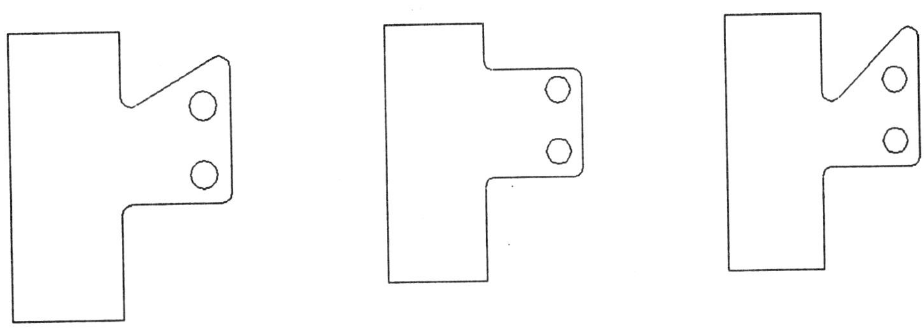

Step 4: Under the Pro/ENGINEER design environment, define 4 datum points for specifying the 2 bearing loads.

Define PNT0 and PNT1

Select the icon of datum point from the toolbar of creating datum features.
Select the **FRONT** datum plane and, while holding down the Ctrl key, select the axis of the upper hole > **OK. PNT0** is defined.
Repeat the above procedure to define **PNT1** by selecting the axis of the lower hole.

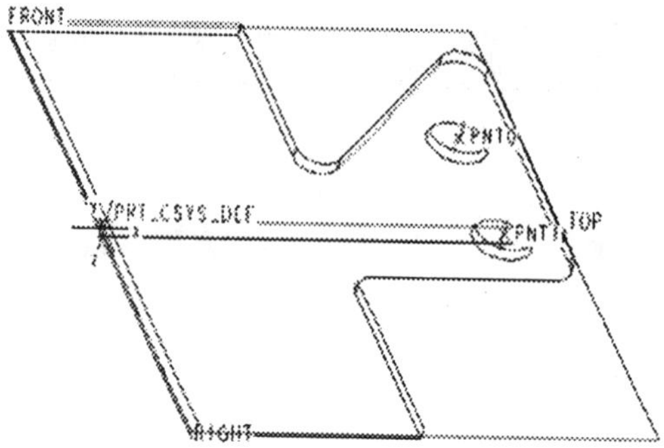

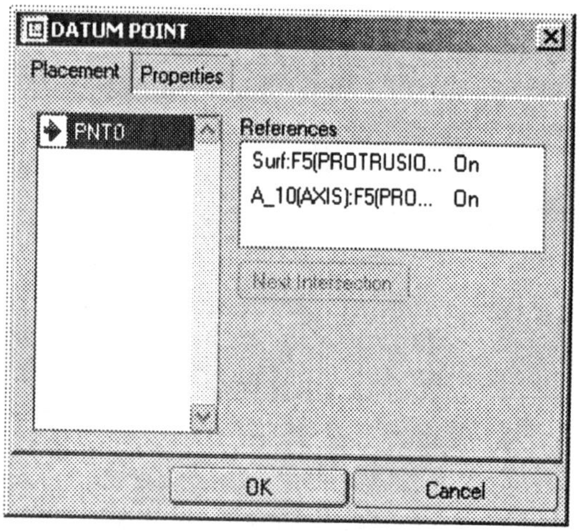

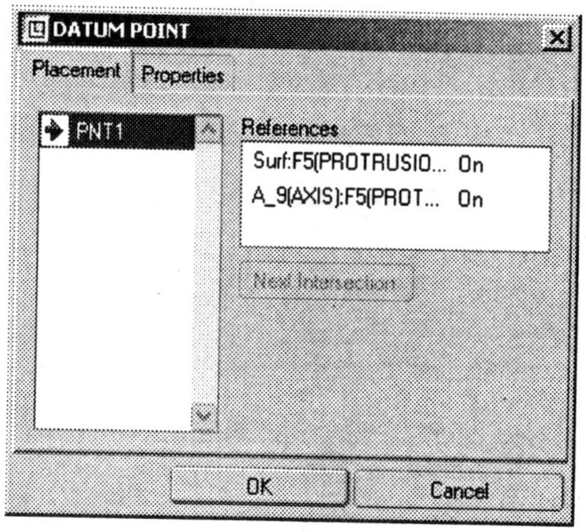

Define PNT2 and PNT3

We first define two coordinate systems.

Select the icon of coordinate system > pick the axis of the upper hole and a vertical edge of the bracket component > re-orient CSO > **OK**

Repeat this procedure to define CS1 by picking the axis of the lower hole.

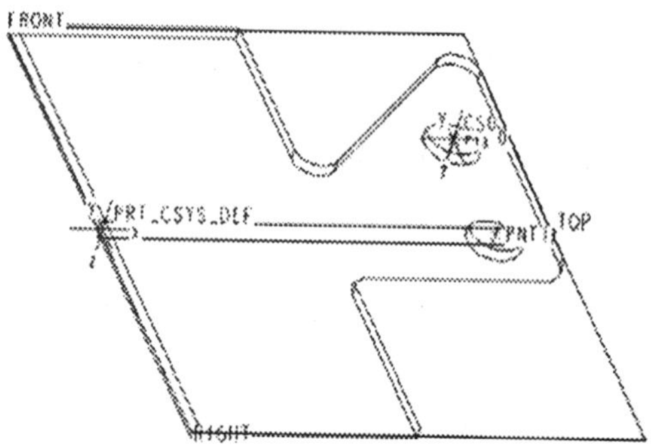

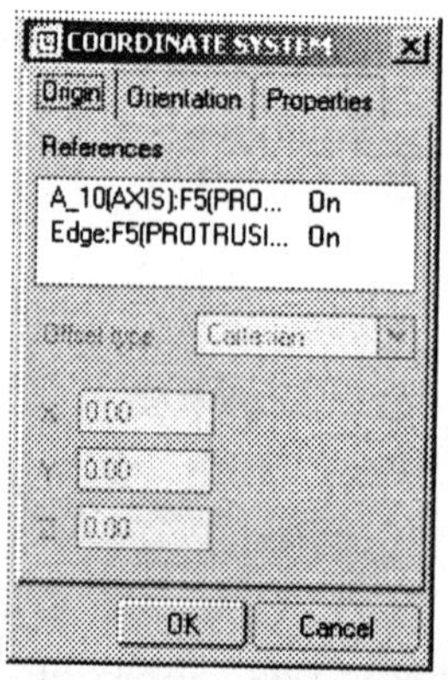

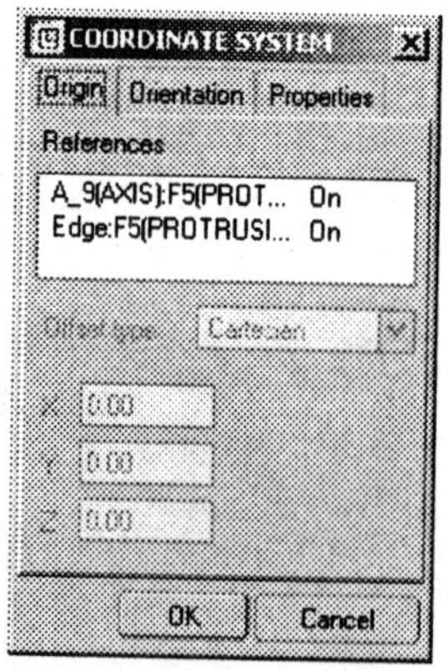

To define **PNT2**, select the icon of datum point > pick PNT0 and CSO > along the x direction, specify the offset distance equal to 10 > OK. **PNT2** is defined.

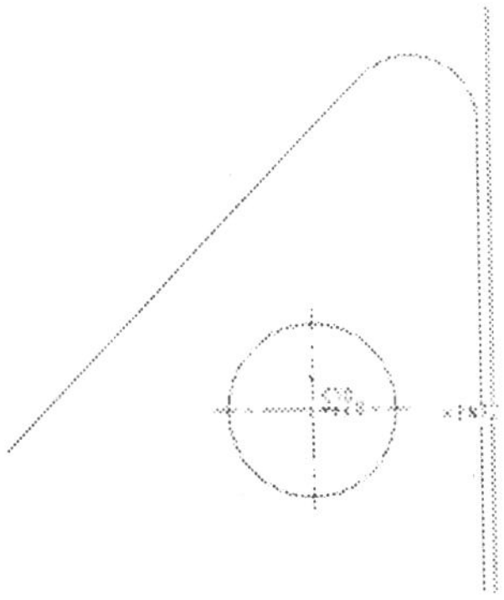

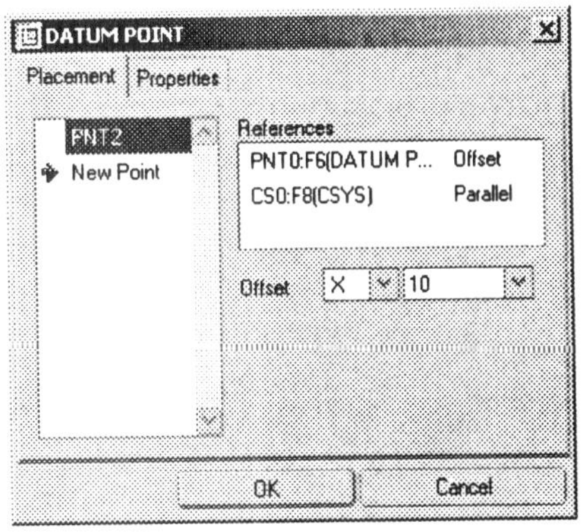

Repeat the above procedure to define **CS1** first, then **PNT3**. The offset value along the x direction is -10.

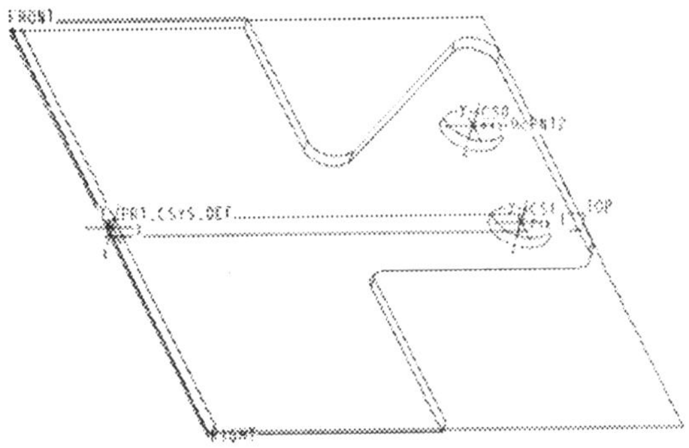

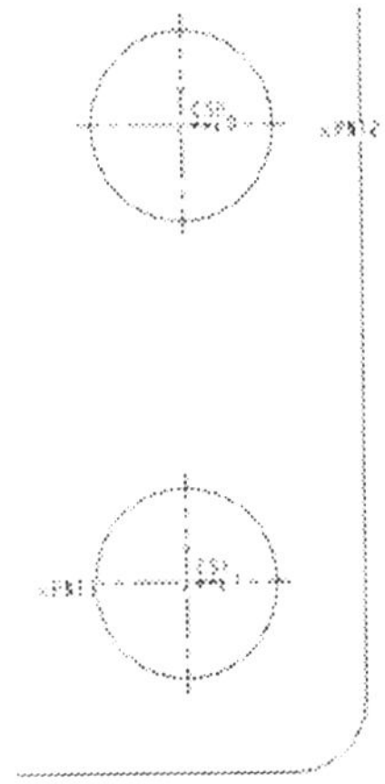

Step 5: Under the Pro/MECHANICA environment, develop a shell model.
Applications > Mechanica > Continue > Structure > OK.

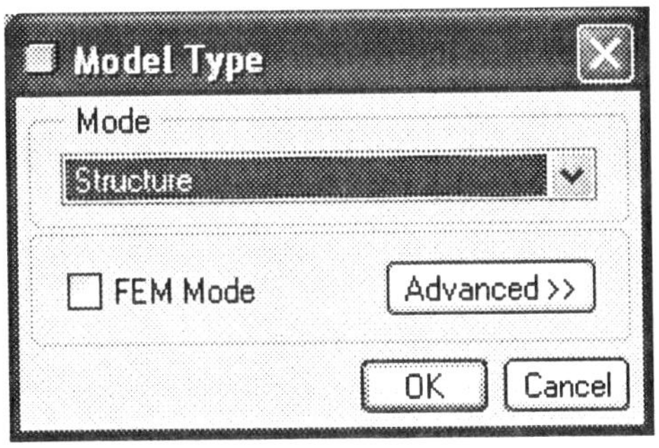

From the toolbar of functions, select the icon of **Materials** > select **STEEL**.

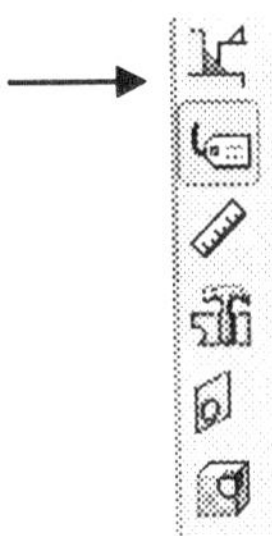

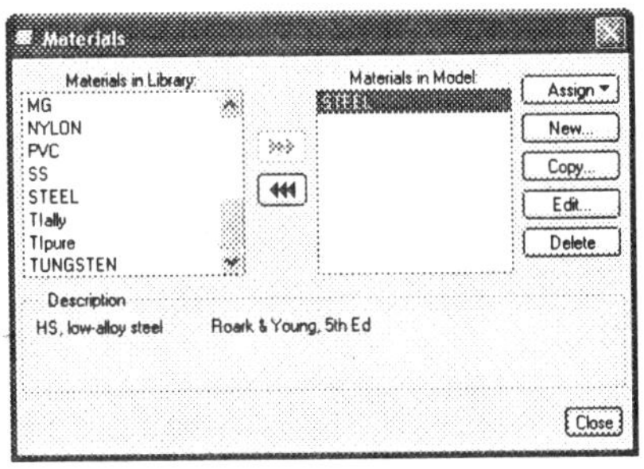

Assign > **Part** > select the bracket structure from the screen display > **OK** > **Close**.

From the toolbar of functions, select the icon of **New Shell Pair** > **Auto Detect** > **Compress** > **Shells Only**. Note that the system does not respond to the request.

Select **New** > **Constant** > pick the front and back surfaces while holding the Ctrl key > **OK**.

Compression > **Shells Only** > **Done**.

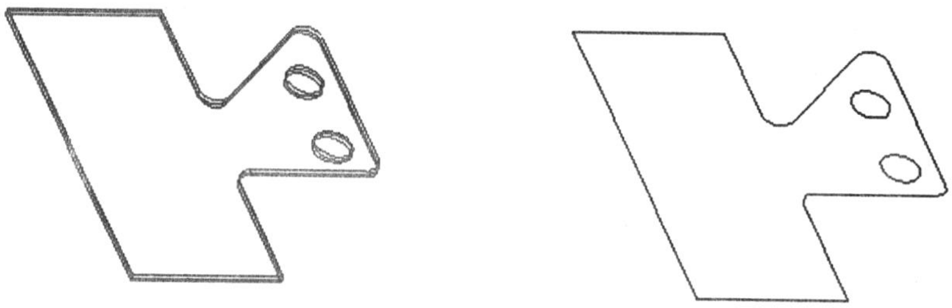

Step 6: Define the constraint condition. Assume that the top and bottom edges are welded.

Model > **Constraints** > **New** > **Edge/Curve** > name the constraint set *weld*, select the front top edge and the front bottom edge, and fix all six degrees of freedom.

From the toolbar of functions, select the icon of **Constraints** > select **Edge(s)/Curve(s)** > pick the top and bottom edges while holding the Ctrl key > **OK**.

Fix all 6 degrees of freedom > **OK**.

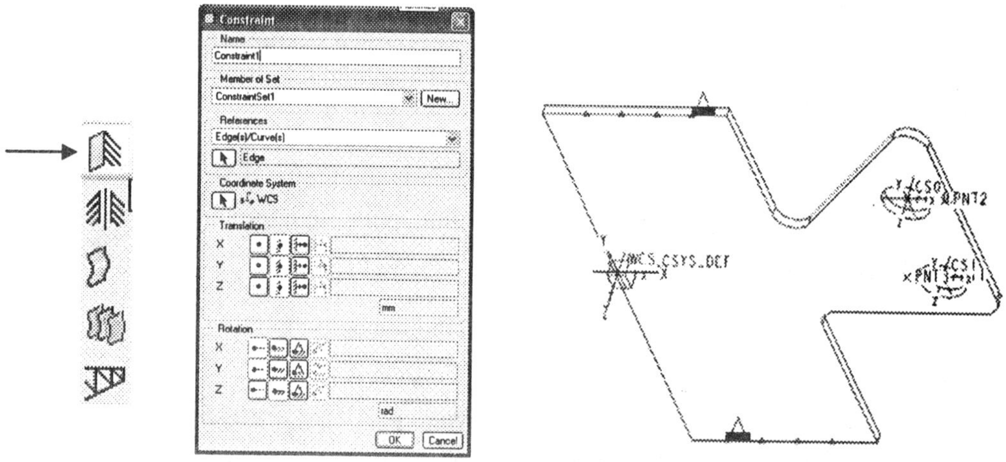

Step 7: Define the load condition. There are two bearing loads. One is acting on the upper hole and the other on the lower hole.

We have already defined four (4) datum points. To define the bearing load acting on the upper hole, select the icon of **bearing load** from the toolbar > **Hole/Pin** > pick the edge on the right side of the upper hole > **OK.** Change the coordinate system to CS0. To specify the direction and magnitude, use **Dir Points & Mag**, select **PNT0** as the starting point and **PNT2** as the ending point. To specify the magnitude, type *1001.25*.

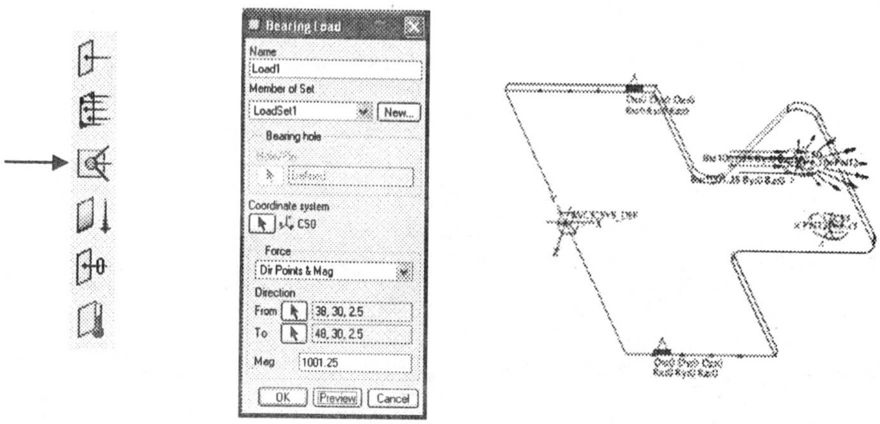

For the bearing load acting on the lower hole, select the icon of **bearing load** from the toolbar.

Select **Hole/Pin** > pick the edge on the left side of the lower hole > **OK.**

To specify the direction, use **Dir Points & Mag**, select **PNT1** as the starting point and **PNT3** as the ending point.

To specify the magnitude, type 1001.25.

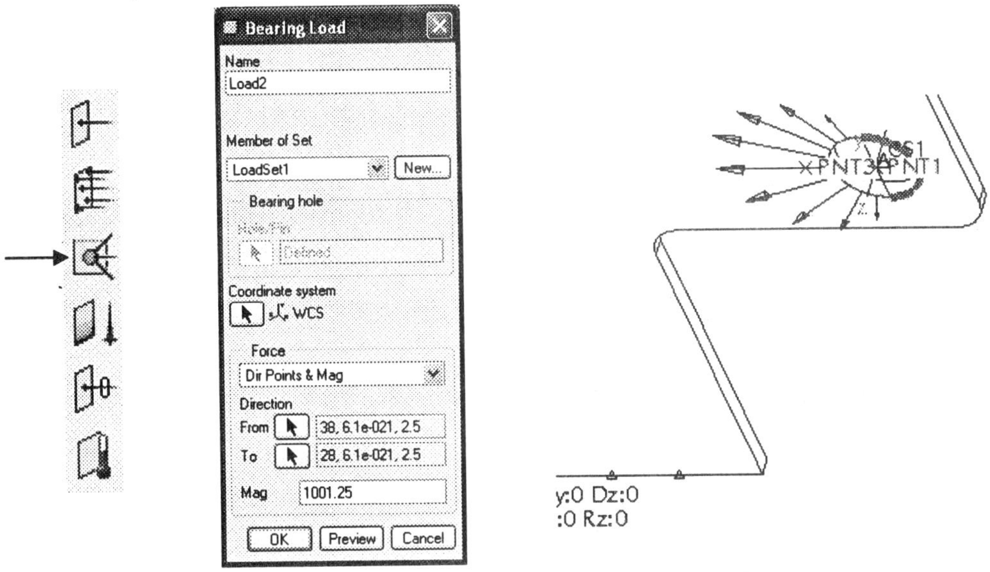

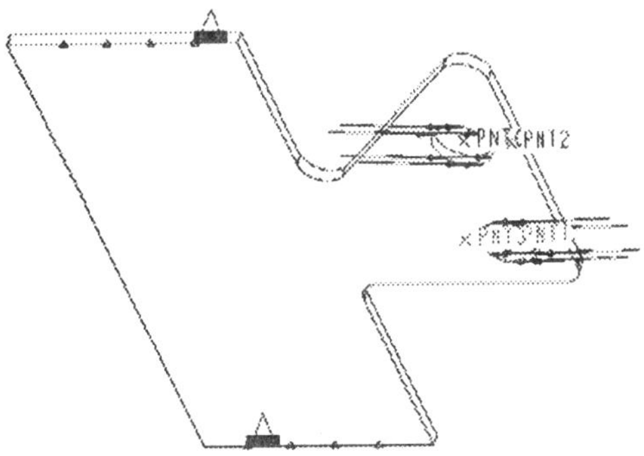

Step 8: Perform a static analysis and run it.

Click the icon of performing a design study from the main toolbar > File > Select New Static. Type bracket_base_run as the folder name and run the program.

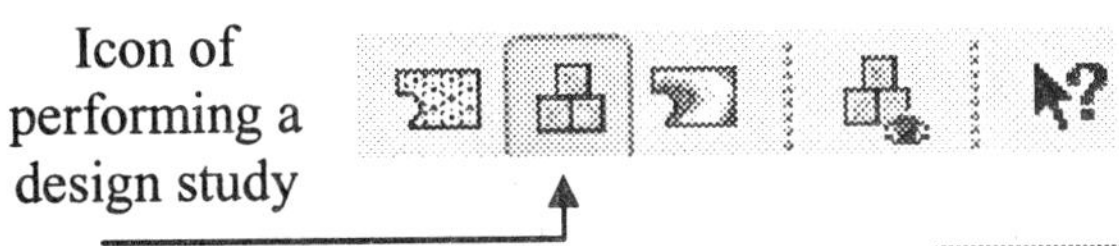

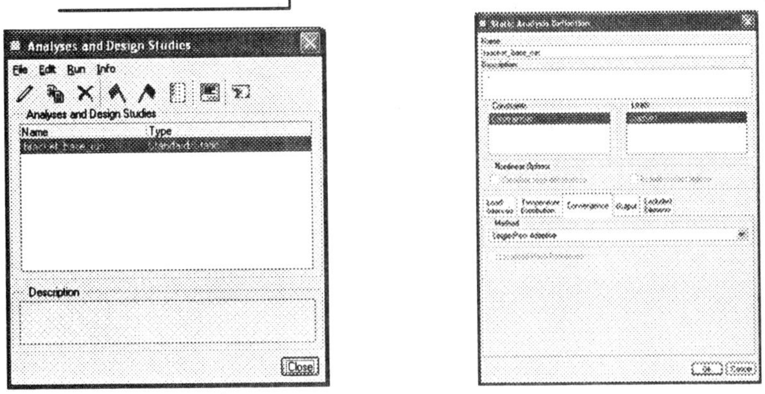

Step 9: Study the evaluation results.

Results > Insert > Result Windows > select bracket_base_run

The von Mises stress distribution is shown below. The maximum von Mises stress is 142 MPa at the location of the upper inner fillet.

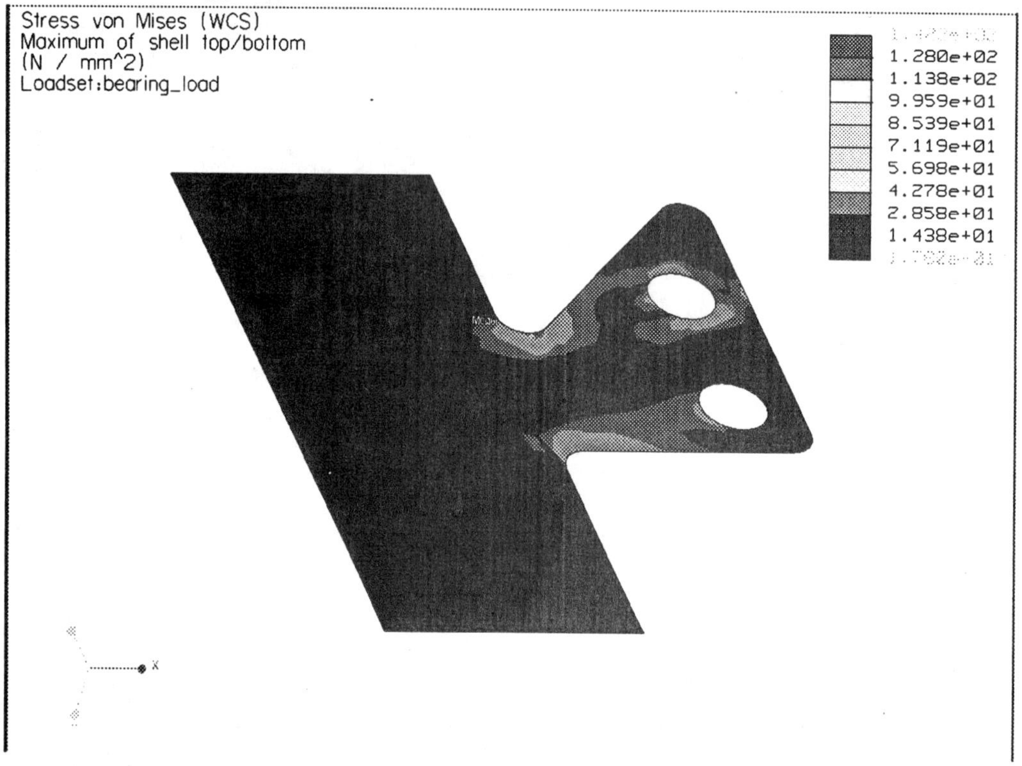

The displacement distribution is shown below. The maximum displacement is 0.05279 mm at the location of the upper tip.

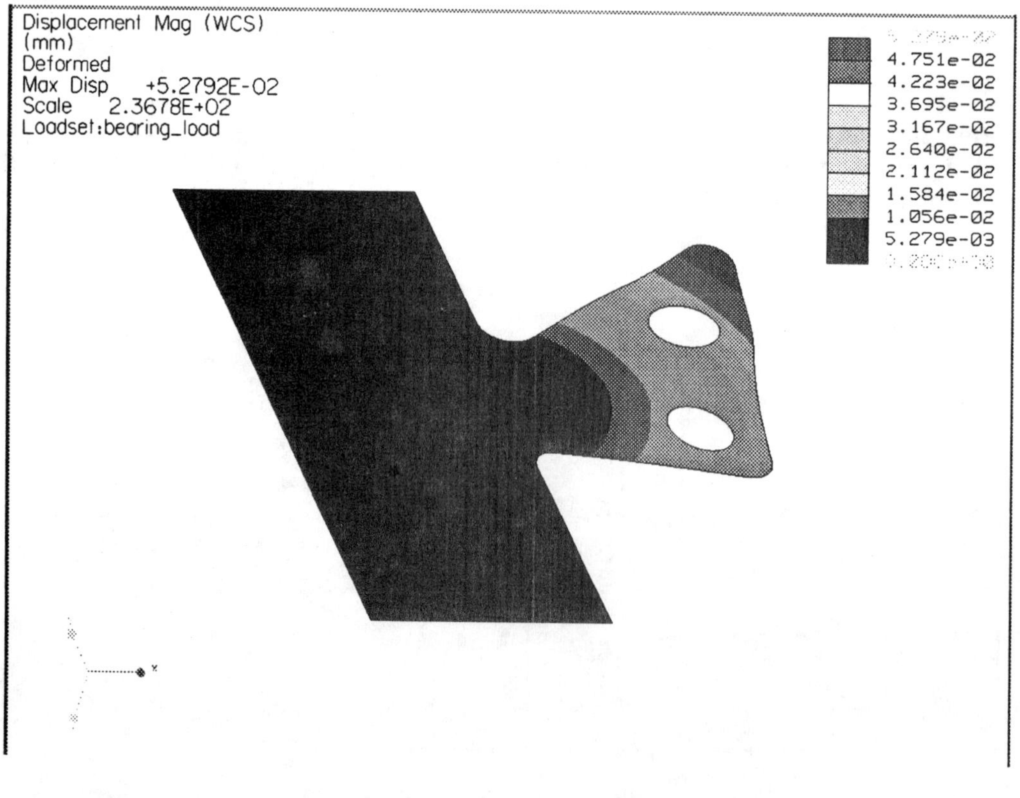

Step 10: Define the design parameters for the optimization.

Because the maximum Von Mises stress is 158 MPa at the current design, which is larger than 90 MPa (we do not use 80 at this moment), a redesign process is needed. The following flow chart outlines the basic steps involved in a re-design process.

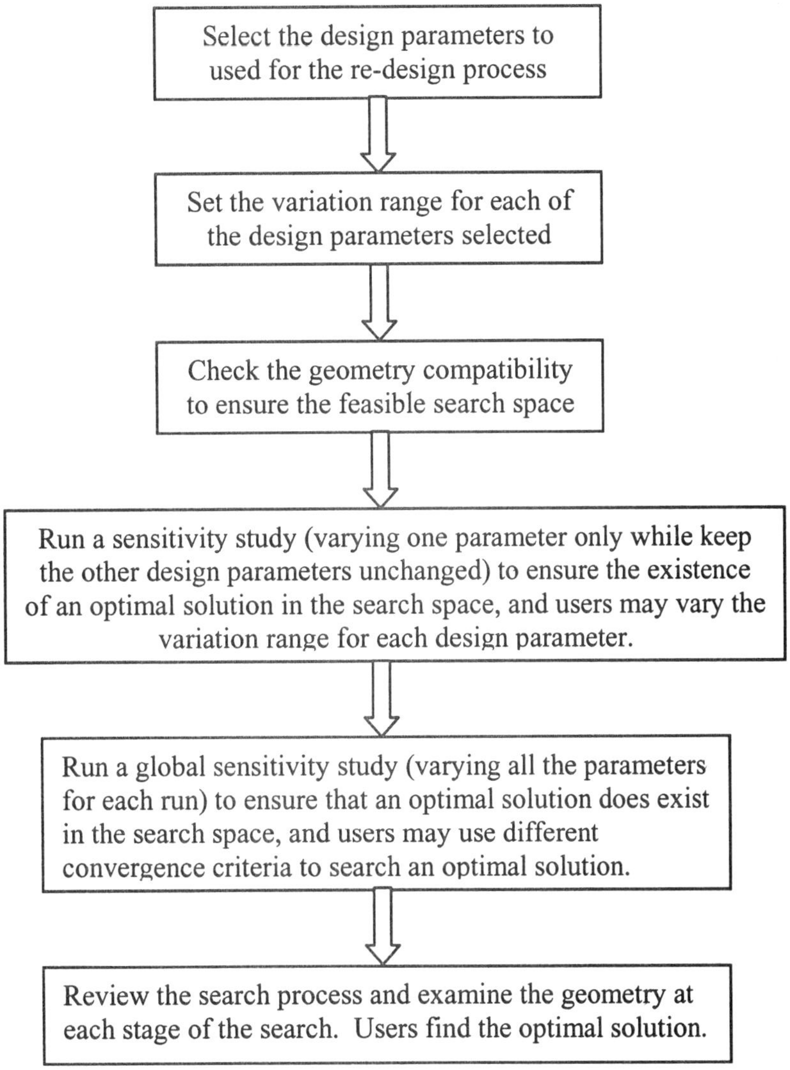

At the end of search, users should be able to obtain a set of solutions, which represent the redesigned bracket components. Users should be able to examine each of these redesigned bracket components through the use of **Optimal History**, and find the optimal design based on the design specifications.

To define a set of design parameters, From the main toolbar, select **Analysis > Dsgn Controls** > **Design Params** > **Create.**

Select **Dimensions** and pick the following three parameters: ang, over_width and inner_fillets. When selecting the parameter of ang, the following page appears. Set the range from 45 to 60.

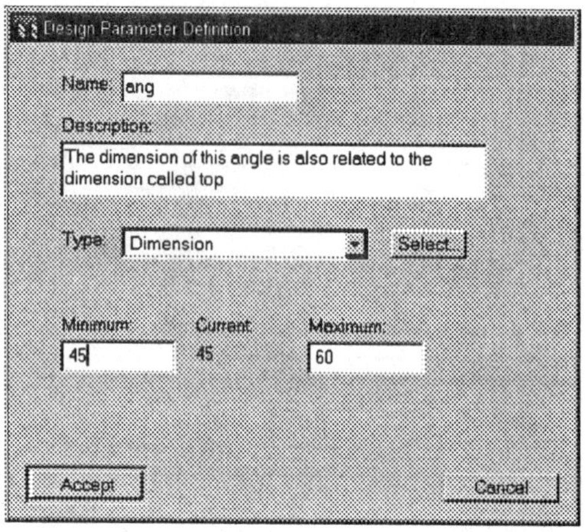

When selecting the parameter of inner_fillets, the following page appears. Set the range from 2.5 to 12.5.

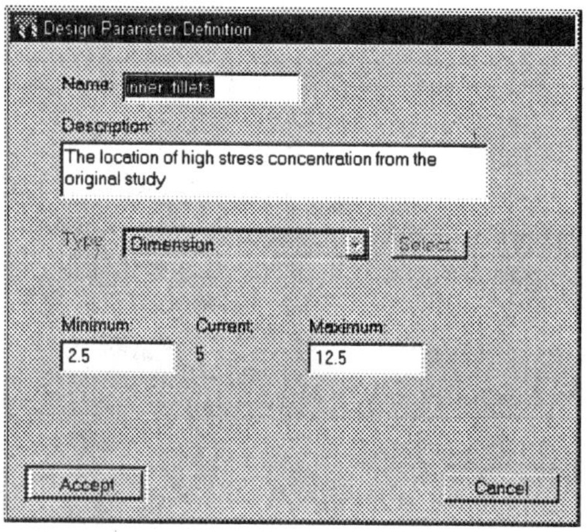

When selecting the parameter of over_width, the following page appears. Set the range from 45 to 55.

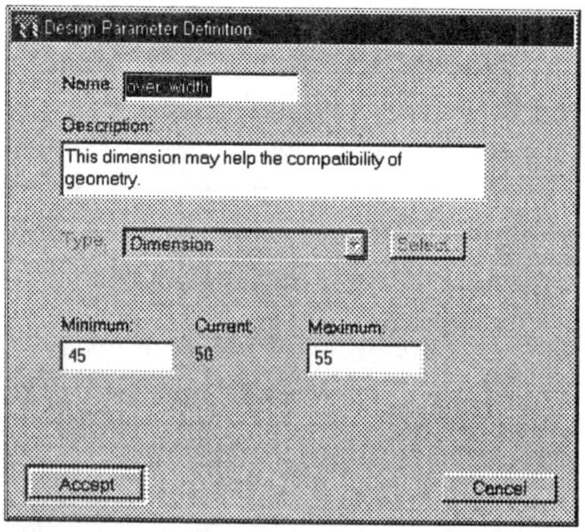

To verify the compatibility of geometrical shape, we can perform shape animation.

Dsgn Controls > Shape Animate > specify the following information on the shape animate window.

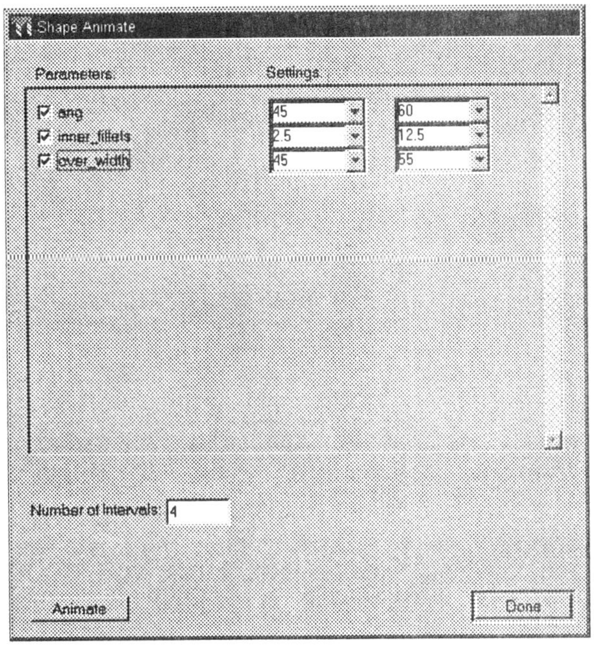

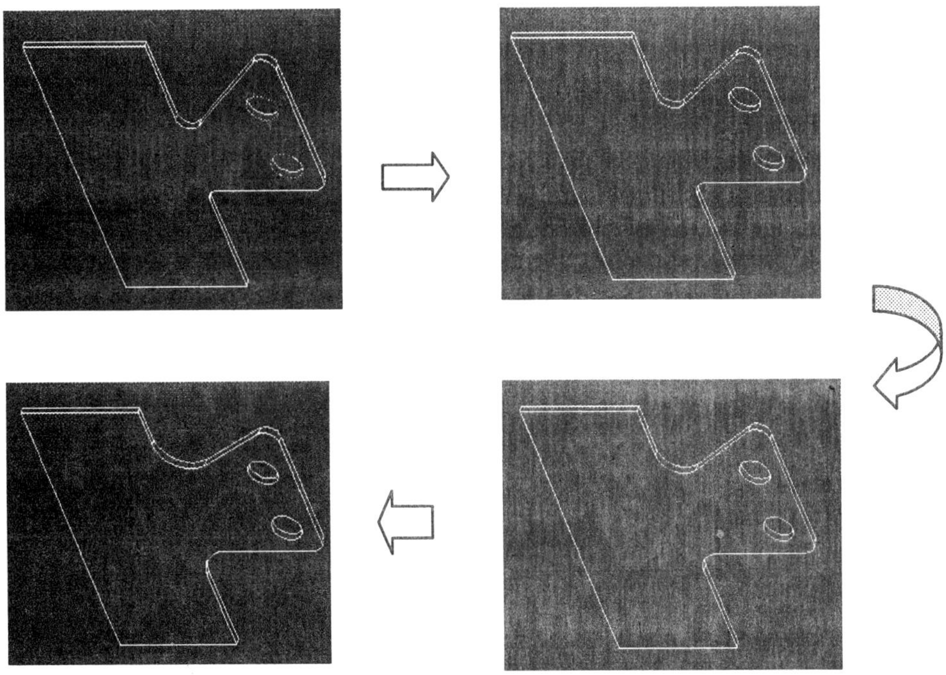

Step 11: Local Sensitivity Study with the Defined Design Parameters

The purpose of carrying out a sensitivity study is to make sure that the constraint conditions used in the process of optimization are meaningful. For example, if we want to set the maximum Von Mises Stress below 80 MPa while searching the redesign with a minimized mass. Are we sure the three (3) ranges set in Step 9 are appropriate, or the search space contains feasible solutions to meet the constrain condition of 80 MPa.

Pick the icon of **performing a design study** > **New Design Study** > type local_sen as the name for the local sensitivity study, select **Local Sensitivity Study** and fill in the following information: ang=45, inner_fillets=5 and over_width=50. Make sure that bracket_base_run is selected > **Accept** > **Run**.

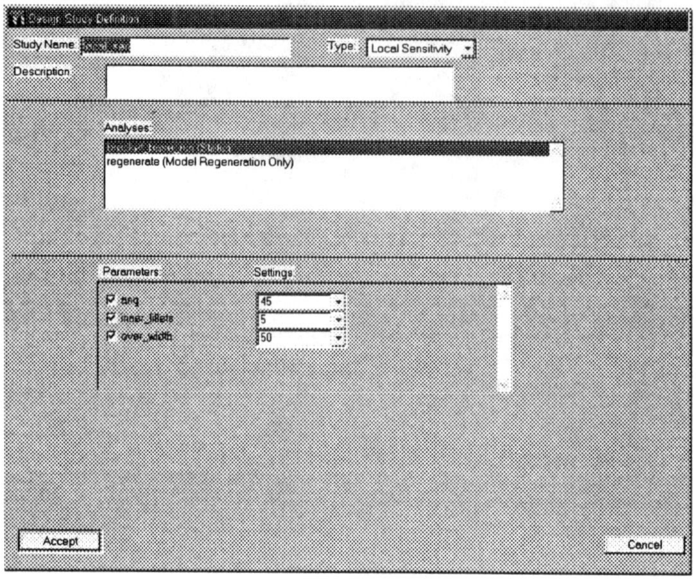

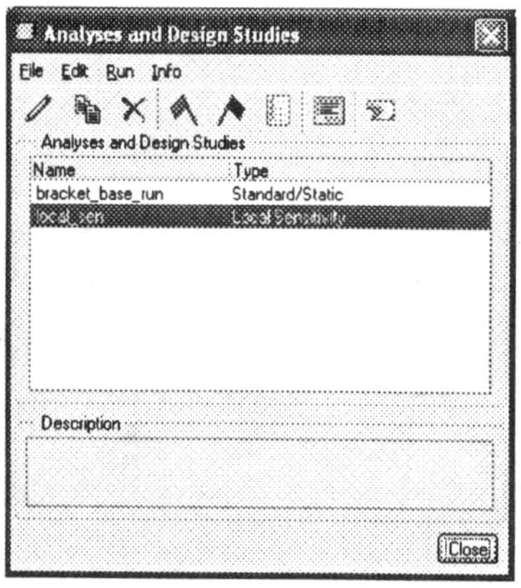

After running the local sensitivity study, plot the maximum von Mises stress vs a design parameter for ang, inner_fillets, and over_width, respectively.

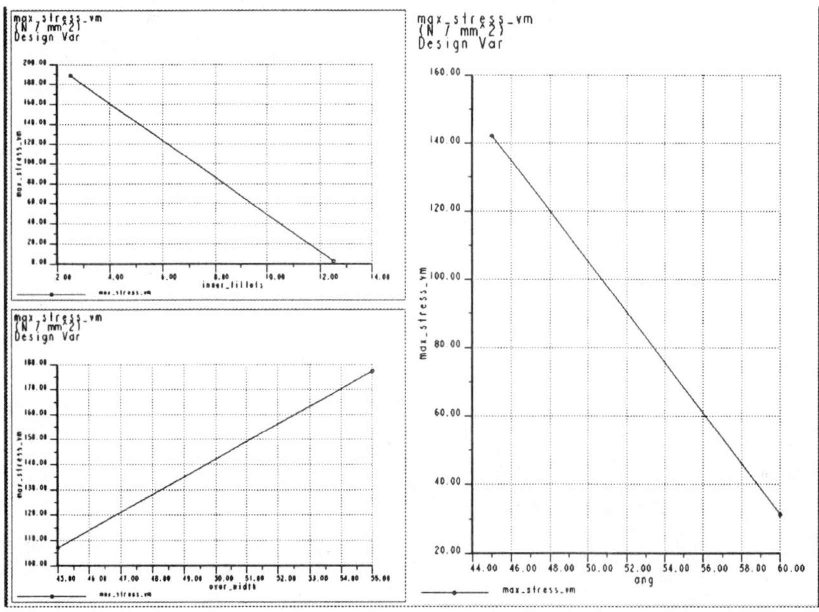

Parameters	Parameter variation ranges	Max VM variation ranges
Ang	45-60	142-30 MPa
Inner_fillets	2.5-12.5	190-0 MPa
Over_width	45-55	105-180 MPa

Note that there is a numerical value of the von Mises stress, which is equal to zero on the list. As we recall that the von Mises stress is always a positive and non-zero value. Is there any thing wrong with the local sensitivity study? The fundamentals underlying the local sensitivity study are to calculate the system response, as defined by the user, to perturbations (a very slightly change: 1% of the design parameter's setting) in the design parameters. Such a calculation comes up with a slope estimated, which is used to linearly interpolate the variation of the system response for the entire range covered by the design parameter. This zero von Mises stress came from this linear interpolation.

It is important to know that the parameter variation ranges for ang and inner_fillets are appropriate because they cover the allowable VM value of 90 MPa. However, the parameter variation related to over_width is questionable because its maximum VM stress variation is between 110 and 200 MPa.

Next, let us review the results for the displacement requirement. The following three plots illustrate the variation of the maximum displacement when each of the three parameters varies within the specified range.

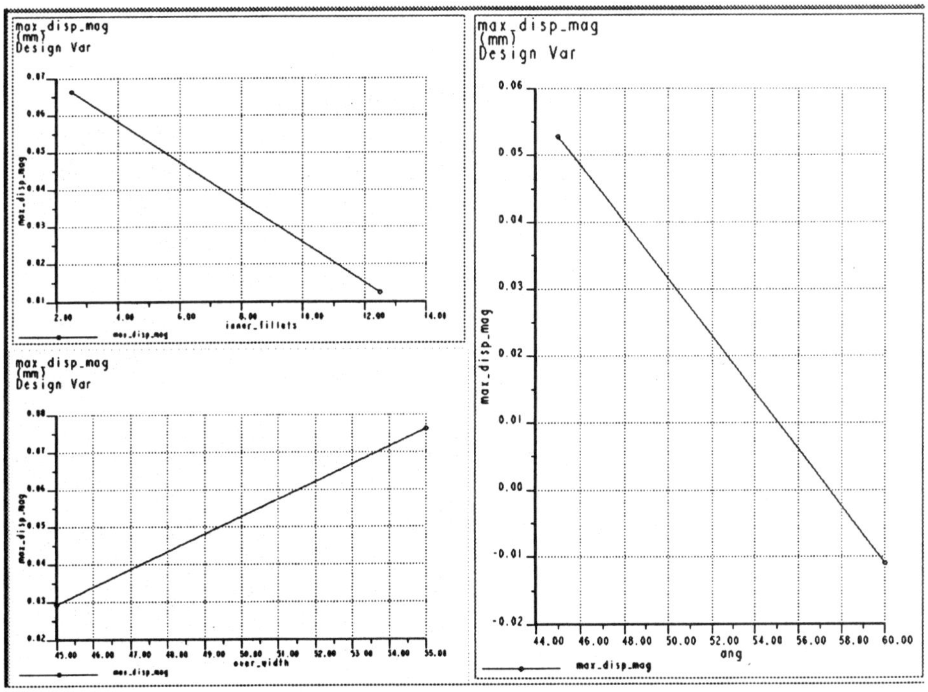

Parameters	Parameter variation ranges	Max Displacement ranges
Ang	45-60	0.055 – (-0.010) mm
Inner_fillets	2.5-12.5	0.015 – 0.065 mm
Over_width	45-55	0.030 – 0.075 mm

Note that there is a numerical value of the displacement, which is negative on the list. Intuitively, it is impossible for the displacement to go up as the angle value increases. As we recall the fundamentals underlying the local sensitivity study, the linear interpolation of the result obtained from perturbation could lead to such a negative value.

The calculation is based on a very slightly change: 1% of the design parameter's setting of the design parameters. Such a calculation comes up with a slope estimated, which is used to linearly interpolate the variation of the system response for the entire range covered by the design parameter. This zero von Mises stress came from this linear interpolation.

At the end of a local sensitivity study, let us make an adjustment to the settings of the three design parameters for the global sensitivity to be carried out.

Parameters	Parameter variation ranges
Ang	45-60
Inner_fillets	5-12.5
Over_width	45-50

Step 12: Global Sensitivity Study with the Adjusted Variation Ranges of the 3 Design Parameters

Pick the icon of **performing a design study** > **New Design Study** > type in *global_sen* as the name for the global sensitivity study, select Global Sensitivity Study and fill in the following information:

ang: 45 – 60
inner_fillets: 5 - 12.5
over_width: 45 - 50

Make sure that bracket_base_run is selected > **Accept** > **Run.**

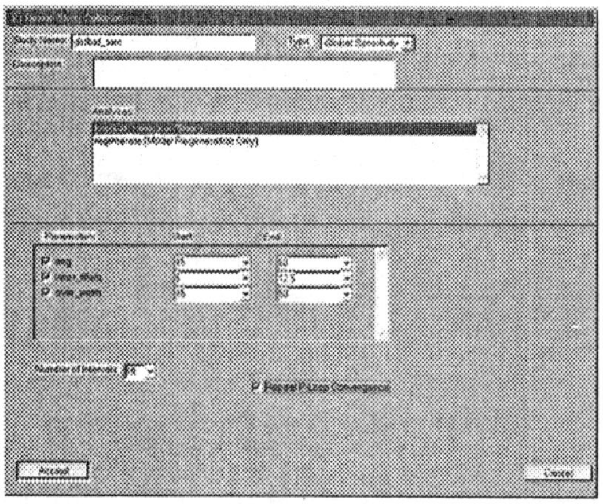

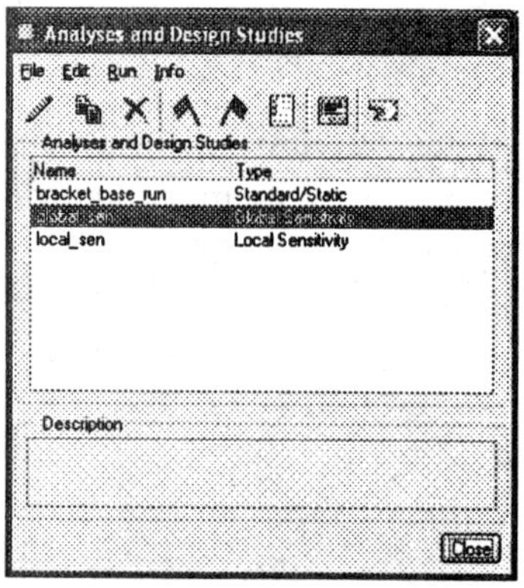

Two graphs are shown below to illustrate the variation of the maximum von Mises stress and the variation of the maximum displacement magnitude as the settings of the three design parameters vary. Note that the fundamentals behind the global sensitivity study are to evaluate the system responses to a set of design parameter settings. Those settings are so selected that the obtained results present a general picture of how the system measure or measures respond to the changing of the settings of the design parameters within the search space.

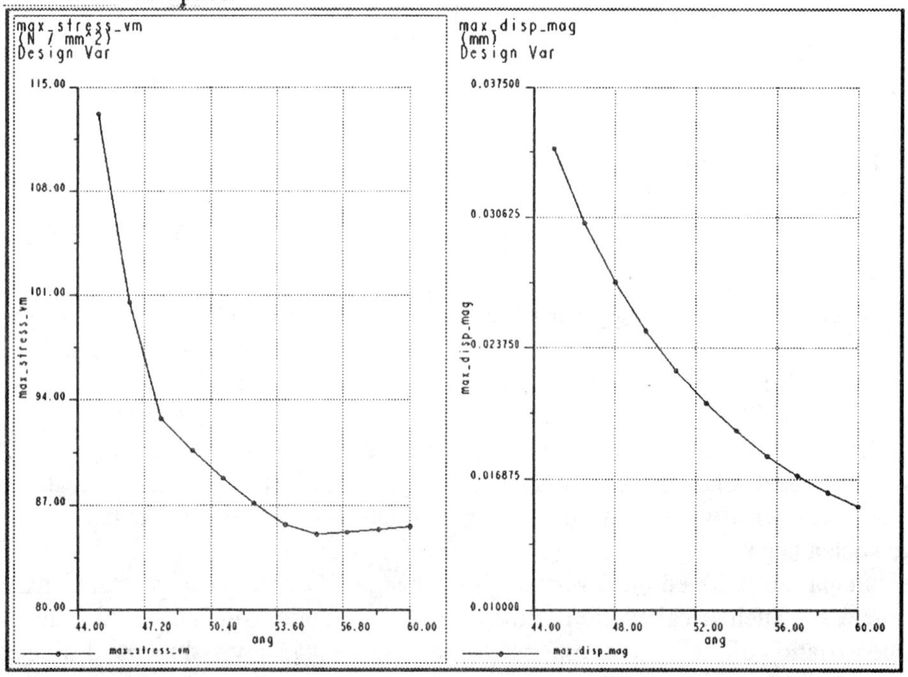

Examining these two curves, the maximum von Mises stress approaches to 85 MPa, and the maximum displacement magnitude approaches to 0.015 mm. Upon completing the global study, we feel comfortable to establish the two constraint conditions. They are

Maximum Von Mises stress <= 85 MPa
Maximum Displacement Magnitude <= 0.020 mm

Step 13: Perform a design study to search for an optimal solution

The objective function is to minimize the total mass of the bracket design. The two constraint conditions are the two limits on the maximum von Mises stress and the maximum displacement.

Minimize: Total Mass of the Bracket Design
Subjected to:
The Maximum Von Mises Stress <= 85 MPa
The Maximum Displacement <= 0.020 mm
Where
45 <= Ang <= 60
5 <= Inner_fillets <= 12.5
45 <= Over_width <= 50

Pick the icon of **performing a design study** > **New Design Study** > select **Optimization** and type in optimal_bracket as the study name. Set up the two constraint conditions and specify the variation ranges of the three design parameters. The convergence criterion is 1%, and the maximum number of runs is set at 20. Run the optimal study.

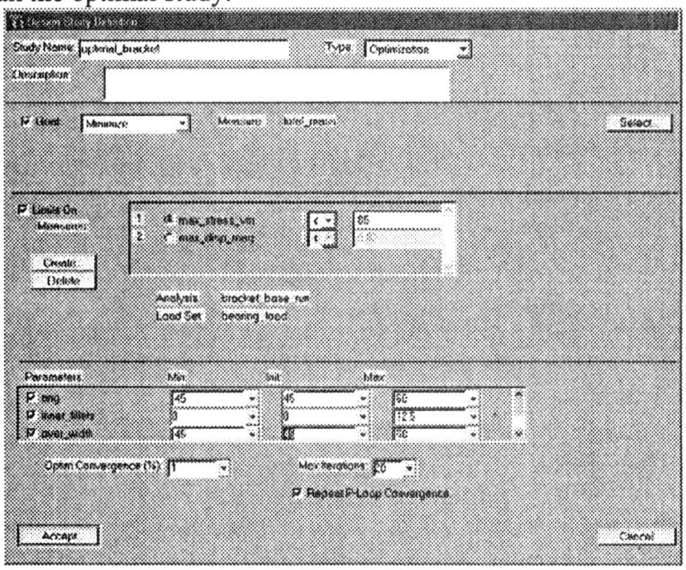

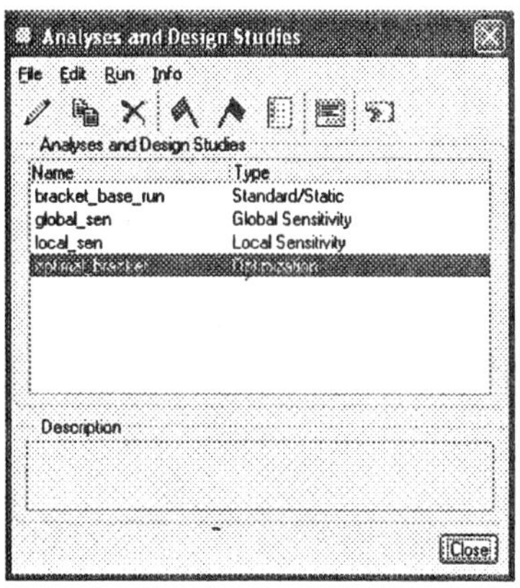

It may take sometime to complete the search for an optimal solution. When the optimization process is completed, plot the graph showing the objective function (total mass in this study) values during the iterations.

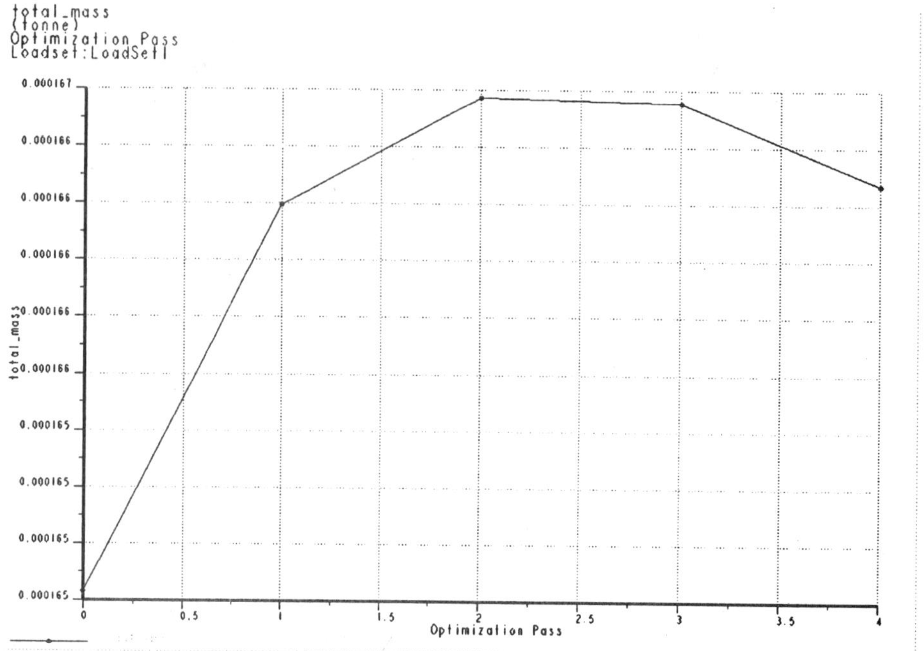

The final result listed in the summary report is shown below, indicating that the total mass at the optimality is 166.27 grams. The von Mises stress constraint condition is in its equality format, or the numerical of its associated slack variable is zero (actually there is a little violation, but within 1%). The displacement constraint still has its potential for reduction.

```
Result of Optimization Iteration 4
Parameters:
    ang                          55.6832
    inner_fillets                 8.3078
    over_width                   45.1037
Goal:  1.6627e-04
Status of Optimization Limits:
1. max_stress_vm          8.5202e+01 <  8.5000e+01 (satisfied)
2. max_disp_mag           1.6338e-02 <  2.0000e-02 (satisfied)
```

Let us examine this minimization process. At the starting points, the three design parameters were set at the following values: ang=45, inner_fillets=8, and over_width=45 with the total mass equal to 164.83 grams. However, the resulting max von Mises stress and max displacement were 96.2 MPa and 0..025 mm. Both values were above their limits. As a result, both the safety and accuracy requirements were violated.

```
Initial Design Status
Parameters:
    ang                               45
    inner_fillets                      8
    over_width                        45
Goal:  1.6483e-04
Status of Optimization Limits:
1. max_stress_vm          9.6199e+01 <  8.5000e+01 (VIOLATED)
2. max_disp_mag           2.5067e-02 <  2.0000e-02 (VIOLATED)
```

In order to meet the two constraint conditions, the search went to a region where the total mass of the bracket was increased. The total mass was increased to 166.16 gram, first, and finally reached 166.47

gram. In the meantime, the maximum von Mises stress reduced from 94.89 MPa to 85.0 MPa. The maximum displacement was reduced from 0.025 mm to 0.018 mm.

	Total Mass (gram)	Max VM Stress (Mpa)	Max Displacement (mm)
Iteration 1	164.8	96.2	0.0251
Iteration 2	166.2	87.3	0.0176
Iteration 3	166.6	85.8	0.016
Iteration 4	166.6	85.8	0.0161
Iteration 5	166.3	85.2	0.0163

Generally speaking, there are three possible reasons for the termination of a search process. The first reason would be a good reason indicating that the optimality has been reached (it was set at 1% in the current study). The second reason would be the confinement of the current feasible space available for the search.

The search was terminated in the current study because the improvement of reducing the total mass was within 1% from iteration 4 and iteration 5. The actual increase is given by

$$Mass\_\mathrm{Re}duction = \frac{166.3 - 166.6}{166.6} = -0.18\%$$

At the optimality, the three design parameter settings were ang = 55.7 degrees, inner_fillets = 8.3 mm, and over_width = 45.1 mm, which was very close to its lower boundary of 45 mm. The maximum von Mises stress was 85.2 MPa and the maximum displacement magnitude was 0.0163 mm.

Have we done this job? Are you satisfied with the results?

Step 14: Perform a second design study to search for a new optimal solution.

For this optimization process, we may take another approach. First, let us relax the constraint condition on the maximum von Mises stress. The new limit is adjusted to 90 MPa. Second, let us lower the limit on the design parameter of Over_Width from 45 mm to 44 mm, and lower the limit on the design parameter of Ang from 60 degrees to 58 degrees. Such adjustments expand the search space. On the other hand, we narrow the range of inner_fillets from 5 mm to 10 mm.

Before using these two limits in the search for optimality, we need to adjust these limits in Design Parameters. Otherwise, the optimality search page will not accept these two new limits. After adjusting them, we run the second optimal search. Note that we narrow the range of inner_fillets. The new range is from 5 mm and 10 mm. The motivation for making these adjustments is to increase the search space and/or shift the search to a space, which includes some new hopes for exploration.

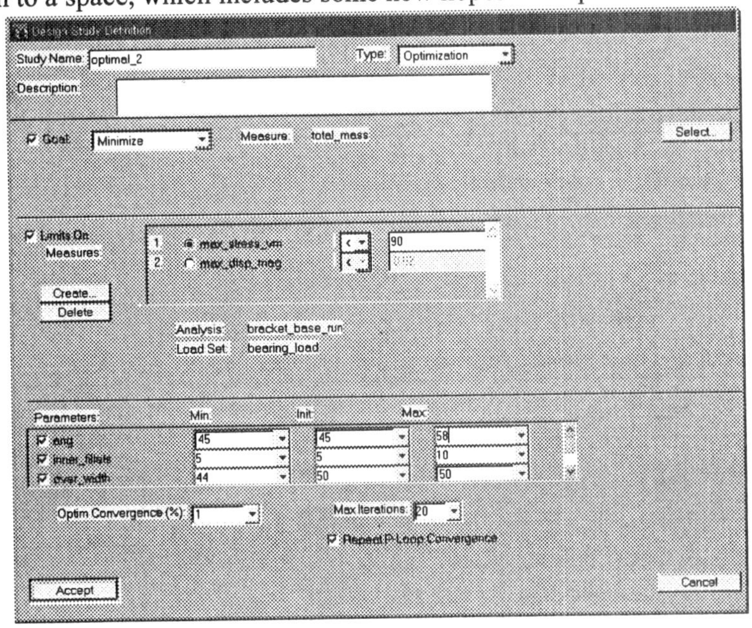

After running the optimization study, the results obtained from the second approach are shown below. It indicates that there are 6 iterations. It also indicates that the total mass at its optimality was 165.20 gram.. Note that this number was much better than the result obtained by the first approach, which was 166.27 gram at its optimality. However, this value is still larger than 164.8 gram, which was the total mass before initiating the re-design process.

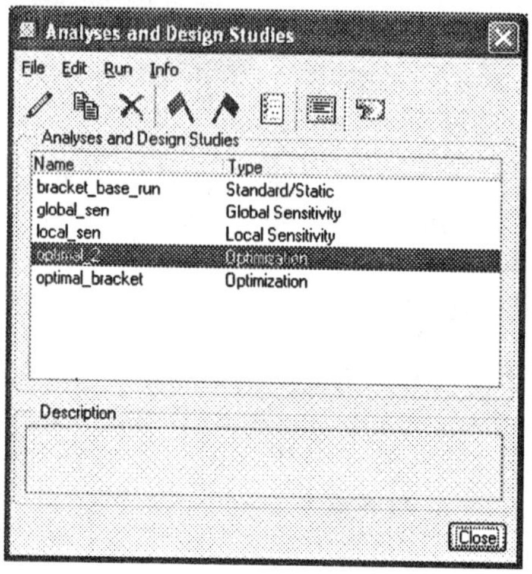

Iteration	angle	inner_fill	over_width	total mass	svm-ax	disp.
1	45.0	5.00	50.0	166.4	146.0	0.053
2	50.2	5.98	48.8	167.5	101.0	0.031
3	54.1	6.85	47.6	167.5	88.0	0.023
4	55.4	7.18	46.9	167.3	86.8	0.020
5	55.7	7.23	46.6	167.0	86.4	0.020
6	57.3	7.52	44.6	165.7	85.4	0.016
7	57.8	7.56	44.0	165.2	84.8	0.015

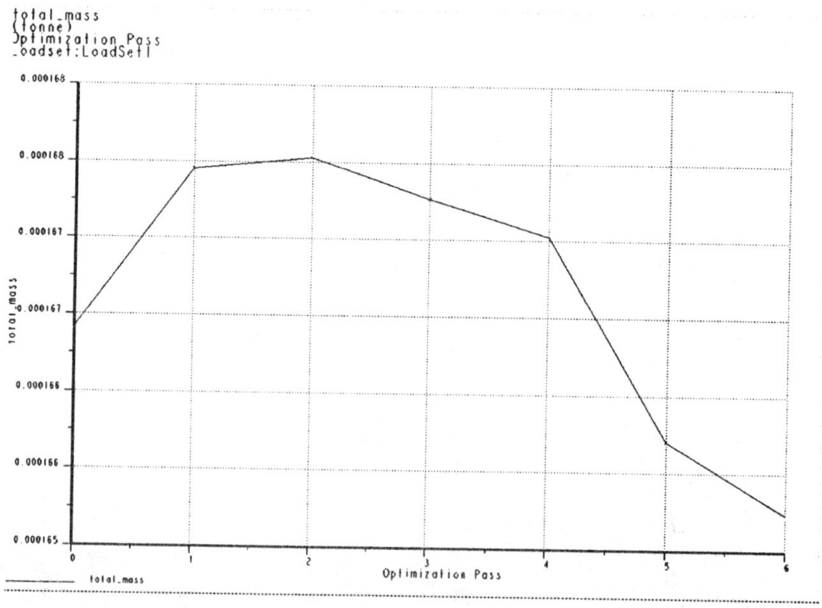

The settings of the three design parameters at its optimality were listed below:

Ang = 57.8 degrees
Inner_fillets = 7.56 mm
Over_width = 44 mm

The two plots shown below are the plot of the maximum VM stress and the plot of the maximum displacement. Examining them carefully, both upper limits were not reached. Therefore, the termination of the optimal search was terminated because the convergence criterion set for the optimality was at the level of 1%, which was reached at iteration 6.

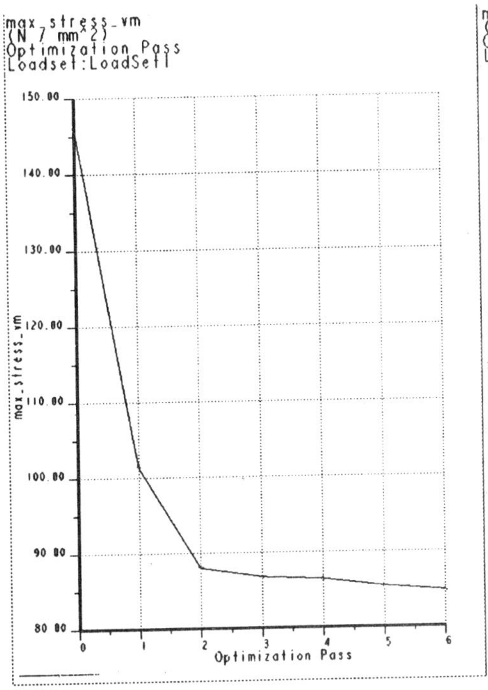

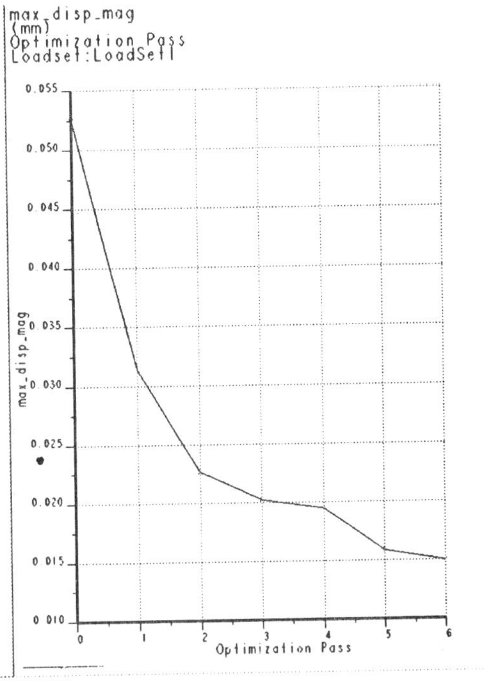

For these reasons, there was a need to continue the search by resetting the convergence criterion set for the optimality at 0.5%. Before carrying out the third round search for optimality, let us further relax the lower limit on the design parameter of Over_Width, from 44 mm to 42 mm. This limit was reached during the previous search. In order to ensure there would be certain search spaces when the new search initiated, relaxing this limit seemed reasonable.

Let us modify the design study by making the following adjustments:

Convergence Criterion for Optimality: 0.5%
Adjust the lower limit on Over_Width from 44 mm to 42 mm.

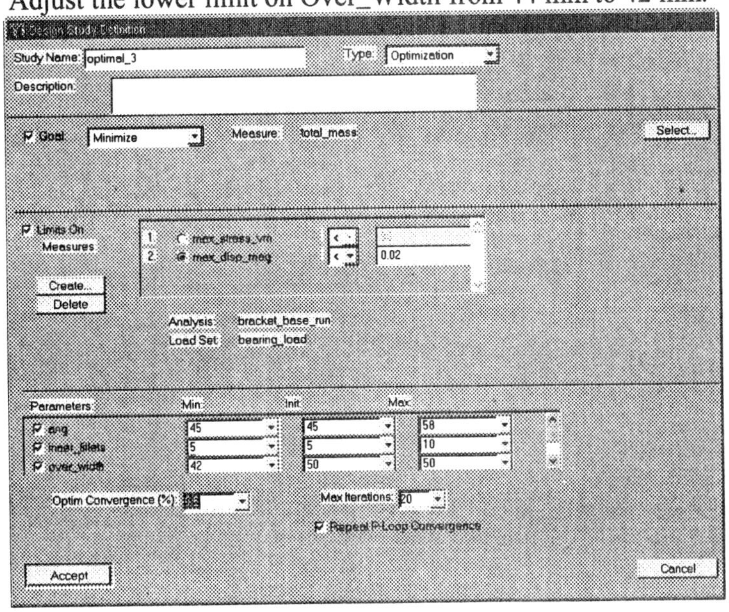

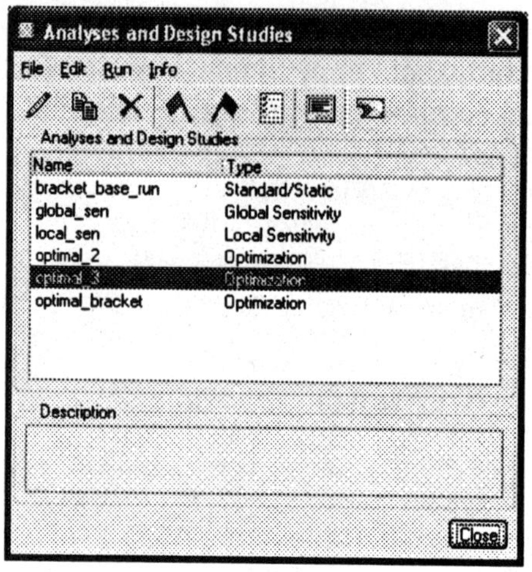

After running the optimization study, the results obtained from the third around are shown below. It indicates that there are 9 iterations. It also indicates that the total mass at its optimality is 162.88 gram. Note that this number was much better than the result obtained from the second round, which was 165.23 gram at its optimality. More important is the fact that this value is much smaller than 164.8 gram, which was the total mass before initiating the re-design process.

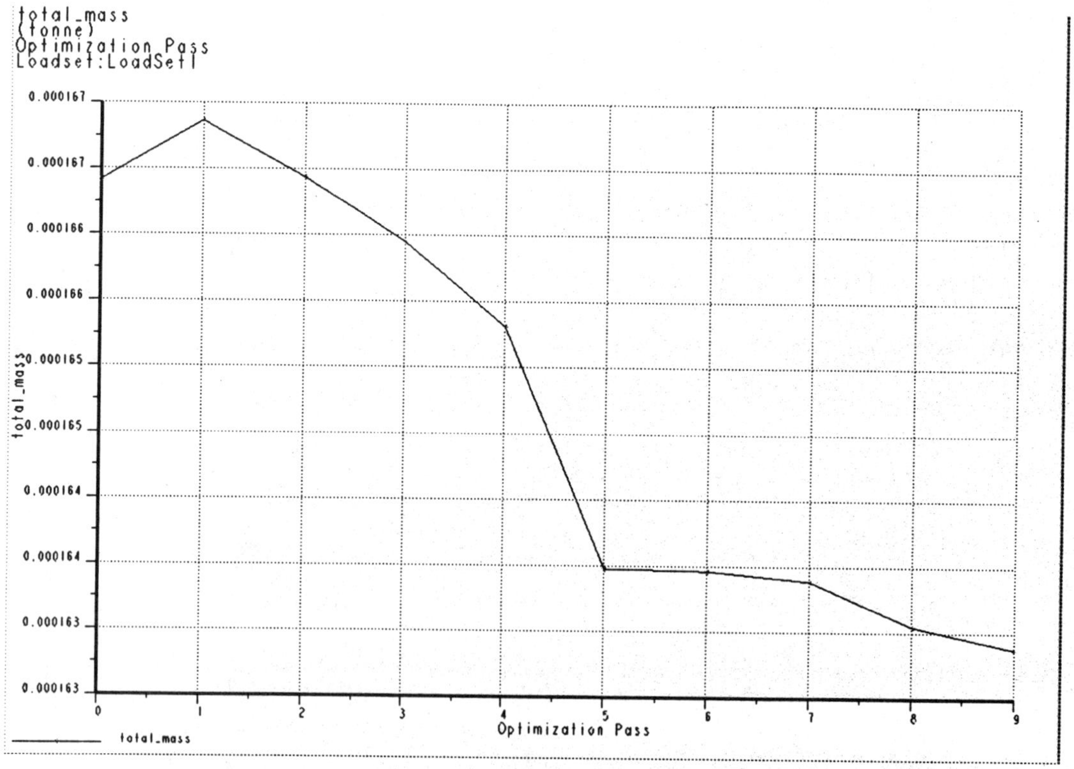


```
Result of Optimization Iteration 9
Parameters:
   ang                          53.6015
   inner_fillets                      5
   over_width                        42
Goal:   1.6288e-04
Status of Optimization Limits:
1. max_stress_vm      8.7197e+01 <  9.0000e+01 (satisfied)
2. max_disp_mag       1.7802e-02 <  2.0000e-02 (satisfied)
```

The settings of the three design parameters at its optimality were listed below:

Ang = 53.6 degrees
Inner_fillets = 5 mm
Over_width = 42 mm

The two plots shown below are the plot of the maximum VM stress and the plot of the maximum displacement. Examining them carefully, the maximum VM stress was 87.2 MPa. Therefore, the upper limit on the maximum VM stress was almost reached. The maximum displacement magnitude was 0.0178 mm. This value was still below the upper limit on the maximum displacement, which was set at 0.020 mm. However, the current value was very close to the limit. Therefore, the optimal search was terminated because the convergence criterion set for the optimality, which was at the level of 0.5%, was met.

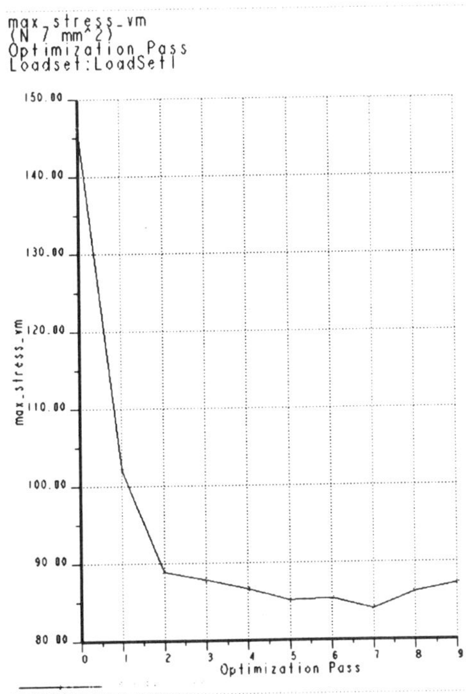

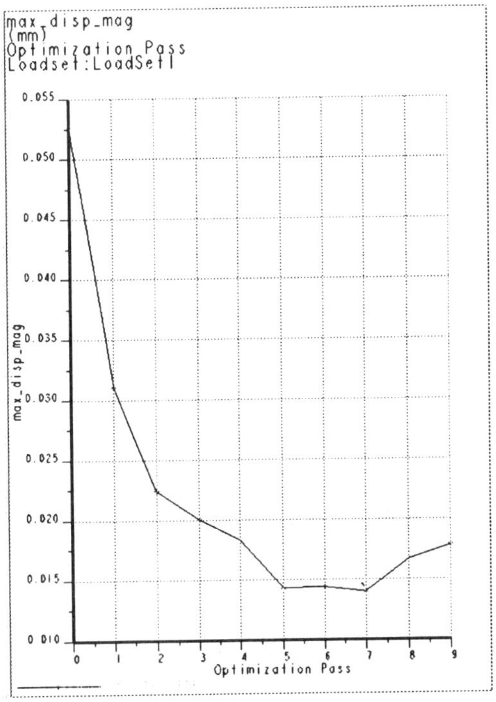

Are we satisfied with the optimal solution this time? We should be satisfied with what we have obtained.

Step 15: plot the geometrical shape of the bracket design at its optimality and compare it with its original design.

In order to display the geometrical shape that the optimality, do the following:

Applications > Mechanica > Continue

From the main toolbar, select **Analysis** > **Dsgn Controls > Optimize Hist > Search Study** > select optimal_3.

Yes to the question on display: Do you want to review the next shape? There should be 9 shapes because the number of the iterations is 9.

Yes to the question on display: Leave the model at the optimized shape?

The geometrical shape at the optimal design is shown below:

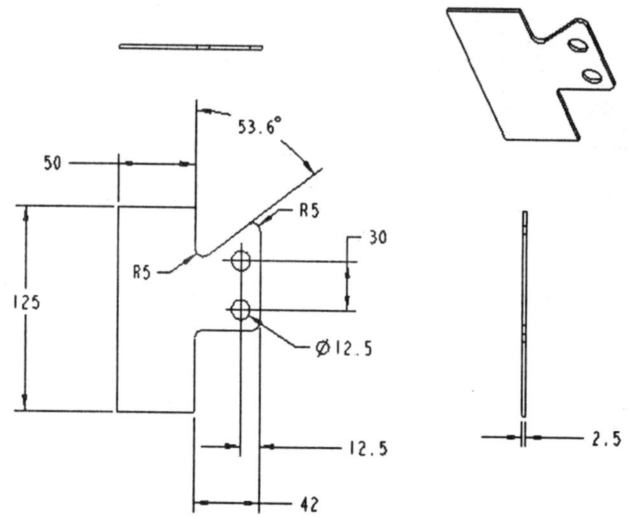

The geometrical shape at its original design is also shown below:

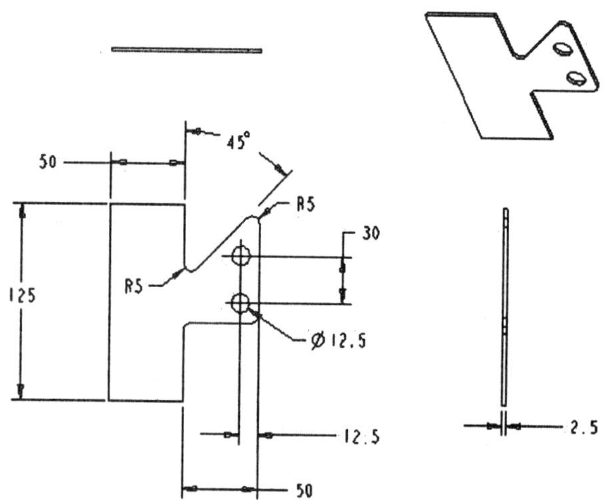

Note the changes of the dimensions of the three design parameters. They are the angle value from 45 degrees to 53.6 degrees, the radius value of the inner fillet remains unchanged (5 mm), and the over_width value from 50 mm to 42 mm.

We note that the design specification requires the maximum von Mises stress remains below 85 MPa, which has not been met. Should we carry out a third round search? In fact, there is no need. Let us re-examine the process of optimization during the second round.

Step 16: Re-examine the process of optimization during the second round search.

When reading the summary report, we notice that at iteration 7, the maximum von Mises stress was 85.54 MPa and the maximum displacement magnitude was 0.01497 mm. The total mass was 163.39, which was less than the total mass before the redesign process.

```
Result of Optimization Iteration 7
Parameters:
   ang                        55.0751
   inner_fillets               6.8367
   over_width                      42
Goal:   1.6339e-04
Status of Optimization Limits:
1. max_stress_vm      8.3958e+01 <  9.0000e+01 (satisfied)
2. max_disp_mag       1.3980e-02 <  2.0000e-02 (satisfied)
```

The settings of the three parameters were

Ang = 55.1 degrees
Inner_fillets = 6.84 mm
Over_width = 42.0 mm

The geometrical shape at the 7th iteration is shown below:

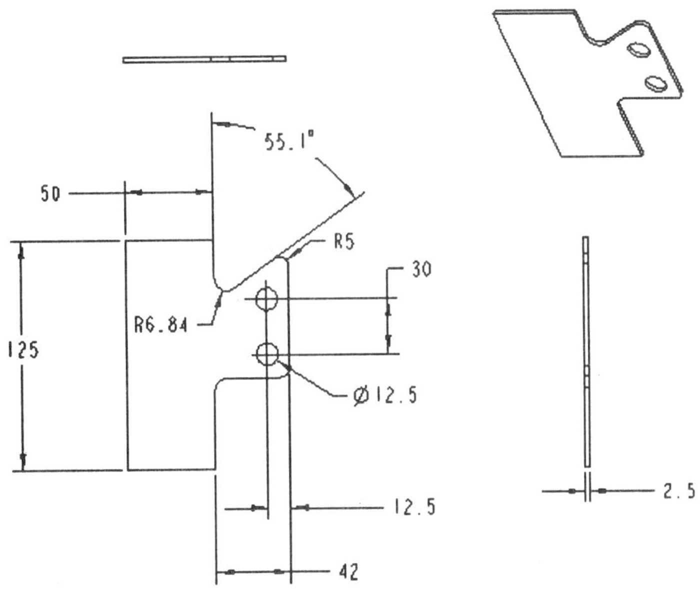

The plot of the VM stress distribution at Iteration 4 is shown below. Note that the maximum value of the von Mises stress is about 85 MPa.

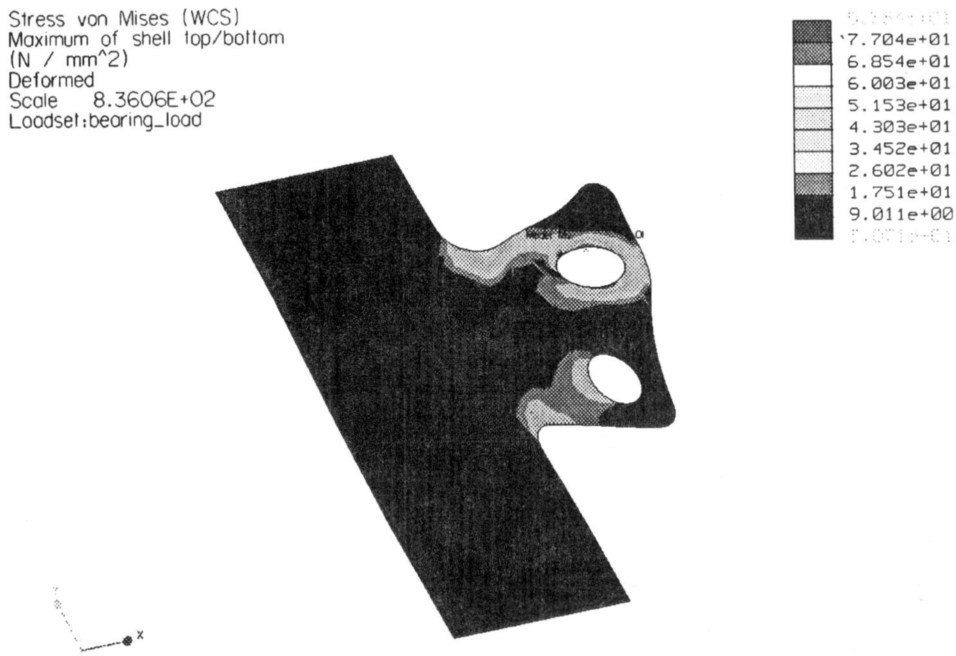

distribution of von Mises stress

The plot of the displacement distribution at Iteration 4 is shown below. Note that the maximum value of the displacement is about 0.01495 mm.

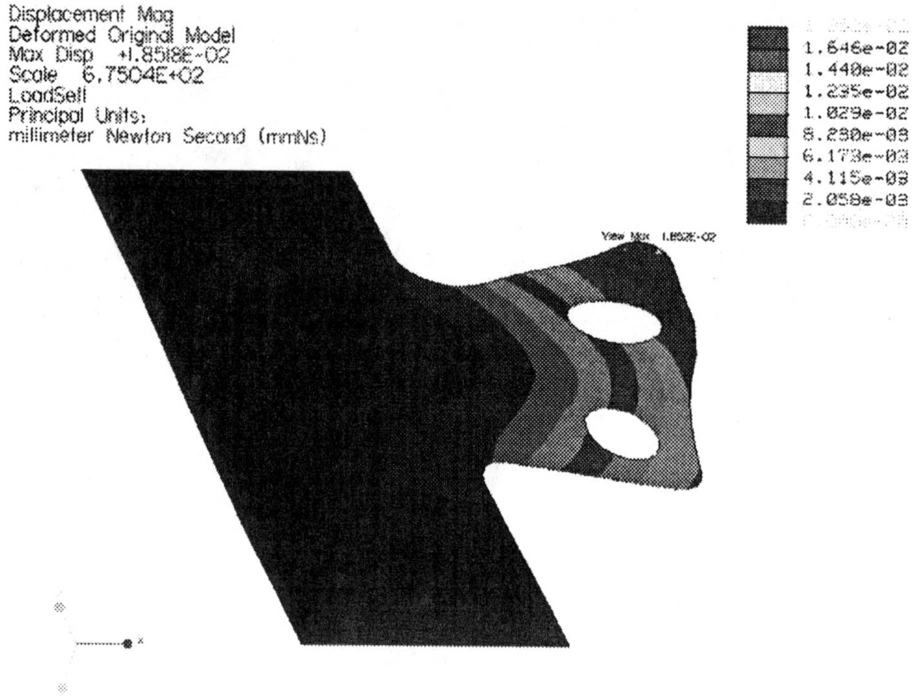

distribution of displacement at Iteration 4

It is very important to note the shift of the location of the maximum von Mises stress from the inner-fillet location to the upper hole location.

Are you amazed at the solution you have come up with? By performing an engineering analysis, the value of the total mass of the notch plate design has been reduced. The maximum M stress has been reduced from 142 MPa to the current 85 MPa, and the maximum value of the displacement has been reduced from 0.0587 mm to 0.01495 mm. Without the assistance from the use of a software system, such as Pro/ENGINEER and Pro/Mechanica, it would be impossible for an engineer designer to carry out such a task with such great an amount of computing requirement.

8.4 References

1. Algor Processor Reference Manual, Algor Interactive Systems, 150 Alpha Drive, Pittsburgh, PA 15238
2. O. Axelsson, Iterative Solution Methods, Cambridge University Press, New York, 1994.
3. K. J. Bathe, Finite Element Procedure, Prentice Hall, Upper Saddle River, New Jersey, 1996.
4. K. Bell, A Refined Triangular Plate Bending Finite Element, Int. J. Numer. Methods Eng., Vol. 1, No. 1 January 1969, pp. 101-122.
5. B. Carnahan, H. A. Luther, and J. O. Wikes, Applied Numerical Methods, Wiley New York, NY, 1969.
6. R. W. Clough, The Finite Element in Plane Stress Analysis, Proceedings of 2nd ASCE Conference on Electronic Computation, Pittsburgh, PA. September 1960.
7. R. D. Cook, D. S. Maikus and M. E. Plesha, Concepts and Applications of Finite Element Analysis, 3rd edition, Wiley, New York, NY, 1989.
8. J. W. Dally and W. F. Riley, Experimental Stress Analysis, McGraw Hill, 1965.
9. J. W. Dally, Design Analysis of Structural Elements, Goodington, New York, NY, 2003.
10. C. S. Desal and J. F. Abel, Introduction to the Finite Element Method, Van Nostrand Reinhold, New York, NY, 1972.

11. J. H. Earle, Graphics for Engineers, AutoCAD Release 13, Addision-Wesley, Reading, Massachusetts, 1996.
12. R. H. Gallagher, Finite Element Analysis Fundamentals, Prentice-Hall, Englewoood Cliffs, NJ, 1975.
13. J. M. Gere and S. P. Timoshenko, Mechanics of Materials, 3rd edition, PWS-KENT Publishing Company, Boston, 1990.
14. W. W. Hager, Applied Numerical Linear Algebra, Prentice-Hall, Englewood Cliffs, NJ, 1988.
15. Holland, The Finite Element Method in Plane Stress Analysis, in the Finite Element Method in Stress Analysis, I. Holland and K. Bell (eds), Tapir Press, Trondheim, Norway, 1969, Chapter 2.
16. Y. Y. Hsieh, Elementary Theory of Structure, 2nd edition, Prentice-Hall, Englewood Cliffs, NJ, 1982.
17. M. A. N. Hendriks, H Jongedijk, J. G. Rots and W. J. E. van Spange (eds), Finite Elements in Engineering and Science, DIANA Computational Mechanics, 1997.
18. K. H. Huebner, D. L. Dewhirst, D. E. Smith, and T. G. Byrom, The Finite Element Method for Engineers, 4th edition, John Wiley & Sons, Inc., 2001.
19. C.S. Krishnamoorthy, Finite Element Analysis, Theory and Programming, 2nd Ed., 1995.
20. D. L. Logan, A First Course in the Finite Element Method Using ALGOR, PWS Publishing, Boston, 1997.
21. H. C. Martin, Introduction to Matrix Methods of Structural Analysis, McGraw-Hill, New York, NY, 1966.
22. R. L. Norton, Machine Design: An Integrated Approach, Prentice Hall, Upper Saddle River, New Jersey, 1996.
23. M. Ortiz, Y. Leroy and A. Needleman, A Finite Element Method for Localized Failure Analysis, Comp. Meth. Applied Mech. Enging, 61, pp. 189-214, 1987.
24. M. Petyl, Introduction to Finite Element Vibration Analysis, Cambridge University Press, 1990.
25. T. H. H. Pian and P. Tong, Basis of Finite Element Methods for Solid Continua, Int. J. Numer. Methods Eng., Vol. 2, No. 7, April 1964, pp. 1333-1336.
26. P. G. Plockner, Symmetry in Structural Mechanics, J. of the Structural Division, American Society of Civil Engineers, Vol. 99, No. ST1, pp. 71-89, 1973.
27. J. N. Reddy, An Introduction to the Finite Element Method, McGraw-Hill Publishing Company, New York, NY, 1984.
28. T. Ross, Finite Element Program in Structural Engineering and Continuum Mechanics, Horwood Publishing Limited, Chichester, England, 1996.
29. T. Ross, Advanced Applied Finite Element Methods, Horwood Publishing Limited, Chichester, England, 1998.
30. L. Stasa, Applied Finite Element Analysis for Engineers, Holt, Rinehart and Winston, New York, NY, 1985, pp. 101-157.
31. Y. Saad, Iterative Methods for Sparse Linear Systems, PWS Publishing Company, Boston, 1996.
32. Soc., Vol. 7, No. 8, pp. 849-861.
33. Y. Tada and G. C. Lee, Finite Element Solution to an Elastic Problem of Beams, Int. J. Numer. Methods Eng., Vol. 2, No. 2, April 1970, pp. 229-241.
34. R. L. Taylor, On Completeness of Shape Functions for Finite Element Analysis, Int. J. Numer., Methods Eng., Vol. 4., No. 1, 1972, pp. 17-22.
35. S. P. Timoshenko and D. H. Young, Theory of Structure, McGraw-Hill, Kogaksha Limited, 1965.
36. M. J. Turner, R. W. Clough, H. C. Martin, and L. J. Topp, Stiffness and Deflection Analysis of Complex Structures, J. of the Aeronautical Sciences, Vol. 23, No. 9, pp. 805-824, Sept. 1956.
37. O. C. Zienkiewicz, The Finite Element Method, third edition, McGraw-Hill, New York, NY, 1977.
38. O. C. Zienkiewicz and R. L. Taylor, The Finite Element Method, 4th edition, Vol. 1, McGraw-Hill (UK),London, 1989.

8.5 Exercises

1. The following drawing shows the design of a notch plate. We have interests on 4 design parameters, which are listed below

Design Parameters	Symbolic Names
Radius of the notch: R2.5	notch_radius
Thickness of the plate: 2.5	thickness
Cut distance: 60	cut_distance
Total height: 30	total_height

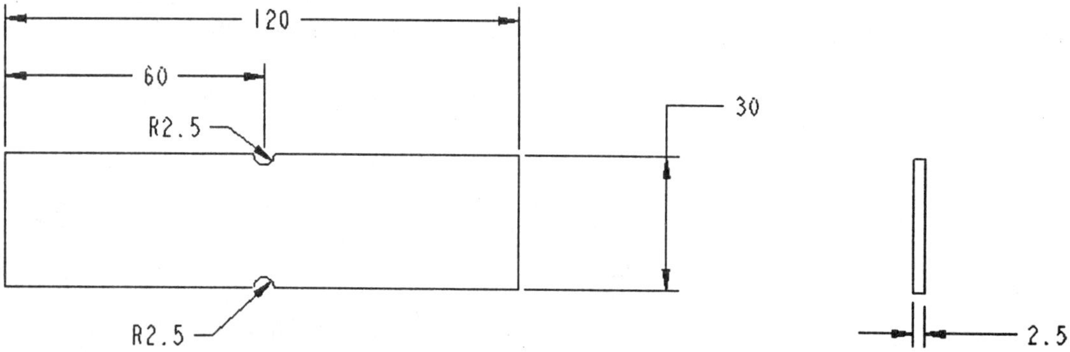

The working condition is that the left end of the notch plate is fixed. The right end of the notch plate is the free end where a load is acting on it. The two force components are 80 N along the x-direction and –200 N along the y-direction.

F_x = 80 Newton F_y = -200 Newton

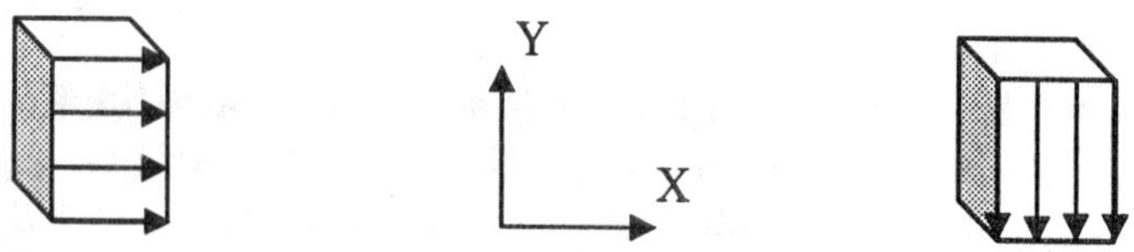

The material of the notch plate is steel. Its yield strength is assumed to be 240 MPa. The safety factor is 3. Therefore, the maximum allowable stress is 80 MPa. The objective of this study is to search for a new design, if necessary, so that the mass of the notch plate can be reduced. In the meantime, the maximum von Mises stress is limited to 80 MPa, and the maximum displacement is limited to 0.1 mm at the upper right corner.

Minimize the total mass of the notch component

Subject to

$$\sigma_{\max\_vm} \leq 80$$

$$\delta_{\max\_deflection} \leq 0.10 \quad \text{(mm)}$$

and $$2.5 \leq notch\_radius \leq 3.5$$

$$2.5 \leq thickness \leq 3.0$$

Units: mm

$$60 \leq cut\_dis\tan ce \leq 70$$

$$2.5 \leq total\_height \leq 3.5$$

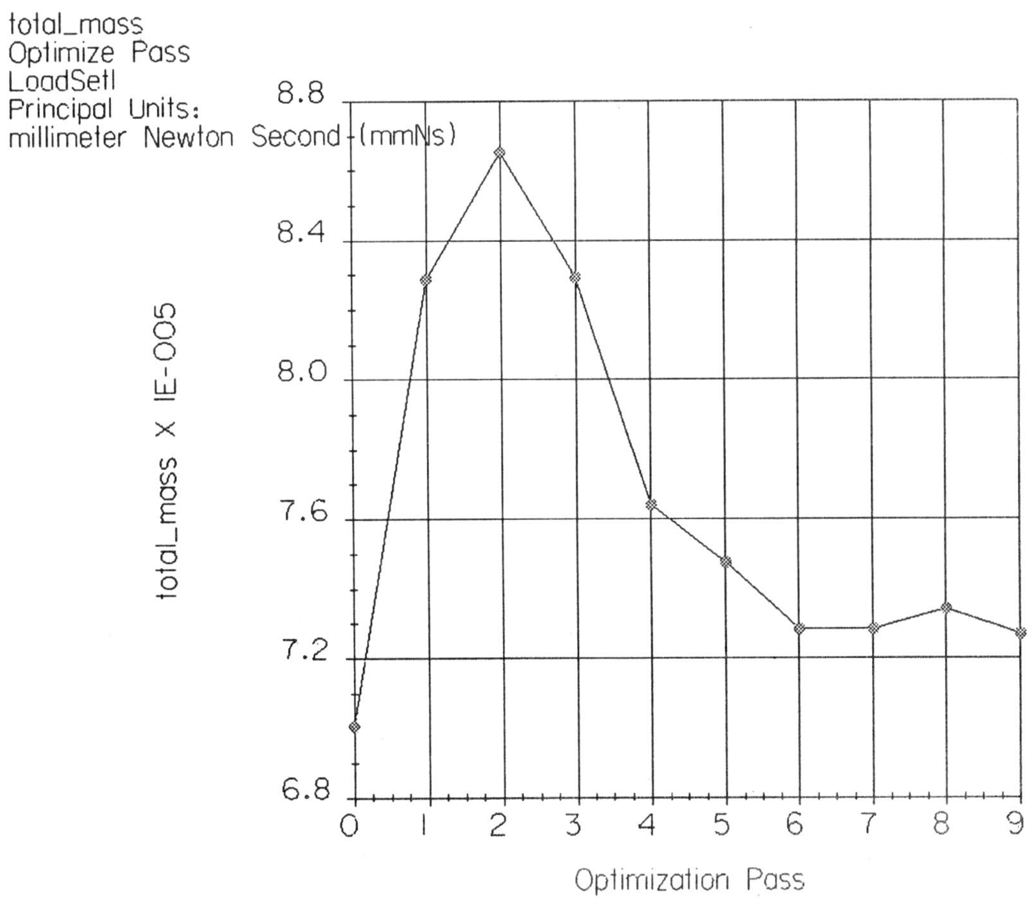

Minimized Total Mass vs Iterations

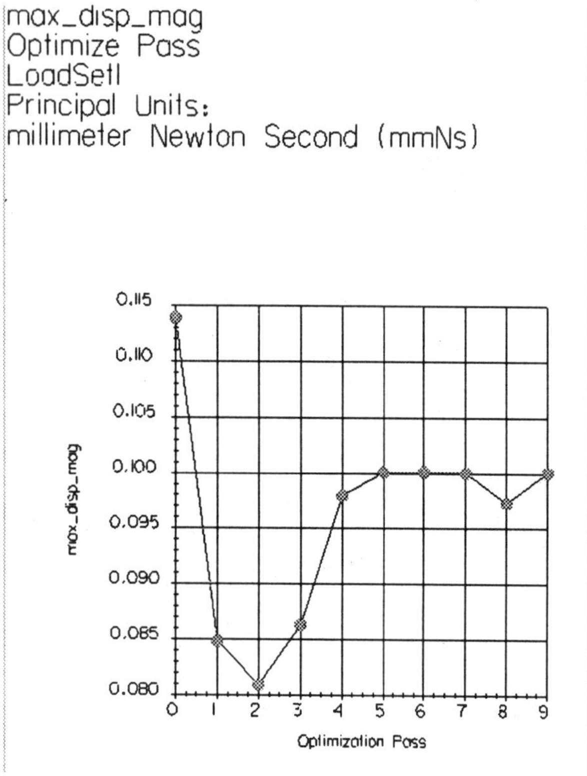

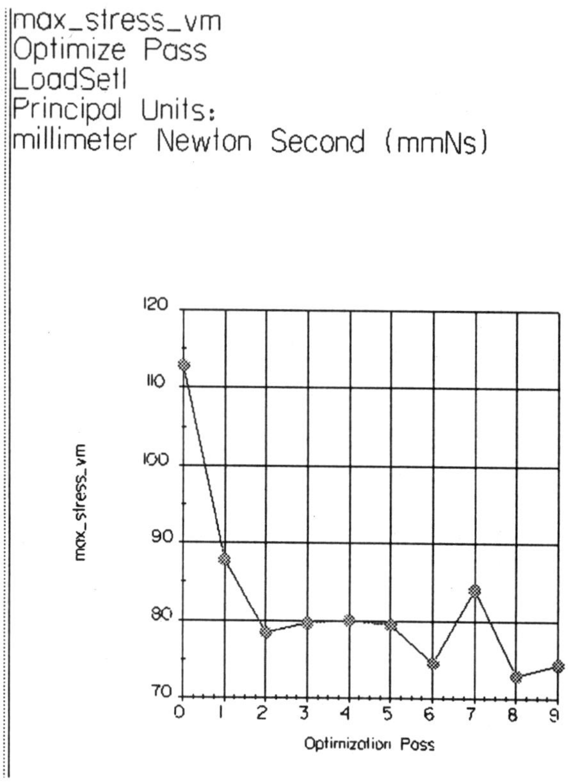

2. In section 8.3, we performed a redesign process for the bracket component. In that study, three design parameters were under consideration. In this exercise, let us perform the redesign process for the bracket component again. Let us add a new design parameter to the set of the design parameters previously selected. The newly added design parameter is the total height of the bracket component. As illustrated in the following figure, the current value of this new design parameter is 125 mm. Let us allow it to vary between 120 mm and 130 mm.

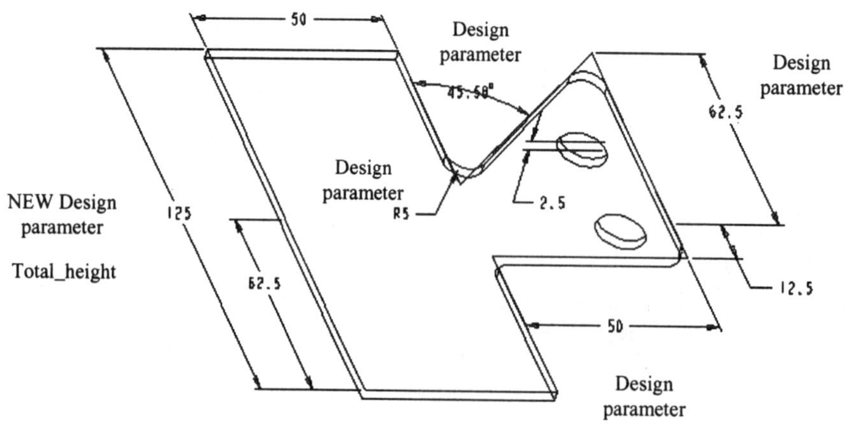

Let us keep the constraint condition, the load condition and the material type unchanged. The mathematical format for this resdesign process is listed below:

Minimize the total mass of the bracket component

Subject to

$$\sigma_{max\_vm} \leq 90 \quad \text{(MPa)}$$

$$\delta_{max} \leq 0.20 \quad \text{(mm)}$$

and

$$45^o \leq angle \leq 60^o$$

$$120 \leq Total\_height \leq 130$$

$$5 \leq inner\_fillets \leq 15$$

$$45 \leq Over\_width \leq 55$$

Units: mm

CHAPTER 9

IDEALIZATION: MODELING WITH TRUSS AND BEAM STRUCTURES AND ALGOR

9.1 Introduction

In Chapter 6, we discussed the idealization of mechanical structures with shells. When using the shell model, we need to detect the middle surface from a pair of surfaces, which are parallel to each other. A shell model is a combination of middle surfaces, namely, an idealization of a 3D mechanical structure. In this chapter, we deal with truss and beam structures through idealization. Instead of using middle surfaces, we will use datum curves to represent the truss and beam structures. The geometrical shape and dimensions are characterized through a process of defining the section area(s). In section 9.2, we use Pro/Mechanica to perform FEA for a simple truss structure.

In this chaper, we also introduce another software system – ALGOR. Four examples will be used to demonstrate the procedure of performing FEA. The first 3 examples are related to truss structures and we use ALGOR V14 to perform FEA. The 4th example is related to a beam structure and we use ALGOR V16, a new release, to perform FEA.

9.2 Application on a Simple Truss Structure

The following figure illustrates a truss structure. It consists of 4 bars. The lengths of these 4 bars are shown in the figure. The three dimensions characterizing the truss structure are 1000, 350 and 250 mm. Regarding the geometrical shape, each bar is a pipe. Its cross section is a hollow circle. The outer diameter is 30 mm and the inner diameter is 26 mm. The load is acting on the free end (PNT2). PNT0 and PNT1 are fixed to the ground.

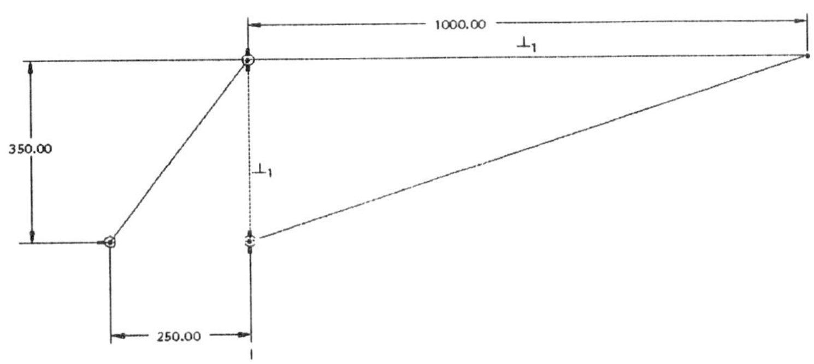

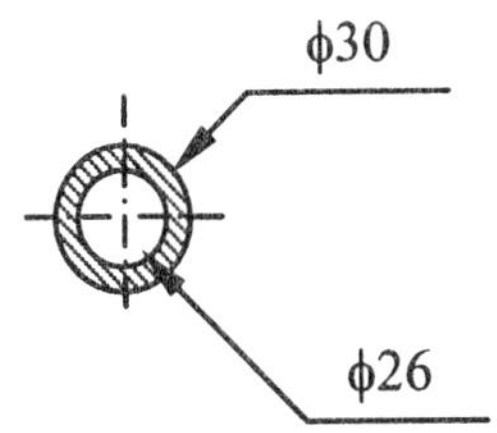

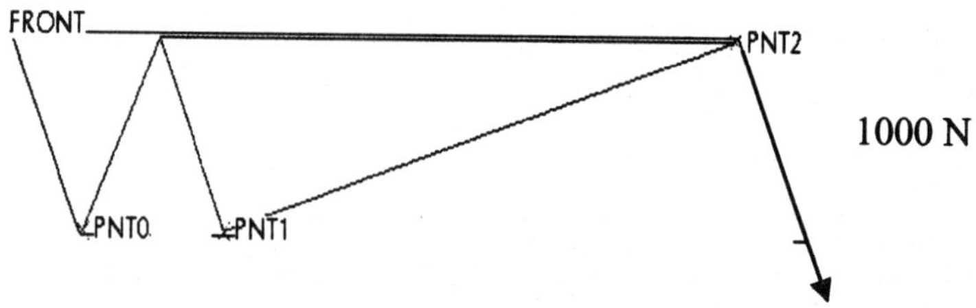

There are 2 objectives of this study. The first objective is to plot the deflection pattern of this truss structure. The second objective is to plot the internal forces induced to each bar when subjected to the load condition.

Step 1: Model the truss structure

From the mail toolbar, select the icon of **Create** a new object. Make sure the **Part** mode is selected. Type *truss_structure* and clear the box of **use default template** > **OK**

Select the unit system to be **mmns_part_solid** > type *truss structure* in **DESCRIPTION** and *student* in **MODELED_BY** > **OK**.

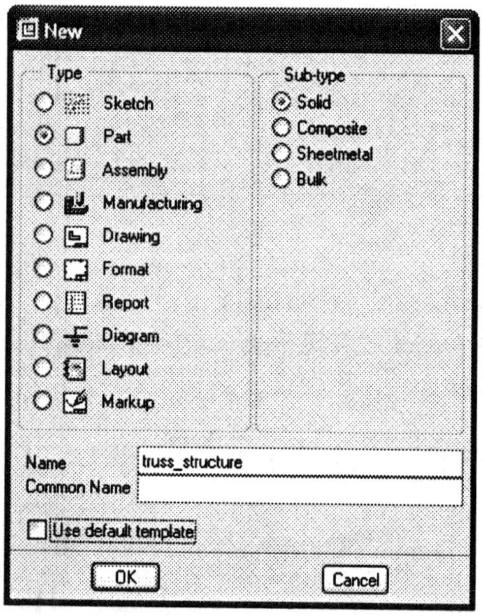

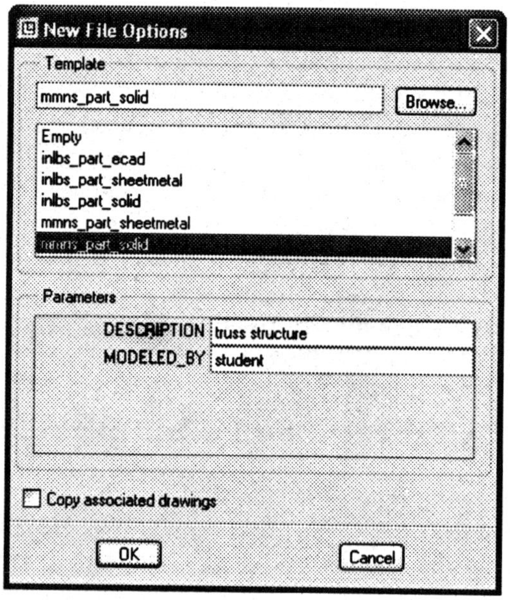

Select the icon of **Sketch Tool** from the toolbar of datum creation.

Pick the **FRONT** datum plane as the sketch plane, and accept the default orientation setting, which uses the **RIGHT** datum plane and position it at the right side > **Sketch**

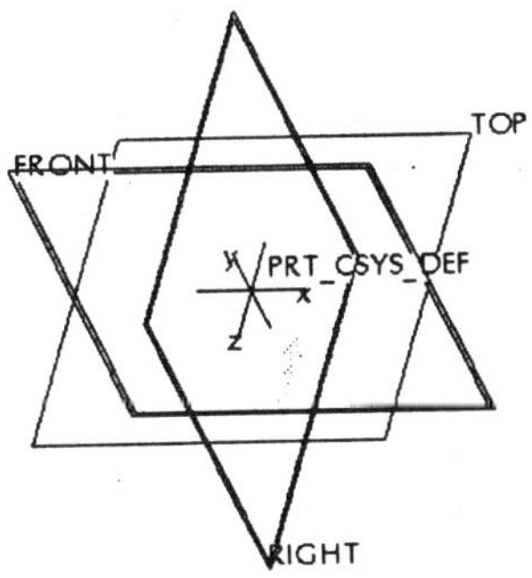

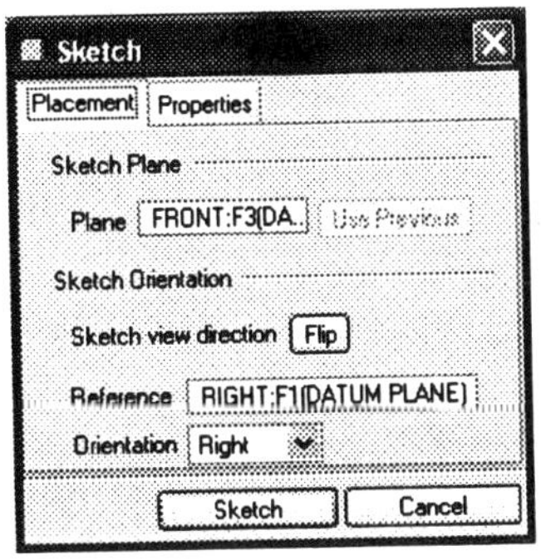

Close the window of **References**. Pick the icon of **Line**, and sketch the following.

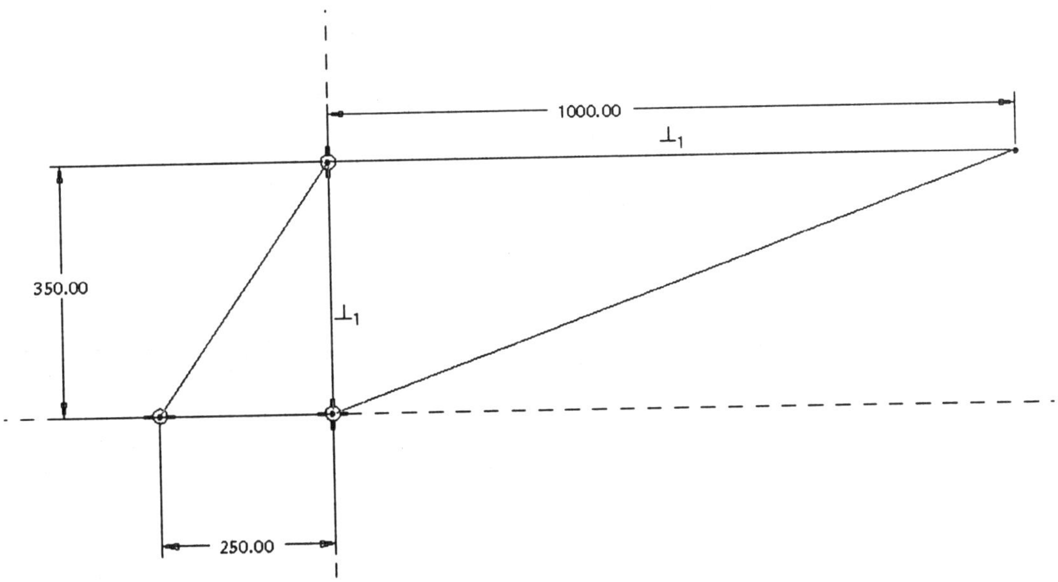

Upon completion, select the icon of **Done** from the toolbar of sketcher. The sketched 4 datum curves (actually, they are straight lines) represent the 4 bars of the truss structure.

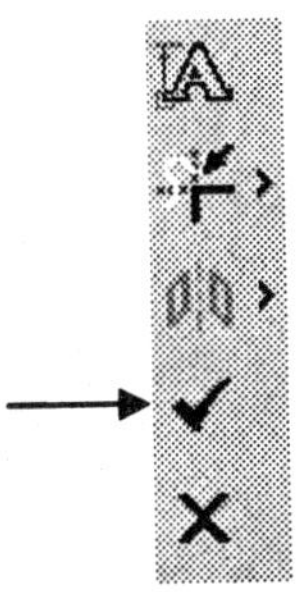

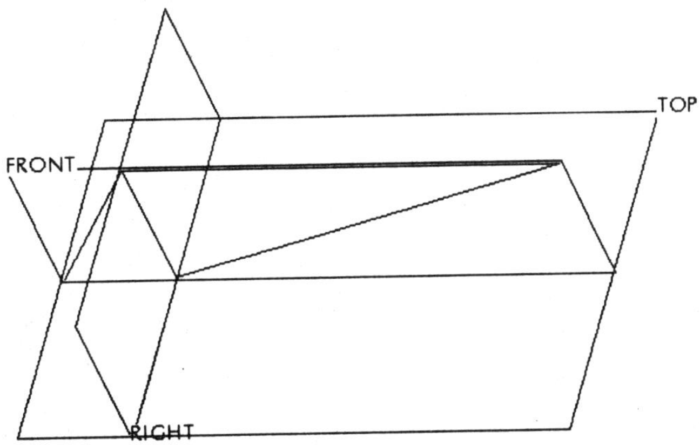

Now let us create three datum points. The first datum point is at the end of the datum curve located at the very left side. Select the icon of datum point from the toolbar of datum feature creation.

Pick the point at the end of the left side of the datum curve on display. Make sure the offset value is set at zero > **OK**. PNT0 is on display.

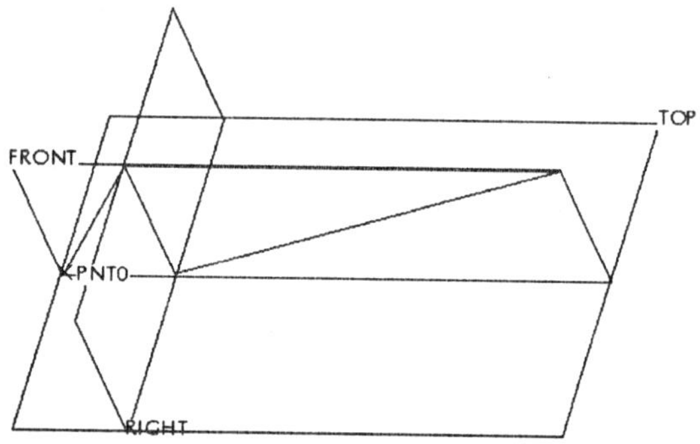

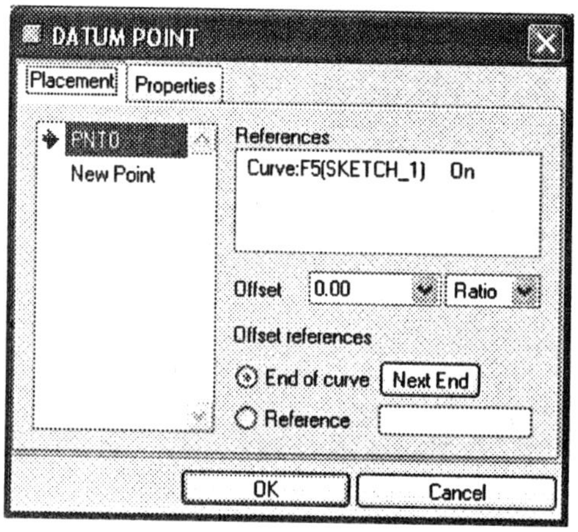

To define the second datum point, select the icon of datum point from the toolbar of datum feature creation. Pick the point at the corner of the datum curve shown below. Make sure the offset value is set at 0.00.

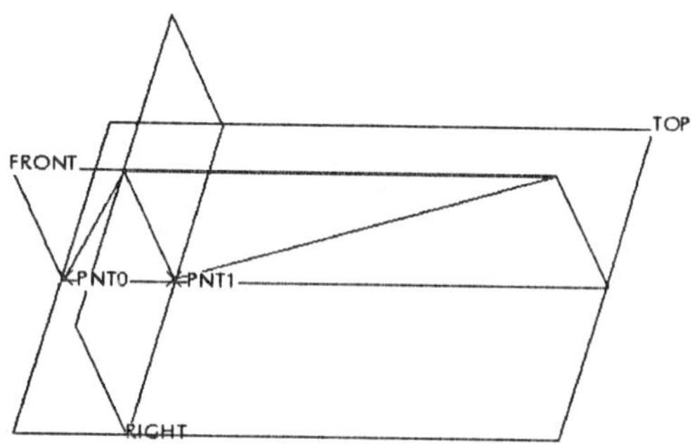

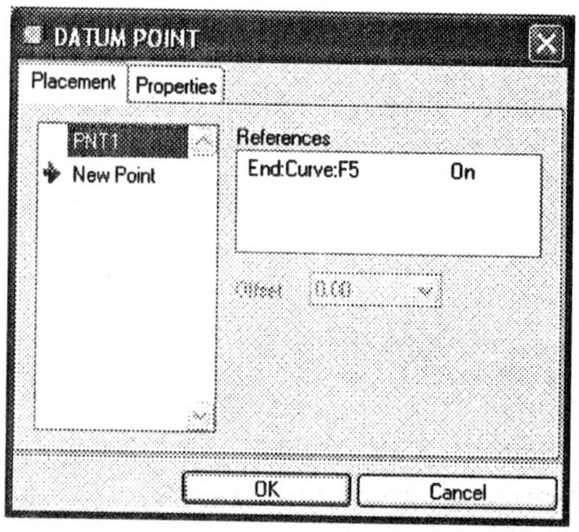

To define the third datum point, select the icon of datum point from the too bar of datum feature creation. Pick the point at the right-upper corner of the datum curve on display. Make sure the offset value is set at 0.00.

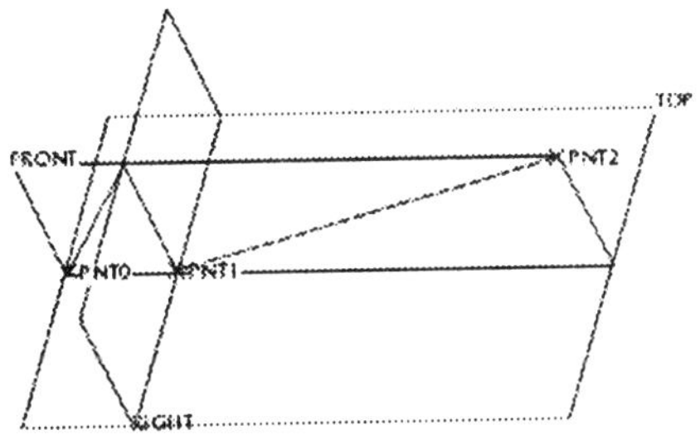

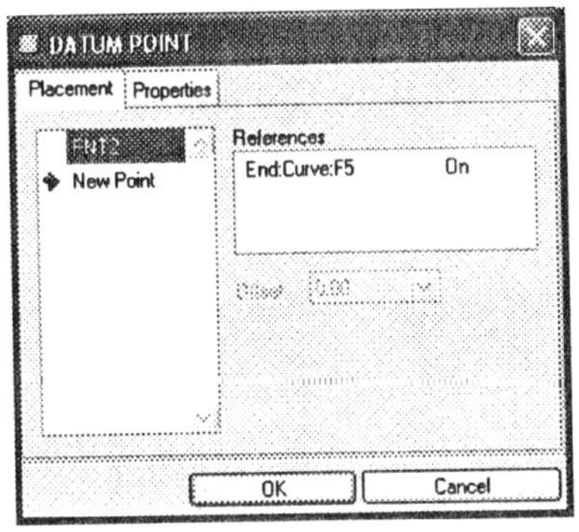

Step 2: under the Pro/Mechanica environment, perform FEA.
Applications > **Mechanica** > **Continue** > **Structure** > **OK.**

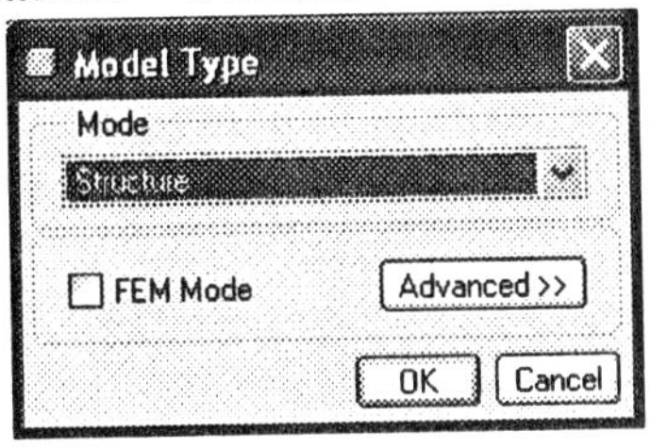

Pick the icon of Beam from the toolbar of functions.

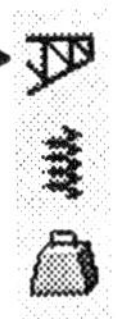

Accept the name: Beam1

Select Edge/Curve

Select Steel

Y vector

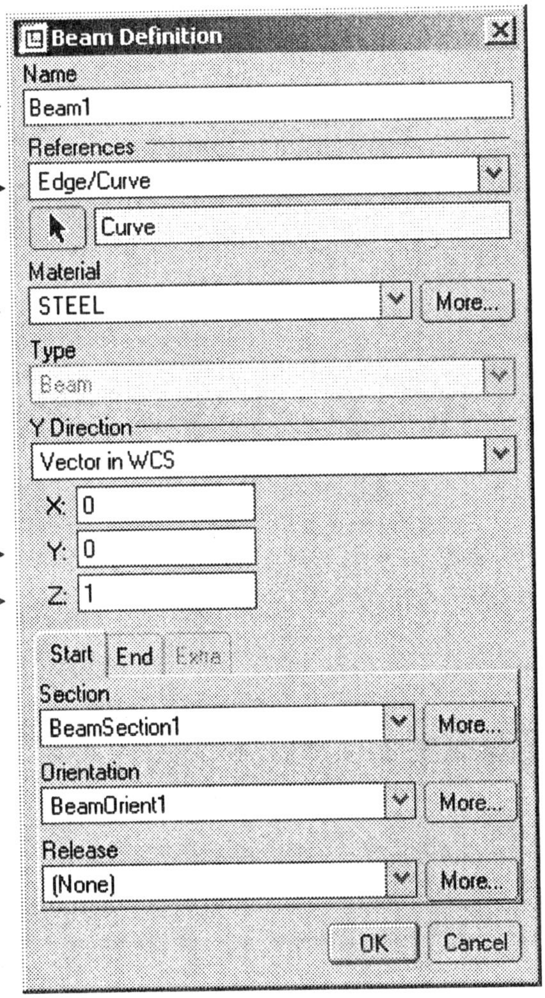

On the appeared Beam Definition window, do the followings:
Select **Edge/Curve** > pick the datum curve.

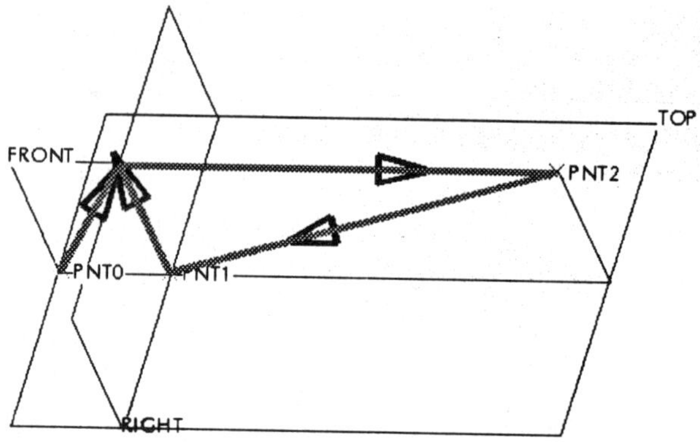

Select More from **Material** > Select **STEEL** > **Close**

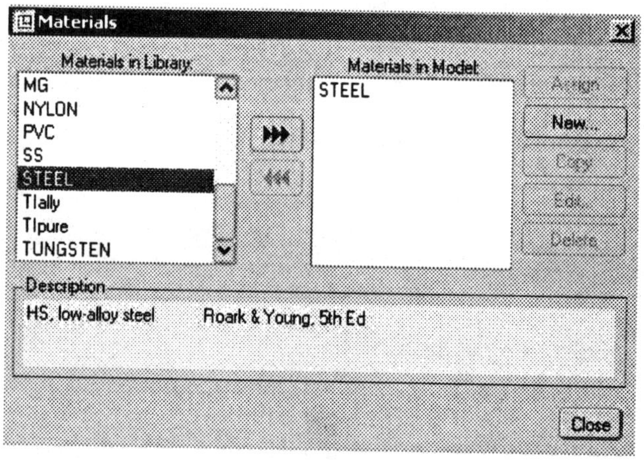

Select More from **Section** > New > select **Hollow circle** > type *30* and *26* as R and Ri, respectively > **OK**.

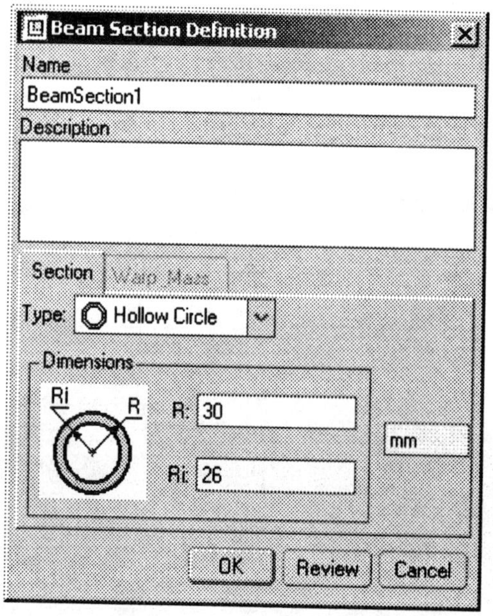

Define the Beam Orientation: click **More** > **New**, and make sure that DY=0 and DZ=0 to complete the process of defining the beam section > **OK > Done/Return.**

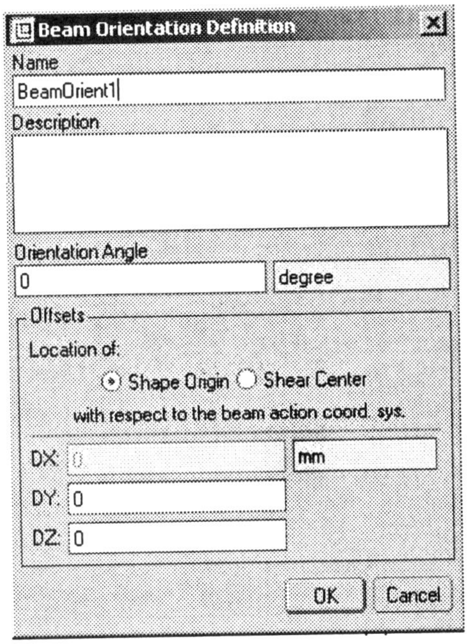

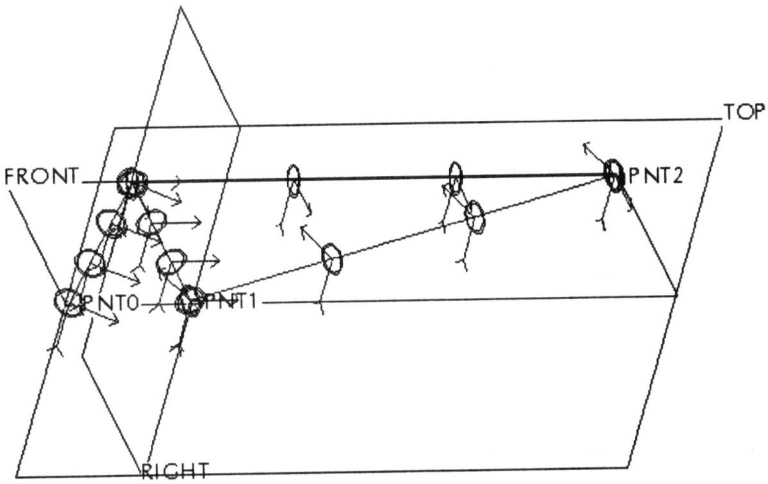

Define the constraint set to present the 2 fixed ends at PNT0 and PNT1. Select the icon of Constraint from the toolbar of functions. Select **Point(s)** > pick **PNT0** > set all 6 degrees of from to fixed > OK.

Repeat the above steps to define the second constraint. Select **Point(s)** > pick **PNT1** > set all 6 degrees of from to fixed > **OK.**

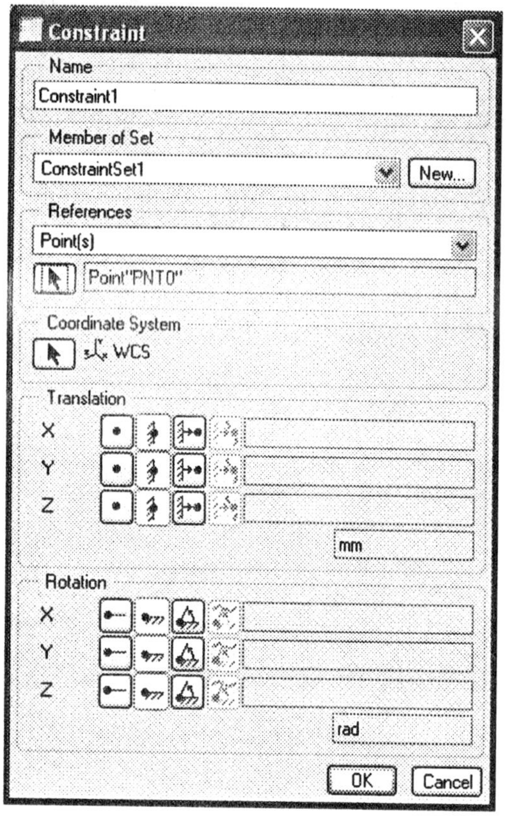

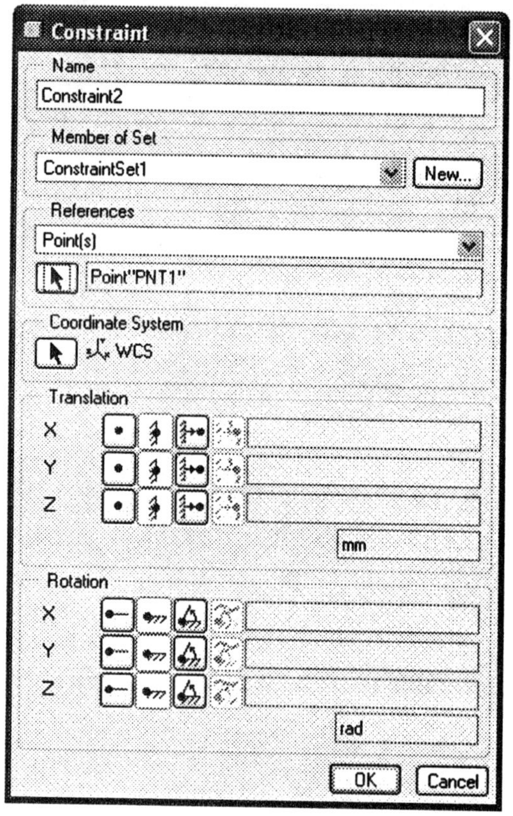

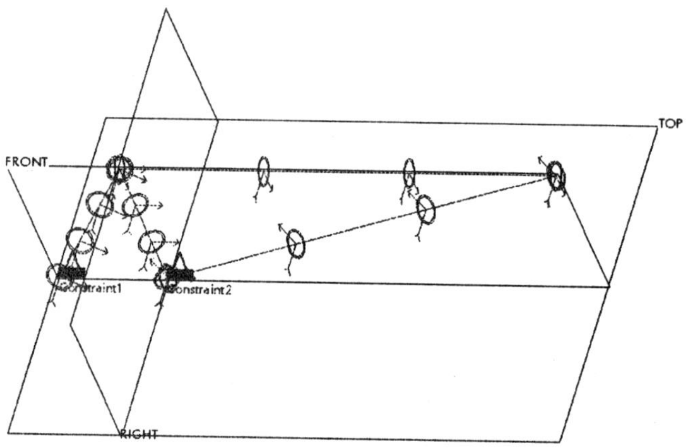

Now let us define the load condition. Select the icon of **New Force/Moment**. Select **Point(s)** > Pick PNT2 > type *–1000* as the magnitude in the Y direction.

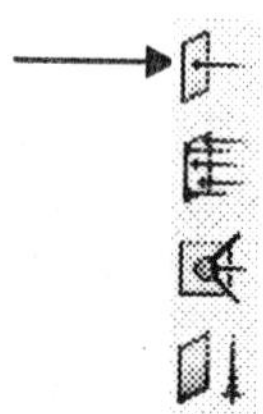

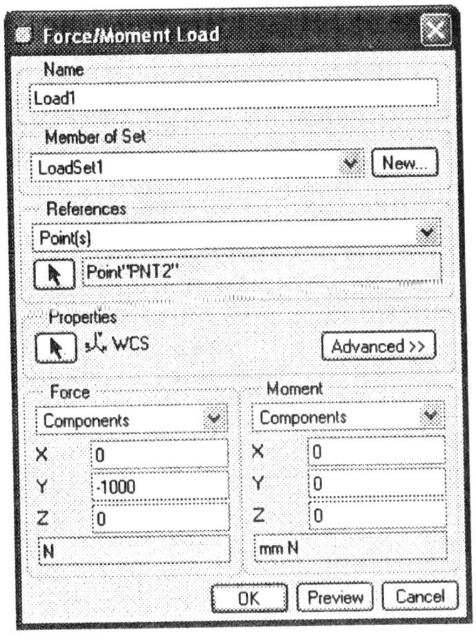

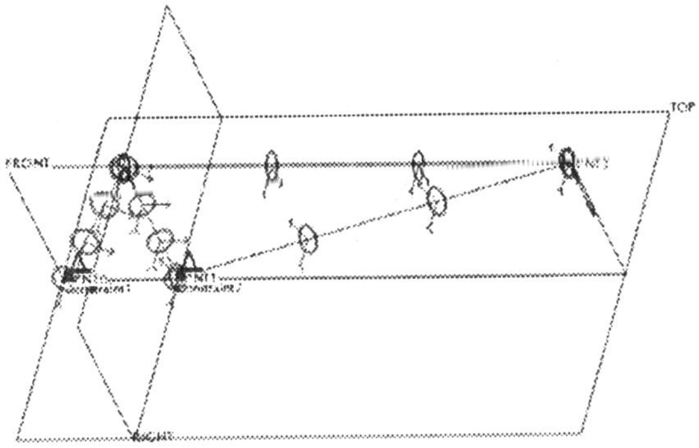

Step 3: Perform a static analysis

Analysis > File > New Static > New > type *truss_structure* as the name of analysis, and **OK** to accept the default setting, which is **Mult-Pass Adaptive > OK.**

Run the program > Close upon completion.

Step 4: perform a static analysis

Click the icon of **performing a design study** from the main toolbar > **File** > Select **New Static.** Type *truss_structure* as the file name > **OK.** Run the static analysis.

Icon of performing a design study

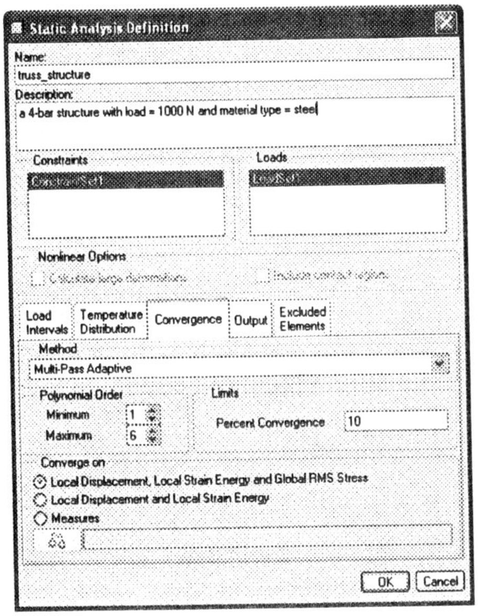

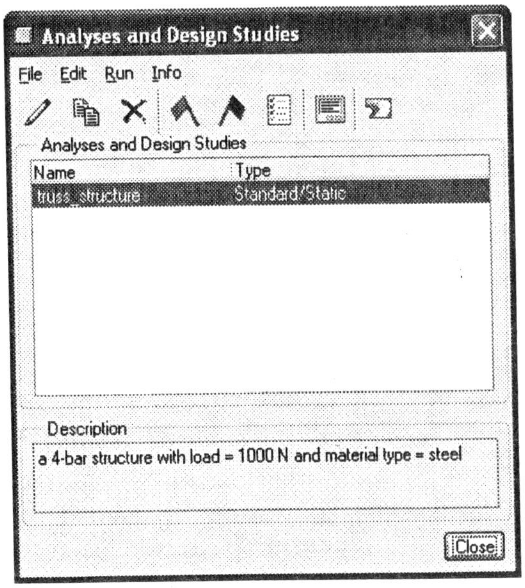

Step 6: Show the results obtained from FEA by plotting the distribution of deflection.

Distribution of the deflection in the XY plane

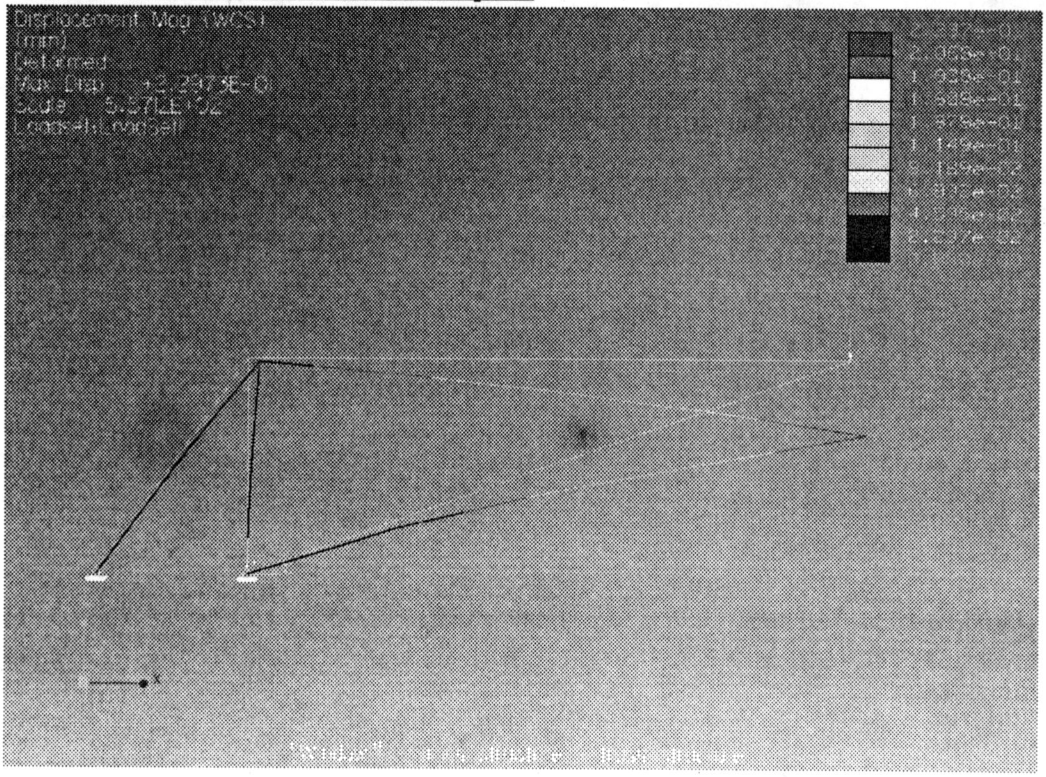

Distribution of the axial forces in individual bars, using GRAPH

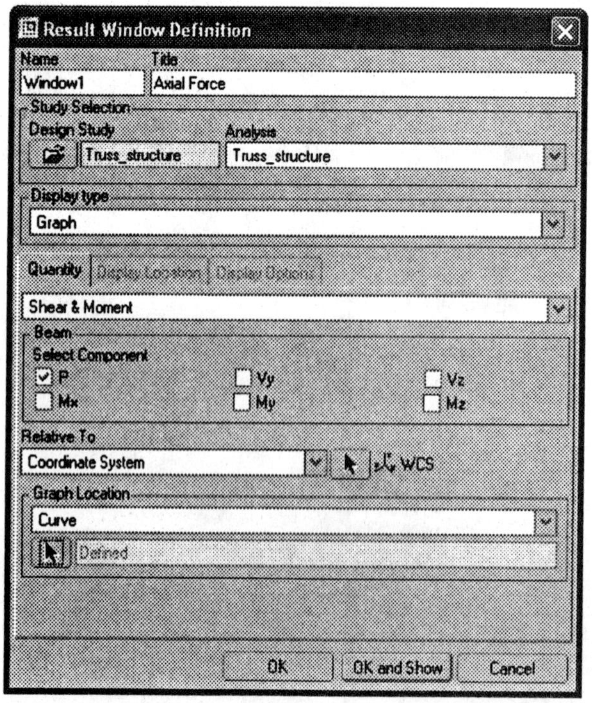

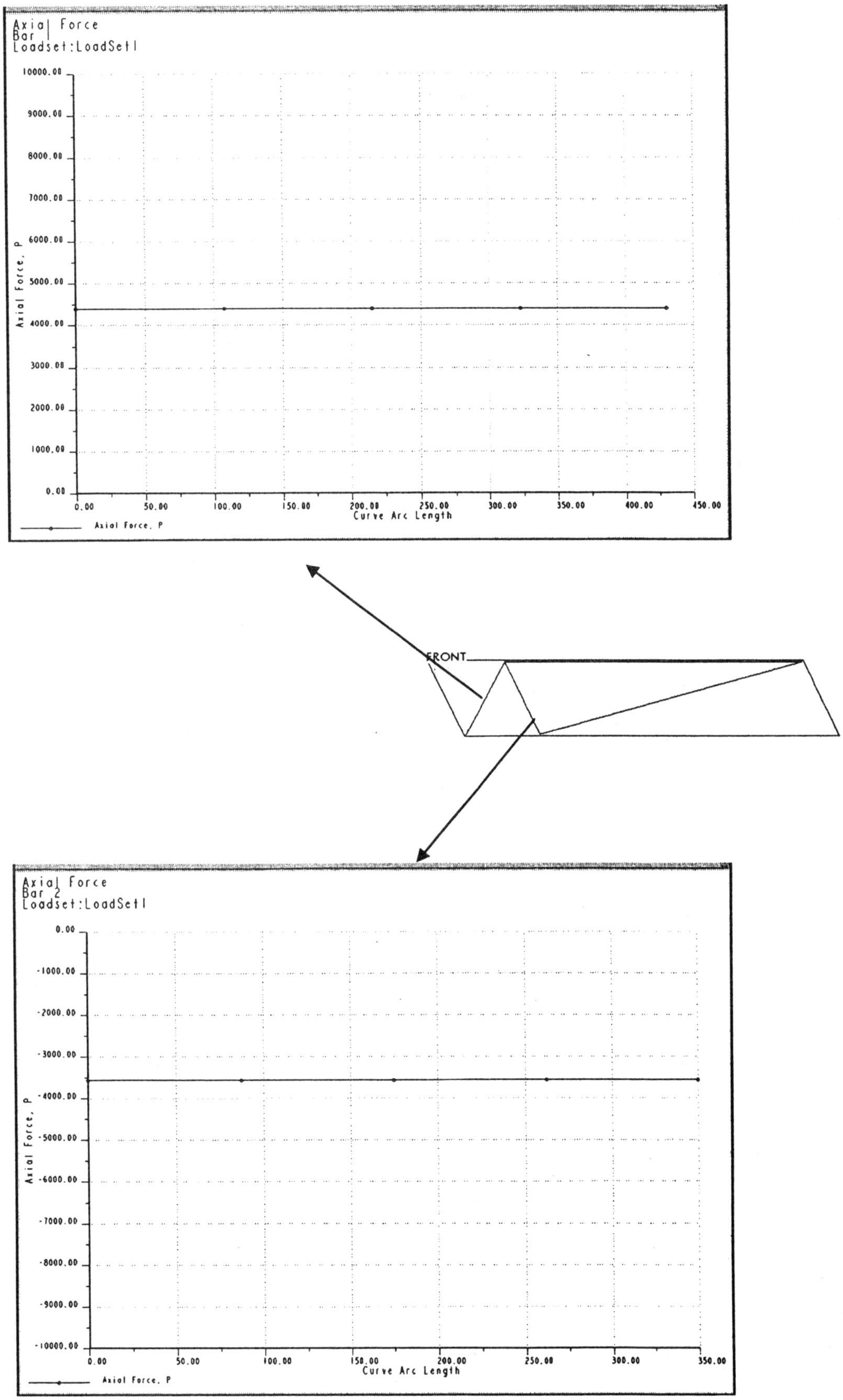
Axial Force
Bar 1
Loadset:LoadSet1
Axial Force, P
Curve Arc Length
Axial Force, P
FRONT
Axial Force
Bar 2
Loadset:LoadSet1
Axial Force, P
Curve Arc Length
Axial Force, P

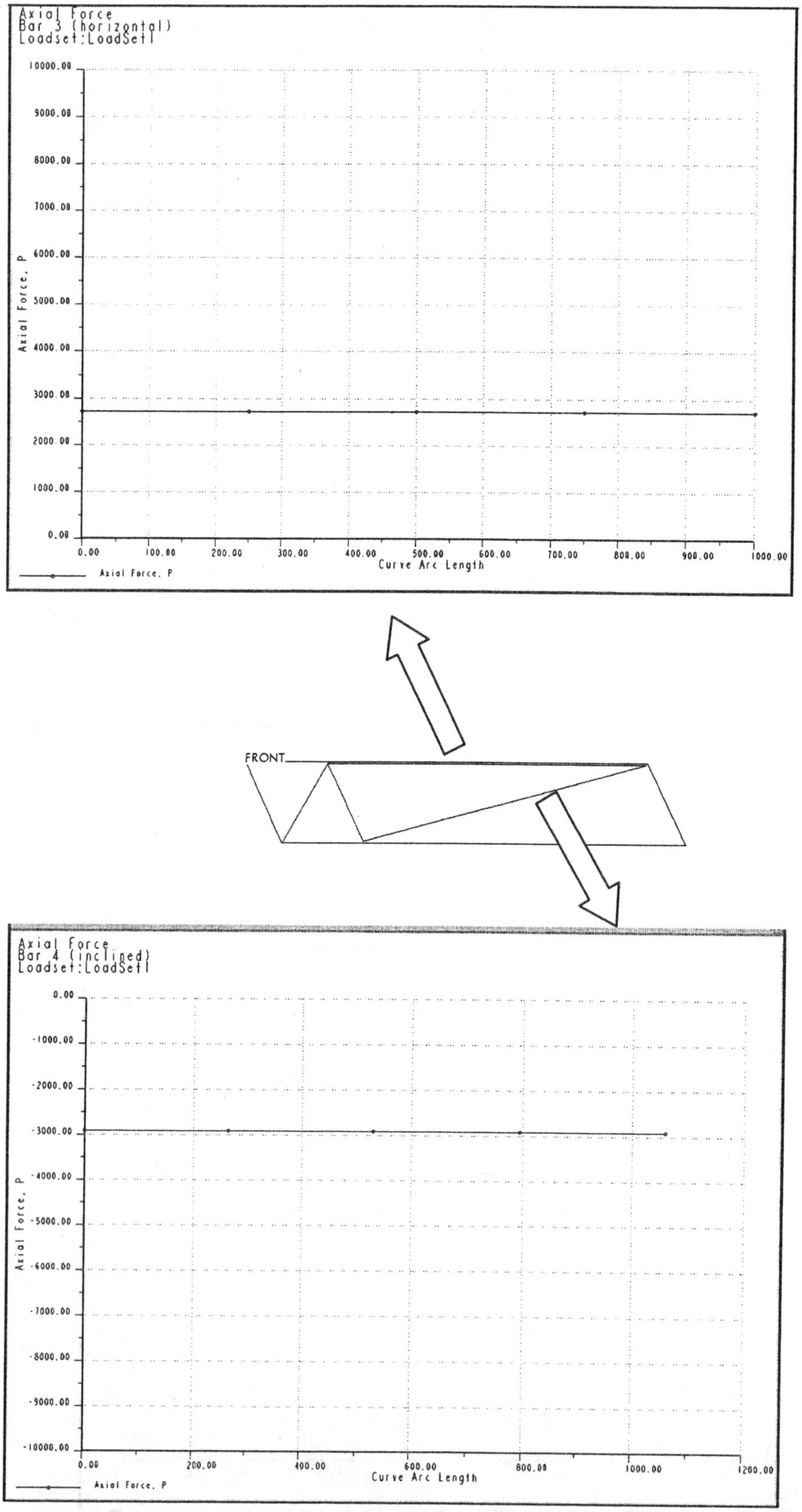
Axial Force
Bar 3 (horizontal)
Loadset:LoadSet1
Axial Force, P
Curve Arc Length
Axial Force, P
FRONT
Axial Force
Bar 4 (inclined)
Loadset:LoadSet1
Axial Force, P
Curve Arc Length
Axial Force, P

9.3 Application on a Simple Supported Beam Structure

A simple supported beam is shown below. The left end of the beam is a pin connection and the right end of the beam is supported by a roller connection. The material type of the beam structure is steel. The load is acting at the middle of the beam length. The magnitude of the load is 100 N and the direction of the load is downward. The cross section of the beam is a rectangle and the two dimensions are 25 mm and 50 mm, respectively.

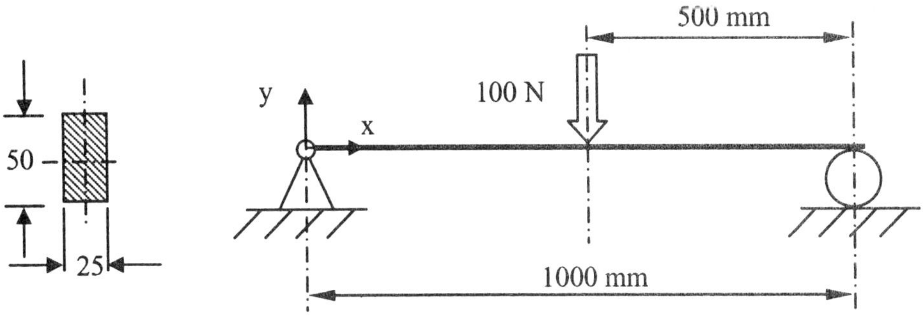

In this example, we will have 3 case studies for three load conditions. In case study 1, the load condition is the external force with the magnitude equal to 100 Newton. In case study 2, we only consider the effect on the deflection of the beam structure due to the gravity force. In case study 3, both loads, namely, the external force and the gravity force, are considered when performing FEA.

	Load 100 N	Gravity Force	Deflection (mm)
Case 1	X		
Case 2		X	
Case 3	X	X	

Step 1: Create a 3D solid model for the simple supported beam with a pin connection at the left side and with a simple support at the right side.

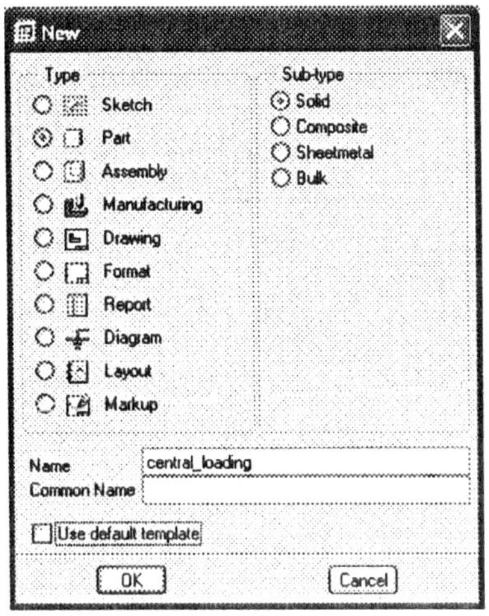

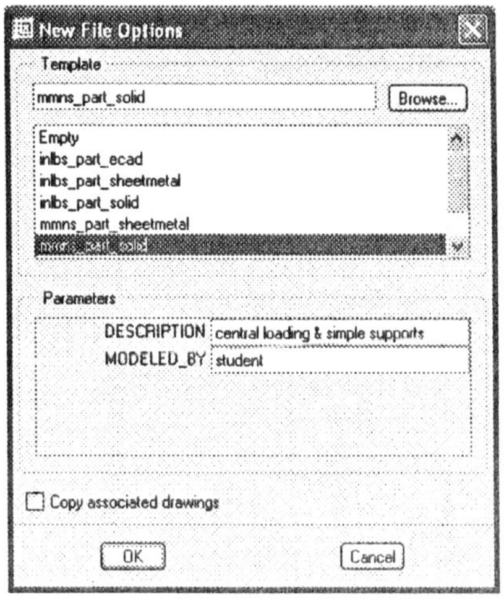

Define a datum curve, which is along the x-axis. The length of the datum curve is 1000 mm and it is symmetric about the y-axis.

Select the icon of **Sketch Tool** from the toolbar of datum creation.

Pick the **FRONT** datum plane as the sketch plane, and accept the default orientation setting, which uses the **RIGHT** datum plane and position it at the right side > **Sketch**.

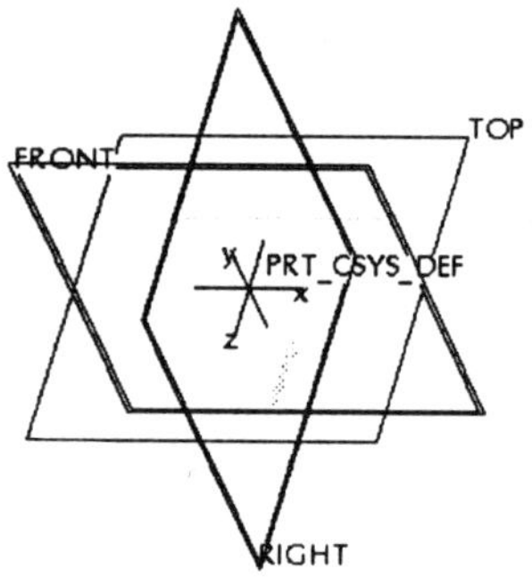

Sketch a centerline along the y-axis, and draw a line along the x-axis. The line is symmetric about the y-axis and the length is 1000 mm. Upon completion, select the icon of **Done**.

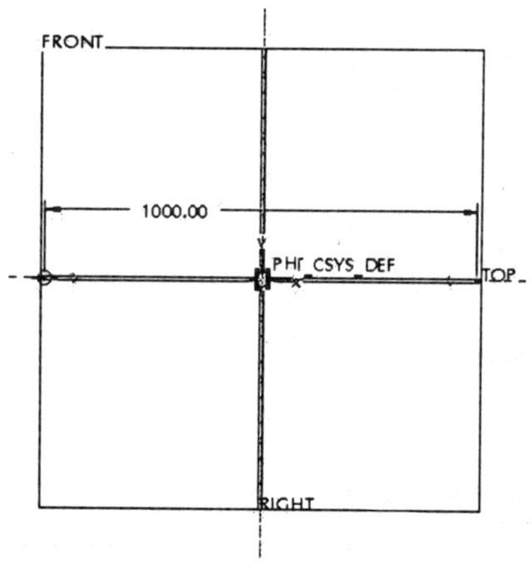

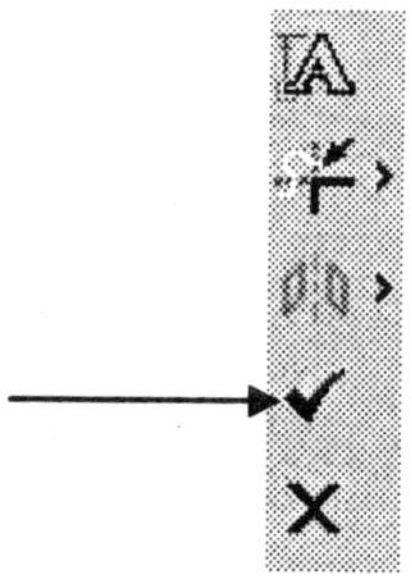

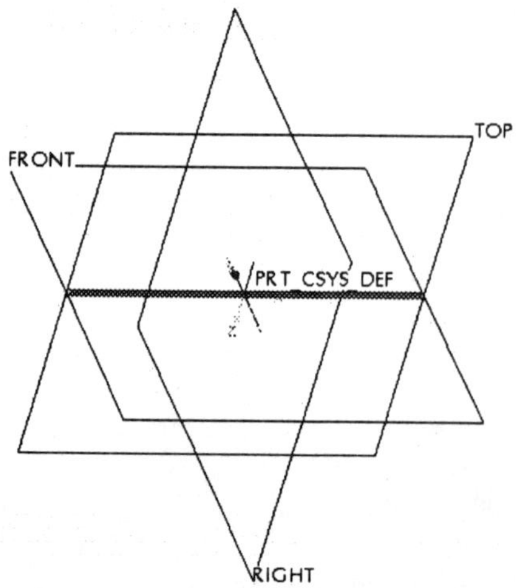

Create three datum points. The first datum point is at the central location of the datum curve. Select the icon of datum point from the toolbar of datum feature creation.

Pick the point at the central location of the datum curve on display. Make sure the offset value is set at 0.50.

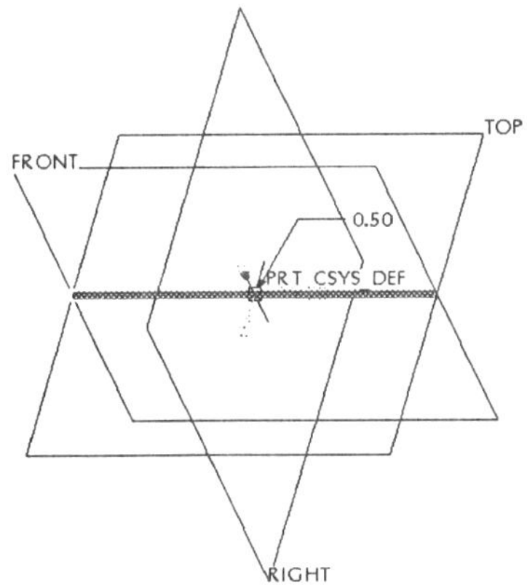

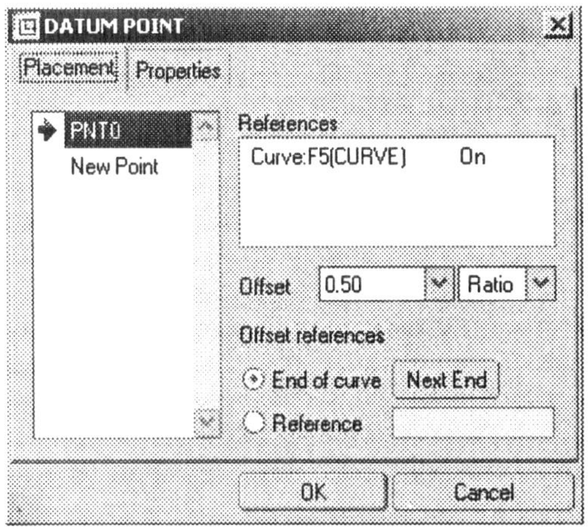

To define the second datum point, select the icon of datum point from the toolbar of datum feature creation. Pick the point at the left end of the datum curve on display. Make sure the offset value is set at 0.00.

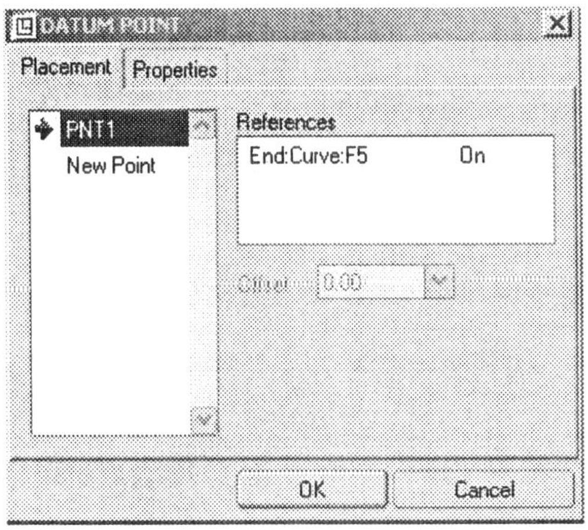

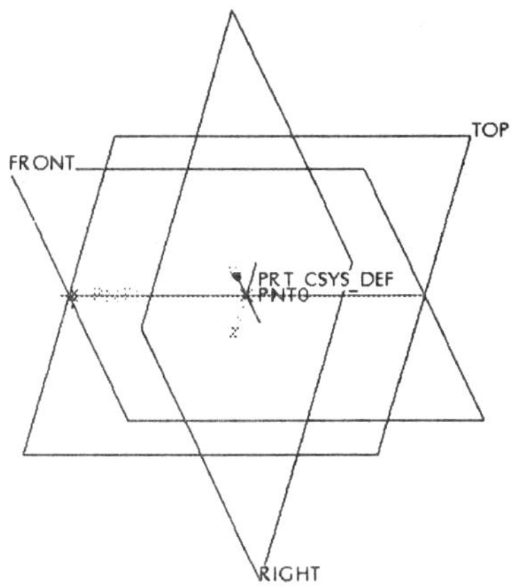

To define the third datum point, select the icon of datum point from the toolbar of datum feature creation. Pick the point at the left end of the datum curve. Make sure the offset value is set at 0.00.

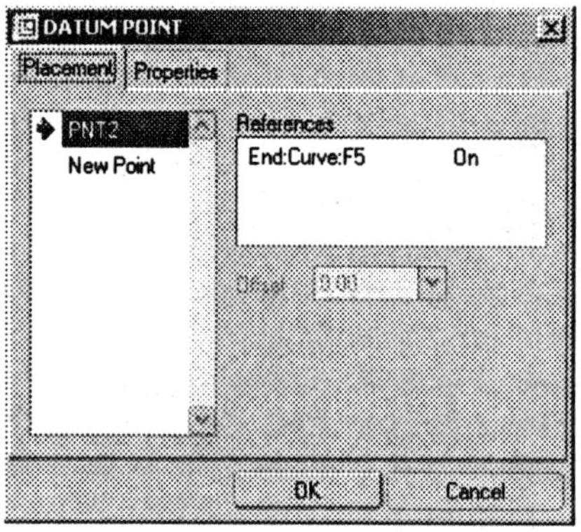

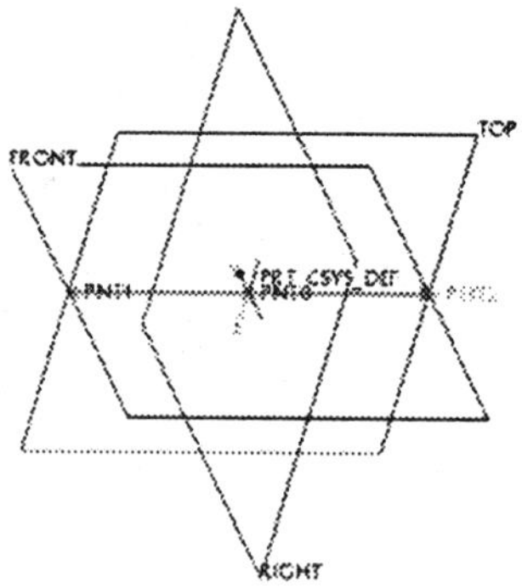

Step 2: Perform FEA under **Pro/Mechanica**
Applications > Mechanica > Continue > Structure > OK.

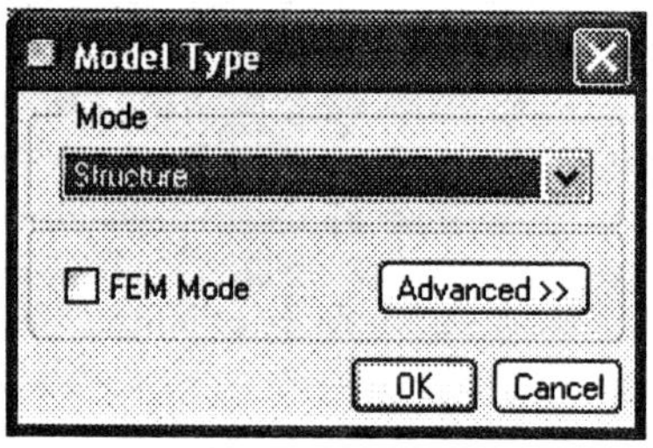

Pick the icon of **Beam** from the toolbar of functions.
On the appeared Beam Definition window, do the followings:

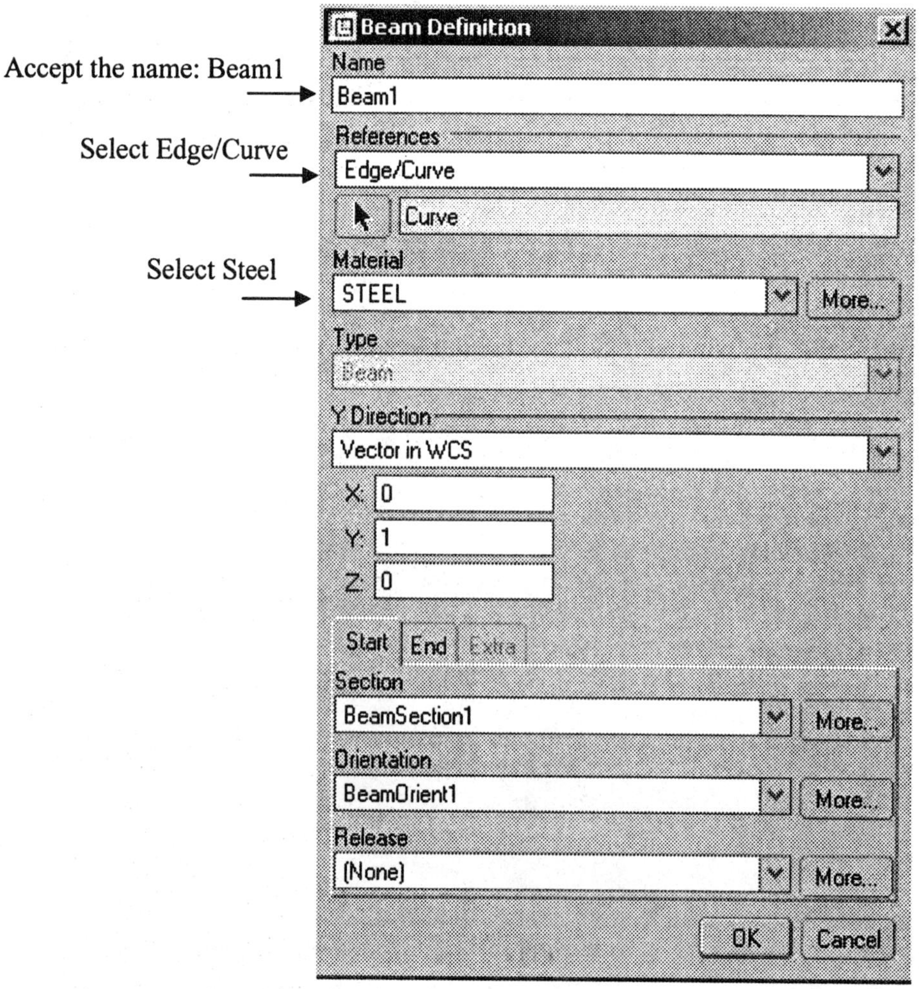

Select More from **Material** > Select **STEEL** > **Close.**

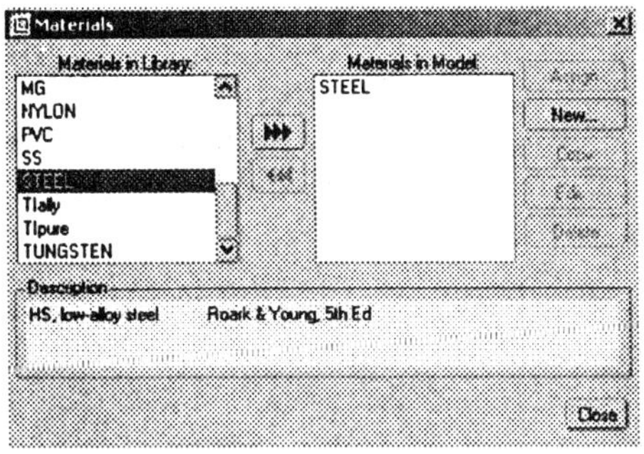

Define the section of the beam: click **More** > **New**, select **Rectangle** and type in 25 and 50.

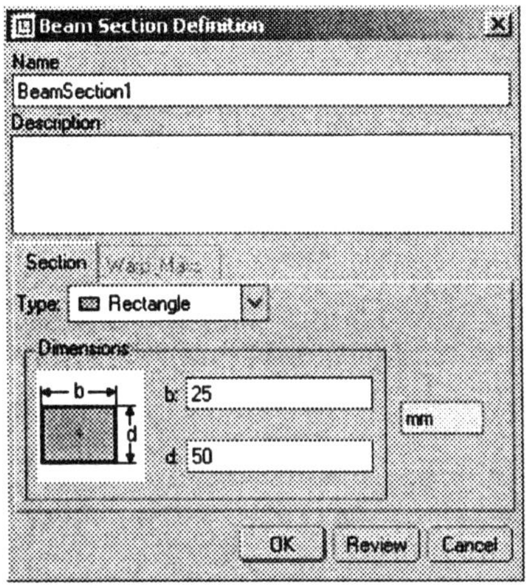

Define the Beam Orientation: click **More** > **New**, and make sure that DY=0 and DZ=0 to complete the process of defining the beam section.

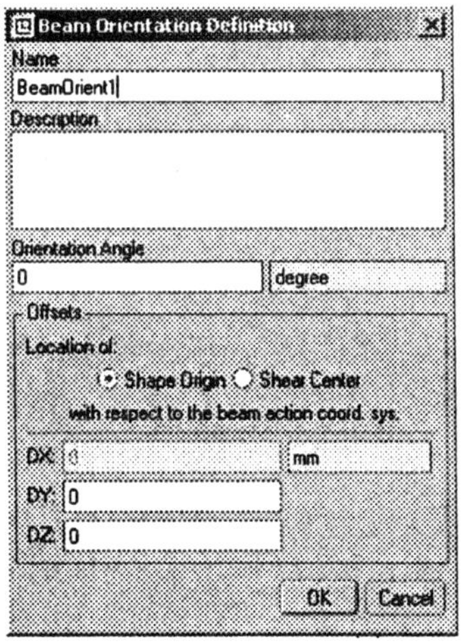

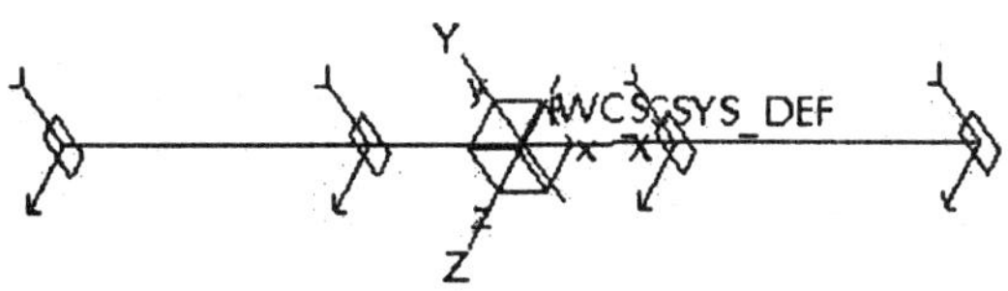

Define a pin connection constraint on the left end of the beam.
Constraints > New > Point > pick **PNT1** > set the rotation about z-axis to free.

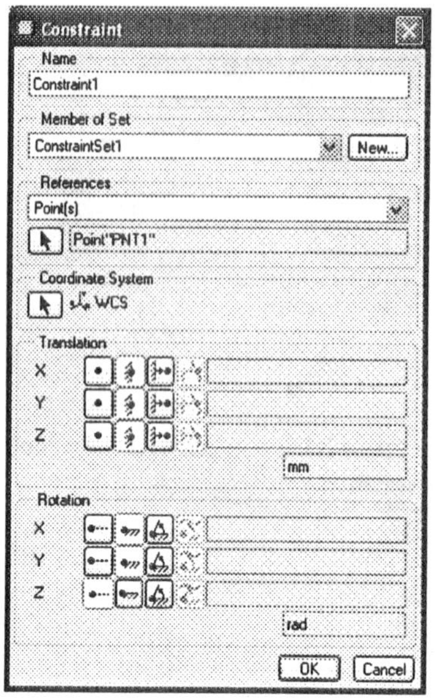

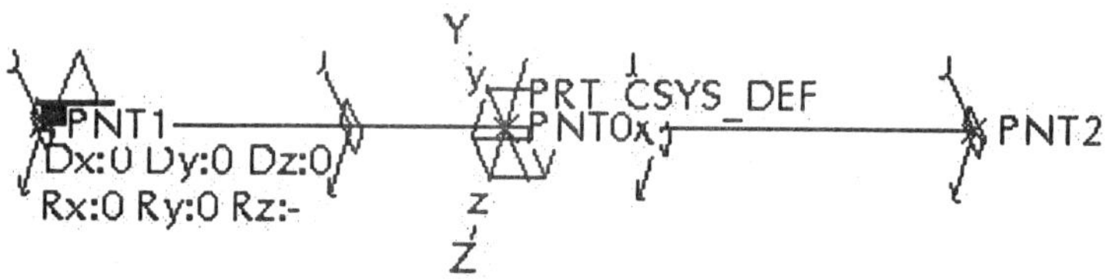

Define a simple supported constraint on the right end of the beam.

Constraints > **Point(s)** > pick **PNT2** > set the translation along x-axis to free and the rotation about z-axis to free.

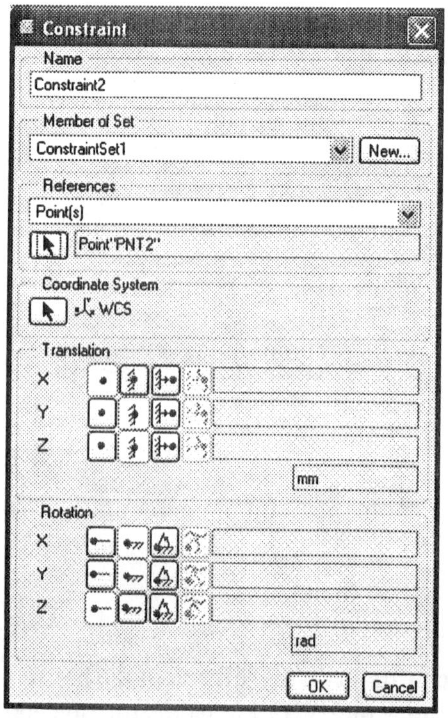

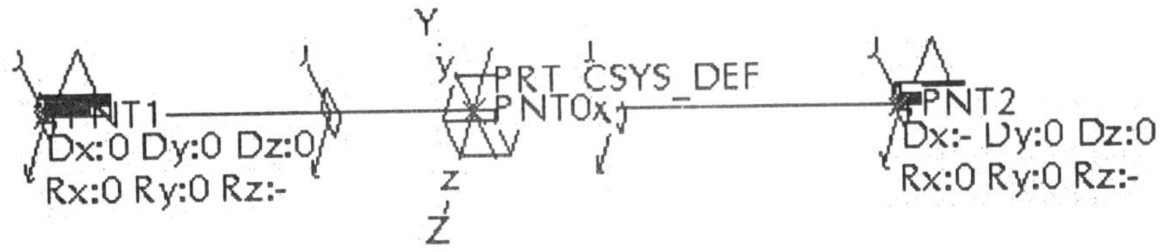

From the toolbar of functions, select the icon of **New Force/Moment** > **Point(s)** > pick PNT0. **Total Load** > **Uniform** > type –100 in the force component **Y** > **OK.**

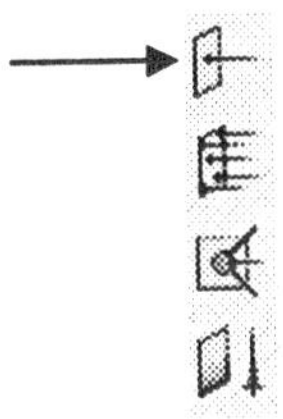

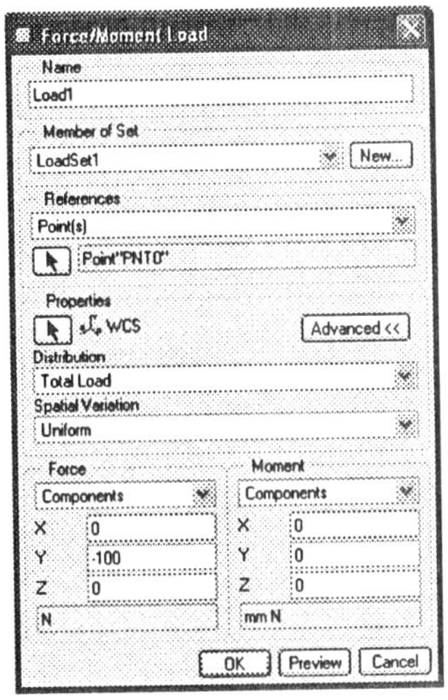

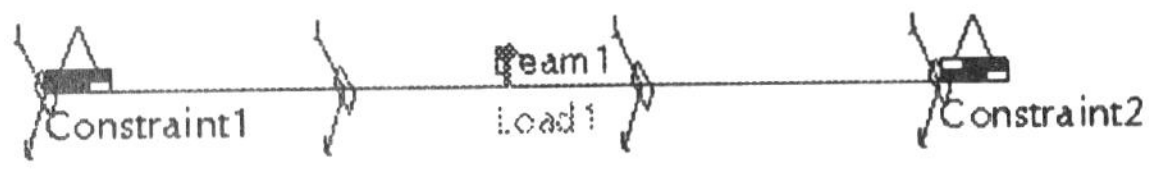

Step 3: Perform a static analysis

Click the icon of **performing a design study** from the main toolbar > **File** > Select **New Static**. Type *simple_support_structure* as the file name > **OK**. Run the static analysis.

Icon of performing a design study

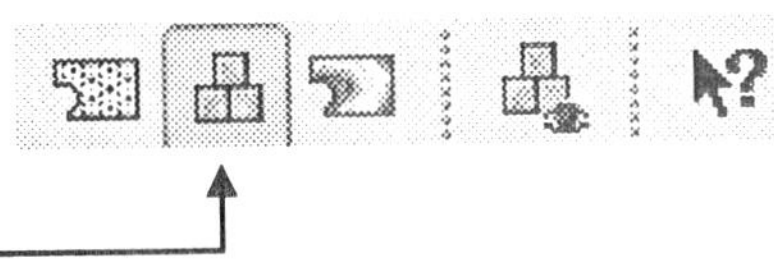

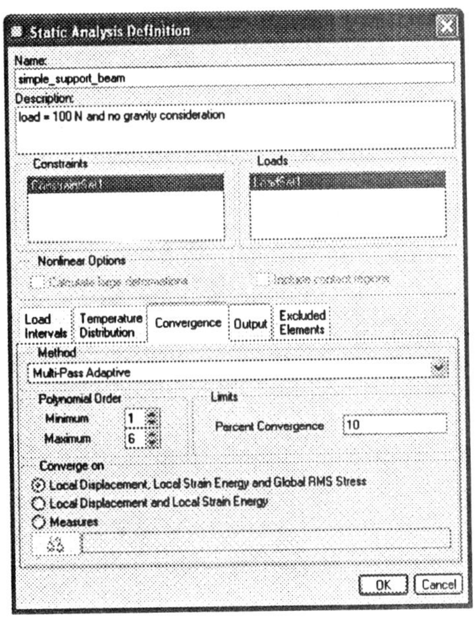

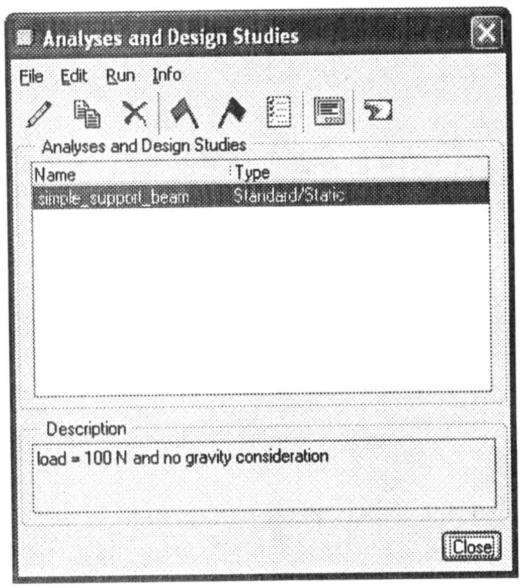

Note: The Percent Convergence is set at 10% (default value).

Results > **Insert** > **Result Window** > type *deflection subjected to 100 N load only* as the title > select the folder called simple_support_beam > Graph > Displacement > Y > select the two segment of the curve > **OK and Show**.

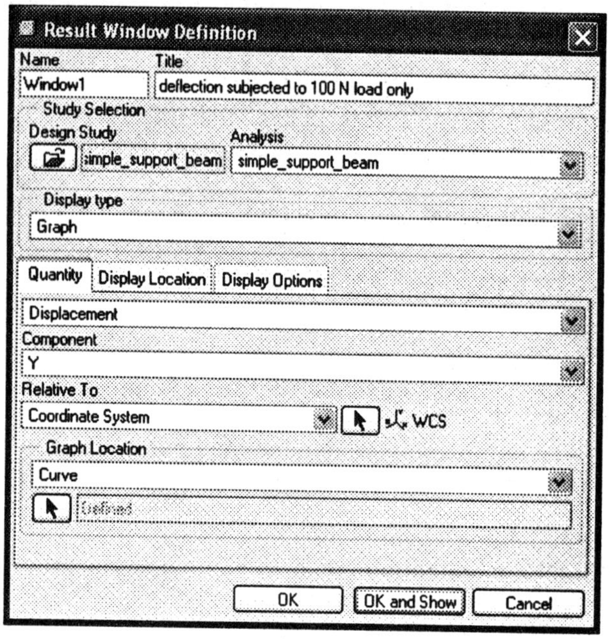

The maximum deflection value at the middle section is 0.040315 mm.

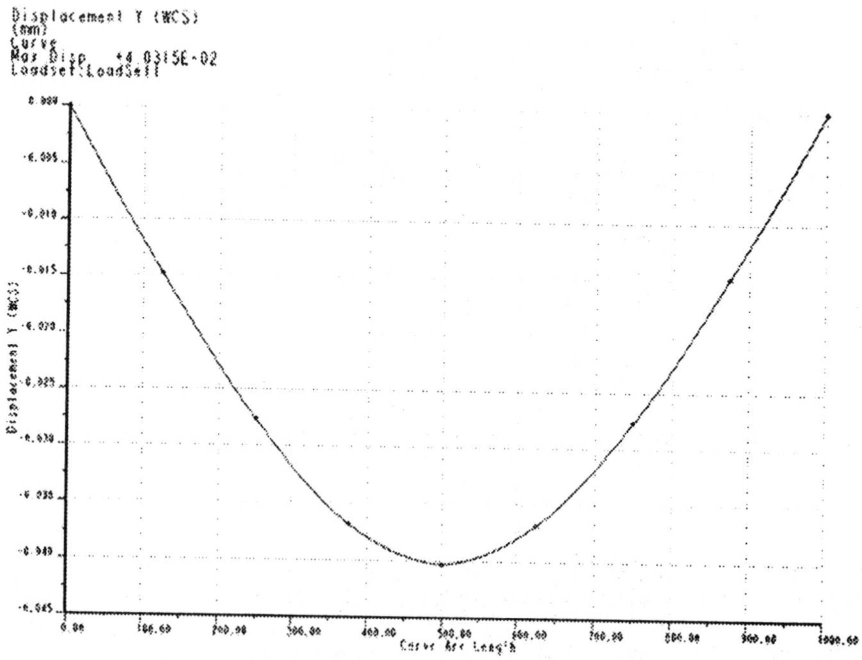

Step 4: Perform a static analysis for the gravity force only.
To run the FEA with the gravity force only, we need to define a second load set.
New Gravity Load and select **New** again > type in –9810 > **Preview > OK.**

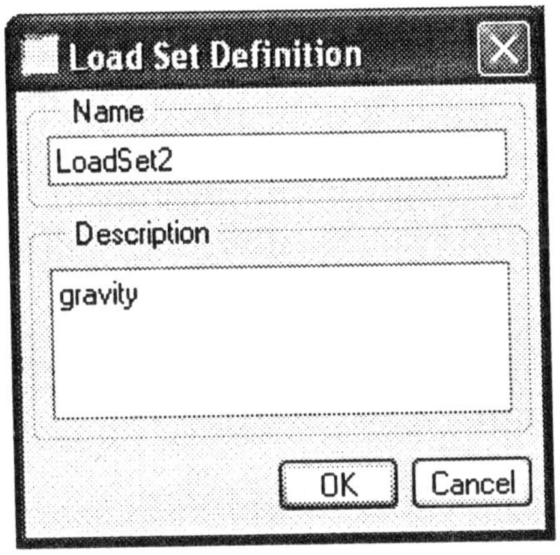

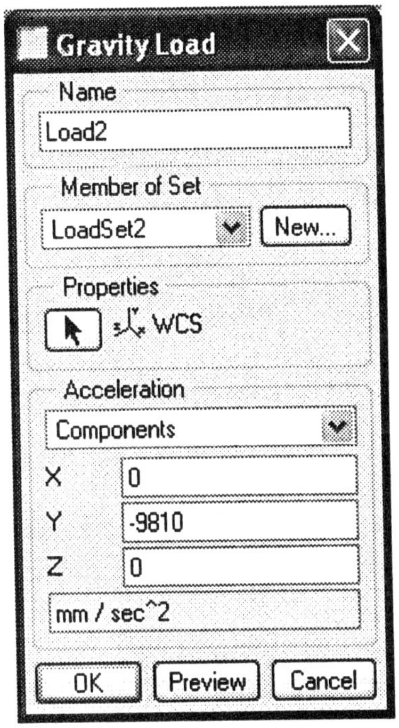

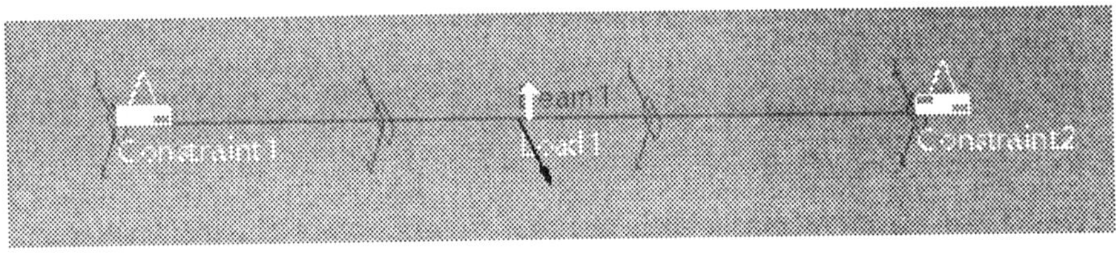

Click the icon of **performing a design study** from the main toolbar > **File** > Select **New Static**. Type *gravity_only* as the file name > **OK**. Run the static analysis.

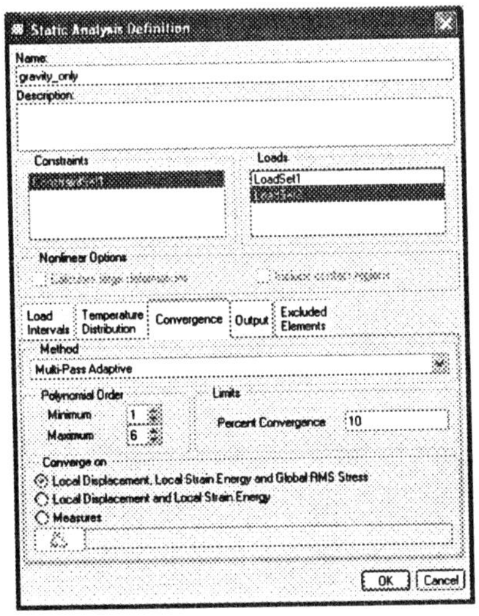

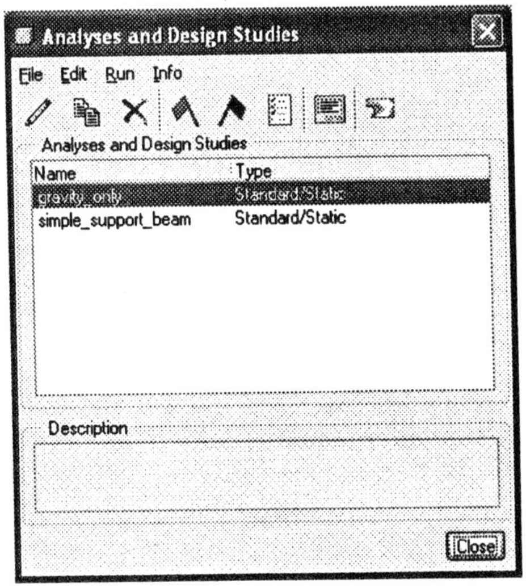

Click the icon of Results > Result Window > type in deflection_gravity_only as the title > select the folder called **gravity_only** > **Graph** > **Displacement** > Y > select the two segment of the curve > **OK and Show**.

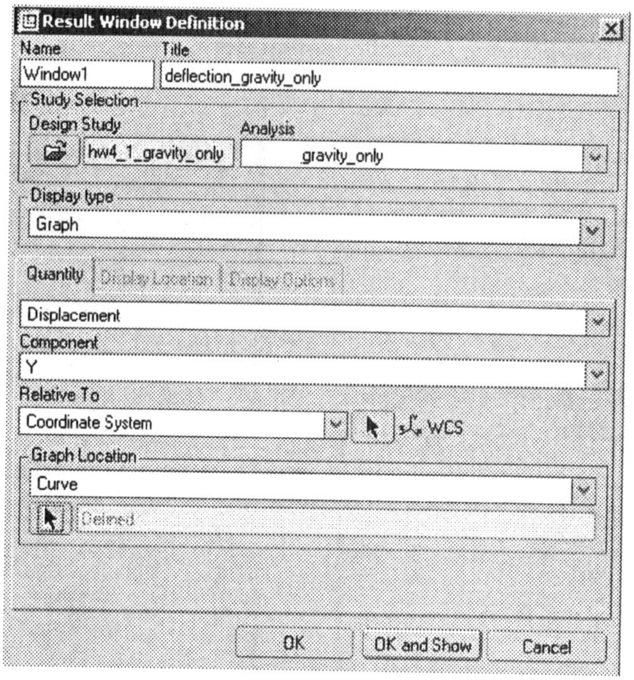

The maximum deflection value at the middle section is 0.024147 mm.

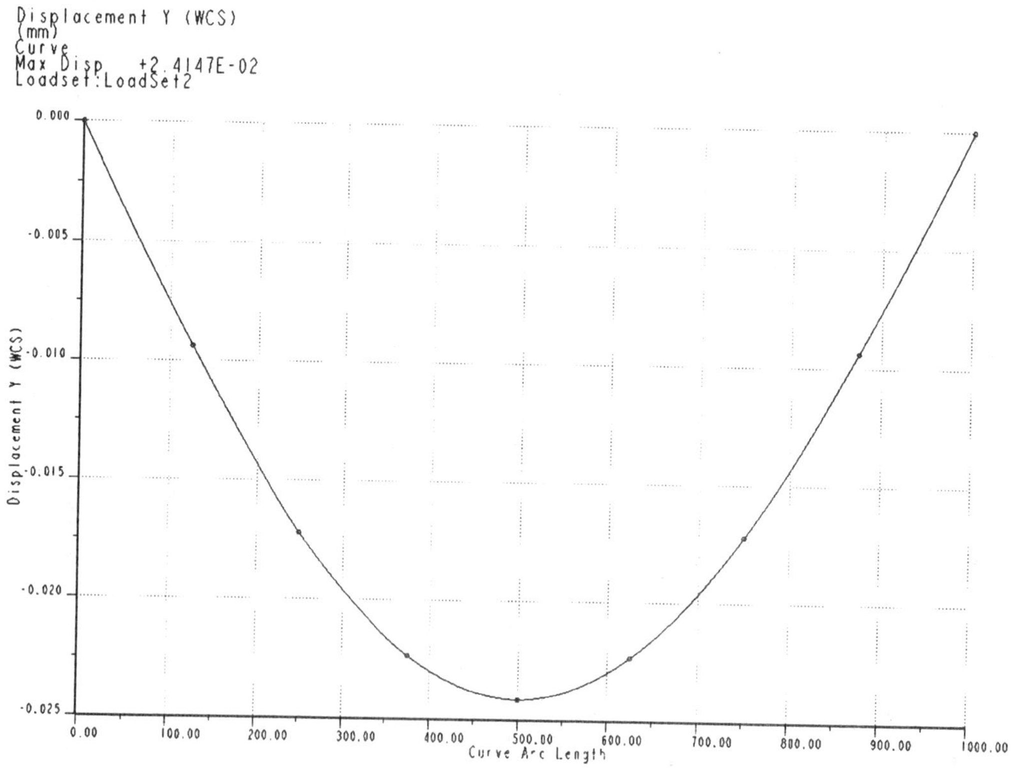

Step 5: Perform a static analysis for the presence of both load and the gravity force.

Click the icon of **performing a design study** from the main toolbar > **File** > Select **New Static.** Type *gravity_only* as the file name > **OK.** Run the static analysis.

Make sure that ConstraintSet1, LoaSset 1 and LoadSet2 are highlighted > **Multi-Pass Adaptive** > **OK** > **Run**

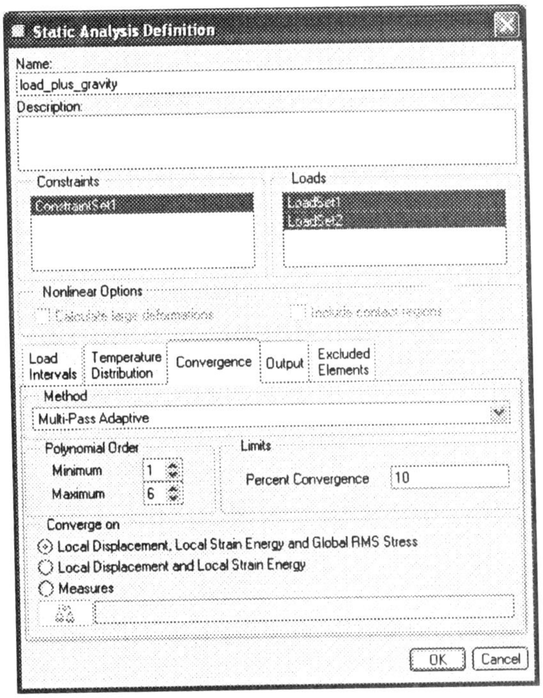

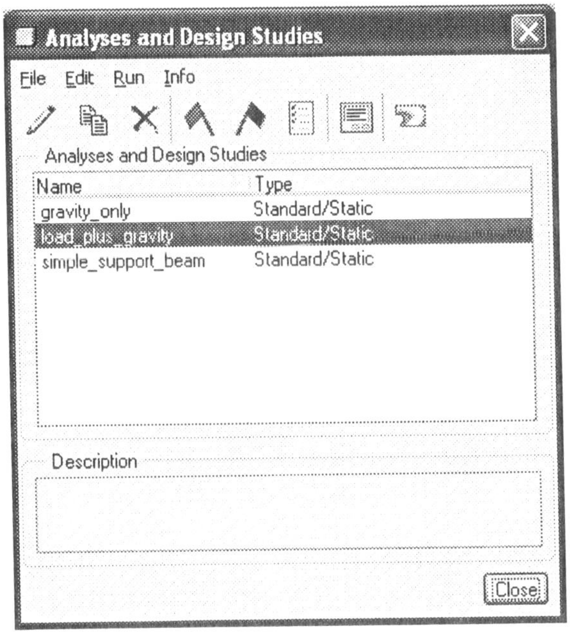

Click the icon of Results > Result Window > type *deflection_gravity_only* as the title > select the folder called load_plus_gravity > Graph > Displacement > Y > select the two segment of the curve > OK and Show.

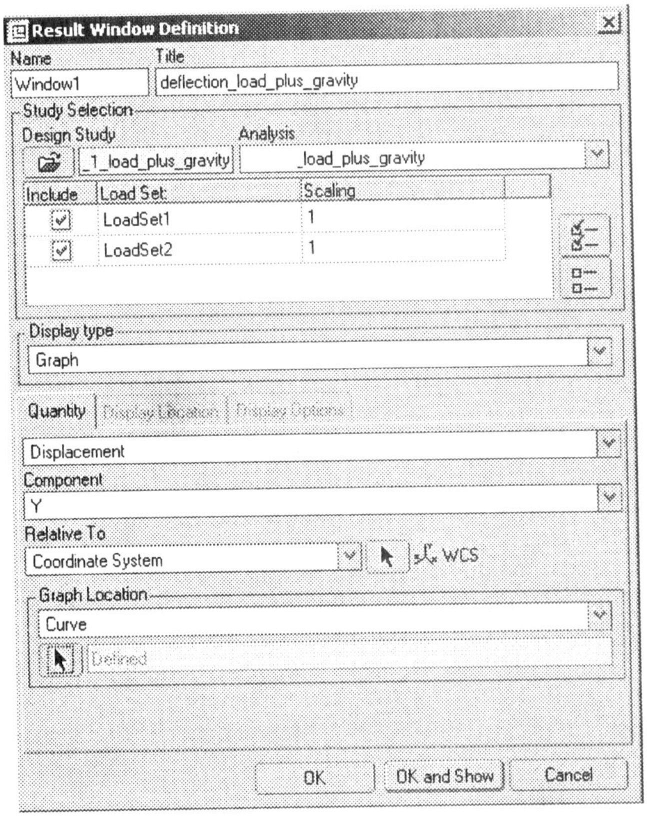

The maximum deflection value at the middle section is 0.06446 mm.

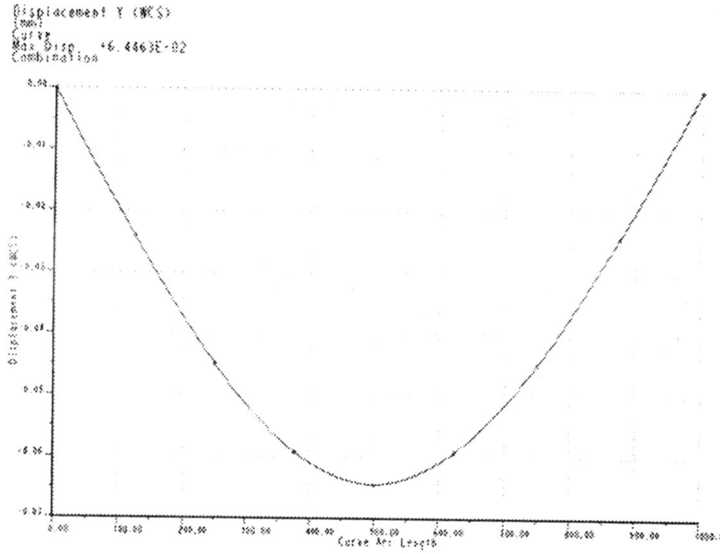

Now let us list these three values obtained from the FEA, and put the three graphs together to gain a clear picture on the gravity force effect on the total deflection of the beam.

	Load 100 N	Gravity Force	Deflection (mm)
Case 1	X		0.040315
Case 2		X	0.024147
Case 3	X	X	0.064460

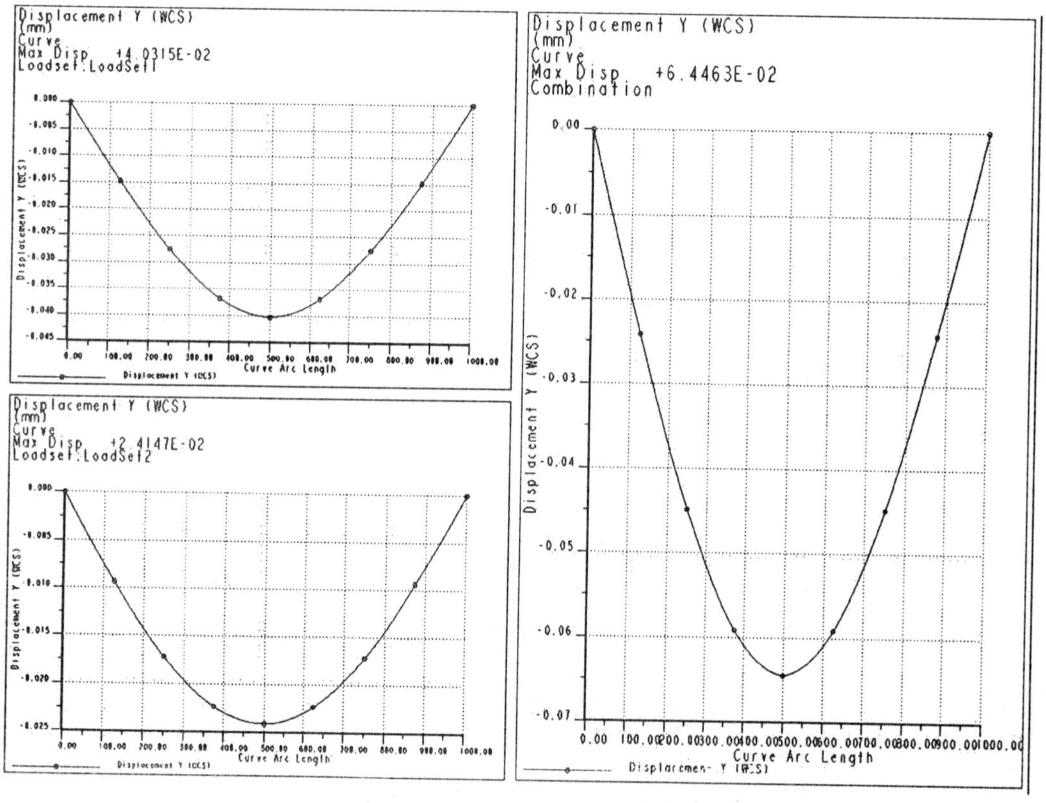

To verify the results obtained from the FEA, let us calculate the maximum deflection of this simple supported beam using an analytical formula. The formula shown below is used by most textbooks published in mechanics of materials,

$$\delta_{\max} = \frac{PL^3}{48EI}$$

Where E = 200 GPa and the moment inertia of I is given by

$$I = \frac{1}{12}bh^3 = \frac{1}{12}(25)(50)^3 \qquad (\text{mm}^4)$$

By substituting the load value of P and the span distanced of the beam structure of L, we have

$$\delta_{\max} = \frac{PL^3}{48EI} = \frac{(100N)(1000mm)^3}{48(200\frac{N}{m^2})(\frac{1}{12})(25mm)(50mm)^3} = 0.04 \quad (\text{mm})$$

This calculated value of 0.04 mm is very closed to the result obtained in case study 1, which was 0.040315. Note that the gravity force was not considered in the study of case 1. Therefore, the calculated value based on the analytical formula does not consider the effect of the gravity force on the deflection of the beam structure. Although the effect of the gravity force on the gravity force was small, which was 0.024147 mm, it should be considered because it accounts for almost 50% of the deflection value due to the external force. As a result, under circumstances where accuracy of the deflection of a beam structure is set to be high, the gravity force should be considered.

9.4 Finite Element Analysis of Truss Structure Using ALGOR V14

In this section, we use the software system of **ALGOR** to perform FEA. The figure illustrates a triangular truss structure with a circular cross section. This truss study will be analyzed under two separate loading conditions. The first load condition is a load of 1000 lbs in the y-direction, and the second load condition is 500 lbs in the x-direction. The left corner of the truss is constrained by a pin connection and the right corner of the truss structure is constrained by a roller connection. This truss structure consists of 2 bars. The cross section area of these 2 bars is a circle. The diameter dimension is 1 inch. The material type of these 2 bars is 6061 T6 aluminum. When performing FEA, consider both the external force and the gravity force, which pulls the truss structure in the negative y-direction.

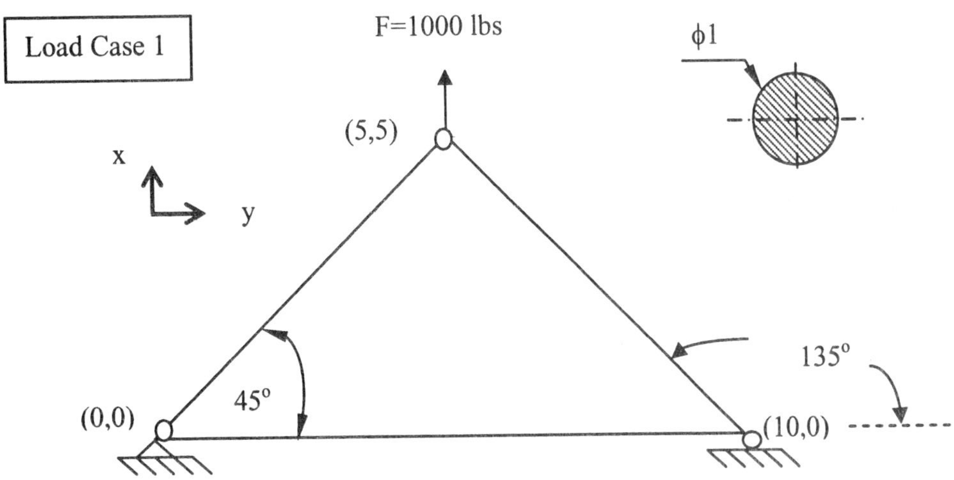

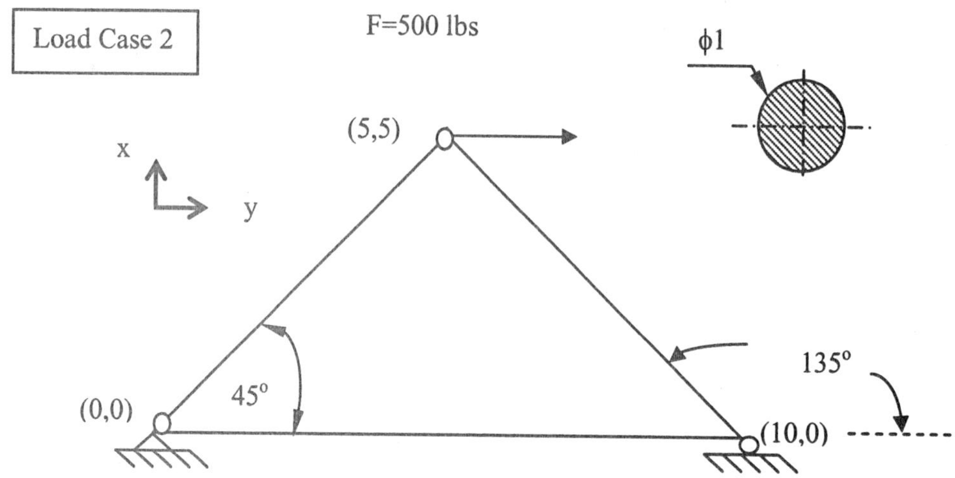

The objective of this study is to find the displacements at each joint of the truss structure, and identify the maximum displacement of the truss structure when it is subjected to each of the 2 loading conditions.

As we discussed in Chapter 1, ALGOR is an engineering program that uses FEA to create computer simulations of mechanical components subjected to different types of loading conditions. In this example, we will give a brief description of features that ALGOR offers for analyzing truss structures. The material covers applying constraints and loads to a structure, selecting materials, and performing analyses.

Step 1: launch the software system: **ALGOR.**

We assume that the software system resides in the folder called **Programs.**

Click the **Start Menu/Programs/ALGOR/FemPro**

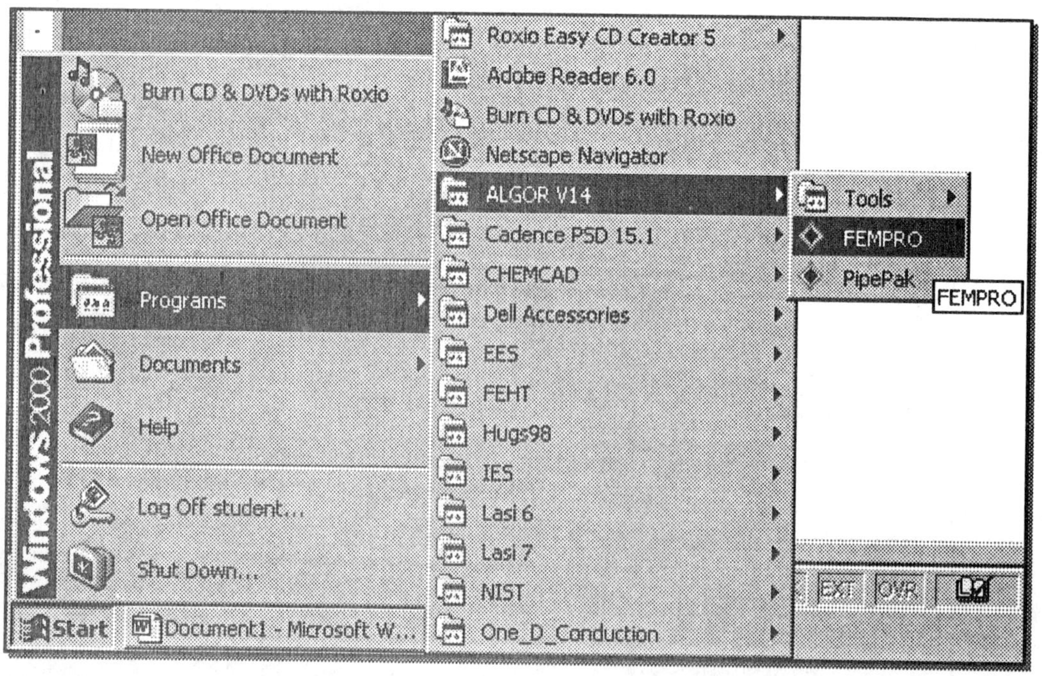

An "**Open**" window will appear. Click **Superdraw** and then click **Open**.

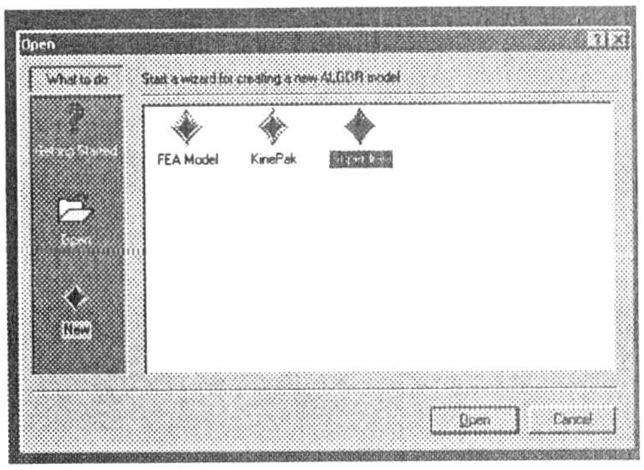

Step 2: Drawing the Truss Structure

From the main toolbar, click **Settings** > **Toolbars** > click **Add FEA Objects** if it is not selected > **Close**.

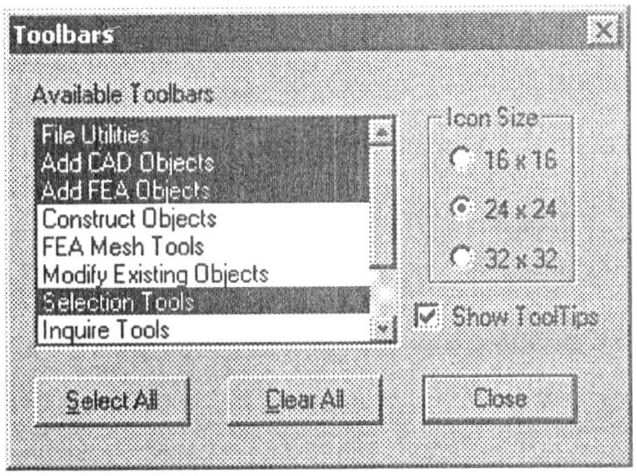

The window on display should look like the picture shown below.

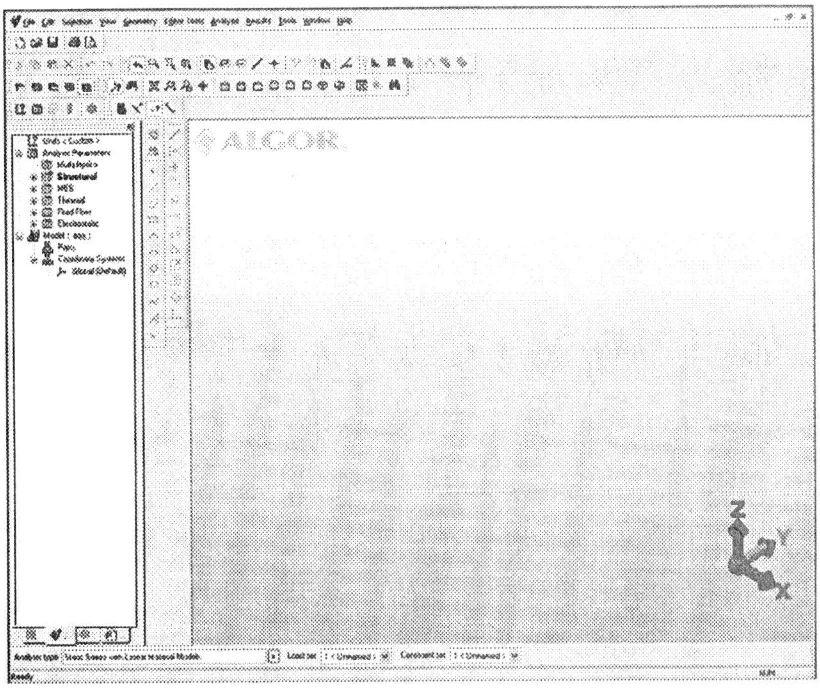

If your window does not look like the one shown below, organize those toolbars on display. The following picture shows a re-organized window. The way to re-organize those toolbars is to your favorite so that you have an easy way to access.

Click the icon of **Line** on the ***Add CAD Objects*** Toolbar.

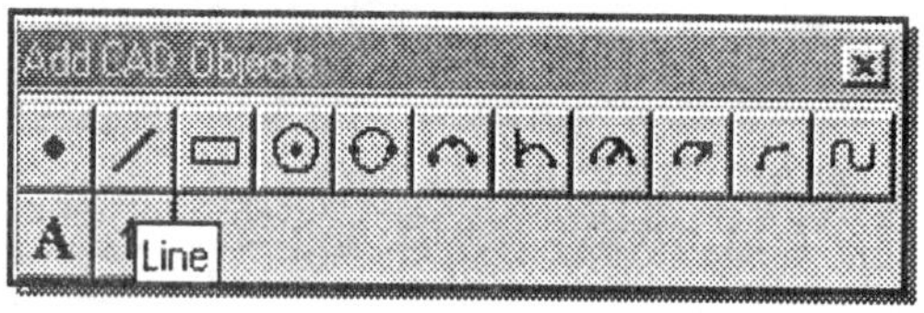

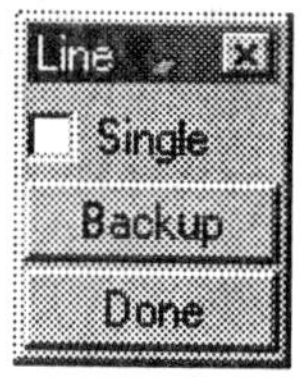

An input window appears at the bottom of the display window (look at the right corner of the screen), type the following coordinates. Note that the unit we are using in this example is inch, which is the default setting of ALGOR. When typing the numerical values, use the **Tab** key to move between the text fields and press the **Enter** key when you complete the data input for the coordinates of the nodes. Make sure that you have two points with (0, 0, 0) so that the trust structure is a closed loop.

X	Y	Z
0	0	0
10	0	0
5	5	0
0	0	0

After you have finished, click ***Done*** on the ***Line*** menu, as shown below. If you make an error, you can click ***Backup*** on the line menu.

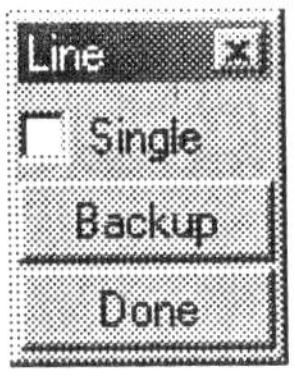

On the View Utilities menu, click on the enclose icon to refit the screen to fit your drawing.

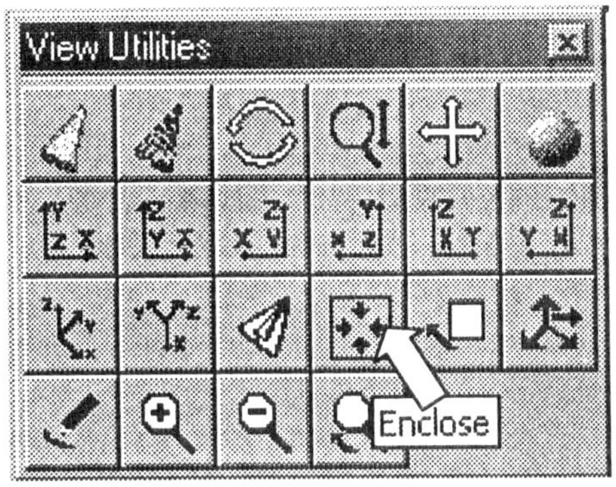

Your screen should look like that pictured below.

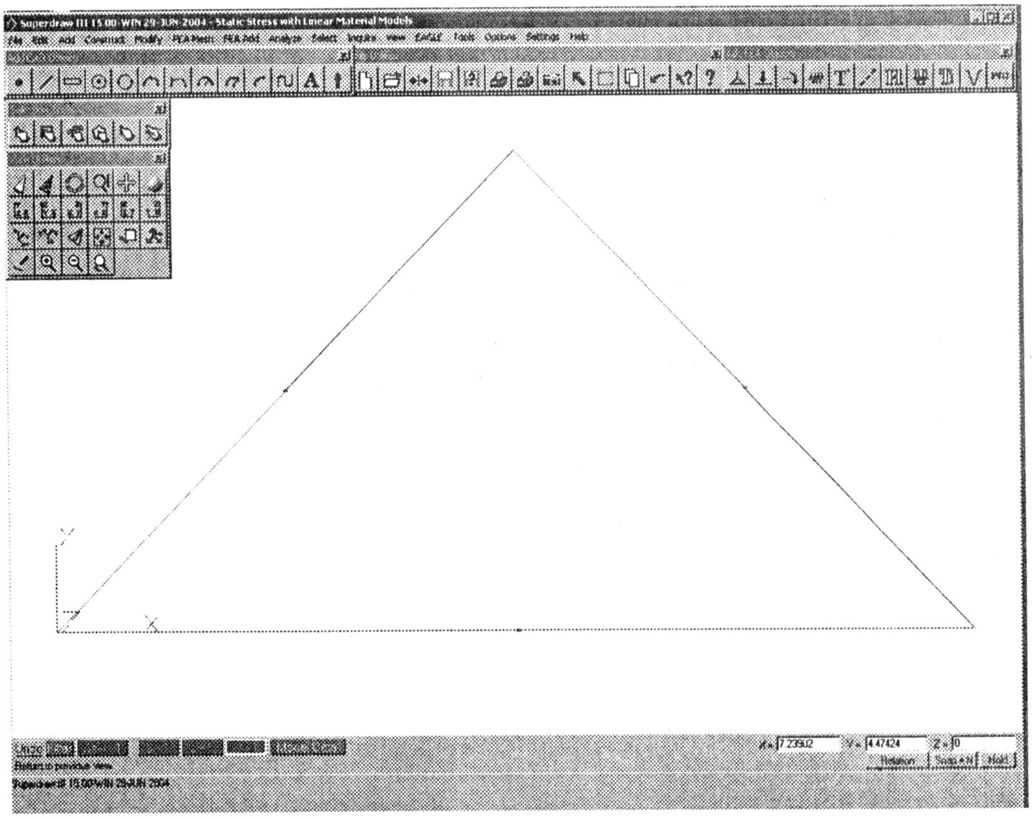

Step 3: Define Loads and Boundary Conditions

Define the Boundary Conditions on the Truss Structure.

On the **Add FEA Objects** window, select the icon of ***Boundary Conditions*** ( ).

A ***Boundary Conditions*** window appears. Click **Change Values** and a new window called **BC Values** appears. In this example, we need to constrain translation of the pin connection at the left end of the truss structure in the x and y-directions. To do so, check the boxes labeled T_x and T_y. Be sure that all of the other boxes are unchecked > Click **Done**.

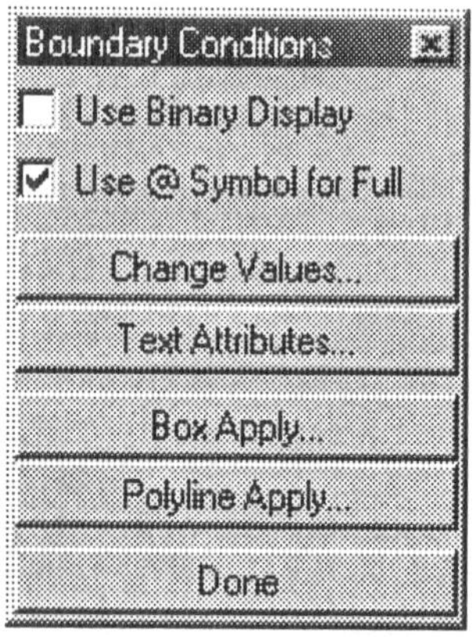

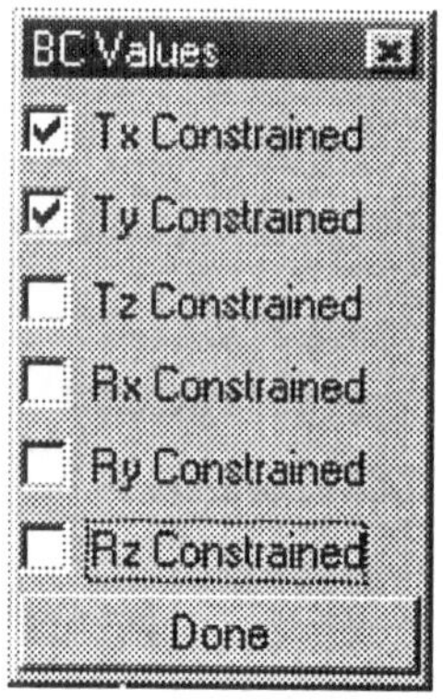

From the Boundary Conditions window, right click the node, which is on the bottom left corner of the triangle. Note that ALGOR will snap a load or constraint to the nearest *node.*

Note that T_{xy} appears on the bottom left corner of the triangle, as is shown in the figure below.

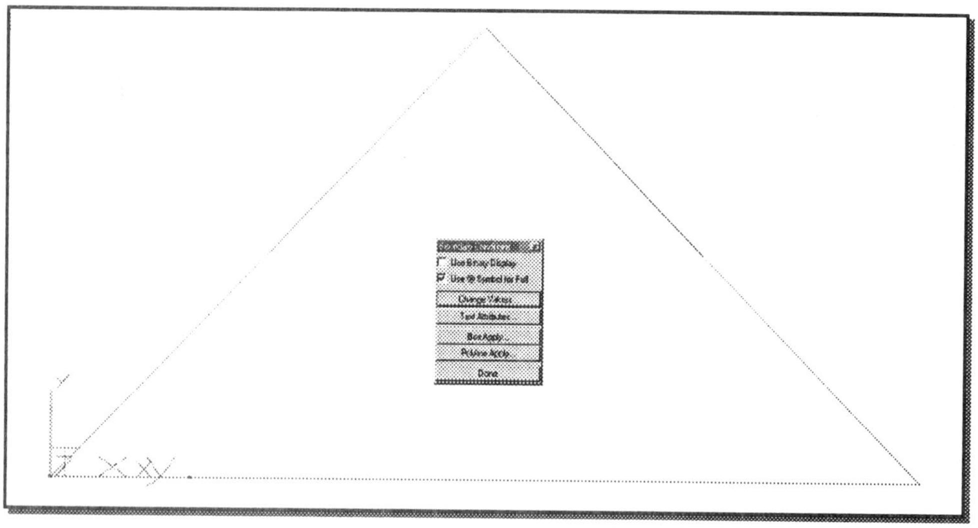

Define the boundary condition at the node, which is the right hand corner of the triangle. Click ***Change Values*** from the ***Boundary Conditions*** window. Because the constraint condition is a roller connection, we need only to constrain translation in the y-direction. Check the box labeled Ty. Be sure that all of the boxes are unchecked.

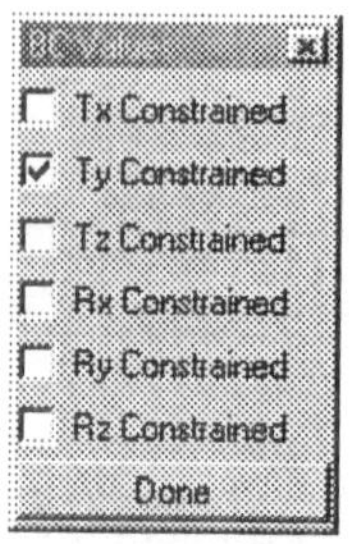

Right-click on the bottom right corner of the triangle to define the second constraint condition. Click the icon of ***Enclose*** from the ***View Utilities*** toolbar, to resize the screen.

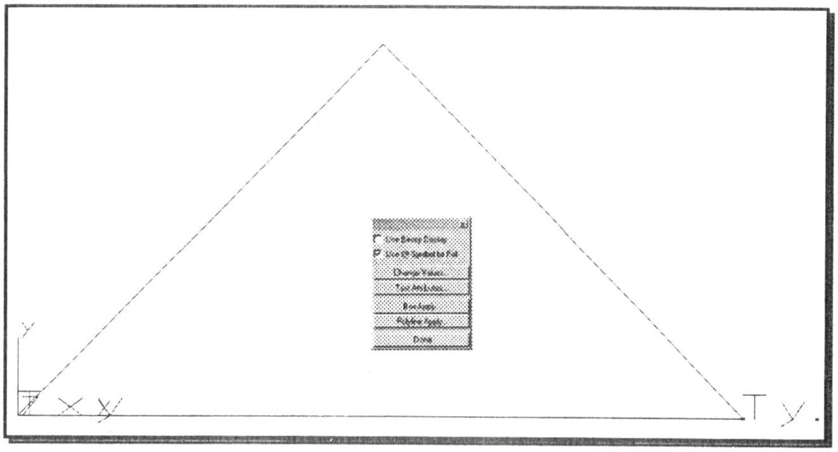

Click the icon of ***Done*** from the Boundary Conditions window.

Now let us define the loads acting on the truss structure. To define Load 1, which is a force acting in the vertical direction and upward.

From the main toolbar, select **FEA Add** > **Structural** > **Nodal Force** > **Vector**, a Vector window appears > click the box of Y Direction.

Click on ***Length***, and Type in 1 > press the Enter key > click **Done** from the Vector window.

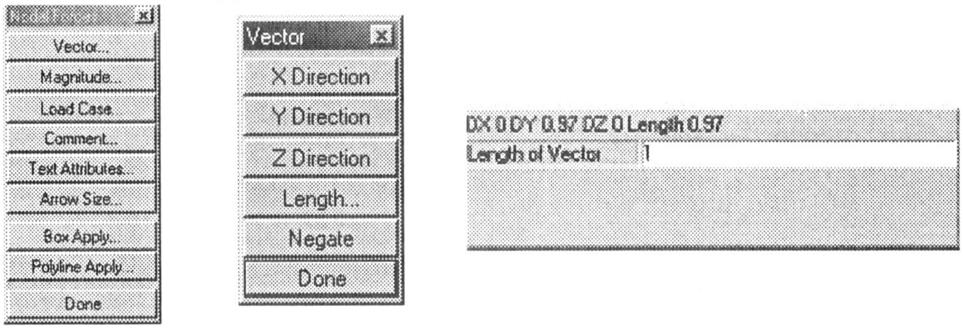

Click the icon of ***Magnitude*** from the ***Nodal Forces*** window.

Type *1000*, and press the **Enter** key.

Right click on the node which is the top vertex of the triangle to specify the location of the load. The following figure illustrates that Load 1 is shown on the screen

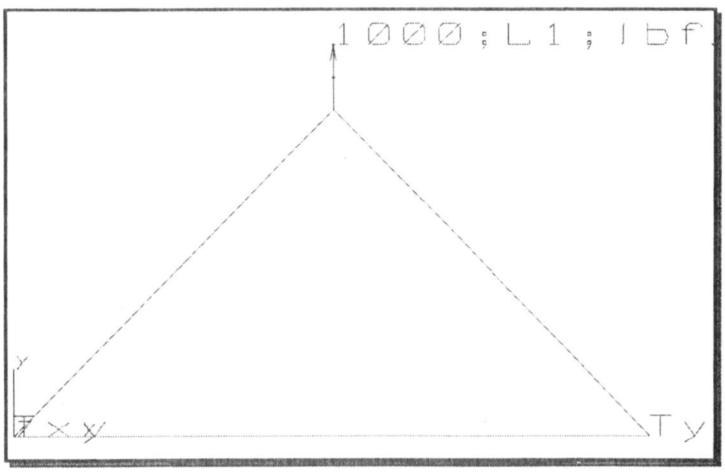

To define Load 2, which is the force acting in the horizontal direction to the right, we following the same steps used to define Load 1.

On the ***Nodal Forces*** window, click ***Load Case*** to define a new load set. Type in **2**, and press **Enter**.

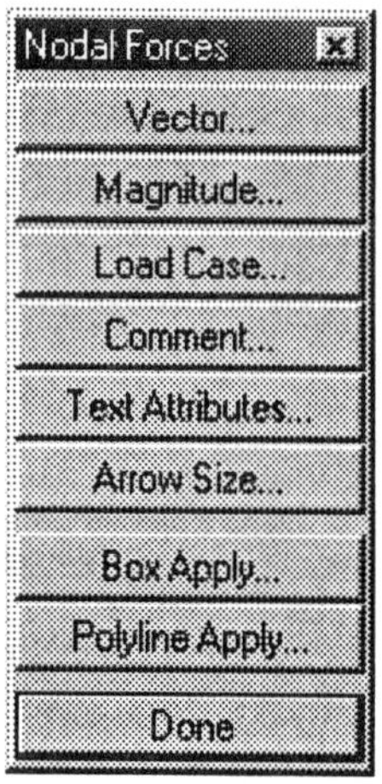

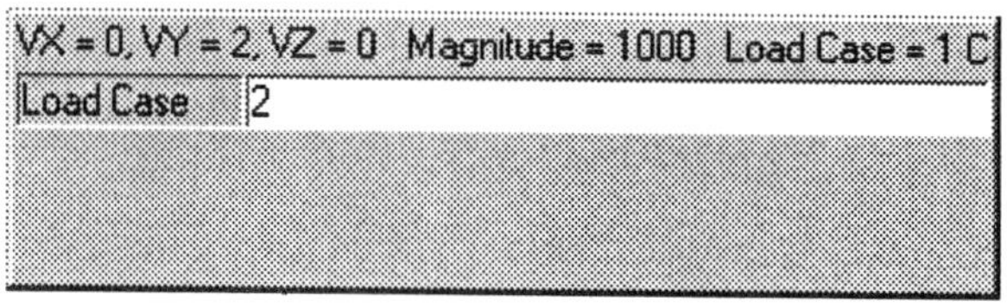

Make sure that users should click ***X-direction*** to define the direction of Load 2 under the **Vector** window:

Click the Nodal Forces button under the add FEA Objects window. Under the Nodal forces menu, click ***Vector***, and then click on the ***X Direction*** button.

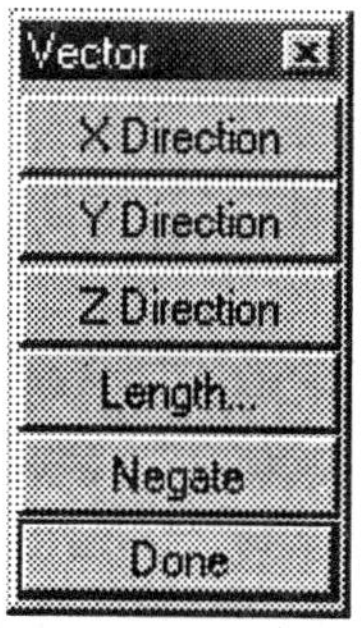

Click ***Length*** and enter **1**. Click **Done**.

Under the Nodal Forces window, click ***Magnitude***. Enter **500** and press Enter. Right click on the apex of the triangle to place the load. Click ***Done***. The figure should look like the one pictured below.

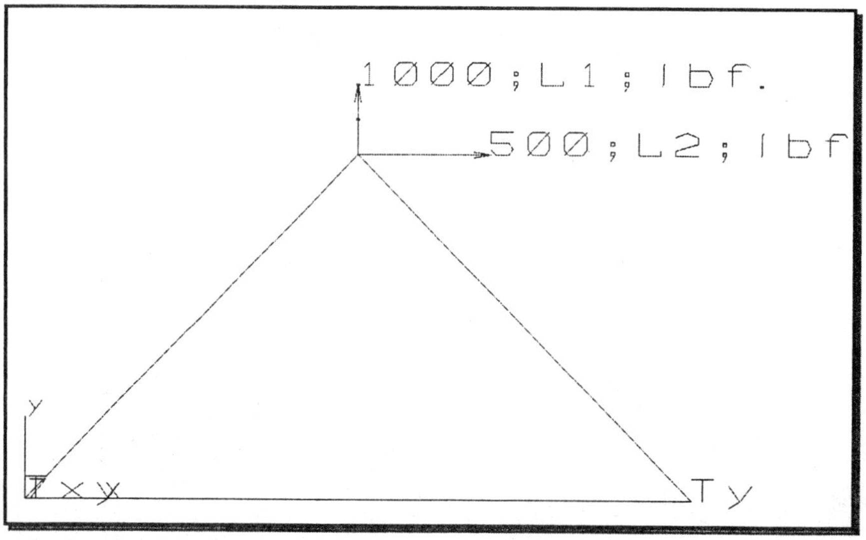

Step 4: **FEA** Analysis using **Model Data Control**

Click on the ***FEA Add*** pull-down menu at the top of the screen. Click ***Structural*** > ***Model Data Control***.

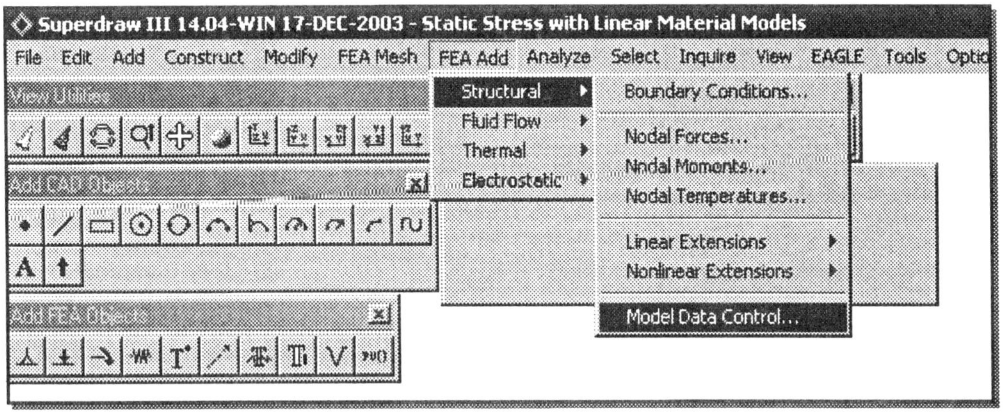

The following window will appear.

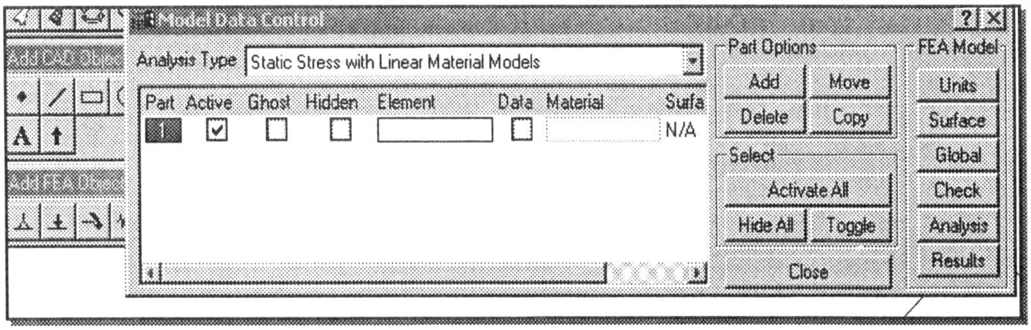

Click on the box below ***Element***. A window will appear in the center of the screen. Click ***OK***.

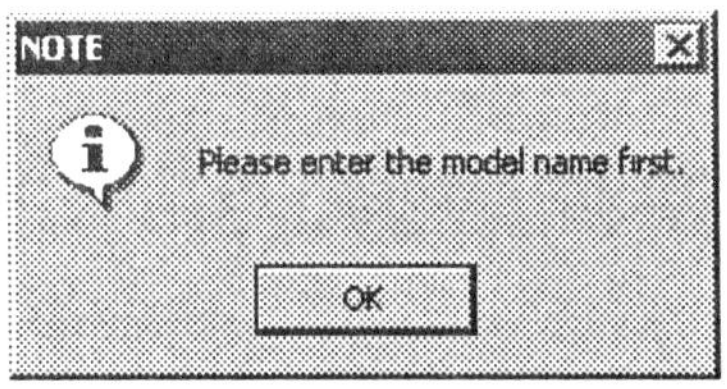

A ***Save As*** window will then appear. Enter ***Truss_structure*** as the filename and click ***Save***. Make sure that you have saved the file before going the steps forwards.

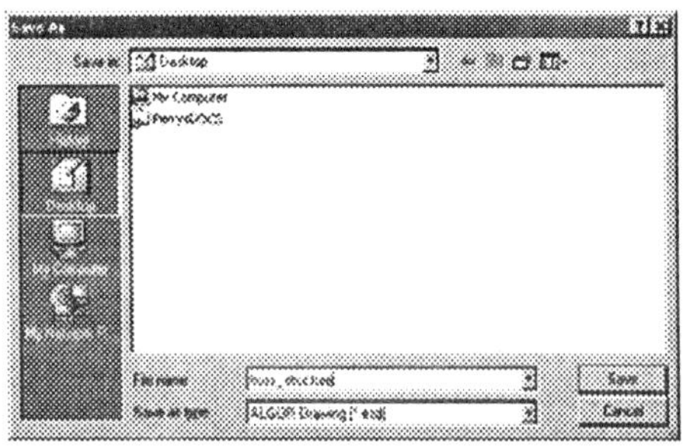

A window entitled Part 1 will appear. Note that Truss is already selected. Click OK to accept the selection.

Truss

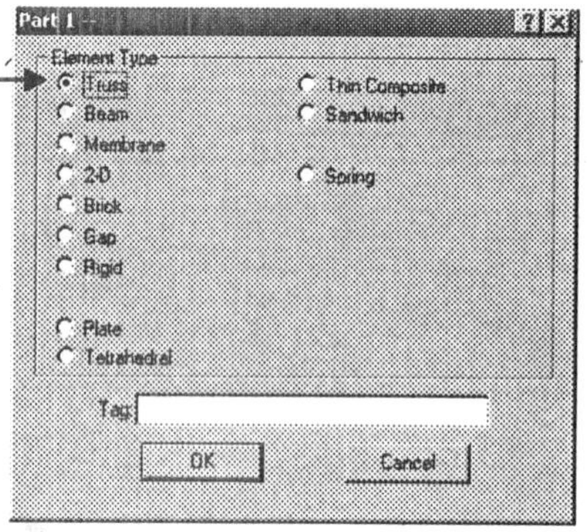

Click the box under ***Data*** on the **Model Data Control** menu. A note will appear asking you to select a units system. Click ***OK***.

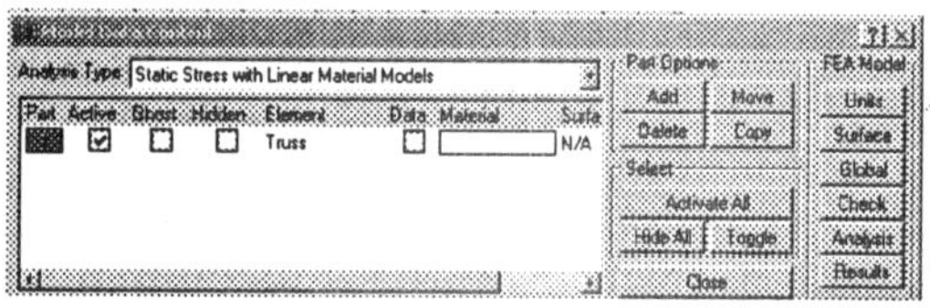

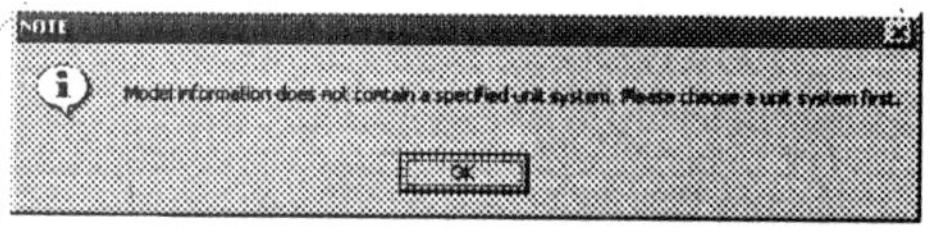

The default unit system is English. Click ***OK*** to accept the English unit system.

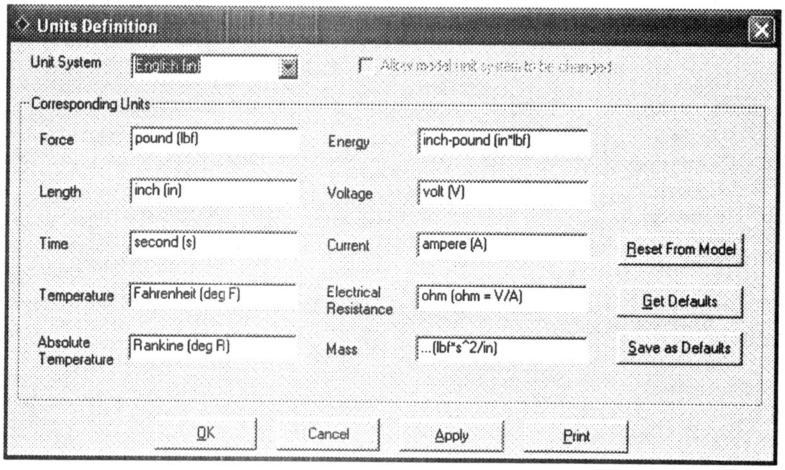

An ***Element Definition*** window will appear. Define the cross sectional area of the truss members as 0.785 in^2 and click ***OK***. Truss members in ALGOR have a circular cross section by default.

$$area = \frac{\pi D^2}{4} = \frac{(3.14)(1^2)}{4} = 0.785 \quad (\text{Inch}^2)$$

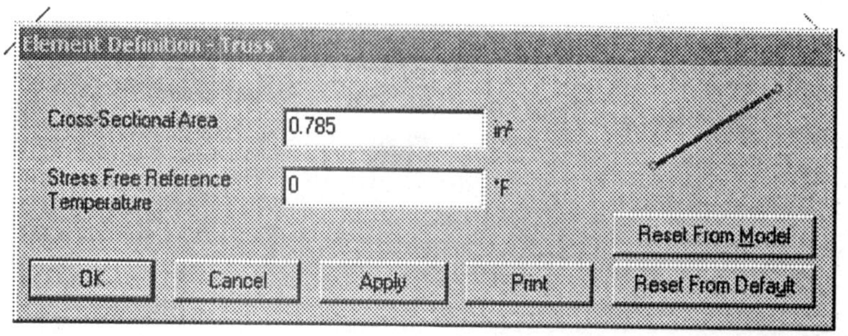

Click on the box under ***Material*** on the ***Model Data Control*** window.

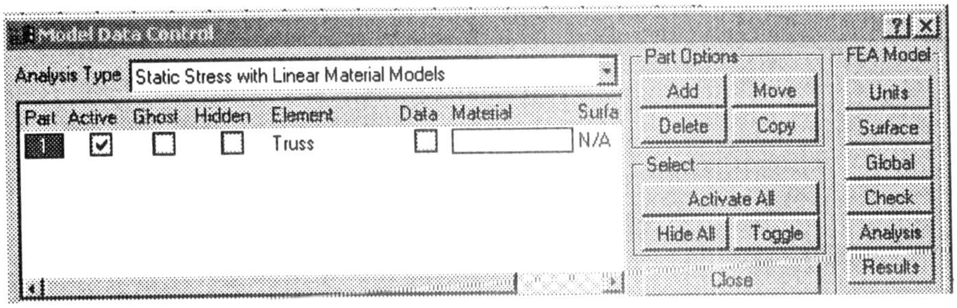

An ***Element Material Selection*** window will appear. Click on Aluminum (6061 T6), and then click ***OK***.

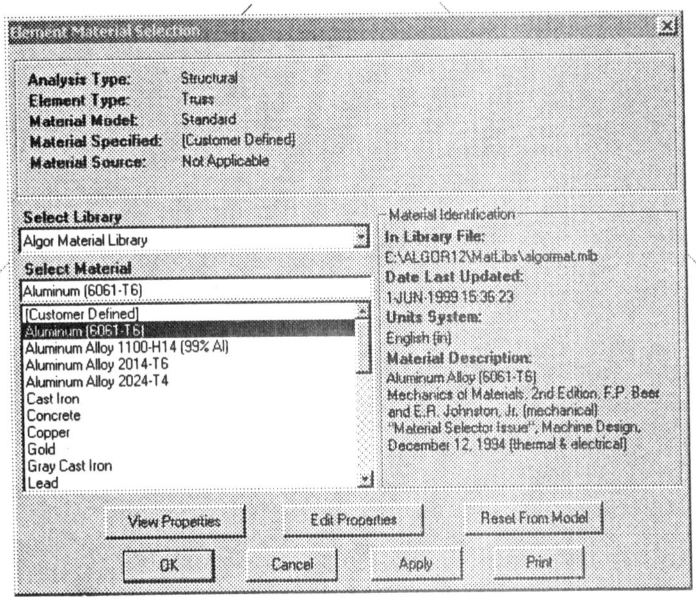

On the Model Data Control window, click Global. A window will appear entitled Analysis ***Parameters***. Ensure that there is a 1 under Accel/Gravity on the Multipliers tab. These include the external loads specified and gravitational forces on the structure.

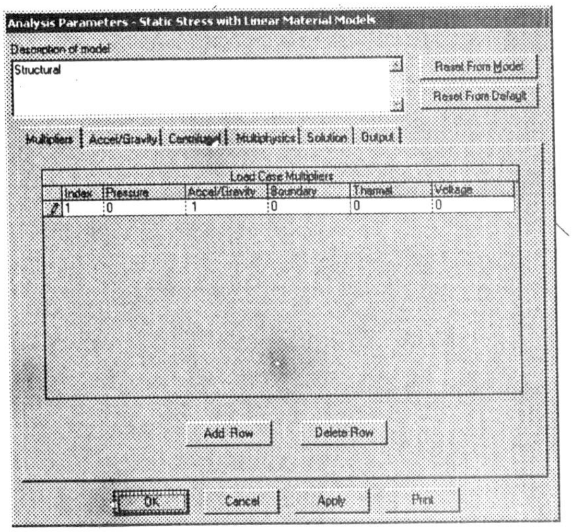

Click on the ***Accel/Gravity*** tab in the ***Analysis Parameters*** menu, and then click on ***Set for Standard Gravity***. Click ***OK***.

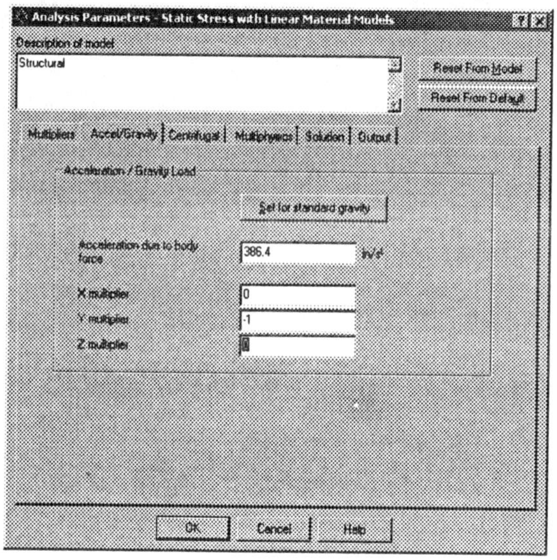

Step 5: Perform an Analysis

From the **Model Data Control** window, click the box of **Analysis** listed on the right side.

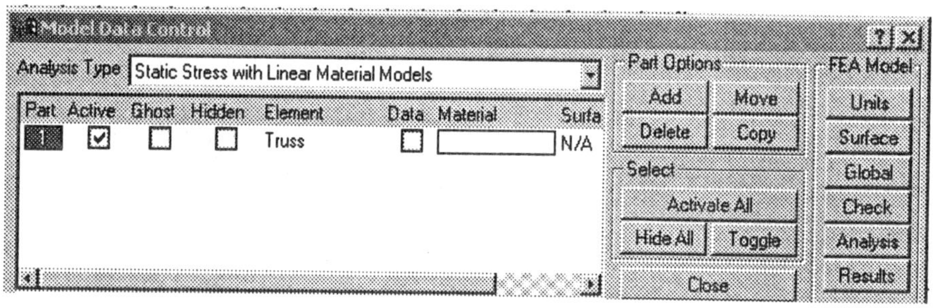

A window entitled Structural – Static Stress with Linear Material Models will appear. Click the box of **Analyze**.

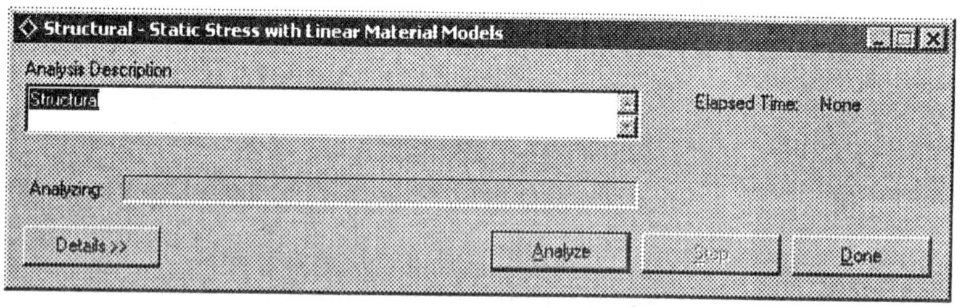

A warning will appear doing analysis; click ***OK***, then click ***Done***.

Step 6: Examine the **Results**

Click ***Results*** from the ***Model Data Control*** window to open ALGOR FEMPRO and display the FEA results.

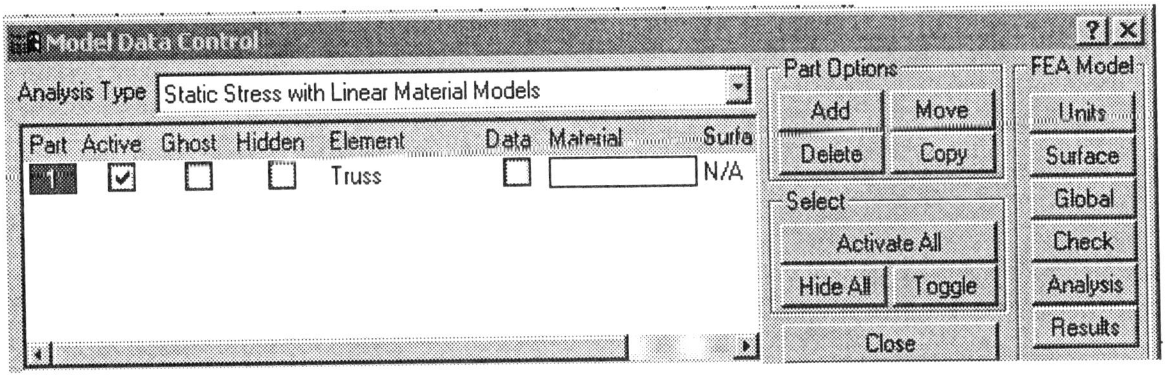

To orient the **view** on display, click **View > Orientation > XY Top**.

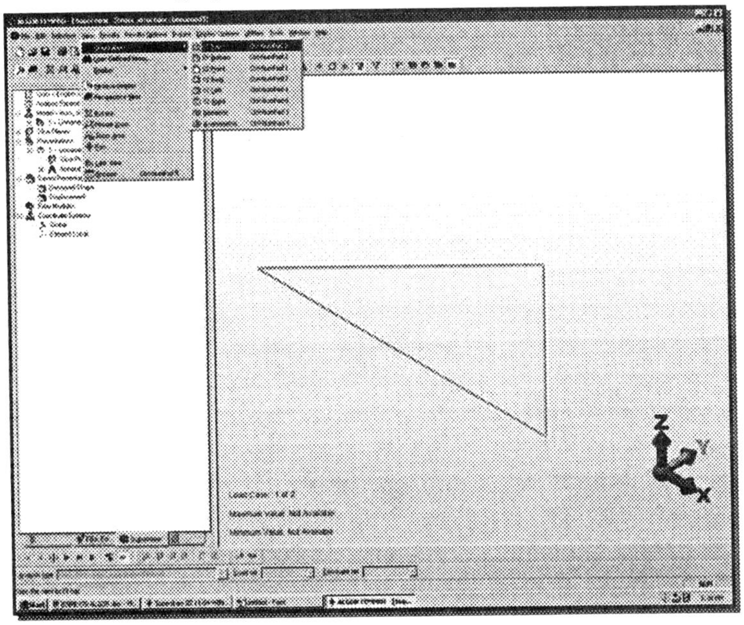

Click on **Results Options > Displaced Model Options** and check the box next to **Show Displaced Model** and **Show Undisplaced Model as Mesh**.

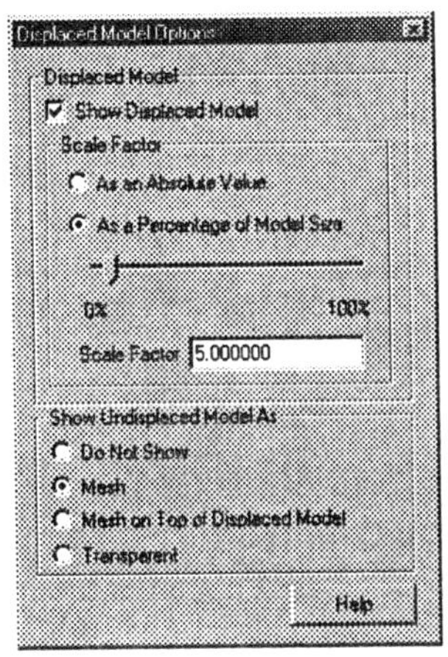

Close the ***Displaced Model Options*** window, and click on *Results>Displacement>Magnitude*, to display the fringe of the displacement magnitude for the truss structure.

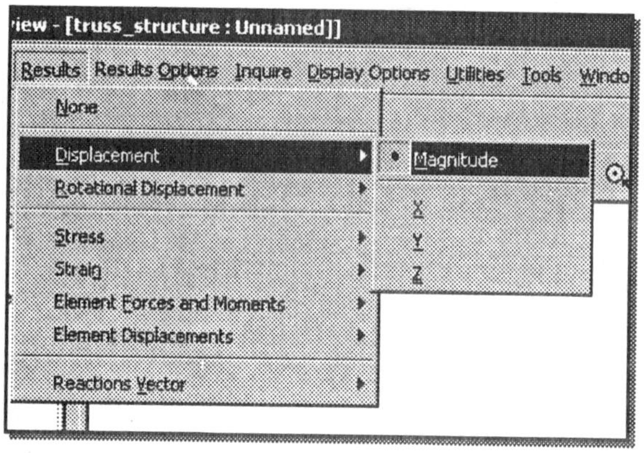

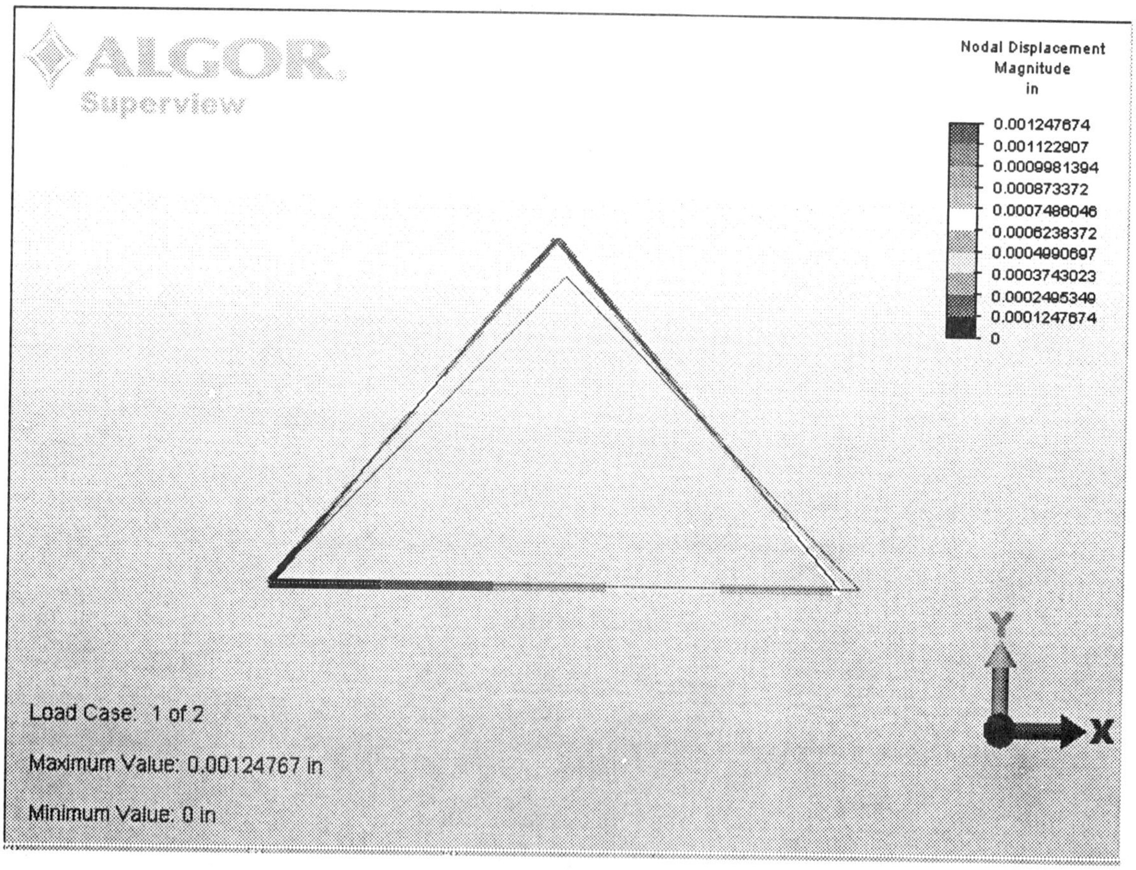

The maximum and minimum values for the displacement magnitudes are 0.000125 inches and 0 inches respectively.

Click on the **right arrow** at the bottom left corner of the screen to display the magnitude of displacement due to **Load 2**. The following figure is obtained.

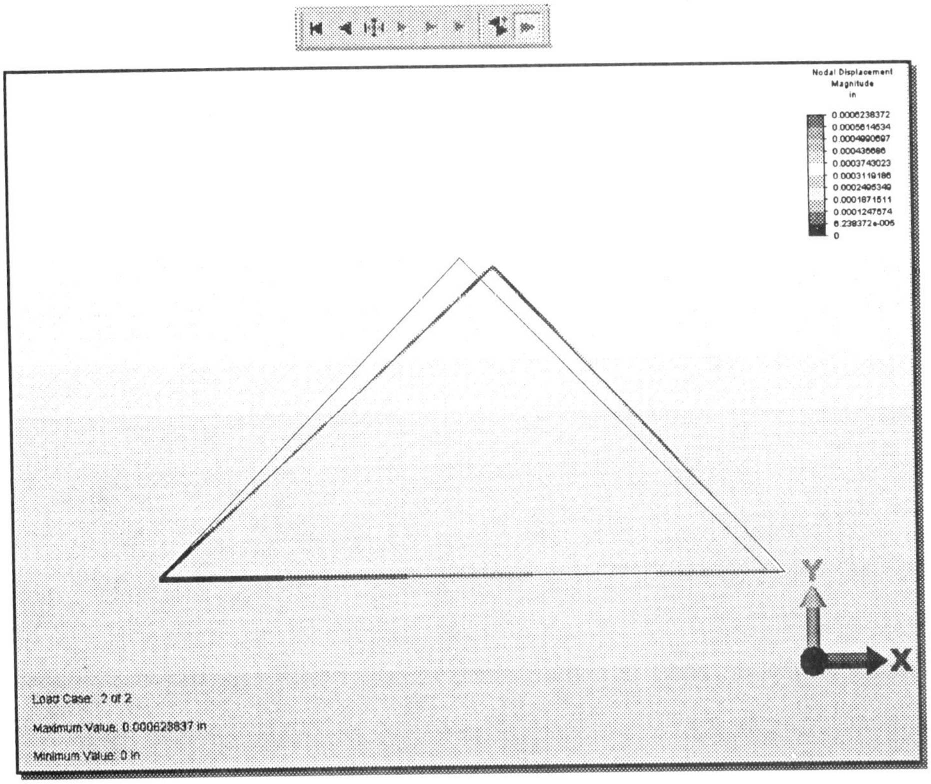

The maximum and minimum values for the displacement magnitudes are 0.000624 inches and 0 inches respectively.

To review the stress distribution, click **Results** > **Stress** > **Beam and Truss** > **Axial Stress**.

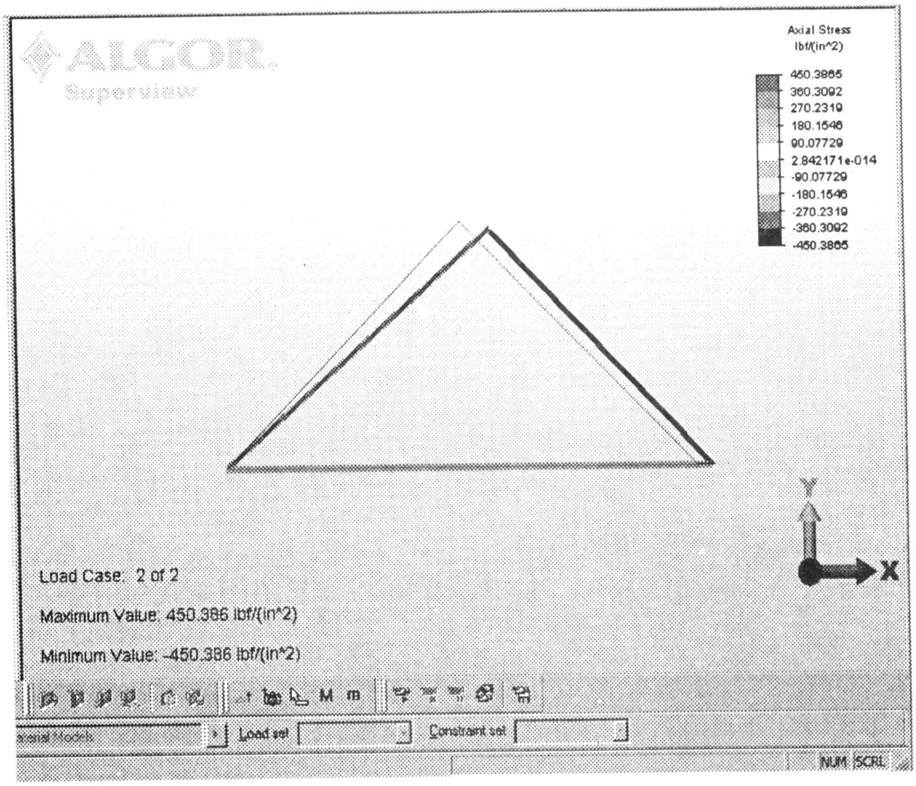

9.5 FEA of a Second Tress Structure Using ALGOR V14

The following figure illustrates a truss structure, which consists of 3 bars. The section area of each bar is circular. The diameter is 100 mm. The geometrical information of the truss structure is characterized by 2 dimensions. The first dimension is the length of the horizontal bar, which is 1000 mm. The second dimension is an angle, which is 30°. The material type is 6061 T6 aluminum.

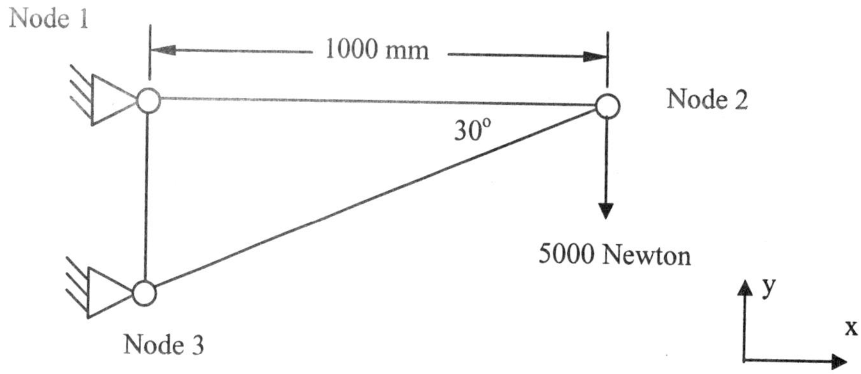

Step 1: launch the software system: **ALGOR**.

We assume that the software system resides in the folder called **Programs**.

Click the **Start Menu/Programs/ALGOR/FemPro**

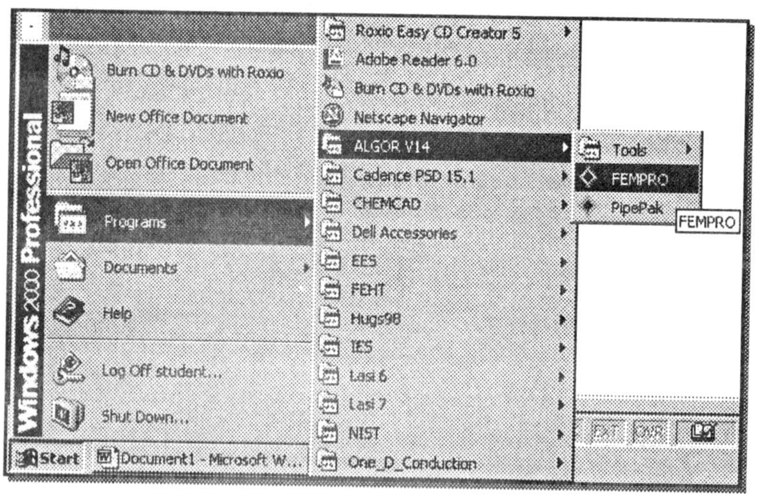

An "**Open**" window will appear. Click **Superdraw** and then click **Open**.

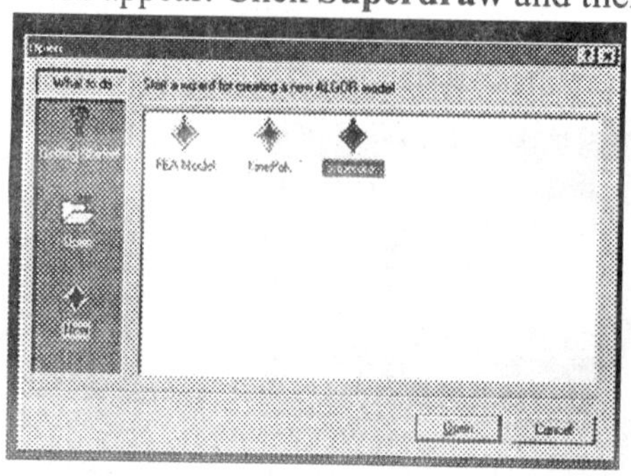

It is important to note that, when launching an updated version of **ALGOR**, say **ALGOR V16**, you may be asked to specify the name of your file, first. Afterwards, you may be asked to select a design scenario from one of the five choices on the list.

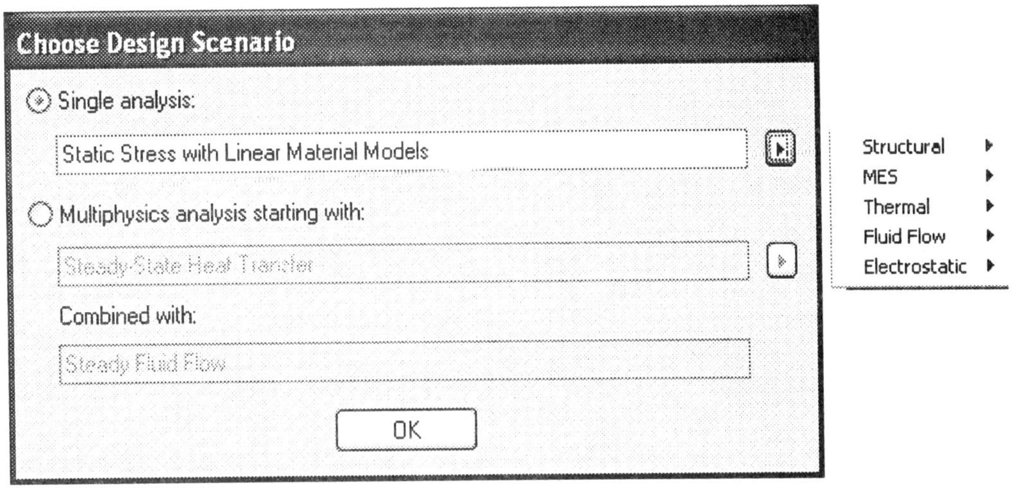

Step 2: Define the unit system before drawing the Truss Structure

The updated version of **ALGOR** requires users to specify the unit system before getting on sketching the truss structure. The unit system window appears immediately after the user has selected a design scenario.

There are five choices, namely, English (ft), English (in), Metric cgs, Metric mks (SI) and Custom. Let us choose Custom and set the force unit to Newton, Length unit to mm, Time unit to second, etc. > OK.

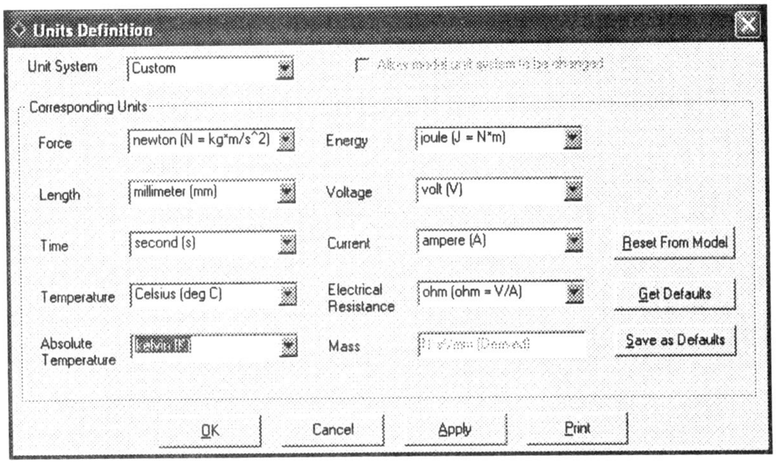

From the main toolbar, click **Settings** > **Toolbars** > click **Add FEA Objects** if it is not selected > **Close**.

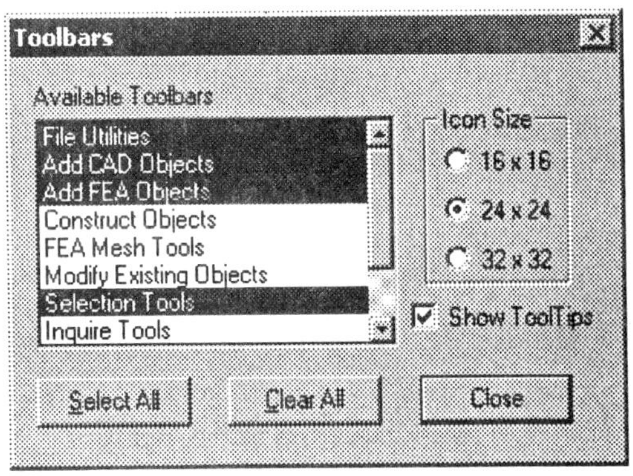

Re-organize those toolbars on display so that you have an easy way to access.
On the **Add CAD Objects** toolbar, click the icon of **Line.**

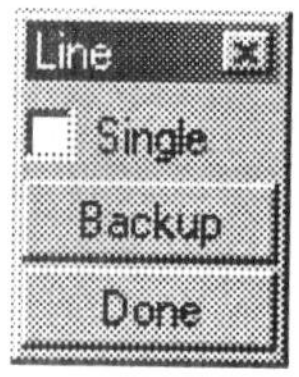

On the bottom of your screen there is now a coordinate box. To move from box to box forward, press the **TAB** key, to move backwards, press both **SHIFT** and **TAB**. When the correct coordinates are in place, press **ENTER**.

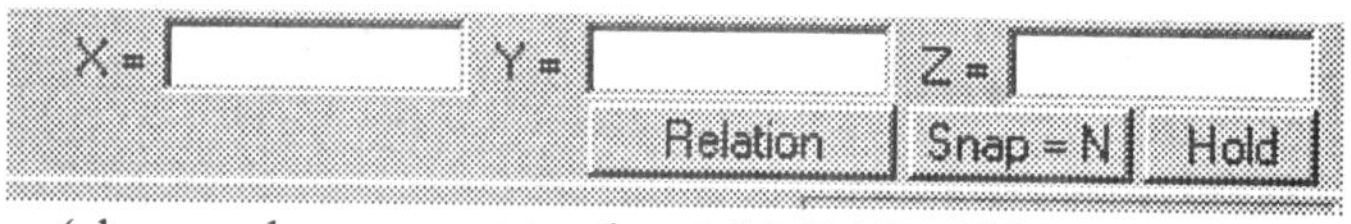

The Line menu (shown above, next to the Add CAD Objects toolbar) is used to correct errors (**Backup**) and to finish line drawing application (**Done**).

For this example, the coordinates are as follows. Note that the coordinates of Node 3 have to be input twice so that a closed form of the truss structure can be formed.

Node	X	Y	Z
3	0	0	0
2	1000	577.35	0
1	0	577.35	0
3	0	0	0

To fit the entire truss in the viewing area, click the button on the **View Utilities** toolbar. You should now have a screen like the one below.

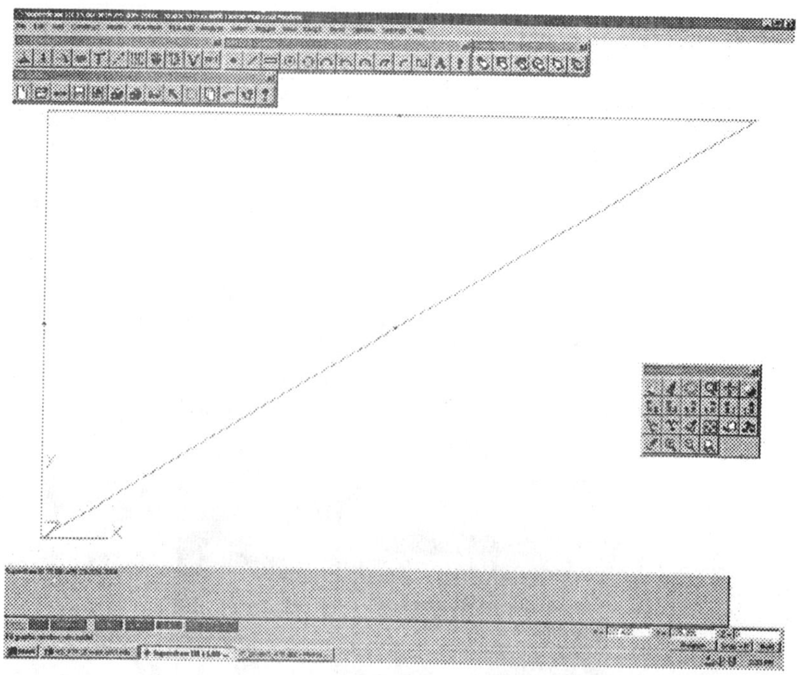

Step 3 – Now we must define the loads and boundary conditions for the truss structure.

Let us first define the boundary conditions. Click the **Boundary** ⊥ **Condition** icon on the **Add FEA Objects** toolbar.

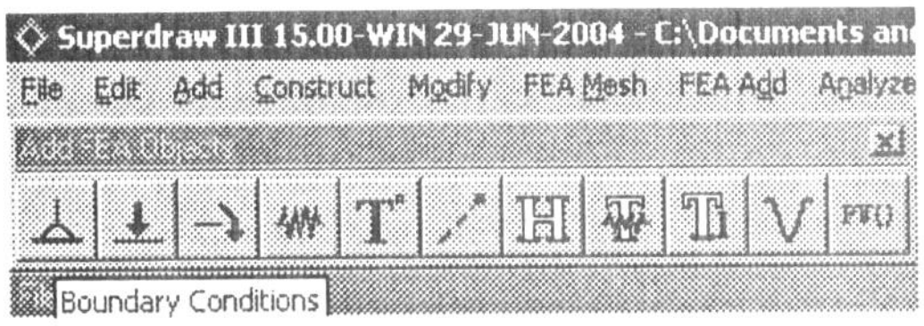

Click **Text Attributes, then **Height** and make sure the text height is at least 12. **Done**. **

Click **Change Values** on **Boundary Conditions** window, as shown below.

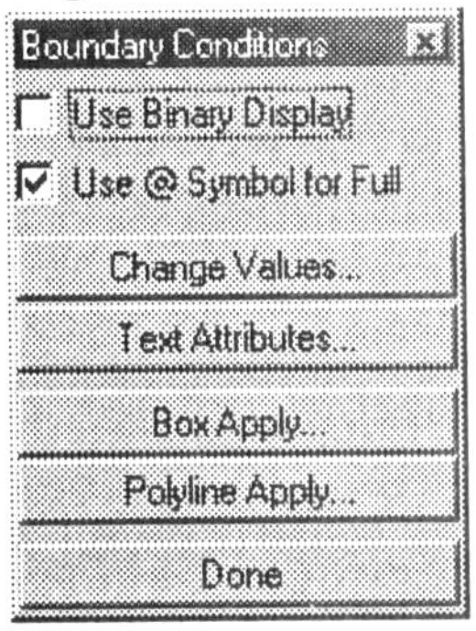

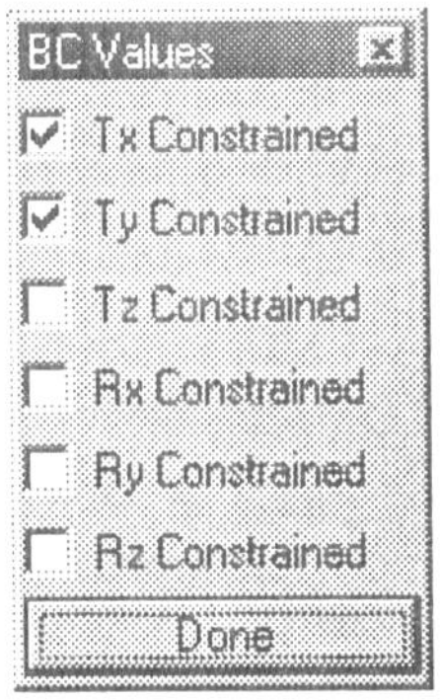

A **BC Value** window will appear. Since the truss structure is constrained in the x & y direction on the wall, choose only the $\mathbf{T_x}$ & $\mathbf{T_y}$ **Constrained** value > **Done**.

Go to drawing screen and click the upper and lower nodes so that the constraint conditions are added.

After defining the constraint conditions, now let us define the 5000 Newton load to the node at the upper right corner.

From the main toolbar, select FEA Add > Structural > Nodal Force > Vector, a vector window will appear.

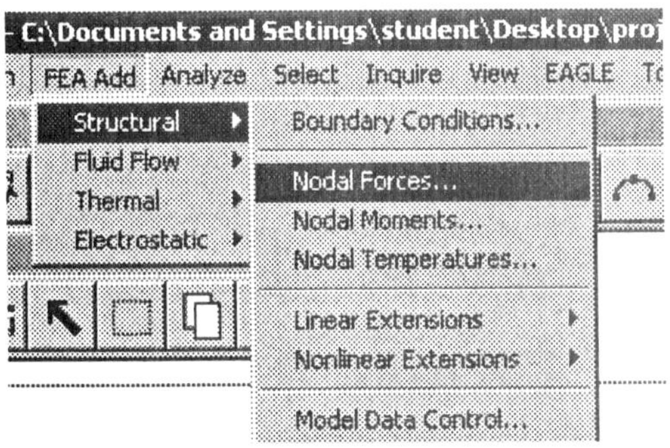

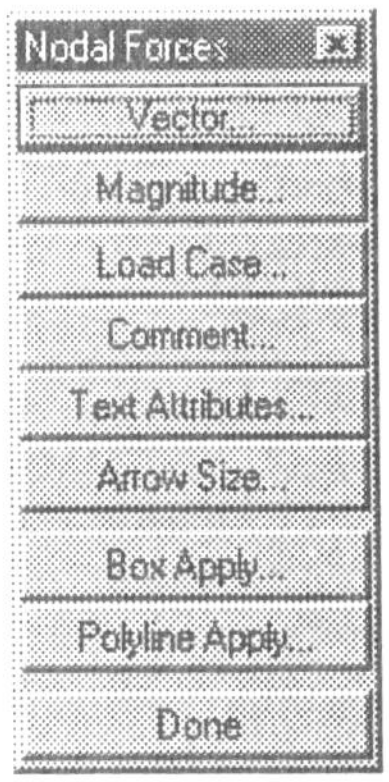

Click on the Y direction.> click Length. Type 100 and Enter > Click **Done** from the Vector window.

Y direction

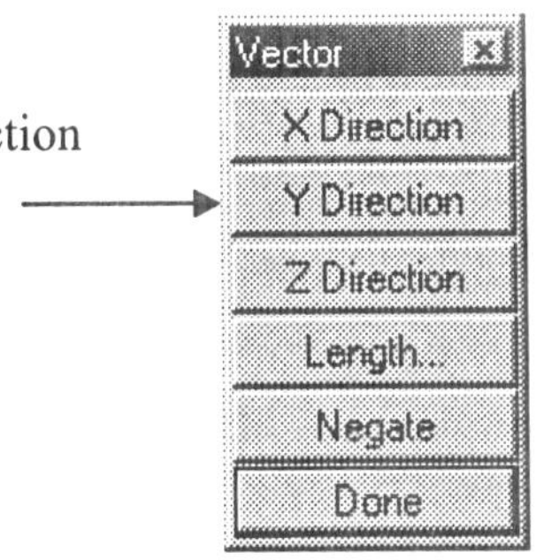

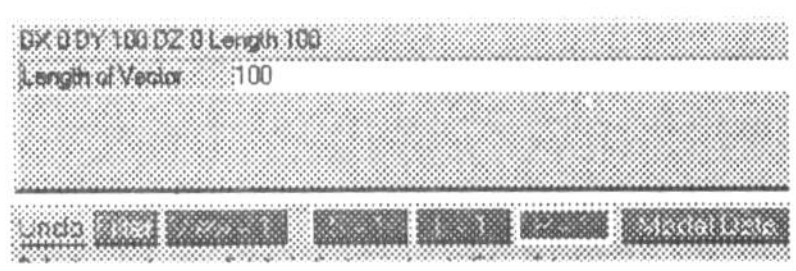

Click the **Magnitude** icon from the **Nodal Forces** window. Type -5000 (since it is in the negative direction) and press Enter. At this moment, users may need to change a text attribute, i.e., height).

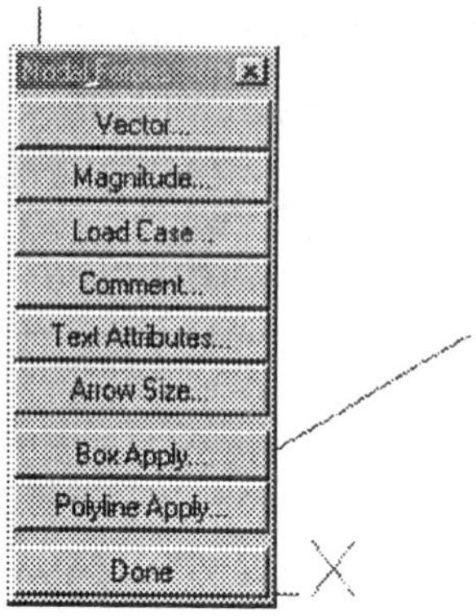

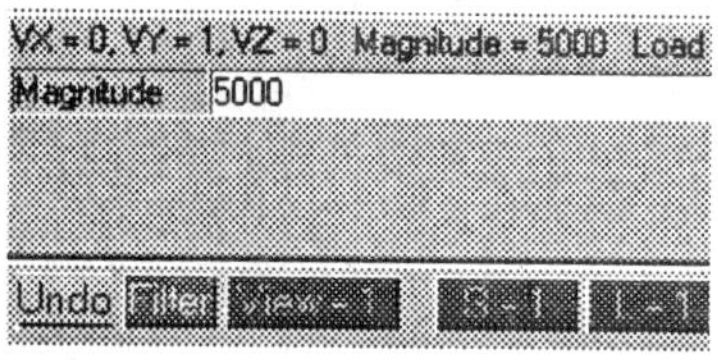

Now click Node 2, this is where the load will be placed. You should now have a screen that looks like this > **Done.**

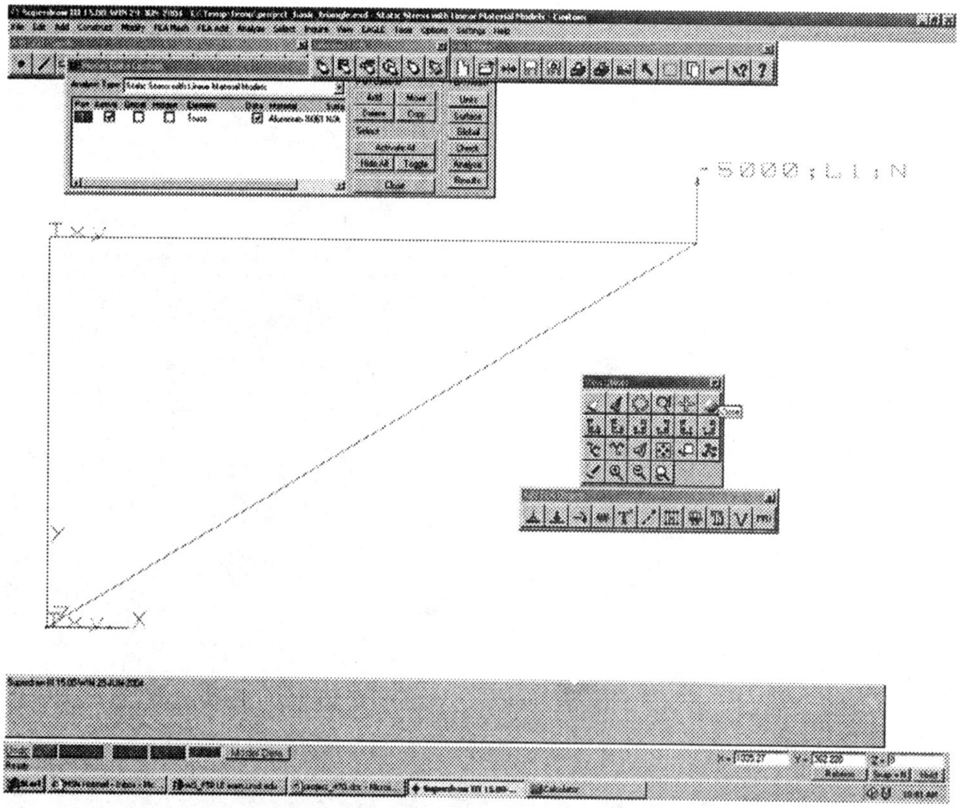

Step 4: Analysis using **Model Date control**
From main toolbar, **Add FEA > Model Date Control.**

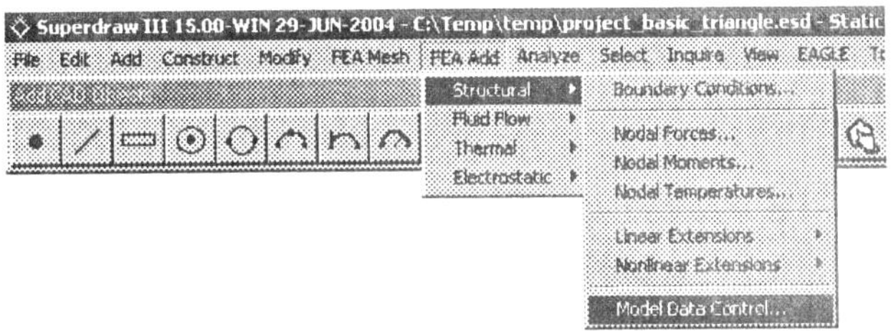

The window called Model Data Control appears.

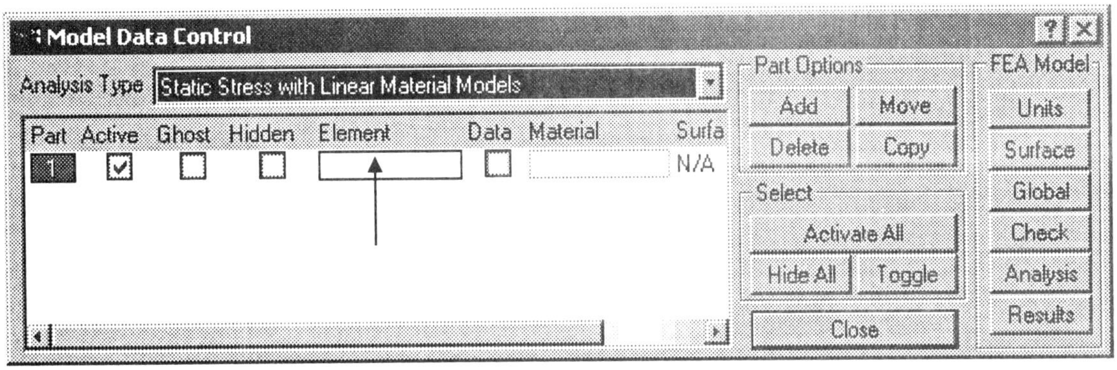

Click the box of **Element**. The following window appears. Click **OK** to accept **Truss.**

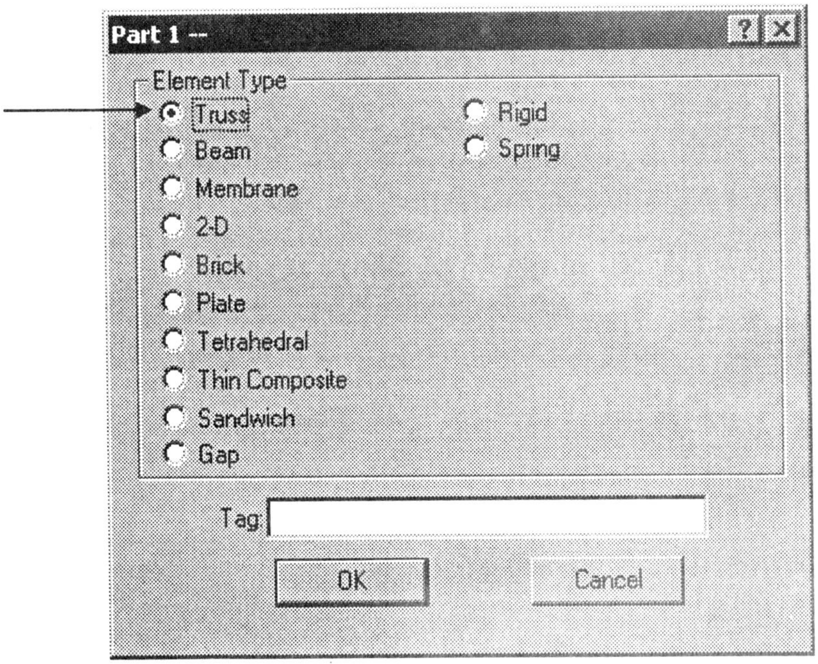

After returning to the *Model Data Control* box, click **Data.** The Element Definition window will appear. After finding the 100 mm diameter cross-sectional area to be 7854.0, enter number into box, as shown below. And click **OK.**

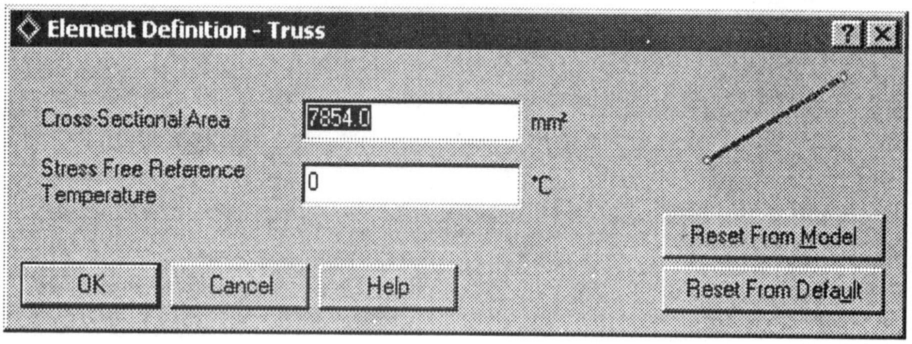

Now click the **Material** box on the *Model Data Control* window. Chose **Aluminum [6061-T6]**.. Click **OK**.

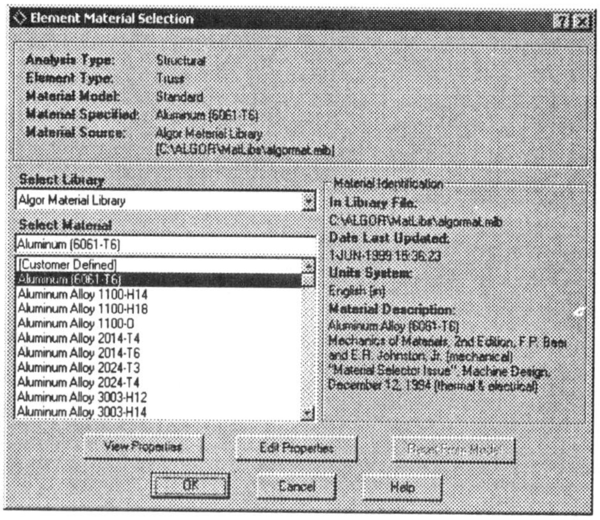

Now click the icon of **Global** on the ***Model Data Control*** window.

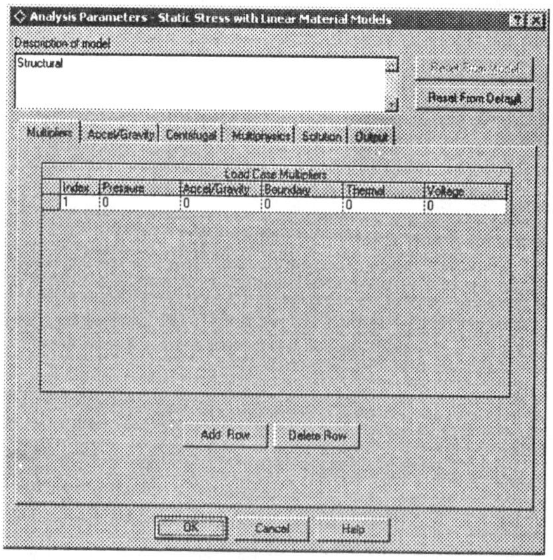

Click the **Accel./Gravity tab** on the *Analysis Parameters* window. Click on **Set for Standard Gravity.** Click **OK**.

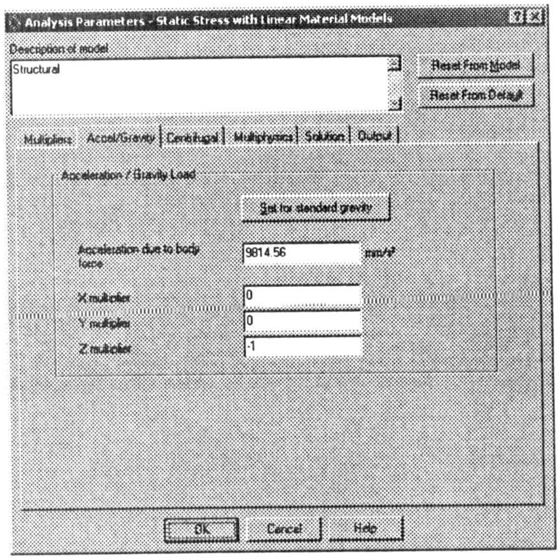

Step 5: Perform **Analysis**

Click on FEA Add from the main toolbar > Structure > Model Data Control. From the *Model Data Control* window – click **Analysis** listed on the right side.

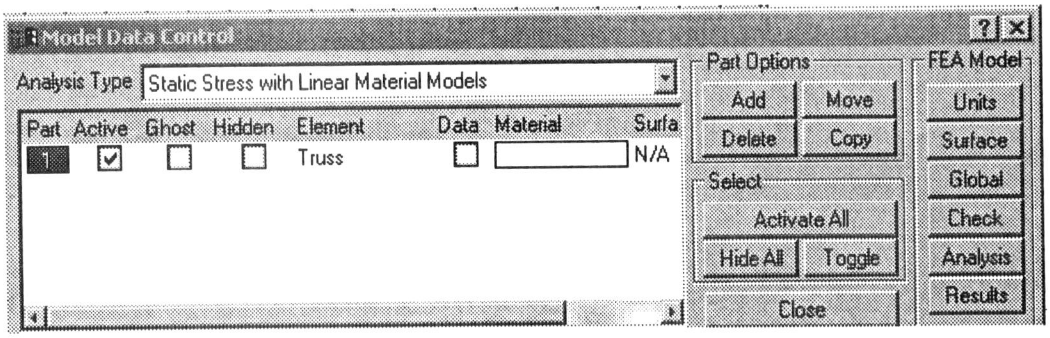

A window called Structural – **Static Stress with Linear Material Models** will appear. Click **Analysis** > **Done**.

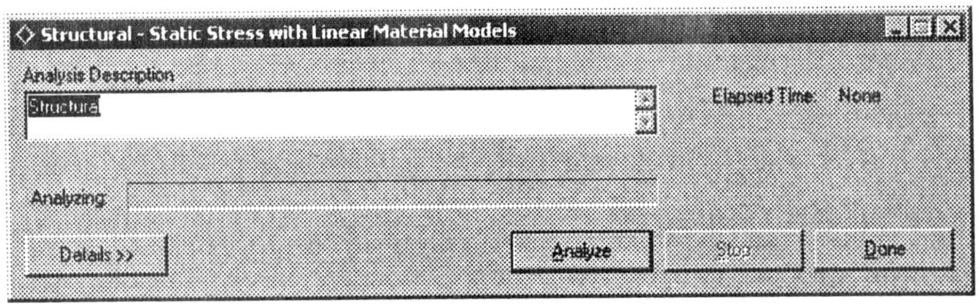

Step 6: Use **Results** to display the results

Click **Results** from the *Model Data Control* window. This will open ALGOR FEMPRO. You will need to orient view. Therefore, click **View > Orient > XY Top.**

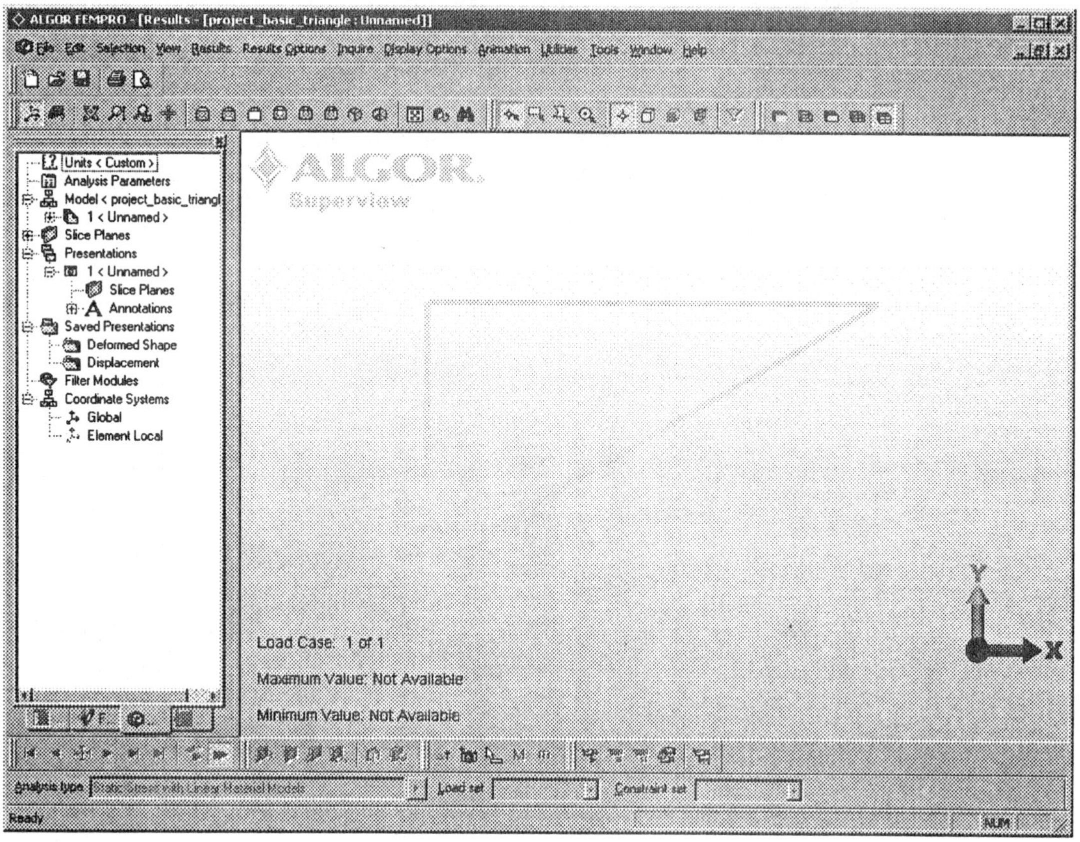

To show display, click **Results Options** from the main toolbar > **Displaced Model Options** > check the box next to *Show Display Model* and *Show Undisplayed Model as* ***Mesh.*** Close the box.

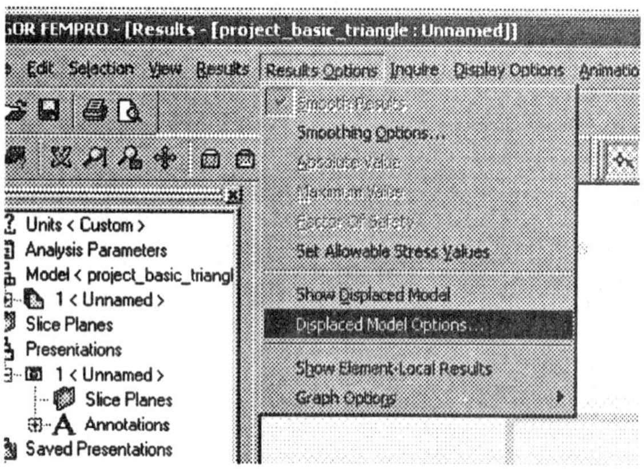

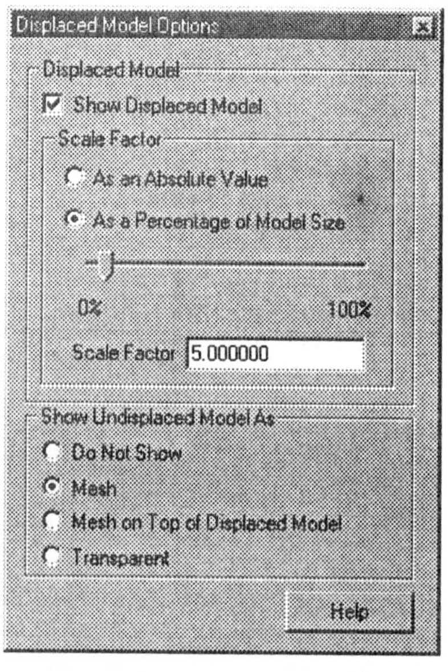

Click **Results** > **Displacement** > **Magnitude** to display the fringe of the displacement magnitude for the truss structure.

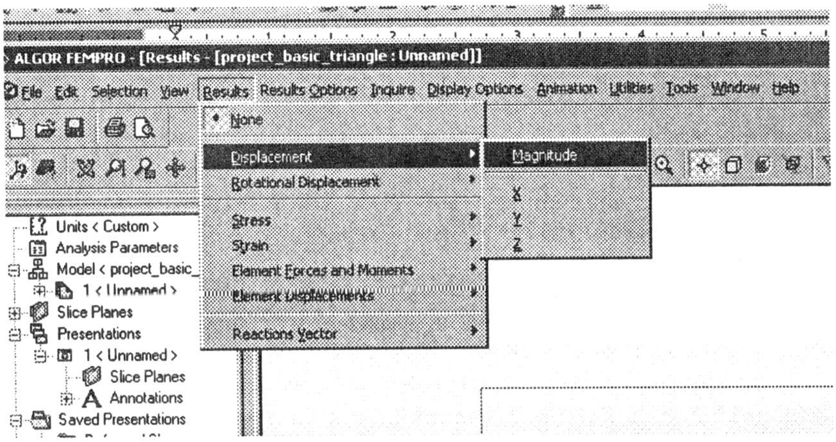

The picture below is the displacement distribution obtained from FEA.

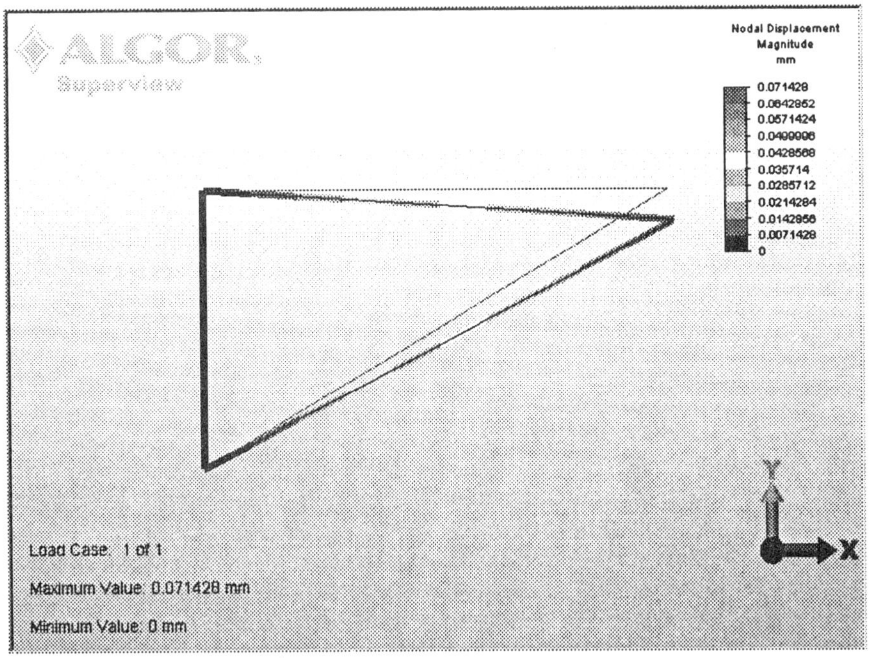

Step 7: Remain in ALGOR FEMPRO and Obtain Results

Now, let's determine the displacement of Node 2 (right node) and the reaction forces of the other two nodes.

First click **Selection** on the main toolbar > **Shape** > **Rectangle**. This will allow you to select the nodes in a rectangle. Left click and hold and drag over structure.

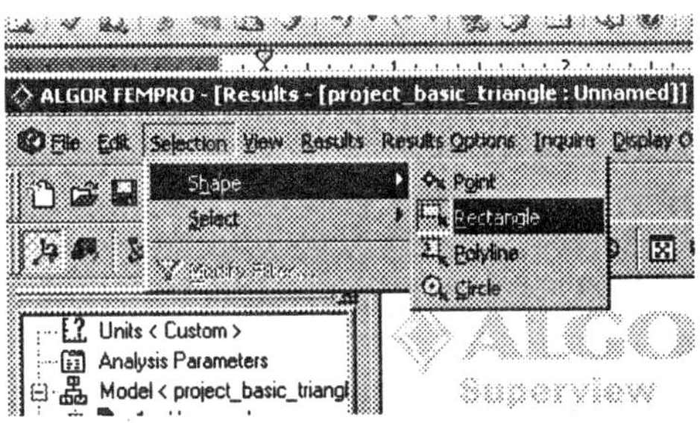

Select all nodes and then right click mouse button.

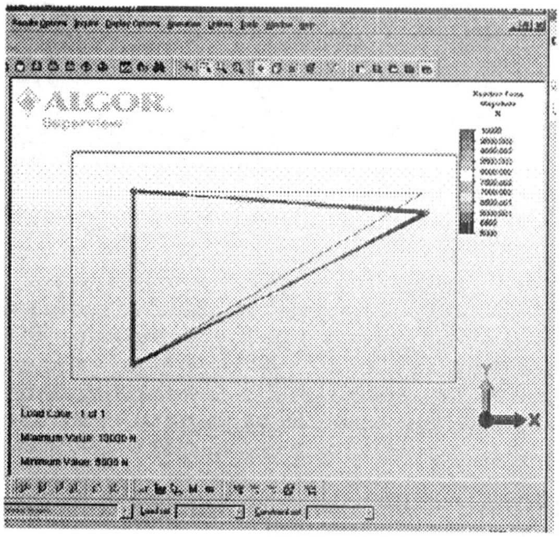

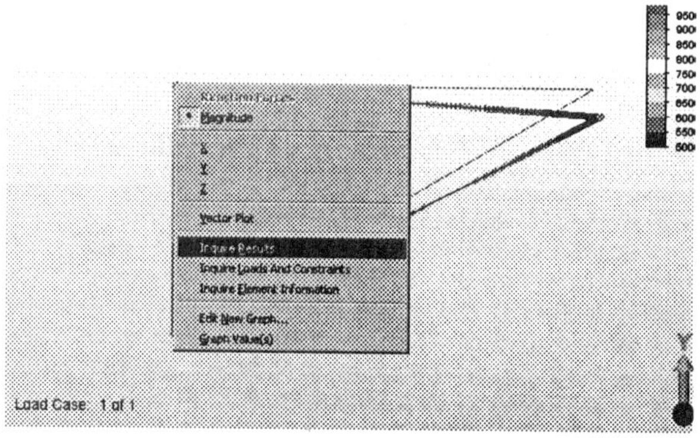

Select **Inquire results.** The following display pops up.

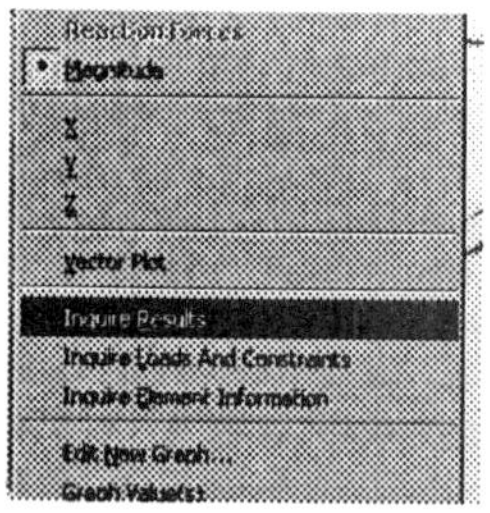

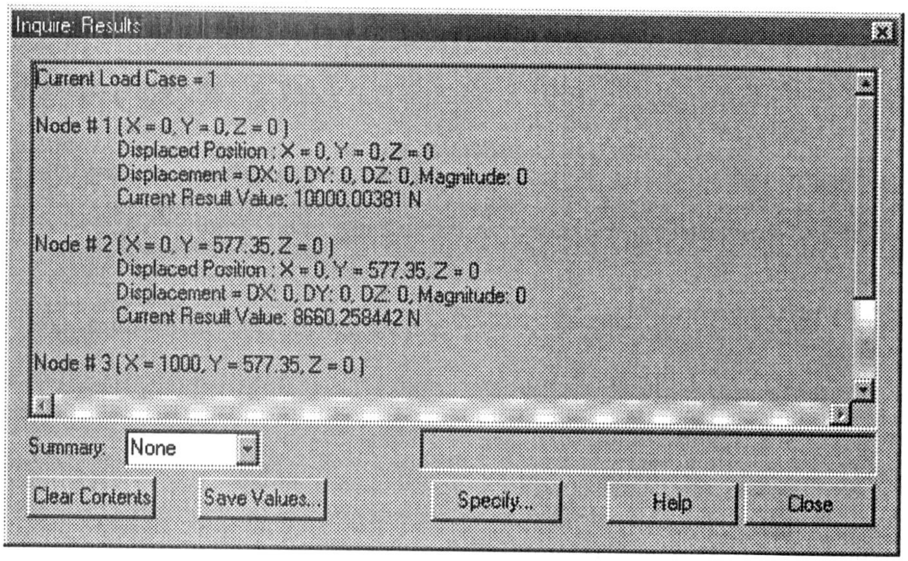

The results are self-explanatory. The *Current Results Value* is the TOTAL reaction force on that node. To obtain the reaction force in the X & Y direction separately, follow the same procedure (after right clicking mouse button). However, instead of Magnitude being highlighted, highlight **X or Y** and then **Inquire Results**.

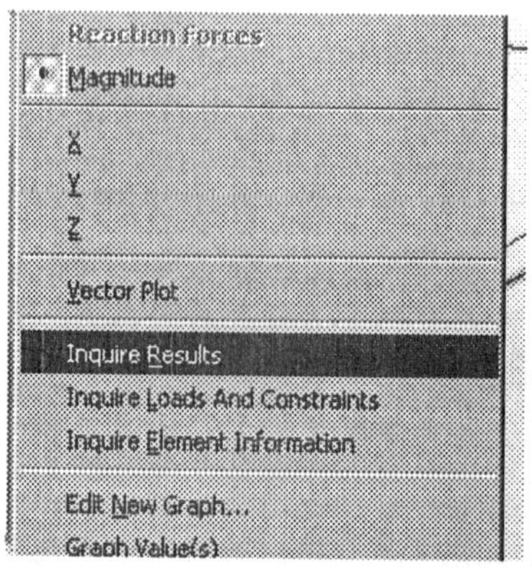

9.6 FEA of a Beam Structure Using ALGOR V16

The following figure illustrates a beam structure, which consists of 3 bars. The two ends of the beam structure, namely, node 1 and node 4, are fixed to the ground. The load condition is external force acting on node 2 in the horizontal direction along the positive x axis. The geometric shape of the cross section is a rectangle. The dimensions of the section area are 10 x 10 mm. The geometrical information of the beam structure is shown in the figure. The material type is 6061 T6 aluminum.

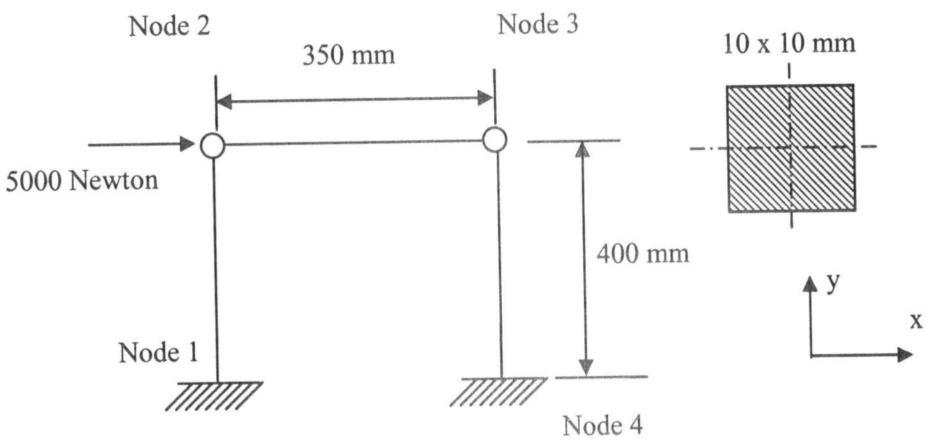

Step 1: launch the software system: **ALGOR V16.**

We assume that the software system resides in the folder called **Programs.**

Click the **Start Menu** > **Programs** > **ALGOR V16** > **FEA Model** > **Open.**

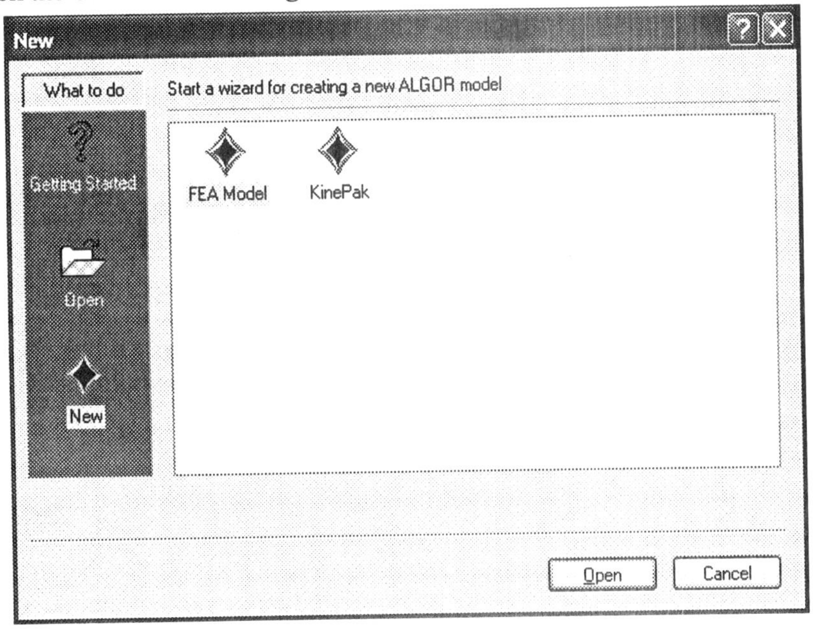

Enter file name *beam* > **Save.**

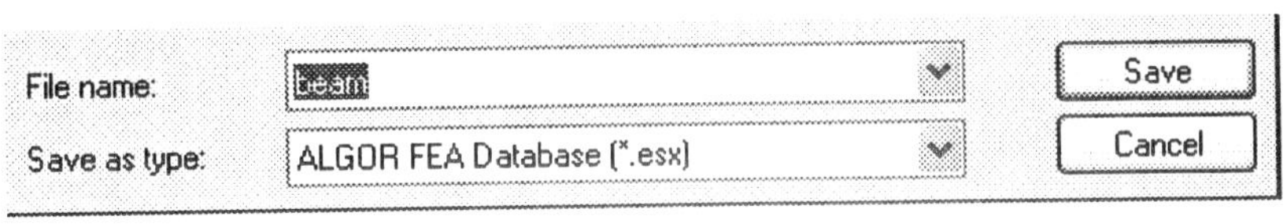

Select *Static Stress with Linear Material Models* as the Design Scenario.

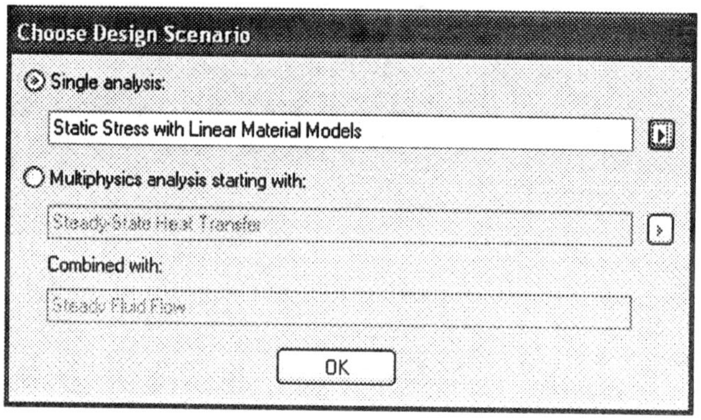

Step 2: Define the unit system before drawing the Truss Structure

ALGOR V16 requires users to specify the unit system before getting on sketching the beam structure. The unit system window appears immediately after the user has selected a design scenario. Let us select **Metric mks (SI)** and set the force unit to Newton, Length unit to meter (m), Time unit to second, etc. > **OK**.

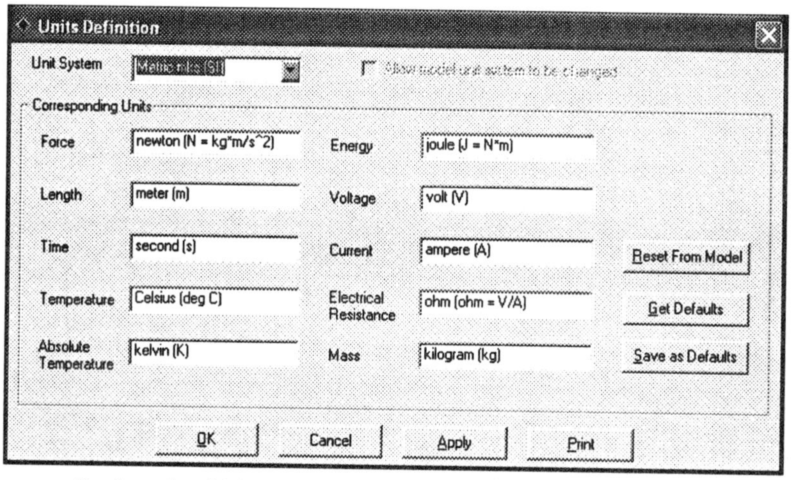

The window on display should look like the picture shown below.

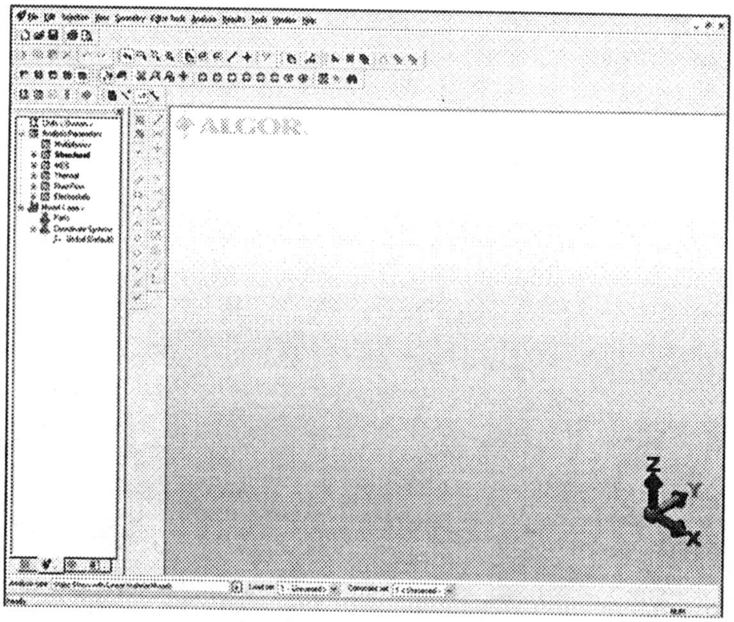

From the main toolbar choose **View > Orientation > XY Top.**

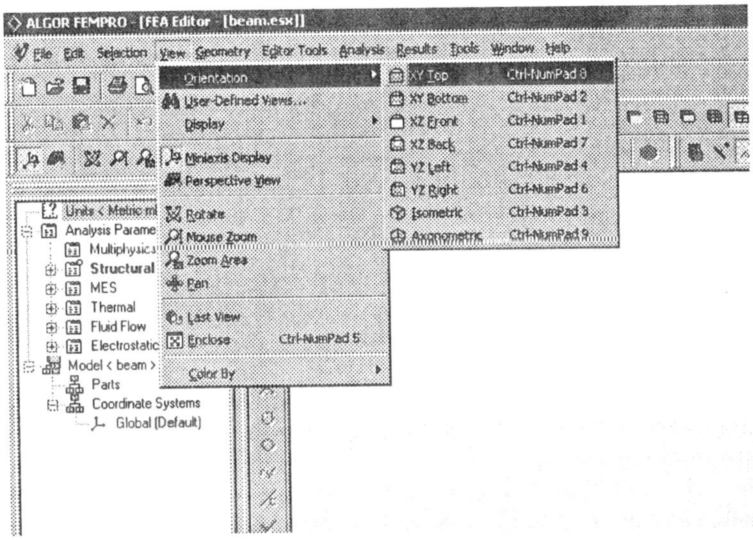

Again, from the main toolbar select **Geometry > Elements > Line.**

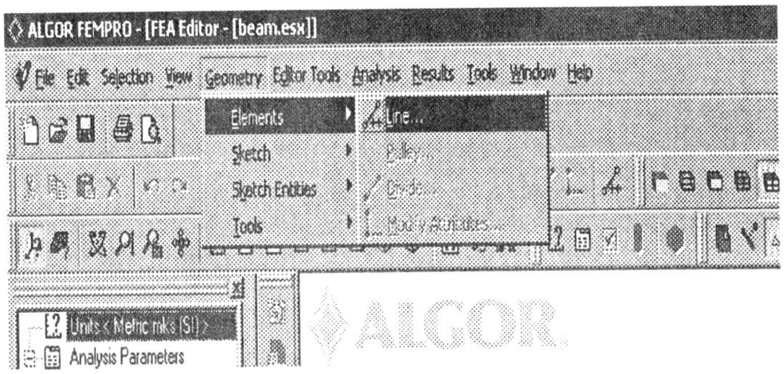

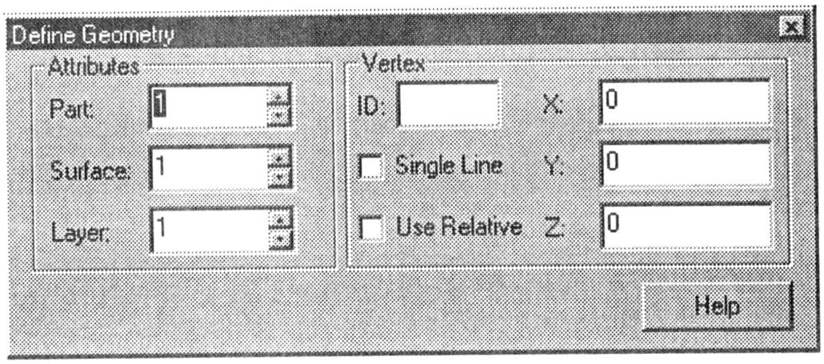

ENTER

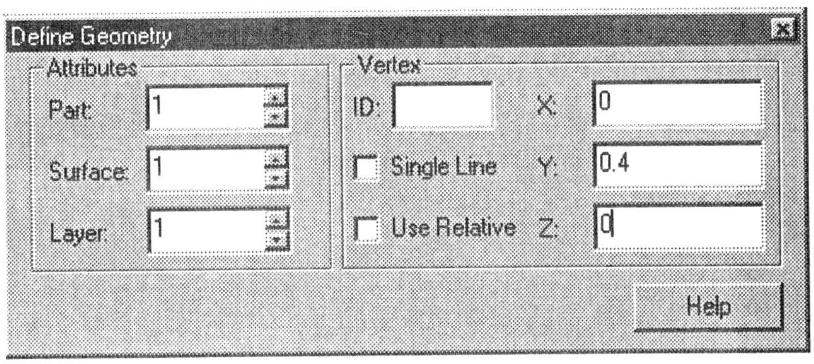

ENTER

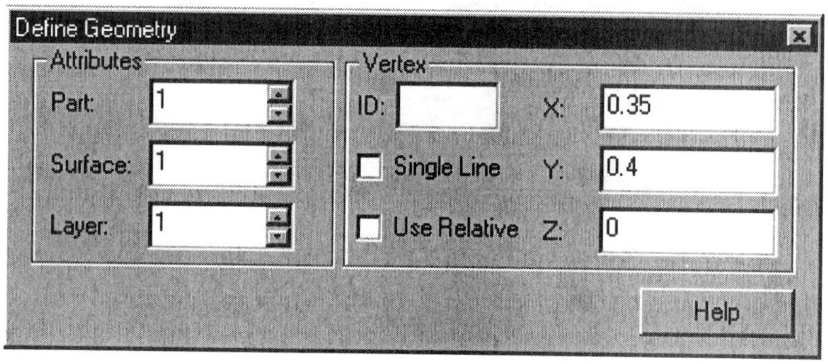

ENTER

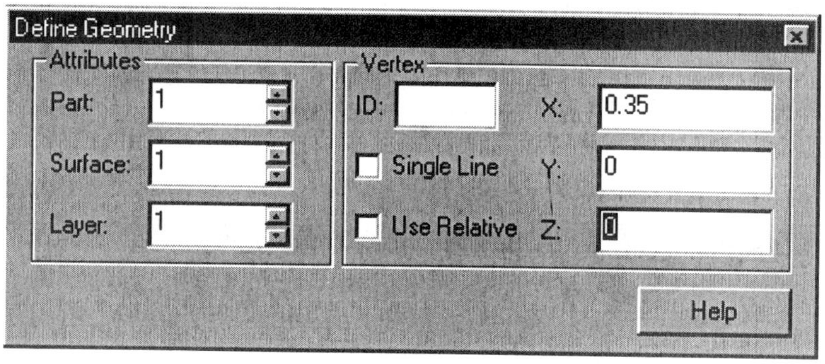

ENTER > Close window to complete the line entries. **View > Enclose.**

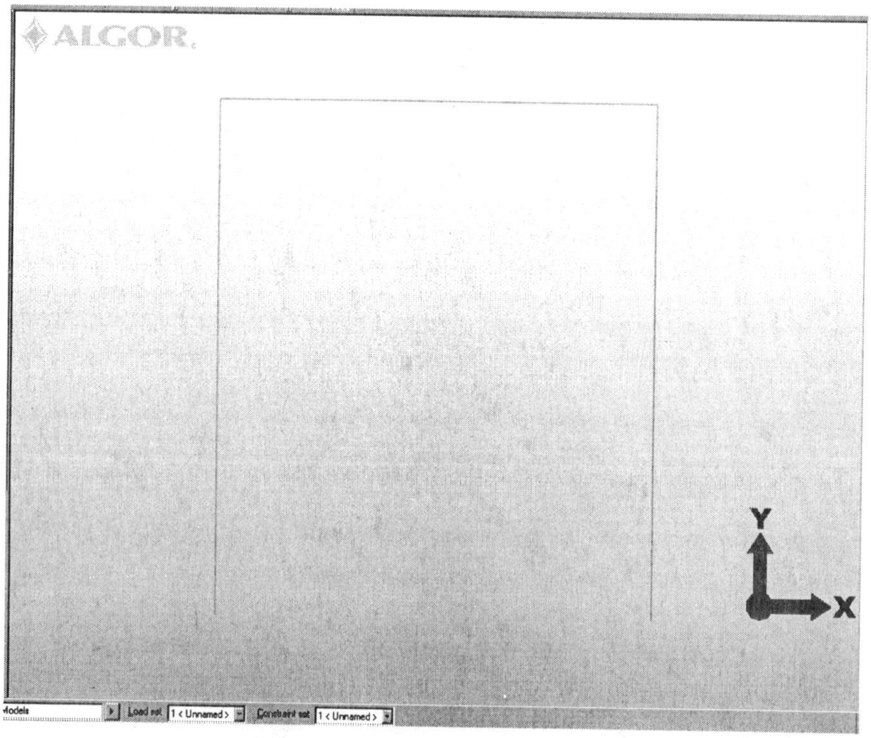

Step 2: Specifying Model Data in FEMPRO
Right click element type in tree view, as shown below.

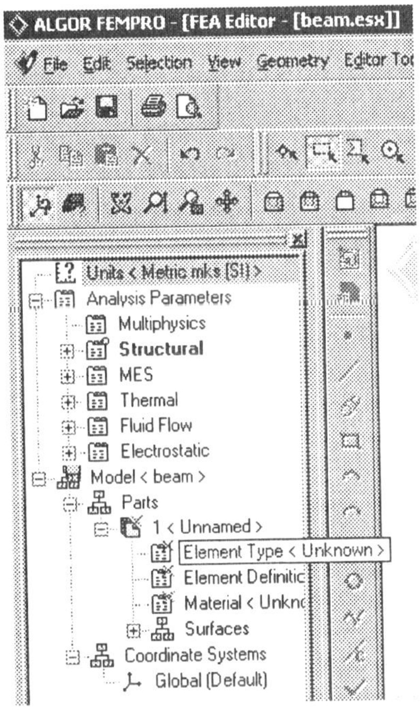

Another list of items appears > Select **beam**.

Now, right click **Element Definition** in the model tree and select **Modify.** Select **Cross-Section Libraries**.

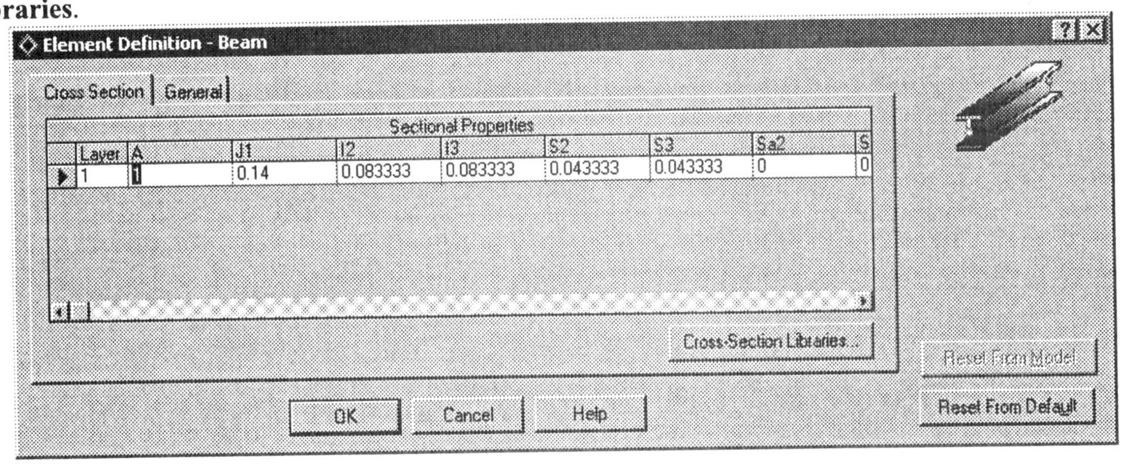

Choose **Rectangular** and define the area as 10 x 10 mm > **OK** > **OK**

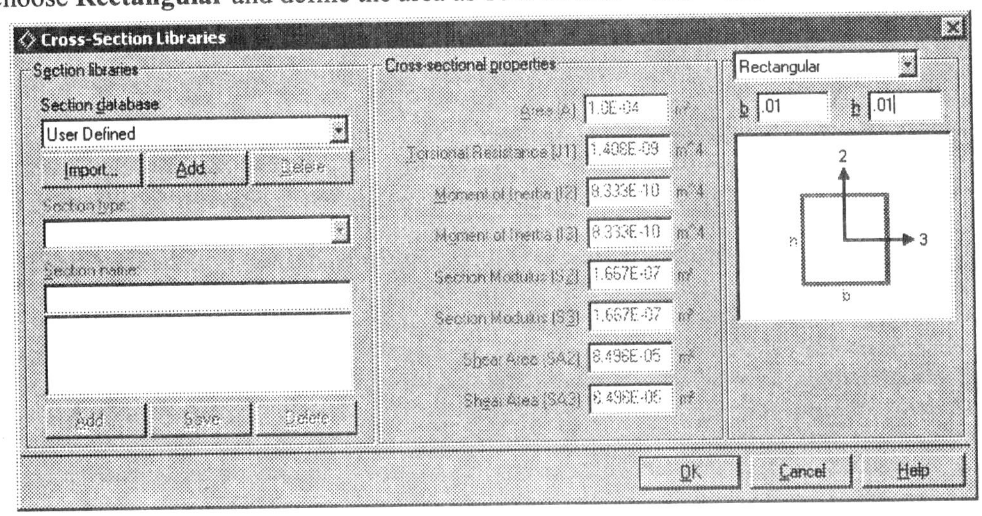

Step 3: Selecting Material Properties

Right click **Material** in the tree view > **Modify** > select the material type: Aluminum 6061-T6 > **OK**.

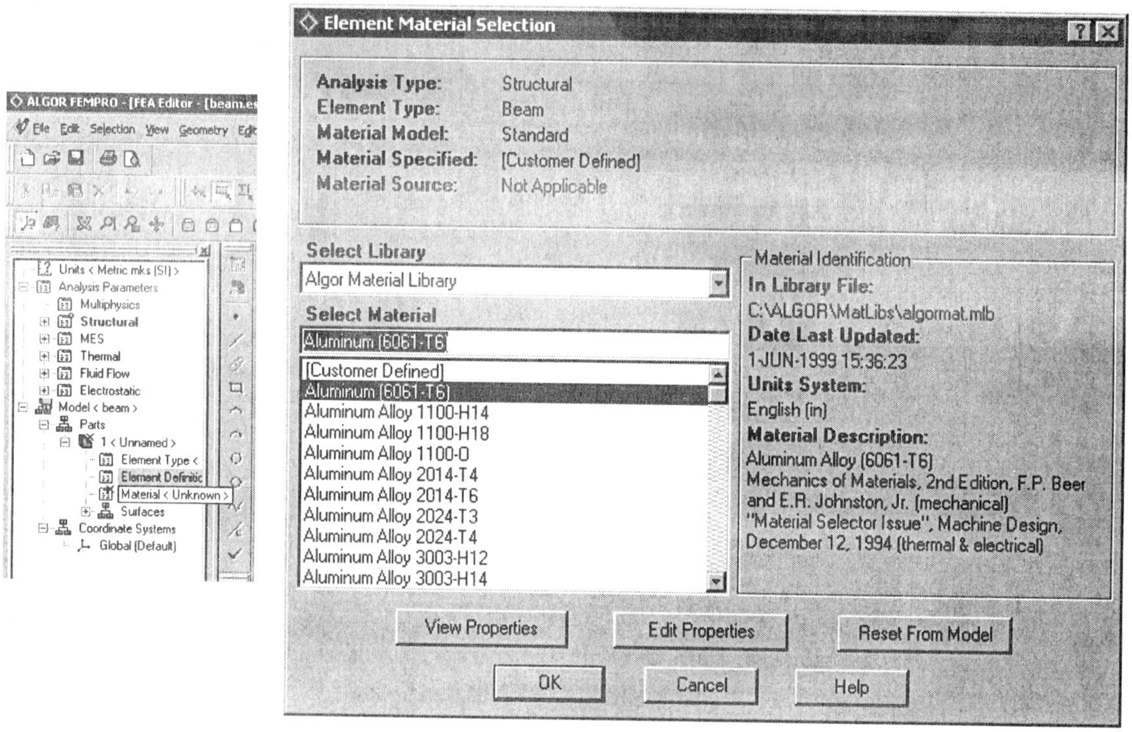

Step 4: Defining Constraints

To select certain parts of the beam follow these steps:

Selection > Shape > Rectangle * This will make the selecting tool a rectangular shape.

Selection > Select > Vertices * This will cause only the vertices to be selected.

Draw a box that includes Node 1 and Node 4, as shown below.

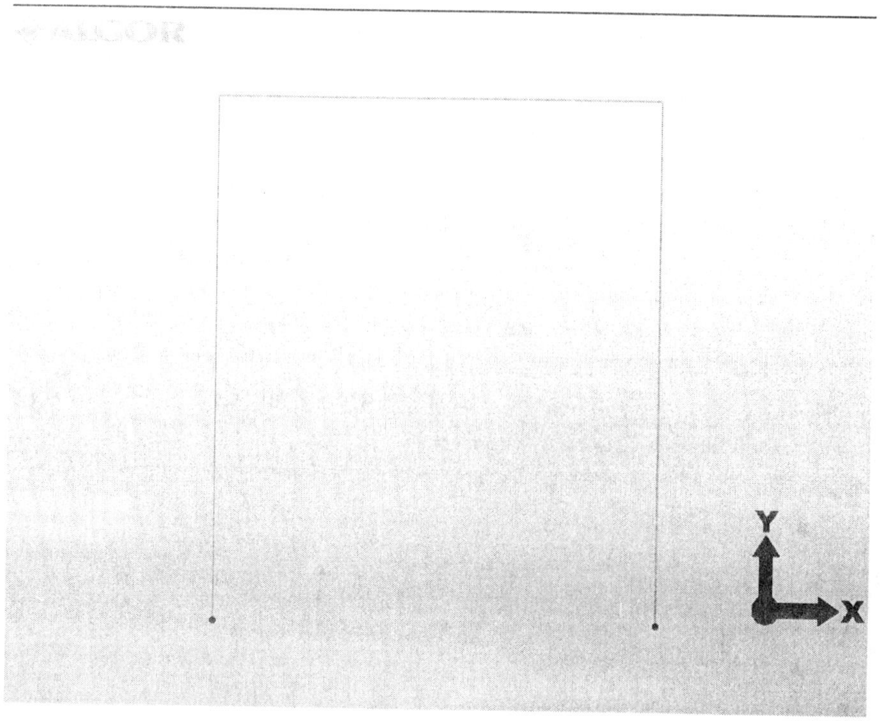

Now, right click the display area > **Add** > **Nodal Boundary Conditions** > **Fixed** > **OK**.

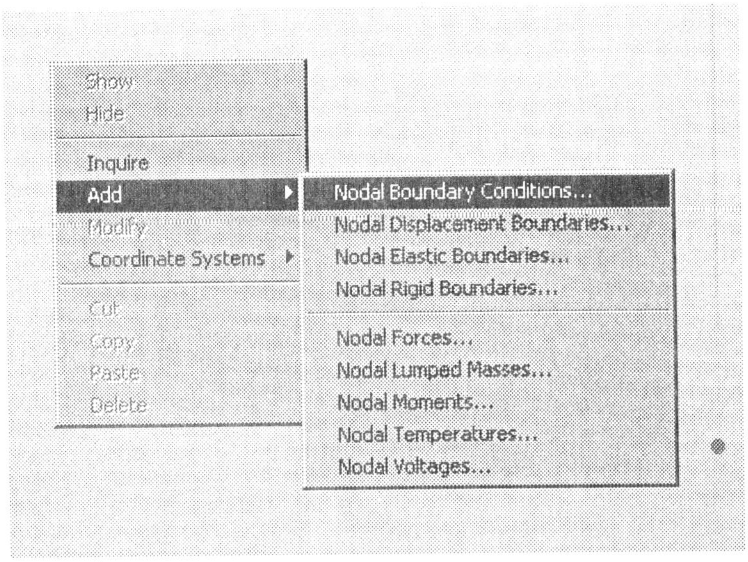

Step 5: Defining Load

Following the instructions above, draw a box around Node 2 to select it.

Right click display area.> **Add** > **Nodal Force.** Tape *5000* as the magnitude > Select **X Direction** > **OK**.

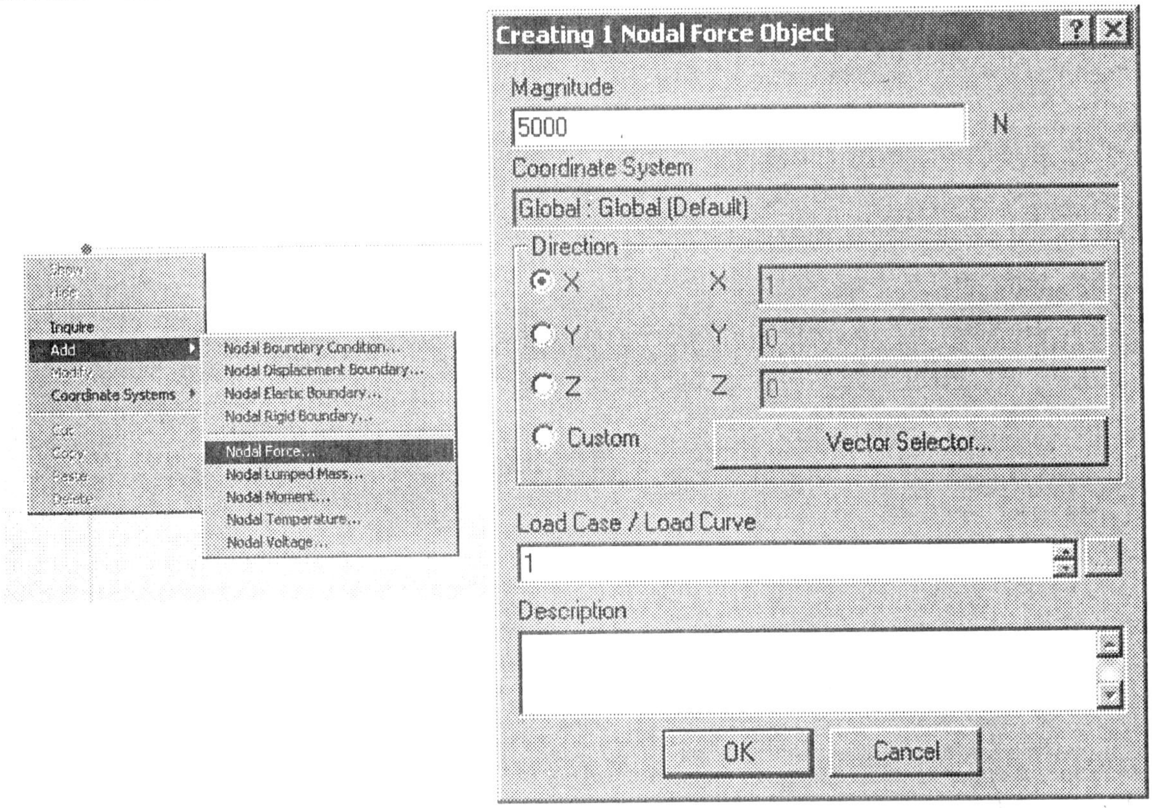

Step 6: Adding Gravity Load

Right click on **Structural** in the tree view > Select **Model Parameters.** Select the tab of **Accel/Gravity**.

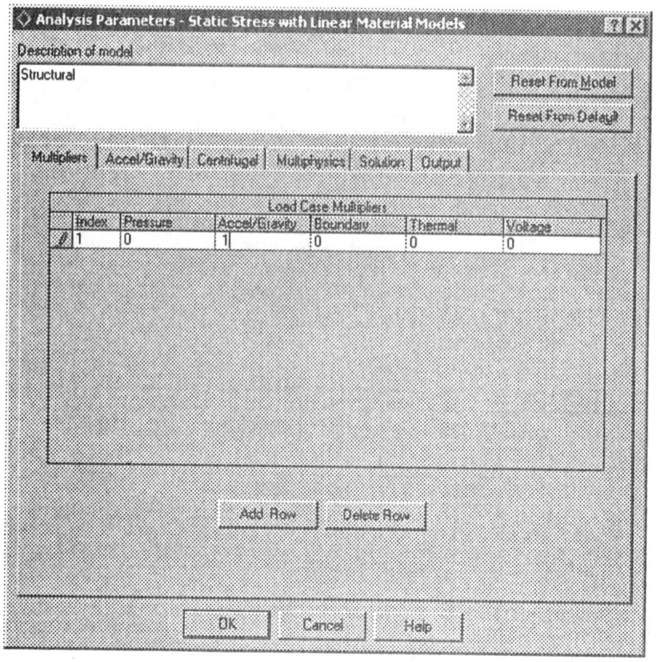

Select **"Set for Standard Gravity"** change **Y multiplier** to **-1** → OK.

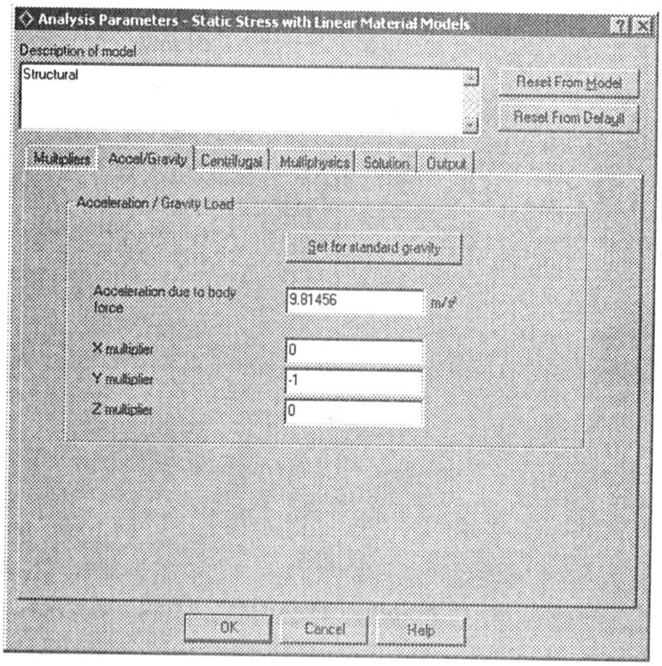

Step 7: Analyzing the model

In the main toolbar, choose **Analysis** > **Check Model.**

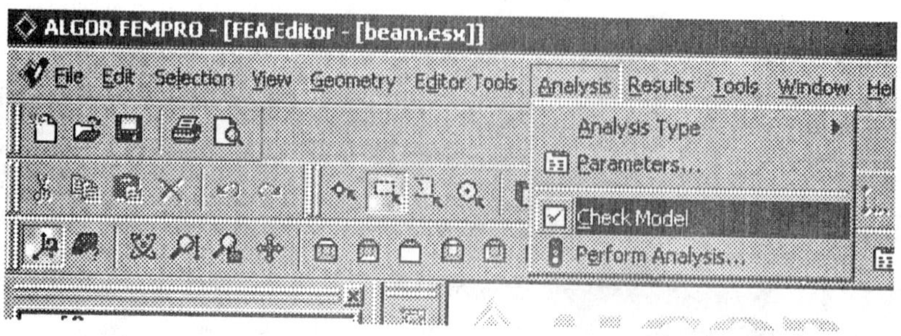

In the main toolbar, select **Tools > FEA editor.**

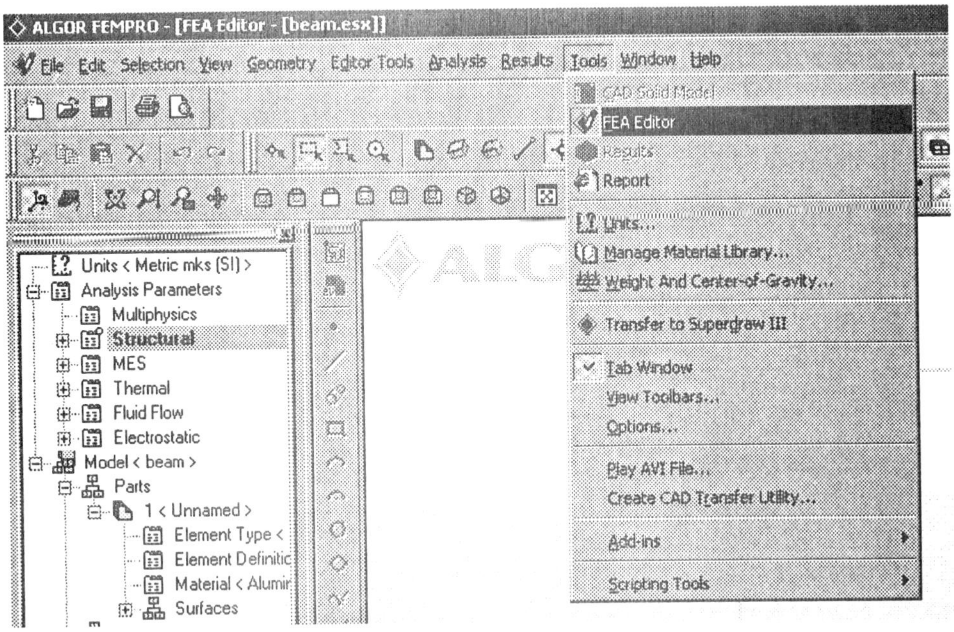

Under the Main toolbar, click **Analysis > Perform Analysis.**

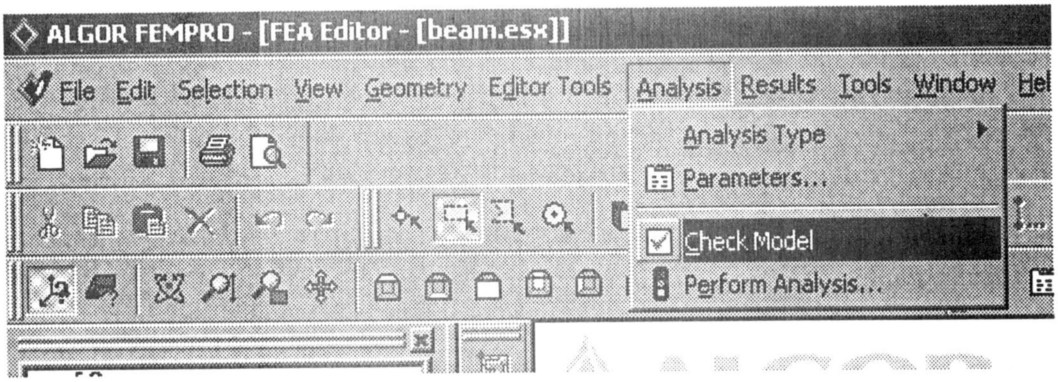

Ignore the warning because we are defining a new model > **OK.**

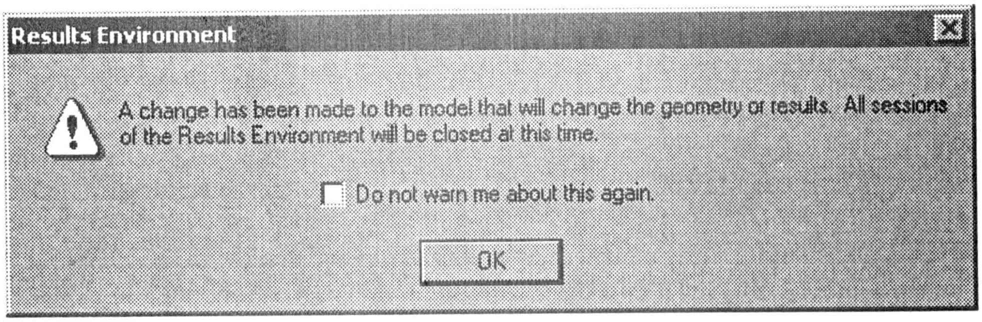

The results window shown below will open after the analysis is complete.

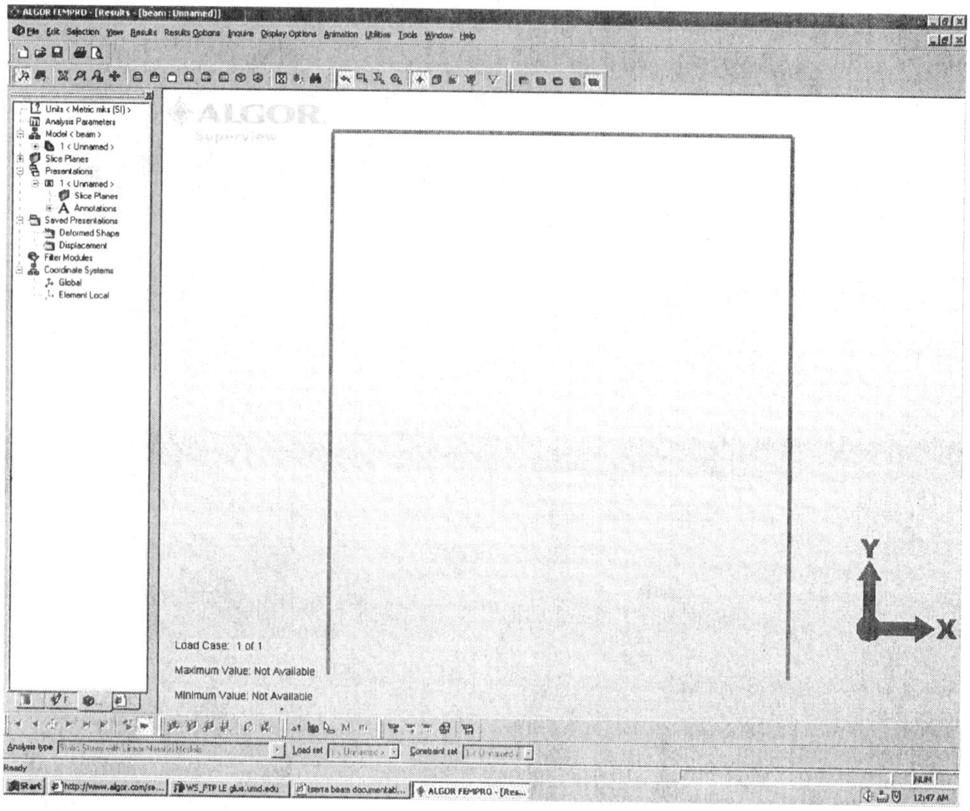

In the main toolbar select **Results Options > Displaced Model Options.**

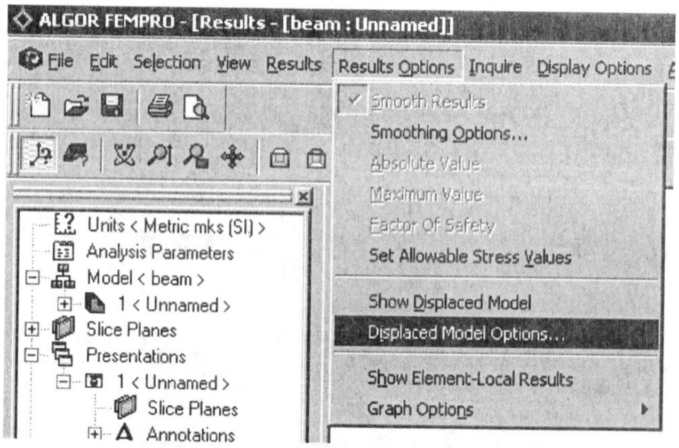

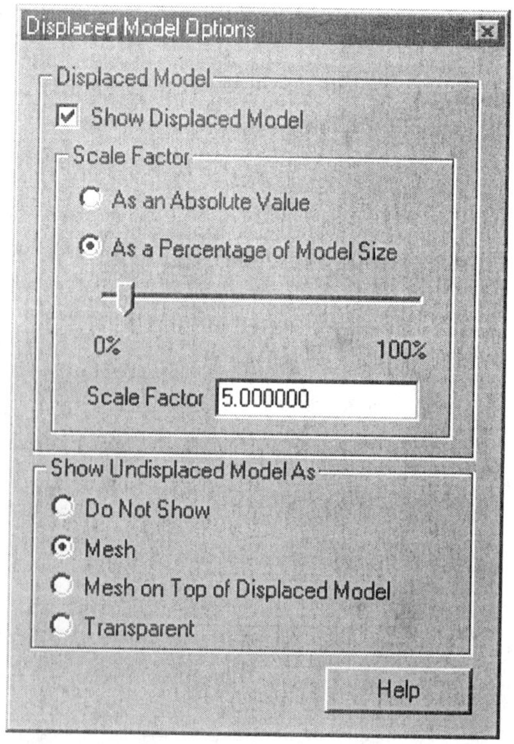

Select "**Show Displaced Model**" select "**Mesh**," you may now close the options window. The following will appear.

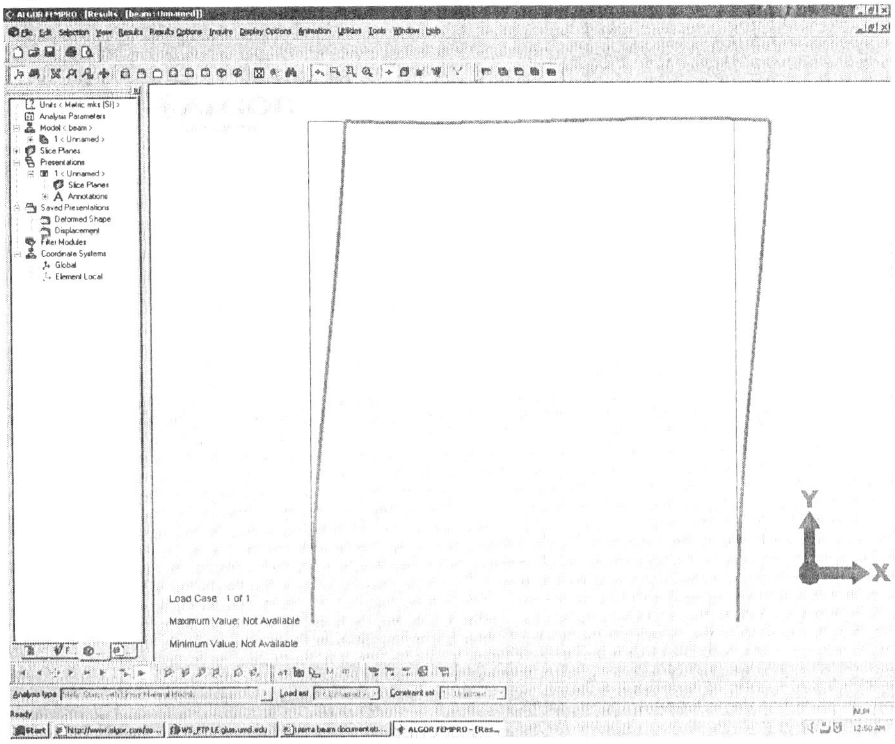

Step 8 - Getting the Results

Under the shown load condition, determine the displacement of Node 2 and Node 3 in the horizontal and vertical directions.

Show the reaction forces by selecting **Results** > **Reactions Vector** > **Reaction Forces** > **Vector Plot**.

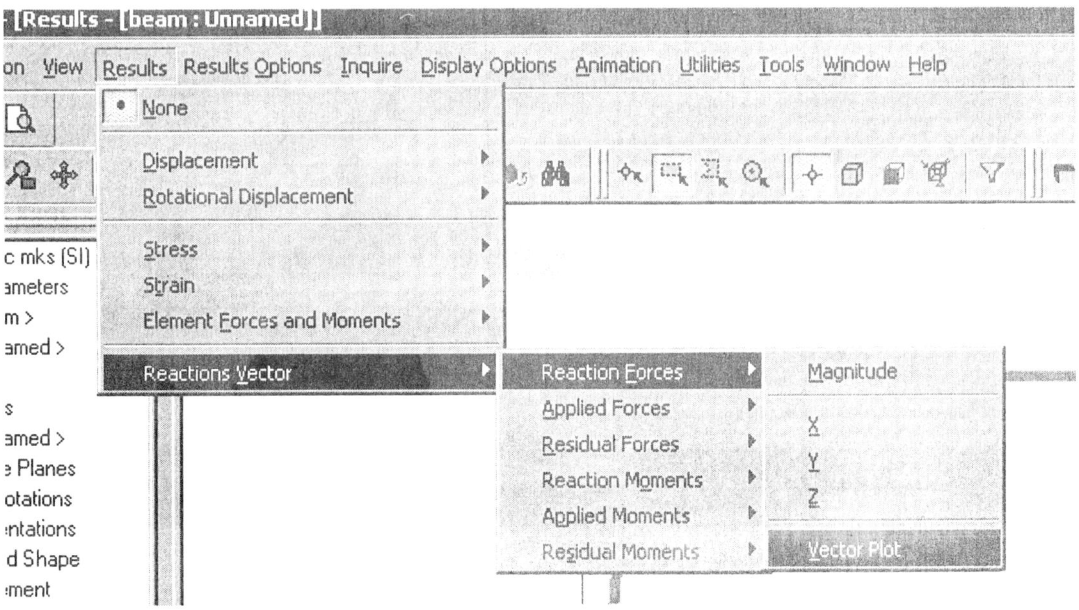

The model should look like this.

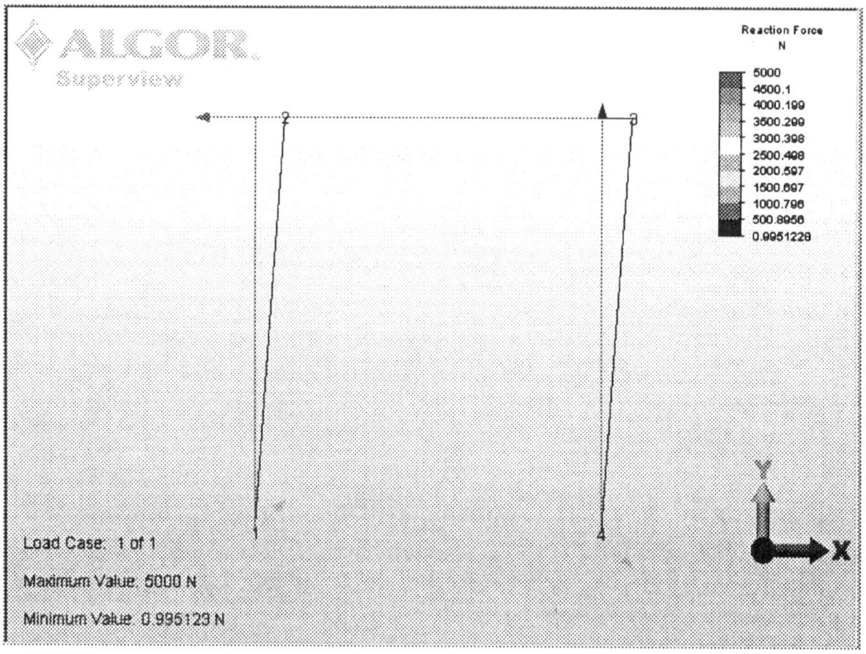

Use the **Rectangle Select Tool** to select all nodes. The nodes will be highlighted.

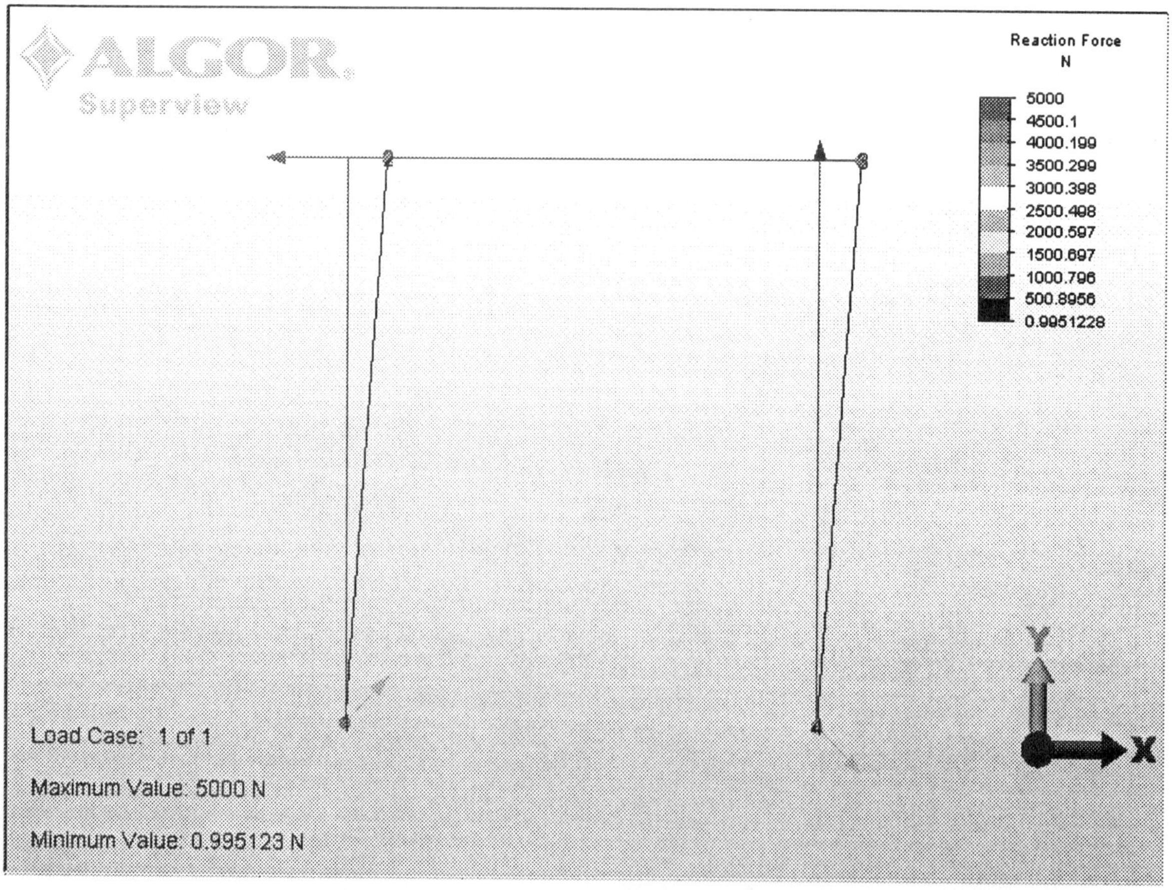

Select **Inquire/ Results**… This will display the displacement and reaction force at each node.

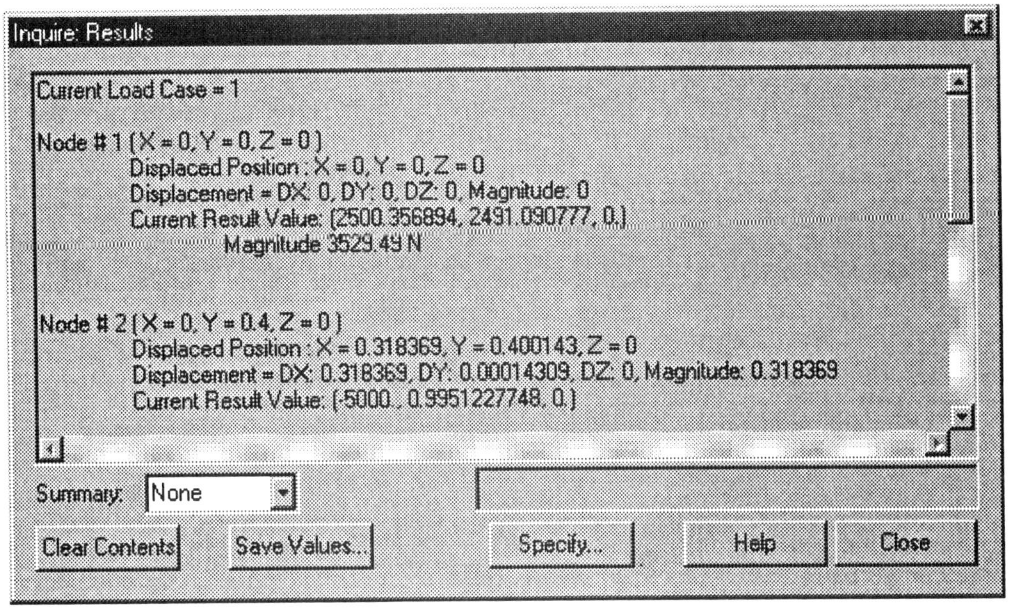

Nodal Displacement

	X displacement (m)	Y displacement (m)
Node 2	0.318369	0.00014309
Node 3	0.318243	-0.000143204

Nodal Reaction Forces

	X force (N)	Y force (N)
Node 1	2500	2491
Node 4	2499	-2493

9.7 References

1. Algor Processor Reference Manual, Algor Interactive Systems, 150 Alpha Drive, Pittsburgh, PA 15238
2. O. Axelsson, Iterative Solution Methods, Cambridge University Press, New York, 1994.
3. K. J. Bathe, Finite Element Procedure, Prentice Hall, Upper Saddle River, New Jersey, 1996.
4. K. Bell, A Refined Triangular Plate Bending Finite Element, Int. J. Numer. Methods Eng., Vol. 1, No. 1 January 1969, pp. 101-122.
5. B. Carnahan, H. A. Luther, and J. O. Wikes, Applied Numerical Methods, Wiley New York, NY, 1969.
6. R. W. Clough, The Finite Element in Plane Stress Analysis, Proceedings of 2nd ASCE Conference on Electronic Computation, Pittsburgh, PA. September 1960.
7. R. D. Cook, D. S. Maikus and M. E. Plesha, Concepts and Applications of Finite Element Analysis, 3rd edition, Wiley, New York, NY, 1989.
8. J. W. Dally and W. F. Riley, Experimental Stress Analysis, McGraw Hill, 1965.

9. J. W. Dally, Design Analysis of Structural Elements, Goodington, New York, NY, 2003.
10. C. S. Desal and J. F. Abel, Introduction to the Finite Element Method, Van Nostrand Reinhold, New York, NY, 1972.
11. J. H. Earle, Graphics for Engineers, AutoCAD Release 13, Addision-Wesley, Reading, Massachusetts, 1996.
12. R. H. Gallagher, Finite Element Analysis Fundamentals, Prentice-Hall, Englewoood Cliffs, NJ, 1975.
13. J. M. Gere and S. P. Timoshenko, Mechanics of Materials, 3rd edition, PWS-KENT Publishing Company, Boston, 1990.
14. W. W. Hager, Applied Numerical Linear Algebra, Prentice-Hall, Englewood Cliffs, NJ, 1988.
15. Holland, The Finite Element Method in Plane Stress Analysis, in the Finite Element Method in Stress Analysis, I. Holland and K. Bell (eds), Tapir Press, Trondheim, Norway, 1969, Chapter 2.
16. Y. Y. Hsieh, Elementary Theory of Structure, 2nd edition, Prentice-Hall, Englewood Cliffs, NJ, 1982.
17. M. A. N. Hendriks, H Jongedijk, J. G. Rots and W. J. E. van Spange (eds), Finite Elements in Engineering and Science, DIANA Computational Mechanics, 1997.
18. K. H. Huebner, D. L. Dewhirst, D. E. Smith, and T. G. Byrom, The Finite Element Method for Engineers, 4th edition, John Wiley & Sons, Inc., 2001.
19. C.S. Krishnamoorthy, Finite Element Analysis, Theory and Programming, 2nd Ed., 1995.
20. D. L. Logan, A First Course in the Finite Element Method Using ALGOR, PWS Publishing, Boston, 1997.
21. H. C. Martin, Introduction to Matrix Methods of Structural Analysis, McGraw-Hill, New York, NY, 1966.
22. R. L. Norton, Machine Design: An Integrated Approach, Prentice Hall, Upper Saddle River, New Jersey, 1996.
23. M. Ortiz, Y. Leroy and A. Needleman, A Finite Element Method for Localized Failure Analysis, Comp. Meth. Applied Mech. Enging, 61, pp. 189-214, 1987.
24. M. Petyl, Introduction to Finite Element Vibration Analysis, Cambridge University Press, 1990.
25. T. H. H. Pian and P. Tong, Basis of Finite Element Methods for Solid Continua, Int. J. Numer. Methods Eng., Vol. 2, No. 7, April 1964, pp. 1333-1336.
26. P. G. Plockner, Symmetry in Structural Mechanics, J. of the Structural Division, American Society of Civil Engineers, Vol. 99, No. ST1, pp. 71-89, 1973.
27. J. N. Reddy, An Introduction to the Finite Element Method, McGraw-Hill Publishing Company, New York, NY, 1984.
28. T. Ross, T. Johns and Emile, Dynamic Buckling of Thin Walled Domes under External Water Pressure, Applied Solid Mechanics 2nd Conference, Ed. Tooth & Spence, Elsevier, pp. 211-224, 1987.
29. T. Ross, Finite Element Program in Structural Engineering and Continuum Mechanics, Horwood Publishing Limited, Chichester, England, 1996.
30. T. Ross, Advanced Applied Finite Element Methods, Horwood Publishing Limited, Chichester, England, 1998.
31. L. Stasa, Applied Finite Element Analysis for Engineers, Holt, Rinehart and Winston, New York, NY, 1985, pp. 101-157.
32. Y. Saad, Iterative Methods for Sparse Linear Systems, PWS Publishing Company, Boston, 1996.
33. P. Silvester, Higher-Order Polynomial Triangular Finite Element for Potential Problems, Int. J. Eng. Soc., Vol. 7, No. 8, pp. 849-861.
34. Y. Tada and G. C. Lee, Finite Element Solution to an Elastic Problem of Beams, Int. J. Numer. Methods Eng., Vol. 2, No. 2, April 1970, pp. 229-241.
35. R. L. Taylor, On Completeness of Shape Functions for Finite Element Analysis, Int. J. Numer., Methods Eng., Vol. 4., No. 1, 1972, pp. 17-22.
36. S. P. Timoshenko and D. H. Young, Theory of Structure, McGraw-Hill, Kogaksha Limited, 1965.
37. M. J. Turner, R. W. Clough, H. C. Martin, and L. J. Topp, Stiffness and Deflection Analysis of Complex Structures, J. of the Aeronautical Sciences, Vol. 23, No. 9, pp. 805-824, Sept. 1956.
38. O. C. Zienkiewicz, The Finite Element Method, third edition, McGraw-Hill, New York, NY, 1977.
39. O. C. Zienkiewicz and R. L. Taylor, The Finite Element Method, 4th edition, Vol. 1, McGraw-Hill (UK),London, 1989.

9.8 Exercises

1. Use Pro/Mechanica to evaluate the deflection at each of the 3 nodes and the reaction force at the pin joint at the left side and the reaction force at the roller support at the right side. Compare these obtained results with the results obtained from performing FEA using **ALGOR**.

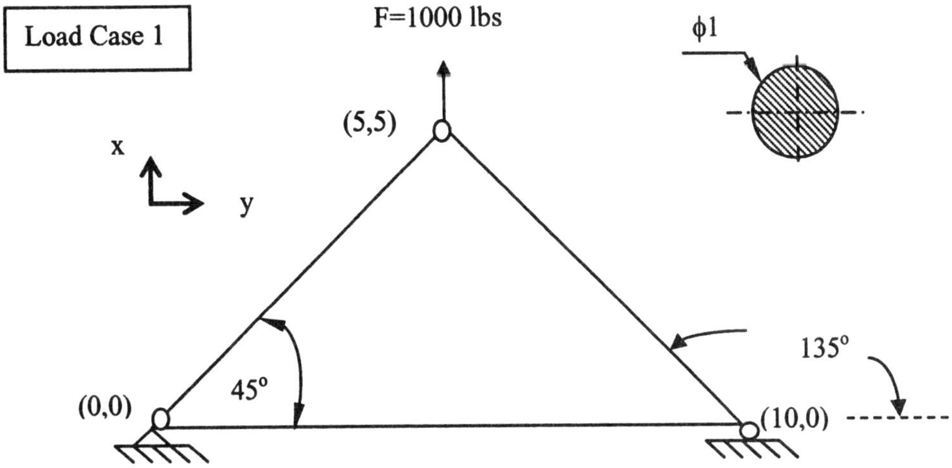

2. Use Pro/Mechanica to evaluate the deflection at each of the 3 nodes and the reaction force at the pin joint at the lèft side and the reaction force at the roller support at the right side. Compare these obtained results with the results obtained from performing FEA using **ALGOR**.

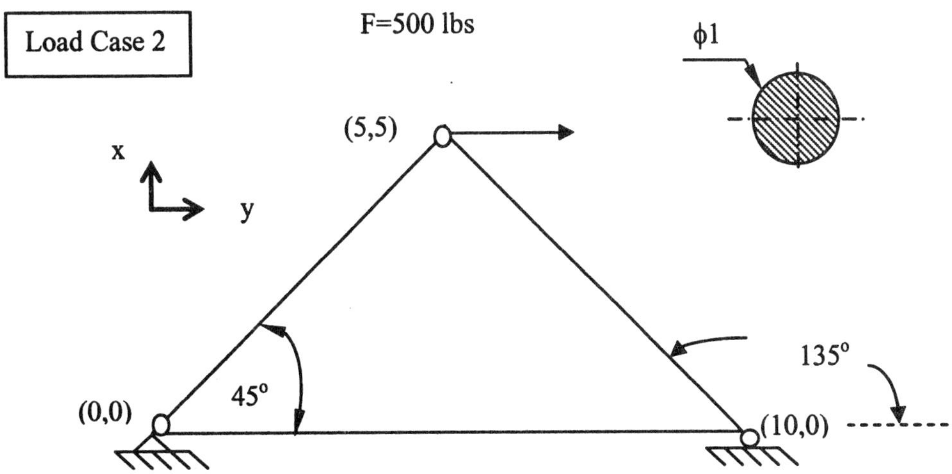

3. Use Pro/Mechanica and ALGOR to evaluate the maximum deflection of the beam under the shown load condition. Assume that the material type is STEEL. Make that the gravity force is considered during the 2 evaluations. Compare the results obtained from both running Pro/Mechanica and running ALGOR with the result obtained from the calculation based on the formula used in a textbook of mechanics of materials.

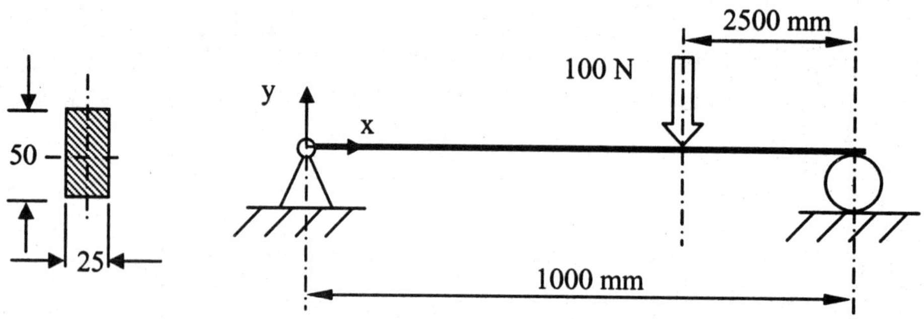

4. Use Pro/Mechanica to evaluate the deflection of node 2, the reaction force components at Node 1 and the reaction force components at Node 3. Compare the obtained results with the results obtained from running ALGOR. Make sure that the two parameters related to the material type are set at the same numerical values as those values used when running ALGOR. Those values should be identified when checking the material properties of 6061 T6 aluminum under the ALGOR software system.

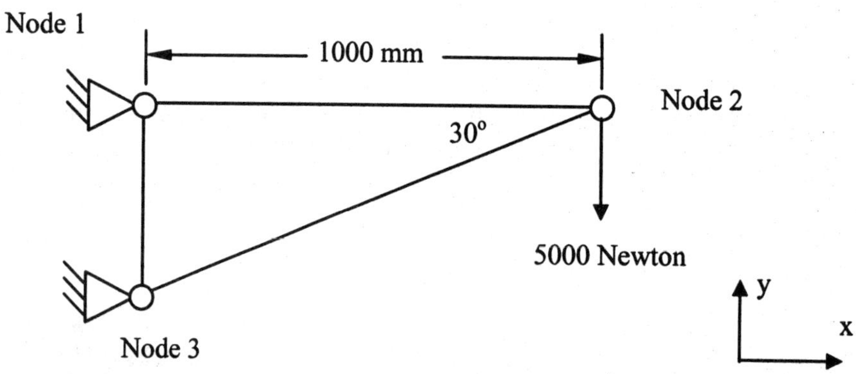

5. Use Pro/Mechanica and ALGOR to evaluate the displacement of node 2, the displacement of node 3, the reaction force components at node 1 and the reaction force compoents at node 3. Compare the obtained results with the results obtained from the calculations carried out in Chapter 1. Set the Young's modulus equal to 300 GPa, and the Poisson's ration equal to 0.27. The section area is a circle. The circular area is 10 mm^2.

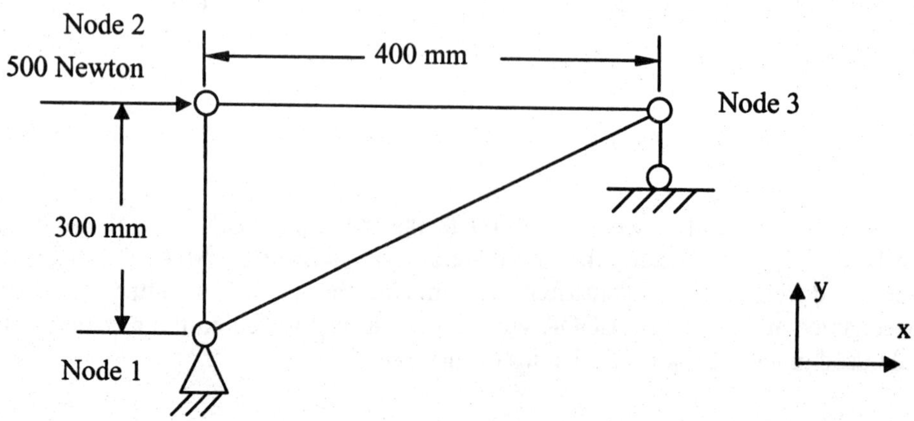

6. The following figure illustrates a truss structure, which consists of 3 bars. The dimension of the section area is 100 mm^2. The geometrical information of the truss structure is shown in the figure. The material type is STEEL. Use Pro/Mechanica and ALGOR to evaluate the displacement components at Node 2, the displacement components at Node 3, the reaction force components at Node 1 and the reaction force components at Node 4.

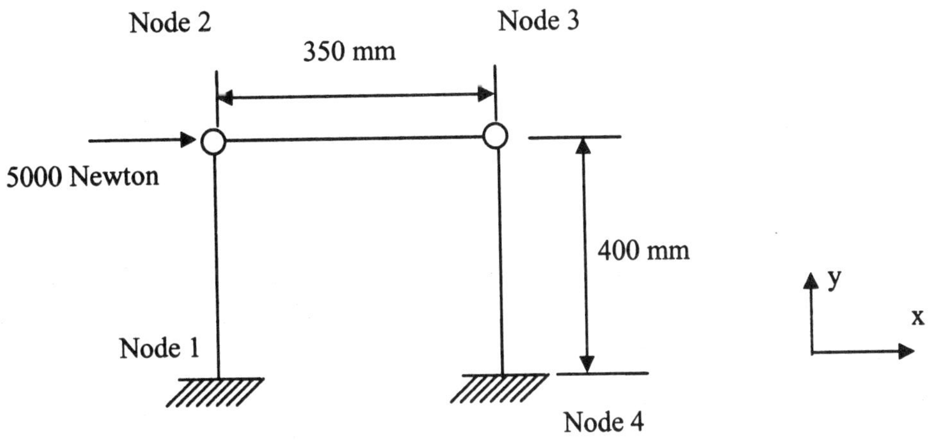

CHAPTER 10

FEA USING ANSYS

10.1 Introduction

In the previous 9 chapters, we have learned two software systems. They are **Pro/Mechanica** and **ALGOR**. In this chapter, we will use **ANSYS** to perform FEA. As discussed in Chapter 1, ANSYS is a simulation software system designed to optimize product development processes. **ANSYS** has been widely used by engineers and designers in industry and academy. The software system focuses on the development of open and flexible solutions that enable users to analyze designs so as to provide a common platform for fast, efficient and cost- conscious product development, from design concept to final-stage testing and validation. The newest release is **ANSYS 9.0**.

The first example in this chapter is to develop a shell model under the **ANSYS** design environment. A procedure to perform FEA using ANSYS is demonstrated step by step. The second example is to work with a 3D solid model. The third example is to work with an assembly. Finally, we use an example of a beam structure to demonstrate an approach to control the size of elements in the stage of creating a geometric model.

10.2 FEA with a Shell Model

The following figure illustrates a plate structure. The dimensions of length and width are 800 and 100 mm, respectively. The thickness of the plate is 10 mm. The left end of the plate is fixed to the wall. The load acts on the edge at the right end, which is set free to move. In terms of the material properties, we assume that the Young's modulus is 30 GPa, and the Poisson's ratio is 0.26. The objective of this study is to obtain the information on the maximum principal stress developed under the load condition.

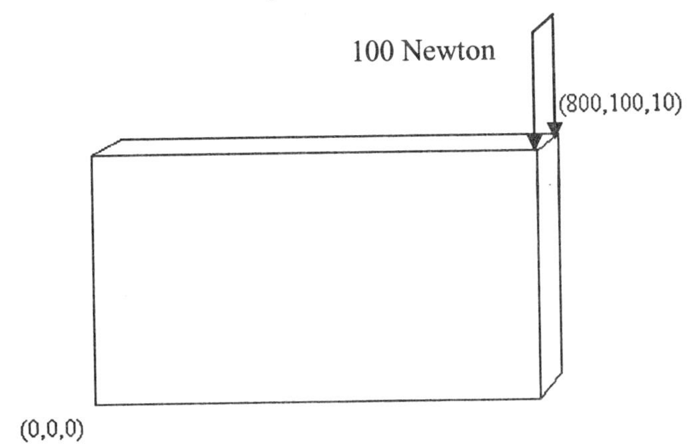

The following flow chart illustrates the general approach of using ANSYS to perform FEA. The approach starts with creating a geometrical model. In this example, we deal with the creation of a shell model of the plate structure. Afterwards, we need to define the type of element. This step distinguishes the ANSYS approach from the other approaches as used in **Pro/Mechanica** or **ALGOR**. A library of elements types serves as a built-in function. Users are able to select the element type, which best fits their need in terms of accuracy and computing efficiency. When the element type is defined, users need to specify the material type. Regarding the mesh generation, a function called **Smart** will automate the mesh generation based on the setting specified by the user. As similar to the other FEA systems, users define load condition(s), constraint condition(s), run the program, examine the obtained solution and plot the results graphically.

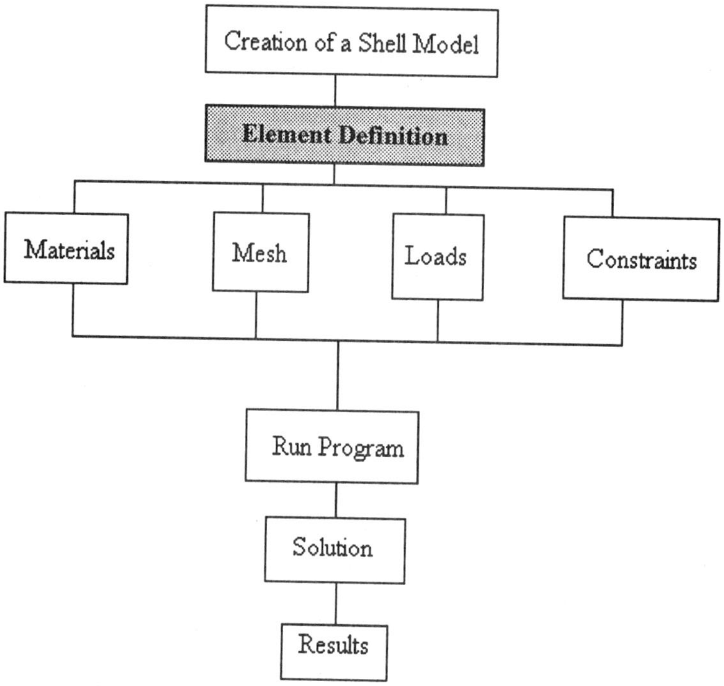

Step 1: Launch the **ANSYS** system.

When starting the ANSYS system, the following window appears first. Select Run **Interactive.**

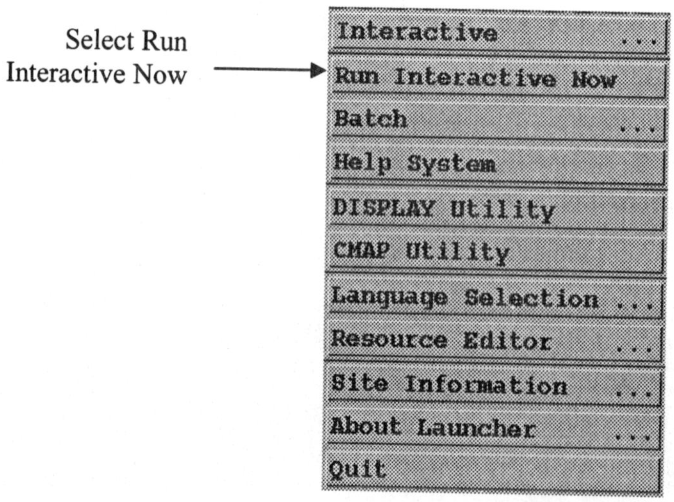

After selecting **Run Interactive Now**, a new window appears, as show below. It is the system default setting window. The displayed information is related to the specifics with the system settings, such as the working directory, the memory size set to run the program, etc. Users may make changes, for example, to crease the size of workspace when running a large scale structure. After checking and/or adjusting the displayed information, select **Run.**

Select Run after checking the system settings

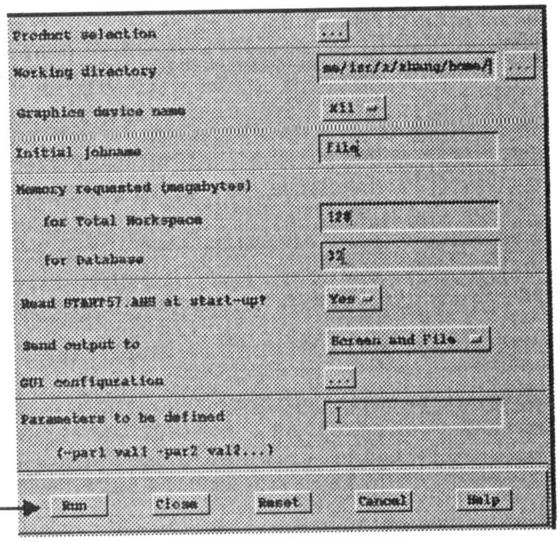

After selecting **Run**, the main window of ANSYS appears.

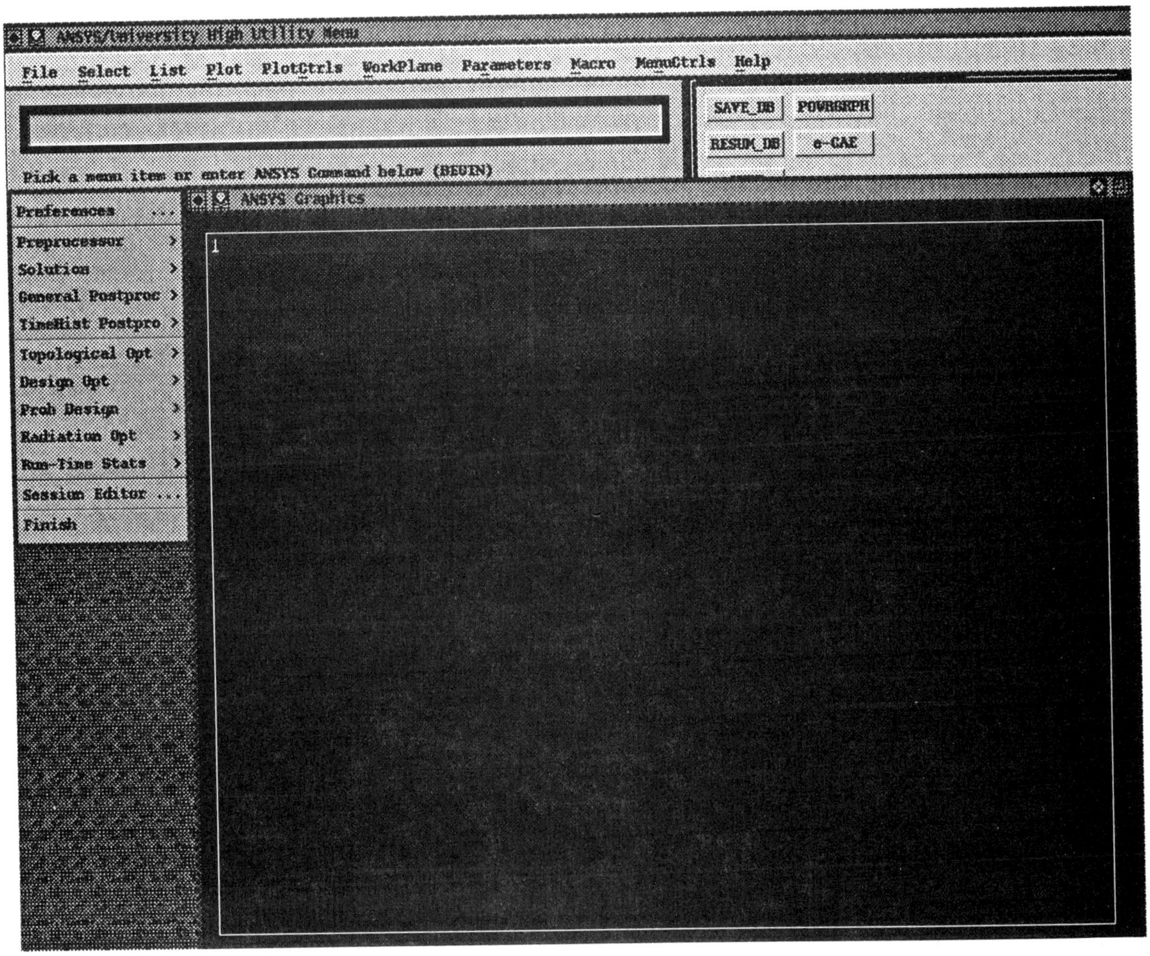

Select Preferences > select Structural > OK.

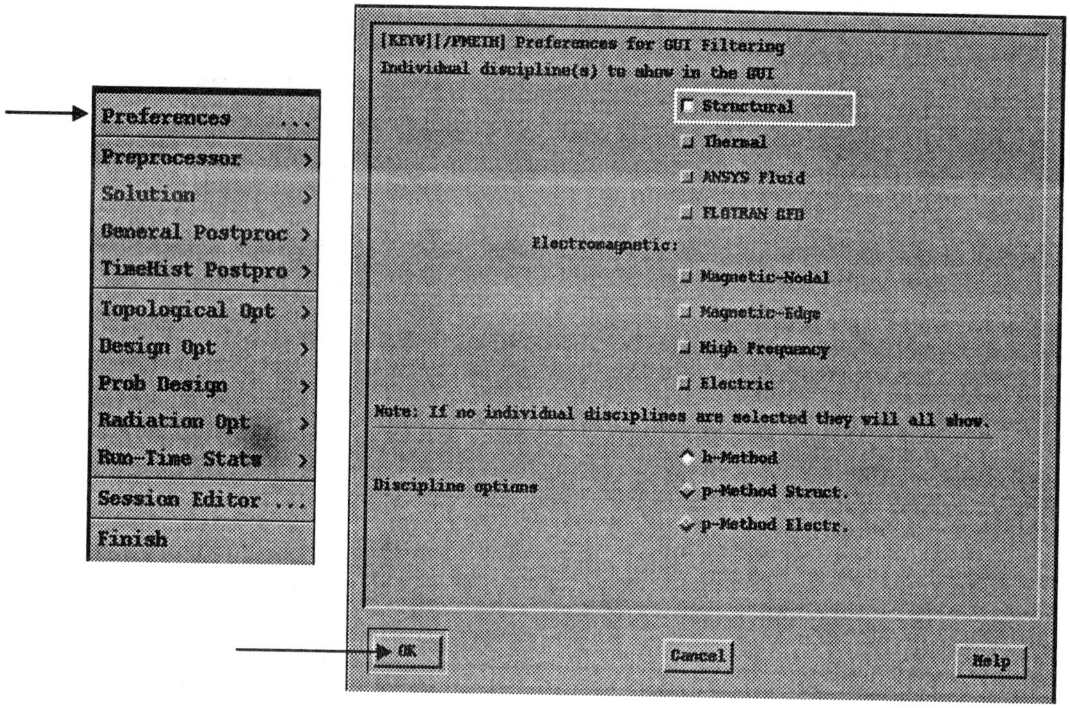

From the main menu, select **Preprocessor**. Then, under modeling, select **Create**. Because we create a 2D geometry, from **Areas**, select **Rectangle > By Dimensions.** In the window for creating the rectangle by dimensions, type 0 and 800 for X1 and X2, and 0 and 100 for Y1 and Y2.

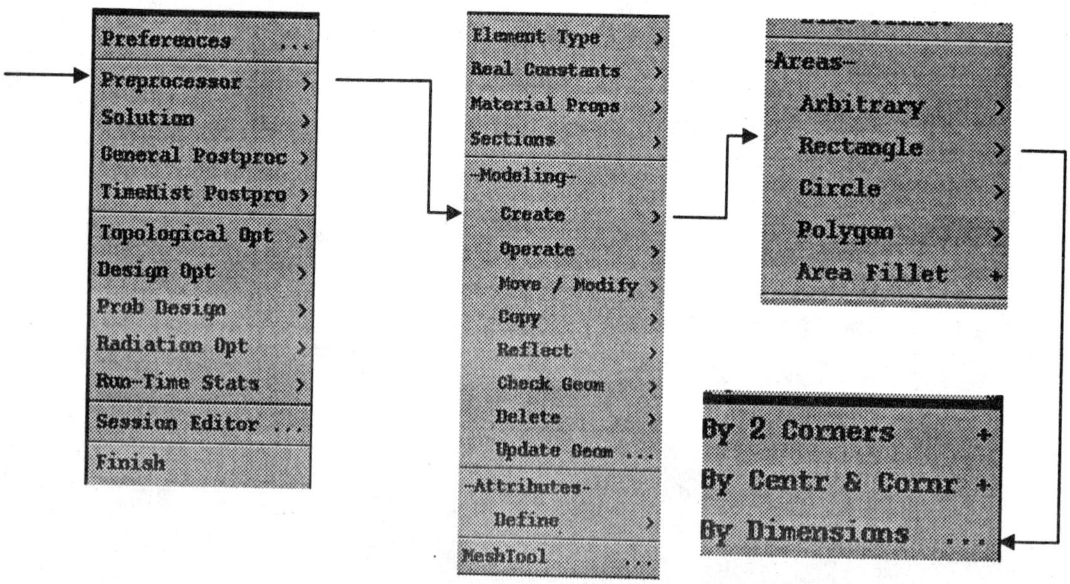

In the window for creating the rectangle by dimensions, type *0* and *0.8* for X1 and X2, and *0* and *0.1* for Y1 and Y2.

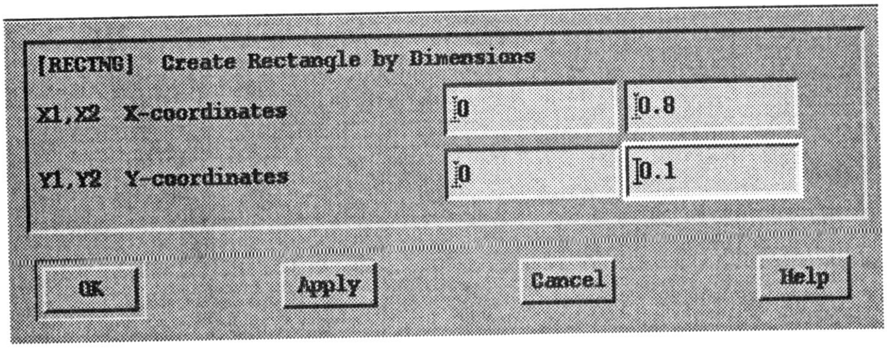

Once the dimensions are defined, click **OK**. This will give you the following shape.

Under the Preprocessor menu, select **Element Type**. Then select **Add/Edit/Delete**, and then **Add**.

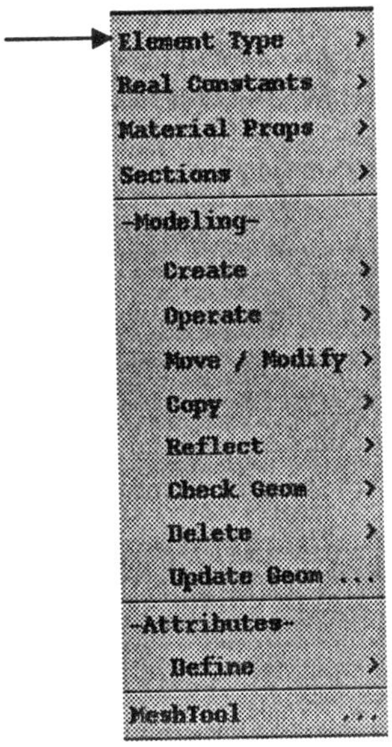

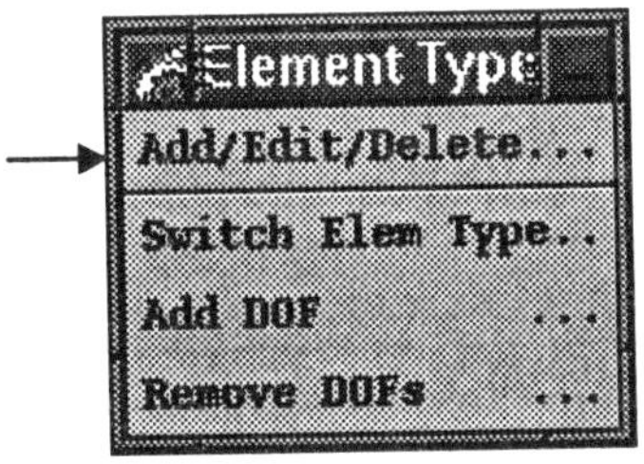

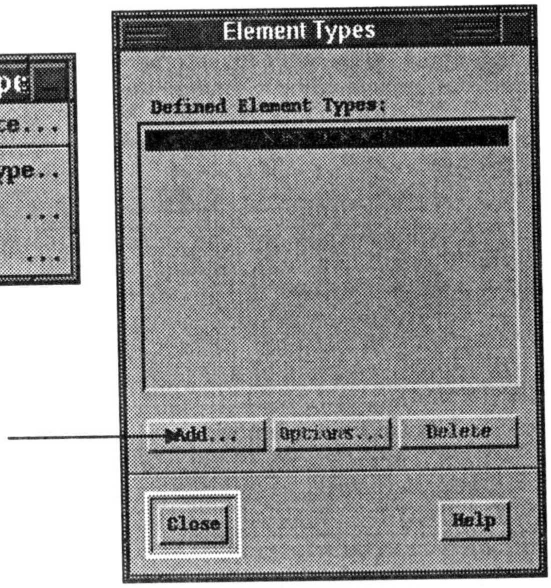

Select **Structural Shell** and choose **Elastic 4node 63**, and then click **OK**.

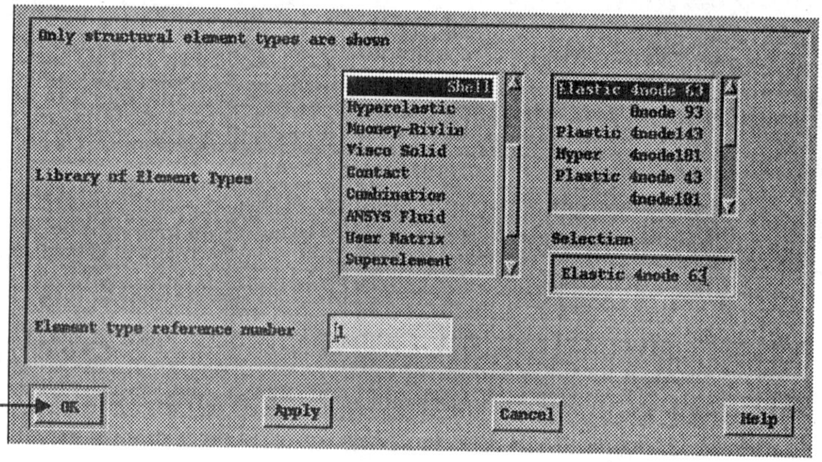

To add a value for the thickness, select **Real Constant** under the **Preprocessor** menu. Then select **Add/Edit/Delete**. Choose **Add** in the **Real Constants** window, and click **OK**.

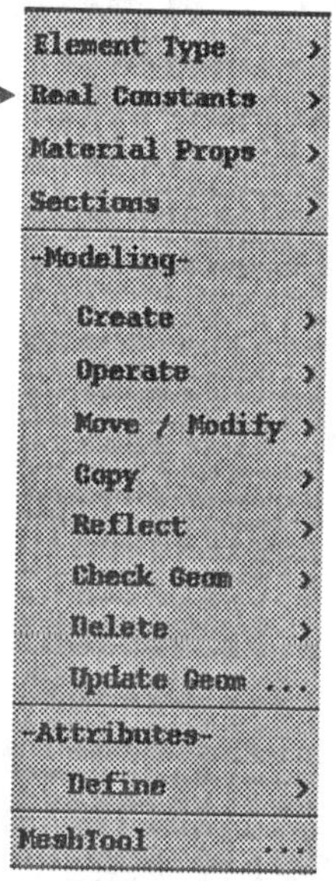

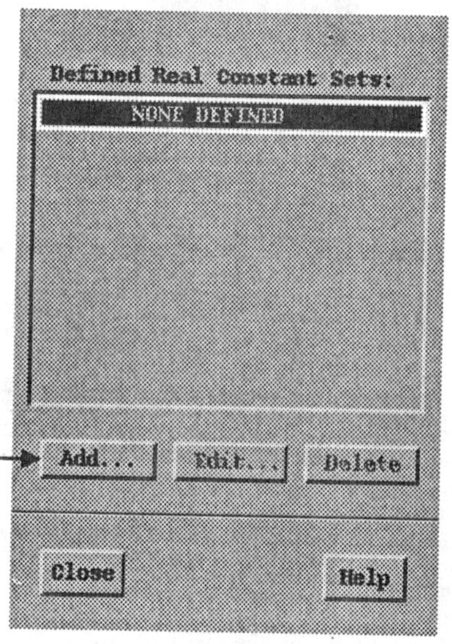

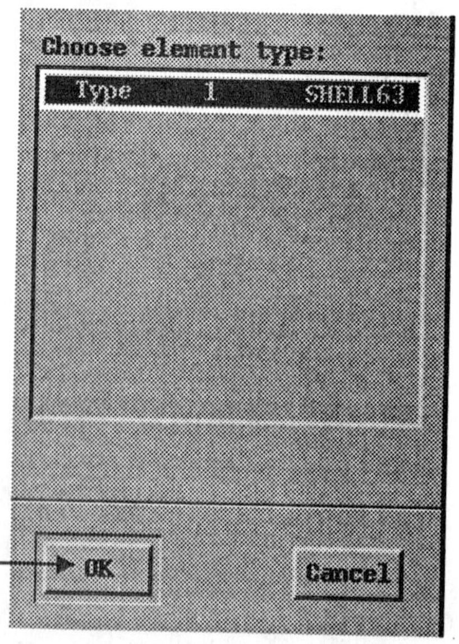

Type *0.1* as the value for the distance between the mid-surface and the top (or bottom) surface. Make sure that you specify 0.05 as the distance value > **OK**. In this way, the position of the midsurface of the shell model is centered with respect to the thickness value of 0.1.

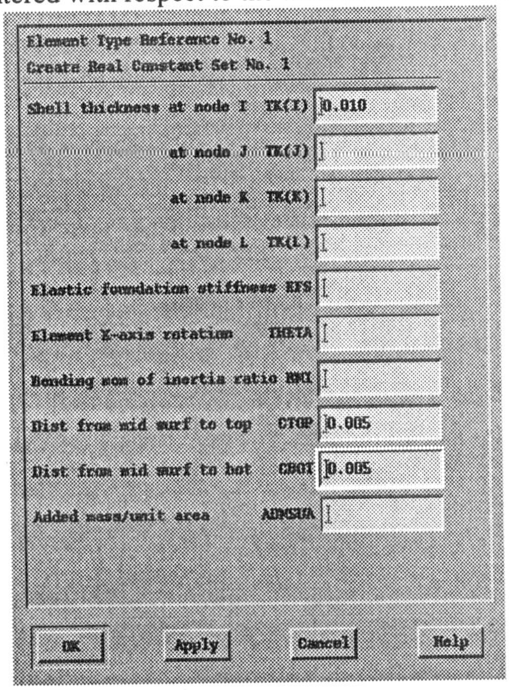

Step 2: Choosing a Material Type

To select the material for the shell model, select **Material Properties** from the Preprocessor menu. Select **Material Model**, and then select **Structural, Linear, Elastic** and **Isotropic** by double clicking on each one.

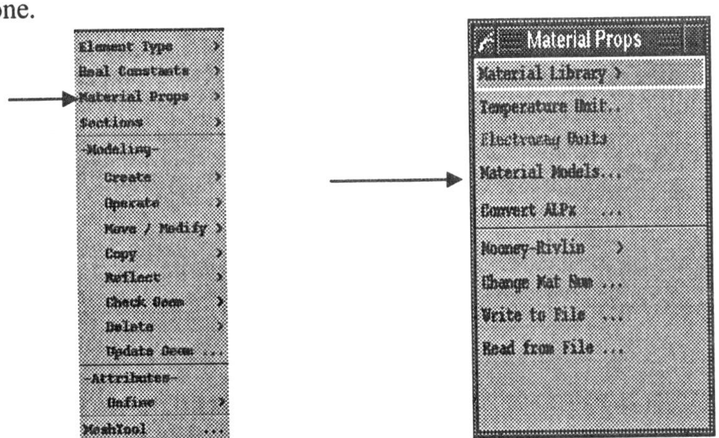

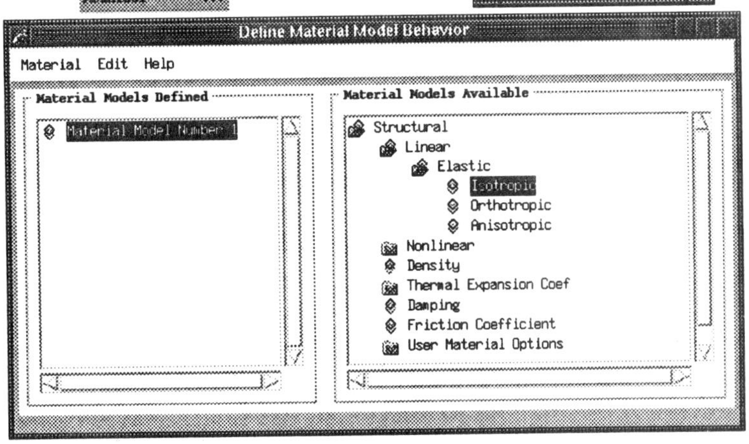

Type 3e10 and 0.26 as the values for Young's Modulus and Poisson's ratio, respectively. Then select OK, and close.

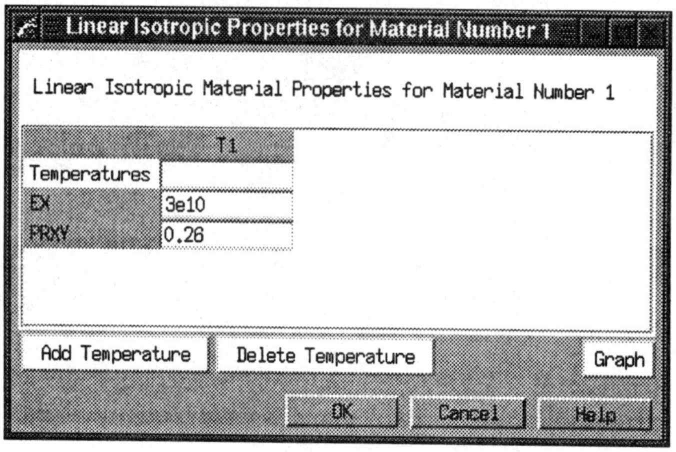

To check the unit system, select **Material Library** under the material properties menu. Then select **Units** and make sure that the **SI** unit system is highlighted. Click **OK.**

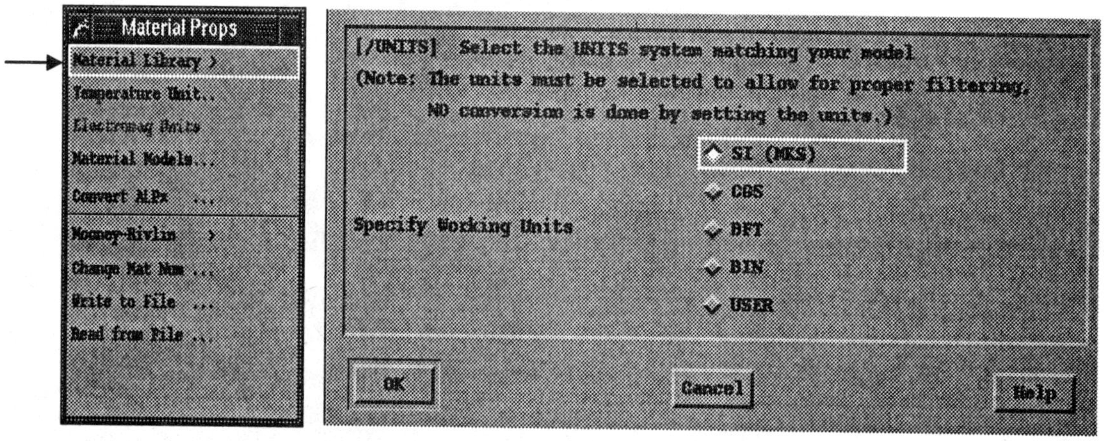

Step 3: Mesh Generation

To create the mesh for the shell model, select **MeshTool** from the **Preprocessor** menu.

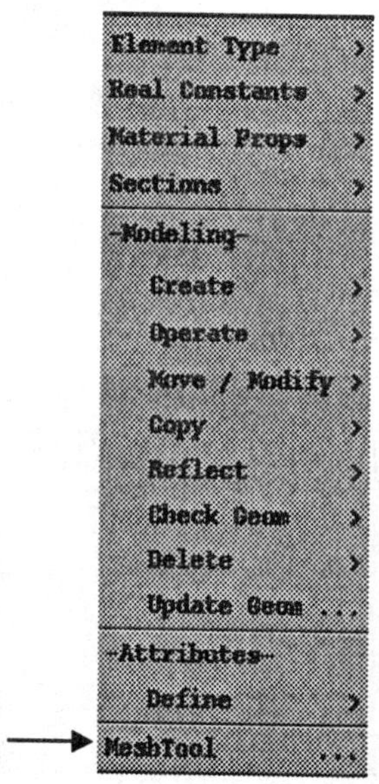

The software system generally picks the size and shape for the defined element type. For example, as shown in the following figure, the system has already picked Quad and Free > click **MESH**. Sometimes, a user may want to control the size and shape by activating Smart Size and adjust the size on his or her own.

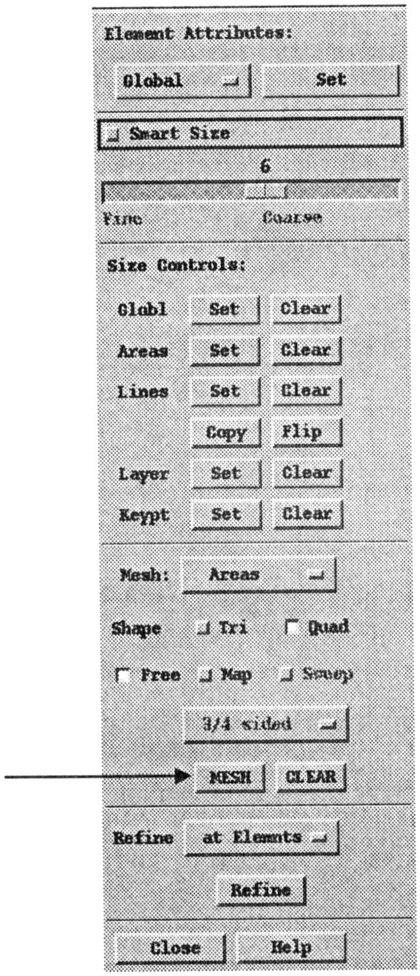

When the selection box comes up, select the area from the display window, and then click **OK**. You may just click the box of Pick All because the maximum area and minimum area are equal to 1, indicating there is only one area.

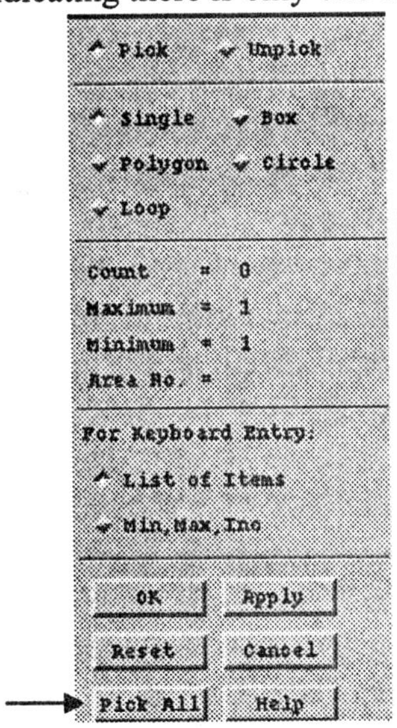

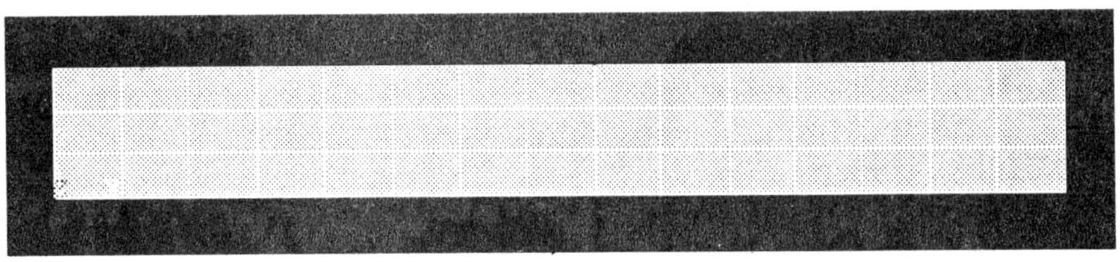

Step 4: Apply the load condition.

To apply the load, on the Preprocessor menu, click **Load**, then **Apply**, **Force/Moment, and On Nodes.** Select the node at the upper right corner, and click **OK**.

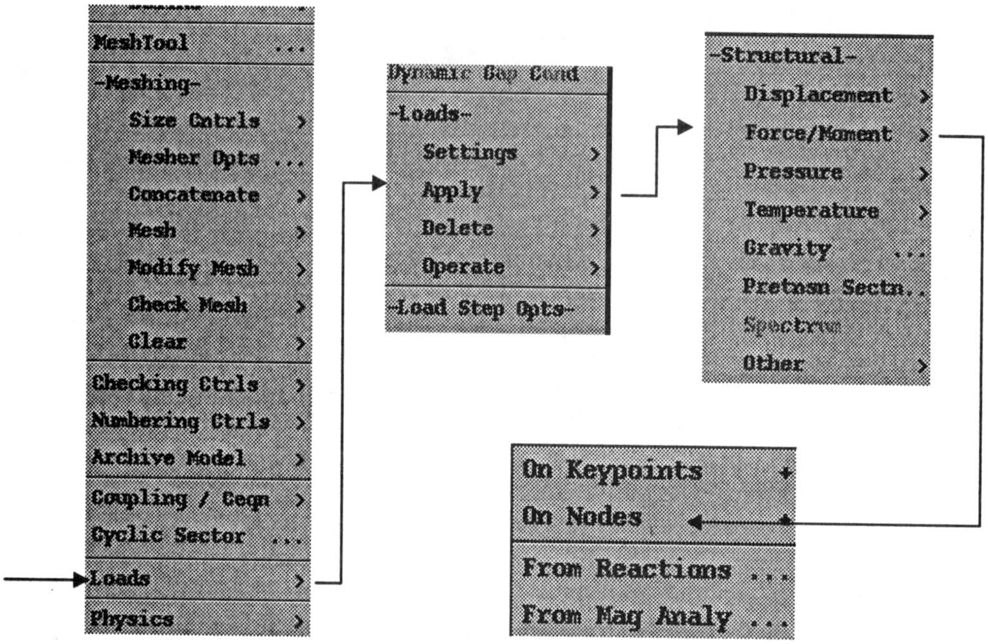

Pick the node at the upper right corner > OK. Users may notice that there are a total of 64 nodes. The picked node is node #17.

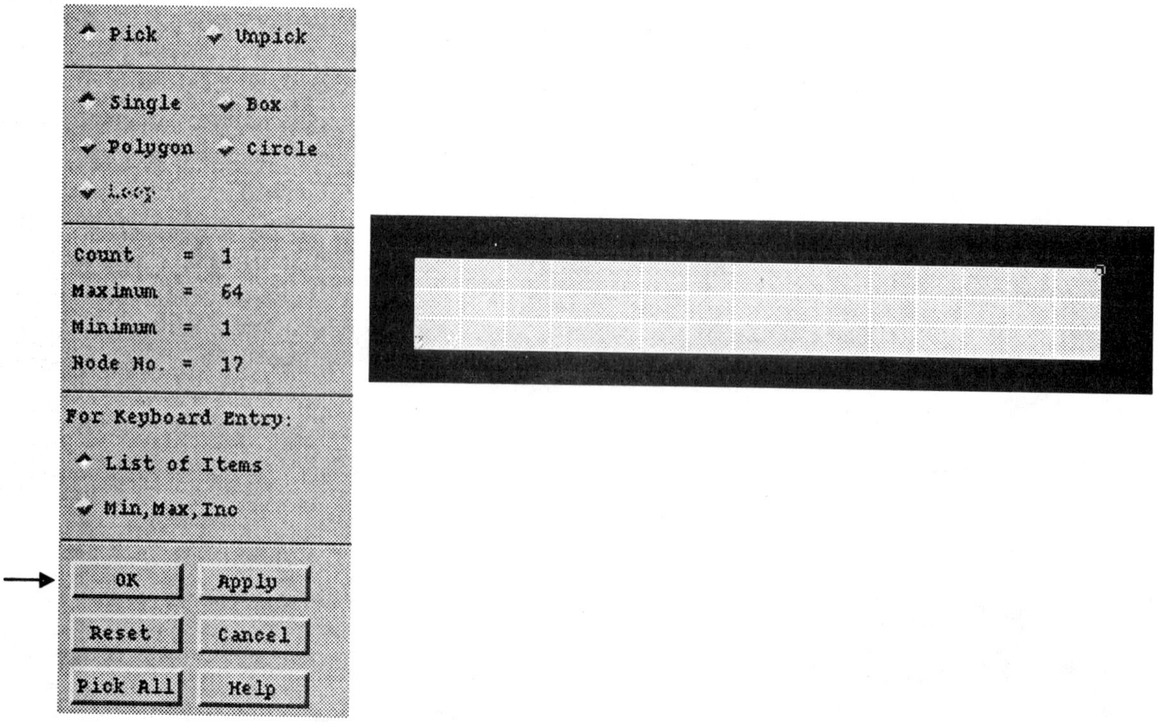

Select FY, and then type –100 as the value for the force.

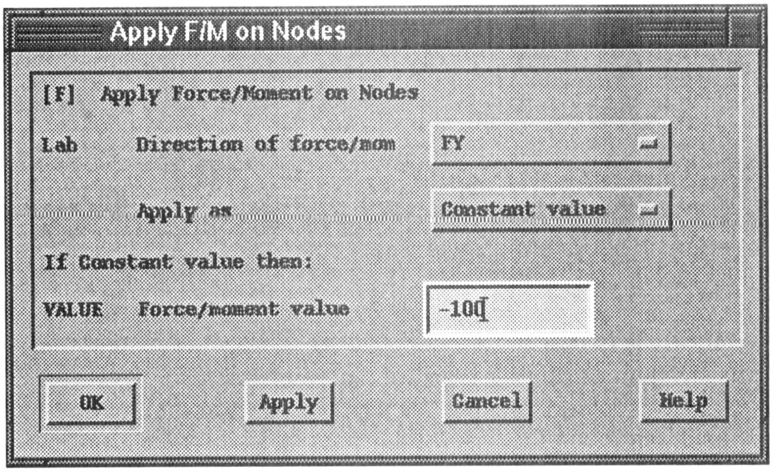

Step 5: Defining the Constraints

To define the constraint conditions, under the Load menu, select **Apply**, **Displacement**, and **On Line**. Pick the left side of the shell model and click **OK**.

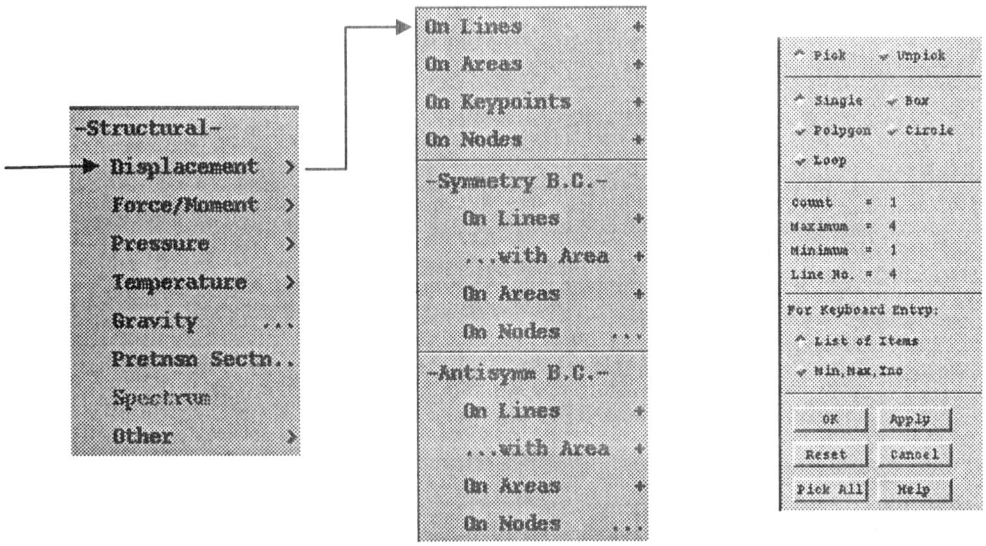

Select **All DOFS**, and type 0 as the value. Then click OK.

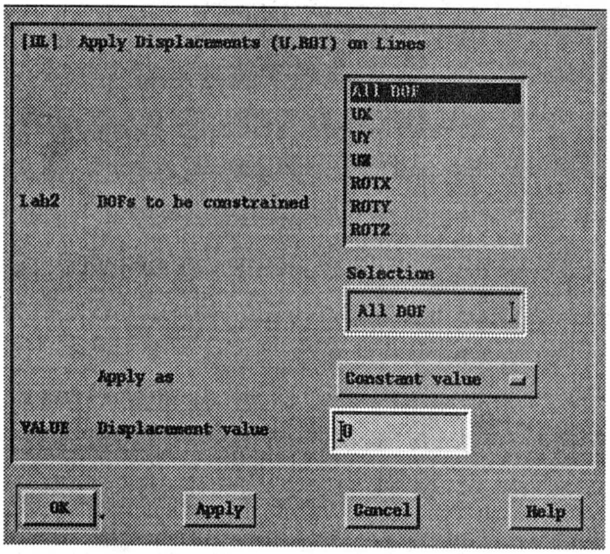

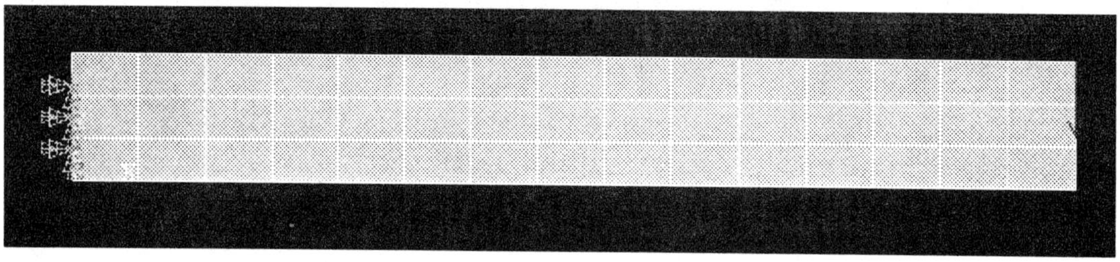

Step 6: Solution

To find the solution, run the program by selecting **Solution** from main ANSYS menu. Choose **Current LS**, and then click **OK**.

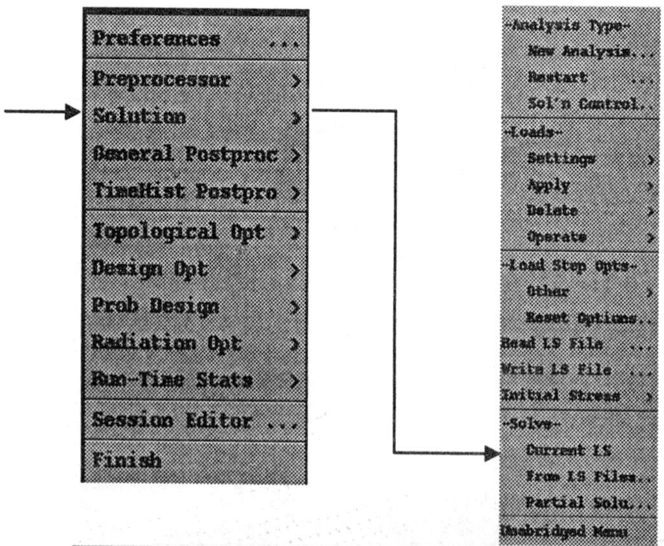

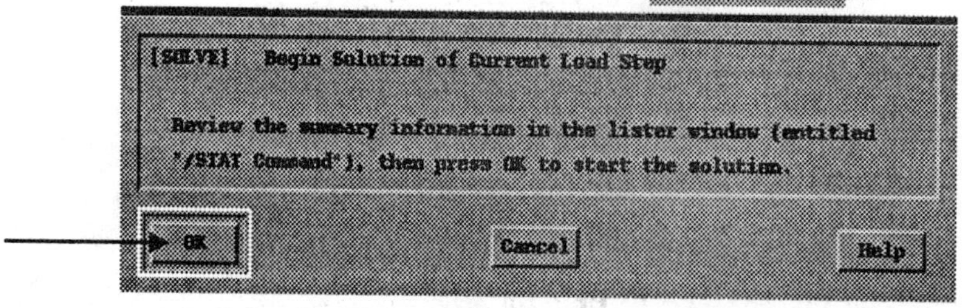

A window will display when the solution is done > **Close**.

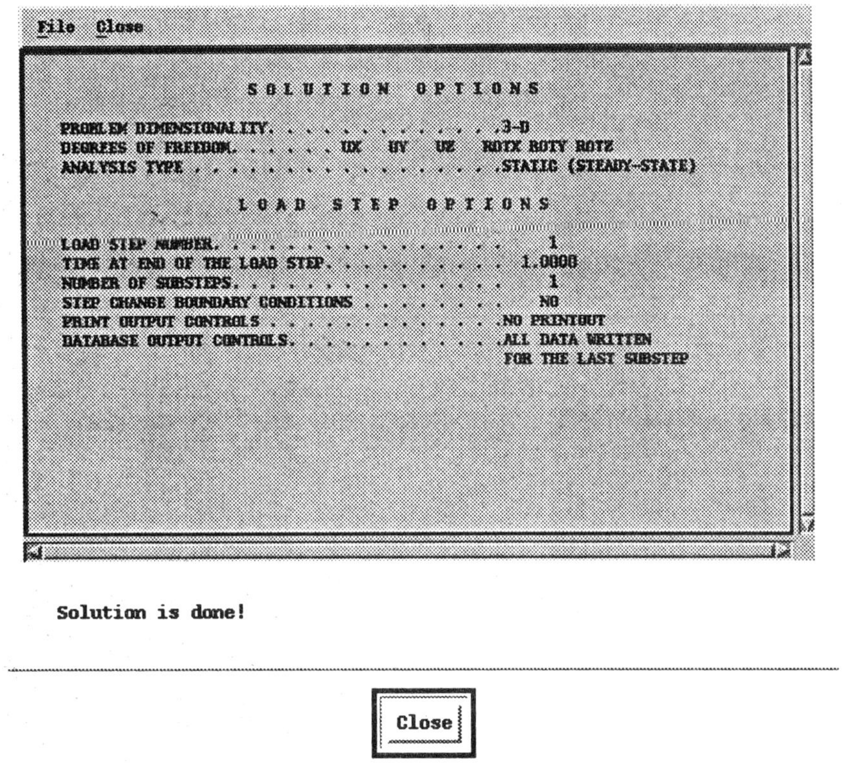

Step 6: Displaying Results
To review the results, select **General Postproc** from the main menu. Then click **Plot Results**, and **Nodal Solution.** .

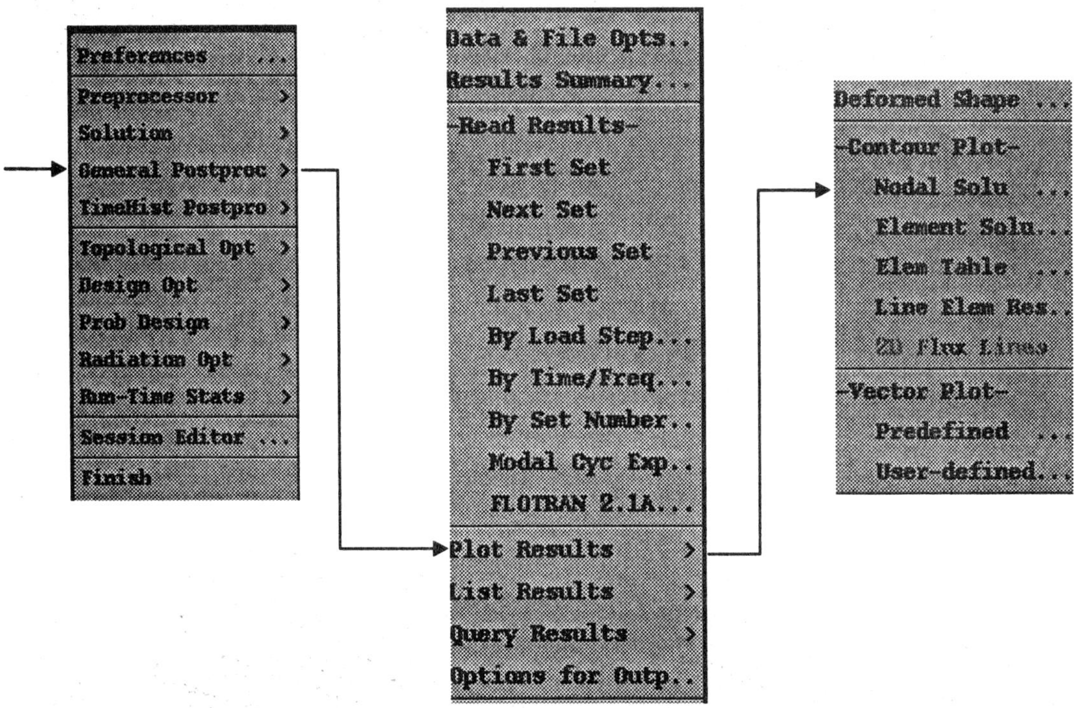

Select **Stress** from the first list, and **1st Principal S1** from the second list. Then click **OK**.

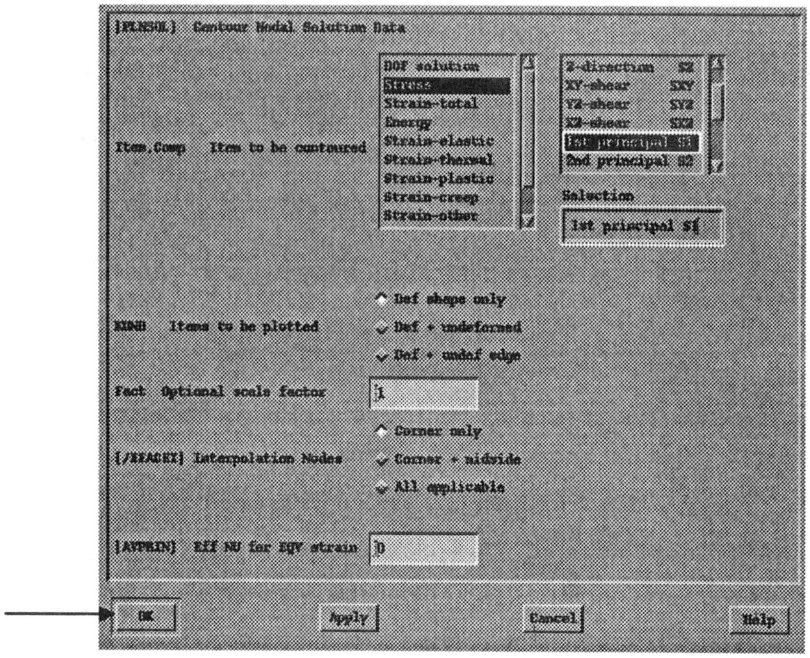

The following will be the result.

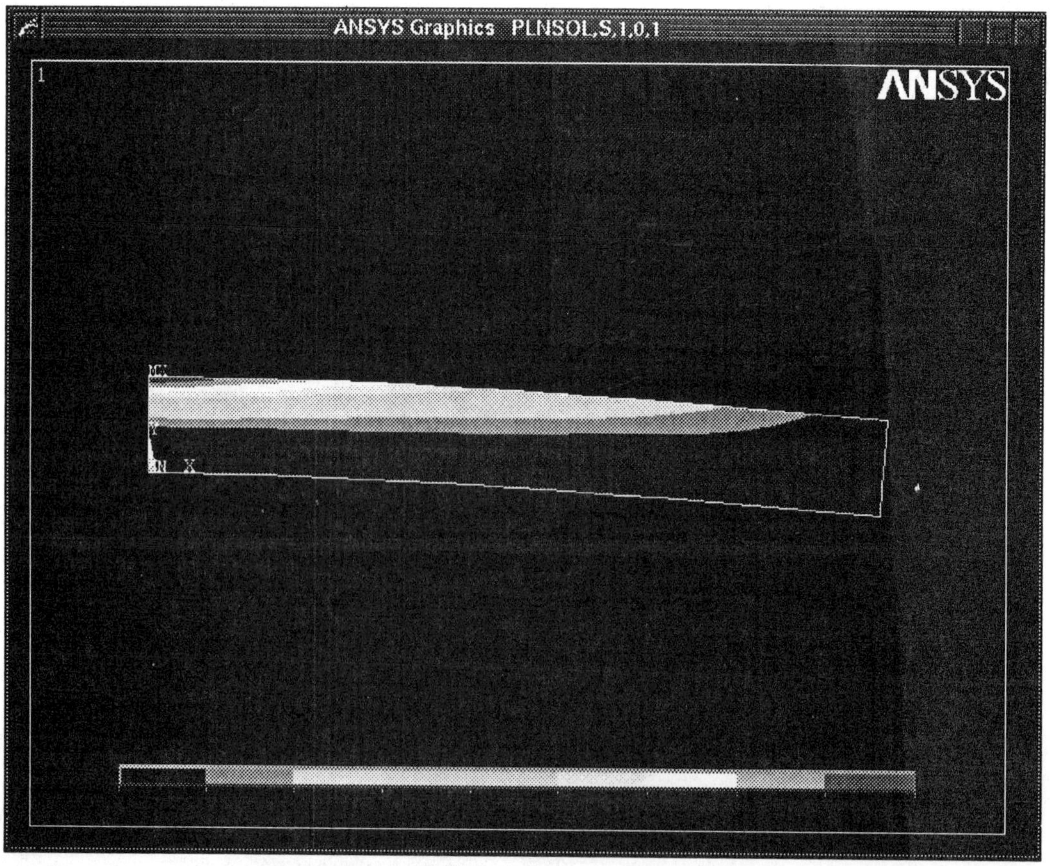

If desired, the results can also be animated. To do this, select **Plot Ctrls** from the upper menu, and click **Animation.** Then select **Deformed Shape** and accept the default settings > **OK**.

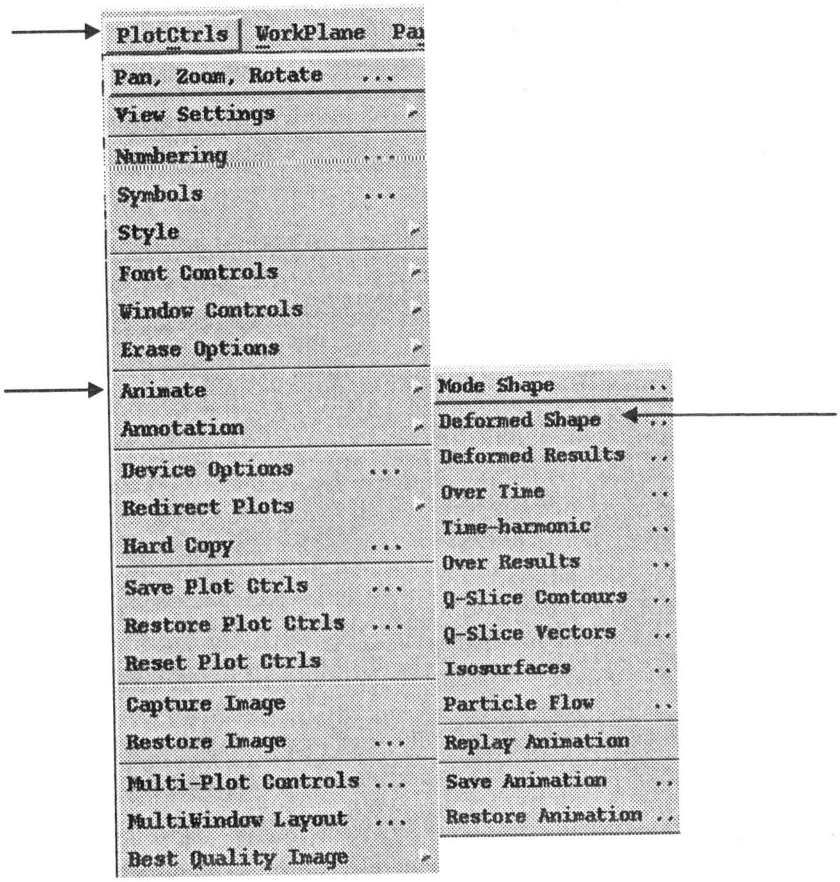

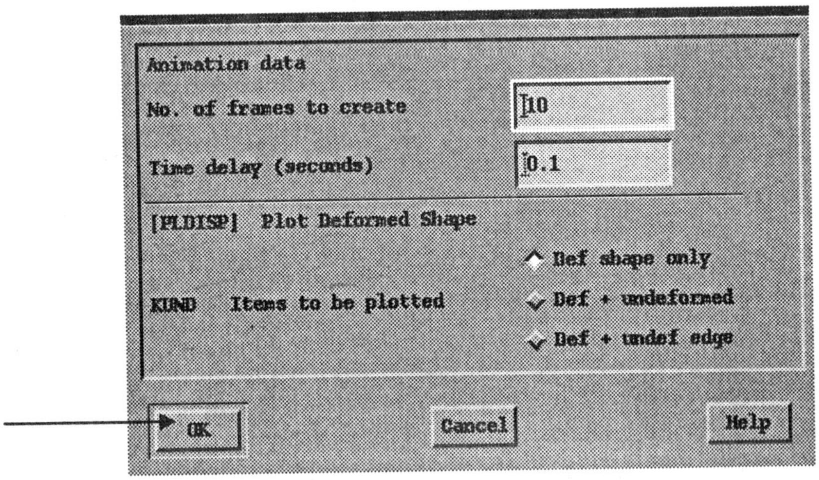

The animation picture from a screen capture is shown on the next page.

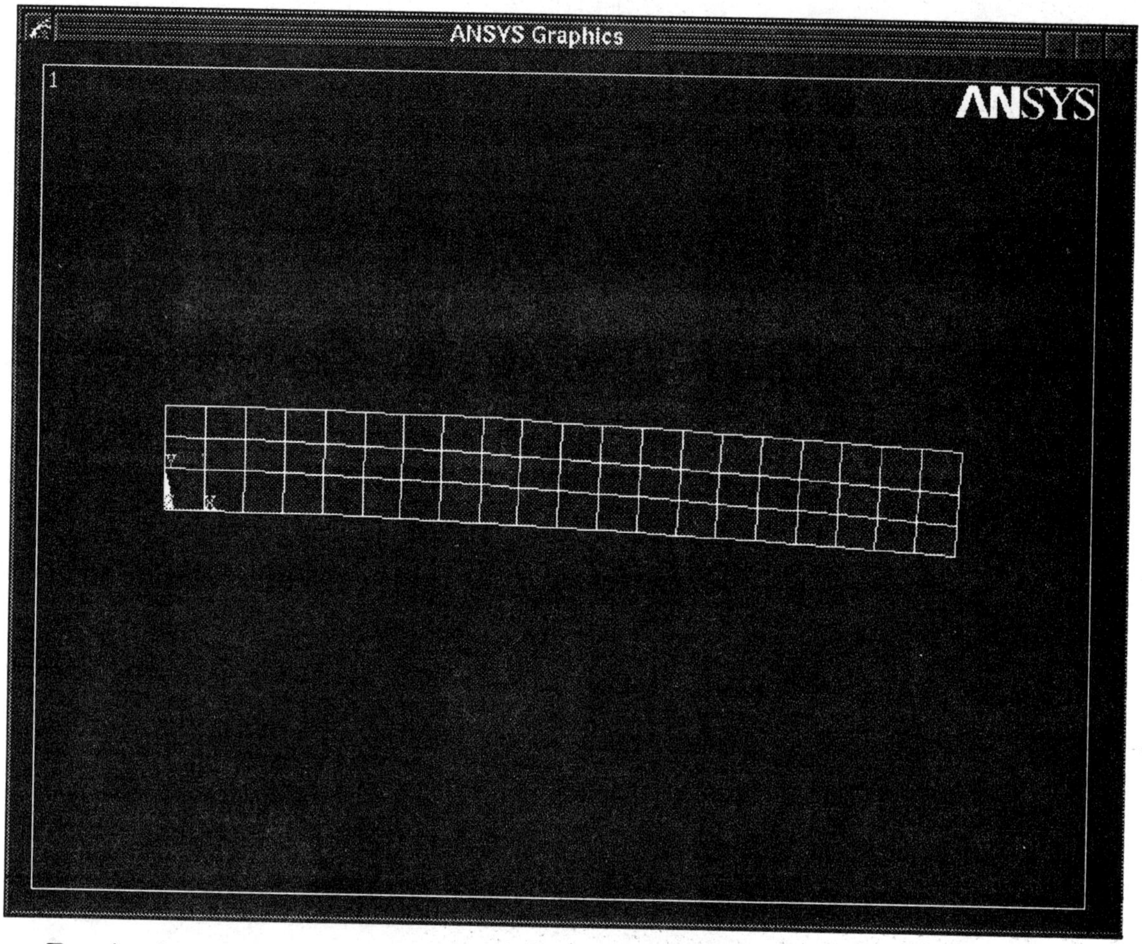

To print the results to an image file, select **PlotCtrls** from the top menu. Select **Hard Copies,** and then choose **Graphics Window**. Then save the file in a .jpg format.

Also, the image can be saved by pressing the “print screen” button on the keyboard and pasting the image into an image-editing program, such as MS Paint.

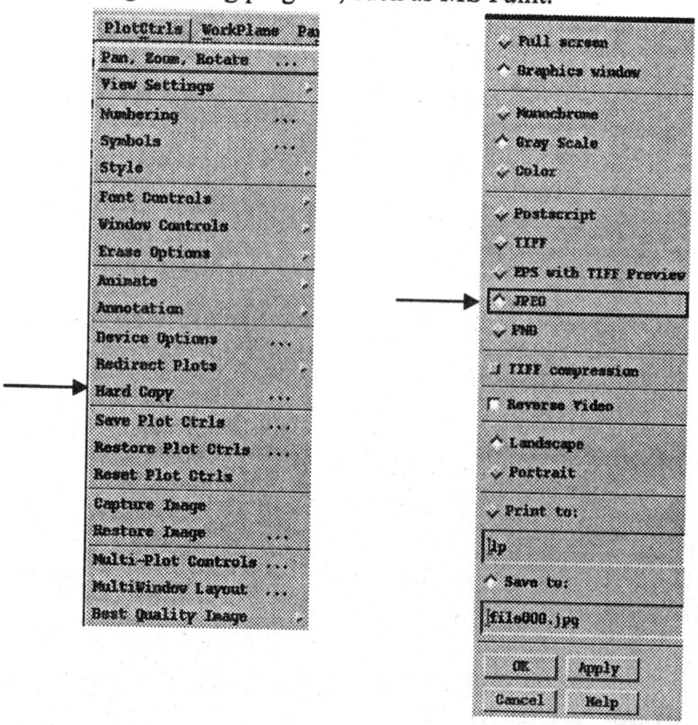

10.3 FEA with a 3D Solid Model

The following figure illustrates a beam structure. The 3 dimensions of the beam structure are 800 x 100 x 50 mm. The left end of the plate is fixed to the wall. The load acts on part of the top surface, as shown. The right end is set free to move. In terms of the material properties, we assume that the Young's modulus is 30 GPa, and the Poisson's ratio is 0.26. The objective of this study is to obtain the information on the maximum principal stress developed under the load condition.

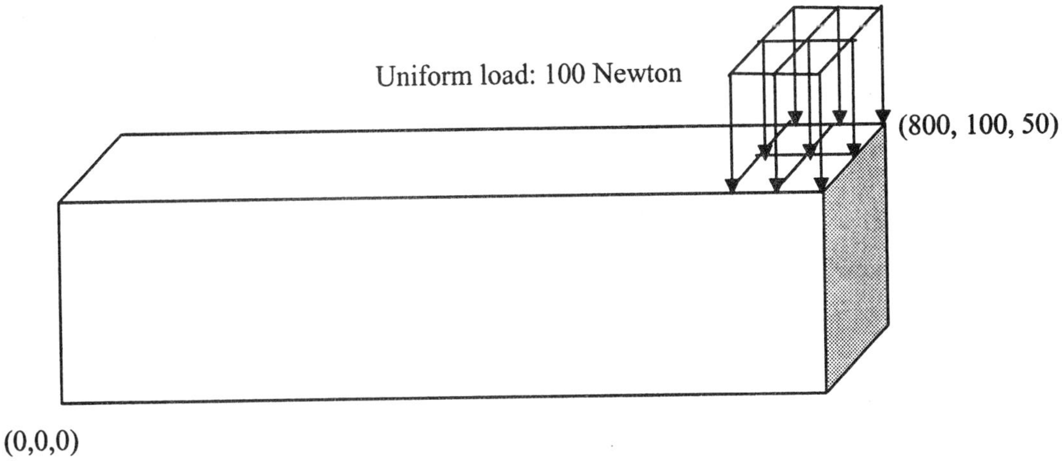

The following flow chart illustrates the general approach of using ANSYS to perform FEA. As discussed in the previous example, we need to define the type of element after creating the 3D solid model for the beam structure.

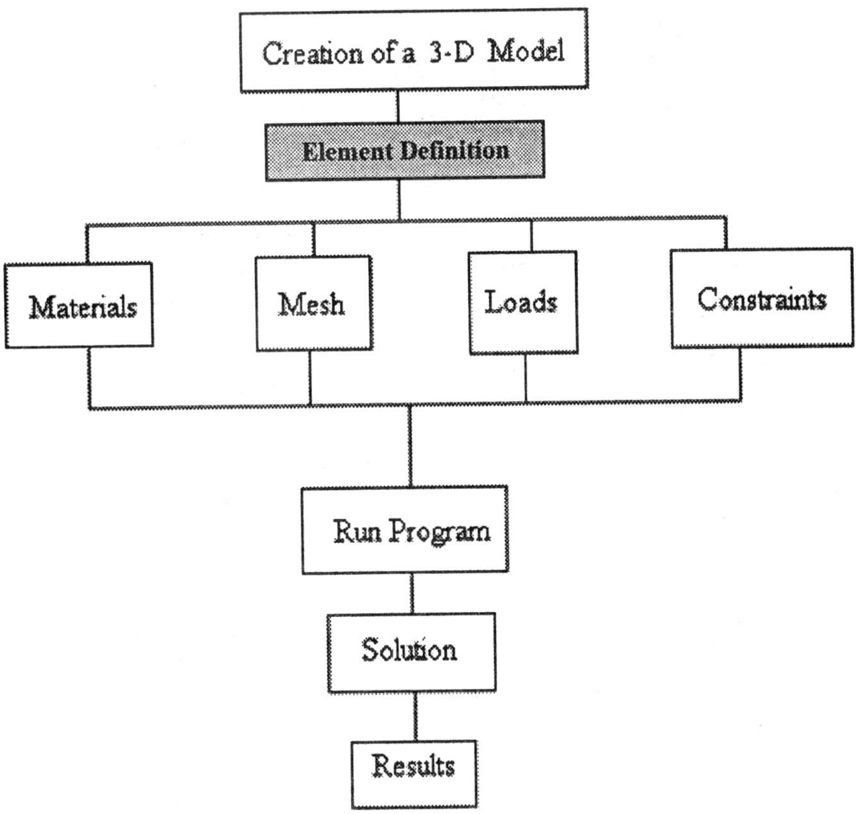

Step 1: Create a 3-D model for analysis.

From the **ANSYS** main menu, select **Preferences**. Make sure that **Structural** is highlighted, and then click **OK**.

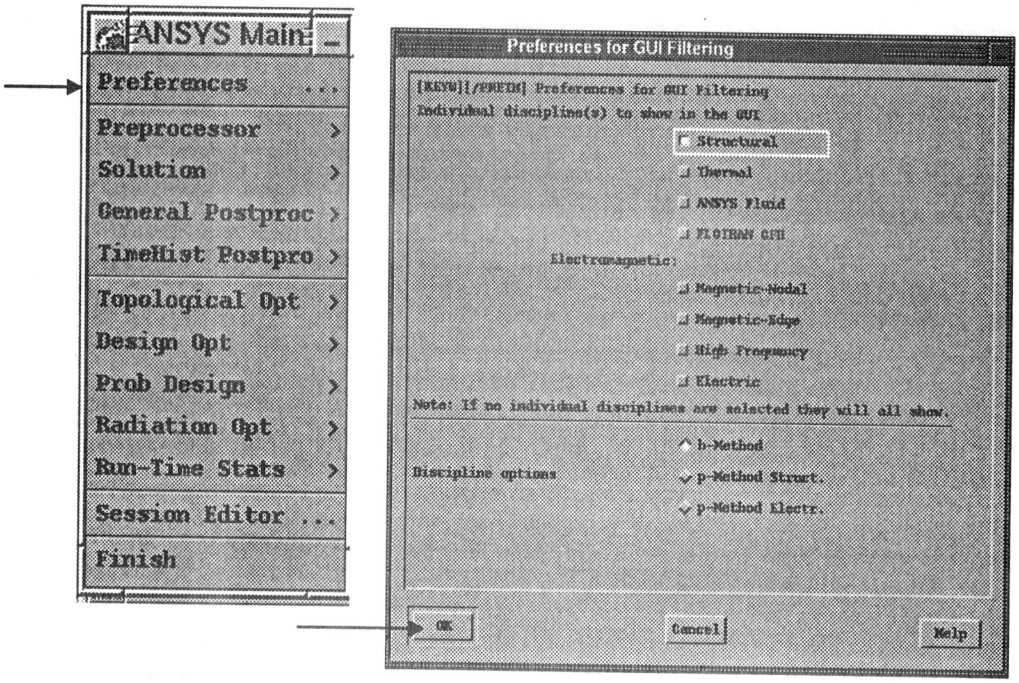

From the main menu, select **Preprocessor**. Then, under modeling, select **Create, Block** under Volumes, and **By Dimensions.** In the window for creating the block by dimensions, type 0 and 0.8 for X1 and X2, 0 and 0.1 for Y1 and Y2, and 0 and 0.05 for Z1 and Z2 > **OK**.

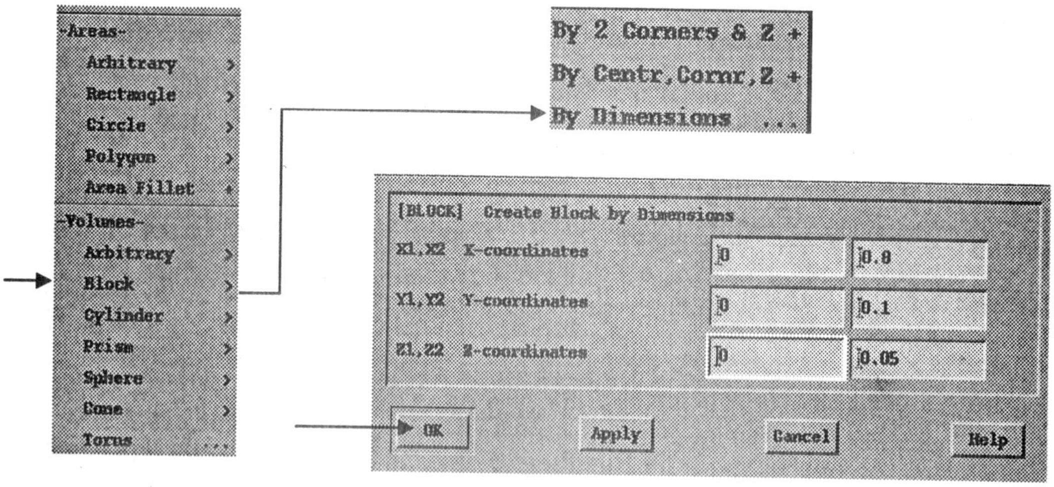

In order to have an isometric view of the created 3D solid model, select PlotCtrls from the main menu > **Pan, Zoom, Rotate** > **Iso** > **Close**.

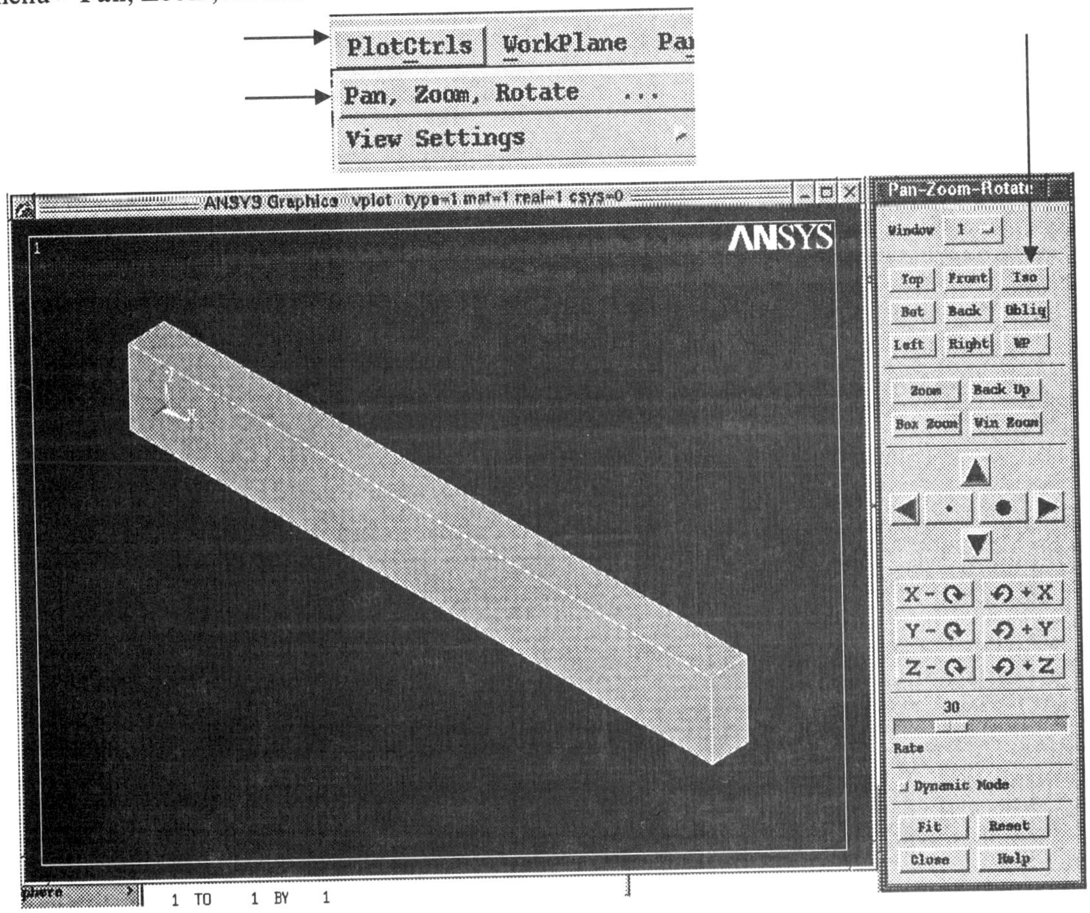

Under the Preprocessor menu, select **Element Type**. Then select **Add/Edit/Delete**, and then **Add**.

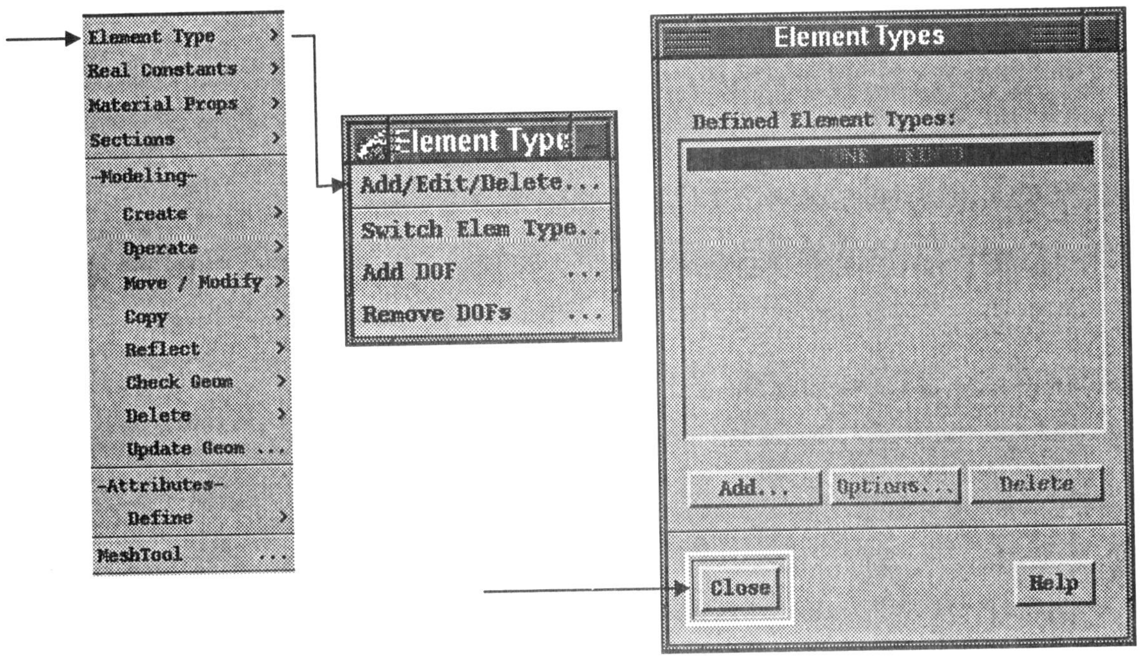

Select **Structural Solid** and choose Brick 8 node 45, and then click OK.

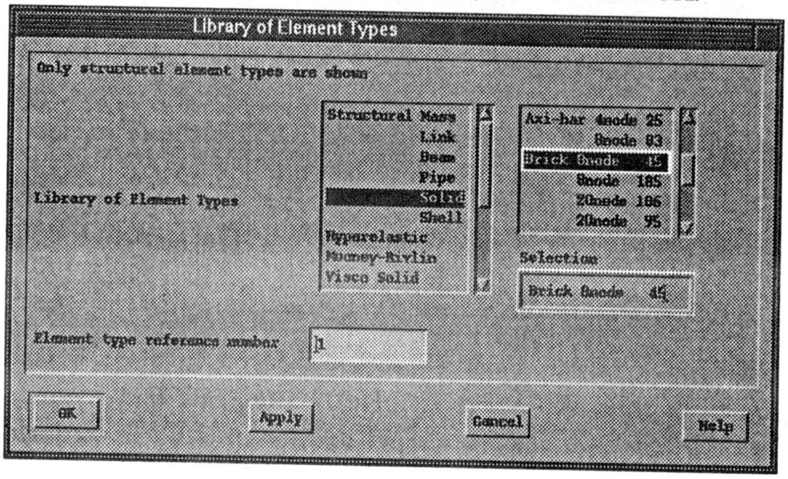

Step 2: Choosing a Material Type

To select the material for the 3D model, select **Material Properties** from the Preprocessor menu. Select **Material Model**, and then select **Structural, Linear, Elastic** and **Isotropic** by double clicking on each one.

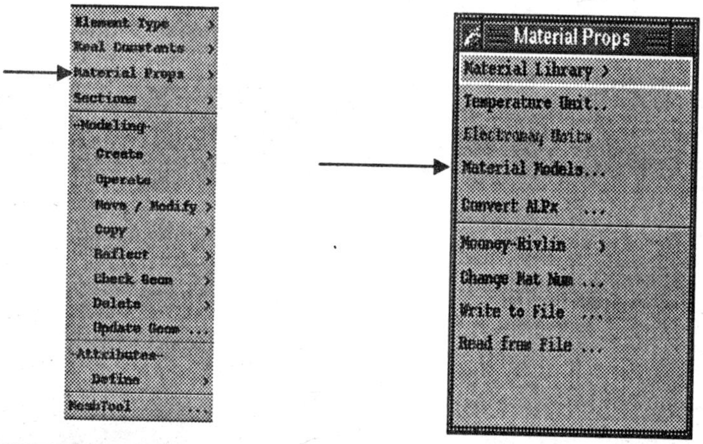

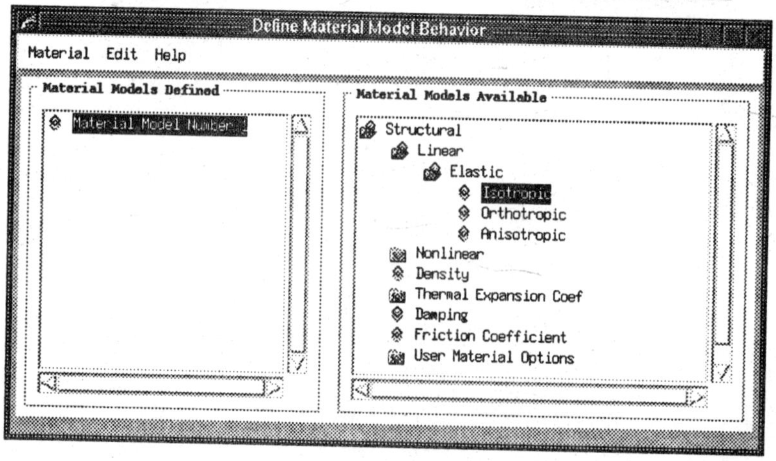

Type 3e10 and 0.26 as the values for Young's Modulus and Poisson's ratio, respectively. Then select OK, and Close.

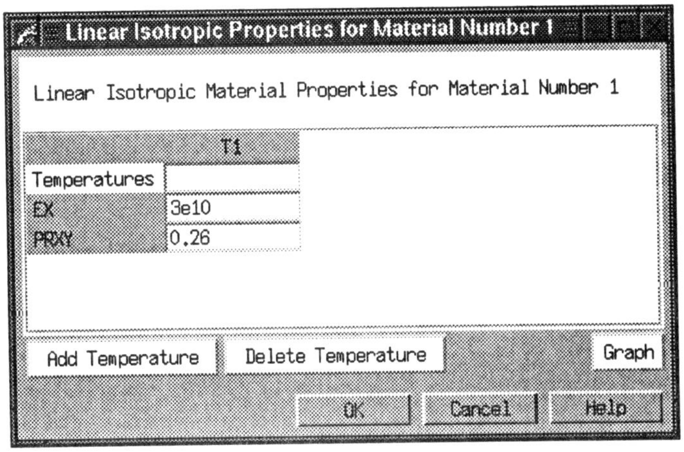

To check the unit system, select **Material Library** under the material properties menu. Then select **Units** and make sure that the **SI** unit system is highlighted. Click **OK**.

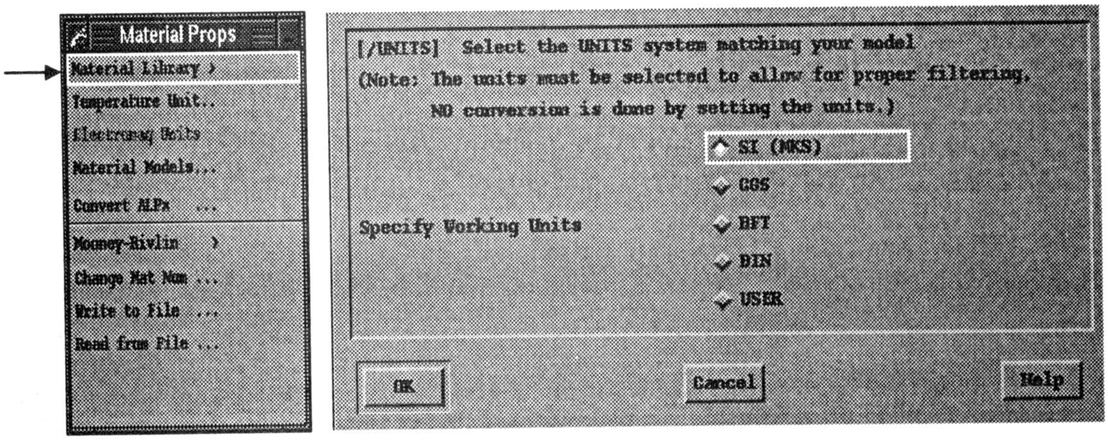

Step 3: Mesh Generation

To create the mesh for the shell model, select **MeshTool** from the **Preprocessor** menu.

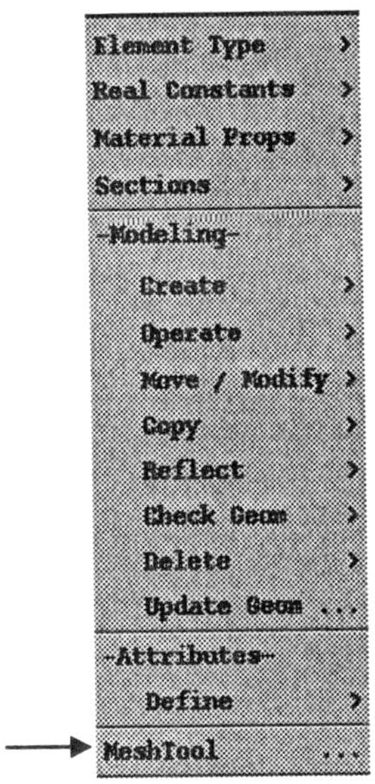

Step 3: Mesh Generation

To create the mesh for the shell model, select **MeshTool** from the Preprocessor menu. Make sure that **Hex** and **Map** are selected. Then click **Mesh**.

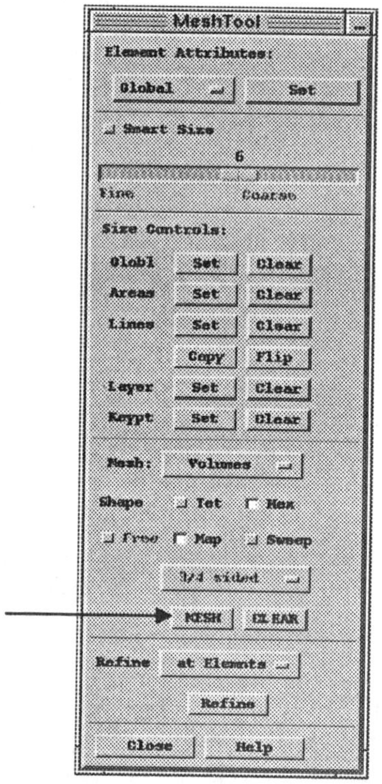

When the selection box comes up, select the volume from the display window, and then click **OK**. You may just click the box of Pick All because the maximum volume and minimum volume are equal to 1, indicating there is only one volume.

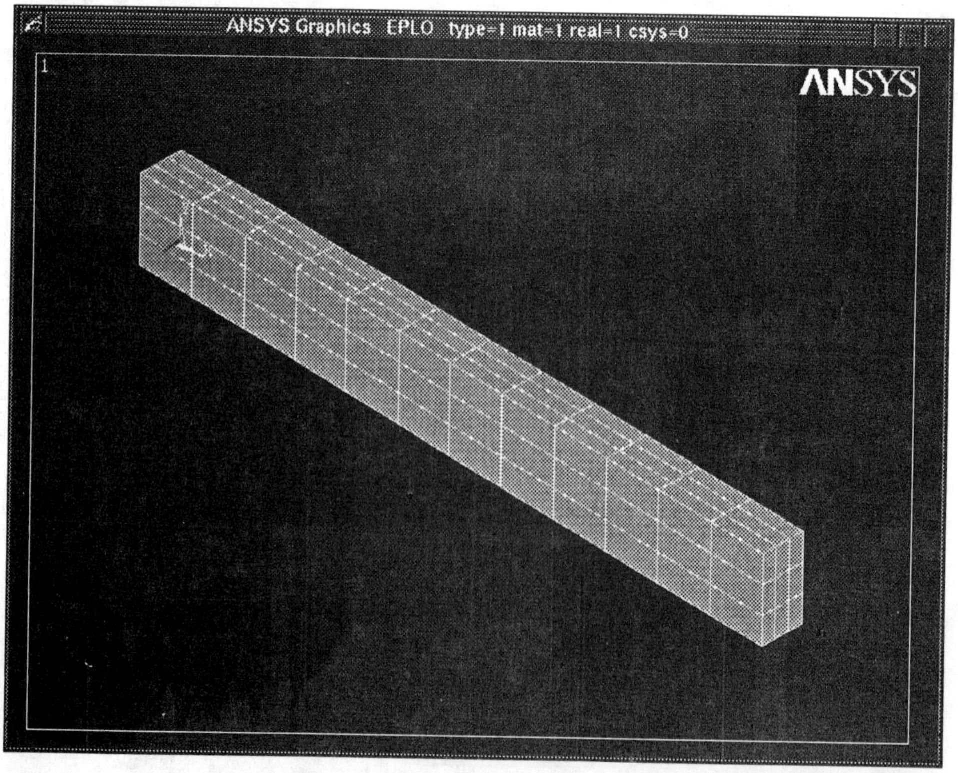

Step 4: Apply the load condition.

To apply the load, on the Preprocessor menu, click **Load**, then **Apply, Force/Moment, and On Nodes**. Select the nodes at the upper right corner, and click **OK**.

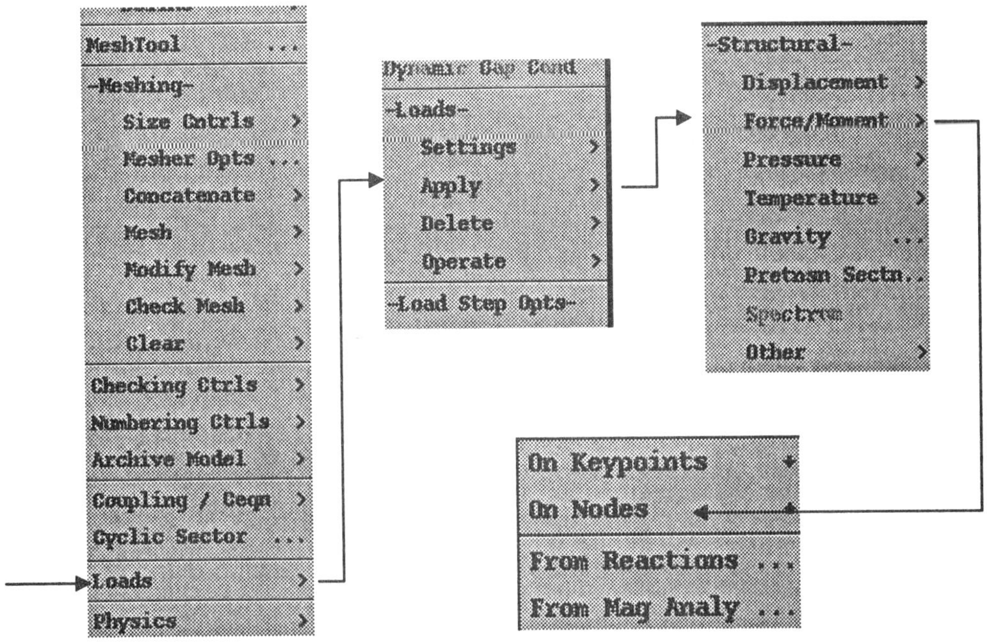

Pick the 12 nodes at the location near the upper right corner > **OK**. Users may notice that there are a total of 208 nodes.

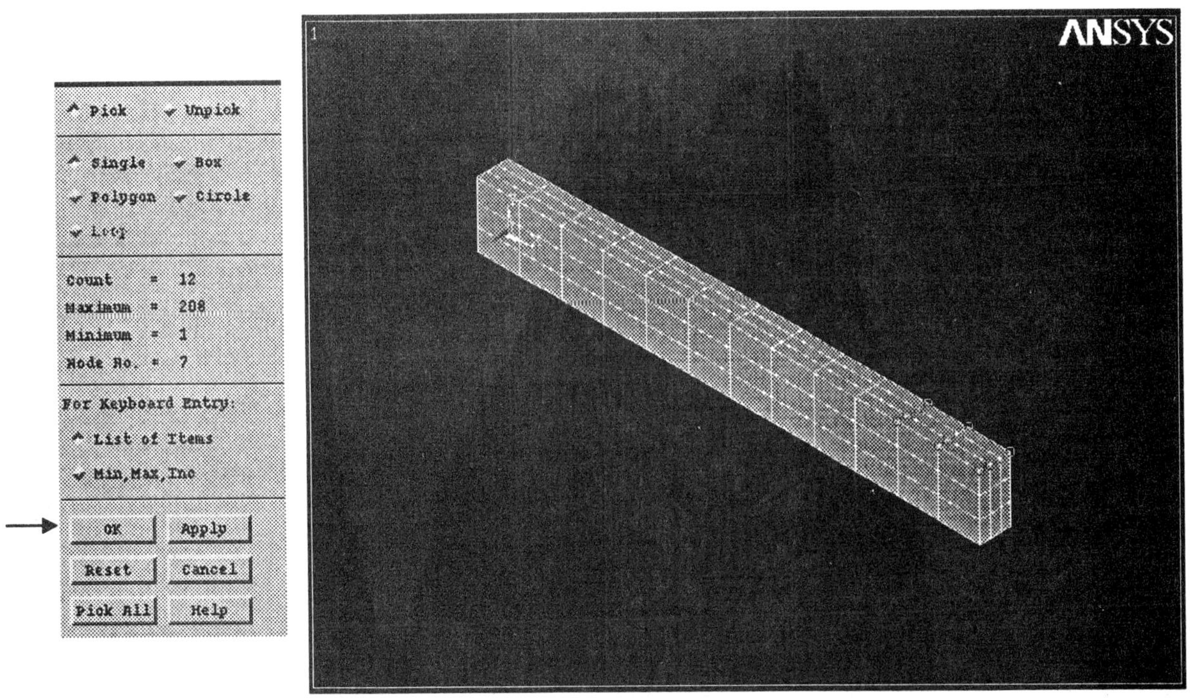

Select FY, and then type –100 as the value for the force.

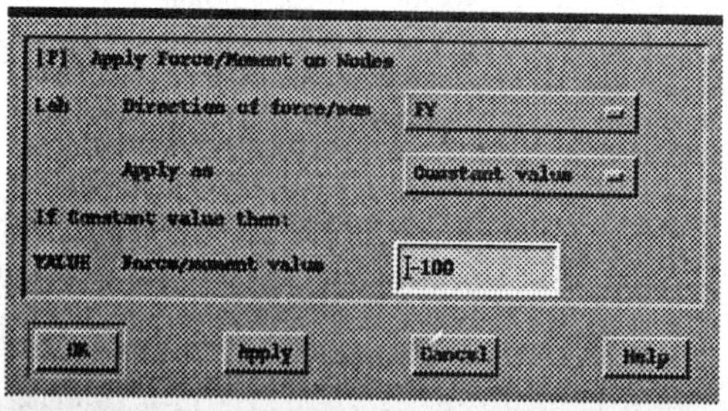

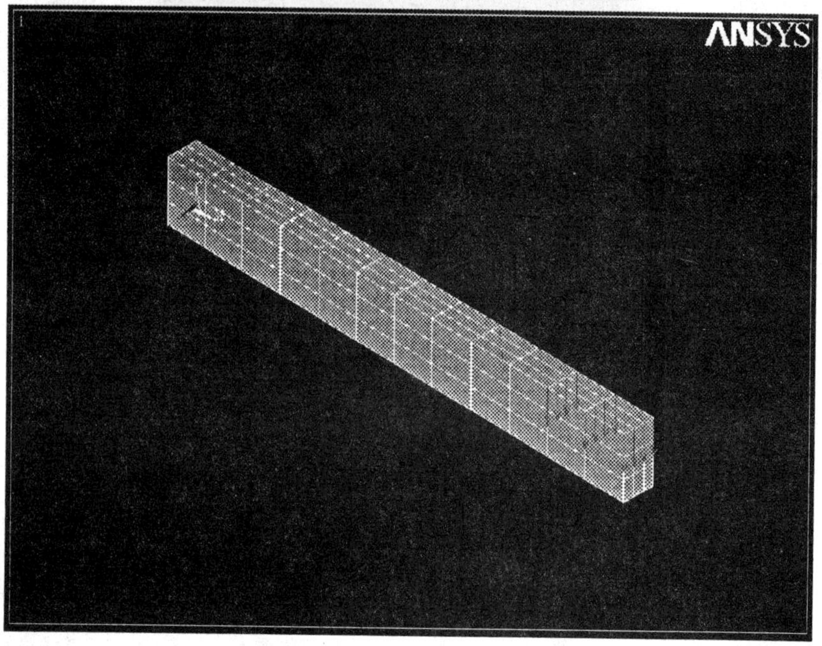

Step 5: Defining the Constraints

To define the constraint conditions, under the Load menu, select **Apply**, **Displacement**, and **On Nodes**. Pick all of the nodes on the left surface of the model and click **OK**. Note that there are 208 nodes. We have picked 16 nodes from them.

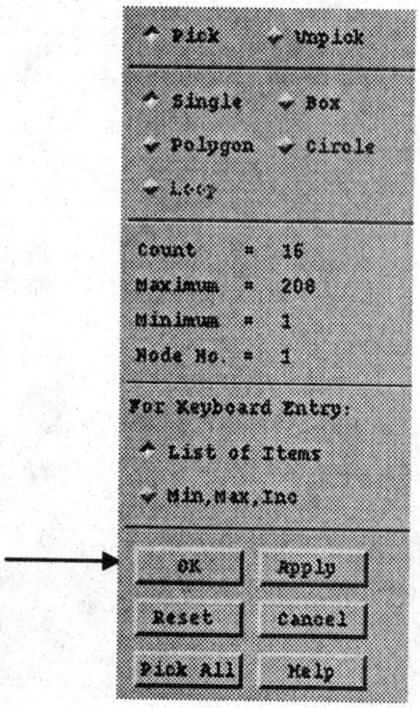

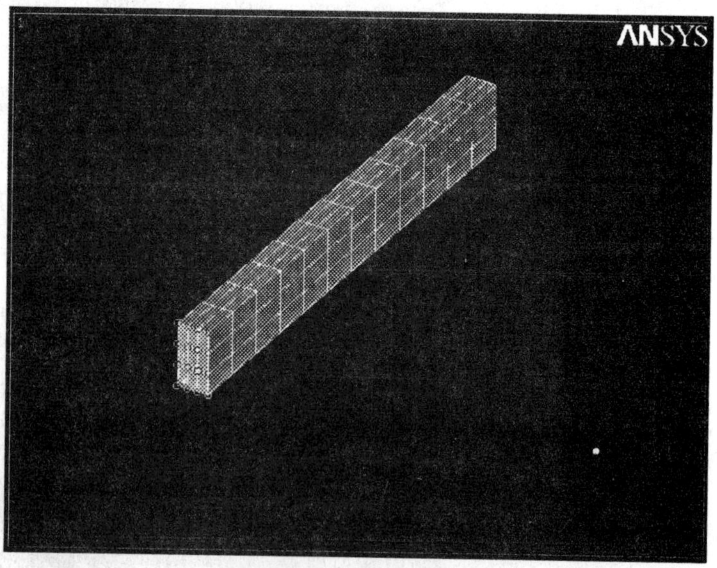

It seems that picking those 16 nodes one by one is not efficient. Let us use the Box function. Set the 3D model to the Front view. Sketch a box at the left side of the rectangle and we are able to pick all 16 nodes by a single sketch.

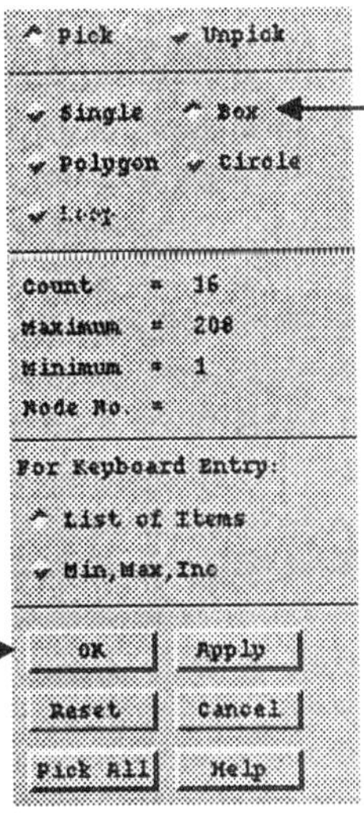

Select **All DOFS**, and type 0 as the value. Then click **OK**.

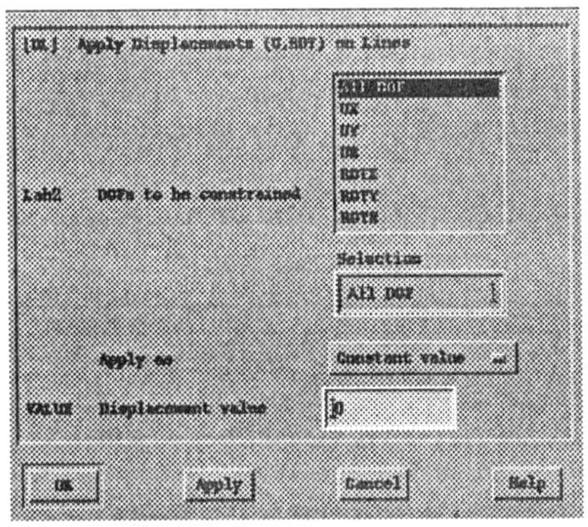

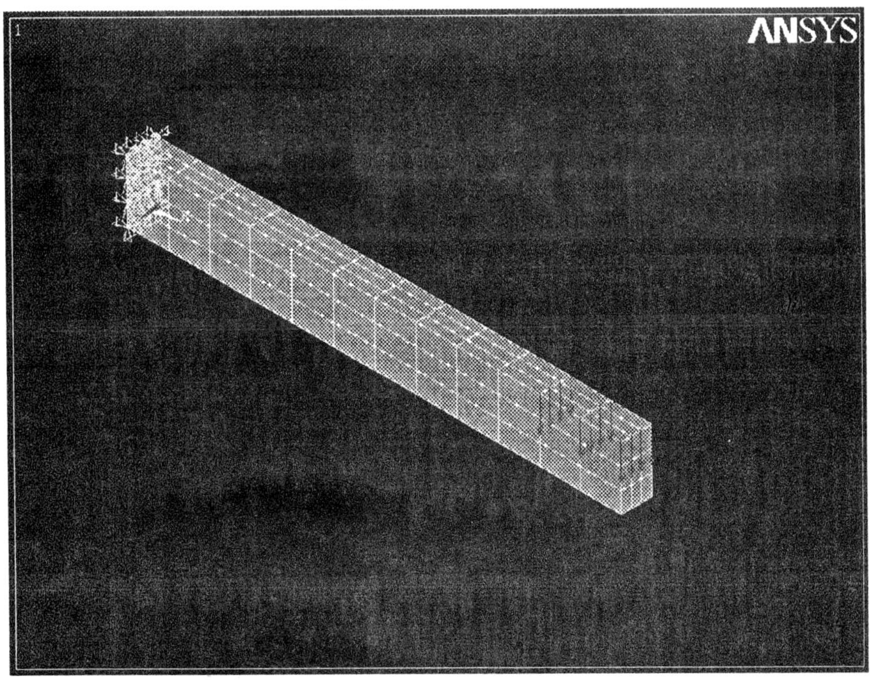

Step 6: Solution

To find the solution, run the program by selecting **Solution** from main ANSYS menu. Choose **Current LS**, and then click **OK**.

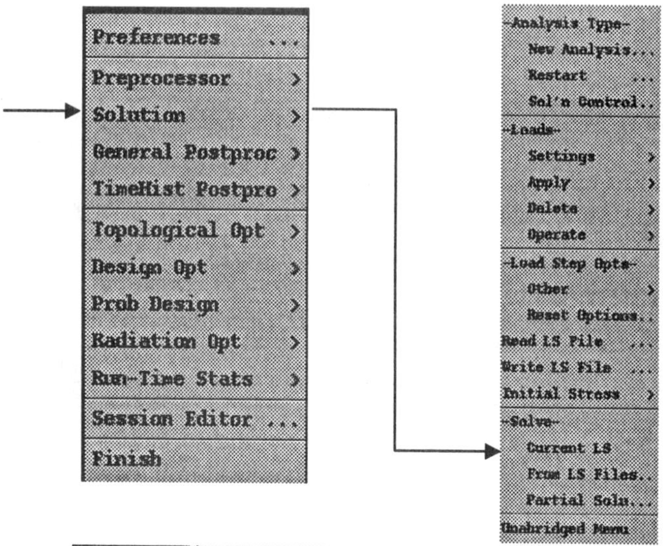

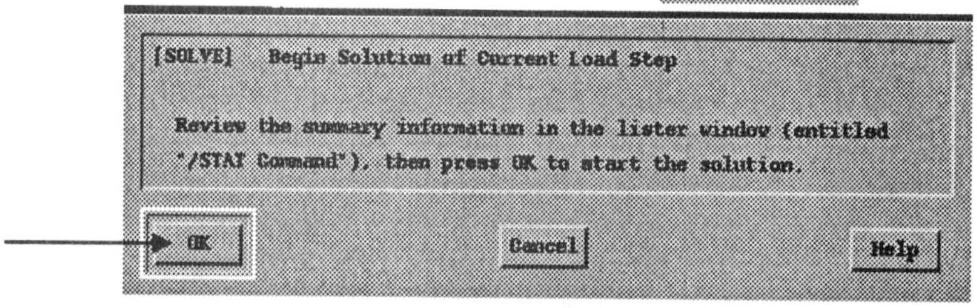

A window will display when the solution is done > **Close**.

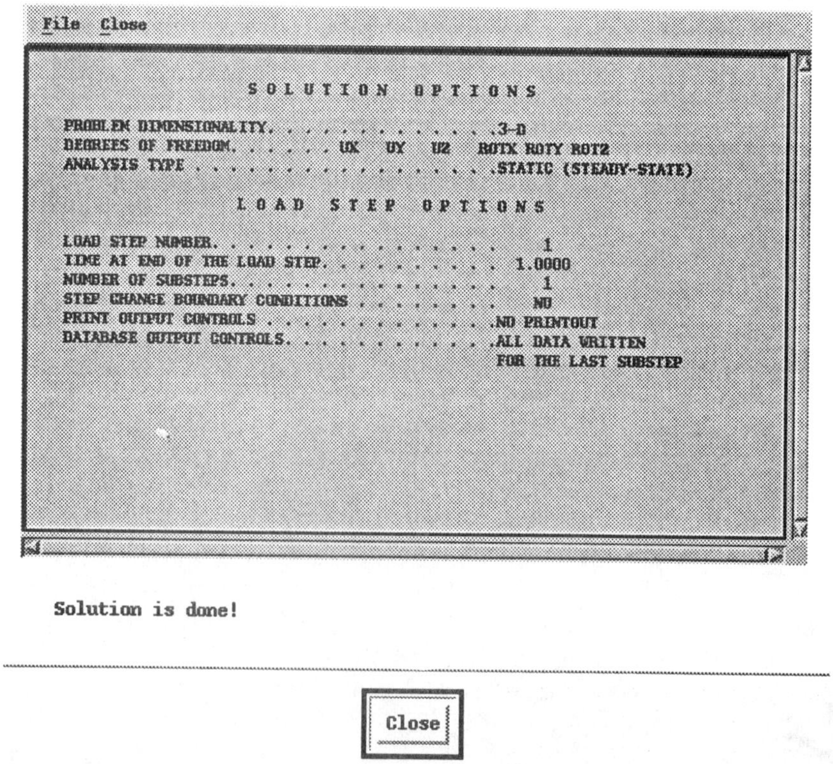

Step 6: Display Results

To review the results, select **General Postproc** from the main menu. Then click **Plot Results,** and **Nodal Solution.**

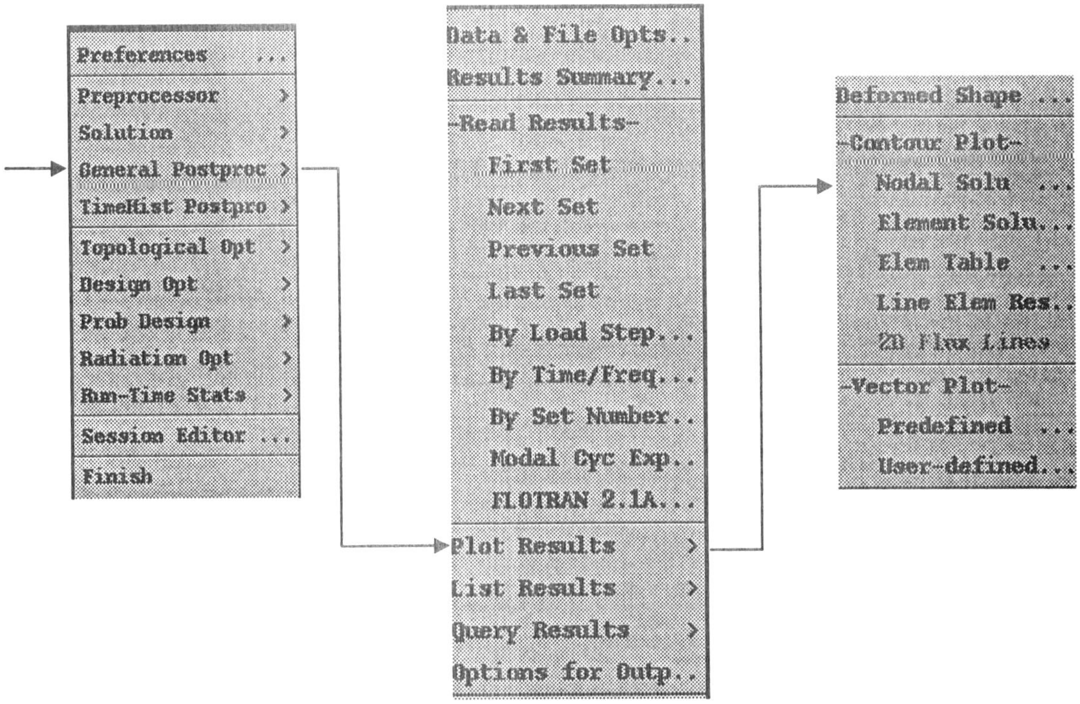

Select **Stress** from the first list, and 1st **Principal S1** from the second list. Then click **OK.**

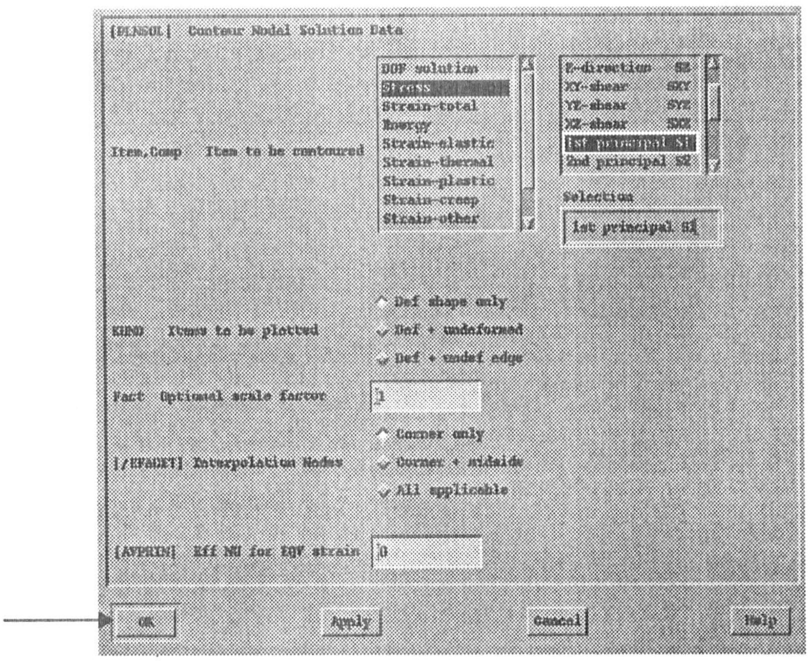

The following will be the result.

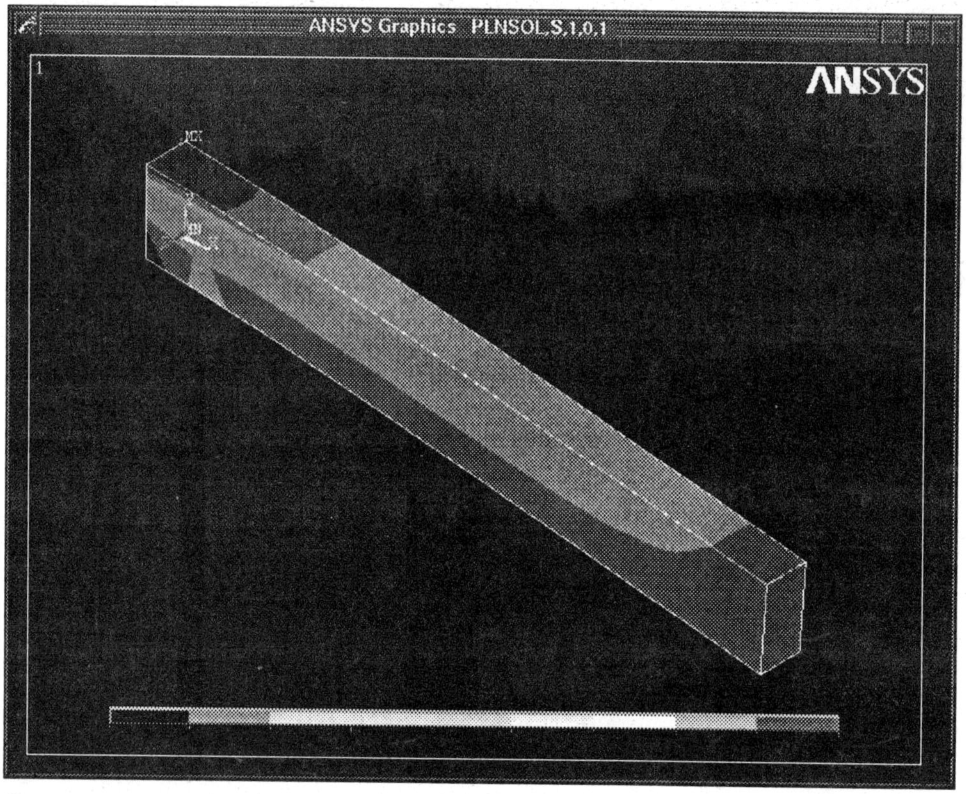

If desired, the results can also be animated. To do this, select **Plot Ctrls** from the upper menu, and click **Animation.** Then select **Deformed Shape** and accept the default settings, following the procedure described in the previous example.

To print the results to an image file, select **PlotCtrls** from the top menu. Select **Hard Copies**, and then choose **Graphics Window**. Then save the file in a .jpg format, following the procedure described in the previous example. Also, the image can be saved by pressing the "print screen" button on the keyboard and pasting the image into an image-editing program, such as MS Paint.

10.4 FEA with an Assembly Structure

The following figure presents an assembly structure. The assembly consists of two parts. They are a block component and a beam component. The block component is assembled on the top of the beam structure. After the assembly, the block component and the beam component are in contact.

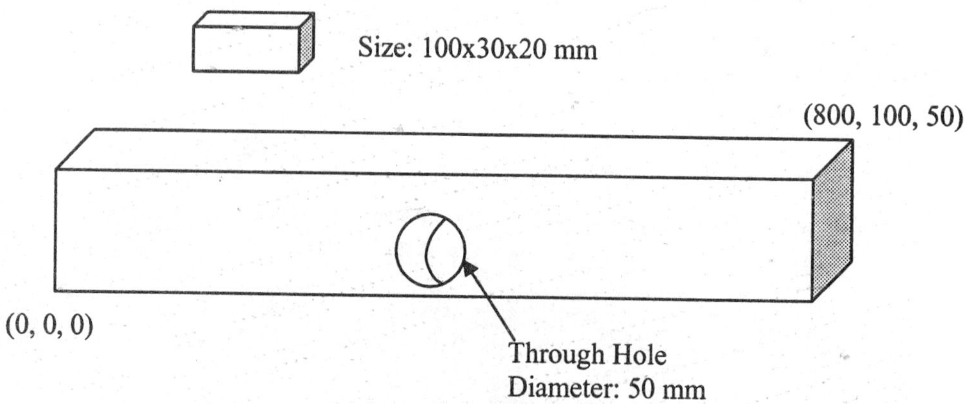

The load acts on the top surface of the block component after the assembly. The load magnitude is 100 Newton in the vertical direction and downward. The material properties of these two components are listed below:

	Young's Modulus (GPa)	Poisson's Ratio
Block Component	3	0.40
Beam Structure	30	0.26

Step1: From the **ANSYS** main menu, select **Preferences**. Make sure that **Structural** is highlighted, and then click **OK**.

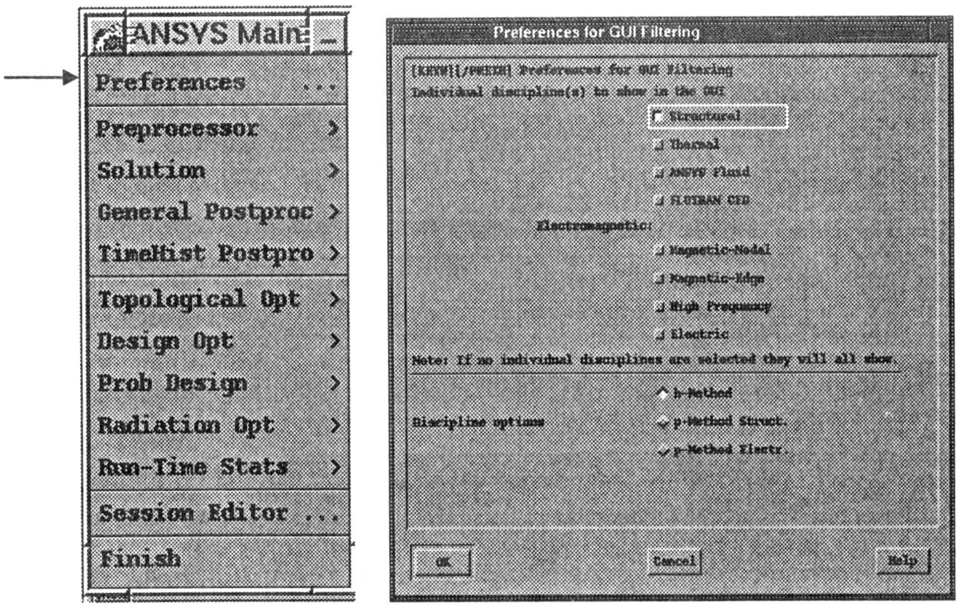

Step 2: From the main menu, select **Preprocessor**. Then, under modeling, select **Create, Block** under Volumes, and **By Dimensions.** In the window for creating the beam structure by dimensions, type 0 and 0.8 for X1 and X2, 0 and 0.1 for Y1 and Y2, and 0 and 0.05 for Z1 and Z2 > **Apply (do not use OK).** Afterwards, type 0.7 and 0.8 for X1 and X2, 0.1 and 0.12 for Y1 and Y2, and 0.01 and 0.04 for Z1 and Z2 > **OK**. The screen will display the 2 3D solid models.

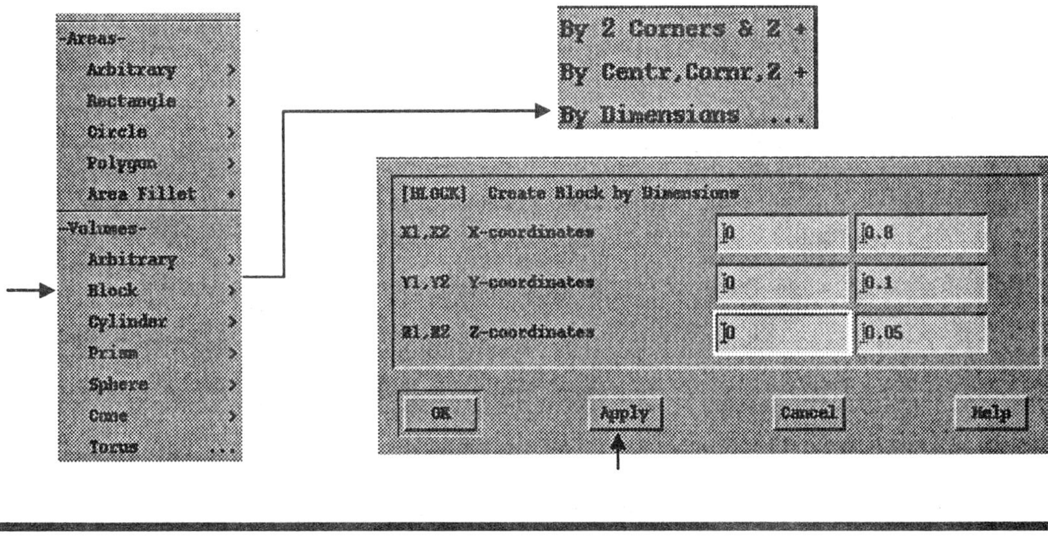

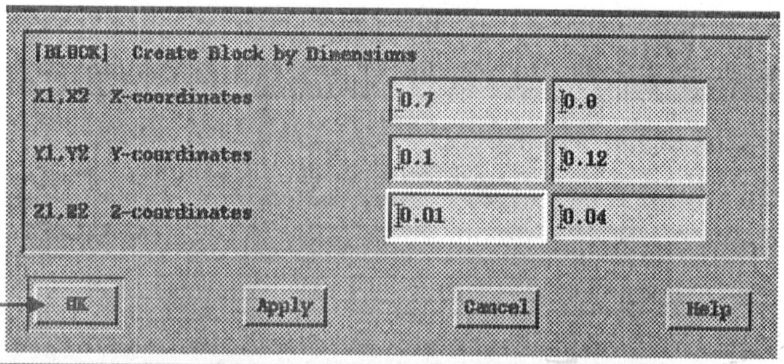

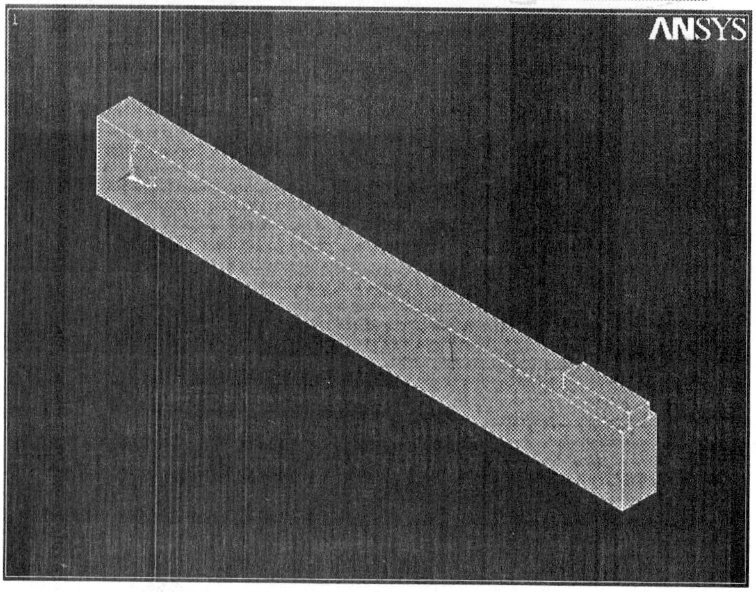

Step 3: To create a hole at the central location of the beam structure, select **Create** > **Cylinder** > **Solid Cylinder**. Type in the Center Location (0.4, 0.05), Radius (0.03) and Depth (0.05) > **OK**.

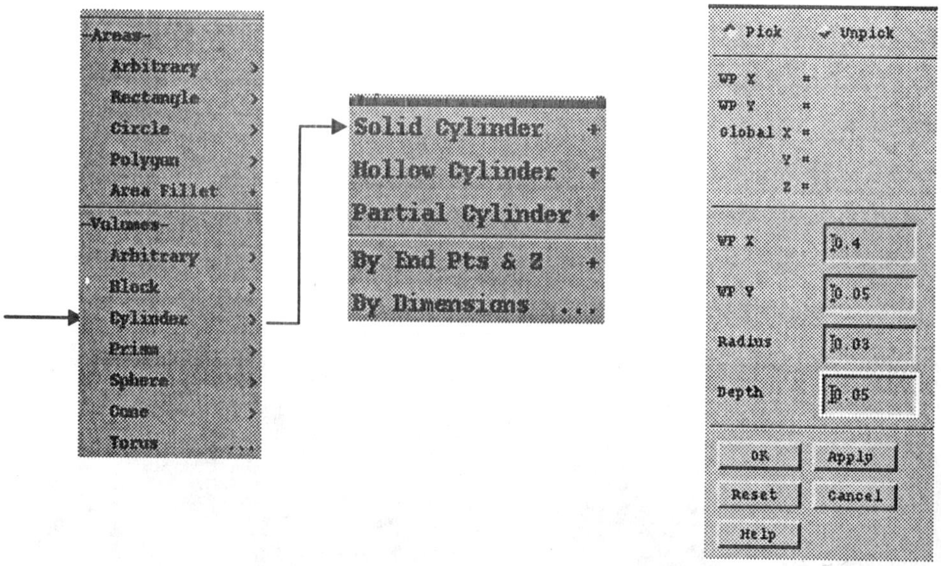

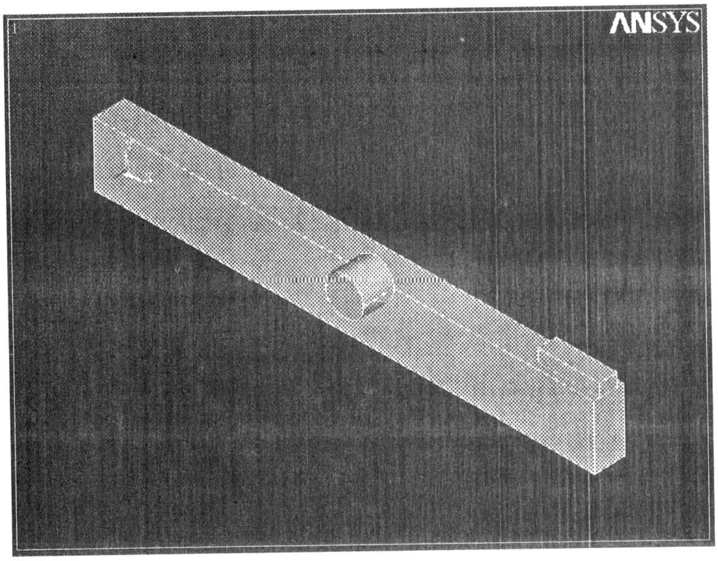

Under **Modeling**, select **Operate** > **Subtract** > **Volume**. Pick the block first. On screen, there is a message indicating that 3 volumes are present. You have picked Volume 1, which is the correct picking. Click **OK** on the message window and click **OK** on the pick up window.

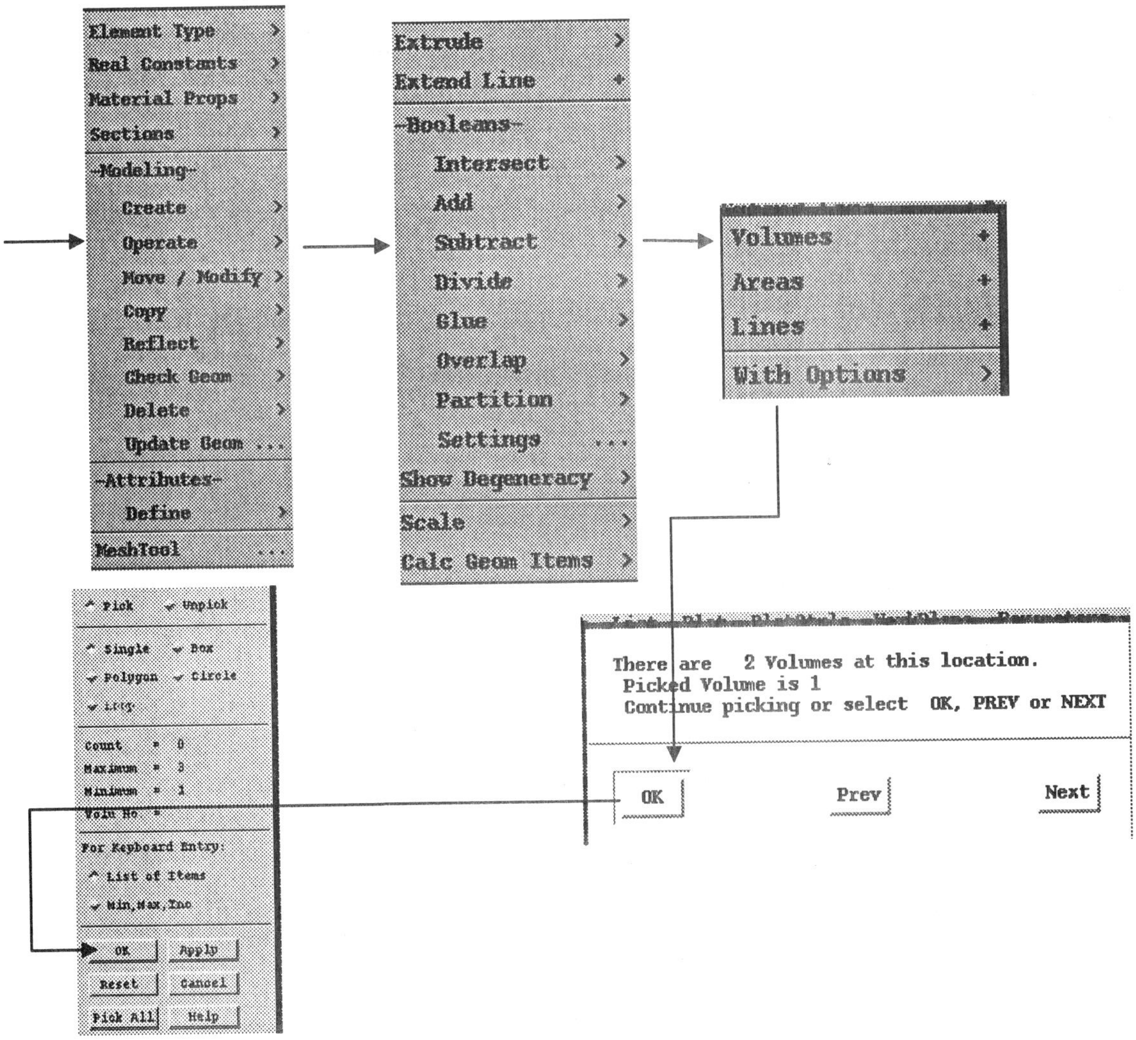

Now pick the solid cylinder from the screen. The message indicates that your pick is Volume 1, which is not correct. Click Next so that the picked volume is Volume 3. Click **OK** on the message window, and click **OK** on the pick up window.

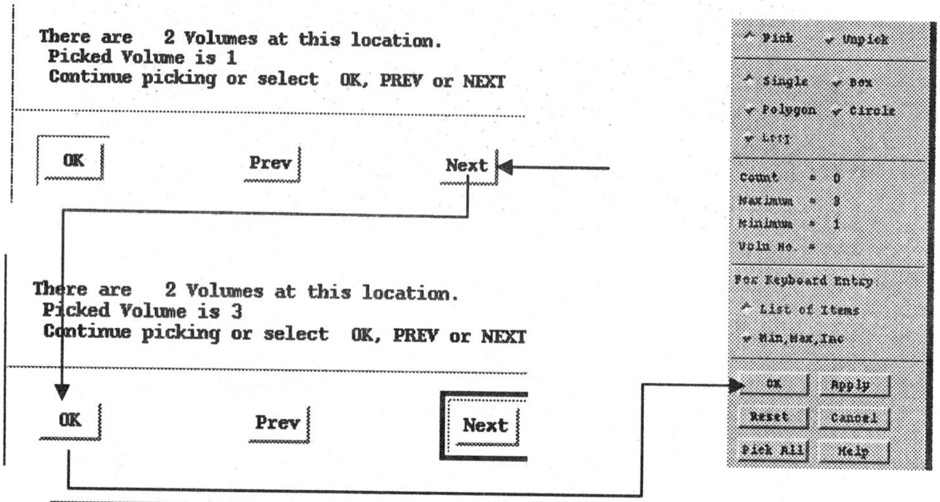

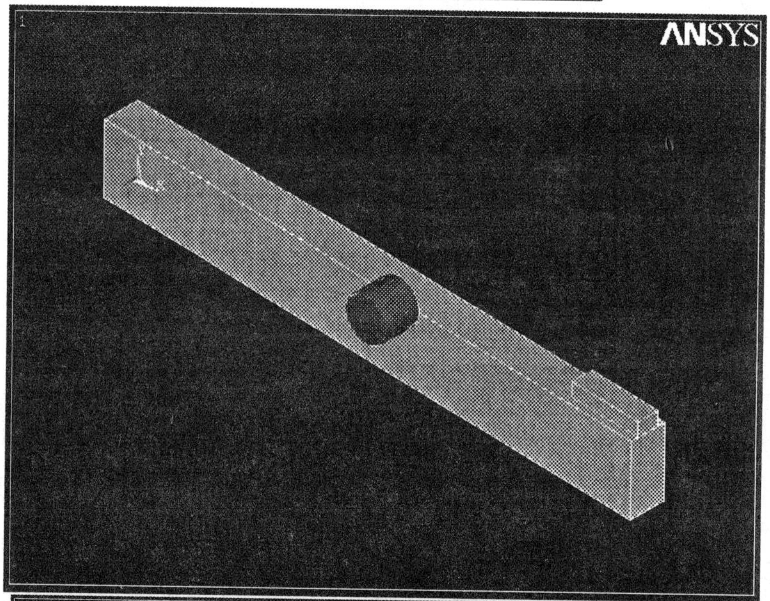

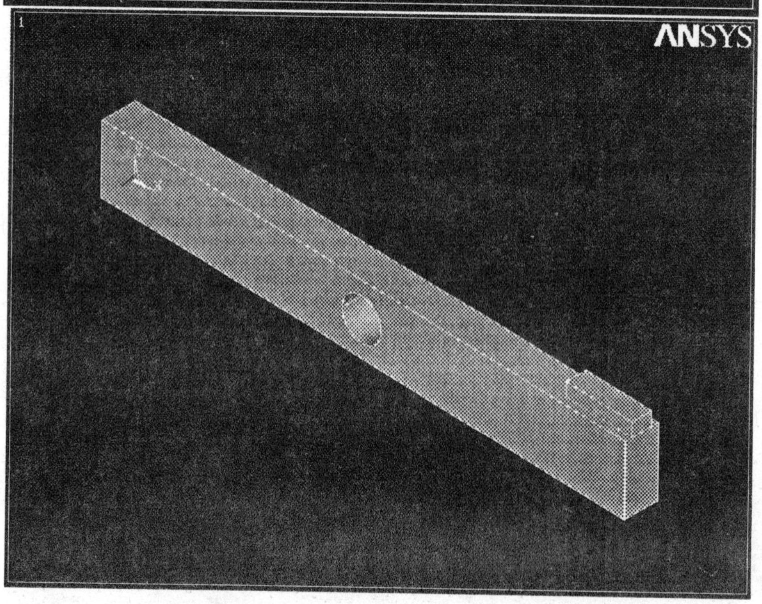

Step 4: Assemble the small block to the beam structure.

Under the Preprocessor menu, select **Operate** > **Glue** > pick both the block component and the beam components > **OK.** For simplicity, a user may just select Pick All. If the user picks the block component from the screen, the volume number is Volume No 4, not Volume No 1. This is because of the subtracting operation just performed. After the substraction, both Volume 1 and Volume 3 became Volume 4. Although this display seems identical to the one shown before, the model at this stage is an assembly model, and the previous model was not an assembly model.

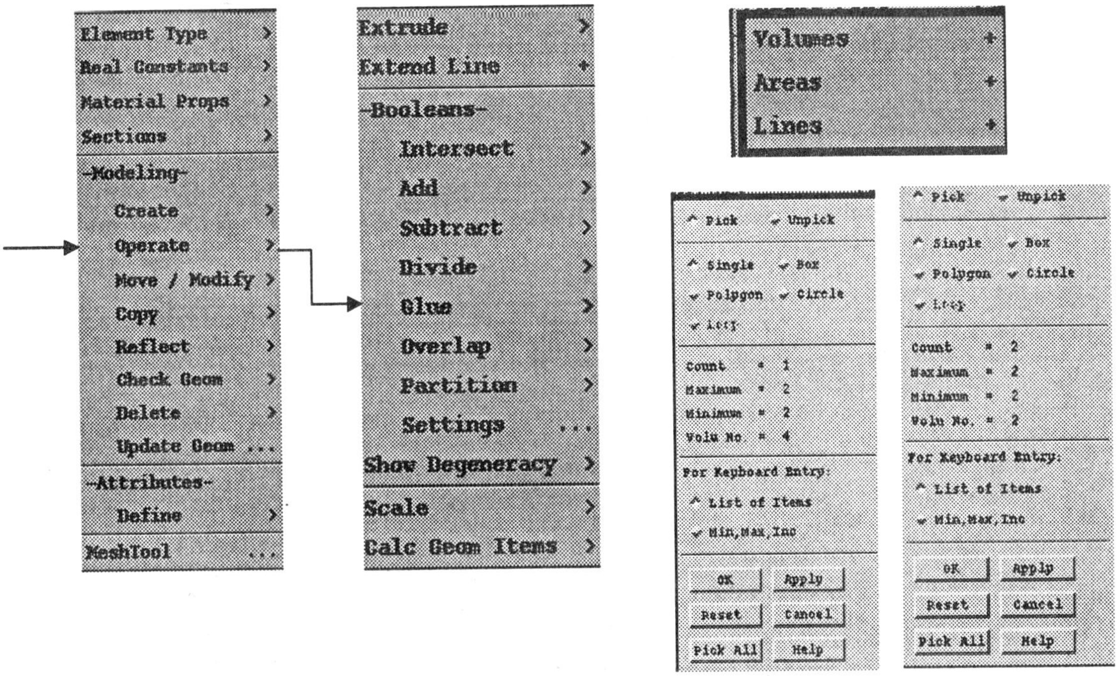

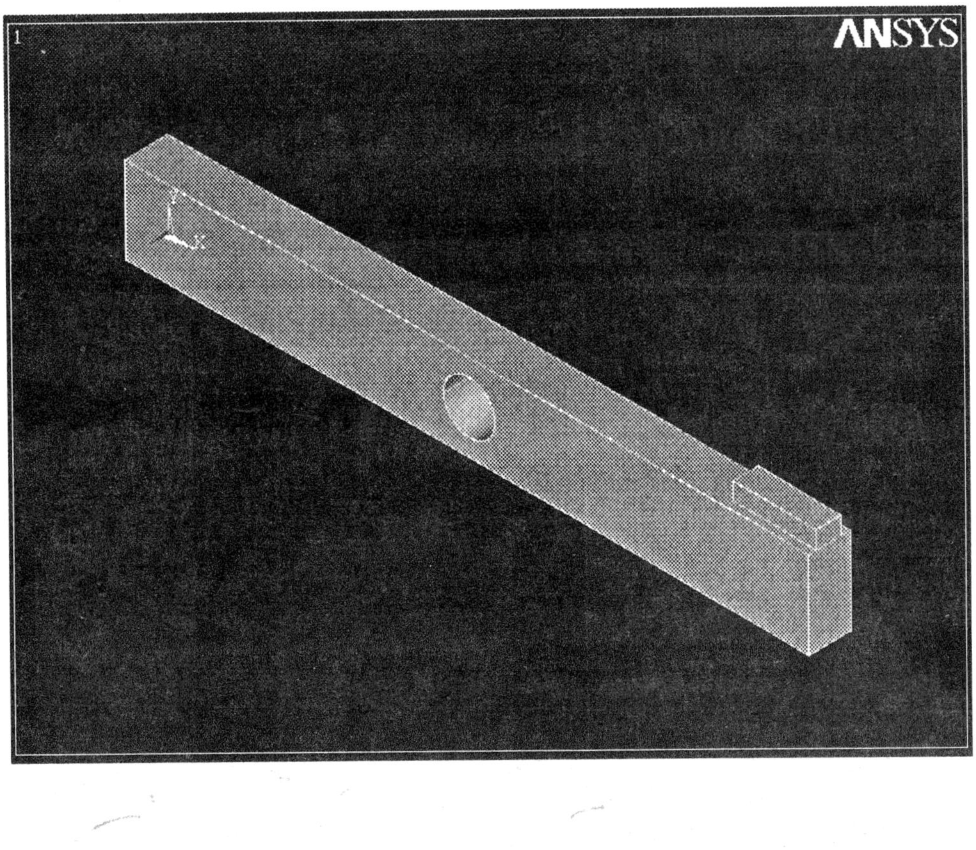

Step 5: Under the Preprocessor menu, select **Element Type**. Then select **Add/Edit/Delete**, and then **Add**.

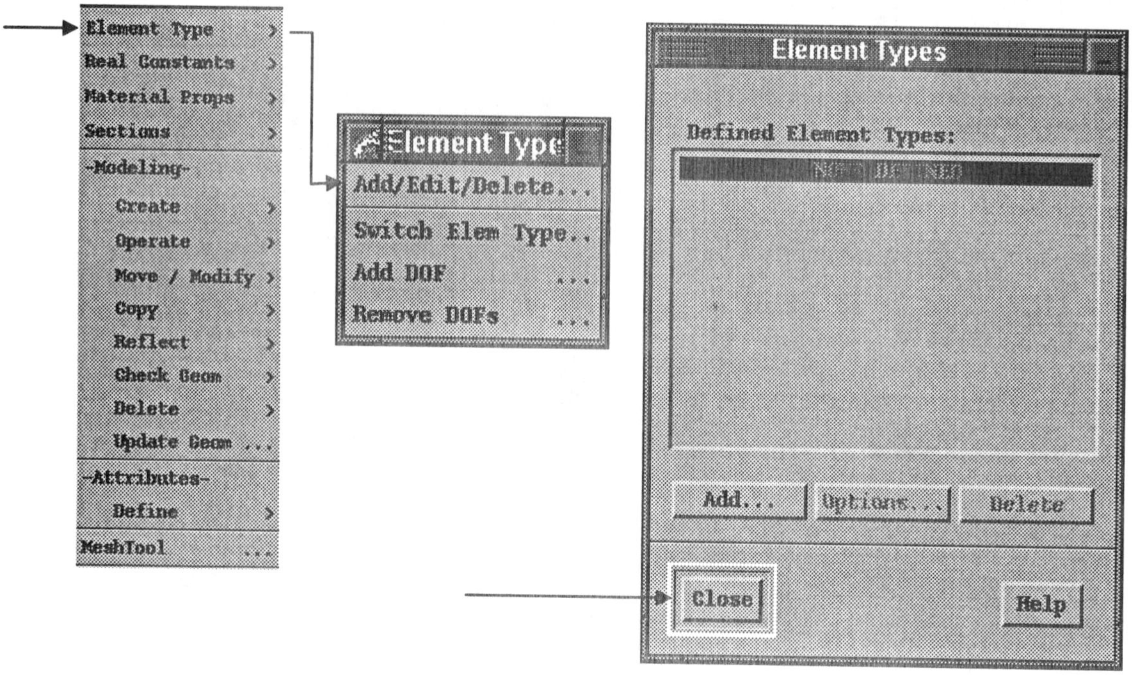

Select **Structural Solid** and choose **Tet 10node 187**, and then click **OK**.

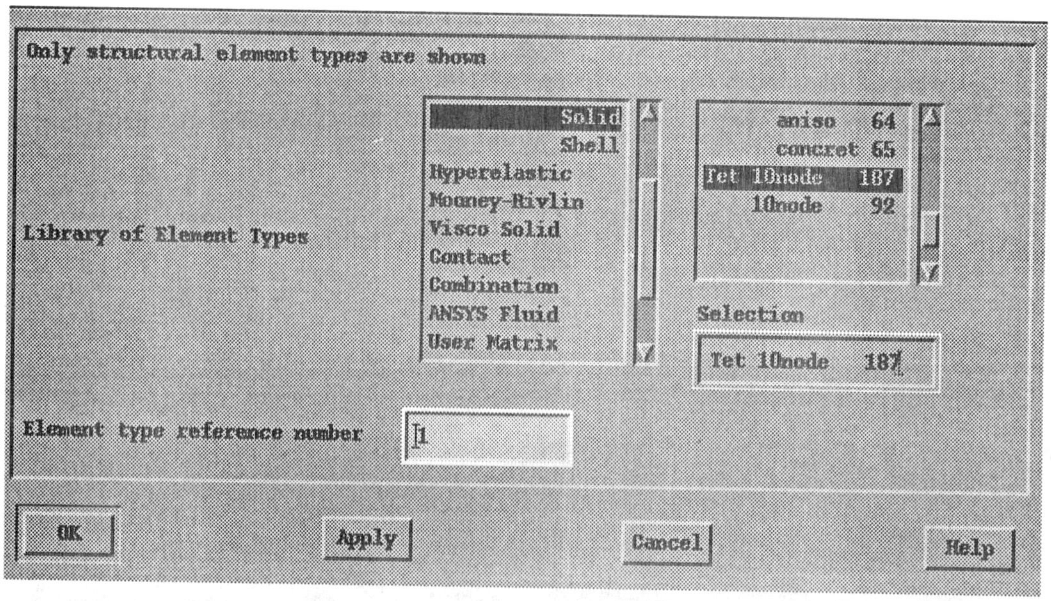

Step 6: Assign material types.

To select the material for the beam model, select **Material Properties** from the Preprocessor menu. Select **Material Model**, and then select **Structural, Linear, Elastic** and **Isotropic** by double clicking on each one.

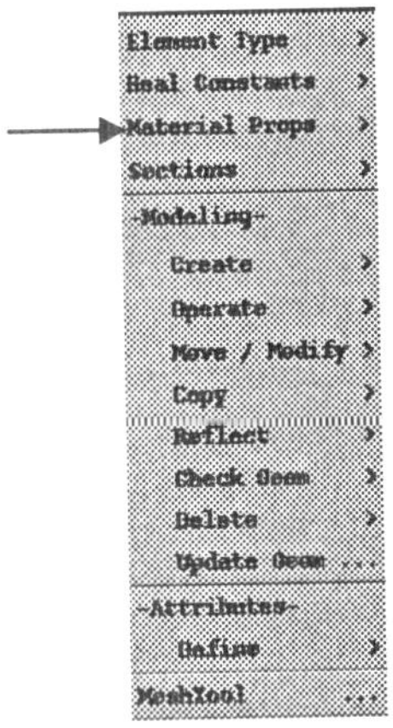

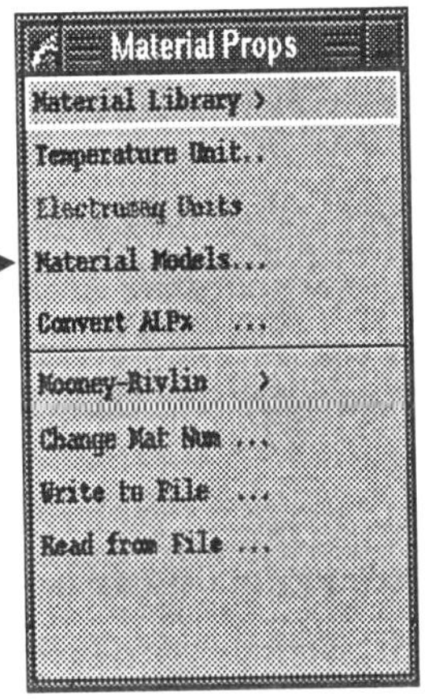

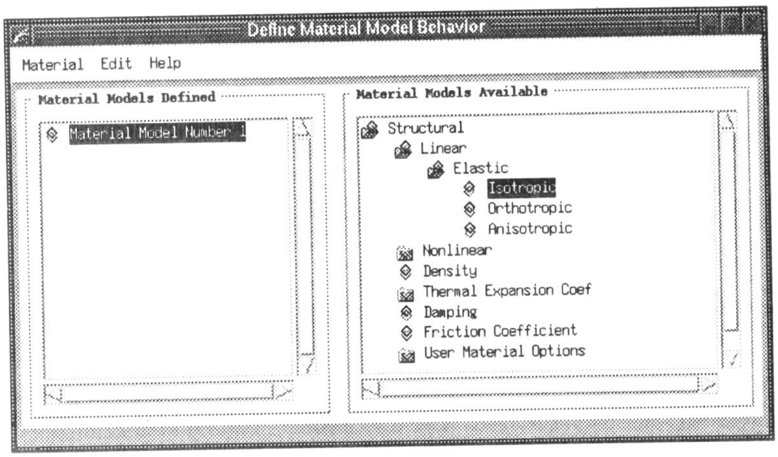

Type 3e10 and 0.26 as the values for Young's Modulus and Poisson's ratio, respectively. Then select OK, and Close.

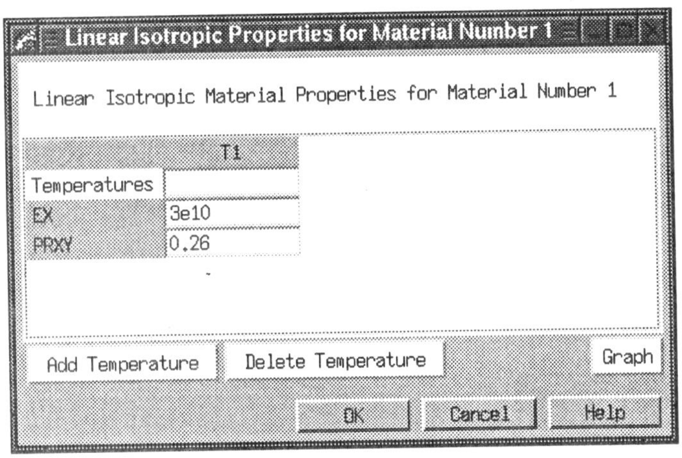

Select **Material Props** > **Material Model** > **Structural** > **Linear** > **Elastic** > **Isotropic**. Type 3e10 for the Young's Modulus and type in .26 as the Poisson's ratio > **OK.**

We also need to define the material properties for the block component, **Material** > **New Model**. A pop up window appears, indicating this is material type 2 > **OK** to accept the assignment for small block > type in 3e9 and 0.40 as the material properties for the block component > **OK** > **Material** > **Exit.**

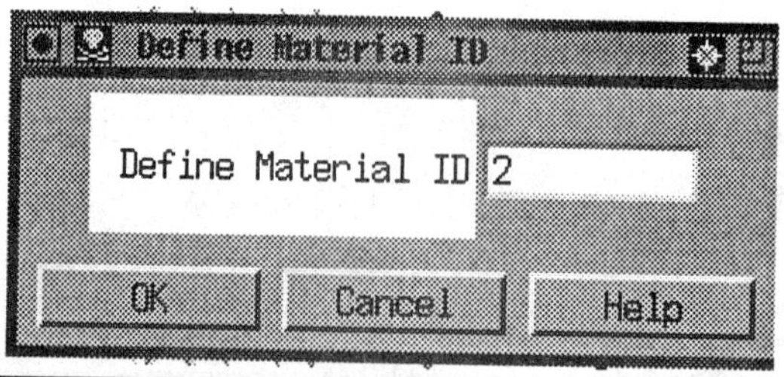

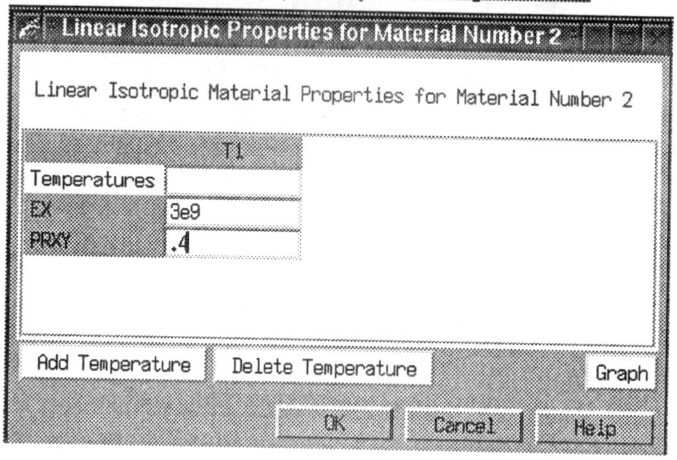

Step 7: Mesh Generation:

To create the mesh for the shell model, select **MeshTool** from the Preprocessor menu. Make sure that **Tet** and **Free** are selected. Then click **Mesh.**

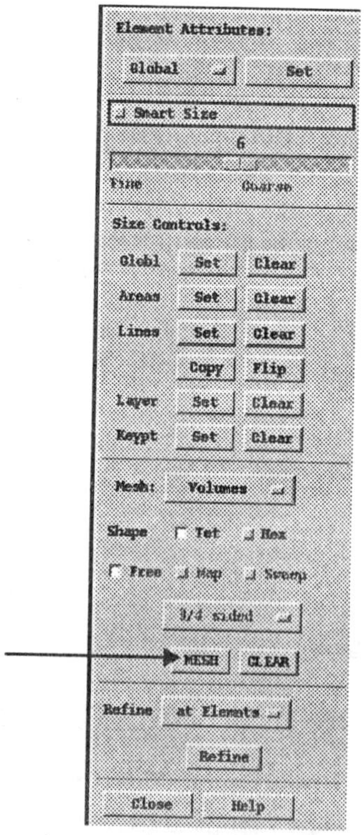

When the selection box comes up, select the volume from the display window, and then click **OK**. You may just click the box of Pick All because there are only 2 volumes.

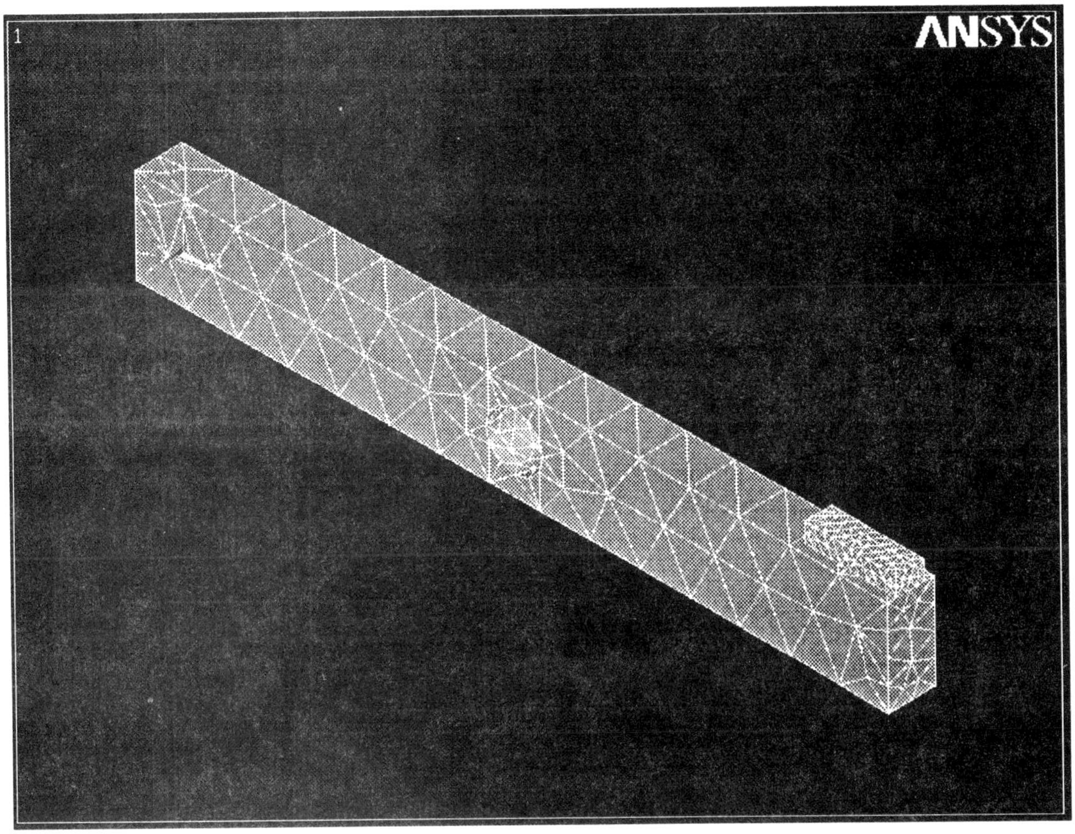

Step 8: Apply the load condition.

To apply the load, on the Preprocessor menu, click **Load**, then **Apply**, **Force/Moment, and On Nodes**. Select the nodes at the top surface of the block component using the box function, and click **OK**.

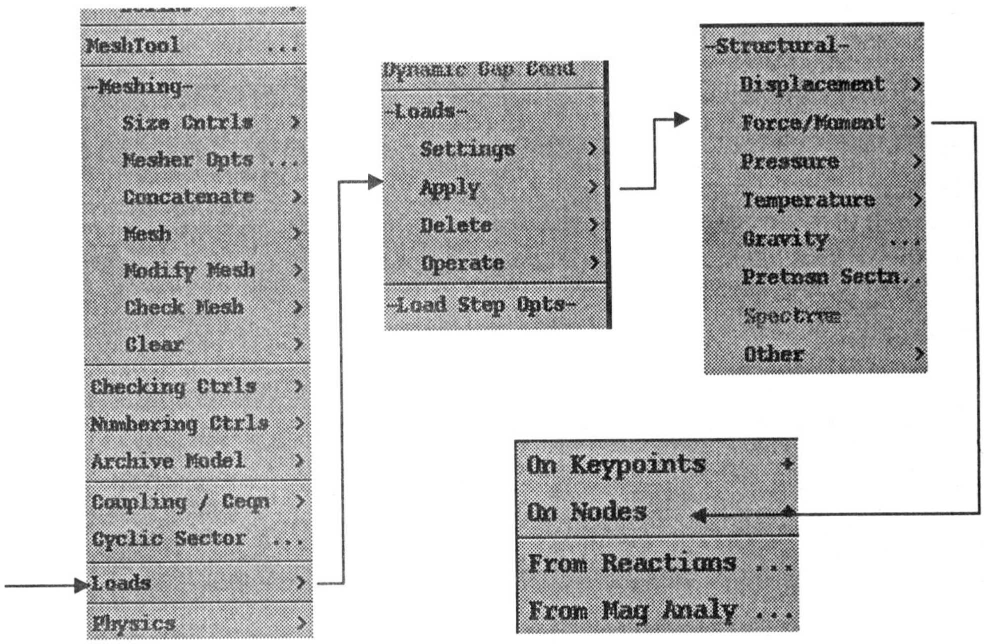

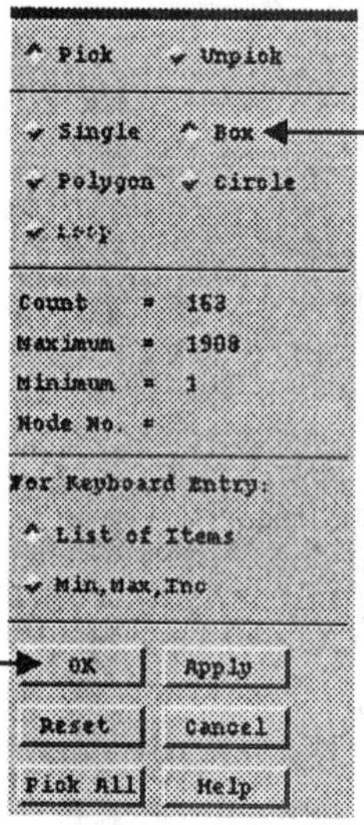

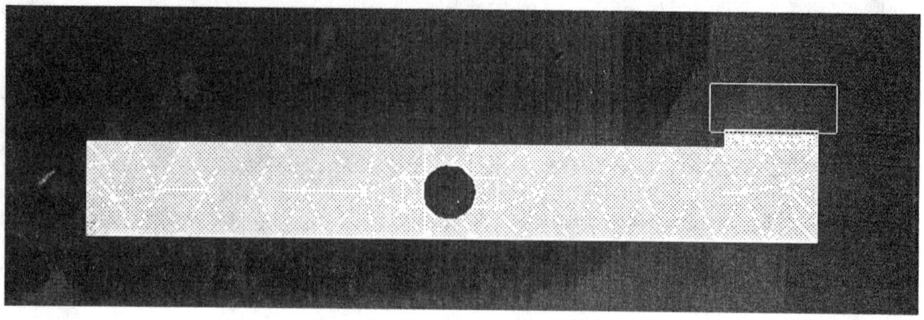

For **Fy**, type in –100 as the force value > **OK.**

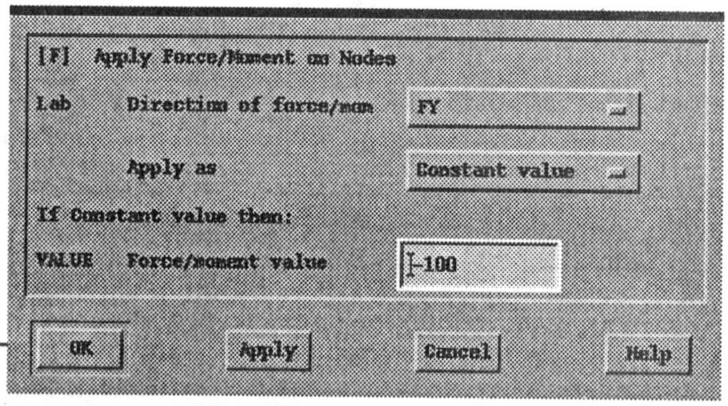

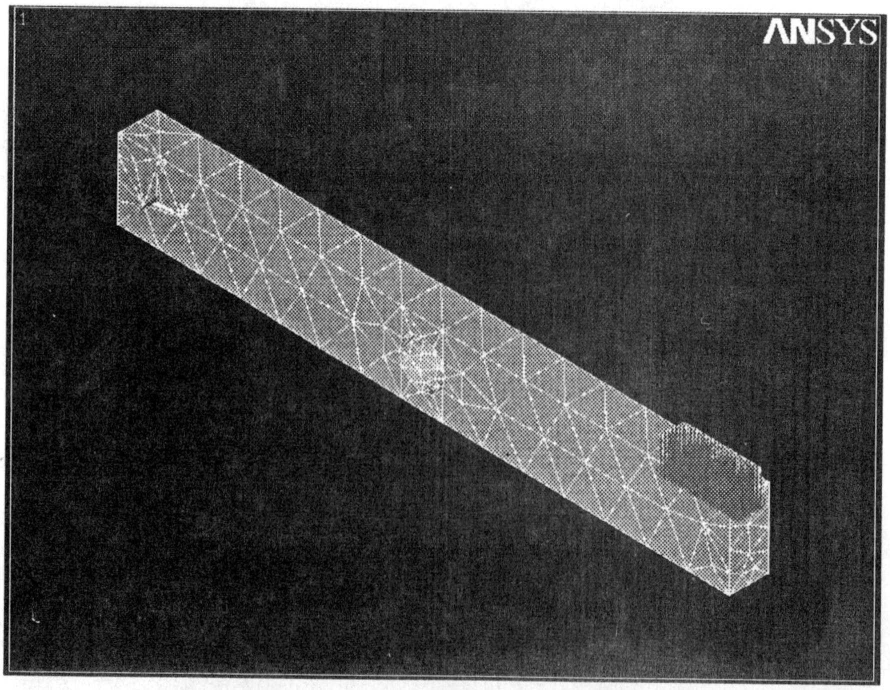

Step 9: Defining the Constraints

To define the constraint conditions, under the Load menu, select **Apply**, **Displacement**, and **On Nodes**. Pick all of the nodes on the left surface of the model using the box function, and set the 3D model to the Front view. Sketch a box at the left side of the rectangle. We are able to pick the 57 nodes from a set of 1308 nodes by a single sketch > **OK**.

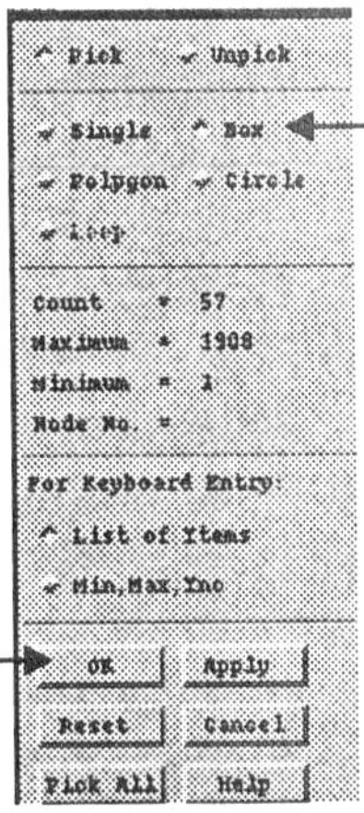

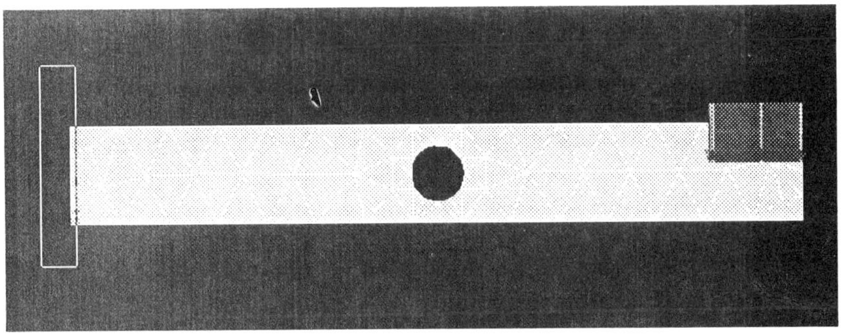

Select **All DOFS**, and type 0 as the value. Then click **OK**.

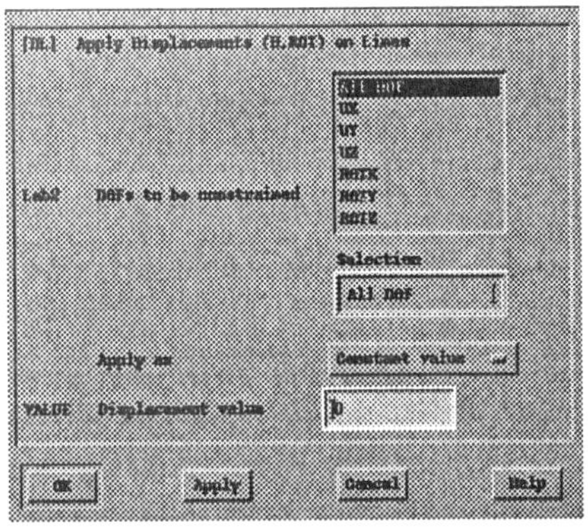

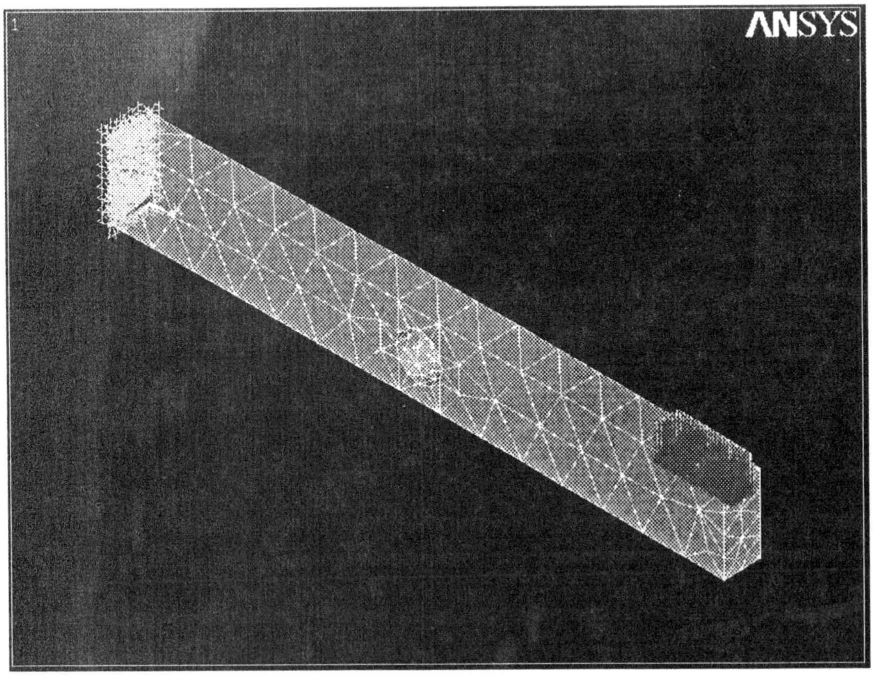

Step 10: Run the program to obtain the solution:

To find the solution, run the program by selecting **Solution** from main **ANSYS** menu. Choose **Current LS**, and then click **OK**.

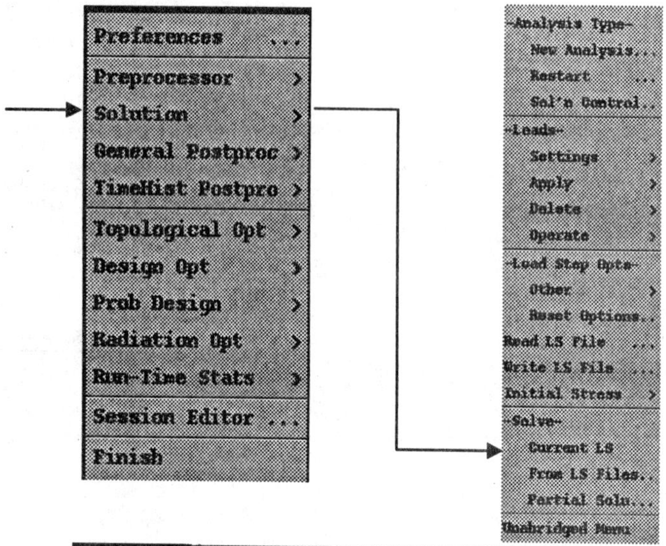

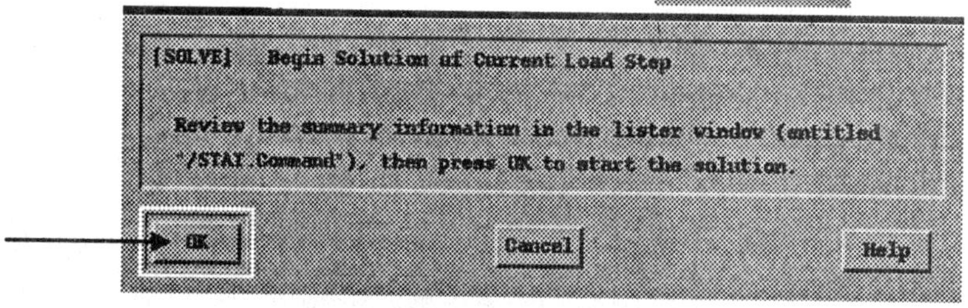

A window will display when the solution is done > **Close**.

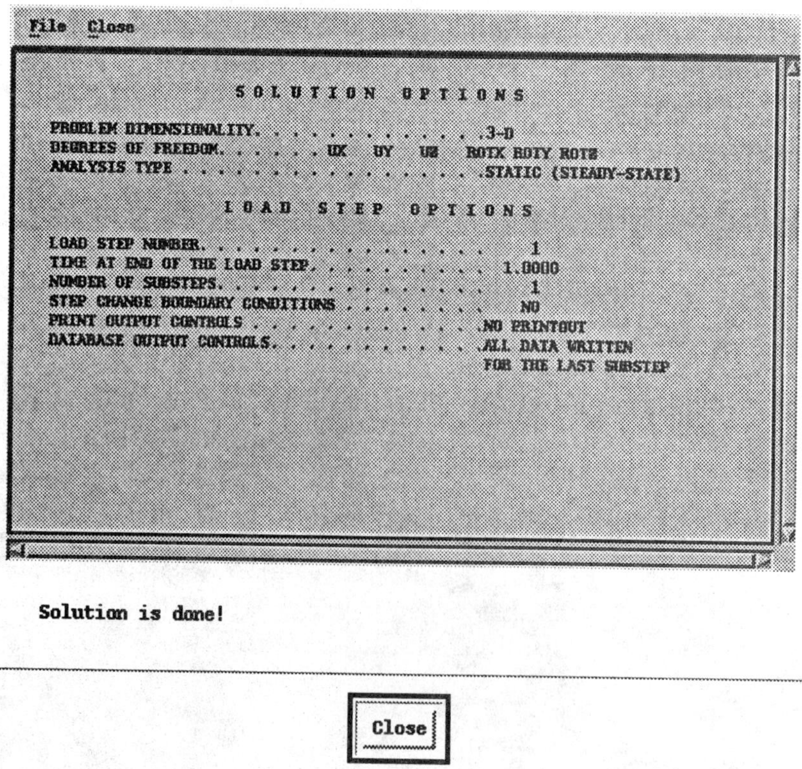

Step 11: Display Results

To review the results, select **General Postproc** from the main menu. Then click **Plot Results,** and **Nodal Solution.**

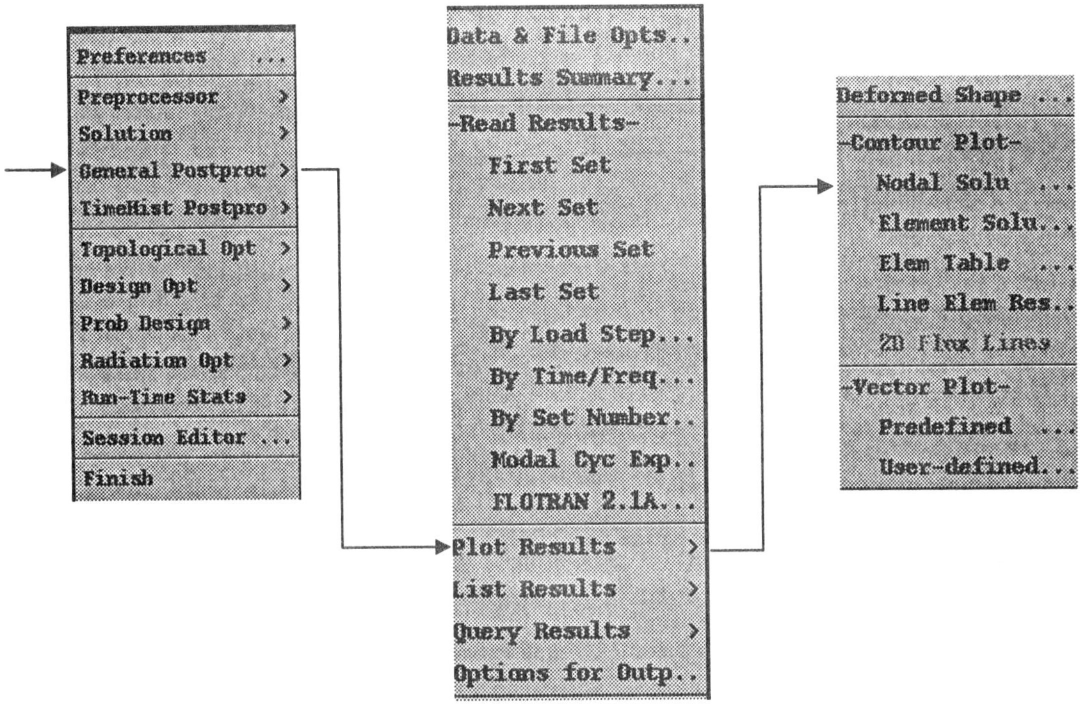

Select **Stress** from the first list, and **1st Principal S1** from the second list. Then click **OK.**

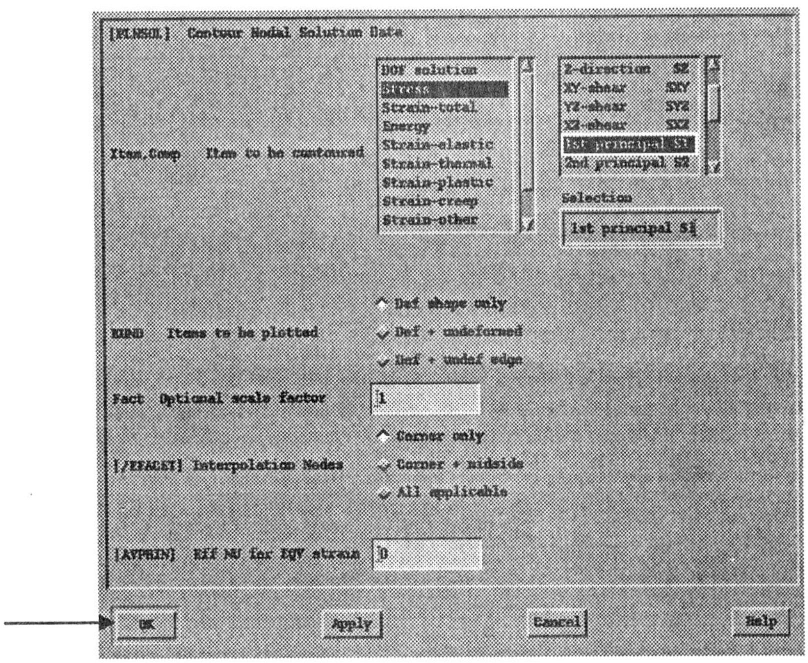

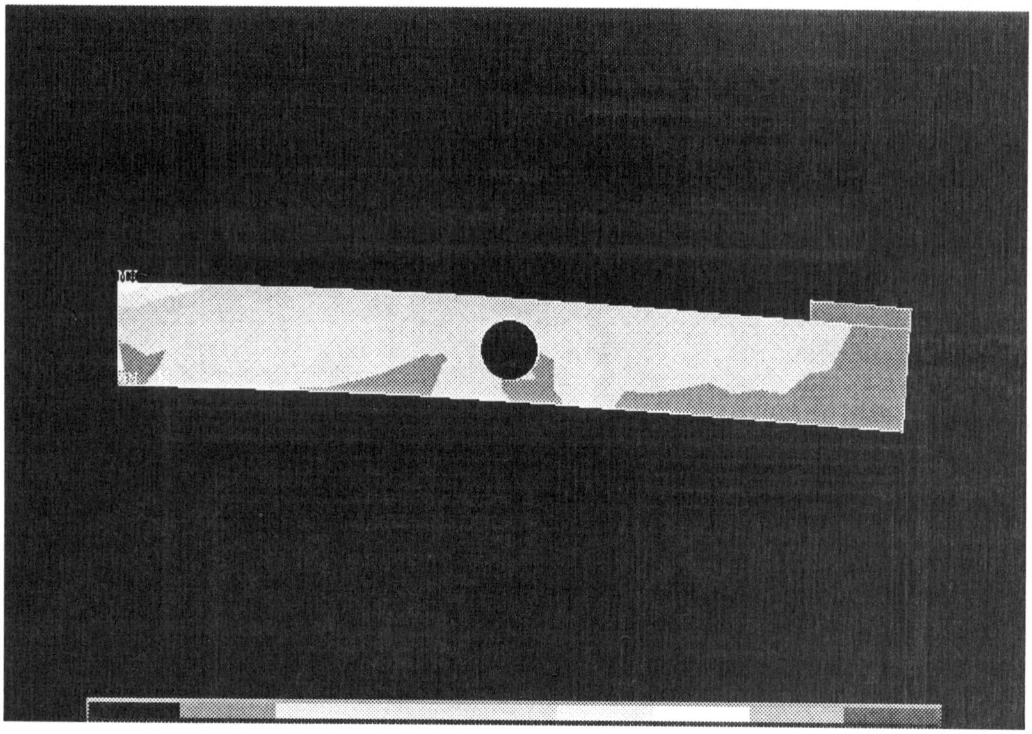

Step 12. Animation:

To animate the results, select **PlotCtrls** > **Animation** > **Deformed Shape**. Accept the default settings, and click **OK**

Step 13: Print the document:

To print the results to an image file, select **PlotCtrls** from the top menu. Select **Hard Copies**, and then choose **Graphics Window**. Then save the file in a .jpg format.

Also, the image can be saved by pressing the "print screen" button on the keyboard and pasting the image into an image-editing program, such as MS Paint.

10.5 Control of the Number of Elements in Modeling

In this section, we present a new procedure to perform FEA. To demonstrate this new procedure, we use an example of a simple support beam. The procedure begins with defining the element type. Afterwards, the process of creating the geometrical model starts. In the process of defining the geometry, we only define two nodes at the two ends of the beam structure. However, we number the node at the left end as node 1 and number the node at the right end as number 81. This implies that we are going to define 80 beam elements over the entire overhand length of the beam structure. This new procedure offers its simplicity in performing FEA.

The following figure illustrates the geometrical shape of the beam structure under consideration. The left end of the beam structure is a pin connection with the ground. The right end of the beam structure is a roller support. The total length of the beam structure is 2000 mm. The section area is a rectangle. The two dimensions are 25 mm and 40 mm, respectively. There are 3 loads acting on the beam. The first load acts on the right node. The magnitude is 4000 Newton and its direction is towards the right, or along the positive x axis direction. The second load is uniformly distributed load acting on the entire over length of the beam structure with the magnitude equal to 2 N/mm. The third load is a concentration load acting on a distance equal to 500 mm with reference to the right end of the beam structure. Pay attention to the assignment of the material properties. We assume that the Young's modulus is 30 MPa and the Poisson's ratio is 0.27. Therefore, the material is easy to be bent in the Y direction. Our objective of this study is to show the deflection pattern and to know the maximum deflection value and the location of the maximum deflection.

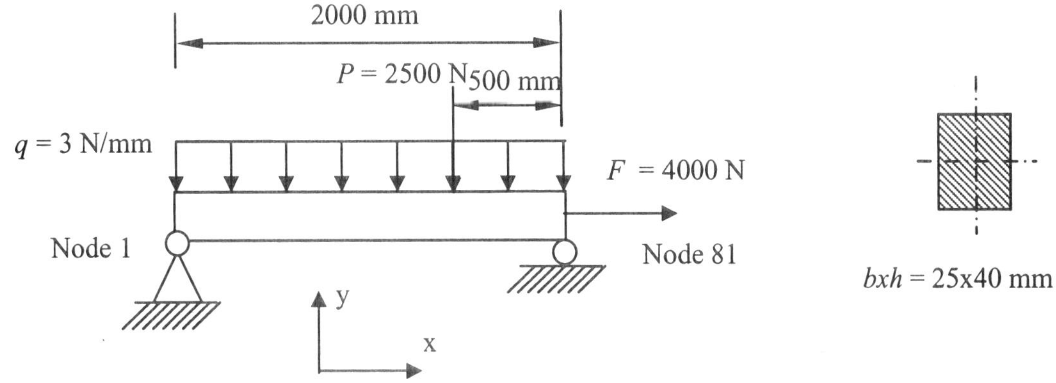

Step 1: Select the type of element that the beam structure will be formed out of.
Under the Preprocessor menu, select **Element Type**. Then select **Add/Edit/Delete**, and then **Add**.

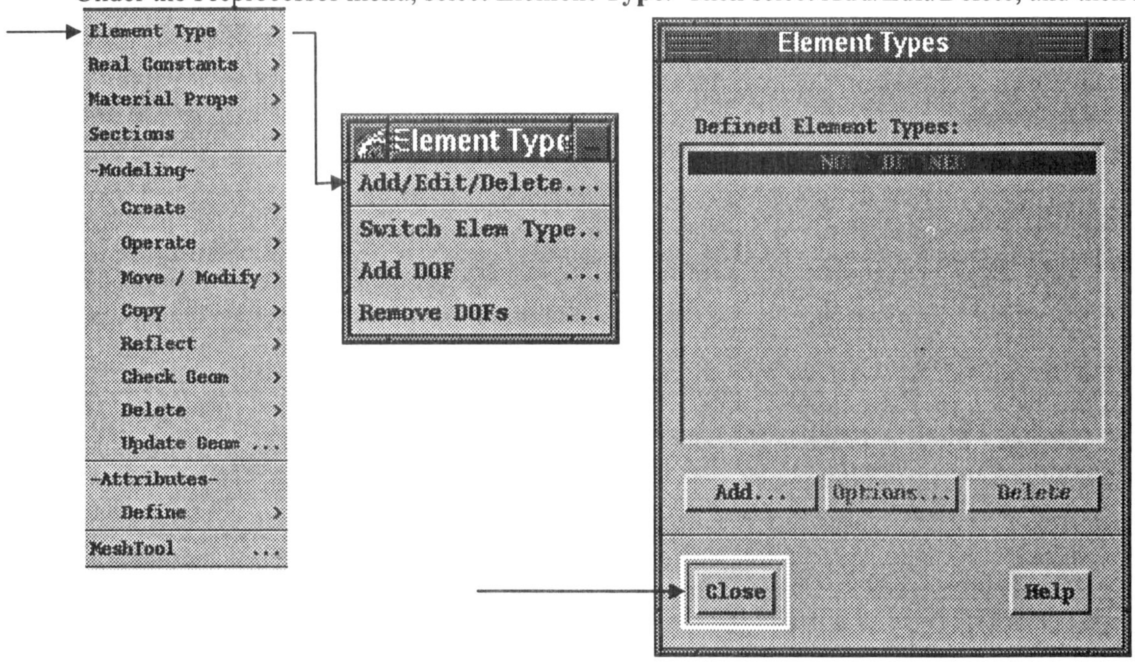

Select **Beam** and choose 3D elastic 4. Click **OK** > **Close** under the Element Type menu and return to the Preprocessor menu.

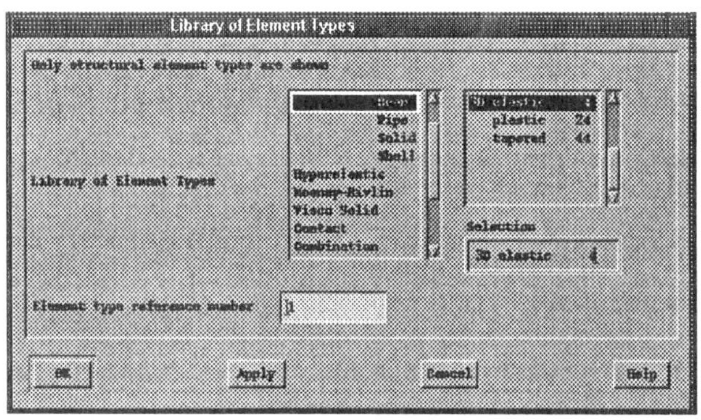

Step 2: Create the element cross-section area.

Under the Preprocessor menu, select **Real Constants**. Then select **Add/Edit/Delete**, and then **Add**. The screen will ask you to select the element type, which the constants will relate to. There should only be one, namely, **Type 1 Beam4**. Select it and click **OK**.

The beam is rectangular in cross-sectional area, with a height of 25 cm (0.25 m) and a width of 40 cm (0.4 m). Therefore, enter *0.1* under area, 0.*0005208* under IZZ, and 0.*0013333* under IYY. Enter 0.4 as the thickness value along z axis and 0.25 as the thickness value along Y axis. Afterwards, click OK and return to the Preprocessor menu.

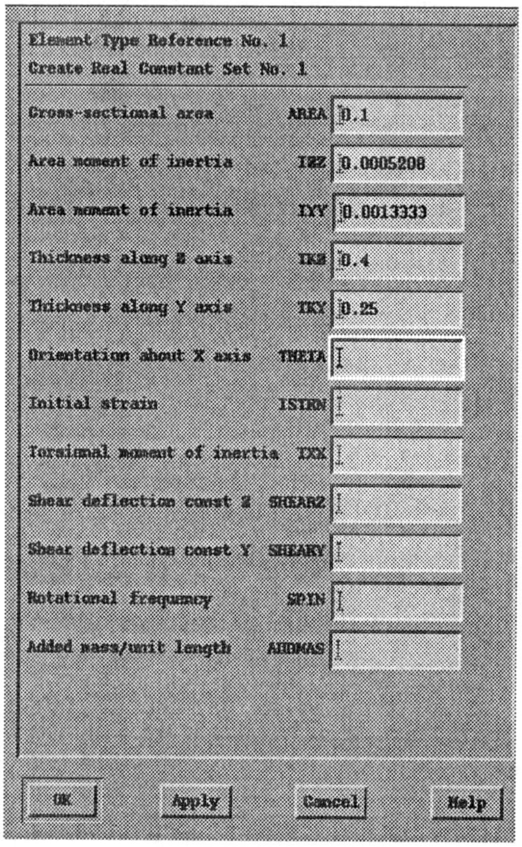

Step 3: Choosing a Material Type

To assign the material for the 3D model, select **Material Properties** from the Preprocessor menu. Select **Material Model,** and then select **Structural, Linear, Elastic** and **Isotropic** by double clicking on each one.

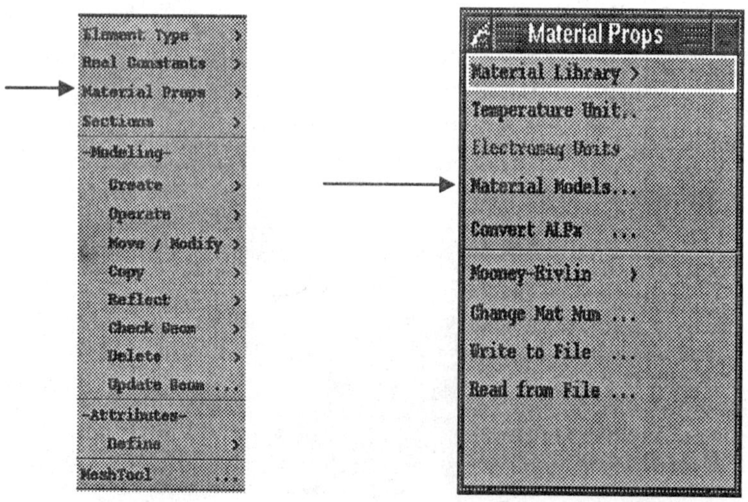

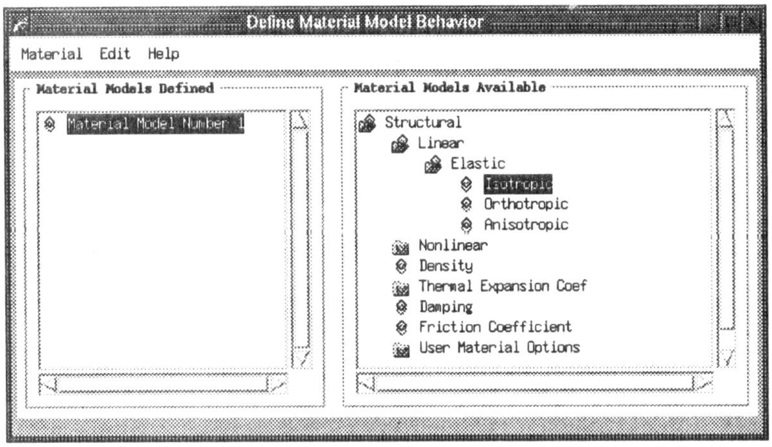

Type 3e10 under EX and 0.26 under PRXY as the values for Young's Modulus and Poisson's ratio, respectively. Then select OK, and Close.

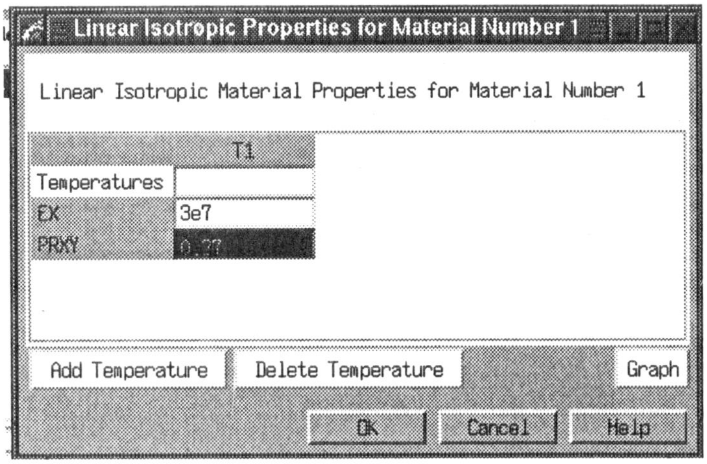

To check the unit system, select **Material Library** under the material properties menu. Then select **Units** and make sure that the **SI** unit system is highlighted. Click **OK**.

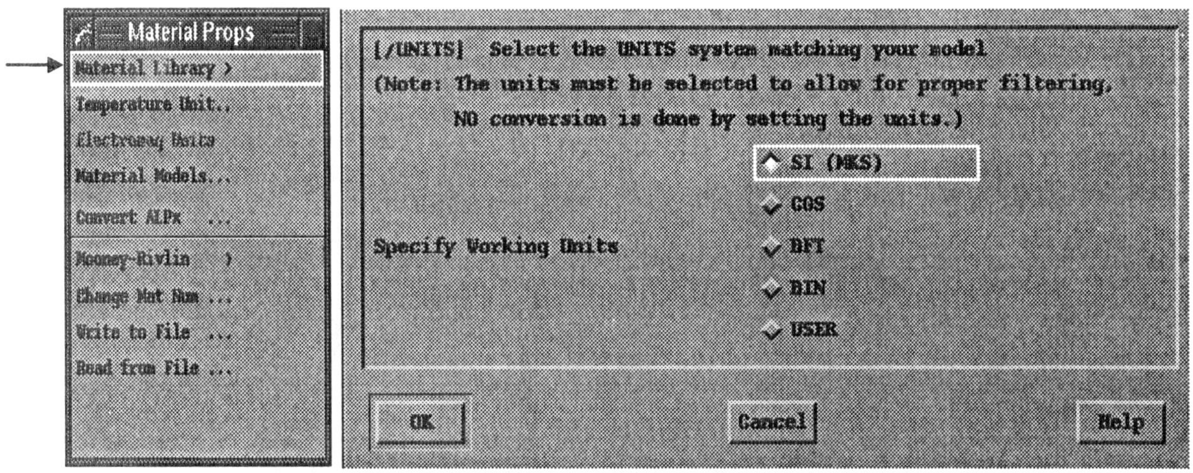

Step 4: Create the geometry for the beam structure.

From the Preprocessor menu and under modeling, select **Create**, select **Nodes**, then **In Active CS**. Create Node 1 at (0, 0, 0) > **Apply** (do not click **OK**).

The beam will be 2 meters in length. It is our intention that the beam will be divided into 80 segments, or beam elements. Create a second node, Node 81, at (2,0,0) > **OK** (do not click **Apply**).

The graphics space displayed on the computer screen should look like the one shown below:

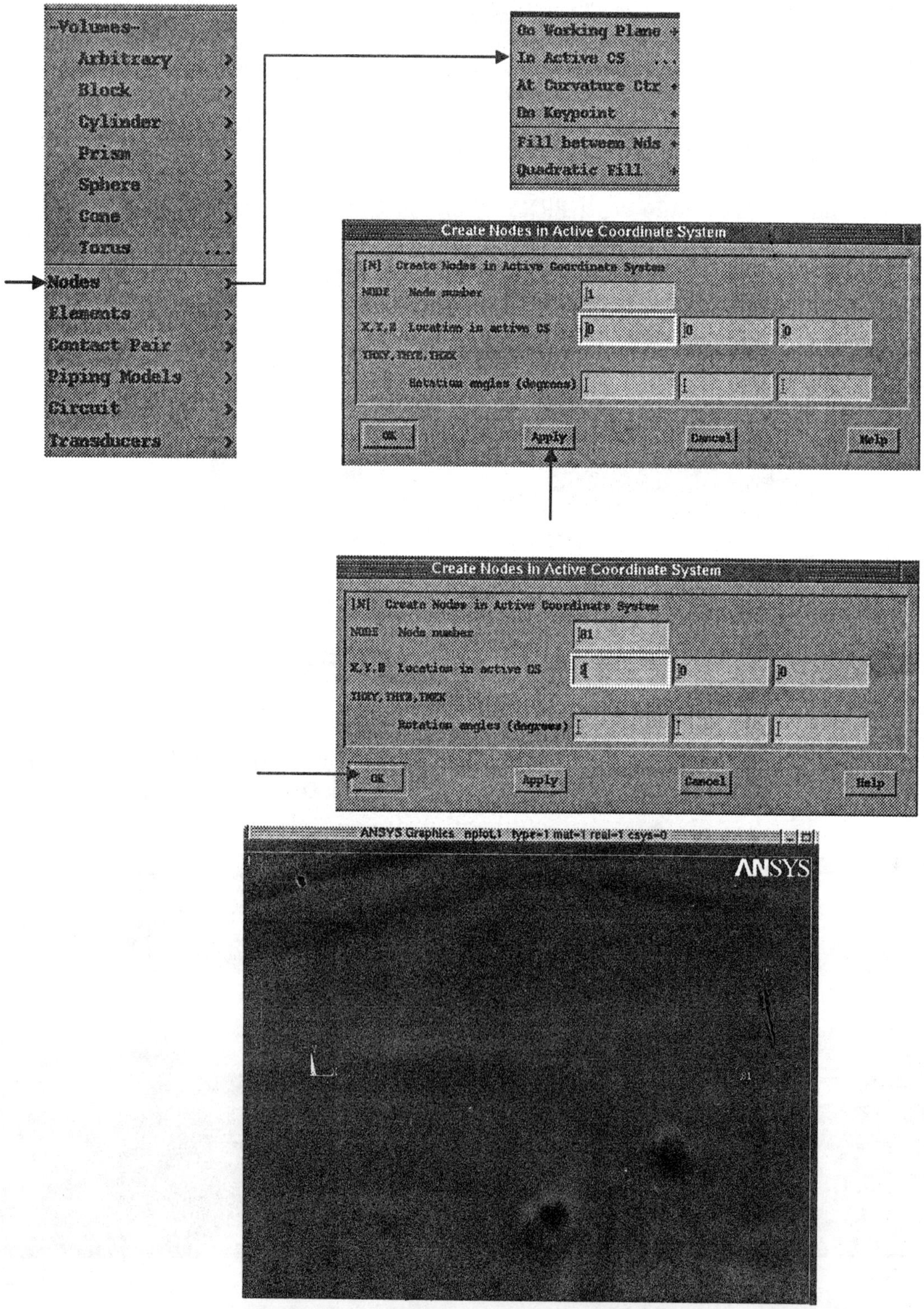

Select **Fill Between Nds** from the **Nodes** Menu. Select nodes 1 and 81 either by clicking on them or typing the numbers into the ANSYS Input window. Click **OK**. Make sure that **NFILL** number is 81, and the **SPACE** value is 1, click **OK**. Return to the Create menu.

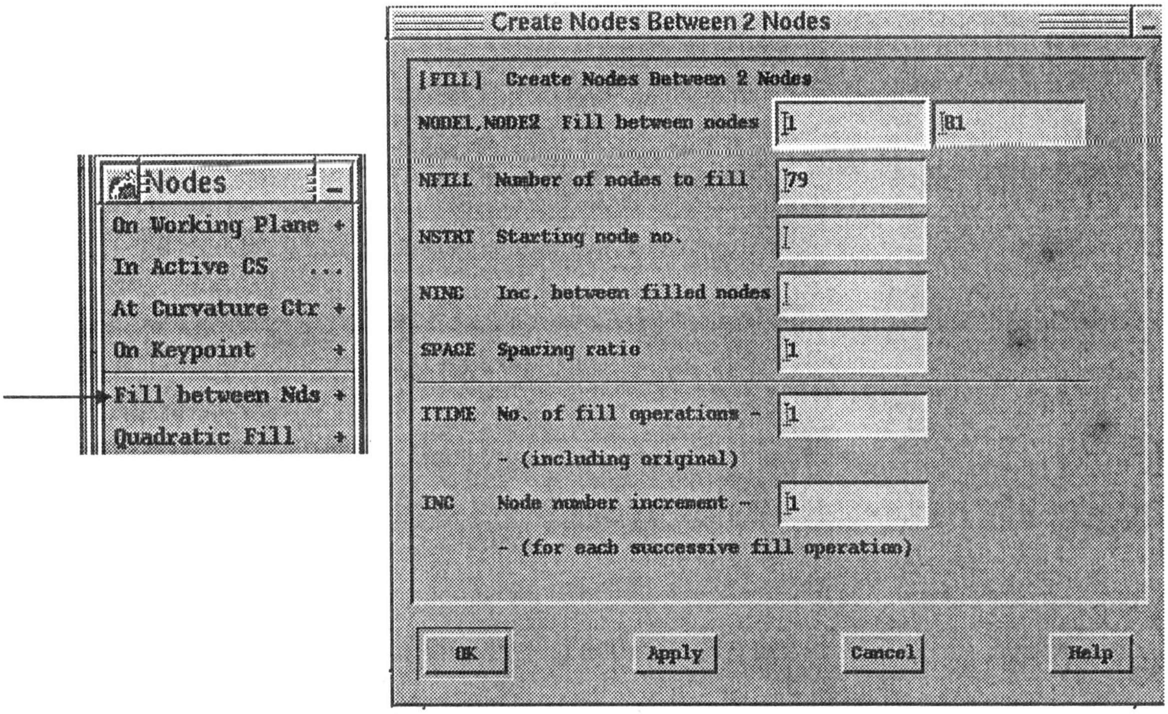

Step 5: Assign a number to each of the element.
Select **Elements** from the Create Menu, and **Thru Nodes** from the **Elements** Menu.
Select nodes 1 and 2 by clicking on them or entering them into the **ANSYS** Input window.

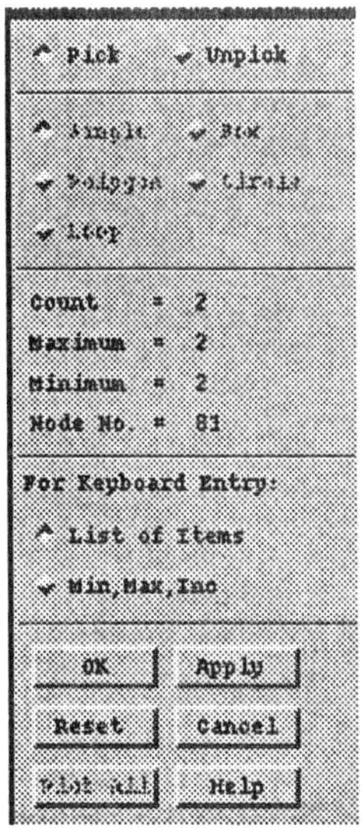

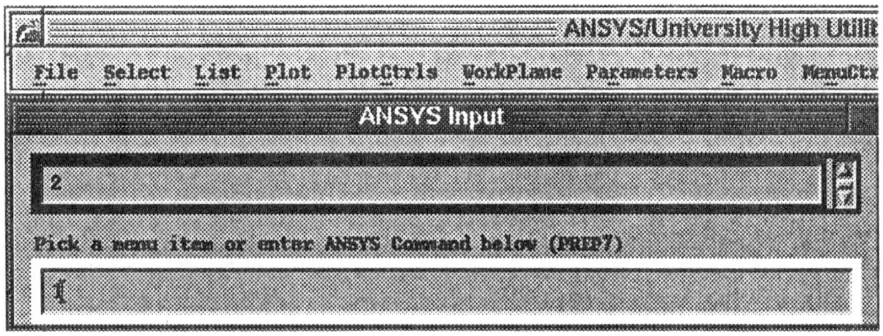

Return to the Preprocessor Menu. Select **Copy** > under **Elements**, select **Auto Numbered** > **Thru Nodes** > Pick node 1 and node 2 > **OK**.

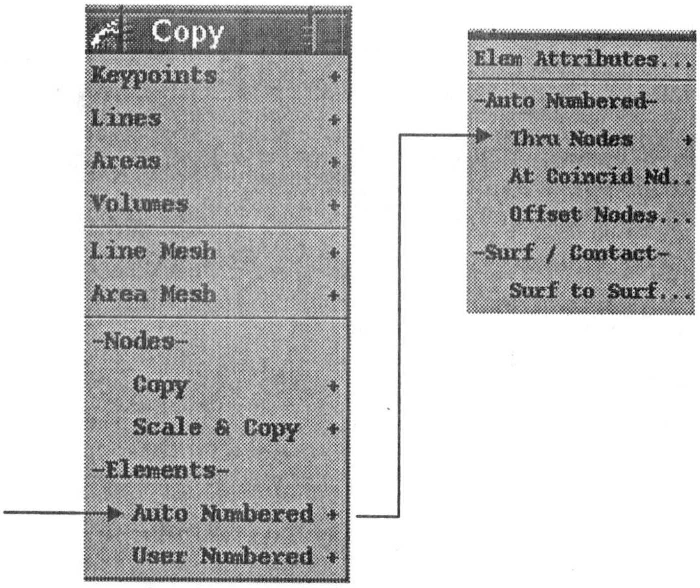

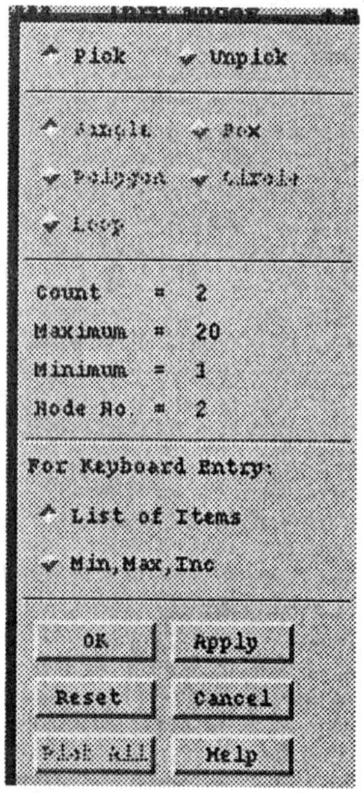

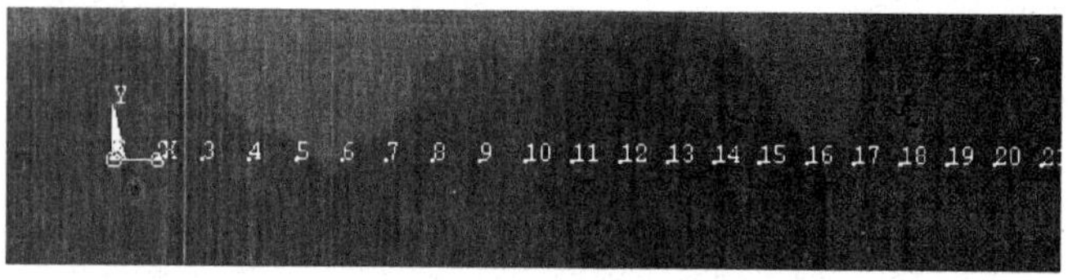

Select the first element, and then click **OK**. Type in *80* under **ITIME**, and check to make sure that **NINC** is 1. Click **OK**, and return to the Preprocessor Menu.

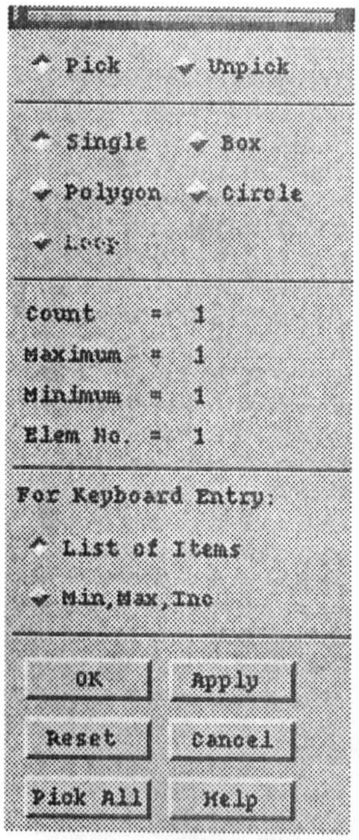

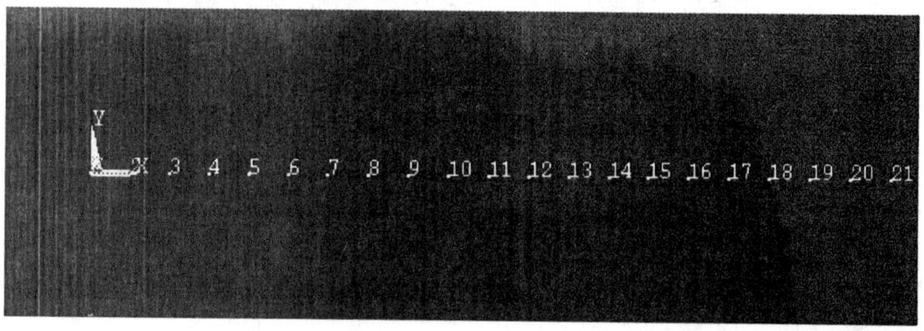

Type in *80* under **ITIME**, and check to make sure that **NINC** is 1. Click **OK**, and return to the Preprocessor Menu.

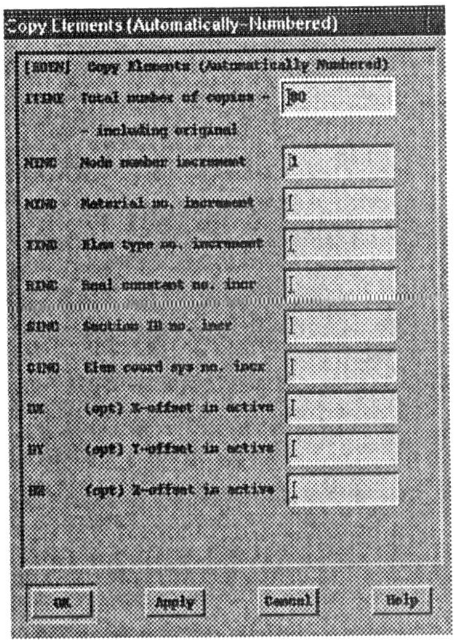

Step 6: define the constraint conditions at the right and left joints.

To define the constraint conditions, under the Load menu, select **Apply**, **Displacement**, and **On Nodes**. Select node 1 and click **OK**. Constrain UX, UY, UZ, ROTX and ROTY, then enter 0 as VALUE. > **OK.**

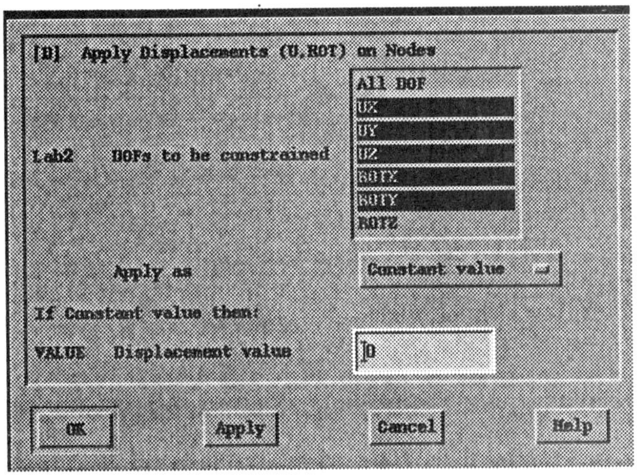

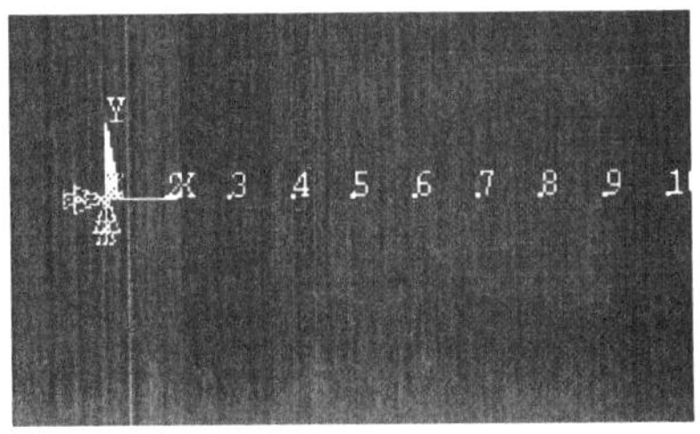

To define the roller support at the right end, click **On Nodes** from the Displacement Menu, and select node 81. Constrain UY, ROTX and ROTY, with a VALUE of 0.

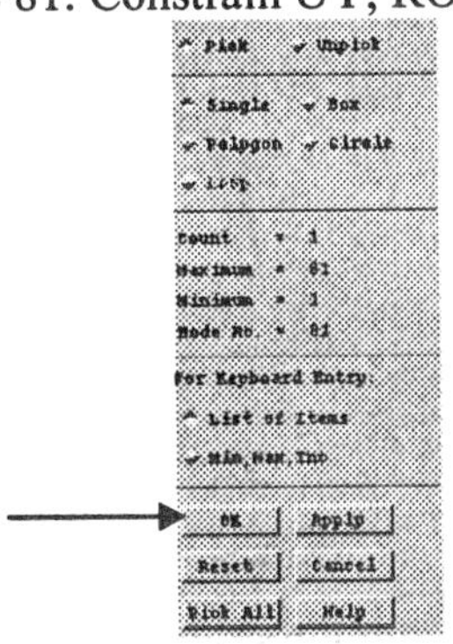

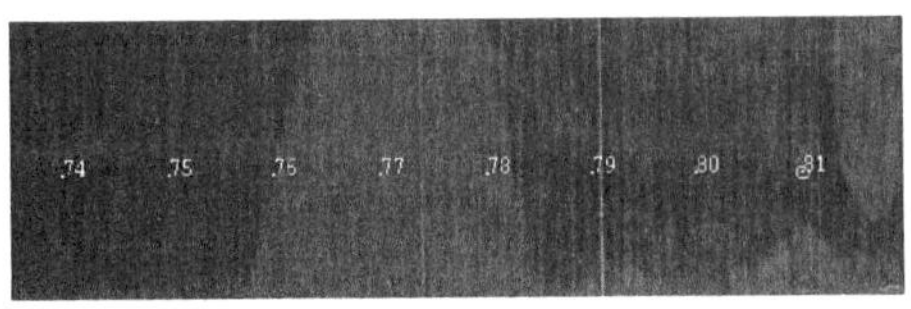

Constrain UY, ROTX and ROTY, with a VALUE of 0.

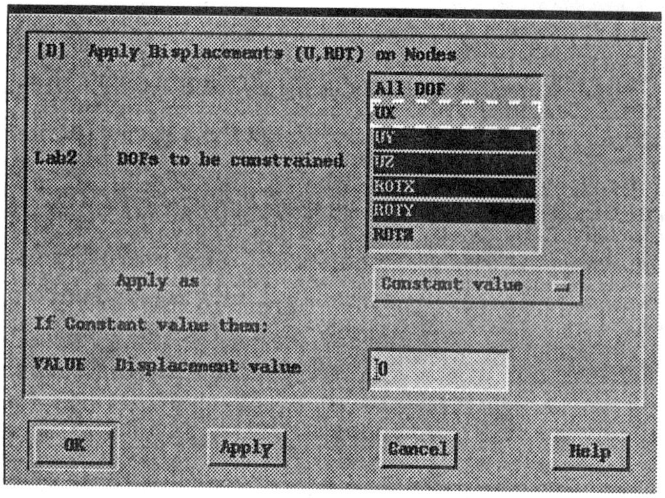

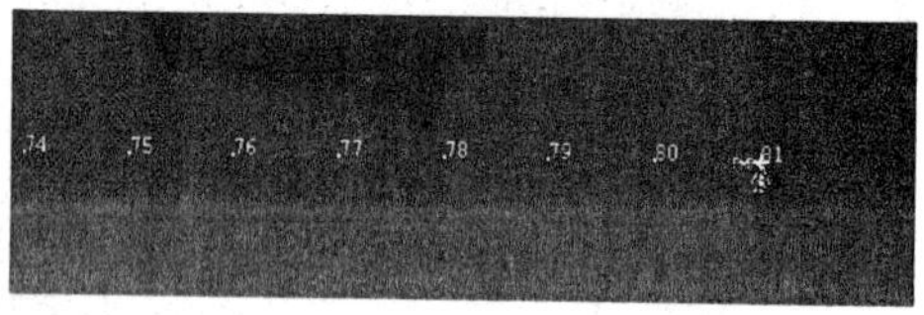

Step 7: Apply the 3 load conditions.

To apply the load, on the Preprocessor menu, click **Load**, then **Apply, Force/Moment, and On Nodes.**

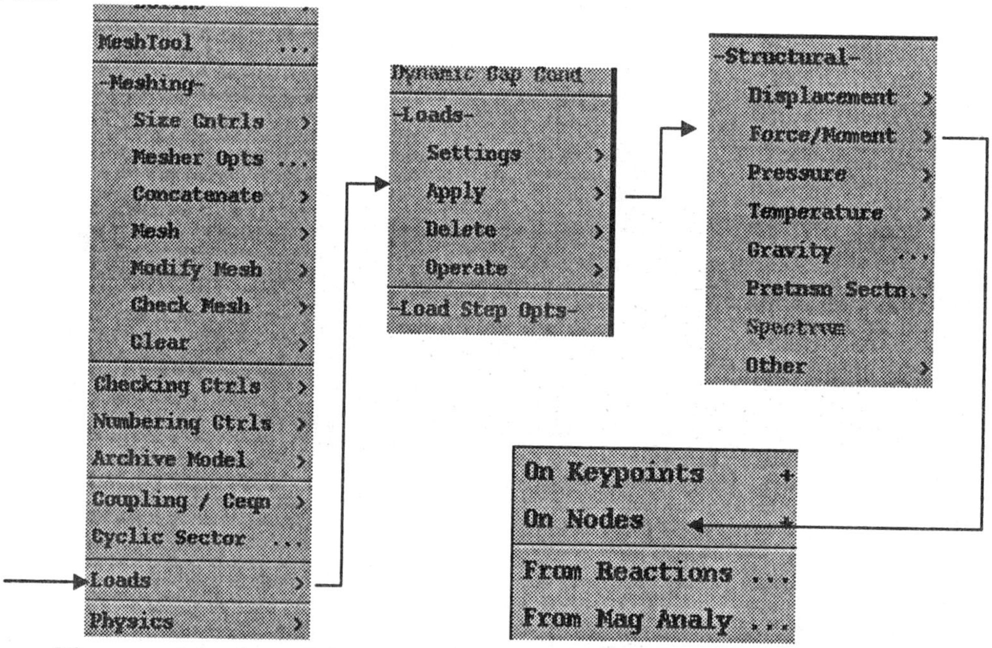

There are 3 load conditions. The first load condition is a 4000 Newton force in the positive x-direction at the right end. Let us select node 81, and click **OK**. The direction of force is FX, and the VALUE should be *4000* > **OK**.

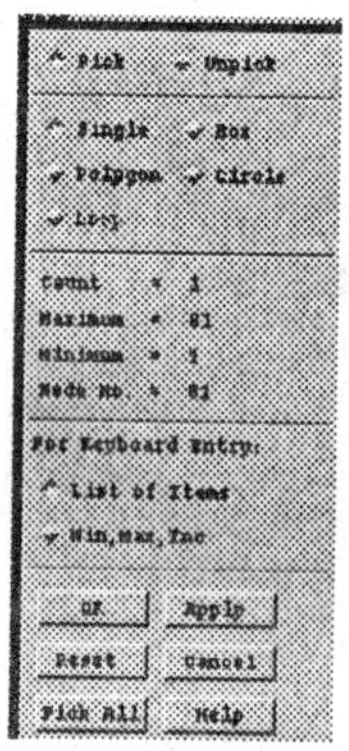

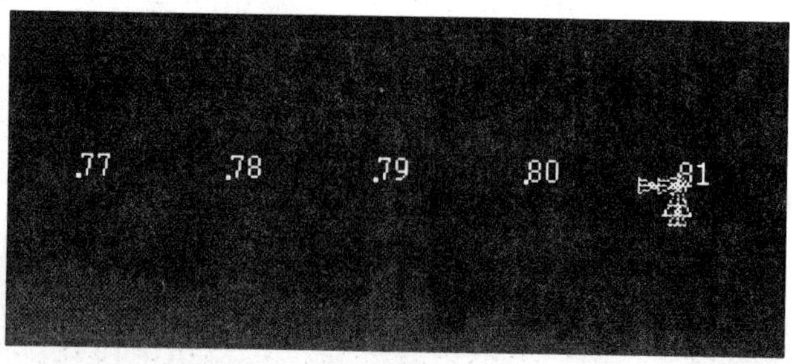

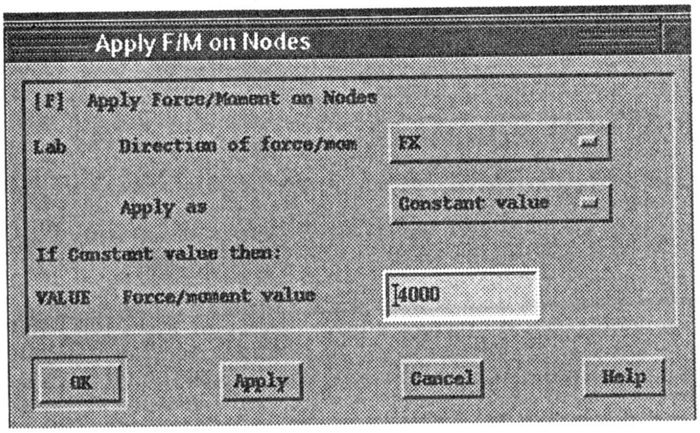

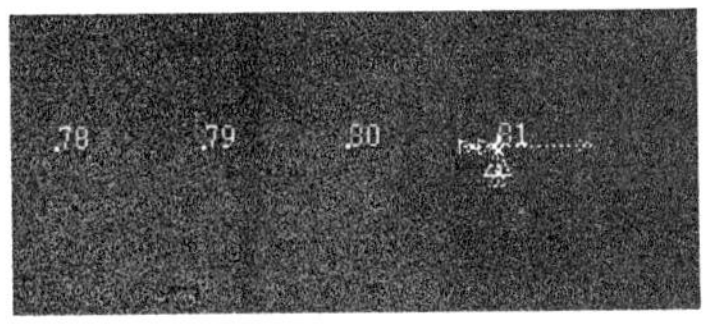

The second load condition is a distributed load of 3 N/mm across the entire beam. To approximate this force, the entire load experienced by the beam structure is 6000 Newton, over the 79 inner nodes. Select **On Nodes** again from the **Forces/Moments** Menu. Select nodes 2-80 using the **box** option.

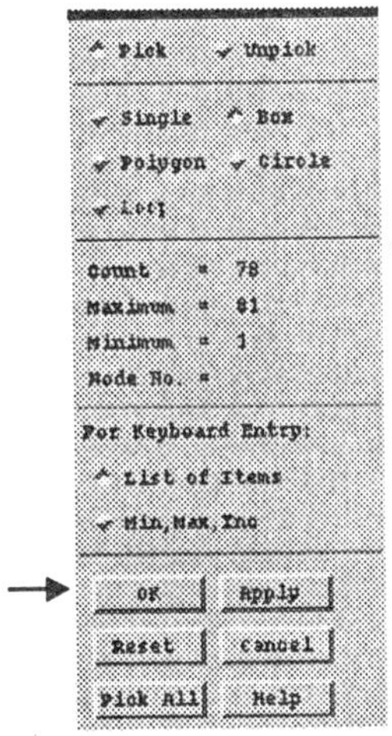

Click **OK**. Under Direction of force, select FY. The force value is -75.95 > **OK**.

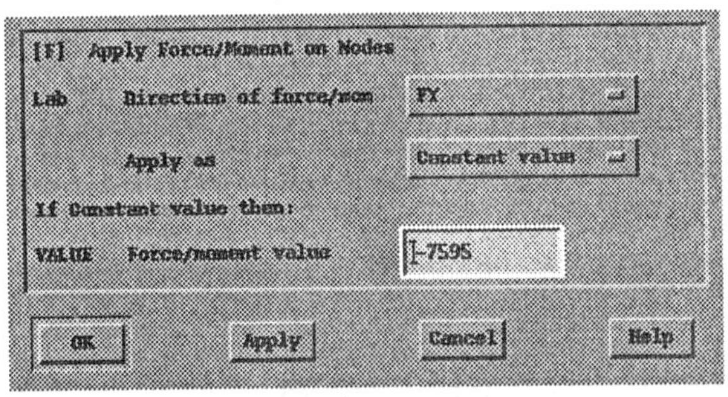

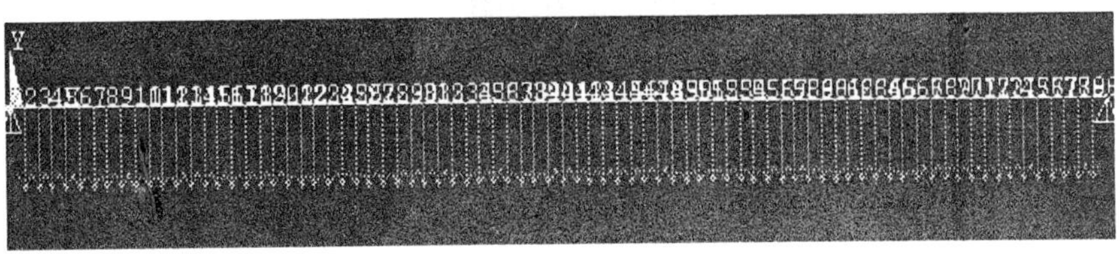

The third or final load is a 2500 Newton point load in the negative y-direction at node 61. Consider the fact that we just added -75.95 Newton on this node. In order to define the third load condition, we need to set the load value, which is -2575.95 Newton, on node 61.

The following figure illustrates the defined 3 load conditions.

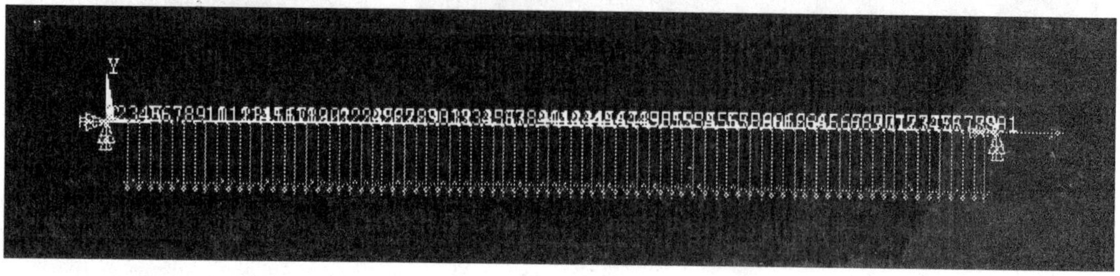

Step 8: Run the program to obtain the solution:

To find the solution, run the program by selecting **Solution** from main ANSYS menu. Choose **Current LS**, and then click **OK**.

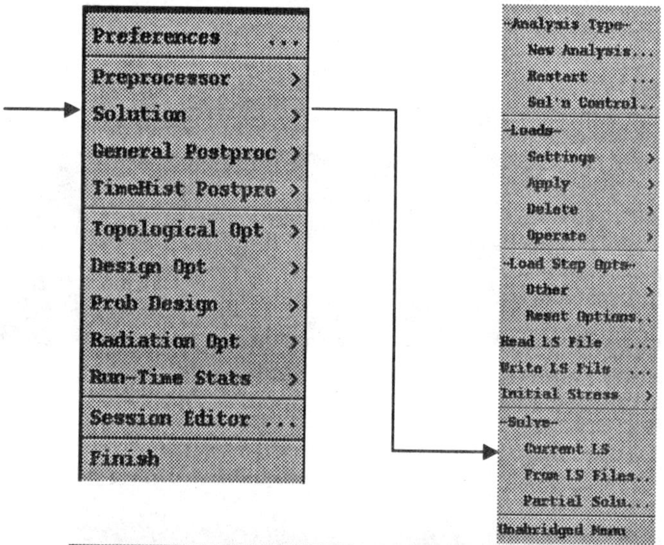

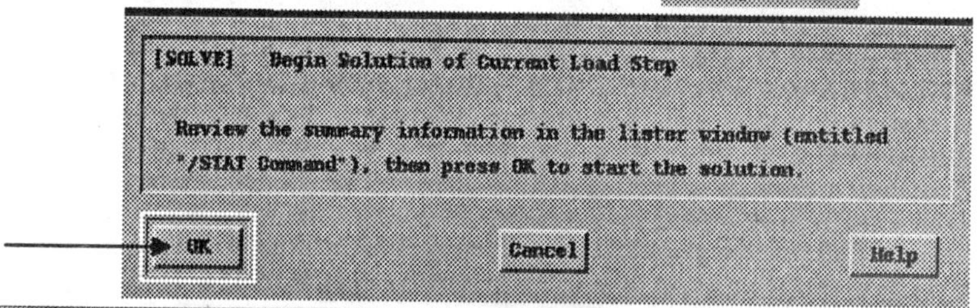

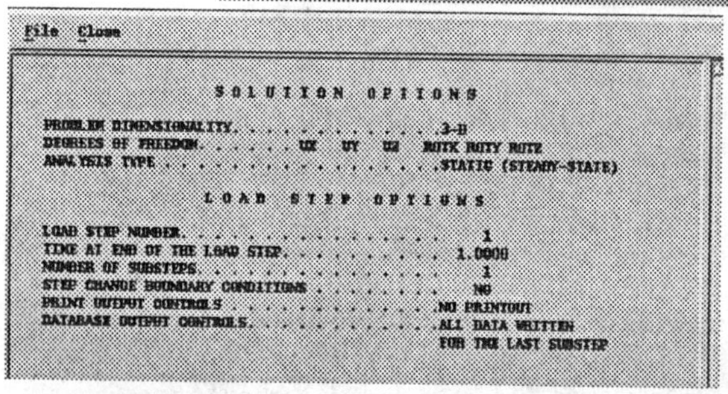

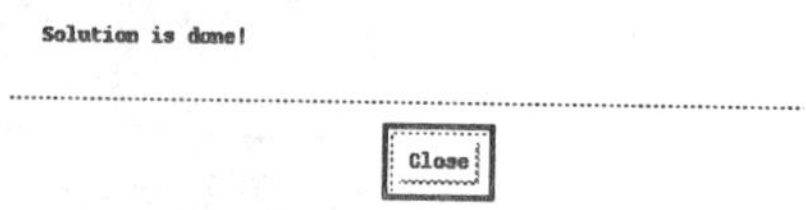

Step 9: Display Results

To review the results, select **General Postproc** from the main menu. Then click **Plot Results**, and **Nodal Solution.**

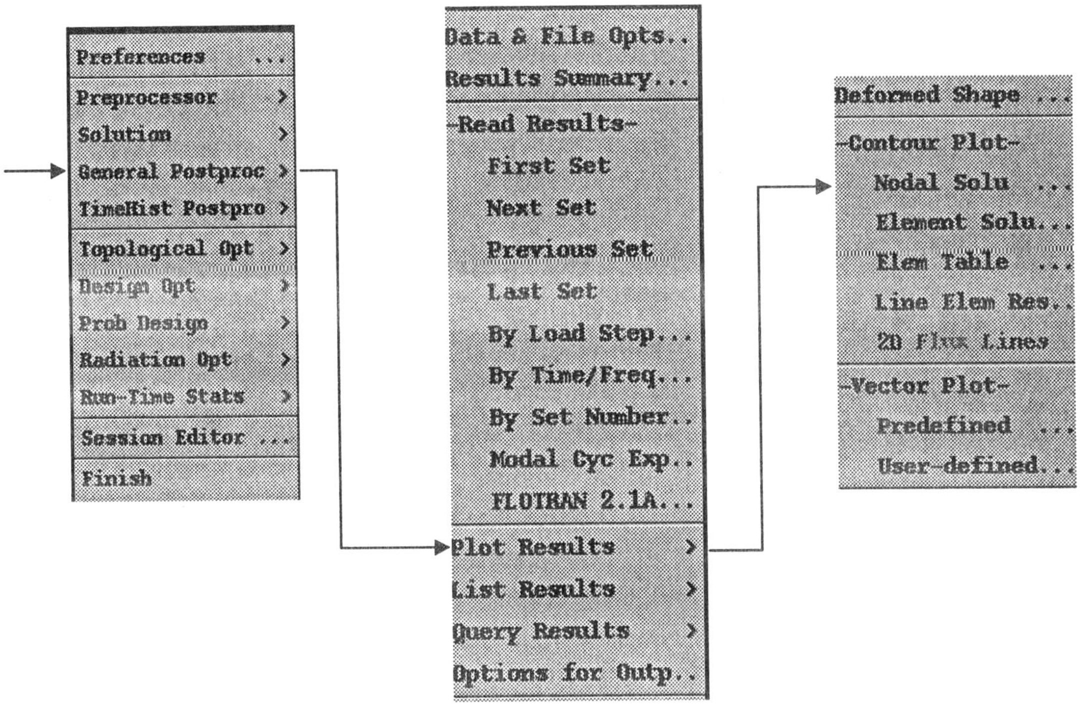

Select **DOF Solution** from the first list, and select the item to be contoured as **Translation UY** > **Def** + **unperformed** > **OK**.

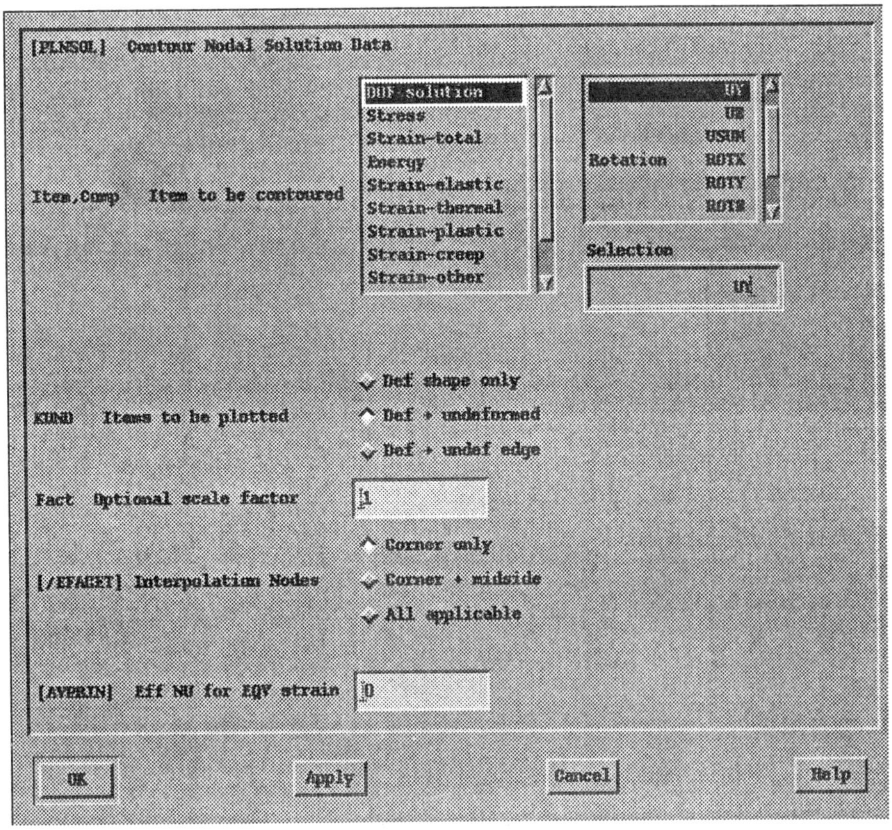

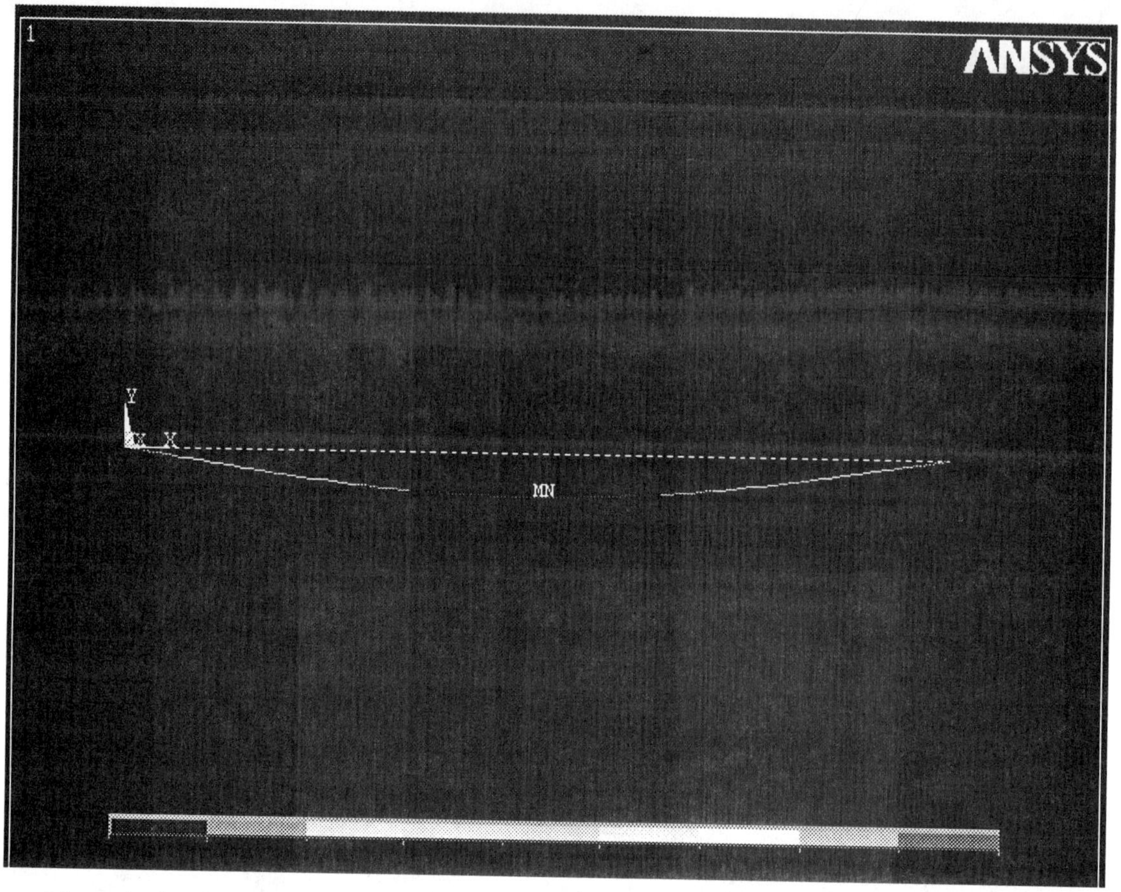

Numerical values for the displacements can be found by selecting **List Results** from the **General Post Processing** menu.

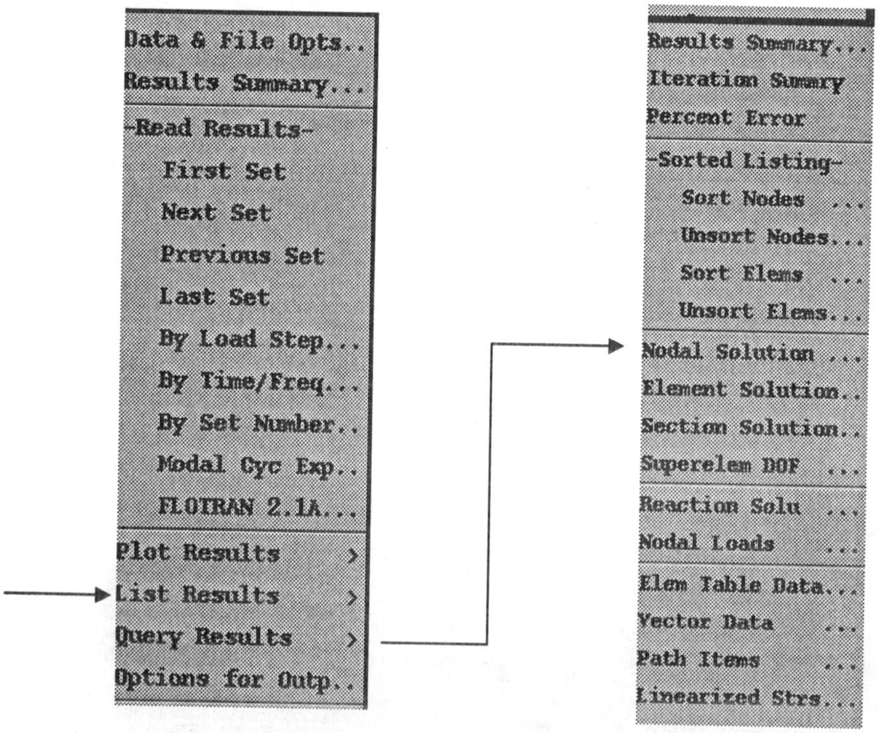

Select **DoF solution** > **Translation UY** > **OK**.

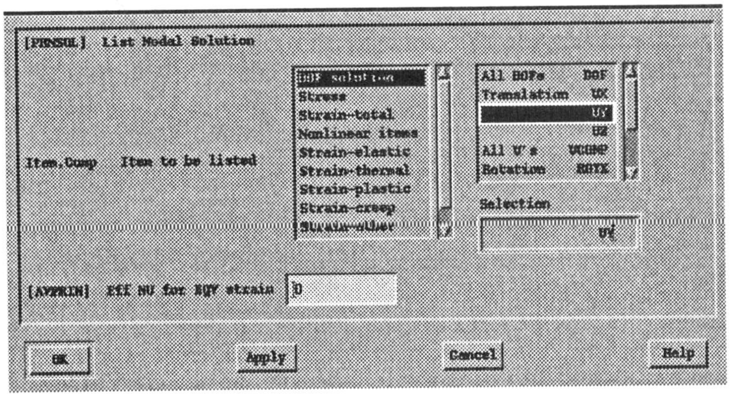
[PRNSOL] List Nodal Solution
Item,Comp Item to be listed
DOF solution
Stress
Strain-total
Nonlinear items
Strain-elastic
Strain-thermal
Strain-plastic
Strain-creep
Strain-other
All DOFs DOF
Translation UX
UY
UZ
All U's UCOMP
Rotation ROTX
Selection
UY
[AVPRIN] Eff NU for EQV strain
0
OK
Apply
Cancel
Help

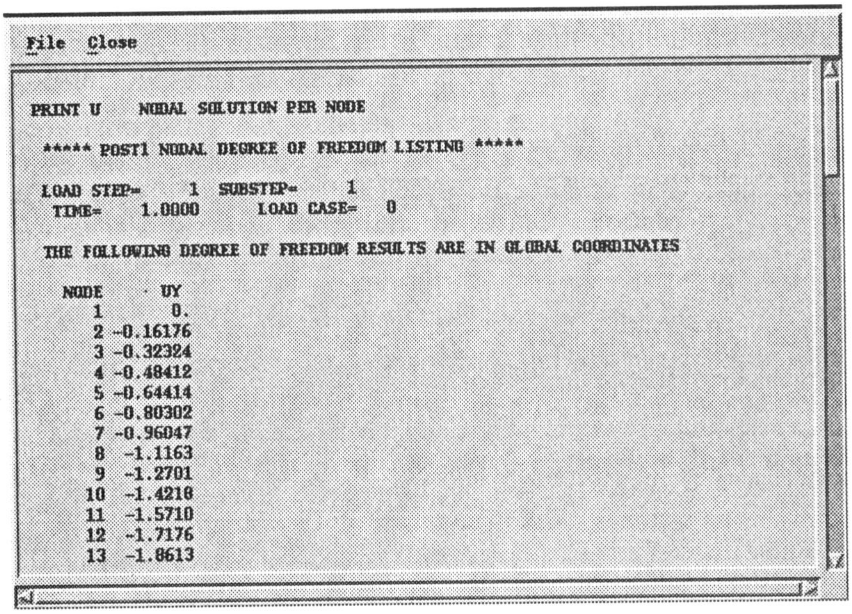
File Close
PRINT U NODAL SOLUTION PER NODE
***** POST1 NODAL DEGREE OF FREEDOM LISTING *****
LOAD STEP= 1 SUBSTEP= 1
TIME= 1.0000 LOAD CASE= 0
THE FOLLOWING DEGREE OF FREEDOM RESULTS ARE IN GLOBAL COORDINATES
NODE UY
1 0.
2 -0.16176
3 -0.32324
4 -0.48412
5 -0.64414
6 -0.80302
7 -0.96047
8 -1.1163
9 -1.2701
10 -1.4218
11 -1.5710
12 -1.7176
13 -1.8613

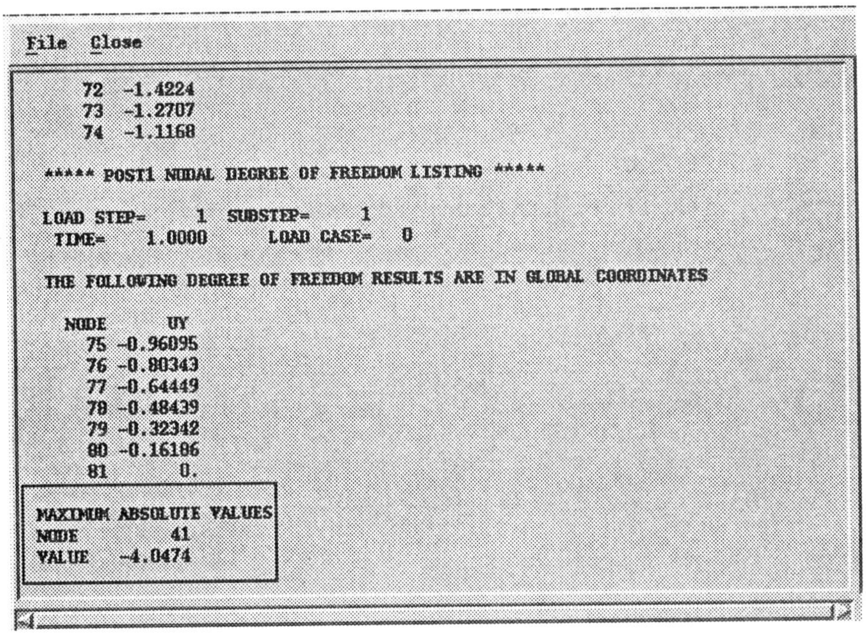
File Close
72 -1.4224
73 -1.2707
74 -1.1168
***** POST1 NODAL DEGREE OF FREEDOM LISTING *****
LOAD STEP= 1 SUBSTEP= 1
TIME= 1.0000 LOAD CASE= 0
THE FOLLOWING DEGREE OF FREEDOM RESULTS ARE IN GLOBAL COORDINATES
NODE UY
75 -0.96095
76 -0.80343
77 -0.64449
78 -0.48439
79 -0.32342
80 -0.16186
81 0.
MAXIMUM ABSOLUTE VALUES
NODE 41
VALUE -4.0474

10.6 References

1. Algor Processor Reference Manual, Algor Interactive Systems, 150 Alpha Drive, Pittsburgh, PA 15238
2. O. Axelsson, Iterative Solution Methods, Cambridge University Press, New York, 1994.
3. K. J. Bathe, Finite Element Procedure, Prentice Hall, Upper Saddle River, New Jersey, 1996.
4. K. Bell, A Refined Triangular Plate Bending Finite Element, Int. J. Numer. Methods Eng., Vol. 1, No. 1 January 1969, pp. 101-122.
5. B. Carnahan, H. A. Luther, and J. O. Wikes, Applied Numerical Methods, Wiley New York, NY, 1969.
6. R. W. Clough, The Finite Element in Plane Stress Analysis, Proceedings of 2nd ASCE Conference on Electronic Computation, Pittsburgh, PA. September 1960.
7. R. D. Cook, D. S. Maikus and M. E. Plesha, Concepts and Applications of Finite Element Analysis, 3rd edition, Wiley, New York, NY, 1989.
8. J. W. Dally and W. F. Riley, Experimental Stress Analysis, McGraw Hill, 1965.
9. J. W. Dally, Design Analysis of Structural Elements, Goodington, New York, NY, 2003.
10. C. S. Desal and J. F. Abel, Introduction to the Finite Element Method, Van Nostrand Reinhold, New York, NY, 1972.
11. J. H. Earle, Graphics for Engineers, AutoCAD Release 13, Addision-Wesley, Reading, Massachusetts, 1996.
12. R. H. Gallagher, Finite Element Analysis Fundamentals, Prentice-Hall, Englewoood Cliffs, NJ, 1975.
13. J. M. Gere and S. P. Timoshenko, Mechanics of Materials, 3rd edition, PWS-KENT Publishing Company, Boston, 1990.
14. W. W. Hager, Applied Numerical Linear Algebra, Prentice-Hall, Englewood Cliffs, NJ, 1988.
15. Holland, The Finite Element Method in Plane Stress Analysis, in the Finite Element Method in Stress Analysis, I. Holland and K. Bell (eds), Tapir Press, Trondheim, Norway, 1969, Chapter 2.
16. Y. Y. Hsieh, Elementary Theory of Structure, 2nd edition, Prentice-Hall, Englewood Cliffs, NJ, 1982.
17. M. A. N. Hendriks, H Jongedijk, J. G. Rots and W. J. E. van Spange (eds), Finite Elements in Engineering and Science, DIANA Computational Mechanics, 1997.
18. K. H. Huebner, D. L. Dewhirst, D. E. Smith, and T. G. Byrom, The Finite Element Method for Engineers, 4th edition, John Wiley & Sons, Inc., 2001.
19. C.S. Krishnamoorthy, Finite Element Analysis, Theory and Programming, 2nd Ed., 1995.
20. D. L. Logan, A First Course in the Finite Element Method Using ALGOR, PWS Publishing, Boston, 1997.
21. H. C. Martin, Introduction to Matrix Methods of Structural Analysis, McGraw-Hill, New York, NY, 1966.
22. M. Ortiz, Y. Leroy and A. Needleman, A Finite Element Method for Localized Failure Analysis, Comp. Meth. Applied Mech. Enging, 61, pp. 189-214, 1987.
23. M. Petyl, Introduction to Finite Element Vibration Analysis, Cambridge University Press, 1990.
24. T. H. H. Pian and P. Tong, Basis of Finite Element Methods for Solid Continua, Int. J. Numer. Methods Eng., Vol. 2, No. 7, April 1964, pp. 1333-1336.
25. P. G. Plockner, Symmetry in Structural Mechanics, J. of the Structural Division, American Society of Civil Engineers, Vol. 99, No. ST1, pp. 71-89, 1973.
26. J. N. Reddy, An Introduction to the Finite Element Method, McGraw-Hill Publishing Company, New York, NY, 1984.
27. T. Ross, Finite Element Program in Structural Engineering and Continuum Mechanics, Horwood Publishing Limited, Chichester, England, 1996.
28. T. Ross, Advanced Applied Finite Element Methods, Horwood Publishing Limited, Chichester, England, 1998.
29. L. Stasa, Applied Finite Element Analysis for Engineers, Holt, Rinehart and Winston, New York, NY, 1985, pp. 101-157.

30. Y. Tada and G. C. Lee, Finite Element Solution to an Elastic Problem of Beams, Int. J. Numer. Methods Eng., Vol. 2, No. 2, April 1970, pp. 229-241.
31. R. L. Taylor, On Completeness of Shape Functions for Finite Element Analysis, Int. J. Numer., Methods Eng., Vol. 4., No. 1, 1972, pp. 17-22.
32. S. P. Timoshenko and D. H. Young, Theory of Structure, McGraw-Hill, Kogaksha Limited, 1965.
33. M. J. Turner, R. W. Clough, H. C. Martin, and L. J. Topp, Stiffness and Deflection Analysis of Complex Structures, J. of the Aeronautical Sciences, Vol. 23, No. 9, pp. 805-824, Sept. 1956.
34. O. C. Zienkiewicz, The Finite Element Method, third edition, McGraw-Hill, New York, NY, 1977.
35. O. C. Zienkiewicz and R. L. Taylor, The Finite Element Method, 4th edition, Vol. 1, McGraw-Hill (UK),London, 1989.

10.7 Exercises

1. Use Pro/Mechanica to evaluate the deflection at each of the 3 nodes and the reaction force at the pin joint at the left side and the reaction force at the roller support at the right side. Compare these obtained results with the results obtained from performing FEA using **ALGOR**.

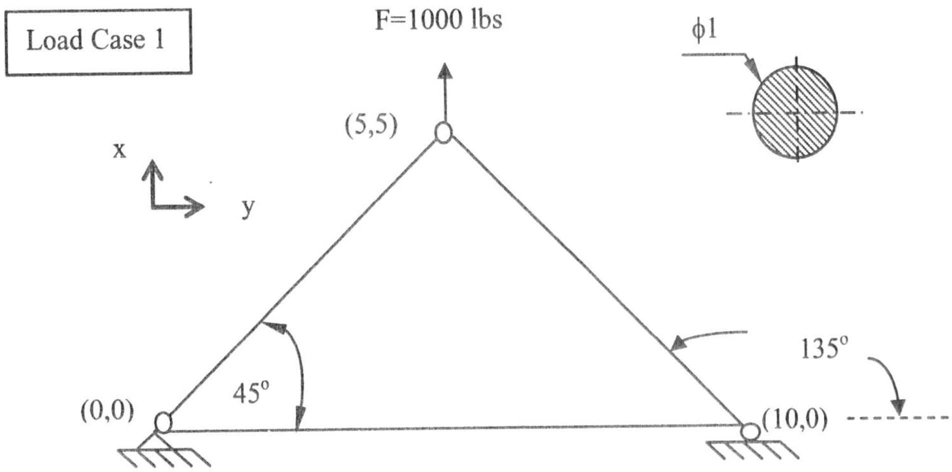

2. Use **ANSYS** to evaluate the deflection at each of the 3 nodes and the reaction force at the pin joint at the left side and the reaction force at the roller support at the right side. Compare these obtained results with the results obtained from performing FEA using **Pro/Mechanica** and **ALGOR** (references to the exercises in Chapter 9).

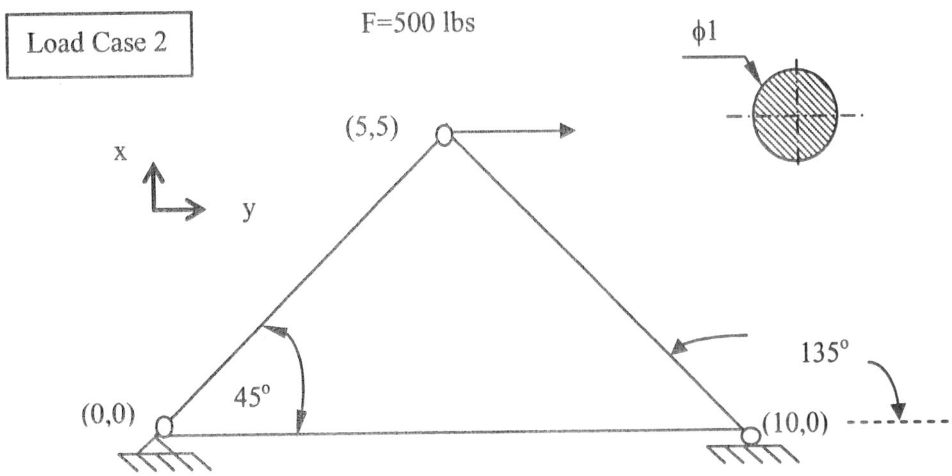

3. Use **ANSYS** to evaluate the maximum deflection of the beam under the shown load condition. Assume that the material type is STEEL. Make that the gravity force is considered during the 2 evaluations. Compare the results obtained from both running Pro/Mechanica and running ALGOR with the result obtained from running Pro/**Mechanica** and **ALGOR** (reference to the exercises in Chapter 9).

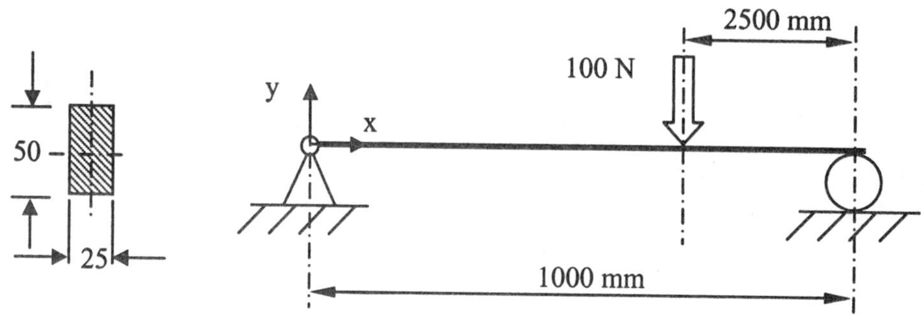

4. Use **ANSYS** to evaluate the deflection of node 2, the reaction force components at Node 1 and the reaction force components at Node 3. Compare the obtained results with the results obtained from running **ALGOR**. Make sure that the two parameters related to the material type are set at the same numerical values as those values used when running **ALGOR** and **Pro/Mechanica** (references to the exercises in Chapter 9). Those values should be identified when checking the material properties of 6061 T6 aluminum under the **ALGOR** software system.

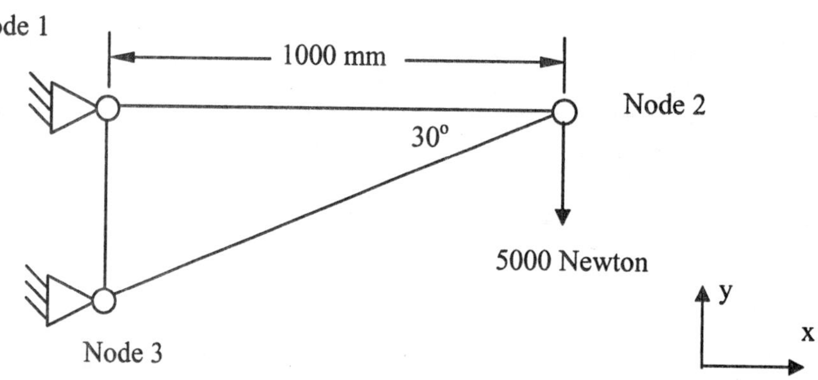

5. Use **ANSYS** to evaluate the displacement of node 2, the displacement of node 3, the reaction force components at node 1 and the reaction force compoents at node 3. Compare the obtained results with the results obtained from the calculations carried out in Chapter 1 and the results obtained from running **Pro/Mechanica** and **ALGOR** (references to the exercises in Chapter 9). Set the Young's modulus equal to 300 GPa, and the Poisson's ration equal to 0.27. The section area is a circle. The circular area is 10 mm^2.

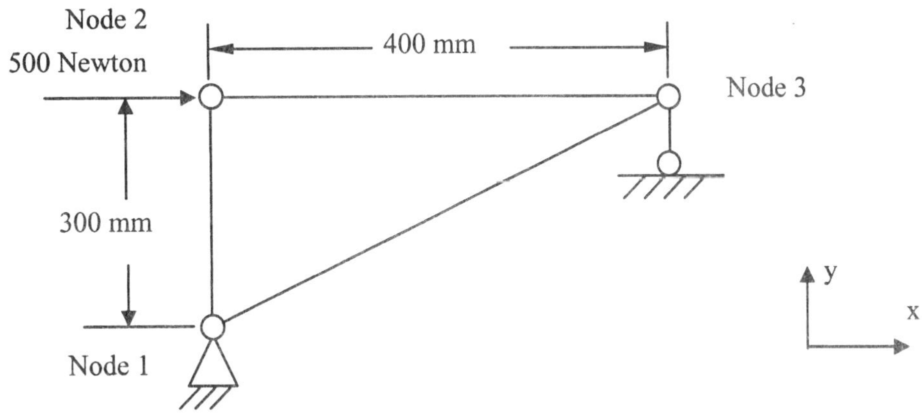

6. The following figure illustrates a truss structure, which consists of 3 bars. The dimension of the section area is 100 mm^2. The geometrical information of the truss structure is shown in the figure. The material type is STEEL. Use **ANSYS** to evaluate the displacement components at Node 2, the displacement components at Node 3, the reaction force components at Node 1 and the reaction force components at Node 4. Compare the results obtained from running **ANSYS** with the results obtained from running **Pro/Mechanica** and **ALGOR** (references to the exercises in Chapter 9).

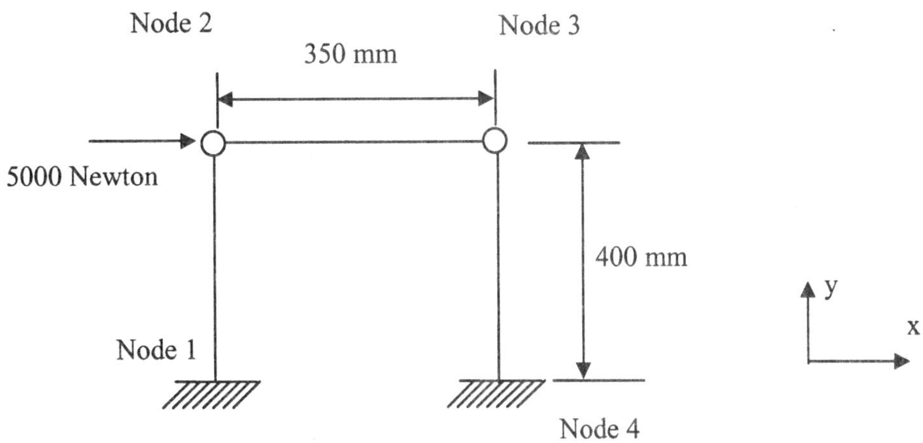

INDEX